T3-AKC-603

RSMeans

Assemblies Cost Data

31st Annual Edition

RSMeans
Construction Publishers & Consultants
63 Smiths Lane
Kingston, MA 02364-0800
(781) 422-5000

Copyright© 2005 by Reed Construction Data, Inc.
All rights reserved.

Printed in the United States of America.
ISSN 0894-4342
ISBN 0-87629-788-2

Senior Editor
Barbara Balboni

Contributing Editors
Christopher Babbitt
Ted Baker
Robert A. Bastoni
John H. Chiang, PE
John Kane
Robert J. Kuchta
Robert C. McNichols
Robert W. Mewis, CCC
Melville J. Mossman, PE
John J. Moylan
Jeannene D. Murphy
Stephen C. Plotner
Michael J. Regan
Eugene R. Spencer
Marshall J. Stetson
Phillip R. Waier, PE

Senior Engineering Operations Manager
John H. Ferguson, PE

Senior Vice President & General Manager
John Ware

Vice President of Sales
John M. Shea

Production Manager
Michael Kokernak

Technical Support
Wayne D. Anderson
Thomas J. Dion
Jonathan Forgit
Mary Lou Geary
Gary L. Hoitt
Paula Reale-Camelio
Kathryn S. Rodriguez
Sheryl A. Rose
Laurie Thom

Book & Cover Design
Norman R. Forgit

 This book is recyclable.

 This book is printed on recycled stock.

 Reed Construction Data®

First Printing

Foreword

RSMeans is a product line of Reed Construction Data, Inc., a leading provider of construction information, products, and services in North America and globally. Reed Construction Data's project information products include more than 100 regional editions, national construction data, sales leads, and local plan rooms in major business centers. Reed Construction Data's PlansDirect provides surveys, plans, and specifications. The First Source suite of products consists of *First Source for Products*, SPEC-DATA™, MANU-SPEC™, CADBlocks, Manufacturer Catalogs, and First Source Exchange (www.firstsourceexchange.com) for the selection of nationally available building products. Reed Construction Data also publishes ProFile, a database of more than 20,000 U.S. architectural firms. RSMeans provides construction cost data, training, and consulting services in print, CD-ROM, and online. Reed Construction Data, headquartered in Atlanta, is owned by Reed Business Information (www.reedconstructiondata.com), a leading provider of critical information and marketing solutions to business professionals in the media, manufacturing, electronics, construction, and retail industries. Its market-leading properties include more than 135 business-to-business publications and over 125 Webzines and Web portals as well as online services, custom publishing, directories, research, and direct-marketing lists. Reed Business Information is a member of the Reed Elsevier plc group (NYSE: RUK and ENL)—a world-leading publisher and information provider operating in the science and medical, legal, education, and business-to-business industry sectors.

Our Mission

Since 1942, RSMeans has been actively engaged in construction cost publishing and consulting throughout North America.

Today, over 60 years after RSMeans began, our primary objective remains the same: to provide you, the construction and facilities professional, with the most current and comprehensive construction cost data possible.

Whether you are a contractor, an owner, an architect, an engineer, a facilities manager, or anyone else who needs a fast and reliable construction cost estimate, you'll find this publication to be a highly useful and necessary tool.

Today, with the constant flow of new construction methods and materials, it's difficult to find the time to look at and evaluate all the different construction cost possibilities. In addition, because labor and material costs keep changing, last year's cost information is not a reliable basis for today's estimate or budget.

That's why so many construction professionals turn to RSMeans. We keep track of the costs for you, along with a wide range of other key information, from city cost indexes . . . to productivity rates . . . to crew composition . . . to contractor's overhead and profit rates.

RSMeans performs these functions by collecting data from all facets of the industry and organizing it in a format that is instantly accessible to you. From the preliminary budget to the detailed unit price estimate, you'll find the data in this book useful for all phases of construction cost determination.

The Staff, the Organization, and Our Services

When you purchase one of RSMeans' publications, you are, in effect, hiring the services of a full-time staff of construction and engineering professionals.

Our thoroughly experienced and highly qualified staff works daily at collecting, analyzing, and disseminating comprehensive cost information for your needs. These staff members have years of practical construction experience and engineering training prior to joining the firm. As a result, you can count on them not only for the cost figures, but also for additional background reference information that will help you create a realistic estimate.

The RSMeans organization is always prepared to help you solve construction problems through its five major divisions: Construction and Cost Data Publishing, Electronic Products and Services, Consulting Services, Insurance Services, and Educational Services.

Besides a full array of construction cost estimating books, RSMeans also publishes a number of other reference works for the construction industry. Subjects include construction estimating and project and business management; special topics such as HVAC, roofing, plumbing, and hazardous waste remediation; and a library of facility management references.

In addition, you can access all of our construction cost data through your computer with *Means CostWorks 2006* CD-ROM, an electronic tool that offers over 50,000 lines of RSMeans detailed construction cost data, along with assembly and whole building cost data. You can also access RSMeans cost information from our Web site at www.rsmeans.com.

What's more, you can increase your knowledge and improve your construction estimating and management performance with an RSMeans Construction Seminar or In-House Training Program. These two-day seminar programs offer unparalleled opportunities for everyone in your organization to get updated on a wide variety of construction-related issues.

RSMeans also is a worldwide provider of construction cost management and analysis services for commercial and government owners and of claims and valuation services for insurers.

In short, RSMeans can provide you with the tools and expertise for constructing accurate and dependable construction estimates and budgets in a variety of ways.

Robert Snow Means Established a Tradition of Quality That Continues Today

Robert Snow Means spent years building RSMeans, making certain he always delivered a quality product.

Today, at RSMeans, we do more than talk about the quality of our data and the usefulness of our books. We stand behind all of our data, from historical cost indexes to construction materials and techniques to current costs.

If you have any questions about our products or services, please call us toll-free at 1-800-334-3509. Our customer service representatives will be happy to assist you. You can also visit our Web site at www.rsmeans.com.

Table of Contents

ASSEMBLIES

SUBSTRUCTURE — A

SHELL — B

INTERIORS — C

SERVICES — D

EQUIPMENT & FURNISHINGS — E

SPECIAL CONSTRUCTION — F

BUILDING SITEWORK — G

REFERENCE INFORMATION

REFERENCE TABLES

COST INDEXES

SQUARE FOOT

REFERENCE AIDS

ESTIMATING FORMS

INDEX

Related RSMeans Products and Services

Barbara Balboni, Senior Editor of this cost data book, suggests the following RSMeans products and services as **companion** information resources to *Assemblies Cost Data*:

Construction Cost Data Books

Building Construction Cost Data 2006
Facilities Construction Cost Data 2006
Square Foot Costs 2006

Reference Books

ADA Compliance Pricing Guide, 2nd Edition
Building Security: Strategies & Costs
Designing & Building with the IBC
Estimating Building Costs
Estimating Handbook, 2nd Edition
Green Building: Project Planning & Estimating
How to Estimate with Means Data and CostWorks
Plan Reading & Material Takeoff
Square Foot & Assemblies Estimating Methods, 3rd Edition

Seminars and In-House Training

Advanced Project Management
Means CostWorks Training
Means Data for Job Order Contracting (JOC)
Plan Reading & Material Takeoff
Scheduling & Project Management
Unit Price Estimating

RSMeans on the Internet

Visit RSMeans at **http://www.rsmeans.com.** The site contains useful *interactive* cost and reference material. Request or download *FREE* estimating software demos. Visit our bookstore for convenient ordering and to learn more about new publications and companion products.

RSMeans Data on CD-ROM

Get the information found in RSMeans traditional cost books on RSMeans' new, easy-to-use CD-ROM, *CostWorks 2006*. *Means CostWorks* users can now enhance their Costlist with the *Means CostWorks Estimator*. For more information see the special full-color brochure inserted in this book.

RSMeans Business Solutions

RSMeans Business Solutions provides construction cost-related services including: Customized Databases, Benchmark Studies, Estimating Audits, Feasibility Studies, Litigation Support, and Predictive Cost Models.

New! RSMeans for Job Order Contracting (JOC)

New methods and "best practices" for estimating and project management to reduce administrative costs for a $147 billion+ renovation market in the U.S. Finish renovation projects faster in schools and universities, on government building projects, and in medical facilities.

* RSMeans Job Order Contracting (JOC) Cost Data
* JOCWorks Software (Basic, Advanced, PRO)
* Consultation in Contracting Methods
* RSMeans Data™ Licensing Program

Construction Costs for Software Applications

Over 25 unit price and assemblies cost databases are available through a number of leading estimating and facilities management software providers (listed below). For more information see the "yellow" pages at the back of this publication.

RSMeansData™ is also available to federal, state, and local government agencies as multi-year, multi-seat licenses.

- 3D International
- 4Clicks-Solutions, LLC
- Aepco, Inc.
- Applied Flow Technology
- ArenaSoft Estimating
- Best Software—Timberline
- BSD – Building Systems Design, Inc.
- CMS – Construction Management Software
- Corecon Technologies, Inc.
- CorVet Systems
- Estimating Systems, Inc.
- Maximus Asset Solutions
- MC^2 – Management Computer Controls
- Shaw Beneco Enterprises, Inc.
- US Cost, Inc.
- VFA–Vanderweil Facility Advisers
- WinEstimator, Inc.

How the Book Is Built: An Overview

A Powerful Construction Tool

You have in your hands one of the most powerful construction tools available today. A successful project is built on the foundation of an accurate and dependable estimate. This book will enable you to construct just such an estimate.

For the casual user the book is designed to be:

- quickly and easily understood so you can get right to your estimate.
- filled with valuable information so you can understand the necessary factors that go into the cost estimate.

For the regular user, the book is designed to be:

- a handy desk reference that can be quickly referred to for key costs.
- a comprehensive, fully reliable source of current construction costs so you'll be prepared to estimate any project.
- a source book for preliminary project cost, product selections, and alternate materials and methods.

To meet all of these requirements we have organized the book into the following clearly defined sections.

How To Use the Book: The Details

This section contains an in-depth explanation of how the book is arranged . . . and how you can use it to determine a reliable construction cost estimate. It includes information about how we develop our cost figures and how to completely prepare your estimate.

Assemblies Section

The cost data in this section has been organized in an "Assemblies" format. These assemblies are the functional elements of a building and are arranged according to the 7 divisions of the UNIFORMAT II classification system. For a complete explanation of a typical "Assemblies" page, see "How To Use the Assemblies Cost Tables."

Reference Section

This section includes information on Reference Numbers, Historical Cost Indexes, City Cost Indexes, Square Foot Costs, Reference Aids, and Estimating Forms and a listing of Abbreviations. It is visually identified by a vertical gray bar on the page edges.

Reference Numbers: Following selected major classifications in the Assemblies Section are "reference numbers" shown in bold squares. These numbers refer you to related information in the Reference Section.

In this section, you'll find reference tables, explanations, and estimating information that support how we develop the unit price data. Also included are alternate pricing methods, technical data, and estimating procedures, along with information on design and economy in construction. You'll also find helpful tips on what to expect and what to avoid when estimating and constructing your project.

It is recommended that you refer to this section if a "reference number" appears within the major classification you are estimating.

Historical Cost Indexes: These indexes provide you with data to adjust construction costs over time. If you know costs for a project completed in the past, you can use these indexes to calculate a rough estimate of what it would cost to construct the same project today.

City Cost Indexes: Costs vary depending on the regional economy. You can adjust the national average costs in this book to 316 major cities throughout the U.S. and Canada by using the data in this section. How-to-use information is included.

Location Factors, to quickly adjust the data to over 930 zip code areas, are included.

Square Foot Costs: In addition to the 7 Systems Divisions there is a S.F. (Square Foot) and C.F. (Cubic Foot) Cost Section. It contains costs for 59 different building types. You can refer to this section to make a rough—and very quick—estimate for the overall cost of a project or its major components.

Reference Aids: This section provides design criteria, code requirements, and many other reference aids useful in designing a building in various geographic locations of the country. Trade-offs by weight, material, thermal conductivity, and fire rating can be developed from this information. Design loads and capacities are based on the requirements of several nationally used building codes. In some cases, local building codes may have more stringent requirements, which would influence the final in-place cost.

Estimating Forms: Detailed estimating forms and checklists are included to be used when estimating on an "Assemblies" basis.

Abbreviations: A listing of the abbreviations used throughout this book, along with the terms they represent, is included.

Index

A comprehensive listing of all terms and subjects in this book will help you quickly find what you need.

The Scope of This Book

This book is designed to be as comprehensive and as easy to use as possible. To that end we have made certain assumptions and limited its scope in three key ways:

1. We have established material prices based on a national average.
2. We have computed labor costs based on a 30-city national average of union wage rates.
3. We have targeted the data for projects of a certain size range.

For a more detailed explanation of how the cost data is developed, see "How To Use the Book: The Details."

Project Size

This book is aimed primarily at commercial and industrial projects costing $1,000,000 and up, or large multi-family housing projects. Costs are primarily for new construction or major renovation of buildings rather than repairs or minor alterations.

With reasonable exercise of judgment the figures can be used for any building work. *However, for civil engineering structures such as bridges, dams, highways, or the like, please refer to RSMeans Heavy Construction Cost Data.*

How to Use the Book: The Details

What's Behind the Numbers? The Development of Cost Data

The staff at RSMeans continuously monitors developments in the construction industry in order to ensure reliable, thorough, and up-to-date cost information.

While *overall* construction costs may vary relative to general economic conditions, price fluctuations within the industry are dependent upon many factors. Individual price variations may, in fact, be opposite to overall economic trends. Therefore, costs are continually monitored and complete updates are published yearly. Also, new items are frequently added in response to changes in materials and methods.

Costs—$ (U.S.)

All costs represent U.S. national averages and are given in U.S. dollars. The RSMeans City Cost Indexes can be used to adjust costs to a particular location. The City Cost Indexes for Canada can be used to adjust U.S. national averages to local costs in Canadian dollars. No exchange rate conversion is necessary.

Material Costs

The RSMeans staff contacts manufacturers, dealers, distributors, and contractors all across the U.S. and Canada to determine national average material costs. If you have access to current material costs for your specific location, you may wish to make adjustments to reflect differences from the national average. Included within material costs are fasteners for a normal installation. RSMeans engineers use manufacturers' recommendations, written specifications, and/or standard construction practice for size and spacing of fasteners. Adjustments to material costs may be required for your specific application or location. Material costs do not include sales tax.

Labor Costs

Labor costs are based on the average of wage rates from 30 major U.S. cities. Rates are determined from labor union agreements or prevailing wages for construction trades for the current year. Rates, along with overhead and profit markups, are listed on the inside back cover of this book.

- If wage rates in your area vary from those used in this book, or if rate increases are expected within a given year, labor costs should be adjusted accordingly.

Labor costs reflect productivity based on actual working conditions. These figures include time spent during a normal workday on tasks other than actual installation, such as material receiving and handling, mobilization at site, site movement, breaks, and cleanup.

Productivity data is developed over an extended period so as not to be influenced by abnormal variations and reflects a typical average.

Equipment Costs

Equipment costs include not only rental, but also operating costs for equipment under normal use. The operating costs include parts and labor for routine servicing such as repair and replacement of pumps, filters, and worn lines. Normal operating expendables, such as fuel, lubricants, tires and electricity (where applicable), are also included. Extraordinary operating expendables with highly variable wear patterns, such as diamond bits and blades, are excluded. These costs are included under materials. Equipment rental rates are obtained from industry sources throughout North America—contractors, suppliers, dealers, manufacturers, and distributors.

General Conditions

Prices given in this book include the Installing Contractor's overhead and profit (O&P). General Conditions, when applicable, should also be added to the Total Cost including O&P. The costs for General Conditions are listed in the Reference Section of this book. General Conditions for the *Installing Contractor* may range from 0% to 10% of the Total Cost including O&P. For the *General* or *Prime Contractor*, costs for General Conditions may range from 5% to 15% of the Total Cost, including O&P, with a figure of 10% as the most typical allowance.

Overhead and Profit

Total Cost, which includes O&P, for the *Installing Contractor* is shown in the last column on the Assemblies Price pages of this book. This figure is the arithmetic sum of the two columns labelled Material and Installation. The value in the Material column is the bare material cost plus 10% for profit. The value in the Installation column is the base labor cost plus total overhead and profit added to the bare equipment cost plus 10% for profit. Details for the calculation of Overhead and Profit on labor are shown on the inside back cover and in the Reference Section of this book. (See the "How To Use the Assemblies Pages" for an example of this calculation.)

Factors Affecting Costs

Costs can vary depending upon a number of variables. Here's how we have handled the main factors affecting costs.

Quality—The prices for materials and the workmanship upon which productivity is based represent sound construction work. They are also in line with U.S. government specifications.

Overtime—We have made no allowance for overtime. If you anticipate premium time or work beyond normal working hours, be sure to make an appropriate adjustment to your labor costs.

Productivity—The productivities used in calculating labor costs are based on working an eight-hour day in daylight hours in moderate temperatures. For work that extends beyond normal work hours or is performed under adverse conditions, productivity may decrease.

Size of Project—The size, scope of work, and type of construction project will have a significant impact on cost. Economies of scale can reduce costs for large projects. Costs can often run higher for small projects. Costs in this book are intended for the size and type of project as previously described in "How the Book Is Built: An Overview." Costs for projects of a significantly different size or type should be adjusted accordingly.

Location—Material prices in this book are for metropolitan areas. However, in dense urban areas, traffic and site storage limitations may increase costs. Beyond a 20-mile radius of large cities, extra trucking or transportation charges may also increase the material costs slightly. On the other hand, lower wage rates may be in effect. Be sure to consider both of these factors when preparing an estimate, particularly if the job site is located in a central city or remote rural location.

In addition, highly specialized subcontract items may require travel and per-diem expenses for mechanics.

Other Factors—
- season of year
- contractor management
- weather conditions
- local union restrictions
- building code requirements
- availability of:
  - adequate energy
  - skilled labor
  - building materials
- owner's special requirements/restrictions
- safety requirements
- environmental considerations

Unpredictable Factors—General business conditions influence "in-place" costs of all items. Substitute materials and construction methods may have to be employed. These may affect the installed cost and/or life cycle costs. Such factors may be difficult to evaluate and cannot necessarily be predicted on the basis of the job's location in a particular section of the country. Thus, where these factors apply, you may find significant, but unavoidable cost variations for which you will have to apply a measure of judgment to your estimate.

Contingencies—Estimates that include an allowance for contingencies have a margin to allow for unforeseen construction difficulties. On alterations or repair jobs, 20% is not an unreasonable allowance to make. If drawings are final and only field contingencies are being considered, 2% to 3% is probably sufficient, and often nothing need be added. As far as the contract is concerned, future changes in plans can be covered by extras. The contractor should consider inflationary price trends and possible material shortages during the course of the job. Escalation factors are dependent upon both economic conditions and the anticipated time between the estimate and actual construction. If drawings are not complete or approved, or if a budget is required before proceeding with a project, it is wise to add 5% to 10%. Contingencies are a matter of judgment.

Rounding of Costs

In general, all unit prices in excess of $5.00 have been rounded to make them easier to use and still maintain adequate precision of the results. The rounding rules we have chosen are in the following table.

Prices from . . .	Rounded to the nearest . . .
$.01 to $5.00	$.01
$5.01 to $20.00	$.05
$20.01 to $100.00	$.50
$100.01 to $300.00	$1.00
$300.01 to $1,000.00	$5.00
$1,000.01 to $10,000.00	$25.00
$10,000.01 to $50,000.00	$100.00
$50,000.01 and above	$500.00

Final Checklist

Estimating can be a straightforward process provided you remember the basics. Here's a checklist of some of the steps you should remember to complete before finalizing your estimate.

Did you remember to . . .
- factor in the City Cost Index for your locale?
- take into consideration which items have been marked up and by how much?
- mark up the entire estimate sufficiently for your purposes?
- read the background information on techniques and technical matters that could impact your project time span and cost?
- include all components of your project in the final estimate?
- double check your figures for accuracy?
- call RSMeans if you have any questions about your estimate or the data you've found in our publications?

Remember, RSMeans stands behind its publications. If you have any questions about your estimate . . . about the costs you've used from our books . . . or even about the technical aspects of the job that may affect your estimate, feel free to call the RSMeans editors at 1-800-334-3509.

Assemblies Estimating

The grouping of several different trades into building components or broad building elements is the "Systems" or "Assemblies" method of estimating. This method allows the estimator or designer to make quick comparisons of systems in various combinations within predetermined guidelines. Systems which are best suited to accommodate budget, code, load, insulation, fireproofing, acoustics, energy considerations, and the owner's special requirements can quickly be determined. This method can also be used to help match existing construction.

In order to understand how a Systems estimate is assembled, it is a good idea to compare a Unit Price estimate with a Systems estimate. In a Unit Price estimate, each item is normally included along the guidelines of the 16-division MasterFormat of the Construction Specifications Institute, Inc. In a Systems estimate, these same items are allocated to one of seven major group elements in the UNIFORMAT II organization of a Systems estimate. Certain items that were formerly grouped into a single trade breakdown must now be allocated among two or more systems.

An example of this difference would be concrete. In a Unit Price estimate, all the concrete items on a job would be priced in the "Concrete" section of the estimate, CSI Division 3. In a Systems estimate, concrete is found in a number of locations. For instance, concrete is used in all of these systems: Division A10, Foundations; Division A20, Basement Construction; Division B10, Superstructure; and Division B20, Exterior Closure.

Conversely, other items that are listed in separate trade breakdowns in a Unit Price estimate are combined into one division in the Systems estimate. For example, interior partitions might include two CSI divisions: Division 6, Wood Stud Wall; and Division 9, Lath, Plaster and Paint. In the UNIFORMAT II Systems Estimate, these items are all combined in Division C, Interior Construction.

This re-allocation of the familiar items from the CSI format may, at first, seem confusing, but once the concept is understood, the resultant increase in estimating speed is well worth the initial familiarization required.

Systems or Assemblies estimating is not a substitute for Unit Price estimating. It is normally done during the earlier conceptual stage before plans have been completed or when preparing a budget. This enables the designer to bring in the project within the owner's budget. During the actual initial design process, the designer will be forced to make important decisions and "trade-offs" for each of the various sytems. Some of the trade-offs can include:

a. Price of each system

b. Appearance, quality and compatibility

c. Story height

d. Clear span

e. Complications and restrictions

f. Thermal characteristics

g. Life-cycle costs

h. Acoustical characteristics

i. Fireproofing characteristics

j. Special owner's requirements in excess of code requirements

k. Code

l. Load

Before starting a Systems Estimate, gather all the information possible pertaining to the project.

Information can be gathered from:

1. Code Requirements (see Reference Section L of this book)
2. Owner's Requirements
3. Preliminary Assumptions
4. Site Inspection and Investigation

Since the Foundation and Substructure design and price are functions of the Superstructure and the site, it is advisable to start the estimate with the Superstructure. Follow this with the Foundation and Substructure, and then the other Systems in the sequence as applicable to your project.

Assemblies Section

Table of Contents

Table of Contents

Table of Contents

Table of Contents

How to Use the Assemblies Cost Tables

The following is a detailed explanation of a sample Assemblies Cost Table. Most Assembly Tables are separated into three parts: 1) an illustration of the system to be estimated; 2) the components and related costs of a typical system; and 3) the costs for similar systems with dimensional and/or size variations. Next to each bold number below is the described item with the appropriate component of the sample entry following in parentheses. In most cases, if the work is to be subcontracted, the general contractor will need to add an additional markup (RSMeans suggests using 10%) to the "Total" figures.

System/Line Numbers (B1010-222-1700)

Each Assemblies Cost Line has been assigned a unique identification number based on the UNIFORMAT II classification system.

UNIFORMAT II Major Group

B1010 **222** **1700**

UNIFORMAT II Level 3

Means Major Classification

Means Individual Line Number

B10 Superstructure

B1010 Floor Construction **1**

2

General: Flat Slab: Solid uniform depth concrete two-way slabs with drop panels at columns and no column capitals.

Design and Pricing Assumptions:
Concrete f'c = 3 KSI, placed by concrete pump.
Reinforcement, fy = 60 KSI.
Forms, four use.
Finish, steel trowel.
Curing, spray on membrane.
Based on 4 bay x 4 bay structure.

System Components	QUANTITY **4**	UNIT	MAT.	INST.	TOTAL **6**
			COST PER S.F.		
SYSTEM B1010 222 1700					
15'X15' BAY 40 PSF S. LOAD, 12" MIN. COL. 6" SLAB, 1-1/2" DROP, 117 PSF					
Forms in place, flat slab with drop panels, to 15' high, 4 uses	.993	S.F.	1.70	4.71	6.41
Forms in place, exterior spandrel, 12" wide, 4 uses	.034	SFCA	.05	.28	.33
Reinforcing in place, elevated slabs #4 to #7	1.588	Lb.	.78	.57	1.35
Concrete ready mix, regular weight, 3000 psi	.513	C.F.	1.82		1.82
Place and vibrate concrete, elevated slab, 6" to 10" pump	.513	C.F.		.55	.55
Finish floor, monolithic steel trowel finish for finish floor	1.000	S.F.		.74	.74
Cure with sprayed membrane curing compound	.010	C.S.F.	.06	.07	.13
3			**8**	**9**	**10**
	TOTAL		4.41	6.92	11.33

5

B1010 222		Cast in Place Flat Slab with Drop Panels						
	BAY SIZE (FT.)	SUPERIMPOSED LOAD (P.S.F.)	MINIMUM COL. SIZE (IN.)	SLAB & DROP (IN.)	TOTAL LOAD (P.S.F.)	COST PER S.F.		
						MAT.	INST.	TOTAL
1700	15 x 15	40	12	6 - 1-1/2	117	4.40	6.90	11.30
1720		75	12	6 - 2-1/2	153	4.49	6.95	11.44
1760	RB1010 -010	125	14	6 - 3-1/2	205	4.70	7.10	11.80
1780		200	16	6 - 4-1/2	281	4.94	7.25	12.19
1840	15 x 20	40	12	6-1/2 - 2	124	4.68	7.05	11.73
1860		75	14	6-1/2 - 4	162	4.89	7.20	12.09
1880	RB1010 -100	125	16	6-1/2 - 5	213	5.20	7.35	12.55
1900		200	18	6-1/2 - 6	293	5.30	7.45	12.75
1960	20 x 20	40	12	7 - 3	132	4.92	7.15	12.07
1980		75	16	7 - 4	168	5.20	7.30	12.50

7

2 Illustration

At the top of most assembly tables is an illustration, a brief description, and the design criteria used to develop the cost.

3 System Components

The components of a typical system are listed separately to show what has been included in the development of the total system price. The table below contains prices for other similar systems with dimensional and/or size variations.

4 Quantity

This is the number of line item units required for one system unit. For example, we assume that it will take 1.588 pounds of reinforcing on a square foot basis.

5 Unit of Measure for Each Item

The abbreviated designation indicates the unit of measure, as defined by industry standards, upon which the price of the component is based. For example, reinforcing is priced by lb. (pound) while concrete is priced by C.F. (cubic foot).

6 Unit of Measure for Each System (cost per S.F.)

Costs shown in the three right-hand columns have been adjusted by the component quantity and unit of measure for the entire system. In this example, "Cost per S.F." is the unit of measure for this system or "assembly."

7 Reference Number Information

RB1010 -100

You'll see reference numbers shown in bold rectangles at the beginning of some sections. These refer to related items in the Reference Section, visually identified by a vertical gray bar on the page edges.

The relation may be: (1) an estimating procedure that should be read before estimating, (2) an alternate pricing method, or (3) technical information.

The "R" designates the Reference Section. The letters and numbers refer to the UNIFORMAT II classification system.

Example: The rectangle number above is directing you to refer to the reference number RB1010-100. This particular reference number shows comparative costs of floor systems.

8 Materials (4.41)

This column contains the Materials Cost of each component. These cost figures are bare costs plus 10% for profit.

9 Installation (6.92)

Installation includes labor and equipment plus the installing contractor's overhead and profit. Equipment costs are the bare rental costs plus 10% for profit. The labor overhead and profit is defined on the inside back cover of this book.

10 Total (11.33)

The figure in this column is the sum of the material and installation costs.

Material Cost	+	Installation Cost	=	Total
$4.41	+	$6.92	=	$11.33

A SUBSTRUCTURE

A1010 Standard Foundations

The Strip Footing System includes: excavation; hand trim; all forms needed for footing placement; forms for 2″ x 6″ keyway (four uses); dowels; and 3,000 p.s.i. concrete.

The footing size required varies for different soils. Soil bearing capacities are listed for 3 KSF and 6 KSF. Depths of the system range from 8″ to 24″. Widths range from 16″ to 96″. Smaller strip footings may not require reinforcement.

Please see the reference section for further design and cost information.

System Components	QUANTITY	UNIT	COST PER L.F. MAT.	COST PER L.F. INST.	COST PER L.F. TOTAL
SYSTEM A1010 110 2500					
STRIP FOOTING, LOAD 5.1KLF, SOIL CAP. 3 KSF, 24″WIDE X12″DEEP, REINF.					
Trench excavation	.148	C.Y.		1.03	1.03
Hand trim	2.000	S.F.		1.46	1.46
Compacted backfill	.074	C.Y.		.23	.23
Formwork, 4 uses	2.000	S.F.	1.88	6.88	8.76
Keyway form, 4 uses	1.000	L.F.	.33	.89	1.22
Reinforcing, fy = 60000 psi	3.000	Lb.	1.47	1.50	2.97
Dowels	2.000	Ea.	1.38	4.34	5.72
Concrete, f'c = 3000 psi	.074	C.Y.	7.07		7.07
Place concrete, direct chute	.074	C.Y.		1.35	1.35
Screed finish	2.000	S.F.		1.12	1.12
TOTAL			12.13	18.80	30.93

A1010 110	Strip Footings		COST PER L.F. MAT.	COST PER L.F. INST.	COST PER L.F. TOTAL
2100	Strip footing, load 2.6KLF, soil capacity 3KSF, 16″wide x 8″deep plain		5.90	11.30	17.20
2300	Load 3.9 KLF, soil capacity, 3 KSF, 24″wide x 8″deep, plain		7.55	12.50	20.05
2500	Load 5.1KLF, soil capacity 3 KSF, 24″wide x 12″deep, reinf.	RA1010 -140	12.15	18.80	30.95
2700	Load 11.1KLF, soil capacity 6 KSF, 24″wide x 12″deep, reinf.		12.15	18.80	30.95
2900	Load 6.8 KLF, soil capacity 3 KSF, 32″wide x 12″deep, reinf.		15.05	20.50	35.55
3100	Load 14.8 KLF, soil capacity 6 KSF, 32″wide x 12″deep, reinf.		15.05	20.50	35.55
3300	Load 9.3 KLF, soil capacity 3 KSF, 40″wide x 12″deep, reinf.		17.90	22.50	40.40
3500	Load 18.4 KLF, soil capacity 6 KSF, 40″wide x 12″deep, reinf.		18.10	22.50	40.60
3700	Load 10.1KLF, soil capacity 3 KSF, 48″wide x 12″deep, reinf.		20	24	44
3900	Load 22.1KLF, soil capacity 6 KSF, 48″wide x 12″deep, reinf.		21.50	25.50	47
4100	Load 11.8KLF, soil capacity 3 KSF, 56″wide x 12″deep, reinf.		24	26.50	50.50
4300	Load 25.8KLF, soil capacity 6 KSF, 56″wide x 12″deep, reinf.		26	29	55
4500	Load 10KLF, soil capacity 3 KSF, 48″wide x 16″deep, reinf.		25.50	28	53.50
4700	Load 22KLF, soil capacity 6 KSF, 48″wide, 16″deep, reinf.		26	28.50	54.50
4900	Load 11.6KLF, soil capacity 3 KSF, 56″wide x 16″deep, reinf.		29	39.50	68.50
5100	Load 25.6KLF, soil capacity 6 KSF, 56″wide x 16″deep, reinf.		31	41	72
5300	Load 13.3KLF, soil capacity 3 KSF, 64″wide x 16″deep, reinf.		34	33.50	67.50
5500	Load 29.3KLF, soil capacity 6 KSF, 64″wide x 16″deep, reinf.		36	36	72
5700	Load 15KLF, soil capacity 3 KSF, 72″wide x 20″deep, reinf.		45	40	85
5900	Load 33KLF, soil capacity 6 KSF, 72″wide x 20″deep, reinf.		48	43	91
6100	Load 18.3KLF, soil capacity 3 KSF, 88″wide x 24″deep, reinf.		64	50.50	114.50
6300	Load 40.3KLF, soil capacity 6 KSF, 88″wide x 24″deep, reinf.		69.50	56	125.50
6500	Load 20KLF, soil capacity 3 KSF, 96″wide x 24″deep, reinf.		70	53.50	123.50
6700	Load 44 KLF, soil capacity 6 KSF, 96″ wide x 24″ deep, reinf.		74	58	132

A
SUBSTRUCTURE

A1010 Standard Foundations

The Spread Footing System includes: excavation; backfill; forms (four uses); all reinforcement; 3,000 p.s.i. concrete (chute placed); and screed finish.

Footing systems are priced per individual unit. The Expanded System Listing at the bottom shows footings that range from 3' square x 12" deep, to 18' square x 52" deep. It is assumed that excavation is done by a truck mounted hydraulic excavator with an operator and oiler.

Backfill is with a dozer, and compaction by air tamp. The excavation and backfill equipment is assumed to operate at 30 C.Y. per hour.

Please see the reference section for further design and cost information.

SUBSTRUCTURE **A**

System Components	QUANTITY	UNIT	COST EACH		
			MAT.	INST.	TOTAL
SYSTEM A1010 210 7100					
SPREAD FOOTINGS, LOAD 25K, SOIL CAPACITY 3 KSF, 3' SQ X 12" DEEP					
Bulk excavation	.590	C.Y.		4.24	4.24
Hand trim	9.000	S.F.		6.57	6.57
Compacted backfill	.260	C.Y.		.79	.79
Formwork, 4 uses	12.000	S.F.	7.80	48.36	56.16
Reinforcing, fy = 60,000 psi	.006	Ton	5.37	5.94	11.31
Dowel or anchor bolt templates	6.000	L.F.	5.52	20.04	25.56
Concrete, f'c = 3,000 psi	.330	C.Y.	31.52		31.52
Place concrete, direct chute	.330	C.Y.		6	6
Screed finish	9.000	S.F.		4.05	4.05
TOTAL			50.21	95.99	146.20

A1010 210	Spread Footings	COST EACH		
		MAT.	INST.	TOTAL
7090	Spread footings, 3000 psi concrete, chute delivered			
7100	Load 25K, soil capacity 3 KSF, 3'-0" sq. x 12" deep	50	96	146
7150	Load 50K, soil capacity 3 KSF, 4'-6" sq. x 12" deep	107	166	273
7200	Load 50K, soil capacity 6 KSF, 3'-6" sq. x 12" deep	50	96	146
7250	Load 75K, soil capacity 3 KSF, 5'-6" sq. x 13" deep	169	235	404
7300	Load 75K, soil capacity 6 KSF, 4'-0" sq. x 12" deep	86.50	142	228.50
7350	Load 100K, soil capacity 3 KSF, 6'-0" sq. x 14" deep	214	282	496
7410	Load 100K, soil capacity 6 KSF, 4'-6" sq. x 15" deep	131	195	326
7450	Load 125K, soil capacity 3 KSF, 7'-0" sq. x 17" deep	340	400	740
7500	Load 125K, soil capacity 6 KSF, 5'-0" sq. x 16" deep	169	234	403
7550	Load 150K, soil capacity 3 KSF 7'-6" sq. x 18" deep	410	475	885
7610	Load 150K, soil capacity 6 KSF, 5'-6" sq. x 18" deep	225	295	520
7650	Load 200K, soil capacity 3 KSF, 8'-6" sq. x 20" deep	585	625	1,210
7700	Load 200K, soil capacity 6 KSF, 6'-0" sq. x 20" deep	295	365	660
7750	Load 300K, soil capacity 3 KSF, 10'-6" sq. x 25" deep	1,075	1,025	2,100
7810	Load 300K, soil capacity 6 KSF, 7'-6" sq. x 25" deep	560	610	1,170
7850	Load 400K, soil capacity 3 KSF, 12'-6" sq. x 28" deep	1,700	1,500	3,200
7900	Load 400K, soil capacity 6 KSF, 8'-6" sq. x 27" deep	775	795	1,570
7950	Load 500K, soil capacity 3 KSF, 14'-0" sq. x 31" deep	2,350	2,000	4,350
8010	Load 500K, soil capacity 6 KSF, 9'-6" sq. x 30" deep	1,075	1,025	2,100

RA1010 -120

A1010 Standard Foundations

A1010 210	Spread Footings	COST EACH		
		MAT.	INST.	TOTAL
8050	Load 600K, soil capacity 3 KSF, 16'-0" sq. x 35" deep	3,450	2,725	6,175
8100	Load 600K, soil capacity 6 KSF, 10'-6" sq. x 33" deep	1,450	1,350	2,800
8150	Load 700K, soil capacity 3 KSF, 17'-0" sq. x 37" deep	4,075	3,100	7,175
8200	Load 700K, soil capacity 6 KSF, 11'-6" sq. x 36" deep	1,850	1,625	3,475
8250	Load 800K, soil capacity 3 KSF, 18'-0" sq. x 39" deep	4,825	3,600	8,425
8300	Load 800K, soil capacity 6 KSF, 12'-0" sq. x 37" deep	2,075	1,775	3,850
8350	Load 900K, soil capacity 3 KSF, 19'-0" sq. x 40" deep	5,600	4,125	9,725
8400	Load 900K, soil capacity 6 KSF, 13'-0" sq. x 39" deep	2,550	2,125	4,675
8450	Load 1000K, soil capacity 3 KSF, 20'-0" sq. x 42" deep	6,450	4,625	11,075
8500	Load 1000K, soil capacity 6 KSF, 13'-6" sq. x 41" deep	2,900	2,375	5,275
8550	Load 1200K, soil capacity 6 KSF, 15'-0" sq. x 48" deep	3,850	3,025	6,875
8600	Load 1400K, soil capacity 6 KSF, 16'-0" sq. x 47" deep	4,700	3,575	8,275
8650	Load 1600K, soil capacity 6 KSF, 18'-0" sq. x 52" deep	6,525	4,750	11,275
8700				

A SUBSTRUCTURE

Important: See the Reference Section for critical supporting data - Reference Numbers and City Cost Indexes

A1010 Standard Foundations

These pile cap systems include excavation with a truck mounted hydraulic excavator, hand trimming, compacted backfill, forms for concrete, templates for dowels or anchor bolts, reinforcing steel and concrete placed and screeded.

Pile embedment is assumed as 6". Design is consistent with the Concrete Reinforcing Steel Institute Handbook f'c = 3000 psi, fy = 60,000.

Please see the reference section for further design and cost information.

System Components	QUANTITY	UNIT	COST EACH		
			MAT.	INST.	TOTAL
SYSTEM A1010 250 5100					
CAP FOR 2 PILES,6'-6"X3'-6"X20",15 TON PILE,8" MIN. COL., 45K COL. LOAD					
Excavation, bulk, hyd excavator, truck mtd. 30" bucket 1/2 CY	2.890	C.Y.		20.78	20.78
Trim sides and bottom of trench, regular soil	23.000	S.F.		16.79	16.79
Dozer backfill & roller compaction	1.500	C.Y.		4.58	4.58
Forms in place pile cap, square or rectangular, 4 uses	33.000	SFCA	29.04	143.88	172.92
Templates for dowels or anchor bolts	8.000	Ea.	7.36	26.72	34.08
Reinforcing in place footings, #8 to #14	.025	Ton	21.13	14.50	35.63
Concrete ready mix, regular weight, 3000 psi	1.400	C.Y.	133.70		133.70
Place and vibrate concrete for pile caps, under 5CY, direct chute	1.400	C.Y.		34.33	34.33
Monolithic screed finish	23.000	S.F.		10.35	10.35
TOTAL			191.23	271.93	463.16

A1010 250		Pile Caps						
	NO. PILES	SIZE FT-IN X FT-IN X IN	PILE CAPACITY (TON)	COLUMN SIZE (IN)	COLUMN LOAD (K)	COST EACH		
						MAT.	INST.	TOTAL
5100	2	6-6x3-6x20	15	8	45	191	272	463
5150	RA1010 -330	26	40	8	155	235	330	565
5200		34	80	11	314	320	425	745
5250		37	120	14	473	340	455	795
5300	3	5-6x5-1x23	15	8	75	227	315	542
5350		28	40	10	232	254	355	609
5400		32	80	14	471	293	405	698
5450		38	120	17	709	340	460	800
5500	4	5-6x5-6x18	15	10	103	256	315	571
5550		30	40	11	308	375	455	830
5600		36	80	16	626	440	530	970
5650		38	120	19	945	465	550	1,015
5700	6	8-6x5-6x18	15	12	156	425	460	885
5750		37	40	14	458	695	670	1,365
5800		40	80	19	936	790	750	1,540
5850		45	120	24	1413	885	830	1,715
5900	8	8-6x7-9x19	15	12	205	645	620	1,265
5950		36	40	16	610	930	810	1,740
6000		44	80	22	1243	1,150	980	2,130
6050		47	120	27	1881	1,250	1,050	2,300

A10 Foundations

A1010   Standard Foundations

A1010 250		Pile Caps						

	NO. PILES	SIZE FT-IN X FT-IN X IN	PILE CAPACITY (TON)	COLUMN SIZE (IN)	COLUMN LOAD (K)	COST EACH		
						MAT.	INST.	TOTAL
6100	10	11-6x7-9x21	15	14	250	910	765	1,675
6150		39	40	17	756	1,400	1,075	2,475
6200		47	80	25	1547	1,700	1,300	3,000
6250		49	120	31	2345	1,800	1,375	3,175
6300	12	11-6x8-6x22	15	15	316	1,150	930	2,080
6350		49	40	19	900	1,825	1,375	3,200
6400		52	80	27	1856	2,025	1,500	3,525
6450		55	120	34	2812	2,200	1,625	3,825
6500	14	11-6x10-9x24	15	16	345	1,475	1,150	2,625
6550		41	40	21	1056	1,975	1,400	3,375
6600		55	80	29	2155	2,575	1,800	4,375
6700	16	11-6x11-6x26	15	18	400	1,800	1,325	3,125
6750		48	40	22	1200	2,425	1,675	4,100
6800		60	80	31	2460	3,025	2,025	5,050
6900	18	13-0x11-6x28	15	20	450	2,050	1,425	3,475
6950		49	40	23	1349	2,825	1,875	4,700
7000		56	80	33	2776	3,325	2,200	5,525
7100	20	14-6x11-6x30	15	20	510	2,500	1,725	4,225
7150		52	40	24	1491	3,400	2,200	5,600

A1010 Standard Foundations

General: Footing drains can be placed either inside or outside of foundation walls depending upon the source of water to be intercepted. If the source of subsurface water is principally from grade or a subsurface stream above the bottom of the footing, outside drains should be used. For high water tables, use inside drains or both inside and outside.

The effectiveness of underdrains depends on good waterproofing. This must be carefully installed and protected during construction.

Costs below include the labor and materials for the pipe and 6″ only of gravel or crushed stone around pipe. Excavation and backfill are not included.

System Components	QUANTITY	UNIT	COST PER L.F.		
			MAT.	INST.	TOTAL
SYSTEM A1010 310 1000					
FOUNDATION UNDERDRAIN, OUTSIDE ONLY, PVC 4″ DIAM.					
PVC pipe 4″ diam. S.D.R. 35	1.000	L.F.	2.78	3.10	5.88
Pipe bedding, graded gravel 3/4″ to 1/2″	.070	C.Y.	2.14	.64	2.78
TOTAL			4.92	3.74	8.66

A1010 310	Foundation Underdrain	COST PER L.F.		
		MAT.	INST.	TOTAL
1000	Foundation underdrain, outside only, PVC, 4″ diameter	4.92	3.74	8.66
1100	6″ diameter	8.60	4.14	12.74
1200	Bituminous fiber, 4″ diameter	4.92	3.74	8.66
1300	6″ diameter	8.60	4.14	12.74
1400	Porous concrete, 6″ diameter	5.70	4.51	10.21
1450	8″ diameter	7	6	13
1500	12″ diameter	12.55	6.90	19.45
1600	Corrugated metal, 16 ga. asphalt coated, 6″ diameter	6.10	7.10	13.20
1650	8″ diameter	8.15	7.50	15.65
1700	10″ diameter	9.90	7.85	17.75
2000	Vitrified clay, C-211, 4″ diameter	4.52	6.65	11.17
2050	6″ diameter	6.70	8.45	15.15
2100	12″ diameter	16.70	10.20	26.90
3000	Outside and inside, PVC, 4″ diameter	9.85	7.50	17.35
3100	6″ diameter	17.20	8.25	25.45
3200	Bituminous fiber, 4″ diameter	9.85	7.50	17.35
3300	6″ diameter	17.20	8.25	25.45
3400	Porous concrete, 6″ diameter	11.35	9	20.35
3450	8″ diameter	13.95	12.05	26
3500	12″ diameter	25	13.85	38.85
3600	Corrugated metal, 16 ga., asphalt coated, 6″ diameter	12.15	14.25	26.40
3650	8″ diameter	16.25	15.05	31.30
3700	10″ diameter	19.85	15.70	35.55
4000	Vitrified clay, C-211, 4″ diameter	9.05	13.30	22.35
4050	6″ diameter	13.35	16.95	30.30
4100	12″ diameter	33.50	20.50	54

General: Apply foundation wall dampproofing over clean concrete giving particular attention to the joint between the wall and the footing. Use care in backfilling to prevent damage to the dampproofing.

Costs for four types of dampproofing are listed below.

System Components			COST PER L.F.		
	QUANTITY	UNIT	MAT.	INST.	TOTAL
SYSTEM A1010 320 1000					
FOUNDATION DAMPPROOFING, BITUMINOUS, 1 COAT, 4' HIGH					
Bituminous asphalt dampproofing brushed on below grade, 1 coat	4.000	S.F.	.40	2.52	2.92
Labor for protection of dampproofing during backfilling	4.000	S.F.		1.02	1.02
TOTAL			.40	3.54	3.94

A1010 320	Foundation Dampproofing	COST PER L.F.		
		MAT.	INST.	TOTAL
1000	Foundation dampproofing, bituminous, 1 coat, 4' high	.40	3.54	3.94
1400	8' high	.80	7.10	7.90
1800	12' high	1.20	10.95	12.15
2000	2 coats, 4' high	.76	4.34	5.10
2400	8' high	1.52	8.70	10.22
2800	12' high	2.28	13.35	15.63
3000	Asphalt with fibers, 1/16" thick, 4' high	.80	4.34	5.14
3400	8' high	1.60	8.70	10.30
3800	12' high	2.40	13.35	15.75
4000	1/8" thick, 4' high	1.40	5.20	6.60
4400	8' high	2.80	10.35	13.15
4800	12' high	4.20	15.90	20.10
5000	Asphalt coated board and mastic, 1/4" thick, 4' high	2.72	4.70	7.42
5400	8' high	5.45	9.40	14.85
5800	12' high	8.15	14.45	22.60
6000	1/2" thick, 4' high	4.08	6.55	10.63
6400	8' high	8.15	13.10	21.25
6800	12' high	12.25	19.95	32.20
7000	Cementitious coating, on walls, 1/8" thick coating, 4' high	6.85	4.80	11.65
7400	8' high	13.70	9.60	23.30
7800	12' high	20.50	14.40	34.90
8000	Cementitious/metallic slurry, 2 coat, 1/4" thick, 2' high	39.50	164	203.50
8400	4' high	79	330	409
8800	6' high	119	490	609

Important: See the Reference Section for critical supporting data - Reference Numbers and City Cost Indexes

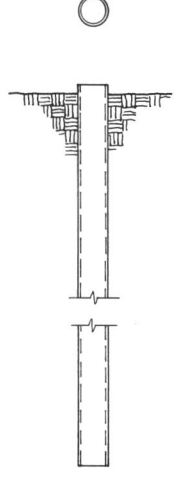

The Cast-in-Place Concrete Pile System includes: a defined number of 4,000 p.s.i. concrete piles with thin-wall, straight-sided, steel shells that have a standard steel plate driving point. An allowance for cutoffs is included.

The Expanded System Listing shows costs per cluster of piles. Clusters range from one pile to twenty piles. Loads vary from 50 Kips to 1,600 Kips. Both end-bearing and friction-type piles are shown.

Please see the reference section for cost of mobilization of the pile driving equipment and other design and cost information.

System Components			COST EACH		
	QUANTITY	UNIT	MAT.	INST.	TOTAL
SYSTEM A1020 110 2220 **CIP SHELL CONCRETE PILE, 25′ LONG, 50K LOAD, END BEARING, 1 PILE**					
Cast in place piles, end bearing, no mobil, 7 Ga. shell, 12″ diam.	27.000	V.L.F.	567	241.11	808.11
Steel pipe pile standard point, 12″ or 14″ diameter pile	1.000	Ea.	53	71	124
Pile cutoff, conc. pile with thin steel shell	1.000	Ea.		12	12
TOTAL			620	324.11	944.11

A1020 110	C.I.P. Concrete Piles	COST EACH		
		MAT.	INST.	TOTAL
2220	CIP shell concrete pile, 25′ long, 50K load, end bearing, 1 pile	620	325	945
2240	100K load, end bearing, 2 pile cluster	1,250	650	1,900
2260	200K load, end bearing, 4 pile cluster	2,475	1,300	3,775
2280	400K load, end bearing, 7 pile cluster	4,350	2,275	6,625
2300	10 pile cluster	6,200	3,250	9,450
2320	800K load, end bearing, 13 pile cluster	12,300	5,925	18,225
2340	17 pile cluster	16,100	7,750	23,850
2360	1200K load, end bearing, 14 pile cluster	13,300	6,375	19,675
2380	19 pile cluster	18,000	8,650	26,650
2400	1600K load, end bearing, 19 pile cluster	18,000	8,650	26,650
2420	50′ long, 50K load, end bearing, 1 pile	1,175	560	1,735
2440	Friction type, 2 pile cluster	2,225	1,100	3,325
2460	3 pile cluster	3,350	1,650	5,000
2480	100K load, end bearing, 2 pile cluster	2,325	1,100	3,425
2500	Friction type, 4 pile cluster	4,450	2,225	6,675
2520	6 pile cluster	6,675	3,350	10,025
2540	200K load, end bearing, 4 pile cluster	4,675	2,225	6,900
2560	Friction type, 8 pile cluster	8,900	4,450	13,350
2580	10 pile cluster	11,100	5,550	16,650
2600	400K load, end bearing, 7 pile cluster	8,150	3,900	12,050
2620	Friction type, 16 pile cluster	17,800	8,900	26,700
2640	19 pile cluster	21,100	10,600	31,700
2660	800K load, end bearing, 14 pile cluster	25,400	10,500	35,900
2680	20 pile cluster	36,400	14,900	51,300
2700	1200K load, end bearing, 15 pile cluster	27,300	11,200	38,500
2720	1600K load, end bearing, 20 pile cluster	36,400	14,900	51,300

RA1020
-100

A1020 Special Foundations

A1020 110	C.I.P. Concrete Piles	COST EACH		
		MAT.	INST.	TOTAL
3740	75' long, 50K load, end bearing, 1 pile	1,800	945	2,745
3760	Friction type, 2 pile cluster	3,450	1,875	5,325
3780	3 pile cluster	5,175	2,825	8,000
3800	100K load, end bearing, 2 pile cluster	3,600	1,875	5,475
3820	Friction type, 3 pile cluster	5,175	2,825	8,000
3840	5 pile cluster	8,625	4,725	13,350
3860	200K load, end bearing, 4 pile cluster	7,200	3,750	10,950
3880	6 pile cluster	10,800	5,650	16,450
3900	Friction type, 6 pile cluster	10,300	5,650	15,950
3910	7 pile cluster	12,100	6,600	18,700
3920	400K load, end bearing, 7 pile cluster	12,600	6,600	19,200
3930	11 pile cluster	19,800	10,400	30,200
3940	Friction type, 12 pile cluster	20,700	11,300	32,000
3950	14 pile cluster	24,100	13,200	37,300
3960	800K load, end bearing, 15 pile cluster	42,000	18,400	60,400
3970	20 pile cluster	56,000	24,500	80,500
3980	1200K load, end bearing, 17 pile cluster	47,600	20,800	68,400
3990				

Important: See the Reference Section for critical supporting data - Reference Numbers and City Cost Indexes

A1020 Special Foundations

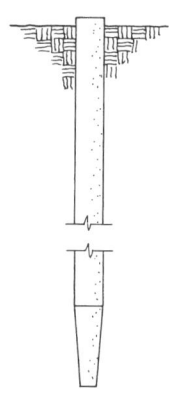

The Precast Concrete Pile System includes: pre-stressed concrete piles; standard steel driving point; and an allowance for cutoffs.

The Expanded System Listing shows costs per cluster of piles. Clusters range from one pile to twenty piles. Loads vary from 50 Kips to 1,600 Kips. Both end-bearing and friction type piles are listed.

Please see the reference section for cost of mobilization of the pile driving equipment and other design and cost information.

System Components	QUANTITY	UNIT	COST EACH MAT.	COST EACH INST.	COST EACH TOTAL
SYSTEM A1020 120 2220					
PRECAST CONCRETE PILE, 50' LONG, 50K LOAD, END BEARING, 1 PILE					
Precast, prestressed conc. piles, 10" square, no mobil.	53.000	V.L.F.	530	403.86	933.86
Steel pipe pile standard point, 8" to 10" diameter	1.000	Ea.	38.25	62.50	100.75
Piling special costs cutoffs concrete piles plain	1.000	Ea.		83	83
TOTAL			568.25	549.36	1,117.61

A1020 120	Precast Concrete Piles		MAT.	INST.	TOTAL
2220	Precast conc pile, 50' long, 50K load, end bearing, 1 pile		570	550	1,120
2240	Friction type, 2 pile cluster		1,450	1,150	2,600
2260	4 pile cluster		2,875	2,300	5,175
2280	100K load, end bearing, 2 pile cluster	RA1020 -100	1,125	1,100	2,225
2300	Friction type, 2 pile cluster		1,450	1,150	2,600
2320	4 pile cluster		2,875	2,300	5,175
2340	7 pile cluster		5,050	4,000	9,050
2360	200K load, end bearing, 3 pile cluster		1,700	1,650	3,350
2380	4 pile cluster		2,275	2,200	4,475
2400	Friction type, 8 pile cluster		5,775	4,550	10,325
2420	9 pile cluster		6,475	5,125	11,600
2440	14 pile cluster		10,100	7,975	18,075
2460	400K load, end bearing, 6 pile cluster		3,400	3,275	6,675
2480	8 pile cluster		4,550	4,400	8,950
2500	Friction type, 14 pile cluster		10,100	7,975	18,075
2520	16 pile cluster		11,500	9,100	20,600
2540	18 pile cluster		13,000	10,300	23,300
2560	800K load, end bearing, 12 pile cluster		7,375	7,150	14,525
2580	16 pile cluster		9,100	8,800	17,900
2600	1200K load, end bearing, 19 pile cluster		30,400	12,800	43,200
2620	20 pile cluster		32,000	13,500	45,500
2640	1600K load, end bearing, 19 pile cluster		30,400	12,800	43,200
4660	100' long, 50K load, end bearing, 1 pile		1,100	945	2,045
4680	Friction type, 1 pile		1,375	975	2,350

A1020 Special Foundations

A1020 120	Precast Concrete Piles	COST EACH		
		MAT.	INST.	TOTAL
4700	2 pile cluster	2,750	1,950	4,700
4720	100K load, end bearing, 2 pile cluster	2,175	1,900	4,075
4740	Friction type, 2 pile cluster	2,750	1,950	4,700
4760	3 pile cluster	4,125	2,925	7,050
4780	4 pile cluster	5,500	3,925	9,425
4800	200K load, end bearing, 3 pile cluster	3,275	2,825	6,100
4820	4 pile cluster	4,350	3,775	8,125
4840	Friction type, 3 pile cluster	4,125	2,925	7,050
4860	5 pile cluster	6,875	4,875	11,750
4880	400K load, end bearing, 6 pile cluster	6,525	5,675	12,200
4900	8 pile cluster	8,700	7,575	16,275
4910	Friction type, 8 pile cluster	11,000	7,800	18,800
4920	10 pile cluster	13,800	9,775	23,575
4930	800K load, end bearing, 13 pile cluster	14,100	12,300	26,400
4940	16 pile cluster	17,400	15,200	32,600
4950	1200K load, end bearing, 19 pile cluster	59,000	22,300	81,300
4960	20 pile cluster	62,000	23,400	85,400
4970	1600K load, end bearing, 19 pile cluster	59,000	22,300	81,300

Important: See the Reference Section for critical supporting data - Reference Numbers and City Cost Indexes

A10 Foundations

A1020 Special Foundations

The Steel Pipe Pile System includes: steel pipe sections filled with 4,000 p.s.i. concrete; a standard steel driving point; splices when required and an allowance for cutoffs.

The Expanded System Listing shows costs per cluster of piles. Clusters range from one pile to twenty piles. Loads vary from 50 Kips to 1,600 Kips. Both end-bearing and friction-type piles are shown.

Please see the reference section for cost of mobilization of the pile driving equipment and other design and cost information.

System Components	QUANTITY	UNIT	COST EACH		
			MAT.	INST.	TOTAL
SYSTEM A1020 130 2220					
CONC. FILL STEEL PIPE PILE, 50′ LONG, 50K LOAD, END BEARING, 1 PILE					
Piles, steel, pipe, conc. filled, 12″ diameter	53.000	V.L.F.	1,298.50	682.11	1,980.61
Steel pipe pile, standard point, for 12″ or 14″ diameter pipe	1.000	Ea.	106	142	248
Pile cut off, concrete pile, thin steel shell	1.000	Ea.		12	12
TOTAL			1,404.50	836.11	2,240.61

A1020 130	Steel Pipe Piles		COST EACH		
			MAT.	INST.	TOTAL
2220	Conc. fill steel pipe pile, 50′ long, 50K load, end bearing, 1 pile		1,400	835	2,235
2240	Friction type, 2 pile cluster		2,800	1,675	4,475
2250	100K load, end bearing, 2 pile cluster	RA1020 -100	2,800	1,675	4,475
2260	3 pile cluster		4,225	2,525	6,750
2300	Friction type, 4 pile cluster		5,625	3,325	8,950
2320	5 pile cluster		7,025	4,175	11,200
2340	10 pile cluster		14,000	8,350	22,350
2360	200K load, end bearing, 3 pile cluster		4,225	2,525	6,750
2380	4 pile cluster		5,625	3,325	8,950
2400	Friction type, 4 pile cluster		5,625	3,325	8,950
2420	8 pile cluster		11,200	6,700	17,900
2440	9 pile cluster		12,600	7,525	20,125
2460	400K load, end bearing, 6 pile cluster		8,425	5,025	13,450
2480	7 pile cluster		9,825	5,850	15,675
2500	Friction type, 9 pile cluster		12,600	7,525	20,125
2520	16 pile cluster		22,500	13,400	35,900
2540	19 pile cluster		26,700	15,900	42,600
2560	800K load, end bearing, 11 pile cluster		15,400	9,200	24,600
2580	14 pile cluster		19,700	11,700	31,400
2600	15 pile cluster		21,100	12,600	33,700
2620	Friction type, 17 pile cluster		23,900	14,300	38,200
2640	1200K load, end bearing, 16 pile cluster		22,500	13,400	35,900
2660	20 pile cluster		28,100	16,700	44,800
2680	1600K load, end bearing, 17 pile cluster		23,900	14,300	38,200
3700	100′ long, 50K load, end bearing, 1 pile		2,775	1,675	4,450
3720	Friction type, 1 pile		2,775	1,675	4,450

A1020 Special Foundations

A1020 130	Steel Pipe Piles	COST EACH		
		MAT.	INST.	TOTAL
3740	2 pile cluster	5,525	3,325	8,850
3760	100K load, end bearing, 2 pile cluster	5,525	3,325	8,850
3780	Friction type, 2 pile cluster	5,525	3,325	8,850
3800	3 pile cluster	8,300	4,975	13,275
3820	200K load, end bearing, 3 pile cluster	8,300	4,975	13,275
3840	4 pile cluster	11,100	6,625	17,725
3860	Friction type, 3 pile cluster	8,300	4,975	13,275
3880	4 pile cluster	11,100	6,625	17,725
3900	400K load, end bearing, 6 pile cluster	16,600	9,950	26,550
3910	7 pile cluster	19,400	11,600	31,000
3920	Friction type, 5 pile cluster	13,800	8,300	22,100
3930	8 pile cluster	22,100	13,300	35,400
3940	800K load, end bearing, 11 pile cluster	30,400	18,200	48,600
3950	14 pile cluster	38,800	23,200	62,000
3960	15 pile cluster	41,500	24,900	66,400
3970	1200K load, end bearing, 16 pile cluster	44,300	26,500	70,800
3980	20 pile cluster	55,500	33,200	88,700
3990	1600K load, end bearing, 17 pile cluster	47,100	28,200	75,300

Important: See the Reference Section for critical supporting data - Reference Numbers and City Cost Indexes

A1020 Special Foundations

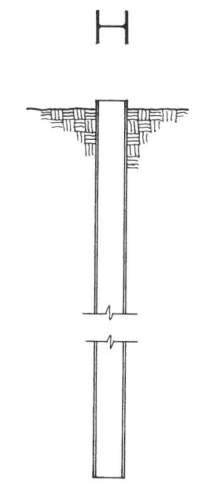

A Steel "H" Pile System includes: steel H sections; heavy duty driving point; splices where applicable and allowance for cutoffs.

The Expanded System Listing shows costs per cluster of piles. Clusters range from one pile to seventeen piles. Loads vary from 50 Kips to 2,000 Kips. All loads for Steel H Pile systems are given in terms of end bearing capacity.

Steel sections range from 10" x 10" to 14" x 14" in the Expanded System Listing. The 14" x 14" steel section is used for all H piles used in applications requiring a working load over 800 Kips.

Please see the reference section for cost of mobilization of the pile driving equipment and other design and cost information.

System Components	QUANTITY	UNIT	COST EACH		
			MAT.	INST.	TOTAL
SYSTEM A1020 140 2220 **STEEL H PILES, 50' LONG, 100K LOAD, END BEARING, 1 PILE**					
Steel H piles 10" x 10", 42 #/L.F.	53.000	V.L.F.	871.85	465.34	1,337.19
Heavy duty point, 10"	1.000	Ea.	142	144	286
Pile cut off, steel pipe or H piles	1.000	Ea.		24	24
TOTAL			1,013.85	633.34	1,647.19

A1020 140	Steel H Piles	COST EACH		
		MAT.	INST.	TOTAL
2220	Steel H piles, 50' long, 100K load, end bearing, 1 pile	1,025	635	1,660
2260	2 pile cluster	2,025	1,275	3,300
2280	200K load, end bearing, 2 pile cluster	2,025	1,275	3,300
2300	3 pile cluster	3,050	1,900	4,950
2320	400K load, end bearing, 3 pile cluster	3,050	1,900	4,950
2340	4 pile cluster	4,050	2,525	6,575
2360	6 pile cluster	6,075	3,800	9,875
2380	800K load, end bearing, 5 pile cluster	5,075	3,175	8,250
2400	7 pile cluster	7,100	4,425	11,525
2420	12 pile cluster	12,200	7,600	19,800
2440	1200K load, end bearing, 8 pile cluster	8,100	5,075	13,175
2460	11 pile cluster	11,200	6,975	18,175
2480	17 pile cluster	17,200	10,800	28,000
2500	1600K load, end bearing, 10 pile cluster	12,700	6,625	19,325
2520	14 pile cluster	17,800	9,300	27,100
2540	2000K load, end bearing, 12 pile cluster	15,200	7,975	23,175
2560	18 pile cluster	22,800	12,000	34,800
2580				
3580	100' long, 50K load, end bearing, 1 pile	3,375	1,450	4,825
3600	100K load, end bearing, 1 pile	3,375	1,450	4,825
3620	2 pile cluster	6,725	2,925	9,650
3640	200K load, end bearing, 2 pile cluster	6,725	2,925	9,650
3660	3 pile cluster	10,100	4,400	14,500
3680	400K load, end bearing, 3 pile cluster	10,100	4,400	14,500
3700	4 pile cluster	13,500	5,875	19,375
3720	6 pile cluster	20,200	8,800	29,000

RA1020 -100

A1020 Special Foundations

A1020 140	Steel H Piles	COST EACH		
		MAT.	INST.	TOTAL
3740	800K load, end bearing, 5 pile cluster	16,800	7,325	24,125
3760	7 pile cluster	23,500	10,300	33,800
3780	12 pile cluster	40,400	17,600	58,000
3800	1200K load, end bearing, 8 pile cluster	26,900	11,700	38,600
3820	11 pile cluster	37,000	16,100	53,100
3840	17 pile cluster	57,000	24,900	81,900
3860	1600K load, end bearing, 10 pile cluster	33,600	14,700	48,300
3880	14 pile cluster	47,100	20,500	67,600
3900	2000K load, end bearing, 12 pile cluster	40,400	17,600	58,000
3920	18 pile cluster	60,500	26,300	86,800

A SUBSTRUCTURE

Important: See the Reference Section for critical supporting data - Reference Numbers and City Cost Indexes

A1020 Special Foundations

The Step Tapered Steel Pile System includes: step tapered piles filled with 4,000 p.s.i. concrete. The cost for splices and pile cutoffs is included.

The Expanded System Listing shows costs per cluster of piles. Clusters range from one pile to twenty-four piles. Both end bearing piles and friction piles are listed. Loads vary from 50 Kips to 1,600 Kips.

Please see the reference section for cost of mobilization of the pile driving equipment and other design and cost information.

System Components	QUANTITY	UNIT	COST EACH		
			MAT.	INST.	TOTAL
SYSTEM A1020 150 1000					
STEEL PILE, STEP TAPERED, 50' LONG, 50K LOAD, END BEARING, 1 PILE					
Steel shell step tapered conc. filled piles, 8" tip 60 ton capacity to 60'	53.000	V.L.F.	498.20	383.19	881.39
Pile cutoff, steel pipe or H piles	1.000	Ea.		24	24
TOTAL			498.20	407.19	905.39

A1020 150	Step-Tapered Steel Piles		COST EACH		
			MAT.	INST.	TOTAL
1000	Steel pile, step tapered, 50' long, 50K load, end bearing, 1 pile		500	405	905
1200	Friction type, 3 pile cluster		1,500	1,225	2,725
1400	100K load, end bearing, 2 pile cluster	RA1020 -100	995	815	1,810
1600	Friction type, 4 pile cluster		2,000	1,625	3,625
1800	200K load, end bearing, 4 pile cluster		2,000	1,625	3,625
2000	Friction type, 6 pile cluster		3,000	2,450	5,450
2200	400K load, end bearing, 7 pile cluster		3,475	2,850	6,325
2400	Friction type, 10 pile cluster		4,975	4,075	9,050
2600	800K load, end bearing, 14 pile cluster		6,975	5,700	12,675
2800	Friction type, 18 pile cluster		8,975	7,325	16,300
3000	1200K load, end bearing, 16 pile cluster		8,950	6,950	15,900
3200	Friction type, 21 pile cluster		11,700	9,150	20,850
3400	1600K load, end bearing, 18 pile cluster		10,100	7,850	17,950
3600	Friction type, 24 pile cluster		13,400	10,500	23,900
5000	100' long, 50K load, end bearing, 1 pile		1,000	810	1,810
5200	Friction type, 2 pile cluster		2,000	1,600	3,600
5400	100K load, end bearing, 2 pile cluster		2,000	1,600	3,600
5600	Friction type, 3 pile cluster		3,025	2,450	5,475
5800	200K load, end bearing, 4 pile cluster		4,025	3,250	7,275
6000	Friction type, 5 pile cluster		5,025	4,050	9,075
6200	400K load, end bearing, 7 pile cluster		7,025	5,650	12,675
6400	Friction type, 8 pile cluster		8,025	6,450	14,475
6600	800K load, end bearing, 15 pile cluster		15,100	12,200	27,300
6800	Friction type, 16 pile cluster		16,100	13,000	29,100
7000	1200K load, end bearing, 17 pile cluster		29,700	21,200	50,900
7200	Friction type, 19 pile cluster		22,300	16,100	38,400
7400	1600K load, end bearing, 20 pile cluster		23,500	16,900	40,400
7600	Friction type, 22 pile cluster		25,800	18,700	44,500

A1020 Special Foundations

The Treated Wood Pile System includes: creosoted wood piles; a standard steel driving point; and an allowance for cutoffs.

The Expanded System Listing shows costs per cluster of piles. Clusters range from three piles to twenty piles. Loads vary from 50 Kips to 400 Kips. Both end-bearing and friction type piles are listed.

Please see the reference section for cost of mobilization of the pile driving equipment and other design and cost information.

System Components	QUANTITY	UNIT	COST EACH		
			MAT.	INST.	TOTAL
SYSTEM A1020 160 2220 **WOOD PILES, 25' LONG, 50K LOAD, END BEARING, 3 PILE CLUSTER**					
Wood piles, treated, 12" butt, 8" tip, up to 30' long	81.000	V.L.F.	862.65	694.17	1,556.82
Point for driving wood piles	3.000	Ea.	66	69	135
Pile cutoff, wood piles	3.000	Ea.		36	36
TOTAL			928.65	799.17	1,727.82

A1020 160	Treated Wood Piles	COST EACH		
		MAT.	INST.	TOTAL
2220	Wood piles, 25' long, 50K load, end bearing, 3 pile cluster	930	800	1,730
2240	Friction type, 3 pile cluster	865	730	1,595
2260	5 pile cluster	1,450	1,225	2,675
2280	100K load, end bearing, 4 pile cluster	1,250	1,075	2,325
2300	5 pile cluster	1,550	1,325	2,875
2320	6 pile cluster	1,850	1,600	3,450
2340	Friction type, 5 pile cluster	1,450	1,225	2,675
2360	6 pile cluster	1,725	1,450	3,175
2380	10 pile cluster	2,875	2,425	5,300
2400	200K load, end bearing, 8 pile cluster	2,475	2,125	4,600
2420	10 pile cluster	3,100	2,675	5,775
2440	12 pile cluster	3,725	3,200	6,925
2460	Friction type, 10 pile cluster	2,875	2,425	5,300
2480	400K load, end bearing, 16 pile cluster	4,950	4,275	9,225
2500	20 pile cluster	6,200	5,325	11,525
4520	50' long, 50K load, end bearing, 3 pile cluster	2,025	1,175	3,200
4540	4 pile cluster	2,700	1,575	4,275
4560	Friction type, 2 pile cluster	1,275	720	1,995
4580	3 pile cluster	1,925	1,075	3,000
4600	100K load, end bearing, 5 pile cluster	3,375	1,950	5,325
4620	8 pile cluster	5,375	3,150	8,525
4640	Friction type, 3 pile cluster	1,925	1,075	3,000
4660	5 pile cluster	3,200	1,775	4,975
4680	200K load, end bearing, 9 pile cluster	6,050	3,525	9,575
4700	10 pile cluster	6,725	3,950	10,675
4720	15 pile cluster	10,100	5,900	16,000
4740	Friction type, 5 pile cluster	3,200	1,775	4,975
4760	6 pile cluster	3,850	2,150	6,000

Note: Row 2280 shows reference box **RA1020 -100**

A
SUBSTRUCTURE

A10 Foundations

A1020 Special Foundations

A1020 160	Treated Wood Piles	COST EACH		
		MAT.	INST.	TOTAL
4780	10 pile cluster	6,400	3,600	10,000
4800	400K load, end bearing, 18 pile cluster	12,100	7,100	19,200
4820	20 pile cluster	13,500	7,875	21,375
4840	Friction type, 9 pile cluster	5,750	3,225	8,975
4860	10 pile cluster	6,400	3,600	10,000
9000	Add for boot for driving tip, each pile	22	16.90	38.90

A1020 Special Foundations

The Grade Beam System includes: excavation with a truck mounted backhoe; hand trim; backfill; forms (four uses); reinforcing steel; and 3,000 p.s.i. concrete placed from chute.

Superimposed loads vary in the listing from 8 Kips per linear foot (KLF) to 50 KLF. In the Expanded System Listing, the span of the beams varies from 15' to 40'. Depth varies from 28" to 52". Width varies from 12" to 28".

Please see the reference section for further design and cost information.

System Components	QUANTITY	UNIT	COST PER L.F.		
			MAT.	INST.	TOTAL
SYSTEM A1020 210 2220					
GRADE BEAM, 15' SPAN, 28" DEEP, 12" WIDE, 8 KLF LOAD					
Excavation, trench, hydraulic backhoe, 3/8 CY bucket	.260	C.Y.		1.80	1.80
Trim sides and bottom of trench, regular soil	2.000	S.F.		1.46	1.46
Backfill, by hand, compaction in 6" layers, using vibrating plate	.170	C.Y.		1.05	1.05
Forms in place, grade beam, 4 uses	4.700	SFCA	2.91	20.02	22.93
Reinforcing in place, beams & girders, #8 to #14	.019	Ton	17.86	14.63	32.49
Concrete ready mix, regular weight, 3000 psi	.090	C.Y.	8.60		8.60
Place and vibrate conc. for grade beam, direct chute	.090	C.Y.		1.31	1.31
TOTAL			29.37	40.27	69.64

A1020 210	Grade Beams	COST PER L.F.		
		MAT.	INST.	TOTAL
2220	Grade beam, 15' span, 28" deep, 12" wide, 8 KLF load	29.50	40	69.50
2240	14" wide, 12 KLF load	30.50	40.50	71
2260	40" deep, 12" wide, 16 KLF load	29.50	49.50	79
2280	20 KLF load	34.50	53.50	88
2300	52" deep, 12" wide, 30 KLF load	39.50	67	106.50
2320	40 KLF load	49	75	124
2340	50 KLF load	58.50	82	140.50
3360	20' span, 28" deep, 12" wide, 2 KLF load	18.10	31	49.10
3380	16" wide, 4 KLF load	24	34.50	58.50
3400	40" deep, 12" wide, 8 KLF load	29.50	49.50	79
3420	12 KLF load	38	56	94
3440	14" wide, 16 KLF load	46.50	62.50	109
3460	52" deep, 12" wide, 20 KLF load	50	75.50	125.50
3480	14" wide, 30 KLF load	69.50	90	159.50
3500	20" wide, 40 KLF load	82	95	177
3520	24" wide, 50 KLF load	102	108	210
4540	30' span, 28" deep, 12" wide, 1 KLF load	19.05	31.50	50.55
4560	14" wide, 2 KLF load	31	44	75
4580	40" deep, 12" wide, 4 KLF load	36.50	52.50	89
4600	18" wide, 8 KLF load	52.50	64.50	117
4620	52" deep, 14" wide, 12 KLF load	67	88	155
4640	20" wide, 16 KLF load	82	95.50	177.50
4660	24" wide, 20 KLF load	102	109	211
4680	36" wide, 30 KLF load	147	136	283
4700	48" wide, 40 KLF load	194	167	361
5720	40' span, 40" deep, 12" wide, 1 KLF load	26	46.50	72.50

Note: Reference box by rows 2260/2280: RA1020 -230

Important: See the Reference Section for critical supporting data - Reference Numbers and City Cost Indexes

A10 Foundations

A1020 Special Foundations

A1020 210	Grade Beams	COST PER L.F.		
		MAT.	INST.	TOTAL
5740	2 KLF load	32.50	52	84.50
5760	52" deep, 12" wide, 4 KLF load	49	75	124
5780	20" wide, 8 KLF load	79	93.50	172.50
5800	28" wide, 12 KLF load	109	112	221
5820	38" wide, 16 KLF load	151	139	290
5840	46" wide, 20 KLF load	198	171	369

A

Caisson Systems are listed for three applications: stable ground, wet ground and soft rock. Concrete used is 3,000 p.s.i. placed from chute. Included are a bell at the bottom of the caisson shaft (if applicable) along with required excavation and disposal of excess excavated material up to two miles from job site.

The Expanded System lists cost per caisson. End-bearing loads vary from 200 Kips to 3,200 Kips. The dimensions of the caissons range from 2' x 50' to 7' x 200'.

Please see the reference section for further design and cost information.

System Components	QUANTITY	UNIT	COST EACH		
			MAT.	INST.	TOTAL
SYSTEM A1020 310 2200					
CAISSON, STABLE GROUND, 3000 PSI CONC., 10KSF BRNG, 200K LOAD, 2'X50'					
Caissons, drilled, to 50', 24" shaft diameter, .116 C.Y./L.F.	50.000	V.L.F.	692.50	1,567.50	2,260
Reinforcing in place, columns, #3 to #7	.060	Ton	56.40	84	140.40
Caisson bell excavation and concrete, 4' diameter .444 CY	1.000	Ea.	42.50	298	340.50
Load & haul excess excavation, 2 miles	6.240	C.Y.		30.52	30.52
TOTAL			791.40	1,980.02	2,771.42

A1020 310	Caissons	COST EACH		
		MAT.	INST.	TOTAL
2200	Caisson, stable ground, 3000 PSI conc, 10KSF brng, 200K load, 2'x50'	790	1,975	2,765
2400	400K load, 2'-6"x50'-0"	1,300	3,175	4,475
2600	800K load, 3'-0"x100'-0"	3,550	7,550	11,100
2800	1200K load, 4'-0"x100'-0"	6,175	9,475	15,650
3000	1600K load, 5'-0"x150'-0"	13,900	14,600	28,500
3200	2400K load, 6'-0"x150'-0"	20,200	19,000	39,200
3400	3200K load, 7'-0"x200'-0"	36,500	27,400	63,900
5000	Wet ground, 3000 PSI conc., 10 KSF brng, 200K load, 2'-0"x50'-0"	690	2,725	3,415
5200	400K load, 2'-6"x50'-0"	1,125	4,650	5,775
5400	800K load, 3'-0"x100'-0"	3,100	12,500	15,600
5600	1200K load, 4'-0"x100'-0"	5,350	17,500	22,850
5800	1600K load, 5'-0"x150'-0"	11,900	37,800	49,700
6000	2400K load, 6'-0"x150'-0"	17,400	46,400	63,800
6200	3200K load, 7'-0"x200'-0"	31,400	74,500	105,900
7800	Soft rock, 3000 PSI conc., 10 KSF brng, 200K load, 2'-0"x50'-0"	690	13,900	14,590
8000	400K load, 2'-6"x50'-0"	1,125	22,600	23,725
8200	800K load, 3'-0"x100'-0"	3,100	59,000	62,100
8400	1200K load, 4'-0"x100'-0"	5,350	86,500	91,850
8600	1600K load, 5'-0"x150'-0"	11,900	178,000	189,900
8800	2400K load, 6'-0"x150'-0"	17,400	211,500	228,900
9000	3200K load, 7'-0"x200'-0"	31,400	337,000	368,400

Note: Between rows 2600/2800 the reference marker **RA1020 -200** appears.

Important: See the Reference Section for critical supporting data - Reference Numbers and City Cost Indexes

A1020 Special Foundations

Pressure Injected Piles are usually uncased up to 25' and cased over 25' depending on soil conditions.

These costs include excavation and hauling of excess materials; steel casing over 25'; reinforcement; 3,000 p.s.i. concrete; plus mobilization and demobilization of equipment for a distance of up to fifty miles to and from the job site.

The Expanded System lists cost per cluster of piles. Clusters range from one pile to eight piles. End-bearing loads range from 50 Kips to 1,600 Kips.

Please see the reference section for further design and cost information.

System Components	QUANTITY	UNIT	COST EACH		
			MAT.	INST.	TOTAL
SYSTEM A1020 710 4200					
PRESSURE INJECTED FOOTING, END BEARING, 50' LONG, 50K LOAD, 1 PILE					
Pressure injected footings, cased, 30-60 ton cap., 12" diameter	50.000	V.L.F.	735	1,011	1,746
Pile cutoff, concrete pile with thin steel shell	1.000	Ea.		12	12
TOTAL			735	1,023	1,758

A1020 710	Pressure Injected Footings		COST EACH		
			MAT.	INST.	TOTAL
2200	Pressure injected footing, end bearing, 25' long, 50K load, 1 pile		340	715	1,055
2400	100K load, 1 pile		340	715	1,055
2600	2 pile cluster		685	1,425	2,110
2800	200K load, 2 pile cluster	RA1020 -100	685	1,425	2,110
3200	400K load, 4 pile cluster		1,375	2,850	4,225
3400	7 pile cluster		2,400	5,000	7,400
3800	1200K load, 6 pile cluster		2,875	5,450	8,325
4000	1600K load, 7 pile cluster		3,350	6,350	9,700
4200	50' long, 50K load, 1 pile		735	1,025	1,760
4400	100K load, 1 pile		1,275	1,025	2,300
4600	2 pile cluster		2,550	2,050	4,600
4800	200K load, 2 pile cluster		2,550	2,050	4,600
5000	4 pile cluster		5,100	4,100	9,200
5200	400K load, 4 pile cluster		5,100	4,100	9,200
5400	8 pile cluster		10,200	8,200	18,400
5600	800K load, 7 pile cluster		8,925	7,150	16,075
5800	1200K load, 6 pile cluster		8,250	6,150	14,400
6000	1600K load, 7 pile cluster		9,625	7,150	16,775

A1030 Slab on Grade

There are four types of Slab on Grade Systems listed: Non-industrial, Light industrial, Industrial and Heavy industrial. Each type is listed two ways: reinforced and non-reinforced. A Slab on Grade system includes three passes with a grader; 6" of compacted gravel fill; polyethylene vapor barrier; 3500 p.s.i. concrete placed by chute; bituminous fibre expansion joint; all necessary edge forms (4 uses); steel trowel finish; and sprayed-on membrane curing compound.

The Expanded System Listing shows costs on a per square foot basis. Thicknesses of the slabs range from 4" to 8". Non-industrial applications are for foot traffic only with negligible abrasion. Light industrial applications are for pneumatic wheels and light abrasion. Industrial applications are for solid rubber wheels and moderate abrasion. Heavy industrial applications are for steel wheels and severe abrasion. All slabs are either shown unreinforced or reinforced with welded wire fabric.

System Components	QUANTITY	UNIT	COST PER S.F.		
			MAT.	INST.	TOTAL
SYSTEM A1030 120 2220					
SLAB ON GRADE, 4" THICK, NON INDUSTRIAL, NON REINFORCED					
Fine grade, 3 passes with grader and roller	.110	S.Y.		.36	.36
Gravel under floor slab, 6" deep, compacted	1.000	S.F.	.18	.23	.41
Polyethylene vapor barrier, standard, .006" thick	1.000	S.F.	.05	.12	.17
Concrete ready mix, regular weight, 3500 psi	.012	C.Y.	1.18		1.18
Place and vibrate concrete for slab on grade, 4" thick, direct chute	.012	C.Y.		.24	.24
Expansion joint, premolded bituminous fiber, 1/2" x 6"	.100	L.F.	.09	.26	.35
Edge forms in place for slab on grade to 6" high, 4 uses	.030	L.F.	.01	.08	.09
Cure with sprayed membrane curing compound	1.000	S.F.	.06	.07	.13
Finishing floor, monolithic steel trowel	1.000	S.F.		.74	.74
TOTAL			1.57	2.10	3.67

A1030 120	Plain & Reinforced		COST PER S.F.		
			MAT.	INST.	TOTAL
2220	Slab on grade, 4" thick, non industrial, non reinforced		1.56	2.09	3.65
2240	Reinforced		1.69	2.39	4.08
2260	Light industrial, non reinforced	RA1030 -200	2.01	2.57	4.58
2280	Reinforced		2.14	2.87	5.01
2300	Industrial, non reinforced		2.41	5.30	7.71
2320	Reinforced		2.54	5.55	8.09
3340	5" thick, non industrial, non reinforced		1.86	2.16	4.02
3360	Reinforced		1.99	2.46	4.45
3380	Light industrial, non reinforced		2.31	2.64	4.95
3400	Reinforced		2.44	2.94	5.38
3420	Heavy industrial, non reinforced		3.18	6.30	9.48
3440	Reinforced		3.27	6.65	9.92
4460	6" thick, non industrial, non reinforced		2.25	2.11	4.36
4480	Reinforced		2.44	2.51	4.95
4500	Light industrial, non reinforced		2.71	2.59	5.30
4520	Reinforced		3.01	3.13	6.14
4540	Heavy industrial, non reinforced		3.58	6.40	9.98
4560	Reinforced		3.77	6.80	10.57
5580	7" thick, non industrial, non reinforced		2.56	2.19	4.75
5600	Reinforced		2.78	2.62	5.40
5620	Light industrial, non reinforced		3.02	2.67	5.69
5640	Reinforced		3.25	3.10	6.35
5660	Heavy industrial, non reinforced		3.89	6.30	10.19
5680	Reinforced		4.08	6.70	10.78

Important: See the Reference Section for critical supporting data - Reference Numbers and City Cost Indexes

A10 Foundations

A1030 Slab on Grade

A1030 120	Plain & Reinforced	COST PER S.F.		
		MAT.	INST.	TOTAL
6700	8″ thick, non industrial, non reinforced	2.85	2.23	5.08
6720	Reinforced	3.04	2.59	5.63
6740	Light industrial, non reinforced	3.32	2.71	6.03
6760	Reinforced	3.51	3.07	6.58
6780	Heavy industrial, non reinforced	4.21	6.40	10.61
6800	Reinforced	4.47	6.75	11.22

A2010 Basement Excavation

1/2 : 1

1 : 1

Line of Excavation

Pricing Assumptions: Two-thirds of excavation is by 2-1/2 C.Y. wheel mounted front end loader and one-third by 1-1/2 C.Y. hydraulic excavator.

Two-mile round trip haul by 12 C.Y. tandem trucks is included for excavation wasted and storage of suitable fill from excavated soil. For excavation in clay, all is wasted and the cost of suitable backfill with two-mile haul is included.

Sand and gravel assumes 15% swell and compaction; common earth assumes 25% swell and 15% compaction; clay assumes 40% swell and 15% compaction (non-clay).

In general, the following items are accounted for in the costs in the table below.

1. Excavation for building or other structure to depth and extent indicated.
2. Backfill compacted in place.
3. Haul of excavated waste.
4. Replacement of unsuitable material with bank run gravel.

Note: Additional excavation and fill beyond this line of general excavation for the building (as required for isolated spread footings, strip footings, etc.) are included in the cost of the appropriate component systems.

System Components				COST PER S.F.		
		QUANTITY	UNIT	MAT.	INST.	TOTAL
SYSTEM A2010 110 2280						
EXCAVATE & FILL, 1000 S.F., 8′ DEEP, SAND, ON SITE STORAGE						
Excavating bulk shovel, 1.5 C.Y. bucket, 150 cy/hr		.262	C.Y.		.40	.40
Excavation, front end loader, 2-1/2 C.Y.		.523	C.Y.		.84	.84
Haul earth, 12 C.Y. dump truck		.341	C.Y.		1.64	1.64
Backfill, dozer bulk push 300′, including compaction		.562	C.Y.		1.71	1.71
	TOTAL				4.59	4.59

A2010 110	Building Excavation & Backfill		COST PER S.F.		
			MAT.	INST.	TOTAL
2220	Excav & fill, 1000 S.F. 4′ sand, gravel, or common earth, on site storage			.78	.78
2240	Off site storage			1.16	1.16
2260	Clay excavation, bank run gravel borrow for backfill		2	1.75	3.75
2280	8′ deep, sand, gravel, or common earth, on site storage	RG1010 -010		4.60	4.60
2300	Off site storage			10.05	10.05
2320	Clay excavation, bank run gravel borrow for backfill		8.10	8.15	16.25
2340	16′ deep, sand, gravel, or common earth, on site storage	RG1030 -020		12.90	12.90
2350	Off site storage			25	25
2360	Clay excavation, bank run gravel borrow for backfill		22.50	21	43.50
3380	4000 S.F.,4′ deep, sand, gravel, or common earth, on site storage	RG1030 -400		.42	.42
3400	Off site storage			.80	.80
3420	Clay excavation, bank run gravel borrow for backfill		1.02	.89	1.91
3440	8′ deep, sand, gravel, or common earth, on site storage			3.21	3.21
3460	Off site storage			5.60	5.60
3480	Clay excavation, bank run gravel borrow for backfill		3.69	4.87	8.56
3500	16′ deep, sand, gravel, or common earth, on site storage			7.90	7.90
3520	Off site storage			15.40	15.40
3540	Clay, excavation, bank run gravel borrow for backfill		9.90	11.60	21.50
4560	10,000 S.F., 4′ deep, sand gravel, or common earth, on site storage			.24	.24
4580	Off site storage			.48	.48
4600	Clay excavation, bank run gravel borrow for backfill		.62	.55	1.17
4620	8′ deep, sand, gravel, or common earth, on site storage			2.77	2.77
4640	Off site storage			4.20	4.20
4660	Clay excavation, bank run gravel borrow for backfill		2.23	3.78	6.01
4680	16′ deep, sand, gravel, or common earth, on site storage			6.40	6.40
4700	Off site storage			10.75	10.75
4720	Clay excavation, bank run gravel borrow for backfill		5.95	8.60	14.55
5740	30,000 S.F., 4′ deep, sand, gravel, or common earth, on site storage			.14	.14

Important: See the Reference Section for critical supporting data - Reference Numbers and City Cost Indexes

A2010 Basement Excavation

A2010 110	Building Excavation & Backfill	COST PER S.F.		
		MAT.	INST.	TOTAL
5760	Off site storage		.28	.28
5780	Clay excavation, bank run gravel borrow for backfill	.36	.32	.68
5800	8' deep, sand & gravel, or common earth, on site storage		2.47	2.47
5820	Off site storage		3.27	3.27
5840	Clay excavation, bank run gravel borrow for backfill	1.26	3.05	4.31
5860	16' deep, sand, gravel, or common earth, on site storage		5.45	5.45
5880	Off site storage		7.85	7.85
5900	Clay excavation, bank run gravel borrow for backfill	3.29	6.70	9.99
6910	100,000 S.F., 4' deep, sand, gravel, or common earth, on site storage		.08	.08
6920	Off site storage		.16	.16
6930	Clay excavation, bank run gravel borrow for backfill	.18	.16	.34
6940	8' deep, sand, gravel, or common earth, on site storage		2.30	2.30
6950	Off site storage		2.73	2.73
6960	Clay excavation, bank run gravel borrow for backfill	.68	2.62	3.30
6970	16' deep, sand, gravel, or common earth, on site storage		4.92	4.92
6980	Off site storage		6.20	6.20
6990	Clay excavation, bank run gravel borrow for backfill	1.78	5.55	7.33

SUBSTRUCTURE

A

A2020 Basement Walls

The Foundation Bearing Wall System includes: forms up to 16' high (four uses); 3,000 p.s.i. concrete placed and vibrated; and form removal with breaking form ties and patching walls. The wall systems list walls from 6" to 16" thick and are designed with minimum reinforcement.

Excavation and backfill are not included.

Please see the reference section for further design and cost information.

System Components			COST PER L.F.		
	QUANTITY	UNIT	MAT.	INST.	TOTAL
SYSTEM A2020 110 1500					
FOUNDATION WALL, CAST IN PLACE, DIRECT CHUTE, 4' HIGH, 6" THICK					
Formwork	8.000	SFCA	3.60	34.40	38
Reinforcing	3.300	Lb.	1.48	1.15	2.63
Unloading & sorting reinforcing	3.300	Lb.		.07	.07
Concrete, 3,000 psi	.074	C.Y.	7.07		7.07
Place concrete, direct chute	.074	C.Y.		1.82	1.82
Finish walls, break ties and patch voids, one side	4.000	S.F.	.12	3	3.12
TOTAL			12.27	40.44	52.71

A2020 110			Walls, Cast in Place						
	WALL HEIGHT (FT.)	PLACING METHOD	CONCRETE (C.Y. per L.F.)	REINFORCING (LBS. per L.F.)	WALL THICKNESS (IN.)	COST PER L.F.			
						MAT.	INST.	TOTAL	
1500	4'	direct chute	.074	3.3	6	12.25	40.50	52.75	
1520			.099	4.8	8	15.30	41.50	56.80	
1540			.123	6.0	10	18.15	42	60.15	
1560			.148	7.2	12	21	43	64	
1580	RA2020-210		.173	8.1	14	24	44	68	
1600			.197	9.44	16	27	45	72	
1700	4'	pumped	.074	3.3	6	12.25	41.50	53.75	
1720			.099	4.8	8	15.30	43	58.30	
1740			.123	6.0	10	18.15	44	62.15	
1760			.148	7.2	12	21	45	66	
1780			.173	8.1	14	24	45.50	69.50	
1800			.197	9.44	16	27	47	74	
3000	6'	direct chute	.111	4.95	6	18.40	60.50	78.90	
3020			.149	7.20	8	23	62.50	85.50	
3040			.184	9.00	10	27	63.50	90.50	
3060			.222	10.8	12	31.50	65	96.50	
3080			.260	12.15	14	36	66	102	
3100			.300	14.39	16	40.50	67.50	108	

A2020 Basement Walls

A2020 110		Walls, Cast in Place					

	WALL HEIGHT (FT.)	PLACING METHOD	CONCRETE (C.Y. per L.F.)	REINFORCING (LBS. per L.F.)	WALL THICKNESS (IN.)	COST PER L.F.		
						MAT.	INST.	TOTAL
3200	6'	pumped	.111	4.95	6	18.40	62	80.40
3220			.149	7.20	8	23	64.50	87.50
3240			.184	9.00	10	27	66	93
3260			.222	10.8	12	31.50	67.50	99
3280			.260	12.15	14	36	69	105
3300			.300	14.39	16	40.50	70.50	111
5000	8'	direct chute	.148	6.6	6	24.50	81	105.50
5020			.199	9.6	8	30.50	83	113.50
5040			.250	12	10	36.50	84.50	121
5060			.296	14.39	12	42	86.50	128.50
5080			.347	16.19	14	44	87	131
5100			.394	19.19	16	53.50	90.50	144
5200	8'	pumped	.148	6.6	6	24.50	82.50	107
5220			1.99	9.6	8	30.50	85.50	116
5240			.250	12	10	36.50	88	124.50
5260			.296	14.39	12	42	90.50	132.50
5280			.347	16.19	14	44	90	134
5300			.394	19.19	16	53.50	94.50	148
6020	10'	direct chute	.248	12	8	38.50	104	142.50
6040			.307	14.99	10	45.50	106	151.50
6060			.370	17.99	12	52.50	108	160.50
6080			.433	20.24	14	59.50	110	169.50
6100			.493	23.99	16	67	112	179
6220	10'	pumped	.248	12	8	38.50	107	145.50
6240			.307	14.99	10	45.50	109	154.50
6260			.370	17.99	12	52.50	113	165.50
6280			.433	20.24	14	59.50	115	174.50
6300			.493	23.99	16	67	117	184
7220	12'	pumped	.298	14.39	8	46	129	175
7240			.369	17.99	10	54.50	132	186.50
7260			.444	21.59	12	63	135	198
7280			.52	24.29	14	71.50	138	209.50
7300			.591	28.79	16	80.50	141	221.50
7420	12'	crane & bucket	.298	14.39	8	46	134	180
7440			.369	17.99	10	54.50	137	191.50
7460			.444	21.59	12	63	143	206
7480			.52	24.29	14	71.50	145	216.50
7500			.591	28.79	16	80.50	152	232.50
8220	14'	pumped	.347	16.79	8	53.50	150	203.50
8240			.43	20.99	10	63.50	153	216.50
8260			.519	25.19	12	74	158	232
8280			.607	28.33	14	83.50	160	243.50
8300			.69	33.59	16	94	165	259
8420	14'	crane & bucket	.347	16.79	8	53.50	156	209.50
8440			.43	20.99	10	63.50	159	222.50
8460			.519	25.19	12	74	166	240
8480			.607	28.33	14	83.50	169	252.50
8500			.69	33.59	16	94	175	269
9220	16'	pumped	.397	19.19	8	61.50	171	232.50
9240			.492	23.99	10	72.50	175	247.50
9260			.593	28.79	12	84.50	181	265.50
9280			.693	32.39	14	95.50	183	278.50
9300			.788	38.38	16	107	189	296

SUBSTRUCTURE

A

A2020 Basement Walls

A2020 110	Walls, Cast in Place						
	WALL HEIGHT (FT.)	PLACING METHOD	CONCRETE (C.Y. per L.F.)	REINFORCING (LBS. per L.F.)	WALL THICKNESS (IN.)	COST PER L.F.		
						MAT.	INST.	TOTAL
9420	16'	crane & bucket	.397	19.19	8	61.50	179	240.50
9440			.492	23.99	10	72.50	183	255.50
9460			.593	28.79	12	84.50	189	273.50
9480			.693	32.39	14	95.50	194	289.50
9500			.788	38.38	16	107	200	307

Important: See the Reference Section for critical supporting data - Reference Numbers and City Cost Indexes

A20 Basement Construction

A2020 Basement Walls

A2020 220	Subdrainage Piping	COST PER L.F.		
		MAT.	INST.	TOTAL
2000	Piping, excavation & backfill excluded, PVC, perforated			
2110	3" diameter	2.78	3.10	5.88
2130	4" diameter	2.78	3.10	5.88
2140	5" diameter	5.85	3.32	9.17
2150	6" diameter	5.85	3.32	9.17
3000	Metal alum. or steel, perforated asphalt coated			
3150	6" diameter	3.34	6.30	9.64
3160	8" diameter	4.77	6.50	11.27
3170	10" diameter	5.95	6.70	12.65
3180	12" diameter	6.70	8.40	15.10
3220	18" diameter	10.05	11.70	21.75
4000	Porous wall concrete			
4130	4" diameter	2.26	3.47	5.73
4150	6" diameter	2.93	3.69	6.62
4160	8" diameter	3.62	5.05	8.67
4180	12" diameter	7.65	5.45	13.10
4200	15" diameter	8.80	6.80	15.60
4220	18" diameter	11.60	9.45	21.05
5000	Vitrified clay C-211, perforated			
5130	4" diameter	2.38	6	8.38
5150	6" diameter	3.93	7.65	11.58
5160	8" diameter	5.60	8.30	13.90
5180	12" diameter	11.80	8.75	20.55

For information about Means Estimating Seminars, see yellow pages 12 and 13 in back of book

SUBSTRUCTURE

A

B SHELL

CONCRETE COLUMNS

General: It is desirable for purposes of consistency and simplicity to maintain constant column sizes throughout the building height. To do this, concrete strength may be varied (higher strength concrete at lower stories and lower strength concrete at upper stories), as well as varying the amount of reinforcing.

The first portion of the table provides probable minimum column sizes with related costs and weights per lineal foot of story height for bottom level columns.

The second portion of the table provides costs by column size for top level columns with minimum code reinforcement. Probable maximum loads for these columns are also given.

How to Use Table:

1. Enter the second portion (minimum reinforcing) of the table with the minimum allowable column size from the selected cast in place floor system.

   If the total load on the column does not exceed the allowable working load shown, use the cost per L.F. multiplied by the length of columns required to obtain the column cost.

2. If the total load on the column exceeds the allowable working load shown in the second portion of the table, enter the first portion of the table with the total load on the column and the minimum allowable column size from the selected cast in place floor system.

Select a cost per L.F. for bottom level columns by total load or minimum allowable column size.

Select a cost per L.F. for top level columns using the column size required for bottom level columns from the second portion of the table.

$$\frac{\text{Btm. + Top Col. Costs/L.F.}}{2} = \text{Avg. Col. Cost/L.F.}$$

Column Cost = Average Col. Cost/L.F. x Length of Cols. Required.

See reference section to determine total loads.

Design and Pricing Assumptions:

Normal wt. concrete, f'c = 4 or 6 KSI, placed by pump.
Steel, fy = 60 KSI, spliced every other level.
Minimum design eccentricity of 0.1t.
Assumed load level depth is 8″ (weights prorated to full story basis).
Gravity loads only (no frame or lateral loads included).

Please see the reference section for further design and cost information.

System Components	QUANTITY	UNIT	COST PER V.L.F.		
			MAT.	INST.	TOTAL
SYSTEM B1010 201 1050					
ROUND TIED COLUMNS, 4 KSI CONCRETE, 100K MAX. LOAD, 10′ STORY, 12″ SIZE					
Forms in place, columns, round fiber tube, 12″ diam 1 use	1.000	L.F.	2.55	11.15	13.70
Reinforcing in place, column ties	1.393	Lb.	3.24	4.56	7.80
Concrete ready mix, regular weight, 4000 psi	.029	C.Y.	2.90		2.90
Placing concrete, incl. vibrating, 12″ sq./round columns, pumped	.029	C.Y.		1.81	1.81
Finish, burlap rub w/grout	3.140	S.F.	.09	2.83	2.92
TOTAL			8.78	20.35	29.13

B1010 201		C.I.P. Column - Round Tied						
	LOAD (KIPS)	STORY HEIGHT (FT.)	COLUMN SIZE (IN.)	COLUMN WEIGHT (P.L.F.)	CONCRETE STRENGTH (PSI)	COST PER V.L.F.		
						MAT.	INST.	TOTAL
1050	100	10	12	110	4000	8.80	20.50	29.30
1060	RB1010-112	12	12	111	4000	8.65	20	28.65
1070		14	12	112	4000	8.65	20	28.65
1080	150	10	12	110	4000	13.55	27	40.55
1090		12	12	111	4000	14.05	28	42.05
1100		14	14	153	4000	11.65	23	34.65
1120	200	10	14	150	4000	14	26.50	40.50
1140		12	14	152	4000	13.90	26.50	40.40

B1010 Floor Construction

B1010 201			C.I.P. Column - Round Tied					

	LOAD (KIPS)	STORY HEIGHT (FT.)	COLUMN SIZE (IN.)	COLUMN WEIGHT (P.L.F.)	CONCRETE STRENGTH (PSI)	COST PER V.L.F.		
						MAT.	INST.	TOTAL
1160	200	14	14	153	4000	18.35	32.50	50.85
1180	300	10	16	194	4000	22.50	37	59.50
1190		12	18	250	4000	19.20	28	47.20
1200		14	18	252	4000	17.85	26.50	44.35
1220	400	10	20	306	4000	22	31	53
1230		12	20	310	4000	24	33.50	57.50
1260		14	20	313	4000	24	33.50	57.50
1280	500	10	22	368	4000	28	36.50	64.50
1300		12	22	372	4000	30	39.50	69.50
1325		14	22	375	4000	33	43.50	76.50
1350	600	10	24	439	4000	32	40.50	72.50
1375		12	24	445	4000	31.50	39.50	71
1400		14	24	448	4000	35	44.50	79.50
1420	700	10	26	517	4000	42.50	49	91.50
1430		12	26	524	4000	41.50	47.50	89
1450		14	26	528	4000	41	46.50	87.50
1460	800	10	28	596	4000	44.50	50.50	95
1480		12	28	604	4000	43.50	49	92.50
1490		14	28	609	4000	43	48.50	91.50
1500	900	10	28	596	4000	53.50	63	116.50
1510		12	28	604	4000	52	61	113
1520		14	28	609	4000	51	60	111
1530	1000	10	30	687	4000	54.50	60	114.50
1540		12	30	695	4000	53.50	58.50	112
1620		14	30	701	4000	52.50	57	109.50
1640	100	10	12	110	6000	8.65	20	28.65
1660		12	12	111	6000	8.60	20	28.60
1680		14	12	112	6000	8.50	19.95	28.45
1700	150	10	12	110	6000	8.65	20	28.65
1710		12	12	111	6000	8.55	20	28.55
1720		14	12	112	6000	7.25	18.20	25.45
1730	200	10	12	110	6000	12.65	26	38.65
1740		12	12	111	6000	13.95	27.50	41.45
1760		14	12	112	6000	18.70	34.50	53.20
1780	300	10	14	150	6000	14.90	27.50	42.40
1790		12	14	152	6000	17.55	31.50	49.05
1800		14	16	153	6000	14	25	39
1810	400	10	16	194	6000	21	35	56
1820		12	16	196	6000	24.50	39.50	64
1830		14	18	252	6000	20	29.50	49.50
1850	500	10	18	247	6000	23	33	56
1870		12	18	250	6000	28	40.50	68.50
1880		14	20	252	6000	22	30.50	52.50
1890	600	10	20	306	6000	22.50	31.50	54
1900		12	20	310	6000	24.50	34.50	59
1905		14	20	313	6000	29	41	70
1910	700	10	22	368	6000	28	36.50	64.50
1915		12	22	372	6000	30.50	39.50	70
1920		14	22	375	6000	33	43	76
1925	800	10	24	439	6000	33.50	42	75.50
1930		12	24	445	6000	32.50	41	73.50
1935		14	24	448	6000	33	41.50	74.50

B1010   Floor Construction

B1010 201		C.I.P. Column - Round Tied						
	LOAD (KIPS)	STORY HEIGHT (FT.)	COLUMN SIZE (IN.)	COLUMN WEIGHT (P.L.F.)	CONCRETE STRENGTH (PSI)	COST PER V.L.F.		
						MAT.	INST.	TOTAL
1940	900	10	24	439	6000	70.50	94.50	165
1945		12	24	445	6000	36	46	82
1970	1000	10	26	517	6000	43	49.50	92.50
1980		12	26	524	6000	42	48	90
1995		14	26	528	6000	41	47	88

B1010 202		C.I.P. Columns, Round Tied - Minimum Reinforcing						
	LOAD (KIPS)	STORY HEIGHT (FT.)	COLUMN SIZE (IN.)	COLUMN WEIGHT (P.L.F.)	CONCRETE STRENGTH (PSI)	COST PER V.L.F.		
						MAT.	INST.	TOTAL
2500	100	10-14	12	107	4000	8.45	19.90	28.35
2510	200	10-14	16	190	4000	14.20	25	39.20
2520	400	10-14	20	295	4000	21	29.50	50.50
2530	600	10-14	24	425	4000	30	37.50	67.50
2540	800	10-14	28	580	4000	40.50	44.50	85
2550	1100	10-14	32	755	4000	53	52.50	105.50
2560	1400	10-14	36	960	4000	62.50	60.50	123
2570								

B

SHELL

Important: See the Reference Section for critical supporting data - Reference Numbers and City Cost Indexes

CONCRETE COLUMNS

General: It is desirable for purposes of consistency and simplicity to maintain constant column sizes throughout the building height. To do this, concrete strength may be varied (higher strength concrete at lower stories and lower strength concrete at upper stories), as well as varying the amount of reinforcing.

The first portion of the table provides probable minimum column sizes with related costs and weights per lineal foot of story height for bottom level columns.

The second portion of the table provides costs by column size for top level columns with minimum code reinforcement. Probable maximum loads for these columns are also given.

How to Use Table:

1. Enter the second portion (minimum reinforcing) of the table with the minimum allowable column size from the selected cast in place floor system.

   If the total load on the column does not exceed the allowable working load shown, use the cost per L.F. multiplied by the length of columns required to obtain the column cost.

2. If the total load on the column exceeds the allowable working load shown in the second portion of the table, enter the first portion of the

table with the total load on the column and the minimum allowable column size from the selected cast in place floor system.

Select a cost per L.F. for bottom level columns by total load or minimum allowable column size.

Select a cost per L.F. for top level columns using the column size required for bottom level columns from the second portion of the table.

$$\frac{\text{Btm.} + \text{Top Col. Costs/L.F.}}{2} = \text{Avg. Col. Cost/L.F}$$

Column Cost = Average Col. Cost/L.F. x Length of Cols. Required.

See reference section in back of book to determine total loads.

Design and Pricing Assumptions:

Normal wt. concrete, f'c = 4 or 6 KSI, placed by pump.
Steel, fy = 60 KSI, spliced every other level.
Minimum design eccentricity of 0.1t.
Assumed load level depth is 8" (weights prorated to full story basis).
Gravity loads only (no frame or lateral loads included).

Please see the reference section for further design and cost information.

System Components			COST PER V.L.F.		
	QUANTITY	UNIT	MAT.	INST.	TOTAL
SYSTEM B1010 203 0640					
SQUARE COLUMNS, 100K LOAD,10' STORY, 10" SQUARE					
Forms in place, columns, plywood, 10" x 10", 4 uses	3.323	SFCA	2.60	24.66	27.26
Chamfer strip,wood, 3/4" wide	4.000	L.F.	.44	3.36	3.80
Reinforcing in place, column ties	1.405	Lb.	2.94	4.14	7.08
Concrete ready mix, regular weight, 4000 psi	.026	C.Y.	2.60		2.60
Placing concrete, incl. vibrating, 12" sq./round columns, pumped	.026	C.Y.		1.61	1.61
Finish, break ties, patch voids, burlap rub w/grout	3.323	S.F.	.10	3	3.10
TOTAL			8.68	36.77	45.45

B1010 203				C.I.P. Column, Square Tied				
	LOAD (KIPS)	STORY HEIGHT (FT.)	COLUMN SIZE (IN.)	COLUMN WEIGHT (P.L.F.)	CONCRETE STRENGTH (PSI)	COST PER V.L.F.		
						MAT.	INST.	TOTAL
0640	100	10	10	96	4000	8.70	37	45.70
0680	RB1010 -112	12	10	97	4000	8.45	36.50	44.95
0700		14	12	142	4000	11.40	44.50	55.90
0710								

B1010 Floor Construction

B1010 203	C.I.P. Column, Square Tied

	LOAD (KIPS)	STORY HEIGHT (FT.)	COLUMN SIZE (IN.)	COLUMN WEIGHT (P.L.F.)	CONCRETE STRENGTH (PSI)	COST PER V.L.F.		
						MAT.	INST.	TOTAL
0740	150	10	10	96	4000	8.70	37	45.70
0780		12	12	142	4000	11.45	44.50	55.95
0800		14	12	143	4000	11.40	44.50	55.90
0840	200	10	12	140	4000	11.55	44.50	56.05
0860		12	12	142	4000	11.45	44.50	55.95
0900		14	14	196	4000	14.75	51.50	66.25
0920	300	10	14	192	4000	15.20	52	67.20
0960		12	14	194	4000	15	51.50	66.50
0980		14	16	253	4000	17.05	56.50	73.55
1020	400	10	16	248	4000	18.90	59	77.90
1060		12	16	251	4000	18.65	58.50	77.15
1080		14	16	253	4000	18.50	58.50	77
1200	500	10	18	315	4000	26	70.50	96.50
1250		12	20	394	4000	26	73.50	99.50
1300		14	20	397	4000	26	73	99
1350	600	10	20	388	4000	31	80.50	111.50
1400		12	20	394	4000	30.50	80	110.50
1600		14	20	397	4000	30.50	79.50	110
1900	700	10	20	388	4000	31.50	83	114.50
2100		12	22	474	4000	31	82.50	113.50
2300		14	22	478	4000	31	82	113
2600	800	10	22	388	4000	46	103	149
2900		12	22	474	4000	46	103	149
3200		14	22	478	4000	45.50	102	147.50
3400	900	10	24	560	4000	45	104	149
3800		12	24	567	4000	44.50	103	147.50
4000		14	24	571	4000	44	102	146
4250	1000	10	24	560	4000	56	119	175
4500		12	26	667	4000	47	108	155
4750		14	26	673	4000	46.50	107	153.50
5600	100	10	10	96	6000	8.85	37	45.85
5800		12	10	97	6000	8.60	36.50	45.10
6000		14	12	142	6000	11.45	44.50	55.95
6200	150	10	10	96	6000	8.65	37	45.65
6400		12	12	98	CONCRETE	11.45	44.50	55.95
6600		14	12	143	6000	11.45	44.50	55.95
6800	200	10	12	140	6000	12.15	45.50	57.65
7000		12	12	142	6000	11.45	44.50	55.95
7100		14	14	196	6000	14.75	51.50	66.25
7300	300	10	14	192	6000	15	51.50	66.50
7500		12	14	194	6000	14.85	51.50	66.35
7600		14	14	196	6000	14.75	51.50	66.25
7700	400	10	14	192	6000	15	51.50	66.50
7800		12	14	194	6000	14.85	51.50	66.35
7900		14	16	253	6000	18.50	58	76.50
8000	500	10	16	248	6000	18.90	59	77.90
8050		12	16	251	6000	18.65	58.50	77.15
8100		14	16	253	6000	18.50	58.50	77
8200	600	10	18	315	6000	23	68	91
8300		12	18	319	6000	23	68	91
8400		14	18	321	6000	22.50	67.50	90

Important: See the Reference Section for critical supporting data - Reference Numbers and City Cost Indexes

B1010 Floor Construction

| B1010 203 | | C.I.P. Column, Square Tied | | | | | | |

	LOAD (KIPS)	STORY HEIGHT (FT.)	COLUMN SIZE (IN.)	COLUMN WEIGHT (P.L.F.)	CONCRETE STRENGTH (PSI)	COST PER V.L.F.		
						MAT.	INST.	TOTAL
8500	700	10	18	315	6000	25.50	71.50	97
8600		12	18	319	6000	25	71	96
8700		14	18	321	6000	24.50	70.50	95
8800	800	10	20	388	6000	26.50	74	100.50
8900		12	20	394	6000	26	73.50	99.50
9000		14	20	397	6000	26	73	99
9100	900	10	20	388	6000	39	92	131
9300		12	20	394	6000	38.50	91	129.50
9600		14	20	397	6000	38	90	128
9800	1000	10	22	469	6000	34	86.50	120.50
9840		12	22	474	6000	33.50	85.50	119
9900		14	22	478	6000	34.50	86.50	121

| B1010 204 | | C.I.P. Column, Square Tied-Minimum Reinforcing | | | | | | |

	LOAD (KIPS)	STORY HEIGHT (FT.)	COLUMN SIZE (IN.)	COLUMN WEIGHT (P.L.F.)	CONCRETE STRENGTH (PSI)	COST PER V.L.F.		
						MAT.	INST.	TOTAL
9913	150	10-14	12	135	4000	11.15	44	55.15
9918	300	10-14	16	240	4000	16.90	56	72.90
9924	500	10-14	20	375	4000	25.50	72.50	98
9930	700	10-14	24	540	4000	37.50	93.50	131
9936	1000	10-14	28	740	4000	47.50	111	158.50
9942	1400	10-14	32	965	4000	62	129	191
9948	1800	10-14	36	1220	4000	75.50	148	223.50
9954	2300	10-14	40	1505	4000	86	162	248

SHELL **B**

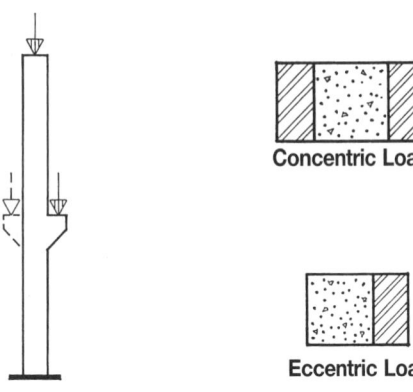

Concentric Load

Eccentric Load

General: Data presented here is for plant produced members transported 50 miles to 100 miles to the site and erected.

Design and pricing assumptions:
Normal wt. concrete, f'c = 5 KSI

Main reinforcement, fy = 60 KSI
Ties, fy = 40 KSI

Minimum design eccentricity, 0.1t.

Concrete encased structural steel haunches are assumed where practical; otherwise galvanized rebar haunches are assumed.

Base plates are integral with columns.

Foundation anchor bolts, nuts and washers are included in price.

System Components	QUANTITY	UNIT	COST PER V.L.F.		
			MAT.	INST.	TOTAL
SYSTEM B1010 206 0560					
PRECAST TIED COLUMN, CONCENTRIC LOADING, 100 K MAX.					
TWO STORY - 10' PER STORY, 5 KSI CONCRETE, 12"X12"					
Precast column, two story-10'/story, 5 KSI conc., 12"x12"	1.000	Ea.	47		47
Anchor bolts, 3/4" diameter x 24" long	.200	Ea.	.72		.72
Steel bearing plates; top, bottom, haunches	3.250	Lb.	3.74		3.74
Erection crew	.013	Hr.		9.92	9.92
TOTAL			51.46	9.92	61.38

B1010 206		Tied, Concentric Loaded Precast Concrete Columns						
	LOAD (KIPS)	STORY HEIGHT (FT.)	COLUMN SIZE (IN.)	COLUMN WEIGHT (P.L.F.)	LOAD LEVELS	COST PER V.L.F.		
						MAT.	INST.	TOTAL
0560	100	10	12x12	164	2	51.50	9.95	61.45
0570		12	12x12	162	2	47.50	8.25	55.75
0580		14	12x12	161	2	47	8.25	55.25
0590	150	10	12x12	166	3	47.50	8.25	55.75
0600		12	12x12	169	3	47	7.45	54.45
0610		14	12x12	162	3	47.50	7.45	54.95
0620	200	10	12x12	168	4	47	9.10	56.10
0630		12	12x12	170	4	47	8.25	55.25
0640		14	12x12	220	4	47	8.25	55.25
0680	300	10	14x14	225	3	64	8.25	72.25
0690		12	14x14	225	3	64.50	7.45	71.95
0700		14	14x14	250	3	64	7.45	71.45
0710	400	10	16x16	255	4	68	9.10	77.10
0720		12	16x16	295	4	67.50	8.25	75.75
0750		14	16x16	305	4	54.50	8.25	62.75
0790	450	10	16x16	320	3	68	8.25	76.25
0800		12	16x16	315	3	68	7.45	75.45
0810		14	16x16	330	3	67.50	7.45	74.95
0820	600	10	18x18	405	4	85.50	9.10	94.60
0830		12	18x18	395	4	85.50	8.25	93.75
0840		14	18x18	410	4	85.50	8.25	93.75
0910	800	10	20x20	495	4	82	9.10	91.10
0920		12	20x20	505	4	82	8.25	90.25
0930		14	20x20	510	4	82	8.25	90.25

B10 Superstructure

B1010 Floor Construction

B1010 206 — Tied, Concentric Loaded Precast Concrete Columns

	LOAD (KIPS)	STORY HEIGHT (FT.)	COLUMN SIZE (IN.)	COLUMN WEIGHT (P.L.F.)	LOAD LEVELS	COST PER V.L.F.		
						MAT.	INST.	TOTAL
0970	900	10	22x22	625	3	99.50	8.25	107.75
0980		12	22x22	610	3	99.50	7.45	106.95
0990		14	22x22	605	3	99	7.45	106.45

B1010 207 — Tied, Eccentric Loaded Precast Concrete Columns

	LOAD (KIPS)	STORY HEIGHT (FT.)	COLUMN SIZE (IN.)	COLUMN WEIGHT (P.L.F.)	LOAD LEVELS	COST PER V.L.F.		
						MAT.	INST.	TOTAL
1130	100	10	12x12	161	2	46.50	9.95	56.45
1140		12	12x12	159	2	46.50	8.25	54.75
1150		14	12x12	159	2	46	8.25	54.25
1160	150	10	12x12	161	3	45.50	8.25	53.75
1170		12	12x12	160	3	45.50	7.45	52.95
1180		14	12x12	159	3	45.50	7.45	52.95
1190	200	10	12x12	161	4	45.50	9.10	54.60
1200		12	12x12	160	4	45	8.25	53.25
1210		14	12x12	177	4	45.50	8.25	53.75
1250	300	10	12x12	185	3	62	8.25	70.25
1260		12	14x14	215	3	62	7.45	69.45
1270		14	14x14	215	3	62.50	7.45	69.95
1280	400	10	14x14	235	4	61.50	9.10	70.60
1290		12	16x16	285	4	61	8.25	69.25
1300		14	16x16	295	4	61.50	8.25	69.75
1360	450	10	14x14	245	3	65.50	8.25	73.75
1370		12	16x16	285	3	65.50	7.45	72.95
1380		14	16x16	290	3	66	7.45	73.45
1390	600	10	18x18	385	4	82	9.10	91.10
1400		12	18x18	380	4	82	8.25	90.25
1410		14	18x18	375	4	82	8.25	90.25
1480	800	10	20x20	490	4	79	9.10	88.10
1490		12	20x20	480	4	79	8.25	87.25
1500		14	20x20	475	4	79	8.25	87.25

SHELL B

B1010 Floor Construction

(A) Wide Flange

(B) Pipe

(C) Pipe, Concrete Filled

(D) Square Tube

(E) Square Tube
Concrete Filled

(F) Rectangular Tube

(G) Rectangular Tube, Concrete Filled

General: The following pages provide data for seven types of steel columns: wide flange, round pipe, round pipe concrete filled, square tube, square tube concrete filled, rectangular tube and rectangular tube concrete filled.

Design Assumptions: Loads are concentric; wide flange and round pipe bearing capacity is for 36 KSI steel. Square and rectangular tubing bearing capacity is for 46 KSI steel.

The effective length factor K=1.1 is used for determining column values in the tables. K=1.1 is within a frequently used range for pinned connections with cross bracing.

How To Use Tables:
a. Steel columns usually extend through two or more stories to minimize splices. Determine floors with splices.
b. Enter Table No. below with load to column at the splice. Use the unsupported height.
c. Determine the column type desired by price or design.

Cost:
a. Multiply number of columns at the desired level by the total height of the column by the cost/VLF.
b. Repeat the above for all tiers.

Please see the reference section for further design and cost information.

B1010 208 — Steel Columns

	LOAD (KIPS)	UNSUPPORTED HEIGHT (FT.)	WEIGHT (P.L.F.)	SIZE (IN.)	TYPE	COST PER V.L.F.		
						MAT.	INST.	TOTAL
1000	25	10	13	4	A	16.50	8.90	25.40
1020	RB1010 -130		7.58	3	B	9.65	8.90	18.55
1040			15	3-1/2	C	11.75	8.90	20.65
1060			6.87	3	D	8.75	8.90	17.65
1080			15	3	E	11.35	8.90	20.25
1100			8.15	4x3	F	10.35	8.90	19.25
1120			20	4x3	G	13.60	8.90	22.50
1200		16	16	5	A	18.80	6.65	25.45
1220			10.79	4	B	12.70	6.65	19.35
1240			36	5-1/2	C	17.75	6.65	24.40
1260			11.97	5	D	14.10	6.65	20.75
1280			36	5	E	18.90	6.65	25.55
1300			11.97	6x4	F	14.10	6.65	20.75
1320			64	8x6	G	27.50	6.65	34.15
1400		20	20	6	A	22.50	6.65	29.15
1420			14.62	5	B	16.25	6.65	22.90
1440			49	6-5/8	C	22	6.65	28.65
1460			11.97	5	D	13.30	6.65	19.95
1480			49	6	E	22	6.65	28.65
1500			14.53	7x5	F	16.15	6.65	22.80
1520			64	8x6	G	26	6.65	32.65
1600	50	10	16	5	A	20.50	8.90	29.40
1620			14.62	5	B	18.55	8.90	27.45
1640			24	4-1/2	C	14.05	8.90	22.95
1660			12.21	4	D	15.50	8.90	24.40
1680			25	4	E	15.85	8.90	24.75
1700			11.97	6x4	F	15.20	8.90	24.10
1720			28	6x3	G	18.05	8.90	26.95

Important: See the Reference Section for critical supporting data - Reference Numbers and City Cost Indexes

B1010 Floor Construction

B1010 208	Steel Columns

	LOAD (KIPS)	UNSUPPORTED HEIGHT (FT.)	WEIGHT (P.L.F.)	SIZE (IN.)	TYPE	COST PER V.L.F.		
						MAT.	INST.	TOTAL
1800	50	16	24	8	A	28	6.65	34.65
1820			18.97	6	B	22.50	6.65	29.15
1840			36	5-1/2	C	17.75	6.65	24.40
1860			14.63	6	D	17.20	6.65	23.85
1880			36	5	E	18.90	6.65	25.55
1900			14.53	7x5	F	17.10	6.65	23.75
1920			64	8x6	G	27.50	6.65	34.15
1940								
2000		20	28	8	A	31	6.65	37.65
2020			18.97	6	B	21	6.65	27.65
2040			49	6-5/8	C	22	6.65	28.65
2060			19.02	6	D	21	6.65	27.65
2080			49	6	E	22	6.65	28.65
2100			22.42	8x6	F	25	6.65	31.65
2120			64	8x6	G	26	6.65	32.65
2200	75	10	20	6	A	25.50	8.90	34.40
2220			18.97	6	B	24	8.90	32.90
2240			36	4-1/2	C	35	8.90	43.90
2260			14.53	6	D	18.45	8.90	27.35
2280			28	4	E	15.80	8.90	24.70
2300			14.33	7x5	F	18.20	8.90	27.10
2320			35	6x4	G	20.50	8.90	29.40
2400		16	31	8	A	36.50	6.65	43.15
2420			28.55	8	B	33.50	6.65	40.15
2440			49	6-5/8	C	23	6.65	29.65
2460			17.08	7	D	20	6.65	26.65
2480			36	5	E	18.90	6.65	25.55
2500			23.34	7x5	F	27.50	6.65	34.15
2520			64	8x6	G	27.50	6.65	34.15
2600		20	31	8	A	34.50	6.65	41.15
2620			28.55	8	B	32	6.65	38.65
2640			81	8-5/8	C	33	6.65	39.65
2660			22.42	7	D	25	6.65	31.65
2680			49	6	E	22	6.65	28.65
2700			22.42	8x6	F	25	6.65	31.65
2720			64	8x6	G	26	6.65	32.65
2800	100	10	24	8	A	30.50	8.90	39.40
2820			28.57	6	B	36.50	8.90	45.40
2840			35	4-1/2	C	35	8.90	43.90
2860			17.08	7	D	21.50	8.90	30.40
2880			36	5	E	20.50	8.90	29.40
2900			19.02	7x5	F	24	8.90	32.90
2920			46	8x4	G	25	8.90	33.90
3000		16	31	8	A	36.50	6.65	43.15
3020			28.55	8	B	33.50	6.65	40.15
3040			56	6-5/8	C	34.50	6.65	41.15
3060			22.42	7	D	26.50	6.65	33.15
3080			49	6	E	23	6.65	29.65
3100			22.42	8x6	F	26.50	6.65	33.15
3120			64	8x6	G	27.50	6.65	34.15

B1010 Floor Construction

| B1010 208 | Steel Columns |

	LOAD (KIPS)	UNSUPPORTED HEIGHT (FT.)	WEIGHT (P.L.F.)	SIZE (IN.)	TYPE	COST PER V.L.F.		
						MAT.	INST.	TOTAL
3200	100	20	40	8	A	44.50	6.65	51.15
3220			28.55	8	B	32	6.65	38.65
3240			81	8-5/8	C	33	6.65	39.65
3260			25.82	8	D	28.50	6.65	35.15
3280			66	7	E	26	6.65	32.65
3300			27.59	8x6	F	30.50	6.65	37.15
3320			70	8x6	G	37.50	6.65	44.15
3400	125	10	31	8	A	39.50	8.90	48.40
3420			28.57	6	B	36.50	8.90	45.40
3440			81	8	C	37.50	8.90	46.40
3460			22.42	7	D	28.50	8.90	37.40
3480			49	6	E	25	8.90	33.90
3500			22.42	8x6	F	28.50	8.90	37.40
3520			64	8x6	G	29.50	8.90	38.40
3600		16	40	8	A	47	6.65	53.65
3620			28.55	8	B	33.50	6.65	40.15
3640			81	8	C	35	6.65	41.65
3660			25.82	8	D	30.50	6.65	37.15
3680			66	7	E	27.50	6.65	34.15
3700			27.59	8x6	F	32.50	6.65	39.15
3720			64	8x6	G	27.50	6.65	34.15
3800		20	48	8	A	53.50	6.65	60.15
3820			40.48	10	B	45	6.65	51.65
3840			81	8	C	33	6.65	39.65
3860			25.82	8	D	28.50	6.65	35.15
3880			66	7	E	26	6.65	32.65
3900			37.59	10x6	F	42	6.65	48.65
3920			60	8x6	G	37.50	6.65	44.15
4000	150	10	35	8	A	44.50	8.90	53.40
4020			40.48	10	B	51.50	8.90	60.40
4040			81	8-5/8	C	37.50	8.90	46.40
4060			25.82	8	D	33	8.90	41.90
4080			66	7	E	29.50	8.90	38.40
4100			27.48	7x5	F	35	8.90	43.90
4120			64	8x6	G	29.50	8.90	38.40
4200		16	45	10	A	53	6.65	59.65
4220			40.48	10	B	47.50	6.65	54.15
4240			81	8-5/8	C	35	6.65	41.65
4260			31.84	8	D	37.50	6.65	44.15
4280			66	7	E	27.50	6.65	34.15
4300			37.69	10x6	F	44.50	6.65	51.15
4320			70	8x6	G	39.50	6.65	46.15
4400		20	49	10	A	54.50	6.65	61.15
4420			40.48	10	B	45	6.65	51.65
4440			123	10-3/4	C	47	6.65	53.65
4460			31.84	8	D	35.50	6.65	42.15
4480			82	8	E	30.50	6.65	37.15
4500			37.69	10x6	F	42	6.65	48.65
4520			86	10x6	G	37	6.65	43.65

Important: See the Reference Section for critical supporting data - Reference Numbers and City Cost Indexes

B1010 Floor Construction

| B1010 208 | | | Steel Columns | | | | |

	LOAD (KIPS)	UNSUPPORTED HEIGHT (FT.)	WEIGHT (P.L.F.)	SIZE (IN.)	TYPE	COST PER V.L.F.		
						MAT.	INST.	TOTAL
4600	200	10	45	10	A	57	8.90	65.90
4620			40.48	10	B	51.50	8.90	60.40
4640			81	8-5/8	C	37.50	8.90	46.40
4660			31.84	8	D	40.50	8.90	49.40
4680			82	8	E	34.50	8.90	43.40
4700			37.69	10x6	F	48	8.90	56.90
4720			70	8x6	G	42.50	8.90	51.40
4800		16	49	10	A	57.50	6.65	64.15
4820			49.56	12	B	58.50	6.65	65.15
4840			123	10-3/4	C	49.50	6.65	56.15
4860			37.60	8	D	44	6.65	50.65
4880			90	8	E	46	6.65	52.65
4900			42.79	12x6	F	50.50	6.65	57.15
4920			85	10x6	G	45.50	6.65	52.15
5000		20	58	12	A	64.50	6.65	71.15
5020			49.56	12	B	55	6.65	61.65
5040			123	10-3/4	C	47	6.65	53.65
5060			40.35	10	D	45	6.65	51.65
5080			90	8	E	43.50	6.65	50.15
5100			47.90	12x8	F	53.50	6.65	60.15
5120			93	10x6	G	55.50	6.65	62.15
5200	300	10	61	14	A	77.50	8.90	86.40
5220			65.42	12	B	83	8.90	91.90
5240			169	12-3/4	C	66	8.90	74.90
5260			47.90	10	D	61	8.90	69.90
5280			90	8	E	49	8.90	57.90
5300			47.90	12x8	F	61	8.90	69.90
5320			86	10x6	G	63.50	8.90	72.40
5400		16	72	12	A	84.50	6.65	91.15
5420			65.42	12	B	77	6.65	83.65
5440			169	12-3/4	C	61.50	6.65	68.15
5460			58.10	12	D	68.50	6.65	75.15
5480			135	10	E	58.50	6.65	65.15
5500			58.10	14x10	F	68.50	6.65	75.15
5600		20	79	12	A	88	6.65	94.65
5620			65.42	12	B	73	6.65	79.65
5640			169	12-3/4	C	58	6.65	64.65
5660			58.10	12	D	64.50	6.65	71.15
5680			135	10	E	55.50	6.65	62.15
5700			58.10	14x10	F	64.50	6.65	71.15
5800	400	10	79	12	A	100	8.90	108.90
5840			178	12-3/4	C	86	8.90	94.90
5860			68.31	14	D	87	8.90	95.90
5880			135	10	E	63	8.90	71.90
5900			62.46	14x10	F	79.50	8.90	88.40
6000		16	87	12	A	102	6.65	108.65
6040			178	12-3/4	C	80	6.65	86.65
6060			68.31	14	D	80.50	6.65	87.15
6080			145	10	E	75.50	6.65	82.15
6100			76.07	14x10	F	89.50	6.65	96.15

B1010 Floor Construction

B1010 208	Steel Columns

	LOAD (KIPS)	UNSUPPORTED HEIGHT (FT.)	WEIGHT (P.L.F.)	SIZE (IN.)	TYPE	COST PER V.L.F.		
						MAT.	INST.	TOTAL
6200	400	20	90	14	A	100	6.65	106.65
6240			178	12-3/4	C	76	6.65	82.65
6260			68.31	14	D	76	6.65	82.65
6280			145	10	E	71.50	6.65	78.15
6300			76.07	14x10	F	84.50	6.65	91.15
6400	500	10	99	14	A	126	8.90	134.90
6460			76.07	12	D	96.50	8.90	105.40
6480			145	10	E	81.50	8.90	90.40
6500			76.07	14x10	F	96.50	8.90	105.40
6600		16	109	14	A	128	6.65	134.65
6660			89.68	14	D	105	6.65	111.65
6700			89.68	16x12	F	105	6.65	111.65
6800		20	120	12	A	134	6.65	140.65
6860			89.68	14	D	100	6.65	106.65
6900			89.68	16x12	F	100	6.65	106.65
7000	600	10	120	12	A	152	8.90	160.90
7060			89.68	14	D	114	8.90	122.90
7100			89.68	16x12	F	114	8.90	122.90
7200		16	132	14	A	155	6.65	161.65
7260			103.30	16	D	121	6.65	127.65
7400		20	132	14	A	147	6.65	153.65
7460			103.30	16	D	115	6.65	121.65
7600	700	10	136	12	A	173	8.90	181.90
7660			103.30	16	D	131	8.90	139.90
7800		16	145	14	A	171	6.65	177.65
7860			103.30	16	D	121	6.65	127.65
8000		20	145	14	A	161	6.65	167.65
8060			103.30	16	D	115	6.65	121.65
8200	800	10	145	14	A	184	8.90	192.90
8300		16	159	14	A	187	6.65	193.65
8400		20	176	14	A	196	6.65	202.65
8800	900	10	159	14	A	202	8.90	210.90
8900		16	176	14	A	207	6.65	213.65
9000		20	193	14	A	215	6.65	221.65
9100	1000	10	176	14	A	224	8.90	232.90
9200		16	193	14	A	227	6.65	233.65
9300		20	211	14	A	235	6.65	241.65

Important: See the Reference Section for critical supporting data - Reference Numbers and City Cost Indexes

B10 Superstructure

B1010 Floor Construction

- Interior Bay
- Exterior Bay
- Corner Bay

Description: Table below lists costs of columns per S.F. of bay for wood columns of various sizes and unsupported heights and the maximum allowable total load per S.F. by bay size.

Design Assumptions: Columns are concentrically loaded and are not subject to bending.

Fiber stress (f) is 1200 psi maximum.

Modulus of elasticity is 1,760,000. Use table to factor load capacity figures for modulus of elasticity other than 1,760,000.

The cost of columns per S.F. of exterior bay is proportional to the area supported. For exterior bays, multiply the costs below by two. For corner bays, multiply the cost by four.

Modulus of Elasticity	Factor
1,210,000 psi	0.69
1,320,000 psi	0.75
1,430,000 psi	0.81
1,540,000 psi	0.87
1,650,000 psi	0.94
1,760,000 psi	1.00

SHELL **B**

B1010 210 — Wood Columns

	NOMINAL COLUMN SIZE (IN.)	BAY SIZE (FT.)	UNSUPPORTED HEIGHT (FT.)	MATERIAL (BF per M.S.F.)	TOTAL LOAD (P.S.F.)	COST PER S.F. MAT.	COST PER S.F. INST.	COST PER S.F. TOTAL
1000	4 x 4	10 x 8	8	133	100	.26	.20	.46
1050			10	167	60	.32	.25	.57
1100			12	200	40	.39	.30	.69
1150			14	233	30	.45	.34	.79
1200		10 x 10	8	106	80	.20	.16	.36
1250			10	133	50	.26	.20	.46
1300			12	160	30	.31	.24	.55
1350			14	187	20	.36	.28	.64
1400		10 x 15	8	71	50	.14	.10	.24
1450			10	88	30	.17	.13	.30
1500			12	107	20	.21	.16	.37
1550			14	124	15	.24	.18	.42
1600		15 x 15	8	47	30	.09	.07	.16
1650			10	59	15	.11	.09	.20
1800		15 x 20	8	35	15	.07	.05	.12
1900								
2000	6 x 6	10 x 15	8	160	230	.38	.22	.60
2050			10	200	210	.47	.27	.74
2100			12	240	140	.56	.32	.88
2150			14	280	100	.66	.38	1.04
2200		15 x 15	8	107	150	.25	.14	.39
2250			10	133	140	.31	.18	.49
2300			12	160	90	.38	.22	.60
2350			14	187	60	.44	.25	.69
2400		15 x 20	8	80	110	.19	.11	.30
2450			10	100	100	.24	.14	.38
2500			12	120	70	.28	.16	.44
2550			14	140	50	.33	.19	.52
2600		20 x 20	8	60	80	.14	.08	.22
2650			10	75	70	.18	.10	.28
2700			12	90	50	.21	.12	.33
2750			14	105	30	.25	.14	.39
2800		20 x 25	8	48	60	.11	.06	.17
2850			10	60	50	.14	.08	.22
2900			12	72	40	.17	.10	.27
2950			14	84	20	.20	.11	.31

B1010 Floor Construction

B1010 210	Wood Columns

	NOMINAL COLUMN SIZE (IN.)	BAY SIZE (FT.)	UNSUPPORTED HEIGHT (FT.)	MATERIAL (BF per M.S.F.)	TOTAL LOAD (P.S.F.)	COST PER S.F.		
						MAT.	INST.	TOTAL
3000	6 x 6	25 x 25	8	38	50	.09	.05	.14
3050			10	48	40	.11	.06	.17
3100			12	58	30	.14	.08	.22
3150			14	67	20	.16	.09	.25
3400	8 x 8	20 x 20	8	107	160	.28	.14	.42
3450			10	133	160	.34	.17	.51
3500			12	160	160	.41	.20	.61
3550			14	187	140	.36	.28	.64
3600		20 x 25	8	85	130	.16	.13	.29
3650			10	107	130	.21	.16	.37
3700			12	128	130	.34	.15	.49
3750			14	149	110	.39	.18	.57
3800		25 x 25	8	68	100	.18	.08	.26
3850			10	85	100	.22	.10	.32
3900			12	102	100	.27	.12	.39
3950			14	119	90	.31	.14	.45
4200	10 x 10	20 x 25	8	133	210	.35	.16	.51
4250			10	167	210	.44	.20	.64
4300			12	200	210	.53	.24	.77
4350			14	233	210	.61	.27	.88
4400		25 x 25	8	107	160	.28	.13	.41
4450			10	133	160	.35	.16	.51
4500			12	160	160	.42	.19	.61
4550			14	187	160	.49	.22	.71
4700	12 x 12	20 x 25	8	192	310	.50	.21	.71
4750			10	240	310	.63	.26	.89
4800			12	288	310	.76	.32	1.08
4850			14	336	310	.88	.37	1.25
4900		25 x 25	8	154	240	.40	.17	.57
4950			10	192	240	.50	.21	.71
5000			12	230	240	.60	.25	.85
5050			14	269	240	.71	.30	1.01

Important: See the Reference Section for critical supporting data - Reference Numbers and City Cost Indexes

B1010 Floor Construction

General: Beams priced in the following table are plant produced prestressed members transported to the site and erected.

Pricing Assumptions: Prices are based upon 10,000 S.F. to 20,000 S.F. projects and 50 mile to 100 mile transport.

Normal steel for connections is included in price. Deduct 20% from prices for Southern states. Add 10% to prices for Western states.

Design Assumptions: Normal weight concrete to 150 lbs./C.F. f'c = 5 KSI. Prestressing Steel: 250 KSI straight strand. Non-Prestressing Steel: fy = 60 KSI.

SHELL B

System Components			QUANTITY	UNIT	COST PER L.F.		
					MAT.	INST.	TOTAL
SYSTEM B1010 213 2200							
RECTANGULAR PRECAST BEAM, 15' SPAN, 12"X16"							
Precast beam, rectangular			1.000	L.F.	43.50		43.50
Delivery, 30 mile radius			.006	Hr.		1.83	1.83
Erection crew			.019	Hr.		15.72	15.72
		TOTAL			43.50	17.55	61.05

B1010 213		Rectangular Precast Beams						
	SPAN (FT.)	SUPERIMPOSED LOAD (K.L.F.)	SIZE W X D (IN.)	BEAM WEIGHT (P.L.F.)	TOTAL LOAD (K.L.F.)	COST PER L.F.		
						MAT.	INST.	TOTAL
2200	15	2.32	12x16	200	2.52	43.50	17.55	61.05
2250		3.80	12x20	250	4.05	65.50	15.75	81.25
2300		5.60	12x24	300	5.90	74	16.55	90.55
2350		7.65	12x28	350	8.00	84.50	16.55	101.05
2400		5.85	18x20	375	6.73	79.50	16.55	96.05
2450		8.26	18x24	450	8.71	88.50	17.35	105.85
2500		11.39	18x28	525	11.91	101	17.35	118.35
2550		18.90	18x36	675	19.58	120	17.35	137.35
2600		10.78	24x24	600	11.38	105	17.35	122.35
2700		15.12	24x28	700	15.82	119	17.35	136.35
2750		25.23	24x36	900	26.13	142	18.20	160.20
2800	20	1.22	12x16	200	1.44	57.50	11.55	69.05
2850		2.03	12x20	250	2.28	65.50	12.45	77.95
2900		3.02	12x24	300	3.32	74	12.45	86.45
2950		4.15	12x28	350	4.50	84.50	12.45	96.95
3000		6.85	12x36	450	7.30	101	13.25	114.25
3050		3.13	18x20	375	3.50	79	12.45	91.45
3100		4.45	18x24	450	4.90	88.50	13.25	101.75
3150		6.18	18x28	525	6.70	75	13.25	88.25
3200		10.33	18x36	675	11.00	120	14.05	134.05
3400		15.48	18x44	825	16.30	141	14.85	155.85
3450		21.63	18x52	975	22.60	164	14.85	178.85
3500		5.80	24x24	600	6.40	105	13.25	118.25
3600		8.20	24x28	700	8.90	106	14.05	120.05
3700		13.80	24x36	900	14.70	121	14.85	135.85
3800		20.70	24x44	1100	21.80	167	20.50	187.50
3850		29.20	24x52	1300	30.50	193	20.50	213.50

B1010 Floor Construction

B1010 213	Rectangular Precast Beams

	SPAN (FT.)	SUPERIMPOSED LOAD (K.L.F.)	SIZE W X D (IN.)	BEAM WEIGHT (P.L.F.)	TOTAL LOAD (K.L.F.)	COST PER L.F.		
						MAT.	INST.	TOTAL
3900	25	1.82	12x24	300	2.12	75.50	9.95	85.45
4000		2.53	12x28	350	2.88	79.50	10.75	90.25
4050		5.18	12x36	450	5.63	102	10.75	112.75
4100		6.55	12x44	550	7.10	118	10.75	128.75
4150		9.08	12x52	650	9.73	138	12.45	150.45
4200		1.86	18x20	375	2.24	80	10.75	90.75
4250		2.69	18x24	450	3.14	90	10.75	100.75
4300		3.76	18x28	525	4.29	76	10.75	86.75
4350		6.37	18x36	675	7.05	122	12.45	134.45
4400		9.60	18x44	872	10.43	144	16.55	160.55
4450		13.49	18x52	975	14.47	145	16.55	161.55
4500		18.40	18x60	1125	19.53	185	16.55	201.55
4600		3.50	24x24	600	4.10	107	12.45	119.45
4700		5.00	24x28	700	5.70	121	12.45	133.45
4800		8.50	24x36	900	9.40	144	16.55	160.55
4900		12.85	24x44	1100	13.95	169	16.55	185.55
5000		18.22	24x52	1300	19.52	206	22.50	228.50
5050		24.48	24x60	1500	26.00	225	22.50	247.50
5100	30	1.65	12x28	350	2.00	86.50	9.10	95.60
5150		2.79	12x36	450	3.24	103	9.10	112.10
5200		4.38	12x44	550	4.93	143	9.95	152.95
5250		6.10	12x52	650	6.75	140	9.95	149.95
5300		8.36	12x60	750	9.11	158	14.05	172.05
5400		1.72	18x24	450	2.17	90.50	9.10	99.60
5450		2.45	18x28	525	2.98	103	9.95	112.95
5750		4.21	18x36	675	4.89	124	14.05	138.05
6000		6.42	18x44	825	7.25	144	14.05	158.05
6250		9.07	18x52	975	10.05	167	14.05	181.05
6500		12.43	18x60	1125	13.56	197	18.20	215.20
6750		2.24	24x24	600	2.84	107	9.95	116.95
7000		3.26	24x28	700	3.96	122	14.05	136.05
7100		5.63	24x36	900	6.53	145	14.05	159.05
7200	35	8.59	24x44	1100	9.69	171	15.75	186.75
7300		12.25	24x52	1300	13.55	197	15.75	212.75
7400		16.54	24x60	1500	18.04	221	15.75	236.75
7500	40	2.23	12x44	550	2.78	122	15.75	137.75
7600		3.15	12x52	650	3.80	142	15.75	157.75
7700		4.38	12x60	750	5.13	163	15.75	178.75
7800		2.08	18x36	675	2.76	124	15.75	139.75
7900		3.25	18x44	825	4.08	152	19.05	171.05
8000		4.68	18x52	975	5.66	175	19.05	194.05
8100		6.50	18x60	1175	7.63	192	19.05	211.05
8200		2.78	24x36	900	3.60	152	19.05	171.05
8300		4.35	24x44	1100	5.45	172	19.05	191.05
8400		6.35	24x52	1300	7.65	199	20.50	219.50
8500		8.65	24x60	1500	10.15	223	20.50	243.50

Important: See the Reference Section for critical supporting data - Reference Numbers and City Cost Indexes

B10 Superstructure

B1010 Floor Construction

B1010 213	Rectangular Precast Beams

	SPAN (FT.)	SUPERIMPOSED LOAD (K.L.F.)	SIZE W X D (IN.)	BEAM WEIGHT (P.L.F.)	TOTAL LOAD (K.L.F.)	COST PER L.F.		
						MAT.	INST.	TOTAL
8600	45	2.35	12x52	650	3.00	144	14.85	158.85
8800		3.29	12x60	750	4.04	166	16.55	182.55
9000		2.39	18x44	825	3.22	149	16.55	165.55
9020		3.49	18x52	975	4.47	173	16.55	189.55
9200		4.90	18x60	1125	6.03	193	19.05	212.05
9250		3.21	24x44	1100	4.31	174	19.05	193.05
9300		4.72	24x52	1300	6.02	197	19.05	216.05
9500		6.52	24x60	1500	8.02	226	19.05	245.05
9600	50	2.64	18x52	975	3.62	175	16.55	191.55
9700		3.75	18x60	1125	4.88	197	16.55	213.55
9750		3.58	24x52	1300	4.88	310	16.55	326.55
9800		5.00	24x60	1500	6.50	229	16.55	245.55
9950	55	3.86	24x60	1500	5.36	230	14.85	244.85

General: Beams priced in the following table are plant produced prestressed members transported to the site and erected.

Pricing Assumptions: Prices are based upon 10,000 S.F. to 20,000 S.F. projects and 50 mile to 100 mile transport.

Normal steel for connections is included in price. Deduct 20% from prices for Southern states. Add 10% to prices for Western states.

Design Assumptions: Normal weight concrete to 150 lbs./C.F. f'c = 5 KSI. Prestressing Steel: 250 KSI straight strand. Non-Prestressing Steel: fy = 60 KSI.

System Components	QUANTITY	UNIT	COST PER L.F.		
			MAT.	INST.	TOTAL
SYSTEM B1010 214 2300					
"T" SHAPED PRECAST BEAM, 15' SPAN, 12"X16"					
Precast beam, "T" shaped, 12"x16"	1.000	L.F.	129		129
Delivery, 30 mile radius	.009	Hr.		1.83	1.83
Erection crew	.019	Hr.		15.72	15.72
TOTAL			129	17.55	146.55

B1010 214	"T" Shaped Precast Beams						
	SPAN (FT.)	SUPERIMPOSED LOAD (K.L.F.)	SIZE W X D (IN.)	BEAM WEIGHT (P.L.F.)	TOTAL LOAD (K.L.F.)	COST PER L.F.		
						MAT.	INST.	TOTAL
2300	15	2.8	12x16	260	3.06	129	17.55	146.55
2350		4.33	12x20	355	4.69	128	16.55	144.55
2400		6.17	12x24	445	6.62	149	17.35	166.35
2500		8.37	12x28	515	8.89	161	17.35	178.35
2550		13.54	12x36	680	14.22	182	17.35	199.35
2600		8.83	18x24	595	9.43	165	17.35	182.35
2650		12.11	18x28	690	12.8	177	17.35	194.35
2700		19.72	18x36	905	24.63	203	18.20	221.20
2800		11.52	24x24	745	12.27	182	17.35	199.35
2900		15.85	24x28	865	16.72	197	18.20	215.20
3000		26.07	24x36	1130	27.20	225	19.85	244.85

B1010   Floor Construction

B1010 214	"T" Shaped Precast Beams

	SPAN (FT.)	SUPERIMPOSED LOAD (K.L.F.)	SIZE W X D (IN.)	BEAM WEIGHT (P.L.F.)	TOTAL LOAD (K.L.F.)	COST PER L.F.		
						MAT.	INST.	TOTAL
3100	20	1.46	12x16	260	1.72	129	12.45	141.45
3200		2.28	12x20	355	2.64	128	12.45	140.45
3300		3.28	12x24	445	3.73	149	13.25	162.25
3400		4.49	12x28	515	5.00	161	13.25	174.25
3500		7.32	12x36	680	8.00	182	14.05	196.05
3600		11.26	12x44	840	12.10	206	14.85	220.85
3700		4.70	18x24	595	5.30	164	13.25	177.25
3800		6.51	18x28	690	7.20	177	14.05	191.05
3900		10.7	18x36	905	11.61	203	14.85	217.85
4300		16.19	18x44	1115	17.31	233	20.50	253.50
4400		22.77	18x52	1330	24.10	259	20.50	279.50
4500		6.15	24x24	745	6.90	182	14.05	196.05
4600		8.54	24x28	865	9.41	197	14.85	211.85
4700		14.17	24x36	1130	15.30	226	20.50	246.50
4800		21.41	24x44	1390	22.80	257	20.50	277.50
4900		30.25	24x52	1655	31.91	296	27.50	323.50
5000	25	2.68	12x28	515	3.2	163	10.75	173.75
5050		4.44	12x36	680	5.12	184	12.45	196.45
5100		6.90	12x44	840	7.74	209	16.55	225.55
5200		9.75	12x52	1005	10.76	234	16.55	250.55
5300		13.43	12x60	1165	14.60	257	16.55	273.55
5350		3.92	18x28	690	4.61	180	12.45	192.45
5400		6.52	18x36	905	7.43	206	16.55	222.55
5500		9.96	18x44	1115	11.08	193	16.55	209.55
5600		14.09	18x52	1330	15.42	268	22.50	290.50
5650		19.39	18x60	1540	20.93	290	22.50	312.50
5700		3.67	24x24	745	4.42	184	12.45	196.45
5750		5.15	24x28	865	6.02	200	16.55	216.55
5800		8.66	24x36	1130	9.79	185	16.55	201.55
5850		13.20	24x44	1390	14.59	266	22.50	288.50
5900		18.76	24x52	1655	20.42	296	22.50	318.50
5950		25.35	24x60	1916	27.27	320	22.50	342.50
6000	30	2.88	12x36	680	3.56	187	14.05	201.05
6100		4.54	12x44	840	5.38	215	14.05	229.05
6200		6.46	12x52	1005	7.47	244	18.20	262.20
6250		8.97	12x60	1165	10.14	266	18.20	284.20
6300		4.25	18x36	905	5.16	207	14.05	221.05
6350		6.57	18x44	1115	7.69	240	18.20	258.20
6400		9.38	18x52	1330	10.71	265	18.20	283.20
6500		13.00	18x60	1540	14.54	294	18.20	312.20
6700		3.31	24x28	865	4.18	201	14.05	215.05
6750		5.67	24x36	1130	6.80	233	18.20	251.20
6800		8.74	24x44	1390	10.13	264	18.20	282.20
6850		12.52	24x52	1655	14.18	251	18.20	269.20
6900		17.00	24x60	1215	18.92	325	18.20	343.20

SHELL B

B1010 Floor Construction

B1010 214	"T" Shaped Precast Beams

	SPAN (FT.)	SUPERIMPOSED LOAD (K.L.F.)	SIZE W X D (IN.)	BEAM WEIGHT (P.L.F.)	TOTAL LOAD (K.L.F.)	COST PER L.F.		
						MAT.	INST.	TOTAL
7000	35	3.11	12x44	840	3.95	211	11.55	222.55
7100		4.48	12x52	1005	5.49	244	15.75	259.75
7200		6.28	12x60	1165	7.45	266	15.75	281.75
7300		4.53	18x44	1115	5.65	240	15.75	255.75
7500		6.54	18x52	1330	7.87	267	15.75	282.75
7600		9.14	18x60	1540	10.68	282	19.85	301.85
7700		3.87	24x36	1130	5.00	193	15.75	208.75
7800		6.05	24x44	1390	7.44	265	15.75	280.75
7900		8.76	24x52	1655	10.42	294	19.85	313.85
8000		12.00	24x60	1915	13.92	330	19.85	349.85
8100	40	3.19	12x52	1005	4.2	250	19.05	269.05
8200		4.53	12x60	1165	5.7	274	19.05	293.05
8300		4.70	18x52	1330	6.03	275	19.05	294.05
8400		6.64	18x60	1540	8.18	305	19.05	324.05
8600		4.31	24x44	1390	5.7	265	19.05	284.05
8700		6.32	24x52	1655	7.98	305	19.05	324.05
8800		8.74	24x60	1915	10.66	340	19.05	359.05
8900	45	3.34	12x60	1165	4.51	277	16.55	293.55
9000		4.92	18x60	1540	6.46	310	16.55	326.55
9250		4.64	24x52	1655	6.30	310	16.55	326.55
9900		6.5	24x60	1915	8.42	345	16.55	361.55
9950	50	4.9	24x60	1915	6.82	345	19.85	364.85

Important: See the Reference Section for critical supporting data - Reference Numbers and City Cost Indexes

B1010 Floor Construction

General: Beams priced in the following table are plant produced prestressed members transported to the site and erected.

Pricing Assumptions: Prices are based upon 10,000 S.F. to 20,000 S.F. projects and 50 mile to 100 mile transport.

Normal steel for connections is included in price. Deduct 20% from prices for Southern states. Add 10% to prices for Western states.

Design Assumptions: Normal weight concrete to 150 lbs./C.F. f'c = 5 KSI. Prestressing Steel: 250 KSI straight strand. Non-Prestressing Steel: fy = 60 KSI.

System Components	QUANTITY	UNIT	COST PER L.F.		
			MAT.	INST.	TOTAL
SYSTEM B1010 215 2250					
"L" SHAPED PRECAST BEAM, 15' SPAN, 12"X16"					
Precast beam, "L" shaped, 12"x16"	1.000	L.F.	95.50		95.50
Delivery, 30 mile radius	.008	Hr.		1.83	1.83
Erection crew	.019	Hr.		15.72	15.72
TOTAL			95.50	17.55	113.05

B1010 215		"L" Shaped Precast Beams						
	SPAN (FT.)	SUPERIMPOSED LOAD (K.L.F.)	SIZE W X D (IN.)	BEAM WEIGHT (P.L.F.)	TOTAL LOAD (K.L.F.)	COST PER L.F.		
						MAT.	INST.	TOTAL
2250	15	2.58	12x16	230	2.81	95.50	17.55	113.05
2300		4.10	12x20	300	4.40	104	16.55	120.55
2400		5.92	12x24	370	6.29	114	16.55	130.55
2450		8.09	12x28	435	8.53	124	16.55	140.55
2500		12.95	12x36	565	13.52	143	17.35	160.35
2600		8.55	18x24	520	9.07	128	17.35	145.35
2650		11.83	18x28	610	12.44	141	17.35	158.35
2700		19.30	18x36	790	20.09	162	17.35	179.35
2750		11.24	24x24	670	11.91	146	17.35	163.35
2800		15.40	24x28	780	16.18	160	17.35	177.35
2850		25.65	24x36	1015	26.67	186	19.85	205.85

B1010 Floor Construction

B1010 215	"L" Shaped Precast Beams

	SPAN (FT.)	SUPERIMPOSED LOAD (K.L.F.)	SIZE W X D (IN.)	BEAM WEIGHT (P.L.F.)	TOTAL LOAD (K.L.F.)	COST PER L.F.		
						MAT.	INST.	TOTAL
2900	20	2.18	12x20	300	2.48	104	12.45	116.45
2950		3.17	12x24	370	3.54	113	12.45	125.45
3000		4.37	12x28	435	4.81	124	13.25	137.25
3100		7.04	12x36	565	7.60	143	13.25	156.25
3150		10.80	12x44	695	11.50	162	14.05	176.05
3200		15.08	12x52	825	15.91	185	14.85	199.85
3250		4.58	18x24	520	5.10	128	13.25	141.25
3300		6.39	18x28	610	7.00	141	14.05	155.05
3350		10.51	18x36	790	11.30	163	14.85	177.85
3400		15.73	18x44	970	16.70	187	14.85	201.85
3550		22.15	18x52	1150	23.30	215	20.50	235.50
3600		6.03	24x24	370	6.40	142	12.45	154.45
3650		8.32	24x28	780	9.10	160	14.85	174.85
3700		13.98	24x36	1015	15.00	186	20.50	206.50
3800		21.05	24x44	1245	22.30	214	20.50	234.50
3900		29.73	24x52	1475	31.21	244	20.50	264.50
4000	25	2.64	12x28	435	3.08	124	9.95	133.95
4100		4.30	12x36	565	4.87	143	10.75	153.75
4200		6.65	12x44	695	7.35	161	10.75	171.75
4250		9.35	12x52	875	10.18	185	12.45	197.45
4300		12.81	12x60	950	13.76	207	12.45	219.45
4350		2.74	18x24	570	3.76	135	16.55	151.55
4400		3.67	18x28	610	4.28	141	10.75	151.75
4450		6.44	18x36	790	7.23	163	12.45	175.45
4500		9.72	18x44	970	10.69	191	16.55	207.55
4600		13.76	18x52	1150	14.91	210	10.75	220.75
4700		18.33	18x60	1330	20.16	244	22.50	266.50
4800		3.62	24x24	370	3.99	145	13.25	158.25
4900		5.04	24x28	780	5.82	162	12.45	174.45
5000		8.58	24x36	1015	9.60	188	16.55	204.55
5100		13.00	24x44	1245	14.25	223	22.50	245.50
5200		18.50	24x52	1475	19.98	251	22.50	273.50
5300	30	2.80	12x36	565	3.37	147	9.95	156.95
5400		4.42	12x44	695	5.12	166	14.05	180.05
5500		6.24	12x52	875	7.07	189	14.05	203.05
5600		8.60	12x60	950	9.55	214	14.05	228.05
5700		2.50	18x28	610	3.11	143	9.95	152.95
5800		4.23	18x36	790	5.02	166	14.05	180.05
5900		6.45	18x44	970	7.42	191	14.05	205.05
6000		9.20	18x52	1150	10.35	222	18.20	240.20
6100		12.67	18x60	1330	14.00	245	18.20	263.20
6200		3.26	24x28	780	4.04	163	14.05	177.05
6300		5.65	24x36	1015	6.67	194	18.20	212.20
6400		8.66	24x44	1245	9.90	221	18.20	239.20

Important: See the Reference Section for critical supporting data - Reference Numbers and City Cost Indexes

B1010 215		"L" Shaped Precast Beams						
	SPAN (FT.)	SUPERIMPOSED LOAD (K.L.F.)	SIZE W X D (IN.)	BEAM WEIGHT (P.L.F.)	TOTAL LOAD (K.L.F.)	COST PER L.F.		
						MAT.	INST.	TOTAL
6500	35	3.00	12x44	695	3.70	168	11.55	179.55
6600		4.27	12x52	825	5.20	191	11.55	202.55
6700		6.00	12x60	950	6.95	198	15.75	213.75
6800		2.90	18x36	790	3.69	168	11.55	179.55
6900		4.48	18x44	970	5.45	197	15.75	212.75
7000		6.45	18x52	1150	7.60	223	15.75	238.75
7100		8.95	18x60	1330	10.28	248	15.75	263.75
7200		3.88	24x36	1015	4.90	196	15.75	211.75
7300		6.03	24x44	1245	7.28	222	15.75	237.75
7400		8.71	24x52	1475	10.19	252	19.85	271.85
7500	40	3.15	12x52	825	3.98	195	19.05	214.05
7600		4.43	12x60	950	5.38	222	19.05	241.05
7700		3.20	18x44	970	4.17	198	19.05	217.05
7800		4.68	18x52	1150	5.83	225	19.05	244.05
7900		6.55	18x60	1330	7.88	251	19.05	270.05
8000		4.43	24x44	1245	5.68	224	19.05	243.05
8100		6.33	24x52	1475	7.81	255	19.05	274.05
8200	45	3.30	12x60	950	4.25	224	16.55	240.55
8300		3.45	18x52	1150	4.60	228	16.55	244.55
8500		4.44	18x60	1330	5.77	253	16.55	269.55
8750		3.16	24x44	1245	4.41	227	16.55	243.55
9000		4.68	24x52	1475	6.16	258	16.55	274.55
9900	50	3.71	18x60	1330	5.04	253	15.75	268.75
9950		3.82	24x52	1475	5.30	257	15.75	272.75

SHELL B

B1010 Floor Construction

Description: Table below lists the cost per L.F. and total uniform load allowable for various size beams at various spans.

Design Assumptions: Fiber strength (f) is 1,000 PSI. Maximum deflection does not exceed 1/360 the span of the beam. Modulus of elasticity (E) is 1,100,000 psi. Anything less will result in excessive deflection in the longer members so that spans must be reduced or member size increased.

The total loads are in pounds per foot.

The span is in feet and is the unsupported clear span.

The costs are based on net quantities and do not include any allowance for waste or overlap.

The member sizes are from 2″ x 6″ to 4″ x 12″ and include one, two and three pieces of each member size to arrive at the costs, spans and loading. If more than a three-piece beam is required, its cost may be calculated directly. A four-piece 2″ x 6″ would cost two times as much as a two-piece 2″ x 6″ beam. The maximum span and total load cannot be doubled, however.

B1010 216	Wood Beams						
	BEAM SIZE (IN.)	TOTAL LOAD (P.L.F.) 6′ SPAN	TOTAL LOAD (P.L.F.) 10′ SPAN	TOTAL LOAD (P.L.F.) 14′ SPAN	TOTAL LOAD (P.L.F.) 18′ SPAN	COST PER L.F.		
						MAT.	INST.	TOTAL
1000	2 - 2 x 6	238				1.30	3.40	4.70
1050	3 - 2 x 6	357	RB1010 -240			1.95	5.10	7.05
1100	2 - 2 x 8	314				1.88	3.93	5.81
1150	3 - 2 x 8	471	232			2.82	5.90	8.72
1200	2 - 2 x 10	400	240			3.23	4.03	7.26
1250	3 - 2 x 10	600	360			4.84	6.05	10.89
1300	2 - 2 x 12	487	292	208		3.52	4.40	7.92
1350	3 - 2 x 12	731	438	313		5.30	6.60	11.90
1400	2 - 2 x 14	574	344	246		4.78	4.60	9.38
1450	3 - 2 x 14	861	516	369	243	7.20	6.90	14.10
1550	2 - 3 x 6	397				4.20	3.83	8.03
1600	3 - 3 x 6	595				6.30	5.75	12.05
1650	1 - 3 x 8	261				2.85	2.20	5.05
1700	2 - 3 x 8	523	258			5.70	4.40	10.10
1750	3 - 3 x 8	785	388			8.55	6.60	15.15
1800	1 - 3 x 10	334	200			3.56	2.46	6.02
1850	2 - 3 x 10	668	400			7.15	4.93	12.08
1900	3 - 3 x 10	1,002	600	293		10.70	7.40	18.10
1950	1 - 3 x 12	406	243			4.35	2.66	7.01
2000	2 - 3 x 12	812	487	348		8.70	5.30	14
2050	3 - 3 x 12	1,218	731	522	248	13.05	7.95	21
2100	1 - 4 x 6	278				2.05	2.20	4.25
2150	2 - 4 x 6	556				4.10	4.40	8.50
2200	1 - 4 x 8	366				2.87	2.63	5.50
2250	2 - 4 x 8	733	362			5.75	5.25	11
2300	1 - 4 x 10	467	280			3.58	2.95	6.53
2350	2 - 4 x 10	935	560			7.15	5.90	13.05
2400	1 - 4 x 12	568	341			4.40	3.22	7.62
2450	2 - 4 x 12	1,137	682		232	8.80	6.45	15.25

B1010 Floor Construction

SHELL **B**

General: Solid concrete slabs of uniform depth reinforced for flexure in one direction and for temperature and shrinkage in the other direction.

Design and Pricing Assumptions:
Concrete f'c = 3 KSI normal weight placed by pump.
Reinforcement fy = 60 KSI.
Deflection≤ span/360.
Forms, four uses hung from steel beams plus edge forms.
Steel trowel finish for finish floor and cure.

System Components			COST PER S.F.		
	QUANTITY	UNIT	MAT.	INST.	TOTAL
SYSTEM B1010 217 2000					
6'-0" SINGLE SPAN, 4" SLAB DEPTH, 40 PSF SUPERIMPOSED LOAD					
Forms in place, floor slab forms hung from steel beams, 4 uses	1.000	S.F.	1.45	4.57	6.02
Edge forms to 6" high on elevated slab, 4 uses	.080	L.F.	.07	1.11	1.18
Reinforcing in place, elevated slabs #4 to #7	1.030	Lb.	.50	.37	.87
Concrete ready mix, regular weight, 3000 psi	.330	C.F.	1.18		1.18
Place and vibrate concrete, elevated slab less than 6", pumped	.330	C.F.		.42	.42
Finishing floor, monolithic steel trowel finish for finish floor	1.000	S.F.		.74	.74
Curing with sprayed membrane curing compound	1.000	S.F.	.06	.07	.13
TOTAL			3.26	7.28	10.54

B1010 217		Cast in Place Slabs, One Way							
	SLAB DESIGN & SPAN (FT.)	SUPERIMPOSED LOAD (P.S.F.)	THICKNESS (IN.)	TOTAL LOAD (P.S.F.)			COST PER S.F.		
							MAT.	INST.	TOTAL
2000	Single 6	40	4	90	RB1010 -010		3.26	7.30	10.56
2100		75	4	125			3.26	7.30	10.56
2200		125	4	175	RB1010 -100		3.26	7.30	10.56
2300		200	4	250			3.44	7.40	10.84
2500	Single 8	40	4	90			3.21	6.45	9.66
2600		75	4	125			3.25	7	10.25
2700		125	4-1/2	181			3.40	7.05	10.45
2800		200	5	262			3.80	7.30	11.10
3000	Single 10	40	4	90			3.37	6.95	10.32
3100		75	4	125			3.37	6.95	10.32
3200		125	5	188			3.78	7.10	10.88
3300		200	7-1/2	293			5.25	8	13.25
3500	Single 15	40	5-1/2	90			4.08	7.10	11.18
3600		75	6-1/2	156			4.67	7.40	12.07
3700		125	7-1/2	219			5.10	7.60	12.70
3800		200	8-1/2	306			5.50	7.80	13.30
4000	Single 20	40	7-1/2	115			5.05	7.45	12.50
4100		75	9	200			5.45	7.40	12.85
4200		125	10	250			6.05	7.75	13.80
4300		200	10	324			6.50	8.10	14.60

B1010 Floor Construction

B1010 217	Cast in Place Slabs, One Way

	SLAB DESIGN & SPAN (FT.)	SUPERIMPOSED LOAD (P.S.F.)	THICKNESS (IN.)	TOTAL LOAD (P.S.F.)		COST PER S.F.		
						MAT.	INST.	TOTAL
4500	Multi 6	40	4	90		3.31	6.90	10.21
4600		75	4	125		3.31	6.90	10.21
4700		125	4	175		3.31	6.90	10.21
4800		200	4	250		3.31	6.90	10.21
5000	Multi 8	40	4	90		3.31	6.75	10.06
5100		75	4	125		3.31	6.75	10.06
5200		125	4	175		3.31	6.75	10.06
5300		200	4	250		3.31	6.75	10.06
5500	Multi 10	40	4	90		3.31	6.65	9.96
5600		75	4	125		3.31	6.70	10.01
5700		125	4	175		3.45	6.80	10.25
5800		200	5	263		3.94	7.05	10.99
6600	Multi 15	40	5	103		3.88	6.90	10.78
6800		75	5	138		4.07	7	11.07
7000		125	6-1/2	206		4.58	7.25	11.83
7100		200	6-1/2	281		4.72	7.35	12.07
7500	Multi 20	40	5-1/2	109		4.49	7.20	11.69
7600		75	6-1/2	156		4.89	7.40	12.29
7800		125	9	238		6.20	8.05	14.25
8000		200	9	313		6.20	8.05	14.25

Important: See the Reference Section for critical supporting data - Reference Numbers and City Cost Indexes

B1010 Floor Construction

General: Solid concrete one-way slab cast monolithically with reinforced concrete support beams and girders.

Design and Pricing Assumptions:
Concrete f'c = 3 KSI, normal weight, placed by concrete pump.
Reinforcement, fy = 60 KSI.
Forms, four use.
Finish, steel trowel.
Curing, spray on membrane.
Based on 4 bay x 4 bay structure.

SHELL **B**

System Components		QUANTITY	UNIT	COST PER S.F.		
				MAT.	INST.	TOTAL
SYSTEM B1010 219 3000						
BM. & SLAB ONE WAY 15'X15'BAY, 40 PSF S.LOAD,12" MIN. COL.						
Forms in place, flat plate to 15' high, 4 uses		.858	S.F.	1.32	3.96	5.28
Forms in place, exterior spandrel, 12" wide, 4 uses		.142	SFCA	.19	1.18	1.37
Forms in place, interior beam. 12" wide, 4 uses		.306	SFCA	.45	2.10	2.55
Reinforcing in place, elevated slabs #4 to #7		1.600	Lb.	.78	.58	1.36
Concrete ready mix, regular weight, 3000 psi		.410	C.F.	1.45		1.45
Place and vibrate concrete, elevated slab less than 6", pump		.410	C.F.		.51	.51
Finish floor, monolithic steel trowel finish for finish floor		1.000	S.F.		.74	.74
Cure with sprayed membrane curing compound		.010	C.S.F.	.06	.07	.13
	TOTAL			4.25	9.14	13.39

B1010 219		Cast in Place Beam & Slab, One Way						

	BAY SIZE (FT.)	SUPERIMPOSED LOAD (P.S.F.)	MINIMUM COL. SIZE (IN.)	SLAB THICKNESS (IN.)	TOTAL LOAD (P.S.F.)	COST PER S.F.		
						MAT.	INST.	TOTAL
3000	15x15	40	12	4	120	4.26	9.10	13.36
3100	RB1010 -010	75	12	4	138	4.33	9.15	13.48
3200		125	12	4	188	4.43	9.25	13.68
3300		200	14	4	266	4.70	9.60	14.30
3600	15x20	40	12	4	102	4.35	9.05	13.40
3700	RB1010 -100	75	12	4	140	4.53	9.30	13.83
3800		125	14	4	192	4.77	9.65	14.42
3900		200	16	4	272	5.25	10.30	15.55
4200	20x20	40	12	5	115	4.70	8.85	13.55
4300		75	14	5	154	5.10	9.55	14.65
4400		125	16	5	206	5.30	10	15.30
4500		200	18	5	287	5.95	10.75	16.70
5000	20x25	40	12	5-1/2	121	4.86	8.85	13.71
5100		75	14	5-1/2	160	5.35	9.60	14.95
5200		125	16	5-1/2	215	5.70	10.15	15.85
5300		200	18	5-1/2	294	6.20	10.95	17.15
5500	25x25	40	12	6	129	5.10	8.70	13.80
5600		75	16	6	171	5.55	9.35	14.90
5700		125	18	6	227	6.50	10.70	17.20
5800		200	2	6	300	7.15	11.45	18.60
6500	25x30	40	14	6-1/2	132	5.20	8.85	14.05
6600		75	16	6-1/2	172	5.65	9.40	15.05
6700		125	18	6-1/2	231	6.60	10.65	17.25
6800		200	20	6-1/2	312	7.20	11.55	18.75

B1010 Floor Construction

B1010 219	Cast in Place Beam & Slab, One Way

	BAY SIZE (FT.)	SUPERIMPOSED LOAD (P.S.F.)	MINIMUM COL. SIZE (IN.)	SLAB THICKNESS (IN.)	TOTAL LOAD (P.S.F.)	COST PER S.F.		
						MAT.	INST.	TOTAL
7000	30x30	40	14	7-1/2	150	6	9.55	15.55
7100		75	18	7-1/2	191	6.60	10.10	16.70
7300		125	20	7-1/2	245	7	10.70	17.70
7400		200	24	7-1/2	328	7.85	11.85	19.70
7500	30x35	40	16	8	158	6.30	9.85	16.15
7600		75	18	8	196	6.75	10.20	16.95
7700		125	22	8	254	7.55	11.30	18.85
7800		200	26	8	332	8.15	11.70	19.85
8000	35x35	40	16	9	169	7.05	10.15	17.20
8200		75	20	9	213	7.60	11.05	18.65
8400		125	24	9	272	8.35	11.55	19.90
8600		200	26	9	355	9.15	12.30	21.45
9000	35x40	40	18	9	174	7.20	10.35	17.55
9300		75	22	9	214	7.75	11.15	18.90
9400		125	26	9	273	8.45	11.60	20.05
9600		200	30	9	355	9.25	12.40	21.65

B SHELL

Important: See the Reference Section for critical supporting data - Reference Numbers and City Cost Indexes

B1010 Floor Construction

General: Solid concrete two-way slab cast monolithically with reinforced concrete support beams and girders.

Design and Pricing Assumptions:
Concrete f'c = 3 KSI, normal weight, placed by concrete pump.
Reinforcement, fy = 60 KSI.
Forms, four use.
Finish, steel trowel.
Curing, spray on membrane.
Based on 4 bay x 4 bay structure.

SHELL **B**

System Components	QUANTITY	UNIT	COST PER S.F. MAT.	COST PER S.F. INST.	COST PER S.F. TOTAL
SYSTEM B1010 220 2000					
15'X 15' BAY, 40 PSF, S. LOAD 12" MIN. COL					
Forms in place, flat plate to 15' high, 4 uses	.894	S.F.	1.38	4.12	5.50
Forms in place, exterior spandrel, 12" wide, 4 uses	.142	SFCA	.19	1.18	1.37
Forms in place, interior beam. 12" wide, 4 uses	.178	SFCA	.26	1.22	1.48
Reinforcing in place, elevated slabs #4 to #7	1.598	Lb.	.78	.58	1.36
Concrete ready mix, regular weight, 3000 psi	.437	C.F.	1.55		1.55
Place and vibrate concrete, elevated slab less than 6", pump	.437	C.F.		.55	.55
Finish floor, monolithic steel trowel finish for finish floor	1.000	S.F.		.74	.74
Cure with sprayed membrane curing compound	.010	C.S.F.	.06	.07	.13
TOTAL			4.22	8.46	12.68

B1010 220   Cast in Place Beam & Slab, Two Way

	BAY SIZE (FT.)	SUPERIMPOSED LOAD (P.S.F.)	MINIMUM COL. SIZE (IN.)	SLAB THICKNESS (IN.)	TOTAL LOAD (P.S.F.)	COST PER S.F. MAT.	COST PER S.F. INST.	COST PER S.F. TOTAL
2000	15x15	40	12	4-1/2	106	4.22	8.45	12.67
2200	RB1010-010	75	12	4-1/2	143	4.33	8.60	12.93
2250		125	12	4-1/2	195	4.63	8.80	13.43
2300		200	14	4-1/2	274	5	9.35	14.35
2400	15 x 20	40	12	5-1/2	120	4.71	8.40	13.11
2600	RB1010-100	75	12	5-1/2	159	5.15	9	14.15
2800		125	14	5-1/2	213	5.55	9.75	15.30
2900		200	16	5-1/2	294	6.05	10.40	16.45
3100	20 x 20	40	12	6	132	5.20	8.55	13.75
3300		75	14	6	167	5.55	9.35	14.90
3400		125	16	6	220	5.75	9.70	15.45
3600		200	18	6	301	6.35	10.35	16.70
4000	20 x 25	40	12	7	141	5.60	8.95	14.55
4300		75	14	7	181	6.35	9.80	16.15
4500		125	16	7	236	6.45	10.05	16.50
4700		200	18	7	317	7.15	10.85	18
5100	25 x 25	40	12	7-1/2	149	5.80	9.05	14.85
5200		75	16	7-1/2	185	6.30	9.70	16
5300		125	18	7-1/2	250	6.85	10.55	17.40
5400		200	20	7-1/2	332	8.10	11.40	19.50
6000	25 x 30	40	14	8-1/2	165	6.40	9.10	15.50
6200		75	16	8-1/2	201	6.85	9.90	16.75
6400		125	18	8-1/2	259	7.50	10.65	18.15

B1010 220		Cast in Place Beam & Slab, Two Way						
	BAY SIZE (FT.)	SUPERIMPOSED LOAD (P.S.F.)	MINIMUM COL. SIZE (IN.)	SLAB THICKNESS (IN.)	TOTAL LOAD (P.S.F.)	COST PER S.F.		
						MAT.	INST.	TOTAL
6600	25 x 30	200	20	8-1/2	341	8.40	11.30	19.70
7000	30 x 30	40	14	9	170	6.95	9.70	16.65
7100		75	18	9	212	7.60	10.45	18.05
7300		125	20	9	267	8.15	10.85	19
7400		200	24	9	351	8.85	11.45	20.30
7600	30 x 35	40	16	10	188	7.60	10.30	17.90
7700		75	18	10	225	8.05	10.65	18.70
8000		125	22	10	282	8.95	11.50	20.45
8200		200	26	10	371	9.65	11.85	21.50
8500	35 x 35	40	16	10-1/2	193	8.15	10.50	18.65
8600		75	20	10-1/2	233	8.55	10.90	19.45
9000		125	24	10-1/2	287	9.50	11.70	21.20
9100		200	26	10-1/2	370	9.85	12.40	22.25
9300	35 x 40	40	18	11-1/2	208	8.80	10.85	19.65
9500		75	22	11-1/2	247	9.20	11.15	20.35
9800		125	24	11-1/2	302	9.75	11.55	21.30
9900		200	26	11-1/2	383	10.50	12.30	22.80

Important: See the Reference Section for critical supporting data - Reference Numbers and City Cost Indexes

B1010 Floor Construction

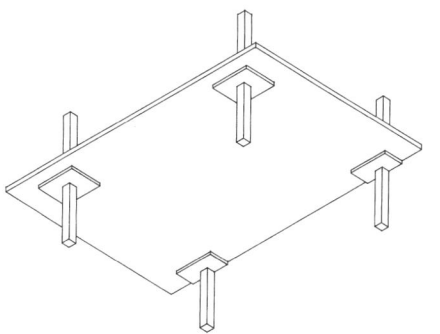

General: Flat Slab: Solid uniform depth concrete two-way slabs with drop panels at columns and no column capitals.

Design and Pricing Assumptions:
Concrete f'c = 3 KSI, placed by concrete pump.
Reinforcement, fy = 60 KSI.
Forms, four use.
Finish, steel trowel.
Curing, spray on membrane.
Based on 4 bay x 4 bay structure.

SHELL **B**

System Components	QUANTITY	UNIT	COST PER S.F.		
			MAT.	INST.	TOTAL
SYSTEM B1010 222 1700					
15'X15' BAY 40 PSF S. LOAD, 12" MIN. COL. 6" SLAB, 1-1/2" DROP, 117 PSF					
Forms in place, flat slab with drop panels, to 15' high, 4 uses	.993	S.F.	1.70	4.71	6.41
Forms in place, exterior spandrel, 12" wide, 4 uses	.034	SFCA	.05	.28	.33
Reinforcing in place, elevated slabs #4 to #7	1.588	Lb.	.78	.57	1.35
Concrete ready mix, regular weight, 3000 psi	.513	C.F.	1.82		1.82
Place and vibrate concrete, elevated slab, 6" to 10" pump	.513	C.F.		.55	.55
Finish floor, monolithic steel trowel finish for finish floor	1.000	S.F.		.74	.74
Cure with sprayed membrane curing compound	.010	C.S.F.	.06	.07	.13
TOTAL			4.41	6.92	11.33

B1010 222		Cast in Place Flat Slab with Drop Panels						
	BAY SIZE (FT.)	SUPERIMPOSED LOAD (P.S.F.)	MINIMUM COL. SIZE (IN.)	SLAB & DROP (IN.)	TOTAL LOAD (P.S.F.)	COST PER S.F.		
						MAT.	INST.	TOTAL
1700	15 x 15	40	12	6 - 1-1/2	117	4.40	6.90	11.30
1720		75	12	6 - 2-1/2	153	4.49	6.95	11.44
1760	RB1010 -010	125	14	6 - 3-1/2	205	4.70	7.10	11.80
1780		200	16	6 - 4-1/2	281	4.94	7.25	12.19
1840	15 x 20	40	12	6-1/2 - 2	124	4.68	7.05	11.73
1860		75	14	6-1/2 - 4	162	4.89	7.20	12.09
1880	RB1010 -100	125	16	6-1/2 - 5	213	5.20	7.35	12.55
1900		200	18	6-1/2 - 6	293	5.30	7.45	12.75
1960	20 x 20	40	12	7 - 3	132	4.92	7.15	12.07
1980		75	16	7 - 4	168	5.20	7.30	12.50
2000		125	18	7 - 6	221	5.75	7.55	13.30
2100		200	20	8 - 6-1/2	309	5.85	7.70	13.55
2300	20 x 25	40	12	8 - 5	147	5.50	7.40	12.90
2400		75	18	8 - 6-1/2	184	5.90	7.65	13.55
2600		125	20	8 - 8	236	6.40	8.05	14.45
2800		200	22	8-1/2 - 8-1/2	323	6.65	8.20	14.85
3200	25 x 25	40	12	8-1/2 - 5-1/2	154	5.75	7.55	13.30
3400		75	18	8-1/2 - 7	191	6.05	7.75	13.80
4000		125	20	8-1/2 - 8-1/2	243	6.50	8.05	14.55
4400		200	24	9 - 8-1/2	329	6.80	8.25	15.05
5000	25 x 30	40	14	9-1/2 - 7	168	6.25	7.75	14
5200		75	18	9-1/2 - 7	203	6.65	8.05	14.70
5600		125	22	9-1/2 - 8	256	6.95	8.25	15.20
5800		200	24	10 - 10	342	7.40	8.50	15.90
6400	30 x 30	40	14	10-1/2 - 7-1/2	182	6.70	7.95	14.65

B10 Superstructure

B1010 Floor Construction

B1010 222	Cast in Place Flat Slab with Drop Panels						

| | BAY SIZE (FT.) | SUPERIMPOSED LOAD (P.S.F.) | MINIMUM COL. SIZE (IN.) | SLAB & DROP (IN.) | TOTAL LOAD (P.S.F.) | COST PER S.F. | | |
						MAT.	INST.	TOTAL
6600	30 x 30	75	18	10-1/2 - 7-1/2	217	7.15	8.30	15.45
6800		125	22	10-1/2 - 9	269	7.50	8.50	16
7000		200	26	11 - 11	359	8	8.85	16.85
7400	30 x 35	40	16	11-1/2 - 9	196	7.30	8.30	15.60
7900		75	20	11-1/2 - 9	231	7.80	8.65	16.45
8000		125	24	11-1/2 - 11	284	8.10	8.85	16.95
9000	35 x 35	40	16	12 - 9	202	7.50	8.35	15.85
9400		75	20	12 - 11	240	8.10	8.75	16.85
9600		125	24	12 - 11	290	8.35	8.95	17.30

Important: See the Reference Section for critical supporting data - Reference Numbers and City Cost Indexes

B1010 Floor Construction

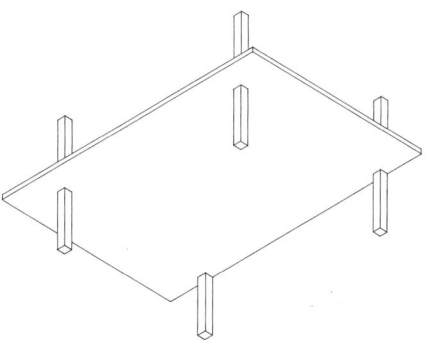

SHELL B

General: Flat Plates: Solid uniform depth concrete two-way slab without drops or interior beams. Primary design limit is shear at columns.

Design and Pricing Assumptions:
Concrete f'c to 4 KSI, placed by concrete pump.
Reinforcement, fy = 60 KSI.
Forms, four use.
Finish, steel trowel.
Curing, spray on membrane.
Based on 4 bay x 4 bay structure.

System Components	QUANTITY	UNIT	COST PER S.F.		
			MAT.	INST.	TOTAL
SYSTEM B1010 223 2000					
15'X15' BAY 40 PSF S. LOAD, 12" MIN. COL.					
Forms in place, flat plate to 15' high, 4 uses	.992	S.F.	1.53	4.57	6.10
Edge forms to 6" high on elevated slab, 4 uses	.065	L.F.	.01	.22	.23
Reinforcing in place, elevated slabs #4 to #7	1.706	Lb.	.84	.61	1.45
Concrete ready mix, regular weight, 3000 psi	.459	C.F.	1.62		1.62
Place and vibrate concrete, elevated slab less than 6", pump	.459	C.F.		.58	.58
Finish floor, monolithic steel trowel finish for finish floor	1.000	S.F.		.74	.74
Cure with sprayed membrane curing compound	.010	C.S.F.	.06	.07	.13
TOTAL			4.06	6.79	10.85

B1010 223		Cast in Place Flat Plate						
	BAY SIZE (FT.)	SUPERIMPOSED LOAD (P.S.F.)	MINIMUM COL. SIZE (IN.)	SLAB THICKNESS (IN.)	TOTAL LOAD (P.S.F.)	COST PER S.F.		
						MAT.	INST.	TOTAL
2000	15 x 15	40	12	5-1/2	109	4.06	6.80	10.86
2200	RB1010 -010	75	14	5-1/2	144	4.09	6.80	10.89
2400		125	20	5-1/2	194	4.25	6.90	11.15
2600		175	22	5-1/2	244	4.33	6.90	11.23
3000	15 x 20	40	14	7	127	4.67	6.90	11.57
3400	RB1010 -100	75	16	7-1/2	169	4.96	7	11.96
3600		125	22	8-1/2	231	5.45	7.20	12.65
3800		175	24	8-1/2	281	5.45	7.20	12.65
4200	20 x 20	40	16	7	127	4.67	6.90	11.57
4400		75	20	7-1/2	175	5	7.05	12.05
4600		125	24	8-1/2	231	5.45	7.20	12.65
5000		175	24	8-1/2	281	5.50	7.25	12.75
5600	20 x 25	40	18	8-1/2	146	5.40	7.20	12.60
6000		75	20	9	188	5.60	7.25	12.85
6400		125	26	9-1/2	244	6	7.50	13.50
6600		175	30	10	300	6.25	7.60	13.85
7000	25 x 25	40	20	9	152	5.60	7.25	12.85
7400		75	24	9-1/2	194	5.90	7.45	13.35
7600		125	30	10	250	6.25	7.60	13.85
8000								

General: Combination of thin concrete slab and monolithic ribs at uniform spacing to reduce dead weight and increase rigidity. The ribs (or joists) are arranged parallel in one direction between supports.

Square end joists simplify forming. Tapered ends can increase span or provide for heavy load.

Costs for multiple span joists are provided in this section. Single span joist costs are not provided here.

Design and Pricing Assumptions:
Concrete f'c = 4 KSI, normal weight placed by concrete pump.
Reinforcement, fy = 60 KSI.
Forms, four use.
　4-1/2″ slab.
　30″ pans, sq. ends (except for shear req.).
　6″ rib thickness.
　Distribution ribs as required.
Finish, steel trowel.
Curing, spray on membrane.
Based on 4 bay x 4 bay structure.

System Components	QUANTITY	UNIT	COST PER S.F.		
			MAT.	INST.	TOTAL
SYSTEM B1010 226 2000					
15′X15′ BAY, 40 PSF, S. LOAD 12″ MIN. COLUMN					
Forms in place, floor slab with 30″ metal pans, 4 use	.905	S.F.	2.85	4.66	7.51
Forms in place, exterior spandrel, 12″ wide, 4 uses	.170	SFCA	.23	1.41	1.64
Forms in place, interior beam. 12″ wide, 4 uses	.095	SFCA	.14	.65	.79
Edge forms, 7″-12″ high on elevated slab, 4 uses	.010	L.F.	.01	.05	.06
Reinforcing in place, elevated slabs #4 to #7	.628	Lb.	.31	.23	.54
Concrete ready mix, regular weight, 4000 psi	.555	C.F.	2.06		2.06
Place and vibrate concrete, elevated slab, 6″ to 10″ pump	.555	C.F.		.59	.59
Finish floor, monolithic steel trowel finish for finish floor	1.000	S.F.		.74	.74
Cure with sprayed membrane curing compound	.010	S.F.	.06	.07	.13
TOTAL			5.66	8.40	14.06

B1010 226			Cast in Place Multispan Joist Slab					
	BAY SIZE (FT.)	SUPERIMPOSED LOAD (P.S.F.)	MINIMUM COL. SIZE (IN.)	RIB DEPTH (IN.)	TOTAL LOAD (P.S.F.)	COST PER S.F.		
						MAT.	INST.	TOTAL
2000	15 x 15	40	12	8	115	5.65	8.40	14.05
2100		75	12	8	150	5.70	8.45	14.15
2200	RB1010 -010	125	12	8	200	5.80	8.55	14.35
2300		200	14	8	275	6	8.85	14.85
2600	15 x 20	40	12	8	115	5.75	8.40	14.15
2800		75	12	8	150	5.90	8.60	14.50
3000	RB1010 -100	125	14	8	200	6.10	9.05	15.15
3300		200	16	8	275	6.40	9.15	15.55
3600	20 x 20	40	12	10	120	5.85	8.25	14.10
3900		75	14	10	155	6.15	8.75	14.90
4000		125	16	10	205	6.20	8.90	15.10
4100		200	18	10	280	6.50	9.30	15.80
4300	20 x 25	40	12	10	120	5.90	8.35	14.25
4400		75	14	10	155	6.15	8.80	14.95
4500		125	16	10	205	6.50	9.25	15.75
4600		200	18	12	280	6.80	9.70	16.50
4700	25 x 25	40	12	12	125	5.95	8.15	14.10
4800		75	16	12	160	6.30	8.60	14.90
4900		125	18	12	210	7	9.50	16.50
5000		200	20	14	291	7.30	9.70	17

Important: See the Reference Section for critical supporting data - Reference Numbers and City Cost Indexes

B1010 Floor Construction

B1010 226	Cast in Place Multispan Joist Slab

	BAY SIZE (FT.)	SUPERIMPOSED LOAD (P.S.F.)	MINIMUM COL. SIZE (IN.)	RIB DEPTH (IN.)	TOTAL LOAD (P.S.F.)	COST PER S.F.		
						MAT.	INST.	TOTAL
5400	25 x 30	40	14	12	125	6.20	8.50	14.70
5600		75	16	12	160	6.40	8.85	15.25
5800		125	18	12	210	6.85	9.45	16.30
6000		200	20	14	291	7.35	9.85	17.20
6200	30 x 30	40	14	14	131	6.50	8.65	15.15
6400		75	18	14	166	6.70	8.95	15.65
6600		125	20	14	216	7.10	9.40	16.50
6700		200	24	16	297	7.60	9.75	17.35
6900	30 x 35	40	16	14	131	6.70	9.05	15.75
7000		75	18	14	166	6.80	9.10	15.90
7100		125	22	14	216	7	9.60	16.60
7200		200	26	16	297	7.70	10.10	17.80
7400	35 x 35	40	16	16	137	6.85	8.95	15.80
7500		75	20	16	172	7.20	9.25	16.45
7600		125	24	16	222	7.25	9.30	16.55
7700		200	26	20	309	7.90	9.90	17.80
8000	35 x 40	40	18	16	137	7.10	9.25	16.35
8100		75	22	16	172	7.40	9.65	17.05
8300		125	26	16	222	7.45	9.50	16.95
8400		200	30	20	309	8.15	10	18.15
8750	40 x 40	40	18	20	149	7.55	9.25	16.80
8800		75	24	20	184	7.80	9.45	17.25
8900		125	26	20	234	8.10	9.85	17.95
9100	40 x 45	40	20	20	149	7.85	9.60	17.45
9500		75	24	20	184	7.95	9.65	17.60
9800		125	28	20	234	8.20	10	18.20

B1010 Floor Construction

General: Waffle slabs are basically flat slabs with hollowed out domes on bottom side to reduce weight. Solid concrete heads at columns function as drops without increasing depth. The concrete ribs function as two-way right angle joist.

Joists are formed with standard sized domes. Thin slabs cover domes and are usually reinforced with welded wire fabric. Ribs have bottom steel and may have stirrups for shear.

Design and Pricing Assumptions:
Concrete f'c = 4 KSI, normal weight placed by concrete pump.
Reinforcement, fy = 60 KSI.
Forms, four use.
  4-1/2″ slab.
  30″ x 30″ voids.
  6″ wide ribs.
  (ribs @ 36″ O.C.).
  Rib depth filler beams as required.
  Solid concrete heads at columns.
Finish, steel trowel.
Curing, spray on membrane.
Based on 4 bay x 4 bay structure.

System Components	QUANTITY	UNIT	MAT.	INST.	TOTAL
SYSTEM B1010 227 3900					
20X20′ BAY, 40 PSF S. LOAD, 12″ MIN. COLUMN					
Formwork, floor slab with 30″ fiberglass domes, 4 uses	1.000	S.F.	5.30	6.15	11.45
Edge forms, 7″-12″ high on elevated slab, 4 uses	.052	SFCA	.07	.27	.34
Forms in place, bulkhead for slab with keyway, 1 use, 3 piece	.010	L.F.	.02	.06	.08
Reinforcing in place, elevated slabs #4 to #7	1.580	Lb.	.77	.57	1.34
Welded wire fabric rolls, 6 x 6 - W4 x W4 (4 x 4) 58 lb./c.s.f	1.000	S.F.	.26	.39	.65
Concrete ready mix, regular weight, 4000 psi	.690	C.F.	2.56		2.56
Place and vibrate concrete, elevated slab, over 10″, pump	.690	C.F.		.76	.76
Finish floor, monolithic steel trowel finish for finish floor	1.000	S.F.		.74	.74
Cure with sprayed membrane curing compound	.010	C.S.F.	.06	.07	.13
TOTAL			9.04	9.01	18.05

B1010 227	Cast in Place Waffle Slab							
	BAY SIZE (FT.)	SUPERIMPOSED LOAD (P.S.F.)	MINIMUM COL. SIZE (IN.)	RIB DEPTH (IN.)	TOTAL LOAD (P.S.F.)	MAT.	INST.	TOTAL
3900	20 x 20	40	12	8	144	9.05	9	18.05
4000		75	12	8	179	9.15	9.10	18.25
4100	RB1010 -010	125	16	8	229	9.30	9.20	18.50
4200		200	18	8	304	9.75	9.50	19.25
4400	20 x 25	40	12	8	146	9.20	9.05	18.25
4500		75	14	8	181	9.35	9.20	18.55
4600	RB1010 -100	125	16	8	231	9.50	9.30	18.80
4700		200	18	8	306	9.85	9.55	19.40
4900	25 x 25	40	12	10	150	9.40	9.15	18.55
5000		75	16	10	185	9.60	9.30	18.90
5300		125	18	10	235	9.85	9.50	19.35
5500		200	20	10	310	10.05	9.65	19.70
5700	25 x 30	40	14	10	154	9.50	9.20	18.70
5800		75	16	10	189	9.75	9.35	19.10
5900		125	18	10	239	9.95	9.50	19.45
6000		200	20	12	329	10.75	9.95	20.70
6400	30 x 30	40	14	12	169	10	9.40	19.40
6500		75	18	12	204	10.20	9.55	19.75
6600		125	20	12	254	10.35	9.65	20
6700		200	24	12	329	11.10	10.15	21.25

Important: See the Reference Section for critical supporting data - Reference Numbers and City Cost Indexes

B1010 Floor Construction

B1010 227	Cast in Place Waffle Slab						

	BAY SIZE (FT.)	SUPERIMPOSED LOAD (P.S.F.)	MINIMUM COL. SIZE (IN.)	RIB DEPTH (IN.)	TOTAL LOAD (P.S.F.)	COST PER S.F.		
						MAT.	INST.	TOTAL
6900	30 x 35	40	16	12	169	10.20	9.50	19.70
7000		75	18	12	204	10.20	9.50	19.70
7100		125	22	12	254	10.60	9.80	20.40
7200		200	26	14	334	11.55	10.35	21.90
7400	35 x 35	40	16	14	174	10.65	9.70	20.35
7500		75	20	14	209	10.85	9.85	20.70
7600		125	24	14	259	11.05	10	21.05
7700		200	26	16	346	11.75	10.50	22.25
8000	35 x 40	40	18	14	176	10.90	9.85	20.75
8300		75	22	14	211	11.20	10.10	21.30
8500		125	26	16	271	11.65	10.30	21.95
8750		200	30	20	372	12.60	10.85	23.45
9200	40 x 40	40	18	14	176	11.20	10.10	21.30
9400		75	24	14	211	11.55	10.35	21.90
9500		125	26	16	271	11.80	10.45	22.25
9700	40 x 45	40	20	16	186	11.55	10.25	21.80
9800		75	24	16	221	11.90	10.50	22.40
9900		125	28	16	271	12.10	10.65	22.75

SHELL B

B1010 Floor Construction

General: Units priced here are for plant produced prestressed members, transported to site and erected.

Normal weight concrete is most frequently used. Lightweight concrete may be used to reduce dead weight.

Structural topping is sometimes used on floors: insulating concrete or rigid insulation on roofs.

Camber and deflection may limit use by depth considerations.

Prices are based upon 10,000 S.F. to 20,000 S.F. projects, and 50 mile to 100 mile transport.

Concrete is f'c = 5 KSI and Steel is fy = 250 or 300 KSI

Note: Deduct from prices 20% for Southern states. Add to prices 10% for Western states.

Description of Table: Enter table at span and load. Most economical sections will generally consist of normal weight concrete without topping. If acceptable, note this price, depth and weight. For topping and/or lightweight concrete, note appropriate data.

Generally used on masonry and concrete bearing or reinforced concrete and steel framed structures.

The solid 4″ slabs are used for light loads and short spans. The 6″ to 12″ thick hollow core units are used for longer spans and heavier loads. Cores may carry utilities.

Topping is used structurally for loads or rigidity and architecturally to level or slope surface.

Camber and deflection and change in direction of spans must be considered (door openings, etc.), especially untopped.

System Components	QUANTITY	UNIT	COST PER S.F.		
			MAT.	INST.	TOTAL
SYSTEM B1010 230 2000					
10′ SPAN, 40 LBS S.F. WORKING LOAD, 2″ TOPPING					
Precast prestressed concrete roof/floor slabs 4″ thick, grouted	1.000	S.F.	4.88	2.75	7.63
Edge forms to 6″ high on elevated slab, 4 uses	.100	L.F.	.02	.33	.35
Welded wire fabric 6 x 6 - W1.4 x W1.4 (10 x 10), 21 lb/csf, 10% lap	.010	C.S.F.	.13	.30	.43
Concrete ready mix, regular weight, 3000 psi	.170	C.F.	.60		.60
Place and vibrate concrete, elevated slab less than 6″, pumped	.170	C.F.		.22	.22
Finishing floor, monolithic steel trowel finish for resilient tile	1.000	S.F.		.68	.68
Curing with sprayed membrane curing compound	.010	C.S.F.	.06	.07	.13
TOTAL			5.69	4.35	10.04

B1010 229		Precast Plank with No Topping						
	SPAN (FT.)	SUPERIMPOSED LOAD (P.S.F.)	TOTAL DEPTH (IN.)	DEAD LOAD (P.S.F.)	TOTAL LOAD (P.S.F.)	COST PER S.F.		
						MAT.	INST.	TOTAL
0720	10	40	4	50	90	4.88	2.75	7.63
0750		75	6	50	125	6.05	2.35	8.40
0770	RB1010 -010	100	6	50	150	6.05	2.35	8.40
0800	15	40	6	50	90	6.05	2.35	8.40
0820		75	6	50	125	6.05	2.35	8.40
0850	RB1010 -100	100	6	50	150	6.05	2.35	8.40
0875	20	40	6	50	90	6.05	2.35	8.40
0900		75	6	50	125	6.05	2.35	8.40
0920		100	6	50	150	6.05	2.35	8.40
0950	25	40	6	50	90	6.05	2.35	8.40
0970		75	8	55	130	6.35	2.06	8.41
1000		100	8	55	155	6.35	2.06	8.41
1200	30	40	8	55	95	6.35	2.06	8.41
1300		75	8	55	130	6.35	2.06	8.41
1400		100	10	70	170	6.70	1.83	8.53
1500	40	40	10	70	110	6.70	1.83	8.53
1600		75	12	70	145	7.60	1.65	9.25

Important: See the Reference Section for critical supporting data - Reference Numbers and City Cost Indexes

B10 Superstructure

B1010 Floor Construction

B1010 229	Precast Plank with No Topping							
	SPAN (FT.)	SUPERIMPOSED LOAD (P.S.F.)	TOTAL DEPTH (IN.)	DEAD LOAD (P.S.F.)	TOTAL LOAD (P.S.F.)	COST PER S.F.		
						MAT.	INST.	TOTAL
1700	45	40	12	70	110	7.60	1.65	9.25

B1010 230	Precast Plank with 2″ Concrete Topping							
	SPAN (FT.)	SUPERIMPOSED LOAD (P.S.F.)	TOTAL DEPTH (IN.)	DEAD LOAD (P.S.F.)	TOTAL LOAD (P.S.F.)	COST PER S.F.		
						MAT.	INST.	TOTAL
2000	10	40	6	75	115	5.70	4.35	10.05
2100		75	8	75	150	6.85	3.95	10.80
2200		100	8	75	175	6.85	3.95	10.80
2500	15	40	8	75	115	6.85	3.95	10.80
2600		75	8	75	150	6.85	3.95	10.80
2700		100	8	75	175	6.85	3.95	10.80
2800	20	40	8	75	115	6.85	3.95	10.80
2900		75	8	75	150	6.85	3.95	10.80
3000		100	8	75	175	6.85	3.95	10.80
3100	25	40	8	75	115	6.85	3.95	10.80
3200		75	8	75	150	6.85	3.95	10.80
3300		100	10	80	180	7.15	3.66	10.81
3400	30	40	10	80	120	7.15	3.66	10.81
3500		75	10	80	155	7.15	3.66	10.81
3600		100	10	80	180	7.15	3.66	10.81
3700	35	40	12	95	135	7.50	3.43	10.93
3800		75	12	95	170	7.50	3.43	10.93
3900		100	14	95	195	8.40	3.25	11.65
4000	40	40	12	95	135	7.50	3.43	10.93
4500		75	14	95	170	8.40	3.25	11.65
5000	45	40	14	95	135	8.40	3.25	11.65

B1010 Floor Construction

Most widely used for moderate span floors and roofs. At shorter spans, they tend to be competitive with hollow core slabs. They are also used as wall panels.

System Components	QUANTITY	UNIT	COST PER S.F.		
			MAT.	INST.	TOTAL
SYSTEM B1010 235 6700					
PRECAST, DOUBLE "T", 2" TOPPING, 30' SPAN, 30 PSF SUP. LOAD, 18" X 8'					
Double "T" beams, reg. wt, 18" x 8' w, 30' span	1.000	S.F.	7.29	1.37	8.66
Edge forms to 6" high on elevated slab, 4 uses	.050	L.F.	.01	.17	.18
Concrete ready mix, regular weight, 3000 psi	.250	C.F.	.89		.89
Place and vibrate concrete, elevated slab less than 6", pumped	.250	C.F.		.32	.32
Finishing floor, monolithic steel trowel finish for finish floor	1.000	S.F.		.74	.74
Curing with sprayed membrane curing compound	.010	C.S.F.	.06	.07	.13
TOTAL			8.25	2.67	10.92

B1010 234		Precast Double "T" Beams with No Topping						
	SPAN (FT.)	SUPERIMPOSED LOAD (P.S.F.)	DBL. "T" SIZE D (IN.) W (FT.)	CONCRETE "T" TYPE	TOTAL LOAD (P.S.F.)	COST PER S.F.		
						MAT.	INST.	TOTAL
1500	30	30	18x8	Reg. Wt.	92	7.30	1.37	8.67
1600		40	18x8	Reg. Wt.	102	7.40	1.78	9.18
1700	RB1010 -010	50	18x8	Reg. Wt	112	7.40	1.78	9.18
1800		75	18x8	Reg. Wt.	137	7.45	1.85	9.30
1900		100	18x8	Reg. Wt.	162	7.45	1.85	9.30
2000	40	30	20x8	Reg. Wt.	87	5.65	1.14	6.79
2100		40	20x8	Reg. Wt.	97	5.70	1.38	7.08
2200	RB1010 -100	50	20x8	Reg. Wt.	107	5.70	1.38	7.08
2300		75	20x8	Reg. Wt.	132	5.75	1.48	7.23
2400		100	20x8	Reg. Wt.	157	5.85	1.82	7.67
2500	50	30	24x8	Reg. Wt.	103	5.95	1.03	6.98
2600		40	24x8	Reg. Wt.	113	6	1.26	7.26
2700		50	24x8	Reg. Wt.	123	6.05	1.34	7.39
2800		75	24x8	Reg. Wt.	148	6.05	1.36	7.41
2900		100	24x8	Reg. Wt.	173	6.15	1.70	7.85
3000	60	30	24x8	Reg. Wt.	82	6.05	1.36	7.41
3100		40	32x10	Reg. Wt.	104	7.05	1.12	8.17
3150		50	32x10	Reg. Wt.	114	7	.97	7.97
3200		75	32x10	Reg. Wt.	139	7.05	1.05	8.10
3250		100	32x10	Reg. Wt.	164	7.10	1.31	8.41
3300	70	30	32x10	Reg. Wt.	94	7.05	1.04	8.09
3350		40	32x10	Reg. Wt.	104	7.05	1.05	8.10
3400		50	32x10	Reg. Wt.	114	7.10	1.31	8.41
3450		75	32x10	Reg. Wt.	139	7.20	1.58	8.78
3500		100	32x10	Reg. Wt.	164	7.35	2.11	9.46
3550	80	30	32x10	Reg. Wt.	94	7.10	1.31	8.41
3600		40	32x10	Reg. Wt.	104	7.30	1.84	9.14
3900		50	32x10	Reg. Wt.	114	7.35	2.10	9.45

Important: See the Reference Section for critical supporting data - Reference Numbers and City Cost Indexes

B10 Superstructure

B1010 Floor Construction

B1010 234 — Precast Double "T" Beams with No Topping

	SPAN (FT.)	SUPERIMPOSED LOAD (P.S.F.)	DBL. "T" SIZE D (IN.) W (FT.)	CONCRETE "T" TYPE	TOTAL LOAD (P.S.F.)	COST PER S.F.		
						MAT.	INST.	TOTAL
4300	50	30	20x8	Lt. Wt.	66	7	1.26	8.26
4400		40	20x8	Lt. Wt.	76	7.50	1.15	8.65
4500		50	20x8	Lt. Wt.	86	7.60	1.38	8.98
4600		75	20x8	Lt. Wt.	111	7.65	1.59	9.24
4700		100	20x8	Lt. Wt.	136	7.75	1.91	9.66
4800	60	30	24x8	Lt. Wt.	70	7.55	1.25	8.80
4900		40	32x10	Lt. Wt.	88	8.25	.87	9.12
5000		50	32x10	Lt. Wt.	98	8.30	1.05	9.35
5200		75	32x10	Lt. Wt.	123	8.35	1.22	9.57
5400		100	32x10	Lt. Wt.	148	8.45	1.48	9.93
5600	70	30	32x10	Lt. Wt.	78	8.25	.87	9.12
5750		40	32x10	Lt. Wt.	88	8.30	1.05	9.35
5900		50	32x10	Lt. Wt.	98	8.35	1.22	9.57
6000		75	32x10	Lt. Wt.	123	8.45	1.48	9.93
6100		100	32x10	Lt. Wt.	148	8.60	2	10.60
6200	80	30	32x10	Lt. Wt.	78	8.35	1.21	9.56
6300		40	32x10	Lt. Wt.	88	8.45	1.47	9.92
6400		50	32x10	Lt. Wt.	98	8.55	1.74	10.29

B1010 235 — Precast Double "T" Beams With 2" Topping

	SPAN (FT.)	SUPERIMPOSED LOAD (P.S.F.)	DBL. "T" SIZE D (IN.) W (FT.)	CONCRETE "T" TYPE	TOTAL LOAD (P.S.F.)	COST PER S.F.		
						MAT.	INST.	TOTAL
6700	30	30	18x8	Reg. Wt.	117	8.25	2.66	10.91
6750		40	18x8	Reg. Wt.	127	8.25	2.66	10.91
6800		50	18x8	Reg. Wt.	137	8.30	2.91	11.21
6900		75	18x8	Reg. Wt.	162	8.30	2.91	11.21
7000		100	18x8	Reg. Wt.	187	8.35	3.01	11.36
7100	40	30	18x8	Reg. Wt.	120	6.50	2.56	9.06
7200		40	20x8	Reg. Wt.	130	6.60	2.44	9.04
7300		50	20x8	Reg. Wt.	140	6.65	2.68	9.33
7400		75	20x8	Reg. Wt.	165	6.70	2.78	9.48
7500		100	20x8	Reg. Wt.	190	6.80	3.12	9.92
7550	50	30	24x8	Reg. Wt.	120	6.95	2.55	9.50
7600		40	24x8	Reg. Wt.	130	7	2.64	9.64
7750		50	24x8	Reg. Wt.	140	7	2.65	9.65
7800		75	24x8	Reg. Wt.	165	7.10	2.99	10.09
7900		100	32x10	Reg. Wt.	189	8.05	2.45	10.50
8000	60	30	32x10	Reg. Wt.	118	7.90	2.08	9.98
8100		40	32x10	Reg. Wt.	129	7.95	2.27	10.22
8200		50	32x10	Reg. Wt.	139	8.05	2.45	10.50
8300		75	32x10	Reg. Wt.	164	8.10	2.61	10.71
8350		100	32x10	Reg. Wt.	189	8.15	2.88	11.03
8400	70	30	32x10	Reg. Wt.	119	8	2.35	10.35
8450		40	32x10	Reg. Wt.	129	8.10	2.61	10.71
8500		50	32x10	Reg. Wt.	139	8.15	2.87	11.02
8550		75	32x10	Reg. Wt.	164	8.30	3.40	11.70
8600	80	30	32x10	Reg. Wt.	119	8.30	3.40	11.70
8800	50	30	20x8	Lt. Wt.	105	8	2.66	10.66
8850		40	24x8	Lt. Wt.	121	8.05	2.78	10.83
8900		50	24x8	Lt. Wt.	131	8.10	2.99	11.09
8950		75	24x8	Lt. Wt.	156	8.10	2.99	11.09
9000		100	24x8	Lt. Wt.	181	8.20	3.32	11.52

SHELL B

75

B1010 Floor Construction

B1010 235	Precast Double "T" Beams With 2" Topping

	SPAN (FT.)	SUPERIMPOSED LOAD (P.S.F.)	DBL. "T" SIZE D (IN.) W (FT.)	CONCRETE "T" TYPE	TOTAL LOAD (P.S.F.)	COST PER S.F.		
						MAT.	INST.	TOTAL
9200	60	30	32x10	Lt. Wt.	103	9.20	2.16	11.36
9300		40	32x10	Lt. Wt.	113	9.30	2.34	11.64
9350		50	32x10	Lt. Wt.	123	9.30	2.34	11.64
9400		75	32x10	Lt. Wt.	148	9.35	2.51	11.86
9450		100	32x10	Lt. Wt.	173	9.40	2.77	12.17
9500	70	30	32x10	Lt. Wt.	103	9.35	2.51	11.86
9550		40	32x10	Lt. Wt.	113	9.35	2.51	11.86
9600		50	32x10	Lt. Wt.	123	9.40	2.77	12.17
9650		75	32x10	Lt. Wt.	148	9.50	3.04	12.54
9700	80	30	32x10	Lt. Wt.	103	9.40	2.77	12.17
9800		40	32x10	Lt. Wt.	113	9.50	3.04	12.54
9900		50	32x10	Lt. Wt.	123	9.55	3.29	12.84

B SHELL

Important: See the Reference Section for critical supporting data - Reference Numbers and City Cost Indexes

B1010 Floor Construction

General: Units priced here are for plant produced prestressed members transported to the site and erected.

System has precast prestressed concrete beams and precast, prestressed hollow core slabs spanning the longer direction when applicable.

Camber and deflection must be considered when using untopped hollow core slabs.

Design and Pricing Assumptions:
Prices are based on 10,000 S.F. to 20,000 S.F. projects and 50 mile to 100 mile transport.

Concrete is f'c 5 KSI and prestressing steel is fy = 250 or 300 KSI.

Note: Deduct from prices 20% for Southern states. Add to prices 10% for Western states.

System Components	QUANTITY	UNIT	COST PER S.F.		
			MAT.	INST.	TOTAL
SYSTEM B1010 236 5200					
20'X20' BAY, 6" DEPTH, 40 PSF S. LOAD, 100PSF TOTAL LOAD					
12" x 20" precast "T" beam, 20' span	.038	L.F.		.83	.83
Installation labor and equipment	.038	L.F.	4.86		4.86
12" x 20" precast "L" beam, 20' span	.025	L.F.		.83	.83
Installation labor and equipment	.025	L.F.	2.60		2.60
Precast prestressed concrete roof/floor slabs 6" deep, grouted	1.000	S.F.	6.05	2.35	8.40
TOTAL			13.51	4.01	17.52

B1010 236	Precast Beam & Plank with No Topping						
	BAY SIZE (FT.)	SUPERIMPOSED LOAD (P.S.F.)	PLANK THICKNESS (IN.)	TOTAL DEPTH (IN.)	TOTAL LOAD (P.S.F.)	COST PER S.F.		

	BAY SIZE (FT.)	SUPERIMPOSED LOAD (P.S.F.)	PLANK THICKNESS (IN.)	TOTAL DEPTH (IN.)	TOTAL LOAD (P.S.F.)	MAT.	INST.	TOTAL
5200	20x20	40	6	20	110	13.50	4	17.50
5400		75	6	24	148	14.30	4	18.30
5600		100	6	24	173	14.30	4	18.30
5800	20x25	40	8	24	113	12.90	3.71	16.61
6000		75	8	24	149	12.90	3.71	16.61
6100		100	8	36	183	13.90	3.71	17.61
6200	25x25	40	8	28	118	13.70	3.71	17.41
6400		75	8	36	158	14.35	3.71	18.06
6500		100	8	36	183	14.35	3.71	18.06
7000	25x30	40	8	36	110	13.05	3.71	16.76
7200		75	10	36	159	13.75	3.48	17.23
7400		100	12	36	188	15.25	3.30	18.55
7600	30x30	40	8	36	121	13.50	3.71	17.21
8000		75	10	44	140	14.20	3.71	17.91
8250		100	12	52	206	16.50	3.30	19.80
8500	30x35	40	12	44	135	14.15	3.30	17.45
8750		75	12	52	176	15.05	3.30	18.35
9000	35x35	40	12	52	141	15.10	3.30	18.40
9250		75	12	60	181	15.85	3.30	19.15
9500	35x40	40	12	52	137	14.40	4.14	18.54
9750	40x40	40	12	60	141	15.30	4.14	19.44

B1010 Floor Construction

General: Beams and hollow core slabs priced here are for plant produced prestressed members transported to the site and erected.

The 2″ structural topping is applied after the beams and hollow core slabs are in place and is reinforced with W.W.F.

Design and Pricing Assumptions:

Prices are based on 10,000 S.F. to 20,000 S.F. projects and 50 mile to 100 mile transport.

Concrete for prestressed members is f'c 5 KSI.

Concrete for topping is f'c 3000 PSI and placed by pump.

Prestressing steel is fy = 250 or 300 KSI.

W.W.F. is 6 x 6 – W1.4 x W1.4 (10 x 10).

Note: Deduct from prices 20% for Southern states. Add to prices 10% for Western states.

System Components	QUANTITY	UNIT	COST PER S.F.		
			MAT.	INST.	TOTAL
SYSTEM B1010 238 4300					
20′X20′ BAY, 6″ PLANK, 40 PSF S. LOAD, 135 PSF TOTAL LOAD					
12″ x 20″ precast "T" beam, 20′ span	.038	L.F.		.83	.83
Installation labor and equipment	.038	L.F.	4.86		4.86
12″ x 20″ precast "L" beam, 20′ span	.025	L.F.		.83	.83
Installation labor and equipment	.025	L.F.	2.60		2.60
Precast prestressed concrete roof/floor slabs 6″ deep, grouted	1.000	S.F.	6.05	2.35	8.40
Edge forms to 6″ high on elevated slab, 4 uses	.050	L.F.	.01	.17	.18
Forms in place, bulkhead for slab with keyway, 1 use, 2 piece	.013	L.F.	.02	.07	.09
Welded wire fabric rolls, 6 x 6 - W1.4 x W1.4 (10 x 10), 21 lb/csf	.010	C.S.F.	.13	.30	.43
Concrete ready mix, regular weight, 3000 psi	.170	C.F.	.60		.60
Place and vibrate concrete, elevated slab less than 6″, pump	.170	C.F.		.22	.22
Finish floor, monolithic steel trowel finish for finish floor	1.000	S.F.		.74	.74
Cure with sprayed membrane curing compound	.010	C.S.F.	.06	.07	.13
TOTAL			14.33	5.58	19.91

B1010 238	Precast Beam & Plank with 2″ Topping						
	BAY SIZE (FT.)	SUPERIMPOSED LOAD (P.S.F.)	PLANK THICKNESS (IN.)	TOTAL DEPTH (IN.)	TOTAL LOAD (P.S.F.)	COST PER S.F.		
						MAT.	INST.	TOTAL
4300	20x20	40	6	22	135	14.35	5.55	19.90
4400	RB1010-010	75	6	24	173	15.15	5.55	20.70
4500		100	6	28	200	15.60	5.55	21.15
4600	20x25	40	6	26	134	13.45	5.55	19
5000	RB1010-100	75	8	30	177	14.25	5.25	19.50
5200		100	8	30	202	14.25	5.25	19.50
5400	25x25	40	6	38	143	14.85	5.50	20.35
5600		75	8	38	183	14.85	5.50	20.35
6000		100	8	46	216	16.30	5.20	21.50
6200	25x30	40	8	38	144	13.85	5.20	19.05
6400		75	10	46	200	15.15	4.97	20.12
6600		100	10	46	225	15.15	4.97	20.12
7000	30x30	40	8	46	150	15.05	5.20	20.25
7200		75	10	54	181	16.10	5.20	21.30
7600		100	10	54	231	16.10	5.20	21.30
7800	30x35	40	10	54	166	14.95	4.94	19.89
8000		75	12	54	200	15.60	4.76	20.36
8200	35x35	40	10	62	170	15.45	4.94	20.39
9300		75	12	62	206	16.65	5.60	22.25
9500	35x40	40	12	62	167	15.95	5.55	21.50
9600	40x40	40	12	62	173	16.75	5.55	22.30

Important: See the Reference Section for critical supporting data - Reference Numbers and City Cost Indexes

B1010 Floor Construction

General: Beams and double tees priced here are for plant produced prestressed members transported to the site and erected.

The 2″ structural topping is applied after the beams and double tees are in place and is reinforced with W.W.F.

Design and Pricing Assumptions:
Prices are based on 10,000 S.F. to 20,000 S.F. projects and 50 mile to 100 mile transport.

Concrete for prestressed members is f'c 5 KSI.

Concrete for topping is f'c 3000 PSI and placed by pump.

Prestressing steel is fy = 250 or 300 KSI.

W.W.F. is 6 x 6 – W1.4 x W1.4 (10x10).

Note: Deduct from prices 20% for Southern states. Add to prices 10% for Western states.

SHELL B

System Components	QUANTITY	UNIT	COST PER S.F.		
			MAT.	INST.	TOTAL
SYSTEM B1010 239 3000					
25′X30′ BAY 38″ DEPTH, 130 PSF T.L., 2″ TOPPING					
12″ x 36″ precast "T" beam, 25′ span	.025	L.F.		.83	.83
Installation labor and equipment	.025	L.F.	4.60		4.60
Precast beam, "L" shaped, 12″ x 20″	.017	L.F.		.83	.83
Installation labor and equipment	.017	L.F.	2.11		2.11
Double T, standard weight, 16″ x 8′ w, 25′ span	.989	S.F.	6.49	1.36	7.85
Edge forms to 6″ high on elevated slab, 4 uses	.037	L.F.	.01	.12	.13
Forms in place, bulkhead for slab with keyway, 1 use, 2 piece	.010	L.F.	.02	.05	.07
Welded wire fabric rolls, 6 x 6 - W1.4 x W1.4 (10 x 10), 21 lb/csf	1.000	S.F.	.13	.30	.43
Concrete ready mix, regular weight, 3000 psi	.170	C.F.	.60		.60
Place and vibrate concrete, elevated slab less than 6″, pumped	.170	C.F.		.22	.22
Finishing floor, monolithic steel trowel finish for finish floor	1.000	S.F.		.74	.74
Curing with sprayed membrane curing compound	.010	S.F.	.06	.07	.13
TOTAL			14.02	4.52	18.54

B1010 239		Precast Double "T" & 2″ Topping on Precast Beams						
	BAY SIZE (FT.)	SUPERIMPOSED LOAD (P.S.F.)	DEPTH (IN.)		TOTAL LOAD (P.S.F.)	COST PER S.F.		
						MAT.	INST.	TOTAL
3000	25x30	40	38		130	14	4.51	18.51
3100	RB1010 -100	75	38		168	14	4.51	18.51
3300		100	46		196	14.65	4.51	19.16
3600	30x30	40	46		150	15.20	4.48	19.68
3750		75	46		174	15.20	4.48	19.68
4000		100	54		203	15.90	4.48	20.38
4100	30x40	40	46		136	12.40	4.25	16.65
4300		75	54		173	12.95	4.25	17.20
4400		100	62		204	13.60	4.25	17.85
4600	30x50	40	54		138	11.85	4.11	15.96
4800		75	54		181	12.35	4.11	16.46
5000		100	54		219	13.60	3.87	17.47
5200	30x60	40	62		151	12.50	3.88	16.38
5400		75	62		192	13.10	3.88	16.98
5600		100	62		215	13.05	3.88	16.93
5800	35x40	40	54		139	13.20	4.24	17.44
6000		75	62		179	14.35	4.12	18.47
6250		100	62		212	14.65	4.12	18.77
6500	35x50	40	62		142	12.60	4.11	16.71
6750		75	62		186	13.15	4.11	17.26
7300		100	62		231	15.20	3.87	19.07

B1010 Floor Construction

B1010 239	Precast Double "T" & 2" Topping on Precast Beams

	BAY SIZE (FT.)	SUPERIMPOSED LOAD (P.S.F.)	DEPTH (IN.)		TOTAL LOAD (P.S.F.)	COST PER S.F.		
						MAT.	INST.	TOTAL
7600	35x60	40	54		154	14.25	3.77	18.02
7750		75	54		179	14.60	3.77	18.37
8000		100	62		224	15.30	3.77	19.07
8250	40x40	40	62		145	14.55	5.05	19.60
8400		75	62		187	15.10	5.05	20.15
8750		100	62		223	16.15	5.05	21.20
9000	40x50	40	62		151	13.55	4.92	18.47
9300		75	62		193	14.30	4.18	18.48
9800	40x60	40	62		164	17.65	4.33	21.98

Important: See the Reference Section for critical supporting data - Reference Numbers and City Cost Indexes

"I'm working on a 1200 space parking structure...the owner wants us to present a steel alternative. Help!

How much will it cost?
How long will it take to construct?"

In today's fast-paced, competitive design and construction industry, success is often a matter of access to the right information at the right time. The AISC Steel Solutions Center makes it easy for you to explore traditional and innovative solutions enabling you to find, compare, select, and specify the right framing system for your building project.

From typical framing studies to total structural steel systems, INCLUDING PROJECT COSTS AND SCHEDULES, we can provide you with up-to-date information for your next building project — FOR FREE!

All you have to do is call 866.ASK.AISC or e-mail solutions@aisc.org.

For a sample of what The Steel Solutions Center can do for you, visit www.aisc.org/solutions.

Steel
SolutionsCenter
Your connection to
ideas + answers

American Institute of Steel Construction: making structural steel the material of choice

B1010   Floor Construction

General: The following table is based upon structural W shape beam and girder framing. Non-composite action is assumed between beams and decking. Deck costs not included.

The deck spans the short direction. The steel beams and girders are fireproofed with sprayed fiber fireproofing.

Design and Pricing Assumptions:
Structural steel is A36, with high strength A325 bolts.

Fireproofing is sprayed fiber (non-asbestos).

Total load includes steel, deck & live load.

Spandrels are assumed the same as interior beams and girders to allow for exterior wall loads and bracing or moment connections. No columns included in price.

See Tables B1010 258 and B1020 125-136 for metal deck costs.

System Components	QUANTITY	UNIT	COST PER S.F.		
			MAT.	INST.	TOTAL
SYSTEM B1010 241 1350 **15'X20' BAY,40 P.S.F. L.L.,12"DEPTH, .535P.S.F. FIREPROOF, 50 PSF T.LOAD**					
Structural steel	3.200	Lb.	3.42	1.12	4.54
Spray mineral fiber/cement for fire proof., 1" thick on beams	.535	S.F.	.25	.43	.68
TOTAL			3.67	1.55	5.22

B1010 241		**W Shape Beams & Girders**						
	BAY SIZE (FT.) BEAM X GIRD	SUPERIMPOSED LOAD (P.S.F.)	STEEL FRAMING DEPTH (IN.)	FIREPROOFING (S.F. PER S.F.)	TOTAL LOAD (P.S.F.)	COST PER S.F.		
						MAT.	INST.	TOTAL
1350	15x20	40	12	.535	50	3.68	1.54	5.22
1400		40	16	.65	90	4.80	1.99	6.79
1450		75	18	.694	125	6.30	2.51	8.81
1500		125	24	.796	175	8.70	3.52	12.22
1550		200	24	.89	263	9.85	3.14	12.99
1600	20x15	40	14	.659	50	3.73	1.65	5.38
1650		40	14	.69	90	5.05	2.09	7.14
1700		75	14	.806	125	6.15	2.53	8.68
1800		125	16	.86	175	7.25	3.05	10.30
1900		200	18	1.00	250	8.60	2.78	11.38
2000	20x20	40	12	.55	50	4.11	1.70	5.81
2050		40	14	.579	90	5.60	2.21	7.81
2100		75	16	.672	125	6.75	2.64	9.39
2150		125	16	.714	175	8.05	3.24	11.29
2200		200	24	.841	263	10.05	3.16	13.21
2300	20x20	40	14	.67	50	4.17	1.80	5.97
2400		40	14	.718	90	5.70	2.32	8.02
2500		75	18	.751	125	6.55	2.63	9.18
2550		125	21	.879	175	8.95	3.66	12.61
2600		200	21	.976	250	10.75	3.43	14.18
2650	20x20	40	14	.746	50	4.20	1.86	6.06
2700		40	14	.839	90	5.75	2.43	8.18
2750		75	18	.894	125	7.25	2.95	10.20
2800		125	21	.959	175	9.65	3.95	13.60
2850		200	21	1.10	250	11.85	3.80	15.65
2900	20x25	40	16	.53	50	4.53	1.83	6.36
2950		40	18	.621	96	7.15	2.74	9.89
3000		75	18	.651	131	8.20	3.11	11.31
3050		125	24	.77	200	10.85	4.24	15.09

B1010 Floor Construction

B1010 241	W Shape Beams & Girders

	BAY SIZE (FT.) BEAM X GIRD	SUPERIMPOSED LOAD (P.S.F.)	STEEL FRAMING DEPTH (IN.)	FIREPROOFING (S.F. PER S.F.)	TOTAL LOAD (P.S.F.)	COST PER S.F. MAT.	INST.	TOTAL
3100	20 x 25	200	27	.855	275	12.35	3.76	16.11
3300	20x25	40	14	.608	50	4.57	1.88	6.45
3350		40	21	.751	90	6.35	2.57	8.92
3400		75	24	.793	125	7.85	3.08	10.93
3450		125	24	.846	175	9.40	3.78	13.18
3500		200	24	.947	256	11.80	3.67	15.47
3550	20x25	40	14	.72	50	5.25	2.18	7.43
3600		40	21	.802	90	6.35	2.60	8.95
3650		75	24	.924	125	8.15	3.26	11.41
3700		125	24	.964	175	9.85	4.02	13.87
3750		200	27	1.09	250	12.50	3.95	16.45
3800	25x20	40	12	.512	50	4.52	1.81	6.33
3850		40	16	.653	90	6.10	2.41	8.51
3900		75	18	.726	125	7.85	3.04	10.89
4000		125	21	.827	175	9.80	3.92	13.72
4100		200	24	.928	250	12	3.71	15.71
4200	25x20	40	12	.65	50	4.59	1.92	6.51
4300		40	18	.702	90	6.75	2.66	9.41
4400		75	21	.829	125	7.90	3.11	11.01
4500		125	24	.914	175	9.65	3.91	13.56
4600		200	24	1.015	250	11.80	3.73	15.53
4700	25x20	40	14	.769	50	4.86	2.09	6.95
4800		40	16	.938	90	7.30	2.99	10.29
4900		75	18	.969	125	8.80	3.50	12.30
5000		125	24	1.136	175	12.30	4.98	17.28
5100		200	24	1.239	250	15.90	4.94	20.84
5200	25x25	40	18	.486	50	4.94	1.93	6.87
5300		40	18	.592	96	7.35	2.78	10.13
5400		75	21	.668	131	8.85	3.33	12.18
5450		125	24	.738	191	11.70	4.51	16.21
5500		200	30	.861	272	13.65	4.09	17.74
5550	25x25	40	18	.597	50	4.77	1.95	6.72
5600		40	18	.704	90	7.60	2.95	10.55
5650		75	21	.777	125	8.70	3.35	12.05
5700		125	24	.865	175	11.10	4.39	15.49
5750		200	27	.96	250	13.45	4.12	17.57
5800	25x25	40	18	.71	50	5.25	2.18	7.43
5850		40	21	.767	90	7.65	2.99	10.64
5900		75	24	.887	125	9.20	3.58	12.78
5950		125	24	.972	175	11.60	4.62	16.22
6000		200	30	1.10	250	14.15	4.39	18.54
6050	25x30	40	24	.547	50	6.25	2.40	8.65
6100		40	24	.629	103	8.85	3.31	12.16
6150		75	30	.726	138	10.60	3.94	14.54
6200		125	30	.751	206	12.55	4.82	17.37
6250		200	33	.868	281	15.10	4.48	19.58
6300	25x30	40	21	.568	50	5.60	2.20	7.80
6350		40	21	.694	90	7.60	2.94	10.54
6400		75	24	.776	125	9.80	3.70	13.50
6450		125	30	.904	175	11.75	4.65	16.40
6500		200	33	1.008	263	14.10	4.32	18.42

Important: See the Reference Section for critical supporting data - Reference Numbers and City Cost Indexes

B1010 Floor Construction

B1010 241	W Shape Beams & Girders

SHELL B

	BAY SIZE (FT.) BEAM X GIRD	SUPERIMPOSED LOAD (P.S.F.)	STEEL FRAMING DEPTH (IN.)	FIREPROOFING (S.F. PER S.F.)	TOTAL LOAD (P.S.F.)	COST PER S.F.		
						MAT.	INST.	TOTAL
6550	25x30	40	16	.632	50	5.85	2.32	8.17
6600		40	21	.76	90	8.05	3.12	11.17
6650		75	24	.857	125	9.60	3.70	13.30
6700		125	30	.983	175	12	4.79	16.79
6750		200	33	1.11	250	15.20	4.67	19.87
6800	30x25	40	16	.532	50	5.40	2.10	7.50
6850		40	21	.672	96	8.25	3.13	11.38
6900		75	24	.702	131	9.75	3.64	13.39
6950		125	27	1.020	175	12.70	4.80	17.50
7000		200	30	1.160	250	16.15	6.05	22.20
7100	30x25	40	18	.569	50	5.60	2.20	7.80
7150		40	24	.740	90	7.85	3.05	10.90
7200		75	24	.787	125	9.80	3.70	13.50
7300		125	24	.874	175	12.20	4.77	16.97
7400		200	30	1.013	250	14.95	4.54	19.49
7450	30x25	40	16	.637	50	5.85	2.33	8.18
7500		40	24	.839	90	8.30	3.26	11.56
7550		75	24	.919	125	10.05	3.88	13.93
7600		125	27	1.02	175	12.70	5.05	17.75
7650		200	30	1.160	250	15.90	4.87	20.77
7700	30x30	40	21	.52	50	6	2.31	8.31
7750		40	24	.629	103	9.30	3.44	12.74
7800		75	30	.715	138	11.05	4.07	15.12
7850		125	36	.822	206	14.50	5.55	20.05
7900		200	36	.878	281	16.15	4.75	20.90
7950	30x30	40	24	.619	50	6.30	2.45	8.75
8000		40	24	.706	90	8.45	3.22	11.67
8020		75	27	.818	125	10	3.80	13.80
8040		125	30	.910	175	12.85	5	17.85
8060		200	33	.999	263	15.80	4.75	20.55
8080	30x30	40	18	.631	50	6.70	2.61	9.31
8100		40	24	.805	90	9.15	3.52	12.67
8120		75	27	.899	125	10.90	4.14	15.04
8150		125	30	1.010	175	13.55	5.30	18.85
8200		200	36	1.148	250	16.10	4.91	21.01
8250	30x35	40	21	.508	50	6.85	2.57	9.42
8300		40	24	.651	109	10.15	3.74	13.89
8350		75	33	.732	150	12.35	4.50	16.85
8400		125	36	.802	225	15.35	5.80	21.15
8450		200	36	.888	300	20	5.80	25.80
8500	30x35	40	24	.554	50	6.05	2.33	8.38
8520		40	24	.655	90	8.85	3.33	12.18
8540		75	30	.751	125	11.05	4.10	15.15
8600		125	33	.845	175	13.45	5.20	18.65
8650		200	36	.936	263	17.65	5.20	22.85
8700	30x35	40	21	.644	50	6.50	2.54	9.04
8720		40	24	.733	90	9.35	3.53	12.88
8740		75	30	.833	125	11.75	4.37	16.12
8760		125	36	.941	175	13.50	5.25	18.75
8780		200	36	1.03	250	17.90	5.30	23.20

B1010 241		W Shape Beams & Girders						
	BAY SIZE (FT.) BEAM X GIRD	SUPERIMPOSED LOAD (P.S.F.)	STEEL FRAMING DEPTH (IN.)	FIREPROOFING (S.F. PER S.F.)	TOTAL LOAD (P.S.F.)	COST PER S.F.		
						MAT.	INST.	TOTAL
8800	35x30	40	24	.540	50	6.65	2.53	9.18
8850		40	30	.670	103	9.95	3.69	13.64
8900		75	33	.748	138	12.10	4.45	16.55
8950		125	36	.824	206	14.95	5.70	20.65
8980		200	36	.874	281	18.05	5.25	23.30
9000	35x30	40	24	.619	50	6.50	2.53	9.03
9050		40	24	.754	90	9.55	3.62	13.17
9100		75	27	.844	125	11.30	4.24	15.54
9200		125	30	.856	175	13.90	5.35	19.25
9250		200	33	.953	263	15.35	4.60	19.95
9300	35x30	40	24	.705	50	6.95	2.73	9.68
9350		40	24	.833	90	9.80	3.75	13.55
9400		75	30	.963	125	12	4.55	16.55
9450		125	33	1.078	175	15.05	5.90	20.95
9500		200	36	1.172	250	18.20	5.45	23.65
9550	35x35	40	27	.560	50	7.10	2.69	9.79
9600		40	36	.706	109	12.95	4.70	17.65
9650		75	36	.750	150	13.40	4.87	18.27
9820		125	36	.797	225	17.90	6.70	24.60
9840		200	36	.914	300	21	6.10	27.10
9860	35x35	40	24	.580	50	6.50	2.50	9
9880		40	30	.705	90	9.95	3.71	13.66
9890		75	33	.794	125	12.15	4.48	16.63
9900		125	36	.878	175	14.95	5.75	20.70
9920		200	36	.950	263	18.50	5.40	23.90
9930	35x35	40	24	.689	50	6.95	2.72	9.67
9940		40	30	.787	90	10.20	3.85	14.05
9960		75	33	.871	125	13.05	4.82	17.87
9970		125	36	.949	175	14.35	5.55	19.90
9980		200	36	1.060	250	19.80	5.80	25.60

Important: See the Reference Section for critical supporting data - Reference Numbers and City Cost Indexes

B1010 Floor Construction

Description: Table below lists costs for light gauge CEE or PUNCHED DOUBLE joists to suit the span and loading with the minimum thickness subfloor required by the joist spacing.

Design Assumptions:
Maximum live load deflection is 1/360 of the clear span.

Maximum total load deflection is 1/240 of the clear span.

Bending strength is 20,000 psi.

8% allowance has been added to framing quantities for overlaps, double joists at openings under partitions, etc.; 5% added to glued & nailed subfloor for waste.

Maximum span is in feet and is the unsupported clear span.

SHELL **B**

System Components	QUANTITY	UNIT	COST PER S.F.		
			MAT.	INST.	TOTAL
SYSTEM B1010 244 1500					
LIGHT GAUGE STL & PLYWOOD FLR SYS,40 PSF S.LOAD,15'SPAN,8"DEPTH,16" O.C.					
Light gauge steel, 8" deep, 16" O.C., 12 ga.	1.870	Lb.	1.65	1.05	2.70
Subfloor plywood CDX 5/8"	1.050	S.F.	.77	.69	1.46
TOTAL			2.42	1.74	4.16

B1010 244		Light Gauge Steel Floor Systems						
	SPAN (FT.)	SUPERIMPOSED LOAD (P.S.F.)	FRAMING DEPTH (IN.)	FRAMING SPAC. (IN.)	TOTAL LOAD (P.S.F.)	COST PER S.F.		
						MAT.	INST.	TOTAL
1500	15	40	8	16	54	2.41	1.74	4.15
1550			8	24	54	2.56	1.78	4.34
1600		65	10	16	80	3.13	2.09	5.22
1650			10	24	80	3.14	2.04	5.18
1700		75	10	16	90	3.13	2.09	5.22
1750			10	24	90	3.14	2.04	5.18
1800		100	10	16	116	4.07	2.64	6.71
1850			10	24	116	3.14	2.04	5.18
1900		125	10	16	141	4.07	2.64	6.71
1950			10	24	141	3.14	2.04	5.18
2500	20	40	8	16	55	2.76	1.97	4.73
2550			8	24	55	2.56	1.78	4.34
2600		65	8	16	80	3.20	2.24	5.44
2650			10	24	80	3.15	2.05	5.20
2700		75	10	16	90	2.96	1.98	4.94
2750			12	24	90	3.50	1.79	5.29
2800		100	10	16	115	2.96	1.98	4.94
2850			10	24	116	4.02	2.33	6.35
2900		125	12	16	142	4.60	2.27	6.87
2950			12	24	141	4.36	2.48	6.84
3500	25	40	10	16	55	3.13	2.09	5.22
3550			10	24	55	3.14	2.04	5.18
3600		65	10	16	81	4.07	2.64	6.71
3650			12	24	81	3.87	1.94	5.81
3700		75	12	16	92	4.59	2.26	6.85
3750			12	24	91	4.36	2.48	6.84
3800		100	12	16	117	5.15	2.49	7.64
3850		125	12	16	143	5.95	3.30	9.25
4500	30	40	12	16	57	4.59	2.26	6.85
4550			12	24	56	3.85	1.94	5.79
4600		65	12	16	82	5.15	2.48	7.63
4650		75	12	16	92	5.15	2.48	7.63

B1010 Floor Construction

Description: Table below lists cost per S.F. for a floor system on bearing walls using open web steel joists, galvanized steel slab form and 2-1/2″ concrete slab reinforced with welded wire fabric.

Design and Pricing Assumptions:
Concrete f'c = 3 KSI placed by pump.
WWF 6 x 6 – W1.4 x W1.4 (10 x 10)
Joists are spaced as shown.
Slab form is 28 gauge galvanized.
Joists costs include appropriate bridging. Deflection is limited to 1/360 of the span. Screeds and steel trowel finish.

Design Loads	Min.	Max.
Joists	3.0 PSF	7.6 PSF
Slab Form	1.0	1.0
2-1/2″ Concrete	27.0	27.0
Ceiling	3.0	3.0
Misc.	9.0	9.4
	43.0 PSF	48.0 PSF

System Components	QUANTITY	UNIT	COST PER S.F. MAT.	INST.	TOTAL
SYSTEM B1010 246 1050					
SPAN 20′ S.LOAD 40 PSF, JOIST SPACING 2′-0″ O.C., 14-1/2″ DEPTH					
Open web joists	2.800	Lb.	2.04	.84	2.88
Slab form, galvanized steel 9/16″ deep, 28 gauge	1.050	S.F.	1.44	.90	2.34
Welded wire fabric rolls, 6 x 6 - W1.4 x W1.4 (10 x 10), 21 lb/csf	1.000	S.F.	.13	.30	.43
Concrete ready mix, regular weight, 3000 psi	.210	C.F.	.74		.74
Place and vibrate concrete, elevated slab less than 6″, pumped	.210	C.F.		.27	.27
Finishing floor, monolithic steel trowel finish for finish floor	1.000	S.F.		.74	.74
Curing with sprayed membrane curing compound	.010	C.S.F.	.06	.07	.13
TOTAL			4.41	3.12	7.53

B1010 246		Deck & Joists on Bearing Walls						
	SPAN (FT.)	SUPERIMPOSED LOAD (P.S.F.)	JOIST SPACING FT. or IN.	DEPTH (IN.)	TOTAL LOAD (P.S.F.)	COST PER S.F. MAT.	INST.	TOTAL
1050	20	40	2-0	14-1/2	83	4.42	3.11	7.53
1070		65	2-0	16-1/2	109	4.81	3.30	8.11
1100		75	2-0	16-1/2	119	4.81	3.30	8.11
1120		100	2-0	18-1/2	145	4.84	3.31	8.15
1150		125	1-9	18-1/2	170	5.90	3.95	9.85
1170	25	40	2-0	18-1/2	84	5.30	3.67	8.97
1200		65	2-0	20-1/2	109	5.50	3.77	9.27
1220		75	2-0	20-1/2	119	5.35	3.52	8.87
1250		100	2-0	22-1/2	145	5.45	3.56	9.01
1270		125	1-9	22-1/2	170	6.25	3.89	10.14
1300	30	40	2-0	22-1/2	84	5.45	3.20	8.65
1320		65	2-0	24-1/2	110	5.95	3.34	9.29
1350		75	2-0	26-1/2	121	6.20	3.40	9.60
1370		100	2-0	26-1/2	146	6.60	3.53	10.13
1400		125	2-0	24-1/2	172	7.15	4.62	11.77
1420	35	40	2-0	26-1/2	85	6.75	4.44	11.19
1450		65	2-0	28-1/2	111	7	4.54	11.54
1470		75	2-0	28-1/2	121	7	4.54	11.54
1500		100	1-11	28-1/2	147	7.60	4.84	12.44
1520		125	1-8	28-1/2	172	8.35	5.20	13.55
1550	5/8″ gyp. fireproof.							
1560	On metal furring, add					.63	2.34	2.97

Important: See the Reference Section for critical supporting data - Reference Numbers and City Cost Indexes

B1010 Floor Construction

Table below lists costs for a floor system on exterior bearing walls and interior columns and beams using open web steel joists, galvanized steel slab form, 2-1/2″ concrete slab reinforced with welded wire fabric.

Design and Pricing Assumptions:
Structural Steel is A36.
Concrete f'c = 3 KSI placed by pump.
WWF 6 x 6 – W1.4 x W1.4 (10 x 10)
Columns are 12′ high.
Building is 4 bays long by 4 bays wide.
Joists are 2′ O.C. ± and span the long direction of the bay.

Joists at columns have bottom chords extended and are connected to columns.

Slab form is 28 gauge galvanized. Column costs in table are for columns to support 1 floor plus roof loading in a 2-story building; however, column costs are from ground floor to 2nd floor only. Joist costs include appropriate bridging. Deflection is limited to 1/360 of the span. Screeds and steel trowel finish.

Design Loads	Min.	Max.
S.S & Joists	4.4 PSF	11.5 PSF
Slab Form	1.0	1.0
2-1/2″ Concrete	27.0	27.0
Ceiling	3.0	3.0
Misc.	7.6	5.5
	43.0 PSF	48.0 PSF

SHELL **B**

System Components			COST PER S.F.		
	QUANTITY	UNIT	MAT.	INST.	TOTAL
SYSTEM B1010 248 1200					
15′X20′ BAY, W.F. STEEL, STEEL JOISTS, SLAB FORM, CONCRETE SLAB					
Structural steel	1.248	Lb.	.98	.33	1.31
Open web joists	3.140	Lb.	2.29	.94	3.23
Slab form, galvanized steel 9/16″ deep, 28 gauge	1.020	S.F.	.98	.61	1.59
Welded wire fabric 6x6 - W1.4 x W1.4 (10 x 10), 21 lb/CSF roll, 10% lap	1.000	S.F.	.13	.30	.43
Concrete ready mix, regular weight, 3000 psi	.210	C.F.	.74		.74
Place and vibrate concrete, elevated slab less than 6″, pumped	.210	C.F.		.27	.27
Finishing floor, monolithic steel trowel finish for finish floor	1.000	S.F.		.74	.74
Curing with sprayed membrane curing compound	.010	C.S.F.	.06	.07	.13
TOTAL			5.18	3.26	8.44

B1010 248	Steel Joists on Beam & Wall							
	BAY SIZE (FT.)	SUPERIMPOSED LOAD (P.S.F.)	DEPTH (IN.)	TOTAL LOAD (P.S.F.)	COLUMN ADD	COST PER S.F.		
						MAT.	INST.	TOTAL
1200	15x20 RB1010-100	40	17	83		5.20	3.25	8.45
1210					columns	.33	.10	.43
1220	15x20	65	19	108		5.35	3.32	8.67
1230					columns	.33	.10	.43
1250	15x20	75	19	119		5.70	3.47	9.17
1260					columns	.39	.13	.52
1270	15x20	100	19	144		6.50	4	10.50
1280					columns	.39	.13	.52
1300	15x20	125	19	170		6.70	4.06	10.76
1310					columns	.52	.17	.69
1350	20x20	40	19	83		5.35	3.31	8.66
1360					columns	.29	.10	.39
1370	20x20	65	23	109		5.95	3.54	9.49
1380					columns	.39	.13	.52
1400	20x20	75	23	119		6.15	3.61	9.76
1410					columns	.39	.13	.52
1420	20x20	100	23	144		6.35	3.67	10.02
1430					columns	.39	.13	.52
1450	20x20	125	23	170		7.65	4.43	12.08
1460					columns	.47	.16	.63

B1010 Floor Construction

B1010 248	Steel Joists on Beam & Wall

	BAY SIZE (FT.)	SUPERIMPOSED LOAD (P.S.F.)	DEPTH (IN.)	TOTAL LOAD (P.S.F.)	COLUMN ADD	COST PER S.F.		
						MAT.	INST.	TOTAL
1500	20x25	40	23	84		6.10	3.82	9.92
1510					columns	.31	.10	.41
1520	20x25	65	26	110		6.60	4.01	10.61
1530					columns	.31	.10	.41
1550	20x25	75	26	120		7.05	4.22	11.27
1560					columns	.37	.13	.50
1570	20x25	100	26	145		6.80	3.89	10.69
1580					columns	.37	.13	.50
1670	20x25	125	29	170		7.75	4.26	12.01
1680					columns	.44	.14	.58
1720	25x25	40	23	84		6.55	3.94	10.49
1730					columns	.30	.10	.40
1750	25x25	65	29	110		6.90	4.08	10.98
1760					columns	.30	.10	.40
1770	25x25	75	26	120		7.15	3.99	11.14
1780					columns	.35	.12	.47
1800	25x25	100	29	145		8.05	4.35	12.40
1810					columns	.35	.12	.47
1820	25x25	125	29	170		8.45	4.49	12.94
1830					columns	.39	.13	.52
1870	25x30	40	29	84		6.70	4.10	10.80
1880					columns	.29	.10	.39
1900	25x30	65	29	110		7.15	3.62	10.77
1910					columns	.29	.10	.39
1920	25x30	75	29	120		7.55	3.74	11.29
1930					columns	.32	.10	.42
1950	25x30	100	29	145		8.25	3.94	12.19
1960					columns	.32	.10	.42
1970	25x30	125	32	170		8.85	5	13.85
1980					columns	.36	.12	.48
2020	30x30	40	29	84		7.05	3.58	10.63
2030					columns	.27	.09	.36
2050	30x30	65	29	110		7.95	3.87	11.82
2060					columns	.27	.09	.36
2070	30x30	75	32	120		8.20	3.94	12.14
2080					columns	.31	.10	.41
2100	30x30	100	35	145		9.05	4.21	13.26
2110					columns	.36	.12	.48
2120	30x30	125	35	172		9.90	5.40	15.30
2130					columns	.43	.14	.57
2170	30x35	40	29	85		7.95	3.86	11.81
2180					columns	.26	.09	.35
2200	30x35	65	29	111		8.90	5.05	13.95
2210					columns	.30	.10	.40
2220	30x35	75	32	121		8.90	5.05	13.95
2230					columns	.30	.10	.40
2250	30x35	100	35	148		9.70	4.39	14.09
2260					columns	.37	.13	.50
2270	30x35	125	38	173		10.90	4.75	15.65
2280					columns	.38	.13	.51
2320	35x35	40	32	85		8.20	3.94	12.14
2330					columns	.26	.09	.35

Important: See the Reference Section for critical supporting data - Reference Numbers and City Cost Indexes

B1010 Floor Construction

B1010 248	Steel Joists on Beam & Wall							
	BAY SIZE (FT.)	SUPERIMPOSED LOAD (P.S.F.)	DEPTH (IN.)	TOTAL LOAD (P.S.F.)	COLUMN ADD	COST PER S.F.		
						MAT.	INST.	TOTAL
2350	35x35	65	35	111		9.40	5.25	14.65
2360					columns	.32	.10	.42
2370	35x35	75	35	121		9.60	5.30	14.90
2380					columns	.32	.10	.42
2400	35x35	100	38	148		10	5.50	15.50
2410					columns	.39	.13	.52
2420	35x35	125	41	173		11.40	6.05	17.45
2430					columns	.40	.13	.53
2460	5/8 gyp. fireproof.							
2475	On metal furring, add					.63	2.34	2.97

B1010   Floor Construction

Table below lists costs for a floor system on steel columns and beams using open web steel joists, galvanized steel slab form, and 2-1/2″ concrete slab reinforced with welded wire fabric.

Design and Pricing Assumptions:
Structural Steel is A36.
Concrete f'c = 3 KSI placed by pump.
WWF 6 x 6 – W1.4 x W1.4 (10 x 10)
Columns are 12′ high.
Building is 4 bays long by 4 bays wide.
Joists are 2′ O.C. ± and span the long direction of the bay.

Joists at columns have bottom chords extended and are connected to columns.

Slab form is 28 gauge galvanized. Column costs in table are for columns to support 1 floor plus roof loading in a 2-story building; however, column costs are from ground floor to 2nd floor only. Joist costs include appropriate bridging. Deflection is limited to 1/360 of the span. Screeds and steel trowel finish.

Design Loads	Min.	Max.
S.S. & Joists	6.3 PSF	15.3 PSF
Slab Form	1.0	1.0
2-1/2″ Concrete	27.0	27.0
Ceiling	3.0	3.0
Misc.	5.7	1.7
	43.0 PSF	48.0 PSF

System Components			COST PER S.F.		
	QUANTITY	UNIT	MAT.	INST.	TOTAL
SYSTEM B1010 250 2350					
15′X20′BAY 40 PSF S. LOAD, 17″ DEPTH, 83 PSF TOTAL LOAD					
Structural steel	1.974	Lb.	2.07	.69	2.76
Open web joists	3.140	Lb.	2.29	.94	3.23
Slab form, galvanized steel 9/16″ deep, 28 gauge	1.020	S.F.	.98	.61	1.59
Welded wire fabric rolls, 6 x 6 - W1.4 x W1.4 (10 x 10), 21 lb/csf	1.000	S.F.	.13	.30	.43
Concrete ready mix, regular weight, 3000 psi	.210	C.F.	.74		.74
Place and vibrate concrete, elevated slab less than 6″, pumped	.210	C.F.		.27	.27
Finishing floor, monolithic steel trowel finish for finish floor	1.000	S.F.		.74	.74
Curing with sprayed membrane curing compound	.010	S.F.	.06	.07	.13
TOTAL			6.27	3.62	9.89

B1010 250		Steel Joists, Beams & Slab on Columns						
	BAY SIZE (FT.)	SUPERIMPOSED LOAD (P.S.F.)	DEPTH (IN.)	TOTAL LOAD (P.S.F.)	COLUMN ADD	COST PER S.F.		
						MAT.	INST.	TOTAL
2350	15x20 RB1010 -100	40	17	83		6.30	3.62	9.92
2400					column	.99	.34	1.33
2450	15x20	65	19	108		6.95	3.84	10.79
2500					column	.99	.34	1.33
2550	15x20	75	19	119		7.25	3.97	11.22
2600					column	1.08	.36	1.44
2650	15x20	100	19	144		7.70	4.12	11.82
2700					column	1.08	.36	1.44
2750	15x20	125	19	170		8.60	4.67	13.27
2800					column	1.44	.48	1.92
2850	20x20	40	19	83		6.80	3.78	10.58
2900					column	.81	.27	1.08
2950	20x20	65	23	109		7.50	4.04	11.54
3000					column	1.08	.36	1.44
3100	20x20	75	26	119		7.95	4.19	12.14
3200					column	1.08	.36	1.44
3400	20x20	100	23	144		8.25	4.29	12.54
3450					column	1.08	.36	1.44
3500	20x20	125	23	170		9.15	4.63	13.78
3600					column	1.30	.43	1.73

B1010 Floor Construction

B1010 250	Steel Joists, Beams & Slab on Columns

	BAY SIZE (FT.)	SUPERIMPOSED LOAD (P.S.F.)	DEPTH (IN.)	TOTAL LOAD (P.S.F.)	COLUMN ADD	COST PER S.F.		
						MAT.	INST.	TOTAL
3700	20x25	40	44	83		7.80	4.35	12.15
3800					column	.87	.29	1.16
3900	20x25	65	26	110		8.50	4.61	13.11
4000					column	.87	.29	1.16
4100	20x25	75	26	120		8.35	4.37	12.72
4200					column	1.04	.35	1.39
4300	20x25	100	26	145		8.85	4.55	13.40
4400					column	1.04	.35	1.39
4500	20x25	125	29	170		9.90	4.95	14.85
4600					column	1.21	.40	1.61
4700	25x25	40	23	84		8.35	4.53	12.88
4800					column	.83	.27	1.10
4900	25x25	65	29	110		8.85	4.72	13.57
5000					column	.83	.27	1.10
5100	25x25	75	26	120		9.25	4.67	13.92
5200					column	.97	.32	1.29
5300	25x25	100	29	145		10.35	5.10	15.45
5400					column	.97	.32	1.29
5500	25x25	125	32	170		10.90	5.30	16.20
5600					column	1.08	.36	1.44
5700	25x30	40	29	84		8.70	4.74	13.44
5800					column	.81	.27	1.08
5900	25x30	65	29	110		9.05	4.93	13.98
6000					column	.81	.27	1.08
6050	25x30	75	29	120		9.80	4.45	14.25
6100					column	.89	.30	1.19
6150	25x30	100	29	145		10.60	4.71	15.31
6200					column	.89	.30	1.19
6250	25x30	125	32	170		11.35	5.85	17.20
6300					column	1.03	.35	1.38
6350	30x30	40	29	84		9.10	4.25	13.35
6400					column	.75	.25	1
6500	30x30	65	29	110		10.35	4.65	15
6600					column	.75	.25	1
6700	30x30	75	32	120		10.55	4.71	15.26
6800					column	.86	.29	1.15
6900	30x30	100	35	145		11.75	5.10	16.85
7000					column	1	.34	1.34
7100	30x30	125	35	172		12.80	6.35	19.15
7200					column	1.11	.37	1.48
7300	30x35	40	29	85		10.30	4.61	14.91
7400					column	.64	.21	.85
7500	30x35	65	29	111		11.45	5.85	17.30
7600					column	.83	.27	1.10
7700	30x35	75	32	121		11.45	5.85	17.30
7800					column	.84	.28	1.12
7900	30x35	100	35	148		12.40	5.25	17.65
8000					column	1.03	.35	1.38
8100	30x35	125	38	173		13.80	5.70	19.50
8200					column	1.05	.35	1.40
8300	35x35	40	32	85		10.55	4.70	15.25
8400					column	.74	.25	.99

SHELL B

91

B1010 Floor Construction

B1010 250		Steel Joists, Beams & Slab on Columns						
	BAY SIZE (FT.)	SUPERIMPOSED LOAD (P.S.F.)	DEPTH (IN.)	TOTAL LOAD (P.S.F.)	COLUMN ADD	COST PER S.F.		
						MAT.	INST.	TOTAL
8500	35x35	65	35	111		12	6.10	18.10
8600					column	.88	.30	1.18
9300	35x35	75	38	121		12.35	6.20	18.55
9400					column	.88	.30	1.18
9500	35x35	100	38	148		13.30	6.55	19.85
9600					column	1.09	.36	1.45
9750	35x35	125	41	173		14.40	5.90	20.30
9800					column	1.11	.37	1.48
9810	5/8 gyp. fireproof.							
9815	On metal furring, add					.63	2.34	2.97

Important: See the Reference Section for critical supporting data - Reference Numbers and City Cost Indexes

B SHELL

B1010 Floor Construction

General: Composite construction of W shape flange beams and concrete slabs is most efficiently used when loads are heavy and spans are moderately long. It is stiffer with less deflection than non-composite construction of similar depth and spans.

In practice, composite construction is typically shallower in depth than non-composite would be.

Design Assumptions:
Steel, fy = 36 KSI
Beams unshored during construction
Deflection limited to span/360
Shear connectors, welded studs
Concrete, f'c = 3 KSI

SHELL **B**

System Components	QUANTITY	UNIT	COST PER S.F.		
			MAT.	INST.	TOTAL
SYSTEM B1010 252 3800					
20X25 BAY, 40 PSF S. LOAD, 4″ THICK SLAB, 20″ TOTAL DEPTH					
Structural steel	3.820	Lb.	4.09	1.34	5.43
Welded shear connectors 3/4″ diameter 3-3/8″ long	.140	Ea.	.07	.21	.28
Forms in place, floor slab forms hung from steel beams, 4 uses	1.000	S.F.	1.45	4.57	6.02
Edge forms to 6″ high on elevated slab, 4 uses	.050	L.F.	.01	.17	.18
Reinforcing in place, elevated slabs #4 to #7	1.190	Lb.	.58	.43	1.01
Concrete ready mix, regular weight, 3000 psi	.330	C.F.	1.17		1.17
Place and vibrate concrete, elevated slab less than 6″, pumped	.330	C.F.		.41	.41
Finishing floor, monolithic steel trowel finish for resilient tile	1.000	S.F.		.68	.68
Curing with sprayed membrane curing compound	.010	S.F.	.06	.07	.13
Spray mineral fiber/cement for fire proof, 1″ thick on beams	.520	S.F.	.24	.42	.66
TOTAL			7.67	8.30	15.97

B1010 252		Composite Beam & Cast in Place Slab						
	BAY SIZE (FT.)	SUPERIMPOSED LOAD (P.S.F.)	SLAB THICKNESS (IN.)	TOTAL DEPTH (FT.-IN.)	TOTAL LOAD (P.S.F.)	COST PER S.F.		
						MAT.	INST.	TOTAL
3800	20x25	40	4	1 - 8	94	7.65	8.30	15.95
3900	RB1010 -100	75	4	1 - 8	130	8.90	8.85	17.75
4000		125	4	1 - 10	181	10.35	9.45	19.80
4100		200	5	2 - 2	272	13.50	10.70	24.20
4200	25x25	40	4-1/2	1 - 8-1/2	99	8.05	8.40	16.45
4300		75	4-1/2	1 - 10-1/2	136	9.70	9.10	18.80
4400		125	5-1/2	2 - 0-1/2	200	11.45	9.90	21.35
4500		200	5-1/2	2 - 2-1/2	278	14.20	11	25.20
4600	25x30	40	4-1/2	1 - 8-1/2	100	8.95	8.80	17.75
4700		75	4-1/2	1 - 10-1/2	136	10.50	9.40	19.90
4800		125	5-1/2	2 - 2-1/2	202	12.90	10.40	23.30
4900		200	5-1/2	2 - 5-1/2	279	15.35	11.40	26.75
5000	30x30	40	4	1 - 8	95	8.95	8.80	17.75
5200		75	4	2 - 1	131	10.35	9.30	19.65
5400		125	4	2 - 4	183	12.80	10.30	23.10
5600		200	5	2 - 10	274	16.35	11.70	28.05

B1010 Floor Construction

B1010 252	Composite Beam & Cast in Place Slab

	BAY SIZE (FT.)	SUPERIMPOSED LOAD (P.S.F.)	SLAB THICKNESS (IN.)	TOTAL DEPTH (FT.-IN.)	TOTAL LOAD (P.S.F.)	COST PER S.F.		
						MAT.	INST.	TOTAL
5800	30x35	40	4	2 - 1	95	9.50	8.95	18.45
6000		75	4	2 - 4	139	11.40	9.70	21.10
6250		125	4	2 - 4	185	14.80	11	25.80
6500		200	5	2 - 11	276	18.10	12.30	30.40
7000	35x30	40	4	2 - 1	95	9.75	9.05	18.80
7200		75	4	2 - 4	139	11.55	9.80	21.35
7400		125	4	2 - 4	185	14.80	11	25.80
7600		200	5	2 - 11	276	18.05	12.30	30.35
8000	35x35	40	4	2 - 1	96	10	9.10	19.10
8250		75	4	2 - 4	133	11.85	9.90	21.75
8500		125	4	2 - 10	185	14.25	10.85	25.10
8750		200	5	3 - 5	276	18.55	12.50	31.05
9000	35x40	40	4	2 - 4	97	11.10	9.55	20.65
9250		75	4	2 - 4	134	13	10.20	23.20
9500		125	4	2 - 7	186	15.75	11.30	27.05
9750		200	5	3 - 5	278	20.50	13.10	33.60

B1010 Floor Construction

Description: Table below lists costs per S.F. for floors using steel beams and girders, composite steel deck, concrete slab reinforced with W.W.F. and sprayed fiber fireproofing (non-asbestos) on the steel beams and girders and on the steel deck.

Design and Pricing Assumptions: Structural Steel is A36, high strength bolted.

Composite steel deck varies from 2"–20 gauge to 3"–16 gauge galvanized. WWF 6 x 6 – W1.4 x W1.4 (10 x 10) Concrete f'c = 3 KSI. Steel trowel finish and cure.

Spandrels are assumed the same weight as interior beams and girders to allow for exterior wall loads and bracing or moment connections.

System Components	QUANTITY	UNIT	COST PER S.F.		
			MAT.	INST.	TOTAL
SYSTEM B1010 254 0540					
W SHAPE BEAMS & DECK, FIREPROOFED, 15'X20', 5" SLAB, 40 PSF LOAD					
Structural steel	4.470	Lb.	4.41	1.44	5.85
Metal decking, non-cellular composite, galv 3" deep, 20 gage	1.050	S.F.	1.97	.85	2.82
Sheet metal edge closure form, 12", w/2 bends, 18 ga, galv	.058	L.F.	.19	.12	.31
Welded wire fabric 6 x 6 - W1.4 x W1.4 (10 x 10), 21 lb/CSF roll, 10% lap	1.000	S.F.	.13	.30	.43
Concrete ready mix, regular weight, 3000 psi	.011	C.Y.	1.05		1.05
Place and vibrate concrete, elevated slab less than 6", pumped	.011	C.Y.		.29	.29
Finishing floor, monolithic steel trowel finish for finish floor	1.000	S.F.		.74	.74
Curing with sprayed membrane curing compound	.010	C.S.F.	.06	.07	.13
Sprayed mineral fiber/cement for fireproof, 1" thick on decks	1.000	S.F.	.70	.96	1.66
Sprayed mineral fiber/cement for fireproof, 1" thick on beams	.615	S.F.	.29	.50	.79
		TOTAL	8.80	5.27	14.07

B1010 254	W Shape, Composite Deck, & Slab							
	BAY SIZE (FT.) BEAM X GIRD	SUPERIMPOSED LOAD (P.S.F.)	SLAB THICKNESS (IN.)	TOTAL DEPTH (FT.-IN.)	TOTAL LOAD (P.S.F.)	COST PER S.F.		
						MAT.	INST.	TOTAL
0540	15x20	40	5	1-7	89	8.80	5.25	14.05
0560	RB1010 -100	75	5	1-9	125	9.65	5.60	15.25
0580		125	5	1-11	176	11.25	6.15	17.40
0600		200	5	2-2	254	13.80	6.95	20.75
0620	20x15	40	4	1-4	89	8.90	5.40	14.30
0640		75	4	1-4	119	9.85	5.60	15.45
0660		125	4	1-6	170	11	6	17
0680		200	4	2-1	247	12.90	6.75	19.65
0700	20x20	40	5	1-9	90	9.55	5.50	15.05
0720		75	5	1-9	126	10.80	5.95	16.75
0740		125	5	1-11	177	12.25	6.45	18.70
0760		200	5	2-5	255	14.70	7.25	21.95
0780	20x25	40	5	1-9	90	9.70	5.45	15.15
0800		75	5	1-9	127	12.10	6.30	18.40
0820		125	5	1-11	180	14.45	6.90	21.35
0840		200	5	2-5	256	16.05	7.80	23.85
0860	25x20	40	5	1-11	90	9.85	5.60	15.45
0880		75	5	2-5	127	11.50	6.20	17.70
0900		125	5	2-5	178	13	6.75	19.75
0920		200	5	2-5	257	16.75	7.95	24.70
0940	25x25	40	5	1-11	91	11	5.85	16.85
0960		75	5	2-5	178	12.65	6.45	19.10
0980		125	5	2-5	181	15.45	7.20	22.65
1000		200	5-1/2	2-8-1/2	263	17.50	8.35	25.85

B1010 Floor Construction

B1010 254	W Shape, Composite Deck, & Slab						
	BAY SIZE (FT.) BEAM X GIRD	SUPERIMPOSED LOAD (P.S.F.)	SLAB THICKNESS (IN.)	TOTAL DEPTH (FT.-IN.)	TOTAL LOAD (P.S.F.)	COST PER S.F.		
						MAT.	INST.	TOTAL
1400	25x30	40	5	2-5	91	10.95	6	16.95
1500		75	5	2-5	128	13.25	6.70	19.95
1600		125	5	2-8	180	15.05	7.40	22.45
1700		200	5	2-11	259	18.75	8.55	27.30
1800	30x25	40	5	2-5	92	11.60	6.30	17.90
1900		75	5	2-5	129	13.65	7	20.65
2000		125	5	2-8	181	15.55	7.70	23.25
2100		200	5-1/2	2-11	200	18.95	8.90	27.85
2200	30x30	40	5	2-2	92	12.10	6.40	18.50
2300		75	5	2-5	129	14.15	7.10	21.25
2400		125	5	2-11	182	16.75	8	24.75
2500		200	5	3-2	263	22.50	10.10	32.60
2600	30x35	40	5	2-5	94	13.20	6.80	20
2700		75	5	2-11	131	15.35	7.50	22.85
2800		125	5	3-2	183	18	8.40	26.40
2900		200	5-1/2	3-5-1/2	268	22	9.70	31.70
3400	35x30	40	5	2-5	93	12.75	6.65	19.40
3500		75	5	2-8	130	15.20	7.50	22.70
3600		125	5	2-11	183	18.05	8.50	26.55
3700		200	5	3-5	262	22	9.90	31.90
3800	35x35	40	5	2-8	94	13.55	6.75	20.30
3900		75	5	2-11	131	15.75	7.55	23.30
4000		125	5	3-5	184	18.85	8.65	27.50
4100		200	5-1/2	3-5-1/2	270	24.50	10.20	34.70
4200	35x40	40	5	2-11	94	14.15	7.10	21.25
4300		75	5	3-2	131	16.40	7.95	24.35
4400		125	5	3-5	184	19.40	8.95	28.35
4500		200	5	3-5-1/2	264	24.50	10.75	35.25

Important: See the Reference Section for critical supporting data - Reference Numbers and City Cost Indexes

B1010 Floor Construction

Description: Table below lists costs ($/S.F.) for a floor system using composite steel beams with welded shear studs, composite steel deck, and light weight concrete slab reinforced with W.W.F. Price includes sprayed fiber fireproofing on steel beams.

Design and Pricing Assumptions:
Structural steel is A36, high strength bolted.
Composite steel deck varies from 22 gauge to 16 gauge, galvanized.

Shear Studs are 3/4".
W.W.F., 6 x 6 – W1.4 x W1.4 (10 x 10)
Concrete f'c = 3 KSI, lightweight.
Steel trowel finish and cure.
Fireproofing is sprayed fiber (non-asbestos).

Spandrels are assumed the same as interior beams and girders to allow for exterior wall loads and bracing or moment connections.

SHELL B

System Components	QUANTITY	UNIT	COST PER S.F.		
			MAT.	INST.	TOTAL
SYSTEM B1010 256 2400					
20X25 BAY, 40 PSF S. LOAD, 5-1/2″ SLAB, 17-1/2″ TOTAL THICKNESS					
Structural steel	4.320	Lb.	4.62	1.52	6.14
Welded shear connectors 3/4″ diameter 4-7/8″ long	.163	Ea.	.10	.26	.36
Metal decking, non-cellular composite, galv. 3″ deep, 22 gauge	1.050	S.F.	1.76	.80	2.56
Sheet metal edge closure form, 12″, w/2 bends, 18 ga, galv	.045	L.F.	.15	.09	.24
Welded wire fabric rolls, 6 x 6 - W1.4 x W1.4 (10 x 10), 21 lb/csf	1.000	S.F.	.13	.30	.43
Concrete ready mix, light weight, 3,000 PSI	.333	C.F.	2.23		2.23
Place and vibrate concrete, elevated slab less than 6″, pumped	.333	C.F.		.42	.42
Finishing floor, monolithic steel trowel finish for finish floor	1.000	S.F.		.74	.74
Curing with sprayed membrane curing compound	.010	C.S.F.	.06	.07	.13
Shores, erect and strip vertical to 10′ high	.020	Ea.		.34	.34
Sprayed mineral fiber/cement for fireproof, 1″ thick on beams	.483	S.F.	.23	.38	.61
TOTAL			9.28	4.92	14.20

B1010 256		Composite Beams, Deck & Slab						
	BAY SIZE (FT.)	SUPERIMPOSED LOAD (P.S.F.)	SLAB THICKNESS (IN.)	TOTAL DEPTH (FT.-IN.)	TOTAL LOAD (P.S.F.)	COST PER S.F.		
						MAT.	INST.	TOTAL
2400	20x25	40	5-1/2	1 - 5-1/2	80	9.30	4.91	14.21
2500	RB1010 -100	75	5-1/2	1 - 9-1/2	115	9.65	4.92	14.57
2750		125	5-1/2	1 - 9-1/2	167	11.75	5.75	17.50
2900		200	6-1/4	1 - 11-1/2	251	13.25	6.20	19.45
3000	25x25	40	5-1/2	1 - 9-1/2	82	9.15	4.68	13.83
3100		75	5-1/2	1 - 11-1/2	118	10.15	4.75	14.90
3200		125	5-1/2	2 - 2-1/2	169	10.60	5.15	15.75
3300		200	6-1/4	2 - 6-1/4	252	14.30	6	20.30
3400	25x30	40	5-1/2	1 - 11-1/2	83	9.35	4.65	14
3600		75	5-1/2	1 - 11-1/2	119	10.05	4.70	14.75
3900		125	5-1/2	1 - 11-1/2	170	11.60	5.30	16.90
4000		200	6-1/4	2 - 6-1/4	252	14.35	6	20.35
4200	30x30	40	5-1/2	1 - 11-1/2	81	9.40	4.79	14.19
4400		75	5-1/2	2 - 2-1/2	116	10.10	5	15.10
4500		125	5-1/2	2 - 5-1/2	168	12.20	5.60	17.80
4700		200	6-1/4	2 - 9-1/4	252	14.60	6.50	21.10
4900	30x35	40	5-1/2	2 - 2-1/2	82	9.80	4.95	14.75
5100		75	5-1/2	2 - 5-1/2	117	10.70	5.05	15.75
5300		125	5-1/2	2 - 5-1/2	169	12.55	5.75	18.30
5500		200	6-1/4	2 - 9-1/4	254	14.75	6.50	21.25
5750	35x35	40	5-1/2	2 - 5-1/2	84	10.45	4.98	15.43
6000		75	5-1/2	2 - 5-1/2	121	11.90	5.30	17.20

B1010 Floor Construction

B1010 256		Composite Beams, Deck & Slab						

	BAY SIZE (FT.)	SUPERIMPOSED LOAD (P.S.F.)	SLAB THICKNESS (IN.)	TOTAL DEPTH (FT.-IN.)	TOTAL LOAD (P.S.F.)	COST PER S.F.		
						MAT.	INST.	TOTAL
7000	35 x 35	125	5-1/2	2 - 8-1/2	170	13.90	6.10	20
7200		200	5-1/2	2 - 11-1/2	254	15.90	6.75	22.65
7400	35x40	40	5-1/2	2 - 5-1/2	85	11.50	5.35	16.85
7600		75	5-1/2	2 - 5-1/2	121	12.45	5.50	17.95
8000		125	5-1/2	2 - 5-1/2	171	14.25	6.20	20.45
9000		200	5-1/2	2 - 11-1/2	255	17.25	7.05	24.30

Important: See the Reference Section for critical supporting data - Reference Numbers and City Cost Indexes

B1010 Floor Construction

How to Use Tables: Enter any table at superimposed load and support spacing for deck.

Cellular decking tends to be stiffer and has the potential for utility integration with deck structure.

System Components	QUANTITY	UNIT	COST PER S.F.		
			MAT.	INST.	TOTAL
SYSTEM B1010 258 0900					
SPAN 6', LOAD 125 PSF, DECK 1-1/2"-22 GA., 4" SLAB					
Metal decking, open, galv., 1-1/2" deep, 22 ga., over 50 sq.	1.050	S.F.	1.22	.50	1.72
Welded wire fabric rolls, 6 x 6 - W1.4 x W1.4 (10 x 10), 21 lb/csf	1.000	S.F.	.13	.30	.43
Edge forms to 6" high on elevated slab, 4 uses	.050	L.F.	.01	.17	.18
Concrete ready mix, regular weight, 3000 psi	.253	C.F.	.86		.86
Place and vibrate concrete, elevated slab less than 6", pumped	.253	C.F.		.24	.24
Finishing floor, monolithic steel trowel finish for finish floor	1.000	S.F.		.74	.74
Curing with sprayed membrane curing compound	.010	C.S.F.	.06	.07	.13
TOTAL			2.28	2.02	4.30

B1010 258		**Metal Deck/Concrete Fill**						
	SUPERIMPOSED LOAD (P.S.F.)	DECK SPAN (FT.)	DECK GAGE DEPTH	SLAB THICKNESS (IN.)	TOTAL LOAD (P.S.F.)	COST PER S.F.		
						MAT.	INST.	TOTAL
0900	125	6	22 1-1/2	4	164	2.28	2.02	4.30
0920		7	20 1-1/2	4	164	2.51	2.13	4.64
0950		8	20 1-1/2	4	165	2.51	2.13	4.64
0970		9	18 1-1/2	4	165	2.93	2.13	5.06
1000		10	18 2	4	165	2.87	2.26	5.13
1020		11	18 3	5	169	3.50	2.43	5.93
1050	150	6	22 1-1/2	4	189	2.29	2.05	4.34
1070		7	22 1-1/2	4	189	2.29	2.05	4.34
1100		8	20 1-1/2	4	190	2.51	2.13	4.64
1120		9	18 1-1/2	4	190	2.93	2.13	5.06
1150		10	18 3	5	190	3.50	2.43	5.93
1170		11	18 3	5-1/2	194	3.50	2.43	5.93
1200	200	6	22 1-1/2	4	239	2.29	2.05	4.34
1220		7	22 1-1/2	4	239	2.29	2.05	4.34
1250		8	20 1-1/2	4	239	2.51	2.13	4.64
1270		9	18 2	4	241	3.35	2.30	5.65
1300		10	16 2	4	240	3.92	2.34	6.26
1320	250	6	22 1-1/2	4	289	2.29	2.05	4.34
1350		7	22 1-1/2	4	289	2.29	2.05	4.34
1370		8	18 2	4	290	3.35	2.30	5.65
1400		9	16 2	4	290	3.92	2.34	6.26

B1010 Floor Construction

B1010 260	Cellular Composite Deck							

	SUPERIMPOSED LOAD (P.S.F.)	DECK SPAN (FT.)	DECK & PLATE GAUGE	SLAB THICKNESS (IN.)	TOTAL LOAD (P.S.F.)	COST PER S.F.		
						MAT.	INST.	TOTAL
1450	150	10	20 20	5	195	8.55	3.40	11.95
1470		11	20 20	5	195	8.55	3.40	11.95
1500		12	20 20	5	195	8.55	3.40	11.95
1520		13	20 20	5	195	8.55	3.40	11.95
1570	200	10	20 20	5	250	8.55	3.40	11.95
1650		11	20 20	5	250	8.55	3.40	11.95
1670		12	20 20	5	250	8.55	3.40	11.95
1700		13	18 20	5	250	10.05	3.44	13.49

B SHELL

Important: See the Reference Section for critical supporting data - Reference Numbers and City Cost Indexes

B1010 Floor Construction

Description: Table below lists the S.F. costs for wood joists and a minimum thickness plywood subfloor.

Design Assumptions: 10% allowance has been added to framing quantities for overlaps, waste, double joists at openings or under partitions, etc. 5% added to subfloor for waste.

SHELL B

System Components	QUANTITY	UNIT	COST PER S.F.		
			MAT.	INST.	TOTAL
SYSTEM B1010 261 2500					
WOOD JOISTS 2"X6", 12" O.C.					
Framing joists, fir, 2"x6"	1.100	B.F.	.72	.78	1.50
Subfloor plywood CDX 1/2"	1.050	S.F.	.71	.62	1.33
TOTAL			1.43	1.40	2.83

B1010 261	Wood Joist	COST PER S.F.		
		MAT.	INST.	TOTAL
2500	Wood joists, 2"x6", 12" O.C.	1.43	1.40	2.83
2550	16" O.C.	1.25	1.21	2.46
2600	24" O.C.	1.30	1.14	2.44
2900	2"x8", 12" O.C.	1.74	1.50	3.24
2950	16" O.C.	1.49	1.29	2.78
3000	24" O.C.	1.47	1.19	2.66
3300	2"x10", 12" O.C.	2.28	1.71	3.99
3350	16" O.C.	1.89	1.44	3.33
3400	24" O.C.	1.72	1.29	3.01
3700	2"x12", 12" O.C.	2.65	1.73	4.38
3750	16" O.C.	2.17	1.45	3.62
3800	24" O.C.	1.91	1.30	3.21
4100	2"x14", 12" O.C.	3.34	1.89	5.23
4150	16" O.C.	2.69	1.57	4.26
4200	24" O.C.	2.27	1.38	3.65
4500	3"x6", 12" O.C.	3.02	1.67	4.69
4550	16" O.C.	2.45	1.41	3.86
4600	24" O.C.	2.11	1.27	3.38
4900	3"x8", 12" O.C.	3.85	1.64	5.49
4950	16" O.C.	3.07	1.39	4.46
5000	24" O.C.	2.51	1.26	3.77
5300	3"x10", 12" O.C.	4.63	1.87	6.50
5350	16" O.C.	3.66	1.56	5.22
5400	24" O.C.	2.91	1.37	4.28
5700	3"x12", 12" O.C.	4.79	1.62	6.41
5750	16" O.C.	4.31	1.83	6.14
5800	24" O.C.	3.34	1.55	4.89
6100	4"x6", 12" O.C.	3.90	1.70	5.60
6150	16" O.C.	3.11	1.43	4.54
6200	24" O.C.	2.54	1.28	3.82

B1010 Floor Construction

Description: Table lists the S.F. costs, total load, and member sizes, for various bay sizes and loading conditions.

Design Assumptions: Dead load = girder, beams, and joist weight plus 3/4" plywood floor.

Maximum deflection is 1/360 of the clear span.

Lumber is stress grade f(w) = 1,800 PSI

System Components	QUANTITY	UNIT	COST PER S.F.		
			MAT.	INST.	TOTAL
SYSTEM B1010 264 2000					
15' X 15' BAY, S. LOAD 40 P.S.F.					
Beams and girders, structural grade, 8" x 12"	.730	B.F.	1.94	.18	2.12
Framing joists fir 4"x12"	.660	B.F.	.73	.32	1.05
Framing joists, 2"x6"	.840	B.F.	.55	.60	1.15
Beam to girder saddles	.510	Lb.	.91	.58	1.49
Column caps	.510	Lb.	.37	.08	.45
Drilling, bolt holes	.510	Lb.		.28	.28
Machine bolts	.510	Lb.	.12	.12	.24
Joist hangers 18 ga.	.213	Ea.	.19	.57	.76
Subfloor plywood CDX 3/4"	1.050	S.F.	.95	.75	1.70
TOTAL			5.76	3.48	9.24

B1010 264		Wood Beam & Joist					
	BAY SIZE (FT.)	SUPERIMPOSED LOAD (P.S.F.)	GIRDER BEAM (IN.)	JOISTS (IN.)	TOTAL LOAD (P.S.F.)	COST PER S.F.		
						MAT.	INST.	TOTAL
2000	15x15	40	8 x 12 4 x 12	2 x 6 @ 16	53	5.75	3.49	9.24
2050	RB1010 -100	75	8 x 16 4 x 16	2 x 8 @ 16	90	7.95	3.81	11.76
2100		125	12 x 16 6 x 16	2 x 8 @ 12	144	12.70	4.89	17.59
2150		200	14 x 22 12 x 16	2 x 10 @ 12	227	24	7.50	31.50
2500	15x20	40	10 x 16 8 x 12	2 x 6 @ 16	58	8.95	3.56	12.51
2550		75	12 x 14 8 x 14	2 x 8 @ 16	96	11.05	4.13	15.18
2600		125	10 x 18 12 x 14	2 x 8 @ 12	152	15.45	5.55	21
2650		200	14 x 20 14 x 16	2 x 10 @ 12	234	21.50	6.80	28.30
3000	20x20	40	10 x 14 10 x 12	2 x 8 @ 16	63	8.85	3.51	12.36
3050		75	12 x 16 8 x 16	2 x 10 @ 16	102	11.20	3.94	15.14
3100		125	14 x 22 12 x 16	2 x 10 @ 12	163	19.35	6.30	25.65

Important: See the Reference Section for critical supporting data - Reference Numbers and City Cost Indexes

B1010 Floor Construction

Description: Table below lists the S.F. costs, total load, and member sizes, for various bay sizes and loading conditions.

Design Assumptions: Dead load = girder, beams, and joist weight plus 3/4″ plywood floor.

Maximum deflection is 1/360 of the clear span.

Lumber is stress grade f(w) = 2,500 PSI

System Components	QUANTITY	UNIT	COST PER S.F.		
			MAT.	INST.	TOTAL
SYSTEM B1010 265 2000					
15′X15′ BAY, 40 PSF L.L., JOISTS 2X6, @ 16″ O.C.					
Laminated framing, straight beams	1.000	B.F.	2.88	.83	3.71
Framing joists,2″x6″	.840	B.F.	.55	.60	1.15
Saddles	3.375	Lb.	.91	.58	1.49
Column caps	3.375	Lb.	.37	.08	.45
Drilling, bolt holes	3.375	Lb.		.26	.26
Machine bolts	3.375	Lb.	.10	.11	.21
Joist & beam hangers 18 ga.	.213	Ea.	.19	.57	.76
Subfloor plywood CDX 3/4″	1.050	S.F.	.95	.75	1.70
TOTAL			5.95	3.78	9.73

B1010 265		Laminated Wood Floor Beams						
	BAY SIZE (FT.)	SUPERIMPOSED LOAD (P.S.F.)	GIRDER (IN.)	BEAM (IN.)	TOTAL LOAD (P.S.F.)	COST PER S.F.		
						MAT.	INST.	TOTAL
2000	15x15	40	5 x 12-3/8	5 x 8-1/4	54	5.95	3.78	9.73
2050		75	5 x 16-1/2	5 x 11	92	7.65	4.37	12.02
2100	RB1010 -100	125	8-5/8 x 15-1/8	8-5/8 x 11	147	11.30	6.15	17.45
2150		200	8-5/8 x 22	8-5/8 x 16-1/2	128	18.45	8.95	27.40
2500	15x20	40	5 x 13-3/4	5 x 11	55	6.15	3.73	9.88
2550		75	5 x 17-3/8	5 x 15-1/8	92	8.35	4.55	12.90
2600		125	8-5/8 x 17-7/8	8-5/8 x 13-3/4	148	12.55	6.55	19.10
2650		200	8-5/8 x 22	8-5/8 x 17-7/8	226	16.65	8.10	24.75
3000	20x20	40	5 x 16-1/2	5 x 13-3/4	55	6.45	3.64	10.09
3050		75	5 x 20-5/8	5 x 16-1/2	92	8.35	4.41	12.76
3100		125	8-5/8 x 22	8-5/8 x 15-1/8	148	12.75	6.10	18.85
3150		200	8-5/8 x 28-7/8	8-5/8 x 20-5/8	229	17.85	8.20	26.05

B1010 Floor Construction

Description: Table below lists the S.F. costs and maximum spans for commonly used wood decking materials for various loading conditions.

Design Assumptions: Total applied load is the superimposed load plus dead load.

Maximum deflection is 1/180 of the clear span which is not suitable if plaster ceilings will be supported by the roof or floor.

Modulus of elasticity (E) and fiber strength (f) are as shown in the table to the right.

No allowance for waste has been included.

Supporting beams or purlins are not included in costs below.

Decking Material Characteristics

Deck Material	Modulus of Elasticity	Fiber Strength
Cedar	1,100,000	1000 psi
Douglas fir	1,760,000	1200 psi
Hemlock	1,600,000	1200 psi
White spruce	1,320,000	1200 psi

B1010 266			Wood Deck					
	NOMINAL THK (IN.)	TYPE WOOD	MAXIMUM SPAN			COST PER S.F.		
			MAXIMUM SPAN 40 P.S.F. LOAD	MAXIMUM SPAN 100 P.S.F.	MAXIMUM SPAN 250 P.S.F. LOAD	MAT.	INST.	TOTAL
1000	2	cedar	6.5	5	3.5	4.93	2.53	7.46
1050		douglas fir	8	6	4	1.35	2.53	3.88
1100		hemlock	7.5	5.5	4	1.76	2.53	4.29
1150		white spruce	7	5	3.5	1.69	2.53	4.22
1200	3	cedar	11	8	6	7.05	2.77	9.82
1250		douglas fir	13	10	7	2.64	2.77	5.41
1300		hemlock	12	9	7	2.64	2.77	5.41
1350		white spruce	11	8	6	2.54	2.77	5.31
1400	4	cedar	15	11	8	9.50	3.54	13.04
1450		douglas fir	18	13	10	3.53	3.54	7.07
1500		hemlock	17	13	9	3.52	3.54	7.06
1550		white spruce	16	12	8	3.39	3.54	6.93
1600	6	cedar	23	17	13	15.50	4.43	19.93
1650		douglas fir	24+	21	15	5.30	4.43	9.73
1700		hemlock	24+	20	15	5.30	4.43	9.73
1750		white spruce	24	18	13	4.22	4.43	8.65

Important: See the Reference Section for critical supporting data - Reference Numbers and City Cost Indexes

B1010 Floor Construction

The table below lists fireproofing costs for steel beams by type, beam size, thickness and fire rating. Weights listed are for the fireproofing material only.

System Components	QUANTITY	UNIT	COST PER L.F.		
			MAT.	INST.	TOTAL
SYSTEM B1010 710 1300					
FIREPROOFING, 5/8" F.R. GYP. BOARD, 12"X 4" BEAM, 2" THICK, 2 HR. F.R.					
Corner bead for drywall, 1-1/4" x 1-1/4", galvanized	.020	C.L.F.	.45	2.54	2.99
L bead for drywall, galvanized	.020	C.L.F.	.35	2.96	3.31
Furring, beams & columns, 3/4" galv. channels, 24" O.C.	2.330	S.F.	.35	4.92	5.27
Drywall on beam, no finish, 2 layers at 5/8" thick	3.000	S.F.	2.22	8.85	11.07
Drywall, taping and finishing joints, add	3.000	S.F.	.12	1.32	1.44
TOTAL			3.49	20.59	24.08

B1010 710		Steel Beam Fireproofing						
	ENCASEMENT SYSTEM	BEAM SIZE (IN.)	THICKNESS (IN.)	FIRE RATING (HRS.)	WEIGHT (P.L.F.)	COST PER L.F.		
						MAT.	INST.	TOTAL
0400	Concrete	12x4	1	1	77	5.50	22	27.50
0450	3000 PSI		1-1/2	2	93	6.40	23.50	29.90
0500			2	3	121	7.15	25.50	32.65
0550		14x5	1	1	100	7.15	26.50	33.65
0600			1-1/2	2	122	8.20	29.50	37.70
0650			2	3	142	9.20	32.50	41.70
0700		16x7	1	1	147	8.40	28.50	36.90
0750			1-1/2	2	169	9.05	29	38.05
0800			2	3	195	10.15	32	42.15
0850		18x7-1/2	1	1	172	9.70	32.50	42.20
0900			1-1/2	2	196	10.85	36	46.85
0950			2	3	225	12.05	39.50	51.55
1000		24x9	1	1	264	13.10	39.50	52.60
1050			1-1/2	2	295	14.20	42	56.20
1100			2	3	328	15.30	44	59.30
1150		30x10-1/2	1	1	366	17.10	50	67.10
1200			1-1/2	2	404	18.50	53	71.50
1250			2	3	449	19.50	56	75.50
1300	5/8" fire rated	12x4	2	2	15	3.49	20.50	23.99
1350	Gypsum board		2-5/8	3	24	4.85	25.50	30.35
1400		14x5	2	2	17	3.74	21.50	25.24
1450			2-5/8	3	27	5.35	27	32.35
1500		16x7	2	2	20	4.19	24	28.19
1550			2-5/8	3	31	5.45	23.50	28.95
1600		18x7-1/2	2	2	22	4.48	25.50	29.98
1650			2-5/8	3	34	6.35	32.50	38.85

B1010 710			Steel Beam Fireproofing					

	ENCASEMENT SYSTEM	BEAM SIZE (IN.)	THICKNESS (IN.)	FIRE RATING (HRS.)	WEIGHT (P.L.F.)	COST PER L.F.		
						MAT.	INST.	TOTAL
1700	5/8" fire rated	24x9	2	2	27	5.50	31.50	37
1750	Gypsum board		2-5/8	3	42	7.85	40	47.85
1800		30x10-1/2	2	2	33	6.50	37	43.50
1850			2-5/8	3	51	9.20	46.50	55.70
1900	Gypsum	12x4	1-1/8	3	18	4.93	15.65	20.58
1950	Plaster on	14x5	1-1/8	3	21	5.55	17.65	23.20
2000	Metal lath	16x7	1-1/8	3	25	6.50	20	26.50
2050		18x7-1/2	1-1/8	3	32	7.30	22.50	29.80
2100		24x9	1-1/8	3	35	9	27.50	36.50
2150		30x10-1/2	1-1/8	3	44	10.80	33	43.80
2200	Perlite plaster	12x4	1-1/8	2	16	4.48	17.35	21.83
2250	On metal lath		1-1/4	3	20	4.74	18.10	22.84
2300			1-1/2	4	22	5.05	19.10	24.15
2350		14x5	1-1/8	2	18	5.70	21.50	27.20
2400			1-1/4	3	23	6	22.50	28.50
2450			1-1/2	4	25	6.60	24	30.60
2500		16x7	1-1/8	2	21	5.45	21.50	26.95
2550			1-1/4	3	26	5.75	22.50	28.25
2600			1-1/2	4	29	6.05	23	29.05
2650		18x7-1/2	1-1/8	2	26	6.60	25	31.60
2700			1-1/4	3	33	6.90	26	32.90
2750			1-1/2	4	36	7.50	27.50	35
2800		24x9	1-1/8	2	30	7.10	28	35.10
2850			1-1/4	3	38	7.35	29	36.35
2900			1-1/2	4	41	7.65	30	37.65
2950		30x10-1/2	1-1/8	2	37	9.05	35	44.05
3000			1-1/4	3	46	9.35	35.50	44.85
3050			1-1/2	4	51	9.90	37.50	47.40
3100	Sprayed Fiber	12x4	5/8	1	12	1.02	1.74	2.76
3150	(non asbestos)		1-1/8	2	21	1.83	3.12	4.95
3200			1-1/4	3	22	2.04	3.47	5.51
3250		14x5	5/8	1	14	1.02	1.74	2.76
3300			1-1/8	2	26	1.83	3.12	4.95
3350			1-1/4	3	29	2.04	3.47	5.51
3400		16x7	5/8	1	18	1.29	2.20	3.49
3450			1-1/8	2	32	2.32	3.95	6.27
3500			1-1/4	3	36	2.58	4.39	6.97
3550		18x7-1/2	5/8	1	20	1.42	2.42	3.84
3600			1-1/8	2	35	2.56	4.36	6.92
3650			1-1/4	3	39	2.83	4.82	7.65
3700		24x9	5/8	1	25	1.80	3.06	4.86
3750			1-1/8	2	45	3.26	5.55	8.81
3800			1-1/4	3	50	3.62	6.15	9.77
3850		30x10-1/2	5/8	1	31	2.18	3.71	5.89
3900			1-1/8	2	55	3.92	6.70	10.62
3950			1-1/4	3	61	4.36	7.45	11.81
4000	On Decking	Flat	1		6	.47	.50	.97
4050	Per S.F.	Corrugated	1		7	.70	.96	1.66
4100		Fluted	1		7	.70	.96	1.66

Important: See the Reference Section for critical supporting data - Reference Numbers and City Cost Indexes

Listed below are costs per V.L.F. for fireproofing by material, column size, thickness and fire rating. Weights listed are for the fireproofing material only.

System Components			COST PER V.L.F.		
	QUANTITY	UNIT	MAT.	INST.	TOTAL
SYSTEM B1010 720 3000					
CONCRETE FIREPROOFING, 8″ STEEL COLUMN, 1″ THICK, 1 HR. FIRE RATING					
Forms in place, columns, plywood, 4 uses	3.330	SFCA	2.70	22.31	25.01
Welded wire fabric, 2 x 2 #14 galv. 21 lb./C.S.F., column wrap	2.700	S.F.	.90	4.32	5.22
Concrete ready mix, regular weight, 3000 psi	.621	C.F.	2.20		2.20
Place and vibrate concrete, 12″ sq./round columns, pumped	.621	C.F.		1.43	1.43
TOTAL			5.80	28.06	33.86

B1010 720		Steel Column Fireproofing						
	ENCASEMENT SYSTEM	COLUMN SIZE (IN.)	THICKNESS (IN.)	FIRE RATING (HRS.)	WEIGHT (P.L.F.)	COST PER V.L.F.		
						MAT.	INST.	TOTAL
3000	Concrete	8	1	1	110	5.80	28.50	34.30
3050			1-1/2	2	133	6.65	31.50	38.15
3100			2	3	145	7.50	34	41.50
3150		10	1	1	145	7.55	35	42.55
3200			1-1/2	2	168	8.40	37.50	45.90
3250			2	3	196	9.30	40	49.30
3300		14	1	1	258	9.50	41	50.50
3350			1-1/2	2	294	10.35	43.50	53.85
3400			2	3	325	11.40	46.50	57.90
3450	Gypsum board	8	1/2	2	8	3.07	17.15	20.22
3500	1/2″ fire rated	10	1/2	2	11	3.25	17.90	21.15
3550	1 layer	14	1/2	2	18	3.35	18.30	21.65
3600	Gypsum board	8	1	3	14	4.43	22	26.43
3650	1/2″ fire rated	10	1	3	17	4.77	23	27.77
3700	2 layers	14	1	3	22	4.96	24	28.96
3750	Gypsum board	8	1-1/2	3	23	6.05	27.50	33.55
3800	1/2″ fire rated	10	1-1/2	3	27	6.85	30.50	37.35
3850	3 layers	14	1-1/2	3	35	7.60	33.50	41.10
3900	Sprayed fiber	8	1-1/2	2	6.3	3.19	5.40	8.59
3950	Direct application		2	3	8.3	4.39	7.50	11.89
4000			2-1/2	4	10.4	5.70	9.70	15.40
4050		10	1-1/2	2	7.9	3.85	6.55	10.40
4100			2	3	10.5	5.25	8.95	14.20
4150			2-1/2	4	13.1	6.80	11.55	18.35

B1010 Floor Construction

B1010 720	Steel Column Fireproofing

	ENCASEMENT SYSTEM	COLUMN SIZE (IN.)	THICKNESS (IN.)	FIRE RATING (HRS.)	WEIGHT (P.L.F.)	COST PER V.L.F.		
						MAT.	INST.	TOTAL
4200	Sprayed fiber	14	1-1/2	2	10.8	4.78	8.15	12.93
4250	Direct application		2	3	14.5	6.50	11.10	17.60
4300			2-1/2	4	18	8.35	14.20	22.55
4350	3/4" gypsum plaster	8	3/4	1	23	6.25	22	28.25
4400	On metal lath	10	3/4	1	28	7.05	25.50	32.55
4450		14	3/4	1	38	8.55	31	39.55
4500	Perlite plaster	8	1	2	18	7	24.50	31.50
4550	On metal lath		1-3/8	3	23	7.65	27	34.65
4600			1-3/4	4	35	9	32.50	41.50
4650	Perlite plaster	10	1	2	21	8	28	36
4700			1-3/8	3	27	9.05	32.50	41.55
4750			1-3/4	4	41	10.15	37	47.15
4800		14	1	2	29	9.95	35.50	45.45
4850			1-3/8	3	35	11.20	40.50	51.70
4900			1-3/4	4	53	11.60	42.50	54.10
4950	1/2 gypsum plaster	8	7/8	1	13	6.85	18.20	25.05
5000	On 3/8" gypsum lath	10	7/8	1	16	8.20	22	30.20
5050		14	7/8	1	21	10.05	26.50	36.55
5100	5/8" gypsum plaster	8	1	1-1/2	20	6.60	20	26.60
5150	On 3/8" gypsum lath	10	1	1-1/2	24	7.85	24	31.85
5200		14	1	1-1/2	33	9.65	29	38.65
5250	1"perlite plaster	8	1-3/8	2	23	6.15	21	27.15
5300	On 3/8" gypsum lath	10	1-3/8	2	28	7	24.50	31.50
5350		14	1-3/8	2	37	8.70	30	38.70
5400	1-3/8" perlite plaster	8	1-3/4	3	27	7.85	24	31.85
5450	On 3/8" gypsum lath	10	1-3/4	3	33	9	27	36
5500		14	1-3/4	3	43	11.35	34.50	45.85
5550	Concrete masonry	8	4-3/4	4	126	8.40	26	34.40
5600	Units 4" thick	10	4-3/4	4	166	10.50	32.50	43
5650	75% solid	14	4-3/4	4	262	12.60	39	51.60

Important: See the Reference Section for critical supporting data - Reference Numbers and City Cost Indexes

B1020 Roof Construction

The table below lists prices per S.F. for roof rafters and sheathing by nominal size and spacing. Sheathing is 5/16″ CDX for 12″ and 16″ spacing and 3/8″ CDX for 24″ spacing.

Factors for Converting Inclined to Horizontal

Roof Slope	Approx. Angle	Factor	Roof Slope	Approx. Angle	Factor
Flat	0°	1.000	12 in 12	45.0°	1.414
1 in 12	4.8°	1.003	13 in 12	47.3°	1.474
2 in 12	9.5°	1.014	14 in 12	49.4°	1.537
3 in 12	14.0°	1.031	15 in 12	51.3°	1.601
4 in 12	18.4°	1.054	16 in 12	53.1°	1.667
5 in 12	22.6°	1.083	17 in 12	54.8°	1.734
6 in 12	26.6°	1.118	18 in 12	56.3°	1.803
7 in 12	30.3°	1.158	19 in 12	57.7°	1.873
8 in 12	33.7°	1.202	20 in 12	59.0°	1.943
9 in 12	36.9°	1.250	21 in 12	60.3°	2.015
10 in 12	39.8°	1.302	22 in 12	61.4°	2.088
11 in 12	42.5°	1.357	23 in 12	62.4°	2.162

System Components		QUANTITY	UNIT	COST PER S.F.		
				MAT.	INST.	TOTAL
SYSTEM B1020 102 2500						
RAFTER 2″X4″, 12″ O.C.						
	Framing joists, fir 2″x 4″	.730	B.F.	.45	.78	1.23
	Sheathing plywood on roof CDX 5/16″	1.050	S.F.	.71	.58	1.29
	TOTAL			1.16	1.36	2.52

B1020 102	Wood/Flat or Pitched	COST PER S.F.		
		MAT.	INST.	TOTAL
2500	Flat rafter, 2″x4″, 12″ O.C.	1.16	1.36	2.52
2550	16″ O.C.	1.05	1.17	2.22
2600	24″ O.C.	.75	1.01	1.76
2900	2″x6″, 12″ O.C.	1.43	1.36	2.79
2950	16″ O.C.	1.25	1.17	2.42
3000	24″ O.C.	.88	1	1.88
3300	2″x8″, 12″ O.C.	1.74	1.46	3.20
3350	16″ O.C.	1.49	1.24	2.73
3400	24″ O.C.	1.05	1.06	2.11
3700	2″x10″, 12″ O.C.	2.28	1.67	3.95
3750	16″ O.C.	1.89	1.40	3.29
3800	24″ O.C.	1.30	1.15	2.45
4100	2″x12″, 12″ O.C.	2.65	1.69	4.34
4150	16″ O.C.	2.17	1.41	3.58
4200	24″ O.C.	1.49	1.16	2.65
4500	2″x14″, 12″ O.C.	3.34	2.47	5.81
4550	16″ O.C.	2.69	2.01	4.70
4600	24″ O.C.	1.85	1.56	3.41
4900	3″x6″, 12″ O.C.	3.02	1.44	4.46
4950	16″ O.C.	2.45	1.22	3.67
5000	24″ O.C.	1.69	1.04	2.73
5300	3″x8″, 12″ O.C.	3.85	1.60	5.45
5350	16″ O.C.	3.07	1.34	4.41
5400	24″ O.C.	2.09	1.12	3.21
5700	3″x10″, 12″ O.C.	4.63	1.83	6.46
5750	16″ O.C.	3.66	1.52	5.18
5800	24″ O.C.	2.49	1.24	3.73
6100	3″x12″, 12″ O.C.	5.50	2.19	7.69
6150	16″ O.C.	4.31	1.79	6.10
6200	24″ O.C.	2.92	1.42	4.34

B1020 Roof Construction

Table below lists the cost per S.F. for a roof system with steel columns, beams, and deck using open web steel joists and 1-1/2" galvanized metal deck. Perimeter of system is supported on bearing walls.

Design and Pricing Assumptions:
Columns are 18' high.
Joists are 5'-0" O.C. and span the long direction of the bay.

Joists at columns have bottom chords extended and are connected to columns. Column costs are not included but are listed separately per S.F. of floor.

Roof deck is 1-1/2", 22 gauge galvanized steel. Joist cost includes appropriate bridging. Deflection is limited to 1/240 of the span. Fireproofing is not included.

Costs/S.F. are based on a building 4 bays long and 4 bays wide.

Design Loads	Min.	Max.
Joists & Beams	3 PSF	5 PSF
Deck	2	2
Insulation	3	3
Roofing	6	6
Misc.	6	6
Total Dead Load	20 PSF	22 PSF

System Components			COST PER S.F.		
	QUANTITY	UNIT	MAT.	INST.	TOTAL
SYSTEM B1020 108 1200					
METAL DECK & STEEL JOISTS,15'X20' BAY,20 PSF S. LOAD,16" DP,40 PSF T.LOAD					
Structural steel	.488	Lb.	.51	.13	.64
Open web joists	1.022	Lb.	.75	.31	1.06
Metal decking, open, galvanized, 1-1/2" deep, 22 gauge	1.050	S.F.	1.22	.50	1.72
TOTAL			2.48	.94	3.42

B1020 108 — Steel Joists, Beams & Deck on Columns & Walls

	BAY SIZE (FT.)	SUPERIMPOSED LOAD (P.S.F.)	DEPTH (IN.)	TOTAL LOAD (P.S.F.)	COLUMN ADD	COST PER S.F.		
						MAT.	INST.	TOTAL
1200	15x20	20	16	40		2.48	.94	3.42
1300					columns	.58	.15	.73
1400		30	16	50		2.70	1.01	3.71
1500					columns	.58	.15	.73
1600		40	18	60		2.93	1.09	4.02
1700					columns	.58	.15	.73
1800	20x20	20	16	40		2.93	1.06	3.99
1900					columns	.44	.12	.56
2000		30	18	50		2.93	1.06	3.99
2100					columns	.44	.12	.56
2200		40	18	60		3.30	1.19	4.49
2300					columns	.44	.12	.56
2400	20x25	20	18	40		2.94	1.10	4.04
2500					columns	.35	.09	.44
2600		30	18	50		3.26	1.27	4.53
2700					columns	.35	.09	.44
2800		40	20	60		3.28	1.21	4.49
2900					columns	.47	.12	.59
3000	25x25	20	18	40		3.17	1.15	4.32
3100					columns	.28	.08	.36
3200		30	22	50		3.59	1.36	4.95
3300					columns	.37	.09	.46
3400		40	20	60		3.78	1.34	5.12
3500					columns	.37	.09	.46

Important: See the Reference Section for critical supporting data - Reference Numbers and City Cost Indexes

B1020   Roof Construction

B1020 108	Steel Joists, Beams & Deck on Columns & Walls

	BAY SIZE (FT.)	SUPERIMPOSED LOAD (P.S.F.)	DEPTH (IN.)	TOTAL LOAD (P.S.F.)	COLUMN ADD	COST PER S.F.		
						MAT.	INST.	TOTAL
3600	25x30	20	22	40		3.38	1.10	4.48
3700					columns	.31	.08	.39
3800		30	20	50		3.67	1.43	5.10
3900					columns	.31	.08	.39
4000		40	25	60		3.99	1.27	5.26
4100					columns	.37	.09	.46
4200	30x30	20	25	42		3.71	1.19	4.90
4300					columns	.26	.06	.32
4400		30	22	52		4.06	1.28	5.34
4500					columns	.31	.08	.39
4600		40	28	62		4.22	1.32	5.54
4700					columns	.31	.08	.39
4800	30x35	20	22	42		3.86	1.23	5.09
4900					columns	.27	.07	.34
5000		30	28	52		4.08	1.29	5.37
5100					columns	.27	.07	.34
5200		40	25	62		4.42	1.38	5.80
5300					columns	.31	.08	.39
5400	35x35	20	28	42		3.88	1.24	5.12
5500					columns	.23	.06	.29
5600		30	25	52		4.72	1.46	6.18
5700					columns	.27	.07	.34
5800		40	28	62		4.78	1.47	6.25
5900					columns	.30	.08	.38

SHELL   B

B1020 Roof Construction

Description: Table below lists the cost per S.F. for a roof system with steel columns, beams, and deck, using open web steel joists and 1-1/2" galvanized metal deck.

Design and Pricing Assumptions:
Columns are 18' high.
Building is 4 bays long by 4 bays wide.
Joists are 5'-0" O.C. and span the long direction of the bay.
Joists at columns have bottom chords extended and are connected to columns. Column costs are not included but are listed separately per S.F. of floor.

Roof deck is 1-1/2", 22 gauge galvanized steel. Joist cost includes appropriate bridging. Deflection is limited to 1/240 of the span. Fireproofing is not included.

Design Loads	Min.		Max.	
Joists & Beams	3	PSF	5	PSF
Deck	2		2	
Insulation	3		3	
Roofing	6		6	
Misc.	6		6	
Total Dead Load	20	PSF	22	PSF

System Components			COST PER S.F.		
	QUANTITY	UNIT	MAT.	INST.	TOTAL
SYSTEM B1020 112 1100					
METAL DECK AND JOISTS,15'X20' BAY,20 PSF S. LOAD					
Structural steel	.954	Lb.	1	.26	1.26
Open web joists	1.260	Lb.	.92	.38	1.30
Metal decking, open, galvanized, 1-1/2" deep, 22 gauge	1.050	S.F.	1.22	.50	1.72
TOTAL			3.14	1.14	4.28

B1020 112		Steel Joists, Beams, & Deck on Columns						
	BAY SIZE (FT.)	SUPERIMPOSED LOAD (P.S.F.)	DEPTH (IN.)	TOTAL LOAD (P.S.F.)	COLUMN ADD	COST PER S.F.		
						MAT.	INST.	TOTAL
1100	15x20	20	16	40		3.14	1.14	4.28
1200					columns	1.62	.41	2.03
1300		30	16	50		3.45	1.22	4.67
1400					columns	1.62	.41	2.03
1500		40	18	60		3.51	1.26	4.77
1600					columns	1.62	.41	2.03
1700	20x20	20	16	40		3.38	1.20	4.58
1800					columns	1.22	.31	1.53
1900		30	18	50		3.73	1.29	5.02
2000					columns	1.22	.31	1.53
2100		40	18	60		4.17	1.43	5.60
2200					columns	1.22	.31	1.53
2300	20x25	20	18	40		3.58	1.28	4.86
2400					columns	.97	.25	1.22
2500		30	18	50		4.02	1.51	5.53
2600					columns	.97	.25	1.22
2700		40	20	60		4.03	1.43	5.46
2800					columns	1.30	.33	1.63
2900	25x25	20	18	40		4.18	1.43	5.61
3000					columns	.78	.20	.98
3100		30	22	50		4.52	1.62	6.14
3200					columns	1.04	.27	1.31
3300		40	20	60		4.84	1.64	6.48
3400					columns	1.04	.27	1.31

Important: See the Reference Section for critical supporting data - Reference Numbers and City Cost Indexes

B1020 Roof Construction

B1020 112	Steel Joists, Beams, & Deck on Columns

	BAY SIZE (FT.)	SUPERIMPOSED LOAD (P.S.F.)	DEPTH (IN.)	TOTAL LOAD (P.S.F.)	COLUMN ADD	COST PER S.F.		
						MAT.	INST.	TOTAL
3500	25x30	20	22	40		4.12	1.29	5.41
3600					columns	.87	.22	1.09
3700		30	20	50		4.57	1.70	6.27
3800					columns	.87	.22	1.09
3900		40	25	60		4.98	1.53	6.51
4000					columns	1.04	.27	1.31
4100	30x30	20	25	42		4.66	1.43	6.09
4200					columns	.72	.18	.90
4300		30	22	52		5.15	1.56	6.71
4400					columns	.87	.22	1.09
4500		40	28	62		5.40	1.63	7.03
4600					columns	.87	.22	1.09
4700	30x35	20	22	42		4.82	1.49	6.31
4800					columns	.74	.19	.93
4900		30	28	52		5.10	1.56	6.66
5000					columns	.74	.19	.93
5100		40	25	62		5.60	1.69	7.29
5200					columns	.87	.22	1.09
5300	35x35	20	28	42		5.10	1.55	6.65
5400					columns	.64	.17	.81
5500		30	25	52		5.70	1.72	7.42
5600					columns	.74	.19	.93
5700		40	28	62		6.15	1.83	7.98
5800					columns	.82	.21	1.03

B1020 Roof Construction

Description: Table below lists cost per S.F. for a roof system using open web steel joists and 1-1/2″ galvanized metal deck. The system is assumed supported on bearing walls or other suitable support. Costs for the supports are not included.

Design and Pricing Assumptions:
Joists are 5′-0″ O.C.
Roof deck is 1-1/2″, 22 gauge galvanized.

System Components	QUANTITY	UNIT	COST PER S.F.		
			MAT.	INST.	TOTAL
SYSTEM B1020 116 1100					
METAL DECK AND JOISTS, 20′ SPAN, 20 PSF S. LOAD					
Open web joists, horiz. bridging T.L. lot, to 30′ span	1.114	Lb.	.81	.34	1.15
Metal decking, open type, galv 1-1/2″ deep	1.050	S.F.	1.22	.50	1.72
TOTAL			2.03	.84	2.87

B1020 116		Steel Joists & Deck on Bearing Walls				COST PER S.F.		
	BAY SIZE (FT.)	SUPERIMPOSED LOAD (P.S.F.)	DEPTH (IN.)	TOTAL LOAD (P.S.F.)		MAT.	INST.	TOTAL
1100	20	20	13-1/2	40		2.03	.84	2.87
1200		30	15-1/2	50		2.08	.85	2.93
1300		40	15-1/2	60		2.23	.92	3.15
1400	25	20	17-1/2	40		2.24	.93	3.17
1500		30	17-1/2	50		2.42	1	3.42
1600		40	19-1/2	60		2.46	1.01	3.47
1700	30	20	19-1/2	40		2.45	1.01	3.46
1800		30	21-1/2	50		2.51	1.03	3.54
1900		40	23-1/2	60		2.71	1.11	3.82
2000	35	20	23-1/2	40		2.69	.93	3.62
2100		30	25-1/2	50		2.77	.96	3.73
2200		40	25-1/2	60		2.95	1.01	3.96
2300	40	20	25-1/2	41		3	1.02	4.02
2400		30	25-1/2	51		3.19	1.08	4.27
2500		40	25-1/2	61		3.31	1.12	4.43
2600	45	20	27-1/2	41		3.35	1.54	4.89
2700		30	31-1/2	51		3.54	1.64	5.18
2800		40	31-1/2	61		3.73	1.73	5.46
2900	50	20	29-1/2	42		3.74	1.73	5.47
3000		30	31-1/2	52		4.09	1.90	5.99
3100		40	31-1/2	62		4.34	2.02	6.36
3200	60	20	37-1/2	42		4.79	1.75	6.54
3300		30	37-1/2	52		5.35	1.94	7.29
3400		40	37-1/2	62		5.35	1.94	7.29
3500	70	20	41-1/2	42		5.30	1.94	7.24
3600		30	41-1/2	52		5.70	2.06	7.76
3700		40	41-1/2	64		6.95	2.50	9.45

B1020 Roof Construction

B1020 116	Steel Joists & Deck on Bearing Walls

	BAY SIZE (FT.)	SUPERIMPOSED LOAD (P.S.F.)	DEPTH (IN.)	TOTAL LOAD (P.S.F.)		COST PER S.F.		
						MAT.	INST.	TOTAL
3800	80	20	45-1/2	44		7.75	2.78	10.53
3900		30	45-1/2	54		7.75	2.78	10.53
4000		40	45-1/2	64		8.65	3.08	11.73
4100	90	20	53-1/2	44		6.30	2.18	8.48
4200		30	53-1/2	54		6.70	2.31	9.01
4300		40	53-1/2	65		8	2.74	10.74
4400	100	20	57-1/2	44		6.70	2.31	9.01
4500		30	57-1/2	54		8	2.74	10.74
4600		40	57-1/2	65		8.85	3.02	11.87
4700	125	20	69-1/2	44		9.45	3.14	12.59
4800		30	69-1/2	56		11.05	3.65	14.70
4900		40	69-1/2	67		12.65	4.16	16.81

SHELL B

B1020 Roof Construction

Description: Table below lists costs for a roof system supported on exterior bearing walls and interior columns. Costs include bracing, joist girders, open web steel joists and 1-1/2" galvanized metal deck.

Design and Pricing Assumptions:
Columns are 18' high.
Joists are 5'-0" O.C.
Joist girders and joists have bottom chords connected to columns. Roof deck

is 1-1/2", 22 gauge galvanized steel. Costs include bridging and bracing. Deflection is limited to 1/240 of the span.

Fireproofing is not included.
Costs/S.F. are based on a building 4 bays long and 4 bays wide.
Costs for bearing walls are not included.

Column costs are not included but are listed separately per S.F. of floor.

System Components	QUANTITY	UNIT	COST PER S.F. MAT.	COST PER S.F. INST.	COST PER S.F. TOTAL
SYSTEM B1020 120 2050					
30' X 30' BAY SIZE, 20 PSF SUPERIMPOSED LOAD					
Open web joist girders, average	.475	Lb.	.31	.12	.43
Open web joists, horiz. bridging, T.L. lots, 30' to 50' span	.067	Lb.	.99	.33	1.32
Cross bracing, rods, shop fabricated, 1" diameter	.062	Lb.	.08	.12	.20
Metal decking, open, galv., 1-1/2" deep, 22 ga., over 50 sq.	1.050	S.F.	1.22	.50	1.72
TOTAL			2.60	1.07	3.67

B1020 120		Steel Joists, Joist Girders & Deck on Columns & Walls						
	BAY SIZE (FT.) GIRD X JOISTS	SUPERIMPOSED LOAD (P.S.F.)	DEPTH (IN.)	TOTAL LOAD (P.S.F.)	COLUMN ADD	COST PER S.F. MAT.	COST PER S.F. INST.	COST PER S.F. TOTAL
2050	30x30	20	17-1/2	40		2.60	1.06	3.66
2100					columns	.31	.08	.39
2150		30	17-1/2	50		2.81	1.14	3.95
2200					columns	.31	.08	.39
2250		40	21-1/2	60		2.96	1.20	4.16
2300					columns	.31	.08	.39
2350	30x35	20	32-1/2	40		2.96	1.20	4.16
2400					columns	.27	.07	.34
2450		30	36-1/2	50		3.10	1.24	4.34
2500					columns	.27	.07	.34
2550		40	36-1/2	60		3.33	1.32	4.65
2600					columns	.31	.08	.39
2650	35x30	20	36-1/2	40		2.74	1.12	3.86
2700					columns	.27	.07	.34
2750		30	36-1/2	50		3.03	1.22	4.25
2800					columns	.27	.07	.34
2850		40	36-1/2	60		4.08	1.57	5.65
2900					columns	.31	.08	.39
3000	35x35	20	36-1/2	40		3.22	1.47	4.69
3050					columns	.23	.06	.29
3100		30	36-1/2	50		3.33	1.51	4.84
3150					columns	.23	.06	.29
3200		40	36-1/2	60		3.64	1.61	5.25
3250					columns	.30	.08	.38

B1020 Roof Construction

B1020 120	Steel Joists, Joist Girders & Deck on Columns & Walls

SHELL B

	BAY SIZE (FT.) GIRD X JOISTS	SUPERIMPOSED LOAD (P.S.F.)	DEPTH (IN.)	TOTAL LOAD (P.S.F.)	COLUMN ADD	COST PER S.F.		
						MAT.	INST.	TOTAL
3300	35x40	20	36-1/2	40		3.29	1.35	4.64
3350					columns	.23	.06	.29
3400		30	36-1/2	50		3.41	1.38	4.79
3450					columns	.26	.07	.33
3500		40	36-1/2	60		3.72	1.50	5.22
3550					columns	.26	.07	.33
3600	40x35	20	40-1/2	40		3.17	1.31	4.48
3650					columns	.23	.06	.29
3700		30	40-1/2	50		3.31	1.36	4.67
3750					columns	.26	.07	.33
3800		40	40-1/2	60		3.90	1.82	5.72
3850					columns	.26	.07	.33
3900	40x40	20	40-1/2	41		3.67	1.72	5.39
3950					columns	.23	.06	.29
4000		30	40-1/2	51		3.80	1.78	5.58
4050					columns	.23	.06	.29
4100		40	40-1/2	61		4.11	1.89	6
4150					columns	.23	.06	.29
4200	40x45	20	40-1/2	41		3.74	1.81	5.55
4250					columns	.20	.05	.25
4300		30	40-1/2	51		4.21	2.02	6.23
4350					columns	.20	.05	.25
4400		40	40-1/2	61		4.67	2.24	6.91
4450					columns	.23	.06	.29
4500	45x40	20	52-1/2	41		3.70	1.79	5.49
4550					columns	.20	.05	.25
4600		30	52-1/2	51		3.66	1.56	5.22
4650					columns	.23	.06	.29
4700		40	52-1/2	61		4.24	2.02	6.26
4750					columns	.23	.06	.29
4800	45x45	20	52-1/2	41		3.52	1.69	5.21
4850					columns	.18	.05	.23
4900		30	52-1/2	51		4.40	2.10	6.50
4950					columns	.20	.05	.25
5000		40	52-1/2	61		4.52	2.15	6.67
5050					columns	.23	.06	.29
5100	45x50	20	52-1/2	41		4.12	1.97	6.09
5150					columns	.16	.04	.20
5200		30	52-1/2	51		4.71	2.24	6.95
5250					columns	.18	.05	.23
5300		40	52-1/2	61		5	2.38	7.38
5350					columns	.23	.06	.29
5400	50x45	20	56-1/2	41		3.99	1.90	5.89
5450					columns	.16	.04	.20
5500		30	56-1/2	51		4.50	2.13	6.63
5550					columns	.18	.05	.23
5600		40	56-1/2	61		5.15	2.43	7.58
5650					columns	.22	.05	.27

B1020 Roof Construction

B1020 120	Steel Joists, Joist Girders & Deck on Columns & Walls							

	BAY SIZE (FT.) GIRD X JOISTS	SUPERIMPOSED LOAD (P.S.F.)	DEPTH (IN.)	TOTAL LOAD (P.S.F.)	COLUMN ADD	COST PER S.F.		
						MAT.	INST.	TOTAL
5700	50x50	20	56-1/2	42		4.49	2.15	6.64
5750					columns	.16	.04	.20
5800		30	56-1/2	53		4.87	2.32	7.19
5850					columns	.21	.05	.26
5900		40	59	64		6.15	2.41	8.56
5950					columns	.23	.06	.29
6000	55x50	20	60-1/2	43		4.56	2.20	6.76
6050					columns	.28	.08	.36
6100		30	64-1/2	54		5.25	2.51	7.76
6150					columns	.35	.09	.44
6200		40	67	65		6.35	2.50	8.85
6250					columns	.35	.09	.44
6300	60x50	20	62-1/2	43		4.53	2.18	6.71
6350					columns	.26	.07	.33
6400		30	68-1/2	54		5.25	2.49	7.74
6450					columns	.32	.08	.40
6500		40	71	65		6.30	2.47	8.77
6550					columns	.34	.09	.43

B SHELL

Important: See the Reference Section for critical supporting data - Reference Numbers and City Cost Indexes

B1020 Roof Construction

Description: Table below lists the cost per S.F. for a roof system supported on columns. Costs include joist girders, open web steel joists and 1-1/2" galvanized metal deck.

Design and Pricing Assumptions:
  Columns are 18' high.
  Joists are 5'-0" O.C.
Joist girders and joists have bottom chords connected to columns. Roof deck is 1-1/2", 22 gauge galvanized steel. Costs include bridging and bracing. Deflection is limited to 1/240 of the span. Fireproofing is not included.

Costs/S.F. are based on a building 4 bays long and 4 bays wide.

Costs for columns are not included, but are listed separately per S.F. of area.

System Components	QUANTITY	UNIT	COST PER S.F. MAT.	INST.	TOTAL
SYSTEM B1020 124 2000 **30'X30' BAY SIZE, 20 PSF S. LOAD, 17-1/2"DEPTH, 40 PSF T. LOAD**					
Open web joist girders, average	.758	Lb.	.50	.18	.68
Open web joists, spans to 30' average	1.580	Lb.	1.15	.48	1.63
Cross bracing, rods, shop fabricated, 1" diameter	.120	Lb.	.15	.22	.37
Metal decking, open, galv., 1-1/2" deep, 22 ga., over 50 sq.	1.050	S.F.	1.22	.50	1.72
TOTAL			3.02	1.38	4.40

B1020 124 — Steel Joists & Joist Girders on Columns

	BAY SIZE (FT.) GIRD X JOISTS	SUPERIMPOSED LOAD (P.S.F.)	DEPTH (IN.)	TOTAL LOAD (P.S.F.)	COLUMN ADD	COST PER S.F. MAT.	INST.	TOTAL
2000	30x30	20	17-1/2	40		3.02	1.39	4.41
2050					columns	.87	.23	1.10
2100		30	17-1/2	50		3.28	1.50	4.78
2150					columns	.87	.23	1.10
2200		40	21-1/2	60		3.37	1.53	4.90
2250					columns	.87	.23	1.10
2300	30x35	20	32-1/2	40		3.19	1.33	4.52
2350					columns	.74	.20	.94
2400		30	36-1/2	50		3.37	1.40	4.77
2450					columns	.74	.20	.94
2500		40	36-1/2	60		3.61	1.48	5.09
2550					columns	.87	.23	1.10
2600	35x30	20	36-1/2	40		3.11	1.40	4.51
2650					columns	.74	.20	.94
2700		30	36-1/2	50		3.45	1.54	4.99
2750					columns	.74	.20	.94
2800		40	36-1/2	60		3.52	1.57	5.09
2850					columns	.87	.23	1.10
2900	35x35	20	36-1/2	40		3.47	1.60	5.07
2950					columns	.64	.17	.81
3000		30	36-1/2	50		3.60	1.65	5.25
3050					columns	.74	.20	.94
3100		40	36-1/2	60		3.93	1.75	5.68
3150					columns	.82	.22	1.04

B1020 Roof Construction

B1020 124	Steel Joists & Joist Girders on Columns

	BAY SIZE (FT.) GIRD X JOISTS	SUPERIMPOSED LOAD (P.S.F.)	DEPTH (IN.)	TOTAL LOAD (P.S.F.)	COLUMN ADD	COST PER S.F.		
						MAT.	INST.	TOTAL
3200	35x40	20	36-1/2	40		3.99	2.03	6.02
3250					columns	.65	.17	.82
3300		30	36-1/2	50		4.12	2.08	6.20
3350					columns	.72	.19	.91
3400		40	36-1/2	60		4.50	2.25	6.75
3450					columns	.72	.19	.91
3500	40x35	20	40-1/2	40		3.68	1.65	5.33
3550					columns	.65	.17	.82
3600		30	40-1/2	50		3.77	1.69	5.46
3650					columns	.72	.19	.91
3700		40	40-1/2	60		4.02	1.77	5.79
3750					columns	.72	.19	.91
3800	40x40	20	40-1/2	41		4.05	2.03	6.08
3850					columns	.63	.17	.80
3900		30	40-1/2	51		4.19	2.08	6.27
3950					columns	.63	.17	.80
4000		40	40-1/2	61		4.55	2.24	6.79
4050					columns	.63	.17	.80
4100	40x45	20	40-1/2	41		4.34	2.28	6.62
4150					columns	.56	.14	.70
4200		30	40-1/2	51		4.72	2.45	7.17
4250					columns	.56	.14	.70
4300		40	40-1/2	61		5.25	2.70	7.95
4350					columns	.63	.17	.80
4400	45x40	20	52-1/2	41		4.21	2.21	6.42
4450					columns	.56	.14	.70
4500		30	52-1/2	51		4.38	2.28	6.66
4550					columns	.56	.14	.70
4600		40	52-1/2	61		4.88	2.50	7.38
4650					columns	.63	.17	.80
4700	45x45	20	52-1/2	41		4.43	2.30	6.73
4750					columns	.50	.13	.63
4800		30	52-1/2	51		4.98	2.54	7.52
4850					columns	.56	.14	.70
4900		40	52-1/2	61		5.10	2.60	7.70
4950					columns	.64	.17	.81
5000	45x50	20	52-1/2	41		4.58	2.35	6.93
5050					columns	.45	.12	.57
5100		30	52-1/2	51		5.25	2.65	7.90
5150					columns	.51	.13	.64
5200		40	52-1/2	61		5.60	2.81	8.41
5250					columns	.58	.16	.74
5300	50x45	20	56-1/2	41		4.46	2.28	6.74
5350					columns	.45	.12	.57
5400		30	56-1/2	51		5.05	2.53	7.58
5450					columns	.51	.13	.64
5500		40	56-1/2	61		5.80	2.87	8.67
5550					columns	.58	.16	.74

Important: See the Reference Section for critical supporting data - Reference Numbers and City Cost Indexes

B1020 Roof Construction

B1020 124	Steel Joists & Joist Girders on Columns						

	BAY SIZE (FT.) GIRD X JOISTS	SUPERIMPOSED LOAD (P.S.F.)	DEPTH (IN.)	TOTAL LOAD (P.S.F.)	COLUMN ADD	COST PER S.F.		
						MAT.	INST.	TOTAL
5600	50x50	20	56-1/2	42		4.99	2.53	7.52
5650					columns	.45	.12	.57
5700		30	56-1/2	53		5.45	2.73	8.18
5750					columns	.58	.16	.74
5800		40	59	64		5.50	2.29	7.79
5850					columns	.64	.17	.81
5900	55x50	20	60-1/2	43		5.05	2.54	7.59
5950					columns	.47	.13	.60
6000		30	64-1/2	54		5.80	2.88	8.68
6050					columns	.58	.16	.74
6100		40	67	65		6.85	2.84	9.69
6150					columns	.58	.16	.74
6200	60x50	20	62-1/2	43		5.05	2.52	7.57
6250					columns	.43	.12	.55
6300		30	68-1/2	54		5.85	2.88	8.73
6350					columns	.53	.14	.67
6400		40	71	65		6.90	2.85	9.75
6450					columns	.57	.16	.73

SHELL B

B1020 Root Construction

NARROW RIB DECK

INTERMEDIATE RIB DECK

WIDE RIB DECK

General: Steel roof deck has the principal advantage of lightweight and speed of erection. It is covered by varying thicknesses of rigid insulation and roofing material.

Design Assumptions: Maximum spans shown in tables are based on construction and maintenance loads as specified by the Steel Deck Institute. Deflection is limited to 1/240 of the span.

Note: Costs shown in tables are based on projects over 5000 S.F. Under 5000 S.F. multiply costs by 1.16. For over four stories or congested site add 50% to the installation costs.

B1020 125 — Narrow Rib, Single Span Steel Deck

	TOTAL LOAD (P.S.F.)	MAX. SPAN (FT. or IN.)	DECK GAUGE	WEIGHT (P.S.F.)	DEPTH (IN.)	COST PER S.F.		
						MAT.	INST.	TOTAL
1250	85	3-10	22	1.80	1-1/2	1.16	.48	1.64
1300	66	4-10	20	2.15	1-1/2	1.37	.56	1.93
1350	60	5-11	18	2.85	1-1/2	1.77	.56	2.33
1360								

B1020 126 — Narrow Rib, Multi Span Steel Deck

	TOTAL LOAD (P.S.F.)	MAX. SPAN (FT.-IN.)	DECK GAUGE	WEIGHT (P.S.F.)	DEPTH (IN.)	COST PER S.F.		
						MAT.	INST.	TOTAL
1500	69	4-9	22	1.80	1-1//2	1.16	.48	1.64
1550	55	5-11	20	2.15	1-1/2	1.37	.56	1.93
1600	55	6-11	18	2.85	1-1/2	1.77	.56	2.33
1610								

B1020 127 — Intermediate Rib, Single Span Steel Deck

	TOTAL LOAD (P.S.F.)	MAX. SPAN (FT.-IN.)	DECK GAUGE	WEIGHT (P.S.F.)	DEPTH (IN.)	COST PER S.F.		
						MAT.	INST.	TOTAL
1750	72	4-6	22	1.75	1-1/2	1.16	.48	1.64
1800	65	5-3	20	2.10	1-1/2	1.37	.56	1.93
1850	64	6-2	18	2.80	1-1/2	1.77	.56	2.33
1860								

Important: See the Reference Section for critical supporting data - Reference Numbers and City Cost Indexes

B1020   Roof Construction

B1020 128 — Intermediate Rib, Multi Span Steel Deck

	TOTAL LOAD (P.S.F.)	MAX. SPAN (FT.-IN.)	DECK GAUGE	WEIGHT (P.S.F.)	DEPTH (IN.)	COST PER S.F.		
						MAT.	INST.	TOTAL
2000	60	5-6	22	1.75	1-1/2	1.16	.48	1.64
2050	57	6-3	20	2.10	1-1/2	1.37	.56	1.93
2100	57	7-4	18	2.80	1-1/2	1.77	.56	2.33
2110								

B1020 129 — Wide Rib, Single Span Steel Deck

	TOTAL LOAD (P.S.F.)	MAX. SPAN (FT.-IN.)	DECK GAUGE	WEIGHT (P.S.F.)	DEPTH (IN.)	COST PER S.F.		
						MAT.	INST.	TOTAL
2250	66	5-6	22	1.80	1-1/2	1.16	.48	1.64
2300	56	6-3	20	2.15	1-1/2	1.37	.56	1.93
2350	46	7-6	18	2.85	1-1/2	1.77	.56	2.33
2375								

B1020 130 — Wide Rib, Multi Span Steel Deck

	TOTAL LOAD (P.S.F.)	MAX. SPAN (FT.-IN.)	DECK GAUGE	WEIGHT (P.S.F.)	DEPTH (IN.)	COST PER S.F.		
						MAT.	INST.	TOTAL
2500	76	6-6	22	1.80	1-1/2	1.16	.48	1.64
2550	64	7-5	20	2.15	1-1/2	1.37	.56	1.93
2600	55	8-10	18	2.80	1-1/2	1.77	.56	2.33
2610								

SHELL B

B1020 Roof Construction

Design & Pricing Assumptions:
Costs in table below are based upon 5000 S.F. or more.
Deflection is limited to 1/240 of the span.

B1020 131	Single Span, Steel Deck						
	TOTAL LOAD (P.S.F.)	DECK SUPPORT SPACING (FT.)	DEPTH (IN.)	DECK GAUGE	WEIGHT (P.S.F.)	COST PER S.F.		
						MAT.	INST.	TOTAL
0850	30	10	3	20	2.5	2.59	.72	3.31
0900		12	3	18	3.0	3.36	.76	4.12
0950		14	3	16	4.0	4.42	.81	5.23
1000		16	4.5	20	3.0	4.16	.90	5.06
1050		18	4.5	18	4.0	5.35	.99	6.34
1100		20	4.5	16	5.0	5.55	1.43	6.98
1200		24	6.0	16	6.0	5.70	1.26	6.96
1300		28	7.5	16	7.0	6.30	1.53	7.83
1350	40	10	3.0	20	2.5	2.59	.72	3.31
1400		12	3.0	18	3.0	3.36	.76	4.12
1450		14	3.0	16	4.0	4.42	.81	5.23
1500		16	4.5	20	3.0	4.16	.90	5.06
1550		18	4.5	18	4.0	5.35	.99	6.34
1600		20	6.0	16	6.0	5.70	1.26	6.96
1700		24	7.5	16	7.0	6.30	1.53	7.83
1750	50	10	3.0	20	2.5	2.59	.72	3.31
1800		12	3.0	18	3.0	3.36	.76	4.12
1850		14	3.0	16	4.0	4.42	.81	5.23
1900		16	4.5	18	4.0	5.35	.99	6.34
1950		18	4.5	16	5.0	5.55	1.43	6.98
2000		20	6.0	16	6.0	5.75	.90	6.65
2100		24	7.5	16	7.0	6.30	1.53	7.83

B1020 132	Double Span, Steel Deck						
	TOTAL LOAD (P.S.F.)	DECK SUPPORT SPACING (FT.)	DEPTH (IN.)	DECK GAUGE	WEIGHT (P.S.F.)	COST PER S.F.		
						MAT.	INST.	TOTAL
2250	30	10	3	20	2.5	2.59	.72	3.31
2300		12	3	20	2.5	2.59	.72	3.31
2350		14	3	18	3.0	3.36	.76	4.12
2400		16	3	16	4.0	4.42	.81	5.23
2450	40	10	3	20	2.5	2.59	.72	3.31
2500		12	3	20	2.5	2.59	.72	3.31
2550		14	3	18	3.0	3.36	.76	4.12
2600		16	3	16	4.0	4.42	.81	5.23
2650	50	10	3	20	2.5	2.59	.72	3.31
2700		12	3	18	3.0	3.36	.76	4.12
2750		14	3	16	4.0	4.42	.81	5.23
2800		16	4.5	20	4.0	4.16	.90	5.06

Important: See the Reference Section for critical supporting data - Reference Numbers and City Cost Indexes

B1020   Roof Construction

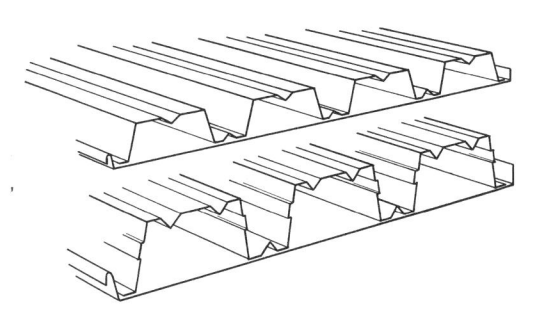

Design & Pricing Assumptions:
Costs in table below are based upon
15,000 S.F. or more.
No utility outlets or modifications included
in cost.
Deflection is limited to 1/240 of the span.

SHELL B

B1020 134	Cellular Steel Deck, Single Span							
	TOTAL LOAD (P.S.F.)	DECK SUPPORT SPACING (FT.)	DEPTH (IN.)	WEIGHT (P.S.F.)		COST PER S.F.		
						MAT.	INST.	TOTAL
1150	30	10	1-5/8	4		7.20	1.68	8.88
1200		12	1-5/8	5		7.50	1.75	9.25
1250		14	3	5		8.55	1.80	10.35
1300		16	3	5		8.55	1.89	10.44
1350		18	3	5		8.55	1.89	10.44
1400		20	3	6		9.60	1.97	11.57
1450	40	8	1-5/8	4		7.20	1.68	8.88
1500		10	1-5/8	5		7.50	1.75	9.25
1550		12	3	5		8.55	1.89	10.44
1600		14	3	5		8.55	1.89	10.44
1650		16	3	5		8.55	1.89	10.44
1700		18	3	6		9.60	1.97	11.57
1750	50	8	1-5/8	4		7.20	1.68	8.88
1800		10	1-5/8	5		7.50	1.75	9.25
1850		12	3	4		8.55	1.80	10.35
1900		14	3	5		8.55	1.89	10.44
1950		16	3	6		9.60	1.97	11.57
1960								

B1020 135	Cellular Steel Deck, Double Span							
	TOTAL LOAD (P.S.F.)	DECK SUPPORT SPACING (FT.)	DEPTH (IN.)	WEIGHT (P.S.F.)		COST PER S.F.		
						MAT.	INST.	TOTAL
2150	30	10	1-5/8	4		7.20	1.68	8.88
2200		12	1-5/8	5		7.50	1.75	9.25
2250		14	1-5/8	5		7.50	1.75	9.25
2300		16	3	5		8.55	1.89	10.44
2350		18	3	5		8.55	1.89	10.44
2400		20	3	6		9.60	1.97	11.57
2450	40	8	1-5/8	4		7.20	1.68	8.88
2500		10	1-5/8	5		7.50	1.75	9.25
2550		12	1-5/8	5		7.50	1.75	9.25
2600		14	3	5		8.55	1.89	10.44
2650		16	3	5		8.55	1.89	10.44
2700		18	3	6		9.60	1.97	11.57

B1020   Roof Construction

B1020 135 — Cellular Steel Deck, Double Span

	TOTAL LOAD (P.S.F.)	DECK SUPPORT SPACING (FT.)	DEPTH (IN.)	WEIGHT (P.S.F.)		COST PER S.F.		
						MAT.	INST.	TOTAL
2750	50	8	1-5/8	4		7.20	1.68	8.88
2800		10	1-5/8	5		7.50	1.75	9.25
2850		12	1-5/8	5		7.50	1.75	9.25
2900		14	3	5		8.55	1.89	10.44
2950		16	3	6		9.60	1.97	11.57
2960								

B1020 136 — Cellular Steel Deck, Triple Span

	TOTAL LOAD (P.S.F.)	DECK SUPPORT SPACING (FT.)	DEPTH (IN.)	WEIGHT (P.S.F.)		COST PER S.F.		
						MAT.	INST.	TOTAL
3200	30	10	1-5/8	4		7.20	1.68	8.88
3210								
3250	40	8	1-5/8	4		7.20	1.68	8.88
3300		10	1-5/8	4		7.20	1.68	8.88
3350		12	1-5/8	5		7.50	1.75	9.25
3400	50	8	1-5/8	4		7.20	1.68	8.88
3450		10	1-5/8	5		7.50	1.75	9.25
3500		12	1-5/8	5		7.50	1.75	9.25

B1020 310 — Canopies

		COST PER S.F.		
		MAT.	INST.	TOTAL
0100	Canopies, wall hung, prefinished aluminum, 8' x 10'	2,075	1,325	3,400
0110	12' x 40'	9,525	2,850	12,375
0120	Canopies, wall hung, canvas awnings, with frame and lettering	72.50	9.85	82.35
0130	Freestanding (walkways), vinyl coated steel, 12' wide	22	6.85	28.85

Important: See the Reference Section for critical supporting data - Reference Numbers and City Cost Indexes

B2010 Exterior Walls

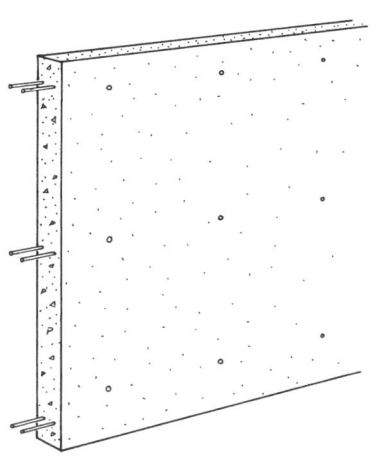

The table below describes a concrete wall system for exterior closure. There are several types of wall finishes priced from plain finish to a finish with 3/4″ rustication strip.

Design Assumptions:
Conc. f'c = 3000 to 5000 psi
Reinf. fy = 60,000 psi

System Components	QUANTITY	UNIT	COST PER S.F.		
			MAT.	INST.	TOTAL
SYSTEM B2010 101 2100					
CONC. WALL, REINFORCED, 8′ HIGH, 6″ THICK, PLAIN FINISH, 3,000 PSI					
Forms in place, wall, job built plyform to 8′ high, 4 uses	2.000	SFCA	2.04	11.90	13.94
Reinforcing in place, walls, #3 to #7	.752	Lb.	.37	.26	.63
Concrete ready mix, regular weight, 3000 psi	.018	C.Y.	1.72		1.72
Place and vibrate concrete, walls 6″ thick, pump	.018	C.Y.		.67	.67
Finish wall, break ties, patch voids	2.000	S.F.	.06	1.50	1.56
TOTAL			4.19	14.33	18.52

B2010 101	Cast In Place Concrete	COST PER S.F.		
		MAT.	INST.	TOTAL
2100	Conc wall reinforced, 8′ high, 6″ thick, plain finish, 3000 PSI	4.19	14.35	18.54
2200	4000 PSI	4.27	14.35	18.62
2300	5000 PSI	4.38	14.35	18.73
2400	Rub concrete 1 side, 3000 PSI	4.19	16.65	20.84
2500	4000 PSI	4.27	16.65	20.92
2600	5000 PSI	4.38	16.65	21.03
2700	Aged wood liner, 3000 PSI	6.50	14.95	21.45
2800	4000 PSI	6.55	14.95	21.50
2900	5000 PSI	6.70	14.95	21.65
3000	Sand blast light 1 side, 3000 PSI	4.46	15.40	19.86
3100	4000 PSI	4.54	15.40	19.94
3300	5000 PSI	4.65	15.40	20.05
3400	Sand blast heavy 1 side, 3000 PSI	4.72	17.40	22.12
3500	4000 PSI	4.80	17.40	22.20
3600	5000 PSI	4.91	17.40	22.31
3700	3/4″ bevel rustication strip, 3000 PSI	4.26	17.20	21.46
3800	4000 PSI	4.34	17.20	21.54
3900	5000 PSI	4.45	17.20	21.65
4000	8″ thick, plain finish, 3000 PSI	4.92	14.65	19.57
4100	4000 PSI	5.05	14.65	19.70
4200	5000 PSI	5.15	14.65	19.80
4300	Rub concrete 1 side, 3000 PSI	4.92	17	21.92
4400	4000 PSI	5.05	17	22.05
4500	5000 PSI	5.15	17	22.15
4550	8″ thick, aged wood liner, 3000 PSI	7.20	15.25	22.45
4600	4000 PSI	7.35	15.25	22.60

B2010 Exterior Walls

B2010 101	Cast In Place Concrete	COST PER S.F.		
		MAT.	INST.	TOTAL
4700	5000 PSI	7.45	15.25	22.70
4750	Sand blast light 1 side, 3000 PSI	5.20	15.70	20.90
4800	4000 PSI	5.30	15.70	21
4900	5000 PSI	5.45	15.70	21.15
5000	Sand blast heavy 1 side, 3000 PSI	5.45	17.75	23.20
5100	4000 PSI	5.55	17.75	23.30
5200	5000 PSI	5.70	17.75	23.45
5300	3/4" bevel rustication strip, 3000 PSI	4.99	17.50	22.49
5400	4000 PSI	5.10	17.50	22.60
5500	5000 PSI	5.25	17.50	22.75
5600	10" thick, plain finish, 3000 PSI	5.60	15	20.60
5700	4000 PSI	5.75	15	20.75
5800	5000 PSI	5.95	15	20.95
5900	Rub concrete 1 side, 3000 PSI	5.60	17.30	22.90
6000	4000 PSI	5.75	17.30	23.05
6100	5000 PSI	5.95	17.30	23.25
6200	Aged wood liner, 3000 PSI	7.90	15.60	23.50
6300	4000 PSI	8.05	15.60	23.65
6400	5000 PSI	8.25	15.60	23.85
6500	Sand blast light 1 side, 3000 PSI	5.90	16.05	21.95
6600	4000 PSI	6	16.05	22.05
6700	5000 PSI	6.20	16.05	22.25
6800	Sand blast heavy 1 side, 3000 PSI	6.15	18.05	24.20
6900	4000 PSI	6.30	18.05	24.35
7000	5000 PSI	6.45	18.05	24.50
7100	3/4" bevel rustication strip, 3000 PSI	5.70	17.85	23.55
7200	4000 PSI	5.85	17.85	23.70
7300	5000 PSI	6	17.85	23.85
7400	12" thick, plain finish, 3000 PSI	6.45	15.35	21.80
7500	4000 PSI	6.60	15.35	21.95
7600	5000 PSI	6.85	15.35	22.20
7700	Rub concrete 1 side, 3000 PSI	6.45	17.65	24.10
7800	4000 PSI	6.60	17.65	24.25
7900	5000 PSI	6.85	17.65	24.50
8000	Aged wood liner, 3000 PSI	8.75	15.95	24.70
8100	4000 PSI	8.90	15.95	24.85
8200	5000 PSI	9.15	15.95	25.10
8300	Sand blast light 1 side, 3000 PSI	6.70	16.40	23.10
8400	4000 PSI	6.90	16.40	23.30
8500	5000 PSI	7.10	16.40	23.50
8600	Sand blast heavy 1 side, 3000 PSI	7	18.45	25.45
8700	4000 PSI	7.15	18.45	25.60
8800	5000 PSI	7.35	18.45	25.80
8900	3/4" bevel rustication strip, 3000 PSI	6.50	18.20	24.70
9000	4000 PSI	6.70	18.20	24.90
9500	5000 PSI	6.90	18.20	25.10

B

SHELL

Important: See the Reference Section for critical supporting data - Reference Numbers and City Cost Indexes

B2010   Exterior Walls

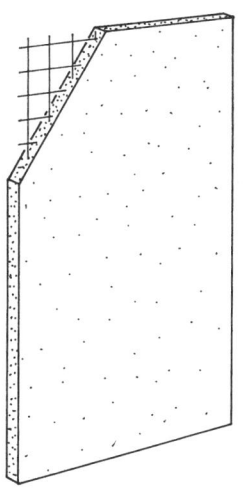

Precast concrete wall panels are either solid or insulated with plain, colored or textured finishes. Transportation is an important cost factor. Prices below are based on delivery within fifty miles of a plant. Engineering data is available from fabricators to assist with construction details. Usual minimum job size for economical use of panels is about 5000 S.F. Small jobs can double the prices below. For large, highly repetitive jobs, deduct up to 15% from the prices below.

B2010 102			**Flat Precast Concrete**					
	THICKNESS (IN.)	PANEL SIZE (FT.)	FINISHES	RIGID INSULATION (IN)	TYPE	COST PER S.F.		
						MAT.	INST.	TOTAL
3000	4	5x18	smooth gray	none	low rise	7.50	6.30	13.80
3050		6x18				6.30	5.25	11.55
3100		8x20				12.70	2.55	15.25
3150		12x20				12	2.41	14.41
3200	6	5x18	smooth gray	2	low rise	8.55	6.75	15.30
3250		6x18				7.30	5.70	13
3300		8x20				13.90	3.07	16.97
3350		12x20				12.60	2.81	15.41
3400	8	5x18	smooth gray	2	low rise	17.90	3.88	21.78
3450		6x18				17	3.69	20.69
3500		8x20				15.50	3.40	18.90
3550		12x20				14.15	3.12	17.27
3600	4	4x8	white face	none	low rise	30	3.65	33.65
3650		8x8				22.50	2.74	25.24
3700		10x10				19.85	2.41	22.26
3750		20x10				17.95	2.18	20.13
3800	5	4x8	white face	none	low rise	31	3.73	34.73
3850		8x8				23	2.81	25.81
3900		10x10				20.50	2.49	22.99
3950		20x20				18.85	2.29	21.14
4000	6	4x8	white face	none	low rise	32	3.86	35.86
4050		8x8				24	2.93	26.93
4100		10x10				21.50	2.59	24.09
4150		20x10				19.60	2.37	21.97
4200	6	4x8	white face	2	low rise	33	4.36	37.36
4250		8x8				25.50	3.43	28.93
4300		10x10				22.50	3.09	25.59
4350		20x10				19.60	2.37	21.97
4400	7	4x8	white face	none	low rise	32.50	3.96	36.46
4450		8x8				25	3.04	28.04
4500		10x10				22.50	2.72	25.22
4550		20x10				20.50	2.49	22.99
4600	7	4x8	white face	2	low rise	34	4.46	38.46
4650		8x8				26	3.54	29.54
4700		10x10				23.50	3.22	26.72

B2010 Exterior Walls

B2010 102 — Flat Precast Concrete

	THICKNESS (IN.)	PANEL SIZE (FT.)	FINISHES	RIGID INSULATION (IN)	TYPE	COST PER S.F.		
						MAT.	INST.	TOTAL
4750	7	20x10				21.50	2.99	24.49
4800	8	4x8	white face	none	low rise	33.50	4.05	37.55
4850		8x8				25.50	3.12	28.62
4900		10x10				23	2.80	25.80
4950		20x10				21.50	2.58	24.08
5000	8	4x8	white face	2	low rise	34.50	4.55	39.05
5050		8x8				27	3.62	30.62
5100		10x10				24	3.30	27.30
5150		20x10				22.50	3.08	25.58

B2010 103 — Fluted Window or Mullion Precast Concrete

	THICKNESS (IN.)	PANEL SIZE (FT.)	FINISHES	RIGID INSULATION (IN)	TYPE	COST PER S.F.		
						MAT.	INST.	TOTAL
5200	4	4x8	smooth gray	none	high rise	17.20	14.40	31.60
5250		8x8				12.30	10.30	22.60
5300		10x10				24	4.87	28.87
5350		20x10				21	4.26	25.26
5400	5	4x8	smooth gray	none	high rise	17.45	14.65	32.10
5450		8x8				12.70	10.65	23.35
5500		10x10				25.50	5.05	30.55
5550		20x10				22.50	4.49	26.99
5600	6	4x8	smooth gray	none	high rise	17.95	15.05	33
5650		8x8				13.15	11.05	24.20
5700		10x10				26	5.20	31.20
5750		20x10				23	4.64	27.64
5800	6	4x8	smooth gray	2	high rise	19.05	15.55	34.60
5850		8x8				14.25	11.55	25.80
5900		10x10				27	5.70	32.70
5950		20x10				24	5.15	29.15
6000	7	4x8	smooth gray	none	high rise	18.35	15.35	33.70
6050		8x8				13.50	11.30	24.80
6100		10x10				27	5.45	32.45
6150		20x10				24	4.79	28.79
6200	7	4x8	smooth gray	2	high rise	19.40	15.85	35.25
6250		8x8				14.60	11.80	26.40
6300		10x10				28	5.95	33.95
6350		20x10				25	5.30	30.30
6400	8	4x8	smooth gray	none	high rise	18.60	15.60	34.20
6450		8x8				13.85	11.65	25.50
6500		10x10				28	5.60	33.60
6550		20x10				25	5	30
6600	8	4x8	smooth gray	2	high rise	19.70	16.10	35.80
6650		8x8				14.95	12.15	27.10
6700		10x10				29	6.10	35.10
6750		20x10				26	5.50	31.50

B2010 104 — Ribbed Precast Concrete

	THICKNESS (IN.)	PANEL SIZE (FT.)	FINISHES	RIGID INSULATION (IN.)	TYPE	COST PER S.F.		
						MAT.	INST.	TOTAL
6800	4	4x8	aggregate	none	high rise	22	14.40	36.40
6850		8x8				16	10.50	26.50
6900		10x10				27	3.82	30.82

Important: See the Reference Section for critical supporting data - Reference Numbers and City Cost Indexes

B2010   Exterior Walls

B2010 104			Ribbed Precast Concrete					

	THICKNESS (IN.)	PANEL SIZE (FT.)	FINISHES	RIGID INSULATION(IN.)	TYPE	COST PER S.F.		
						MAT.	INST.	TOTAL
6950	4	20x10				24	3.37	27.37
7000	5	4x8	aggregate	none	high rise	22.50	14.65	37.15
7050		8x8				16.45	10.85	27.30
7100		10x10				28	3.97	31.97
7150		20x10				25	3.54	28.54
7200	6	4x8	aggregate	none	high rise	22.50	14.95	37.45
7250		8x8				16.90	11.10	28
7300		10x10				29	4.09	33.09
7350		20x10				26	3.67	29.67
7400	6	4x8	aggregate	2	high rise	24	15.45	39.45
7450		8x8				18	11.60	29.60
7500		10x10				30	4.59	34.59
7550		20x10				27	4.17	31.17
7600	7	4x8	aggregate	none	high rise	23.50	15.30	38.80
7650		8x8				17.30	11.40	28.70
7700		10x10				30	4.22	34.22
7750		20x10				26.50	3.78	30.28
7800	7	4x8	aggregate	2	high rise	24.50	15.80	40.30
7850		8x8				18.40	11.90	30.30
7900		10x10				31	4.72	35.72
7950		20x10				28	4.28	32.28
8000	8	4x8	aggregate	none	high rise	23.50	15.55	39.05
8050		8x8				17.85	11.70	29.55
8100		10x10				31	4.37	35.37
8150		20x10				28	3.94	31.94
8200	8	4x8	aggregate	2	high rise	24.50	16.05	40.55
8250		8x8				18.95	12.20	31.15
8300		10x10				32	4.87	36.87
8350		20x10				29	4.44	33.44

B2010 105			Precast Concrete Specialties					

	TYPE	SIZE				COST PER L.F.		
						MAT.	INST.	TOTAL
8400	Coping, precast	6"wide				16.50	10.45	26.95
8450	Stock units	10" wide				11.55	11.20	22.75
8460		12" wide				22	12.05	34.05
8480		14" wide				15.95	13.05	29
8500	Window sills	6" wide				10.25	11.20	21.45
8550	Precast	10" wide				13.75	13.05	26.80
8600		14" wide				14.85	15.65	30.50
8610								

The advantage of tilt up construction is in the low cost of forms and placing of concrete and reinforcing. Tilt up has been used for several types of buildings, including warehouses, stores, offices, and schools. The panels are cast in forms on the ground, or floor slab. Most jobs use 5-1/2" thick solid reinforced concrete panels.

Design Assumptions:
Conc. f'c = 3000 psi
Reinf. fy = 60,000

System Components	QUANTITY	UNIT	COST PER S.F. MAT.	COST PER S.F. INST.	COST PER S.F. TOTAL
SYSTEM B2010 106 3200					
TILT-UP PANELS, 20'X25', BROOM FINISH, 5-1/2" THICK, 3000 PSI					
Apply liquid bond release agent	500.000	S.F.	5	55	60
Edge forms in place for slab on grade	120.000	L.F.	38.40	333.60	372
Reinforcing in place	.350	Ton	313.25	316.75	630
Footings, form braces, steel	1.000	Set	480.20		480.20
Slab lifting inserts	1.000	Set	34.80		34.80
Framing, less than 4" angles	1.000	Set	59.85	662.15	722
Concrete ready mix, regular weight, 3000 psi	8.550	C.Y.	816.53		816.53
Place and vibrate concrete for slab on grade, 4" thick, direct chute	8.550	C.Y.		169.98	169.98
Finish floor, monolithic broom finish	500.000	S.F.		320	320
Cure with curing compound, sprayed	500.000	S.F.	30.50	36	66.50
Erection crew	.058	Day		520.55	520.55
TOTAL	500.000		1,778.53	2,414.03	4,192.56
Cost per S.F.			3.56	4.83	8.39

B2010 106	Tilt-Up Concrete Panel	COST PER S.F. MAT.	COST PER S.F. INST.	COST PER S.F. TOTAL
3200	Tilt up conc panels, broom finish, 5-1/2" thick, 3000 PSI	3.56	4.83	8.39
3250	5000 PSI	3.65	4.73	8.38
3300	6" thick, 3000 PSI	3.93	4.94	8.88
3350	5000 PSI	4.03	4.85	8.89
3400	7-1/2" thick, 3000 PSI	5.01	5.13	10.14
3450	5000 PSI	5.16	5.04	10.20
3500	8" thick, 3000 PSI	5.39	5.26	10.64
3550	5000 PSI	5.56	5.16	10.72
3700	Steel trowel finish, 5-1/2" thick, 3000 PSI	3.56	4.92	8.49
3750	5000 PSI	3.65	4.83	8.48
3800	6" thick, 3000 PSI	3.93	5.04	8.98
3850	5000 PSI	4.03	4.95	8.99
3900	7-1/2" thick, 3000 PSI	5.01	5.23	10.24
3950	5000 PSI	5.16	5.14	10.30
4000	8" thick, 3000 PSI	5.39	5.36	10.74
4050	5000 PSI	5.56	5.26	10.82
4200	Exp. aggregate finish, 5-1/2" thick, 3000 PSI	3.83	4.92	8.76
4250	5000 PSI	3.92	4.83	8.75
4300	6" thick, 3000 PSI	4.20	5.04	9.25
4350	5000 PSI	4.30	4.95	9.26
4400	7-1/2" thick, 3000 PSI	5.28	5.23	10.51
4450	5000 PSI	5.43	5.14	10.57

Important: See the Reference Section for critical supporting data - Reference Numbers and City Cost Indexes

B2010 Exterior Walls

B2010 106	Tilt-Up Concrete Panel	COST PER S.F.		
		MAT.	INST.	TOTAL
4500	8" thick, 3000 PSI	5.65	5.36	11.01
4550	5000 PSI	5.83	5.27	11.09
4600	Exposed aggregate & vert. rustication 5-1/2" thick, 3000 PSI	5.35	8.15	13.51
4650	5000 PSI	5.44	8.06	13.50
4700	6" thick, 3000 PSI	5.72	8.27	14
4750	5000 PSI	5.82	8.18	14.01
4800	7-1/2" thick, 3000 PSI	6.80	8.46	15.26
4850	5000 PSI	6.95	8.37	15.32
4900	8" thick, 3000 PSI	7.17	8.59	15.76
4950	5000 PSI	7.35	8.50	15.84
5000	Vertical rib & light sandblast, 5-1/2" thick, 3000 PSI	5.87	5.87	11.75
5050	5000 PSI	5.96	5.78	11.74
5100	6" thick, 3000 PSI	6.24	5.99	12.24
5150	5000 PSI	6.34	5.90	12.25
5200	7-1/2" thick, 3000 PSI	7.32	6.18	13.50
5250	5000 PSI	7.47	6.09	13.56
5300	8" thick, 3000 PSI	7.69	6.31	14
5350	5000 PSI	7.87	6.22	14.08
6000	Broom finish w/2" polystyrene insulation, 6" thick, 3000 PSI	3.39	5.94	9.33
6050	5000 PSI	3.52	5.94	9.46
6100	Broom finish 2" fiberplank insulation, 6" thick, 3000 PSI	3.96	5.89	9.85
6150	5000 PSI	4.09	5.89	9.98
6200	Exposed aggregate w/2"polystyrene insulation, 6" thick, 3000 PSI	3.57	6.08	9.65
6250	5000 PSI	3.70	6.08	9.78
6300	Exposed aggregate 2" fiberplank insulation, 6" thick, 3000 PSI	4.14	6.03	10.17
6350	5000 PSI	4.27	6.03	10.30

SHELL B

Exterior concrete block walls are defined in the following terms; structural reinforcement, weight, percent solid, size, strength and insulation. Within each of these categories, two to four variations are shown. No costs are included for brick shelf or relieving angles.

System Components		QUANTITY	UNIT	COST PER S.F.		
				MAT.	INST.	TOTAL
SYSTEM B2010 109 1400						
UNREINFORCED CONCRETE BLOCK WALL, 8″ X 8″ X 16″, PERLITE CORE FILL						
Concrete block wall, 8″ thick		1.000	S.F.	2.04	5.35	7.39
Perlite insulation		1.000	S.F.	.48	.33	.81
Horizontal joint reinforcing, alternate courses		.800	S.F.	.10	.14	.24
Control joint		.050	L.F.	.06	.06	.12
	TOTAL			2.68	5.88	8.56

B2010 109 — Concrete Block Wall - Regular Weight

	TYPE	SIZE (IN.)	STRENGTH (P.S.I.)	CORE FILL		COST PER S.F.		
						MAT.	INST.	TOTAL
1200	Hollow	4x8x16	2,000	none		1.39	4.85	6.24
1250			4,500	none		1.16	4.85	6.01
1300	RB2010 -200	6x8x16	2,000	perlite		2.34	5.50	7.84
1310				styrofoam		3.13	5.20	8.33
1340				none		2.01	5.20	7.21
1350			4,500	perlite		1.82	5.50	7.32
1360				styrofoam		2.61	5.20	7.81
1390				none		1.49	5.20	6.69
1400		8x8x16	2,000	perlite		2.68	5.90	8.58
1410				styrofoam		3.32	5.55	8.87
1440				none		2.20	5.55	7.75
1450			4,500	perlite		2.24	5.90	8.14
1460				styrofoam		2.88	5.55	8.43
1490				none		1.76	5.55	7.31
1500		12x8x16	2,000	perlite		3.93	7.80	11.73
1510				styrofoam		4.53	7.15	11.68
1540				none		3.14	7.15	10.29
1550			4,500	perlite		3.18	7.80	10.98
1560				styrofoam		3.78	7.15	10.93
1590				none		2.39	7.15	9.54
2000	75% solid	4x8x16	2,000	none		1.41	4.90	6.31
2050			4,500	none		1.30	4.90	6.20

B2010 Exterior Walls

B2010 109	Concrete Block Wall - Regular Weight							

	TYPE	SIZE (IN.)	STRENGTH (P.S.I.)	CORE FILL		COST PER S.F.		
						MAT.	INST.	TOTAL
2100	75% solid	6x8x16	2,000	perlite		1.79	5.40	7.19
2140				none		1.62	5.25	6.87
2150			4,500	perlite		2.19	5.40	7.59
2190				none		2.02	5.25	7.27
2200		8x8x16	2,000	perlite		2.67	5.80	8.47
2240				none		2.43	5.65	8.08
2250			4,500	perlite		2.97	5.80	8.77
2290				none		2.73	5.65	8.38
2300		12x8x16	2,000	perlite		3.29	7.55	10.84
2340				none		2.90	7.25	10.15
2350			4,500	perlite		4.43	7.55	11.98
2390				none		4.04	7.25	11.29
2500	Solid	4x8x16	2,000	none		1.70	5	6.70
2550			4,500	none		1.30	4.96	6.26
2600		6x8x16	2,000	none		2.08	5.40	7.48
2650			4,500	none		2.02	5.30	7.32
2700		8x8x16	2,000	none		2.88	5.80	8.68
2750			4,500	none		2.73	5.70	8.43
2800		12x8x16	2,000	none		4.34	7.50	11.84
2850			4,500	none		4.04	7.35	11.39

B2010 110	Concrete Block Wall - Lightweight							

	TYPE	SIZE (IN.)	WEIGHT (P.C.F.)	CORE FILL		COST PER S.F.		
						MAT.	INST.	TOTAL
3100	Hollow	8x4x16	105	perlite		1.67	5.60	7.27
3110				styrofoam		2.31	5.25	7.56
3140				none		1.19	5.25	6.44
3150			85	perlite		3.87	5.45	9.32
3160				styrofoam		4.51	5.10	9.61
3190				none		3.39	5.10	8.49
3200		4x8x16	105	none		1.61	4.74	6.35
3250			85	none		1.48	4.64	6.12
3300		6x8x16	105	perlite		2.50	5.35	7.85
3310				styrofoam		3.29	5.10	8.39
3340				none		2.17	5.10	7.27
3350			85	perlite		2.24	5.20	7.44
3360				styrofoam		3.03	4.96	7.99
3390				none		1.91	4.96	6.87
3400		8x8x16	105	perlite		3.13	5.80	8.93
3410				styrofoam		3.77	5.45	9.22
3440				none		2.65	5.45	8.10
3450			85	perlite		2.57	5.65	8.22
3460				styrofoam		3.21	5.30	8.51
3490				none		2.09	5.30	7.39
3500		12x8x16	105	perlite		4.28	7.60	11.88
3510				styrofoam		4.88	6.95	11.83
3540				none		3.49	6.95	10.44
3550			85	perlite		3.63	7.45	11.08
3560				styrofoam		4.23	6.80	11.03
3590				none		2.84	6.80	9.64
3600		4x8x24	105	none		1.22	5.25	6.47
3650			85	none		2.32	4.45	6.77
3690	For stacked bond add						.40	.40

B2010 Exterior Walls

B2010 110				Concrete Block Wall - Lightweight				

B
SHELL

	TYPE	SIZE (IN.)	WEIGHT (P.C.F.)	CORE FILL		COST PER S.F.		
						MAT.	INST.	TOTAL
3700	Hollow	6x8x24	105	perlite		1.94	5.80	7.74
3710				styrofoam		2.73	5.50	8.23
3740				none		1.61	5.50	7.11
3750			85	perlite		3.49	4.90	8.39
3760				styrofoam		4.28	4.64	8.92
3790				none		3.16	4.64	7.80
3800		8x8x24	105	perlite		2.47	6.20	8.67
3810				styrofoam		3.11	5.85	8.96
3840				none		1.99	5.85	7.84
3850			85	perlite		4.22	5.25	9.47
3860				styrofoam		4.86	4.93	9.79
3890				none		3.74	4.93	8.67
3900		12x8x24	105	perlite		3.42	7.35	10.77
3910				styrofoam		4.02	6.70	10.72
3940				none		2.63	6.70	9.33
3950			85	perlite		5.70	7.15	12.85
3960				styrofoam		6.30	6.50	12.80
3990				none		4.92	6.50	11.42
4000	75% solid	4x8x16	105	none		1.71	4.79	6.50
4050			85	none		2.54	4.69	7.23
4100		6x8x16	105	perlite		2.08	5.25	7.33
4140				none		1.91	5.15	7.06
4150			85	perlite		3.10	5.15	8.25
4190				none		2.93	5	7.93
4200		8x8x16	105	perlite		2.75	5.65	8.40
4240				none		2.51	5.50	8.01
4250			85	perlite		4.25	5.50	9.75
4290				none		4.01	5.35	9.36
4300		12x8x16	105	perlite		4.67	7.35	12.02
4340				none		4.28	7.05	11.33
4350			85	perlite		5.75	7.15	12.90
4390				none		5.35	6.85	12.20
4500	Solid	4x8x16	105	none		2.22	4.96	7.18
4550			85	none		2.03	4.74	6.77
4600		6x8x16	105	none		2.56	5.30	7.86
4650			85	none		3.18	5.10	8.28
4700		8x8x16	105	none		3.36	5.70	9.06
4750			85	none		4.26	5.45	9.71
4800		12x8x16	105	none		5.55	7.35	12.90
4850			85	none		6.45	6.95	13.40
4900	For stacked bond, add						.40	.40

B2010 111				Reinforced Concrete Block Wall - Regular Weight				

	TYPE	SIZE (IN.)	STRENGTH (P.S.I.)	VERT. REINF & GROUT SPACING		COST PER S.F.		
						MAT.	INST.	TOTAL
5200	Hollow	4x8x16	2,000	#4 @ 48"		1.52	5.40	6.92
5250			4,500	#4 @ 48"		1.29	5.40	6.69
5300		6x8x16	2,000	#4 @ 48"		2.23	5.80	8.03
5330				#5 @ 32"		2.41	6	8.41
5340				#5 @ 16"		2.80	6.75	9.55
5350			4,500	#4 @ 28"		1.71	5.80	7.51
5380				#5 @ 32"		1.89	6	7.89
5390				#5 @ 16"		2.28	6.75	9.03

Important: See the Reference Section for critical supporting data - Reference Numbers and City Cost Indexes

B2010 Exterior Walls

B2010 111	Reinforced Concrete Block Wall - Regular Weight							
	TYPE	SIZE (IN.)	STRENGTH (P.S.I.)	VERT. REINF & GROUT SPACING		COST PER S.F.		
						MAT.	INST.	TOTAL
5400	Hollow	8x8x16	2,000	#4 @ 48"		2.47	6.10	8.57
5430				#5 @ 32"		2.66	6.45	9.11
5440				#5 @ 16"		3.11	7.35	10.46
5450		8x8x16	4,500	#4 @ 48"		2.01	6.20	8.21
5480				#5 @ 32"		2.22	6.45	8.67
5490				#5 @ 16"		2.67	7.35	10.02
5500		12x8x16	2,000	#4 @ 48"		3.51	7.80	11.31
5530				#5 @ 32"		3.76	8.10	11.86
5540				#5 @ 16"		4.38	9.10	13.48
5550			4,500	#4 @ 48"		2.76	7.80	10.56
5580				#5 @ 32"		3.01	8.10	11.11
5590				#5 @ 16"		3.63	9.10	12.73
6100	75% solid	6x8x16	2,000	#4 @ 48"		1.74	5.75	7.49
6130				#5 @ 32"		1.88	5.90	7.78
6140				#5 @ 16"		2.13	6.55	8.68
6150			4,500	#4 @ 48"		2.14	5.75	7.89
6180				#5 @ 32"		2.28	5.90	8.18
6190				#5 @ 16"		2.53	6.55	9.08
6200		8x8x16	2,000	#4 @ 48"		2.56	6.20	8.76
6230				#5 @ 32"		2.70	6.40	9.10
6240				#5 @ 16"		2.96	7.15	10.11
6250			4,500	#4 @ 48"		2.86	6.20	9.06
6280				#5 @ 32"		3	6.40	9.40
6290				#5 @ 16"		3.26	7.15	10.41
6300		12x8x16	2,000	#4 @ 48"		3.11	7.80	10.91
6330				#5 @ 32"		3.29	8	11.29
6340				#5 @ 16"		3.68	8.75	12.43
6350			4,500	#4 @ 48"		4.25	7.80	12.05
6380				#4 @ 32"		4.43	8	12.43
6390				#5 @ 16"		4.82	8.75	13.57
6500	Solid-double	2-4x8x16	2,000	#4 @ 48" E.W.		4.13	11.35	15.48
6530	Wythe			#5 @ 16" E.W.		4.76	12.10	16.86
6550			4,500	#4 @ 48" E.W.		3.33	11.25	14.58
6580				#5 @ 16" E.W.		3.96	11.95	15.91
6600		2-6x8x16	2,000	#4 @ 48" E.W.		4.88	12.20	17.08
6630				#5 @ 16" E.W.		5.50	12.90	18.40
6650			4,000	#4 @ 48" E.W.		4.76	12	16.76
6680				#5 @ 16" E.W.		5.40	12.70	18.10

B2010 112	Reinforced Concrete Block Wall - Lightweight							
	TYPE	SIZE (IN.)	WEIGHT (P.C.F.)	VERT REINF. & GROUT SPACING		COST PER S.F.		
						MAT.	INST.	TOTAL
7100	Hollow	8x4x16	105	#4 @ 48"		1.44	5.90	7.34
7130				#5 @ 32"		1.65	6.15	7.80
7140				#5 @ 16"		2.10	7.05	9.15
7150			85	#4 @ 48"		3.64	5.75	9.39
7180				#5 @ 32"		3.85	6	9.85
7190				#5 @ 16"		4.30	6.95	11.25
7200		4x8x16	105	#4 @ 48"		1.74	5.30	7.04
7250			85	#4 @ 48"		1.61	5.20	6.81

| B2010 112 | | | Reinforced Concrete Block Wall - Lightweight | | | | |

	TYPE	SIZE (IN.)	WEIGHT (P.C.F.)	VERT REINF. & GROUT SPACING		COST PER S.F.		
						MAT.	INST.	TOTAL
7300	Hollow	6x8x16	105	#4 @ 48"		2.39	5.65	8.04
7330				#5 @ 32"		2.57	5.85	8.42
7340				#5 @ 16"		2.96	6.60	9.56
7350			85	#4 @ 48"		2.13	5.55	7.68
7380				#5 @ 32"		2.31	5.75	8.06
7390				#5 @ 16"		2.70	6.50	9.20
7400		8x8x16	105	#4 @ 48"		2.90	6.10	9
7430				#5 @ 32"		3.11	6.35	9.46
7440				#5 @ 16"		3.56	7.25	10.81
7450		8x8x16	85	#4 @ 48"		2.34	5.95	8.29
7480				#5 @ 32"		2.55	6.20	8.75
7490				#5 @ 16"		3	7.10	10.10
7510		12x8x16	105	#4 @ 48"		3.86	7.60	11.46
7530				#5 @ 32"		4.11	7.90	12.01
7540				#5 @ 16"		4.73	8.90	13.63
7550			85	#4 @ 48"		3.21	7.45	10.66
7580				#5 @ 32"		3.46	7.75	11.21
7590				#5 @ 16"		4.08	8.75	12.83
7600		4x8x24	105	#4 @ 48"		1.35	5.80	7.15
7650			85	#4 @ 48"		2.45	4.99	7.44
7700		6x8x24	105	#4 @ 48"		1.83	6.10	7.93
7730				#5 @ 32"		2.01	6.30	8.31
7740				#5 @ 16"		2.40	7.05	9.45
7750			85	#4 @ 48"		3.38	5.20	8.58
7780				#5 @ 32"		3.56	5.40	8.96
7790				#5 @ 16"		3.95	6.20	10.15
7800		8x8x24	105	#4 @ 48"		2.24	6.50	8.74
7840				#5 @ 16"		2.90	7.65	10.55
7850			85	#4 @ 48"		3.99	5.60	9.59
7880				#5 @ 32"		4.20	5.85	10.05
7890				#5 @ 16"		4.65	6.75	11.40
7900		12x8x24	105	#4 @ 48"		3	7.35	10.35
7930				#5 @ 32"		3.25	7.65	10.90
7940				#5 @ 16"		3.87	8.65	12.52
7950			85	#4 @ 48"		5.30	7.15	12.45
7980				#5 @ 32"		5.55	7.45	13
7990				#5 @ 16"		6.15	8.45	14.60
8100	75% solid	6x8x16	105	#4 @ 48"		2.03	5.60	7.63
8130				#5 @ 32"		2.17	5.80	7.97
8140				#5 @ 16"		2.42	6.40	8.82
8150			85	#4 @ 48"		3.05	5.50	8.55
8180				#5 @ 32"		3.19	5.65	8.84
8190				#5 @ 16"		3.44	6.30	9.74
8200		8x8x16	105	#4 @ 48"		2.64	6.05	8.69
8230				#5 @ 32"		2.78	6.25	9.03
8240				#5 @ 16"		3.04	7	10.04
8250			85	#4 @ 48"		4.14	5.90	10.04
8280				#5 @ 32"		4.28	6.10	10.38
8290				#5 @ 16"		4.54	6.85	11.39

Important: See the Reference Section for critical supporting data - Reference Numbers and City Cost Indexes

B2010 Exterior Walls

B2010 112	Reinforced Concrete Block Wall - Lightweight

	TYPE	SIZE (IN.)	WEIGHT (P.C.F.)	VERT REINF. & GROUT SPACING		COST PER S.F.		
						MAT.	INST.	TOTAL
8300	75% solid	12x8x16	105	#4 @ 48"		4.49	7.60	12.09
8330				#5 @ 32"		4.67	7.80	12.47
8340				#5 @ 16"		5.05	8.55	13.60
8350			85	#4 @ 48"		5.60	7.40	13
8380				#5 @ 32"		5.75	7.60	13.35
8390				#5 @ 16"		6.15	8.35	14.50
8500	Solid-double	2-4x8x16	105	#4 @ 48"		5.15	11.25	16.40
8530	Wythe		105	#5 @ 16"		5.80	11.95	17.75
8550			85	#4 @ 48"		4.79	10.80	15.59
8580				#5 @ 16"		5.40	11.50	16.90
8600		2-6x8x16	105	#4 @ 48"		5.85	12	17.85
8630				#5 @ 16"		6.45	12.70	19.15
8650			85	#4 @ 48"		7.10	11.50	18.60
8680				#5 @ 16"		7.70	12.25	19.95
8900	For stacked bond add						.40	.40
8910								

B2010 Exterior Walls

Exterior split ribbed block walls are defined in the following terms; structural reinforcement, weight, percent solid, size, number of ribs and insulation. Within each of these categories two to four variations are shown. No costs are included for brick shelf or relieving angles. Costs include control joints every 20' and horizontal reinforcing.

System Components			COST PER S.F.		
	QUANTITY	UNIT	MAT.	INST.	TOTAL
SYSTEM B2010 113 1430					
UNREINFORCED SPLIT RIB BLOCK WALL, 8"X8"X16", 8 RIBS(HEX), PERLITE FILL					
Split ribbed block wall, 8" thick	1.000	S.F.	4.27	6.60	10.87
Perlite insulation	1.000	S.F.	.48	.33	.81
Horizontal joint reinforcing, alternate courses	.800	L.F.	.10	.14	.24
Control joint	.050	L.F.	.06	.06	.12
TOTAL			4.91	7.13	12.04

B2010 113 — Split Ribbed Block Wall - Regular Weight

	TYPE	SIZE (IN.)	RIBS	CORE FILL		COST PER S.F.		
						MAT.	INST.	TOTAL
1220	Hollow	4x8x16	4	none		2.44	6	8.44
1250			8	none		2.84	6	8.84
1280			16	none		2.32	6.05	8.37
1330	RB2010 -200		4	none		3.84	6.65	10.49
1340				styrofoam		4.63	6.35	10.98
1350				none		3.51	6.35	9.86
1360			16	perlite		3.40	6.75	10.15
1370				styrofoam		4.19	6.45	10.64
1380				none		3.07	6.45	9.52
1430		8x8x16	8	perlite		4.91	7.15	12.06
1440				styrofoam		5.55	6.80	12.35
1450				none		4.43	6.80	11.23
1460			16	perlite		4.06	7.25	11.31
1470				styrofoam		4.70	6.90	11.60
1480				none		3.58	6.90	10.48
1530		12x8x16	8	perlite		6.15	9.45	15.60
1540				styrofoam		6.75	8.80	15.55
1550				none		5.35	8.80	14.15
1560			16	perlite		5.20	9.60	14.80
1570				styrofoam		5.80	8.95	14.75
1580				none		4.39	8.95	13.34

Important: See the Reference Section for critical supporting data - Reference Numbers and City Cost Indexes

B2010 Exterior Walls

| B2010 113 | Split Ribbed Block Wall - Regular Weight | | | | | | | |

	TYPE	SIZE (IN.)	RIBS	CORE FILL		COST PER S.F.		
						MAT.	INST.	TOTAL
2120	75% solid	4x8x16	4	none		3.06	6.05	9.11
2150			8	none		3.62	6.05	9.67
2180			16	none		2.93	6.15	9.08
2230		6x8x16	8	perlite		4.67	6.60	11.27
2250				none		4.50	6.45	10.95
2260			16	perlite		4.08	6.70	10.78
2280				none		3.91	6.55	10.46
2330		8x8x16	8	perlite		5.90	7.05	12.95
2350				none		5.65	6.90	12.55
2360			16	perlite		4.77	7.15	11.92
2380				none		4.53	7	11.53
2430		12x8x16	8	perlite		7.15	9.25	16.40
2450				none		6.75	8.95	15.70
2460			16	perlite		5.95	9.40	15.35
2480				none		5.55	9.10	14.65
2520	Solid	4x8x16	4	none		3.73	6.15	9.88
2550			8	none		4.36	6.15	10.51
2580			16	none		3.52	6.25	9.77
2620		6x8x16	8	none		5.45	6.55	12
2650			16	none		4.72	6.65	11.37
2680		8x8x16	8	none		6.90	7	13.90
2720			16	none		5.45	7.15	12.60
2750		12x8x16	8	none		8.20	9.10	17.30
2780.			16	none		6.70	9.30	16

| B2010 114 | Split Ribbed Block Wall - Lightweight | | | | | | | |

	TYPE	SIZE (IN.)	RIBS	CORE FILL		COST PER S.F.		
						MAT.	INST.	TOTAL
3250	Hollow	4x8x16	8	none		3.03	5.80	8.83
3280			16	none		2.53	5.90	8.43
3330		6x8x16	8	perlite		4.08	6.45	10.53
3340				styrofoam		4.87	6.15	11.02
3350				none		3.75	6.15	9.90
3360			16	perlite		3.65	6.55	10.20
3370				styrofoam		4.44	6.25	10.69
3380				none		3.32	6.25	9.57
3430		8x8x16	8	perlite		5.70	6.95	12.65
3440				styrofoam		6.35	6.60	12.95
3450				none		5.20	6.60	11.80
3460			16	perlite		4.35	7.05	11.40
3470				styrofoam		4.99	6.70	11.69
3480				none		3.87	6.70	10.57
3530		12x8x16	8	perlite		6.75	9.15	15.90
3540				styrofoam		7.35	8.50	15.85
3550				none		5.95	8.50	14.45
3560			16	perlite		5.55	9.30	14.85
3570				styrofoam		6.15	8.65	14.80
3580				none		4.74	8.65	13.39
4150	75% solid	4x8x16	8	none		3.90	5.90	9.80
4180			16	none		3.17	6	9.17

B2010 Exterior Walls

B2010 114	Split Ribbed Block Wall - Lightweight							
	TYPE	SIZE (IN.)	RIBS	CORE FILL		COST PER S.F.		
						MAT.	INST.	TOTAL
4230	75% solid	6x8x16	8	perlite		4.92	6.40	11.32
4250				none		4.75	6.25	11
4260			16	perlite		4.39	6.50	10.89
4280				none		4.22	6.35	10.57
4330		8x8x16	8	perlite		6.90	6.85	13.75
4350				none		6.65	6.70	13.35
4360			16	perlite		5.15	6.95	12.10
4380		8x8x16	16	none		4.91	6.80	11.71
4430	75% solid	12x8x16	8	perlite		8.05	8.95	17
4450				none		7.65	8.65	16.30
4460			16	perlite		6.35	9.10	15.45
4480				none		5.95	8.80	14.75
4550	Solid	4x8x16	8	none		4.71	6	10.71
4580			16	none		3.85	6.05	9.90
4650		6x8x16	8	none		5.75	6.35	12.10
4680			16	none		5.10	6.45	11.55
4750		8x8x16	8	none		8.05	6.80	14.85
4780			16	none		5.95	6.90	12.85
4850		12x8x16	8	none		9.25	8.80	18.05
4880			16	none		7.25	8.95	16.20

B2010 115	Reinforced Split Ribbed Block Wall - Regular Weight							
	TYPE	SIZE (IN.)	RIBS	VERT. REINF. & GROUT SPACING		COST PER S.F.		
						MAT.	INST.	TOTAL
5200	Hollow	4x8x16	4	#4 @ 48"		2.57	6.55	9.12
5230			8	#4 @ 48"		2.97	6.55	9.52
5260			16	#4 @ 48"		2.45	6.60	9.05
5330		6x8x16	8	#4 @ 48"		3.73	6.95	10.68
5340				#5 @ 32"		3.91	7.15	11.06
5350				#5 @ 16"		4.30	7.90	12.20
5360			16	#4 @ 48"		3.29	7.05	10.34
5370				#5 @ 32"		3.47	7.25	10.72
5380				#5 @ 16"		3.86	8	11.86
5430		8x8x16	8	#4 @ 48"		4.68	7.45	12.13
5440				#5 @ 32"		4.89	7.70	12.59
5450				#5 @ 16"		5.35	8.60	13.95
5460			16	#4 @ 48"		3.83	7.55	11.38
5470				#5 @ 32"		4.04	7.80	11.84
5480				#5 @ 16"		4.49	8.70	13.19
5530		12x8x16	8	#4 @ 48"		5.75	9.45	15.20
5540				#5 @ 32"		6	9.75	15.75
5550				#5 @ 16"		6.60	10.75	17.35
5560			16	#4 @ 48"		4.76	9.60	14.36
5570				#5 @ 32"		5	9.90	14.90
5580				#5 @ 16"		5.65	10.90	16.55
6230	75% solid	6x8x16	8	#4 @ 48"		4.62	6.95	11.57
6240				#5 @ 32"		4.76	7.10	11.86
6250				#5 @ 16"		5	7.75	12.75
6260			16	#4 @ 48"		4.03	7.05	11.08
6270				#5 @ 32"		4.17	7.20	11.37
6280				#5 @ 16"		4.42	7.85	12.27

B2010 Exterior Walls

| B2010 115 | Reinforced Split Ribbed Block Wall - Regular Weight | | | | | | | |

	TYPE	SIZE (IN.)	RIBS	VERT. REINF. & GROUT SPACING		COST PER S.F.		
						MAT.	INST.	TOTAL
6330	75% solid	8x8x16	8	#4 @ 48"		5.80	7.45	13.25
6340				#5 @ 32"		5.95	7.65	13.60
6350				#5 @ 16"		6.20	8.40	14.60
6360			16	#4 @ 48"		4.66	7.55	12.21
6370				#5 @ 32"		4.80	7.75	12.55
6380				#5 @ 16"		5.05	8.50	13.55
6430		12x8x16	8	#4 @ 48"		7	9.50	16.50
6440				#5 @ 32"		7.15	9.70	16.85
6450				#5 @ 16"		7.55	10.45	18
6460			16	#4 @ 48"		5.80	9.65	15.45
6470				#5 @ 32"		5.95	9.85	15.80
6480				#5 @ 16"		6.35	10.60	16.95
6500	Solid-double	2-4x8x16	4	#4 @ 48"		4.52	7.45	11.97
6520	Wythe			#5 @ 16"		4.83	7.80	12.63
6530			8	#4 @ 48"		5.15	7.45	12.60
6550				#5 @ 16"		5.45	7.80	13.25
6560			16	#4 @ 48"		4.31	7.55	11.86
6580				#5 @ 16"		4.62	7.90	12.52
6630		2-6x8x16	8	#4 @ 48"		6.25	7.85	14.10
6650				#5 @ 16"		6.55	8.25	14.80
6660			16	#4 @ 48"		5.50	7.95	13.45
6680				#5 @ 16"		5.80	8.35	14.15

| B2010 116 | Reinforced Split Ribbed Block Wall - Lightweight | | | | | | | |

	TYPE	SIZE (IN.)	RIBS	VERT. REINF. & GROUT SPACING		COST PER S.F.		
						MAT.	INST.	TOTAL
7230	Hollow	4x8x16	8	#4 @ 48"		3.16	6.35	9.51
7260			16	#4 @ 48"		2.66	6.45	9.11
7330		6x8x16	8	#4 @ 48"		3.97	6.75	10.72
7340				#5 @ 32"		4.15	6.95	11.10
7350				#5 @ 16"		4.54	7.70	12.24
7360			16	#4 @ 48"		3.54	6.85	10.39
7370				#5 @ 32"		3.72	7.05	10.77
7380				#5 @ 16"		4.11	7.80	11.91
7430		8x8x16	8	#4 @ 48"		5.45	7.25	12.70
7440				#5 @ 32"		5.65	7.50	13.15
7450				#5 @ 16"		6.10	8.40	14.50
7460			16	#4 @ 48"		4.12	7.35	11.47
7470				#5 @ 32"		4.33	7.60	11.93
7480				#5 @ 16"		4.78	8.50	13.28
7530		12x8x16	8	#4 @ 48"		6.35	9.15	15.50
7540				#5 @ 32"		6.60	9.45	16.05
7550				#5 @ 16"		7.20	10.45	17.65
7560			16	#4 @ 48"		5.10	9.30	14.40
7570				#5 @ 32"		5.35	9.60	14.95
7580				#5 @ 16"		6	10.60	16.60
8230	75% solid	6x8x16	8	#4 @ 48"		4.87	6.75	11.62
8240				#5 @ 32"		5	6.90	11.90
8250				#5 @ 16"		5.25	7.55	12.80
8260			16	#4 @ 48"		4.34	6.85	11.19
8270				#5 @ 32"		4.48	7	11.48
8280				#5 @ 16"		4.73	7.65	12.38

B2010 Exterior Walls

B2010 116		Reinforced Split Ribbed Block Wall - Lightweight			

	TYPE	SIZE (IN.)	RIBS	VERT. REINF. & GROUT SPACING		COST PER S.F.		
						MAT.	INST.	TOTAL
8330	75% solid	8x8x16	8	#4 @ 48"		6.80	7.25	14.05
8340				#5 @ 32"		6.95	7.45	14.40
8350				#5 @ 16"		7.20	8.20	15.40
8360			16	#4 @ 48"		5.05	7.35	12.40
8370				#5 @ 32"		5.20	7.55	12.75
8380				#5 @ 16"		5.45	8.30	13.75
8430		12x8x16	8	#4 @ 48"		7.90	9.20	17.10
8440				#5 @ 32"		8.05	9.40	17.45
8450				#5 @ 16"		8.45	10.15	18.60
8460			16	#4 @ 48"		6.20	9.35	15.55
8470				#5 @ 32"		6.35	9.55	15.90
8480				#5 @ 16"		6.75	10.30	17.05
8530	Solid double	2-4x8x16	8	#4 @ 48" E.W		5.60	7.50	13.10
8550	Wythe			#5 @ 16" E.W.		6.20	8.25	14.45
8560			16	#4 @ 48" E.W.		4.72	7.55	12.27
8580				#5 @ 16" E.W.		5.35	8.30	13.65
8630		2-6x8x16	8	#4 @ 48" E.W.		6.60	7.90	14.50
8650				#5 @ 16" E.W.		7.25	8.60	15.85
8660			16	#4 @ 48" E.W.		5.95	8	13.95
8680				#5 @ 16" E.W.		6.60	8.70	15.30
8900	For stacked bond, add						.40	.40

Important: See the Reference Section for critical supporting data - Reference Numbers and City Cost Indexes

B2010   Exterior Walls

Exterior split face block walls are defined in the following terms; structural reinforcement, weight, percent solid, size, scores and insulation. Within each of these categories two to four variations are shown. No costs are included for brick shelf or relieving angles. Costs include control joints every 20′ and horizontal reinforcing.

System Components	QUANTITY	UNIT	COST PER S.F.		
			MAT.	INST.	TOTAL
SYSTEM B2010 117 1600					
UNREINFORCED SPLIT FACE BLOCK WALL, 8″X8″X16″, 0 SCORES, PERLITE FILL					
Split face block wall, 8″ thick	1.000	S.F.	3.60	6.80	10.40
Perlite insulation	1.000	S.F.	.48	.33	.81
Horizontal joint reinforcing, alternate course	.800	L.F.	.10	.14	.24
Control joint	.050	L.F.	.06	.06	.12
TOTAL			4.24	7.33	11.57

B2010 117		Split Face Block Wall - Regular Weight					COST PER S.F.		
	TYPE	SIZE (IN.)	SCORES	CORE FILL			MAT.	INST.	TOTAL
1200	Hollow	8x4x16	0	perlite			7.20	7.85	15.05
1210				styrofoam			7.85	7.50	15.35
1240				none			6.70	7.50	14.20
1250	RB2010 -200		1	perlite			7.70	7.85	15.55
1260				styrofoam			8.35	7.50	15.85
1290				none			7.20	7.50	14.70
1300		12x4x16	0	perlite			9.35	8.95	18.30
1310				styrofoam			9.95	8.30	18.25
1340				none			8.55	8.30	16.85
1350			1	perlite			9.85	8.95	18.80
1360				styrofoam			10.45	8.30	18.75
1390				none			9.05	8.30	17.35
1400		4x8x16	0	none			2.44	5.90	8.34
1450			1	none			2.75	6	8.75
1500		6x8x16	0	perlite			3.49	6.85	10.34
1510				styrofoam			4.28	6.55	10.83
1540				none			3.16	6.55	9.71
1550			1	perlite			3.81	6.95	10.76
1560				styrofoam			4.60	6.65	11.25
1590				none			3.48	6.65	10.13

B2010 Exterior Walls

B2010 117	Split Face Block Wall - Regular Weight							

B SHELL

	TYPE	SIZE (IN.)	SCORES	CORE FILL		COST PER S.F.		
						MAT.	INST.	TOTAL
1600	Hollow	8x8x16	0	perlite		4.24	7.35	11.59
1610				styrofoam		4.88	7	11.88
1640				none		3.76	7	10.76
1650		8x8x16	1	perlite		4.49	7.50	11.99
1660				styrofoam		5.15	7.15	12.30
1690				none		4.01	7.15	11.16
1700		12x8x16	0	perlite		5.45	9.60	15.05
1710				styrofoam		6.05	8.95	15
1740				none		4.65	8.95	13.60
1750			1	perlite		5.70	9.75	15.45
1760				styrofoam		6.30	9.10	15.40
1790				none		4.91	9.10	14.01
1800	75% solid	8x4x16	0	perlite		8.70	7.80	16.50
1840				none		8.45	7.65	16.10
1850			1	perlite		9.25	7.80	17.05
1890				none		9	7.65	16.65
1900		12x4x16	0	perlite		11.25	8.75	20
1940				none		10.85	8.45	19.30
1950			1	perlite		11.70	8.75	20.45
1990				none		11.30	8.45	19.75
2000		4x8x16	0	none		3.06	6	9.06
2050			1	none		3.32	6.05	9.37
2100		6x8x16	0	perlite		4.24	7.90	12.14
2140				none		4.07	7.75	11.82
2150			1	perlite		4.50	6.90	11.40
2190				none		4.33	6.75	11.08
2200		8x8x16	0	perlite		4.99	7.30	12.29
2240				none		4.75	7.15	11.90
2250			1	perlite		5.25	7.40	12.65
2290				none		5	7.25	12.25
2300		12x8x16	0	perlite		6.25	9.40	15.65
2340				none		5.85	9.10	14.95
2350			1	perlite		6.50	9.60	16.10
2390				none		6.10	9.30	15.40
2400	Solid	8x4x16	0	none		10.35	7.80	18.15
2450			1	none		10.85	7.80	18.65
2500		12x4x16	0	none		13.10	8.65	21.75
2550			1	none		13.60	10.05	23.65
2600		4x8x16	0	none		3.73	6.05	9.78
2650			1	none		3.98	6.15	10.13
2700		6x8x16	0	none		4.96	6.75	11.71
2750			1	none		5.20	6.85	12.05
2800		8x8x16	0	none		5.70	7.25	12.95
2850			1	none		5.95	7.40	13.35
2900		12x8x16	0	none		7.05	9.30	16.35
2950			1	none		7.30	9.45	16.75

B2010 118	Split Face Block Wall - Lightweight							

	TYPE	SIZE (IN.)	SCORES	CORE FILL		COST PER S.F.		
						MAT.	INST.	TOTAL
3200	Hollow	8x4x16	0	perlite		7.45	8.80	16.25
3210				styrofoam		8.10	8.45	16.55
3240				none		6.95	8.45	15.40

Important: See the Reference Section for critical supporting data - Reference Numbers and City Cost Indexes

B2010 Exterior Walls

B2010 118	Split Face Block Wall - Lightweight

	TYPE	SIZE (IN.)	SCORES	CORE FILL		COST PER S.F.		
						MAT.	INST.	TOTAL
3250	Hollow		1	perlite		8	7.60	15.60
3260				styrofoam		8.65	7.25	15.90
3290				none		7.50	7.25	14.75
3300		12x4x16	0	perlite		9.75	8.65	18.40
3310				styrofoam		10.35	8	18.35
3340				none		8.95	8	16.95
3350		12x4x16	1	perlite		10.25	9.95	20.20
3360				styrofoam		10.85	9.30	20.15
3390				none		9.45	9.30	18.75
3400		4x8x16	0	none		2.65	5.75	8.40
3450			1	none		2.88	5.80	8.68
3500		6x8x16	0	perlite		3.88	6.65	10.53
3510				styrofoam		4.67	6.35	11.02
3540				none		3.55	6.35	9.90
3550			1	perlite		4.12	6.75	10.87
3560				styrofoam		4.91	6.45	11.36
3590				none		3.79	6.45	10.24
3600		8x8x16	0	perlite		4.56	7.15	11.71
3610				styrofoam		5.20	6.80	12
3640				none		4.08	6.80	10.88
3650			1	perlite		4.82	7.25	12.07
3660				styrofoam		5.45	6.90	12.35
3690				none		4.34	6.90	11.24
3700		12x8x16	0	perlite		5.75	9.30	15.05
3710				styrofoam		6.35	8.65	15
3740				none		4.98	8.65	13.63
3750			1	perlite		6	9.45	15.45
3760				styrofoam		6.60	8.80	15.40
3790				none		5.20	8.80	14
3800	75% solid	8x4x16	0	perlite		9.25	7.80	17.05
3840				none		9	7.65	16.65
3850			1	perlite		9.60	7.55	17.15
3890				none		9.35	7.40	16.75
3900		12x4x16	0	perlite		11.70	8.75	20.45
3940				none		11.30	8.45	19.75
3950			1	perlite		12.25	9.75	22
3990				none		11.85	9.45	21.30
4000		4x8x16	0	none		3.33	5.80	9.13
4050			1	none		3.61	5.90	9.51
4100		6x8x16	0	perlite		4.65	6.60	11.25
4140				none		4.48	6.45	10.93
4150			1	perlite		4.92	6.70	11.62
4190				none		4.75	6.55	11.30
4200		8x8x16	0	perlite		5.45	7.05	12.50
4240				none		5.20	6.90	12.10
4250			1	perlite		5.65	7.15	12.80
4290				none		5.40	7	12.40
4300		12x8x16	0	perlite		6.65	9.10	15.75
4340				none		6.25	8.80	15.05
4350			1	perlite		6.90	9.25	16.15
4390				none		6.50	8.95	15.45

B2010 Exterior Walls

B2010 118				Split Face Block Wall - Lightweight				

	TYPE	SIZE (IN.)	SCORES	CORE FILL		COST PER S.F.		
						MAT.	INST.	TOTAL
4400	Solid	8x4x16	0	none		10.70	7.50	18.20
4450			1	none		11.25	7.50	18.75
4500		12x4x16	0	none		13.75	8.30	22.05
4550			1	none		14.30	9.65	23.95
4600		4x8x16	0	none		4.03	5.90	9.93
4650			1	none		4.30	6	10.30
4700		6x8x16	0	none		5.45	6.55	12
4750			1	none		5.70	6.65	12.35
4800		8x8x16	0	none		6.30	7	13.30
4850			1	none		6.50	7.15	13.65
4900		12x8x16	0	none		7.55	8.95	16.50
4950			1	none		7.80	9.10	16.90

B2010 119				Reinforced Split Face Block Wall - Regular Weight				

	TYPE	SIZE (IN.)	SCORES	VERT. REINF. & GROUT SPACING		COST PER S.F.		
						MAT.	INST.	TOTAL
5200	Hollow	8x4x16	0	#4 @ 48"		6.95	8.15	15.10
5210				#5 @ 32"		7.15	8.40	15.55
5240				#5 @ 16"		7.60	9.30	16.90
5250			1	#4 @ 48"		7.45	8.15	15.60
5260				#5 @ 32"		7.65	8.40	16.05
5290				#5 @ 16"		8.10	9.30	17.40
5300		12x4x16	0	#4 @ 48"		8.95	8.95	17.90
5310				#5 @ 32"		9.20	9.25	18.45
5340				#5 @ 16"		9.80	10.25	20.05
5350			1	#4 @ 48"		9.45	8.95	18.40
5360				#5 @ 32"		9.70	9.25	18.95
5390				#5 @ 16"		10.30	10.25	20.55
5400		4x8x16	0	#4 @ 48"		2.57	6.45	9.02
5450			1	#4 @ 48"		2.88	6.55	9.43
5500		6x8x16	0	#4 @ 48"		3.38	7.15	10.53
5510				#5 @ 32"		3.56	7.35	10.91
5540				#5 @ 16"		3.95	8.10	12.05
5550			1	#4 @ 48"		3.70	7.25	10.95
5560				#5 @ 32"		3.88	7.45	11.33
5590				#5 @ 16"		4.27	8.20	12.47
5600		8x8x16	0	#4 @ 48"		4.01	7.65	11.66
5610				#5 @ 32"		4.22	7.90	12.12
5640				#5 @ 16"		4.36	8.35	12.71
5650			1	#4 @ 48"		6.25	10.65	16.90
5660				#5 @ 32"		4.47	8.05	12.52
5690				#5 @ 16"		4.92	8.95	13.87
5700		12x8x16	0	#4 @ 48"		5	9.60	14.60
5710				#5 @ 32"		5.25	9.90	15.15
5740				#5 @ 16"		5.90	10.90	16.80
5750			1	#4 @ 48"		5.30	9.75	15.05
5760				#5 @ 32"		5.55	10.05	15.60
5790				#5 @ 16"		6.15	11.05	17.20

B20 Exterior Enclosure

B2010 Exterior Walls

B2010 119 — Reinforced Split Face Block Wall - Regular Weight

	TYPE	SIZE (IN.)	SCORES	VERT. REINF. & GROUT SPACING		COST PER S.F.		
						MAT.	INST.	TOTAL
6000	75% solid	8x4x16	0	#4 @ 48"		8.60	8.20	16.80
6010				#5 @ 32"		8.75	8.40	17.15
6040				#5 @ 16"		9	9.15	18.15
6050			1	#4 @ 48"		9.15	8.20	17.35
6060				#5 @ 32"		9.30	8.40	17.70
6090				#5 @ 16"		9.55	9.15	18.70
6100		12x4x16	0	#4 @ 48"		11.10	9	20.10
6110				#5 @ 32"		11.25	9.20	20.45
6140				#5 @ 16"		11.60	9.80	21.40
6150			1	#4 @ 48"		11.55	9	20.55
6160				#5 @ 32"		11.70	9.20	20.90
6190				#5 @ 16"		12.10	9.95	22.05
6200		6x8x16	0	#4 @ 48"		4.19	8.25	12.44
6210				#5 @ 32"		4.33	8.40	12.73
6240				#5 @ 16"		4.58	9.05	13.63
6250			1	#4 @ 48"		4.45	7.25	11.70
6260				#5 @ 32"		4.59	7.40	11.99
6290				#5 @ 16"		4.84	8.05	12.89
6300		8x8x16	0	#4 @ 48"		4.88	7.70	12.58
6310				#5 @ 32"		5	7.90	12.90
6340				#5 @ 16"		5.30	8.65	13.95
6350			1	#4 @ 48"		5.15	7.80	12.95
6360				#5 @ 32"		5.25	8	13.25
6390				#5 @ 16"		5.55	8.75	14.30
6400	75% solid	12x8x16	0	#4 @ 48"		6.10	9.65	15.75
6410				#5 @ 32"		6.25	9.85	16.10
6440				#5 @ 16"		6.65	10.60	17.25
6450			1	#4 @ 48"		6.35	9.85	16.20
6460				#5 @ 32"		6.50	10.05	16.55
6490				#5 @ 16"		6.90	10.80	17.70
6700	Solid-double Wythe	2-4x8x16	0	#4 @ 48" E.W.		8.20	13.45	21.65
6710				#5 @ 32" E.W.		8.40	13.60	22
6740				#5 @ 16" E.W.		8.80	14.20	23
6750			1	#4 @ 48" E.W.		8.70	13.65	22.35
6760				#5 @ 32" E.W.		8.90	13.80	22.70
6790				#5 @ 16" E.W.		9.30	14.40	23.70
6800		2-6x8x16	0	#4 @ 48" E.W.		10.65	14.90	25.55
6810				#5 @ 32" E.W.		10.85	15.05	25.90
6840				#5 @ 16" E.W.		11.25	15.60	26.85
6850			1	#4 @ 48" E.W.		11.10	15.10	26.20
6860				#5 @ 32" E.W.		11.35	15.25	26.60
6890				#5 @ 16" E.W.		11.75	15.80	27.55

B2010 120 — Reinforced Split Face Block Wall - Lightweight

	TYPE	SIZE (IN.)	SCORES	VERT. REINF. & GROUT SPACING		COST PER S.F.		
						MAT.	INST.	TOTAL
7200	Hollow	8x4x16	0	#4 @ 48"		7.20	9.10	16.30
7210				#5 @ 32"		7.40	9.35	16.75
7240				#5 @ 16"		7.85	10.25	18.10
7250			1	#4 @ 48"		7.75	7.90	15.65
7260				#5 @ 32"		7.95	8.15	16.10
7290				#5 @ 16"		8.40	9.05	17.45

149

| B2010 120 | | | | Reinforced Split Face Block Wall - Lightweight | | | | |

	TYPE	SIZE (IN.)	SCORES	VERT. REINF. & GROUT SPACING		COST PER S.F.		
						MAT.	INST.	TOTAL
7300	Hollow	12x4x16	0	#4 @ 48"		9.35	8.65	18
7310				#5 @ 32"		9.60	8.95	18.55
7340				#5 @ 16"		10.20	9.95	20.15
7350			1	#4 @ 48"		9.85	9.95	19.80
7360				#5 @ 32"		10.10	10.25	20.35
7390				#5 @ 16"		10.70	11.25	21.95
7400		4x8x16	0	#4 @ 48"		2.78	6.30	9.08
7450			1	#4 @ 48"		3.01	6.35	9.36
7500		6x8x16	0	#4 @ 48"		3.77	6.95	10.72
7510				#5 @ 32"		3.95	7.15	11.10
7540				#5 @ 16"		4.34	7.90	12.24
7550			1	#4 @ 48"		4.01	7.05	11.06
7560				#5 @ 32"		4.19	7.25	11.44
7590				#5 @ 16"		4.58	8	12.58
7600		8x8x16	0	#4 @ 48"		4.33	7.45	11.78
7610				#5 @ 32"		4.54	7.70	12.24
7640				#5 @ 16"		4.99	8.60	13.59
7650			1	#4 @ 48"		4.59	7.55	12.14
7660				#5 @ 32"		4.80	7.80	12.60
7690				#5 @ 16"		5.25	8.70	13.95
7700		12x8x16	0	#4 @ 48"		5.35	9.30	14.65
7710				#5 @ 32"		5.60	9.60	15.20
7740		12x8x16	0	#5 @ 16"		6.20	10.60	16.80
7750			1	#4 @ 48"		5.60	9.45	15.05
7760				#5 @ 16"		5.85	9.75	15.60
7790				#5 @ 16"		6.45	10.75	17.20
8000	75% solid	8x4x16	0	#4 @ 48"		9.15	8.20	17.35
8010				#5 @ 32"		9.30	8.40	17.70
8040				#5 @ 16"		9.55	9.15	18.70
8050			1	#4 @ 48"		9.50	7.95	17.45
8060				#5 @ 32"		9.65	8.15	17.80
8090				#5 @ 16"		9.90	8.90	18.80
8100		12x4x16	0	#4 @ 48"		11.55	9	20.55
8110				#5 @ 32"		11.70	9.20	20.90
8140				#5 @ 16"		12.10	9.95	22.05
8150			1	#4 @ 48"		12.10	10	22.10
8160				#5 @ 32"		12.25	10.20	22.45
8190				#5 @ 16"		12.65	10.95	23.60
8200		6x8x16	0	#4 @ 48"		4.60	6.95	11.55
8210				#5 @ 32"		4.74	7.10	11.84
8240				#5 @ 16"		4.99	7.75	12.74
8250			1	#4 @ 48"		4.87	7.05	11.92
8260				#5 @ 32"		5	7.20	12.20
8290				#5 @ 16"		5.25	7.85	13.10
8300		8x8x16	0	#4 @ 48"		5.35	7.45	12.80
8310				#5 @ 32"		5.50	7.65	13.15
8340				#5 @ 16"		5.75	8.40	14.15
8350			1	#4 @ 48"		5.55	7.55	13.10
8360				#5 @ 32"		5.70	7.75	13.45
8390				#5 @ 16"		5.95	8.50	14.45

Important: See the Reference Section for critical supporting data - Reference Numbers and City Cost Indexes

B2010 Exterior Walls

B2010 120	Reinforced Split Face Block Wall - Lightweight

	TYPE	SIZE (IN.)	SCORES	VERT. REINF. & GROUT SPACING		COST PER S.F.		
						MAT.	INST.	TOTAL
8400	75% solid	12x8x16	0	#4 @ 48"		6.50	9.35	15.85
8410				#5 @ 32"		6.65	9.55	16.20
8440				#5 @ 16"		7.05	10.30	17.35
8450			1	#4 @ 48"		6.75	9.50	16.25
8460				#5 @ 32"		6.90	9.70	16.60
8490				#5 @ 16"		7.30	10.45	17.75
8700	Solid-double	2-4x8x16	0	#4 @ 48" E.W.		8.80	13.15	21.95
8710	Wythe			#5 @ 32" E.W.		9	13.30	22.30
8740				#5 @ 16" E.W.		9.40	13.90	23.30
8750			1	#4 @ 48" E.W.		9.35	13.35	22.70
8760				#5 @ 32" E.W.		9.55	13.50	23.05
8790				#5 @ 16" E.W.		9.95	14.10	24.05
8800		2-6x8x16	0	#4 @ 48" E.W.		11.60	14.50	26.10
8810				#5 @ 32" E.W.		11.85	14.65	26.50
8840				#5 @ 16" E.W.		12.25	15.20	27.45
8850			1	#4 @ 48" E.W.		12.10	14.70	26.80
8860				#5 @ 32" E.W.		12.35	14.85	27.20
8890				#5 @ 16" E.W.		12.75	15.40	28.15

Exterior ground face block walls are defined in the following terms; structural reinforcement, weight, percent solid, size, scores and insulation. Within each of these categories two to four variations are shown. No costs are included for brick shelf or relieving angles. Costs include control joints every 20' and horizontal reinforcing.

System Components	QUANTITY	UNIT	COST PER S.F.		
			MAT.	INST.	TOTAL
SYSTEM B2010 121 1600					
UNREINF. GROUND FACE BLOCK WALL, 8"X8"X16", 0 SCORES, PERLITE FILL					
Ground face block wall, 8" thick	1.000	S.F.	6.15	6.95	13.10
Perlite insulation	1.000	S.F.	.48	.33	.81
Horizontal joint reinforcing, alternate course	.800	L.F.	.10	.14	.24
Control joint	.050	L.F.	.06	.06	.12
TOTAL			6.79	7.48	14.27

B2010 121		Ground Face Block Wall				COST PER S.F.		
	TYPE	SIZE (IN.)	SCORES	CORE FILL		MAT.	INST.	TOTAL
1200	Hollow	4x8x16	0	none		4.38	6	10.38
1250			1	none		4.45	6.05	10.50
1300			2 to 5	none		4.57	6.15	10.72
1400	RB2010 -200	6x8x16	0	perlite		5.70	6.95	12.65
1410				styrofoam		6.45	6.65	13.10
1440				none		5.35	6.65	12
1450			1	perlite		5.75	7.05	12.80
1460				styrofoam		6.50	6.75	13.25
1490				none		5.40	6.75	12.15
1500			2 to 5	perlite		5.90	7.15	13.05
1510				styrofoam		6.65	6.85	13.50
1540				none		5.55	6.85	12.40
1600		8x8x16	0	perlite		6.80	7.50	14.30
1610				styrofoam		7.45	7.15	14.60
1640				none		6.30	7.15	13.45
1650			1	perlite		6.90	7.60	14.50
1660				styrofoam		7.55	7.25	14.80
1690				none		6.40	7.25	13.65
1700			2 to 5	perlite		7	7.75	14.75
1710				styrofoam		7.65	7.40	15.05
1740				none		6.50	7.40	13.90

Important: See the Reference Section for critical supporting data - Reference Numbers and City Cost Indexes

B20 Exterior Enclosure

B2010 Exterior Walls

B2010 121 — Ground Face Block Wall

	TYPE	SIZE (IN.)	SCORES	CORE FILL		COST PER S.F.		
						MAT.	INST.	TOTAL
1800	Hollow	12x8x16	0	perlite		8.95	9.75	18.70
1810				styrofoam		9.55	9.10	18.65
1840				none		8.15	9.10	17.25
1850			1	perlite		9.05	9.95	19
1860				styrofoam		9.65	9.30	18.95
1890				none		8.25	9.30	17.55
1900		12x8x16	2 to 5	perlite		9.10	10.10	19.20
1910				styrofoam		9.70	9.45	19.15
1940				none		8.30	9.45	17.75
2200	75% solid	4x8x16	0	none		5.60	6.05	11.65
2250			1	none		5.70	6.15	11.85
2300			2 to 5	none		5.80	6.25	12.05
2400		6x8x16	0	perlite		7	6.90	13.90
2440				none		6.85	6.75	13.60
2450			1	perlite		7.10	7	14.10
2490				none		6.95	6.85	13.80
2500			2 to 5	perlite		7.20	7.10	14.30
2540				none		7.05	6.95	14
2600		8x8x16	0	perlite		8.35	7.40	15.75
2640				none		8.10	7.25	15.35
2650			1	perlite		8.40	7.55	15.95
2690				none		8.15	7.40	15.55
2700			2 to 5	perlite		8.50	7.65	16.15
2740				none		8.25	7.50	15.75
2800		12x8x16	0	perlite		10.90	9.60	20.50
2840				none		10.50	9.30	19.80
2850			1	perlite		10.95	9.75	20.70
2890				none		10.55	9.45	20
2900			2 to 5	perlite		11.05	9.95	21
2940				none		10.65	9.65	20.30
3200	Solid	4x8x16	0	none		6.85	6.15	13
3250			1	none		6.95	6.25	13.20
3300			2 to 5	none		7.05	6.35	13.40
3400		6x8x16	0	none		8.30	6.85	15.15
3450			1	none		8.40	6.95	15.35
3500			2 to 5	none		8.45	7.10	15.55
3600		8x8x16	0	none		9.90	7.40	17.30
3650			1	none		9.95	7.50	17.45
3700			2 to 5	none		10.05	7.65	17.70
3800		12x8x16	0	none		12.75	9.45	22.20
3850			1	none		12.85	9.65	22.50
3900			2 to 5	none		12.95	9.85	22.80

B2010 122 — Reinforced Ground Face Block Wall

	TYPE	SIZE (IN.)	SCORES	VERT. REINF. & GROUT SPACING		COST PER S.F.		
						MAT.	INST.	TOTAL
5200	Hollow	4x8x16	0	#4 @ 48"		4.51	6.55	11.06
5250			1	#4 @ 48"		4.58	6.60	11.18
5300			2 to 5	#4 @ 48"		4.70	6.70	11.40
5400		6x8x16	0	#4 @ 48"		5.55	7.25	12.80
5420				#5 @ 32"		5.75	7.45	13.20
5430				#5 @ 16"		6.15	8.20	14.35

B2010 Exterior Walls

| B2010 122 | | | | Reinforced Ground Face Block Wall | | | | |

B SHELL

	TYPE	SIZE (IN.)	SCORES	VERT. REINF. & GROUT SPACING		COST PER S.F.		
						MAT.	INST.	TOTAL
5450	Hollow		1	#4 @ 48"		5.60	7.35	12.95
5470				#5 @ 32"		5.80	7.55	13.35
5480				#5 @ 16"		6.20	8.30	14.50
5500			2 to 5	#4 @ 48"		5.75	7.45	13.20
5520				#5 @ 32"		5.95	7.65	13.60
5530				#5 @ 16"		6.35	8.40	14.75
5600		8x8x16	0	#4 @ 48"		6.55	7.80	14.35
5620				#5 @ 32"		6.75	8.05	14.80
5630				#5 @ 16"		7.20	8.95	16.15
5650			1	#4 @ 48"		6.65	7.90	14.55
5670				#5 @ 32"		6.85	8.15	15
5680				#5 @ 16"		7.30	9.05	16.35
5700			2 to 5	#4 @ 48"		6.75	8.05	14.80
5720				#5 @ 32"		6.95	8.30	15.25
5730				#5 @ 16"		7.40	9.20	16.60
5800		12x8x16	0	#4 @ 48"		8.55	9.75	18.30
5820				#5 @ 32"		8.80	10.05	18.85
5830				#5 @ 16"		9.40	11.05	20.45
5850			1	#4 @ 48"		8.65	9.95	18.60
5870				#5 @ 32"		8.90	10.25	19.15
5880				#5 @ 16"		9.50	11.25	20.75
5900			2 to 5	#4 @ 48"		8.70	10.10	18.80
5920				#5 @ 32"		8.95	10.40	19.35
5930				#5 @ 16"		9.55	11.40	20.95
6200	75% solid	6x8x16	0	#4 @ 48"		7	7.15	14.15
6220				#5 @ 32"		7.15	7.30	14.45
6230				#5 @ 16"		7.45	7.85	15.30
6250			1	#4 @ 48"		7.10	7.25	14.35
6270				#5 @ 32"		7.25	7.40	14.65
6280				#5 @ 16"		7.55	7.95	15.50
6300			2 to 5	#4 @ 48"		7.20	7.35	14.55
6330				#5 @ 16"		7.65	8.05	15.70
6400		8x8x16	0	#4 @ 48"		8.30	7.65	15.95
6420				#5 @ 32"		8.45	7.90	16.35
6430				#5 @ 16"		8.75	8.45	17.20
6450			1	#4 @ 48"		8.35	7.80	16.15
6470				#5 @ 32"		8.50	8.05	16.55
6480				#5 @ 16"		8.80	8.60	17.40
6500			2 to 5	#4 @ 48"		8.45	7.90	16.35
6520				#5 @ 32"		8.60	8.15	16.75
6530				#5 @ 16"		8.90	8.70	17.60
6600		12x8x16	0	#4 @ 48"		10.75	9.70	20.45
6620				#5 @ 32"		10.95	9.95	20.90
6630				#5 @ 16"		11.35	10.55	21.90
6650			1	#4 @ 48"		10.80	9.85	20.65
6670				#5 @ 32"		11	10.10	21.10
6680				#5 @ 16"		11.40	10.70	22.10
6700			2 to 5	#4 @ 48"		10.90	10.05	20.95
6720				#5 @ 32"		11.10	10.30	21.40
6730				#5 @ 16"		11.50	10.90	22.40

Important: See the Reference Section for critical supporting data - Reference Numbers and City Cost Indexes

B20 Exterior Enclosure

B2010 Exterior Walls

| B2010 122 | Reinforced Ground Face Block Wall |

	TYPE	SIZE (IN.)	SCORES	VERT. REINF. & GROUT SPACING		COST PER S.F.		
						MAT.	INST.	TOTAL
6800	Solid-double	2-4x8x16	0	#4 @ 48" E.W.		14.35	13.55	27.90
6830	Wythe			#5 @ 16" E.W.		14.95	14.15	29.10
6850			1	#4 @ 48" E.W.		14.55	13.75	28.30
6880				#5 @ 16" E.W.		15.15	14.35	29.50
6900			2 to 5	#4 @ 48" E.W.		14.75	13.95	28.70
6930				#5 @ 16" E.W.		15.35	14.55	29.90
7000		2-6x8x16	0	#4 @ 48" E.W.		17.25	14.95	32.20
7030				#5 @ 16" E.W.		17.85	15.55	33.40
7050			1	#4 @ 48" E.W.		17.45	15.15	32.60
7080				#5 @ 16" E.W.		18.05	15.75	33.80
7100			2 to 5	#4 @ 48" E.W.		17.55	15.45	33
7130				#5 @ 16" E.W.		18.15	16.05	34.20

Exterior miscellaneous block walls are defined in the following terms: structural reinforcement, finish texture, percent solid, size, weight and insulation. Within each of these categories two to four variations are shown. No costs are included for brick shelf or relieving angles. Costs include control joints every 20' and horizontal reinforcing.

System Components	QUANTITY	UNIT	COST PER S.F. MAT.	COST PER S.F. INST.	COST PER S.F. TOTAL
SYSTEM B2010 123 2700					
UNREINFORCED HEXAGONAL BLOCK WALL, 8"X8"X16", 125PCF, PERLITE FILL					
Hexagonal block partition, 8" thick	1.000	S.F.	3.78	6.95	10.73
Perlite insulation	1.000	S.F.	.48	.33	.81
Horizontal joint reinf., alternate courses	.800	L.F.	.10	.14	.24
Control joint	.050	L.F.	.06	.06	.12
TOTAL			4.42	7.48	11.90

B2010 123 — Miscellaneous Block Wall

	TYPE	SIZE (IN.)	WEIGHT (P.C.F.)	CORE FILL	COST PER S.F. MAT.	COST PER S.F. INST.	COST PER S.F. TOTAL
1100	Deep groove-hollow	4x8x16	125	none	2.92	6	8.92
1150			105	none	3.06	5.80	8.86
1200	RB2010-200	6x8x16	125	perlite	4.08	6.65	10.73
1210				styrofoam	4.87	6.35	11.22
1240				none	3.75	6.35	10.10
1250			105	perlite	4.24	6.10	10.34
1260				styrofoam	5.05	5.80	10.85
1290				none	3.91	5.80	9.71
1300		8x8x16	125	perlite	5.30	7.25	12.55
1310				styrofoam	5.90	6.90	12.80
1340				none	4.80	6.90	11.70
1350			105	perlite	5.80	7.05	12.85
1360				styrofoam	6.45	6.70	13.15
1390				none	5.30	6.70	12
1400	Deep groove-75% solid	4x8x16	125	none	3.74	6.05	9.79
1450			105	none	3.90	5.90	9.80
1500		6x8x16	125	perlite	4.92	6.60	11.52
1540				none	4.75	6.45	11.20
1550			105	perlite	5.15	6.40	11.55
1590				none	4.98	6.25	11.23

Important: See the Reference Section for critical supporting data - Reference Numbers and City Cost Indexes

| B2010 123 | | | | Miscellaneous Block Wall | | | | |

	TYPE	SIZE (IN.)	WEIGHT (P.C.F.)	CORE FILL		COST PER S.F.		
						MAT.	INST.	TOTAL
1600	Deep groove	8x8x16	125	perlite		6.35	7.15	13.50
1640	75% solid			none		6.10	7	13.10
1650			105	perlite		7	6.95	13.95
1690				none		6.75	6.80	13.55
1700		4x8x16	125	none		4.51	6.15	10.66
1750			105	none		4.74	6	10.74
1800		6x8x16	125	none		5.75	6.55	12.30
1850			105	none		6.05	6.35	12.40
1900		8x8x16	125	none		7.45	7.15	14.60
1950			105	none		8.15	6.90	15.05
2100	Fluted	4x8x16	125	none		2.93	6	8.93
2150			105	none		3.02	5.80	8.82
2200		6x8x16	125	perlite		3.91	6.65	10.56
2210				styrofoam		4.70	6.35	11.05
2240				none		3.58	6.35	9.93
2300		8x8x16	125	perlite		4.78	7.25	12.03
2310				styrofoam		5.40	6.90	12.30
2340				none		4.30	6.90	11.20
2500	Hex-hollow	4x8x16	125	none		2.12	6	8.12
2550			105	none		2.26	5.80	8.06
2600		6x8x16	125	perlite		3.29	6.95	10.24
2610				styrofoam		4.08	6.65	10.73
2640				none		2.96	6.65	9.61
2650			105	perlite		3.53	6.75	10.28
2660				styrofoam		4.32	6.45	10.77
2690				none		3.20	6.45	9.65
2700		8x8x16	125	perlite		4.42	7.50	11.92
2710				styrofoam		5.05	7.15	12.20
2740				none		3.94	7.15	11.09
2750			105	perlite		4.71	7.25	11.96
2760				styrofoam		5.35	6.90	12.25
2790				none		4.23	6.90	11.13
2800	Hex-solid	4x8x16	125	none		2.87	6.15	9.02
2850			105	none		3.11	6	9.11
3100	Slump block	4x4x16	125	none		3.90	4.92	8.82
3200		6x4x16	125	perlite		5.80	5.35	11.15
3210				styrofoam		6.55	5.05	11.60
3240				none		5.45	5.05	10.50
3300		8x4x16	125	perlite		6.35	5.60	11.95
3310				styrofoam		7	5.25	12.25
3340				none		5.85	5.25	11.10
3400		12x4x16	125	perlite		12.15	6.90	19.05
3410				styrofoam		12.70	6.25	18.95
3440				none		11.35	6.25	17.60
3600		6x6x16	125	perlite		5.20	5.50	10.70
3610				styrofoam		6	5.20	11.20
3640				none		4.88	5.20	10.08
3700		8x6x16	125	perlite		7.05	5.75	12.80
3710				styrofoam		7.70	5.40	13.10
3740				none		6.55	5.40	11.95
3800		12x6x16	125	perlite		11.10	7.45	18.55
3810				styrofoam		11.65	6.80	18.45
3840				none		10.30	6.80	17.10

SHELL B

| B2010 124 | | | | Reinforced Misc. Block Walls | | | | |

TYPE	SIZE (IN.)	WEIGHT (P.C.F.)	VERT. REINF. & GROUT SPACING		COST PER S.F.		
					MAT.	INST.	TOTAL
5100 Deep groove-hollow	4x8x16	125	#4 @ 48"		3.05	6.55	9.60
5150		105	#4 @ 48"		3.19	6.35	9.54
5200	6x8x16	125	#4 @ 48"		3.97	6.95	10.92
5220			#5 @ 32"		4.15	7.15	11.30
5230			#5 @ 16"		4.54	7.90	12.44
5250		105	#4 @ 48"		4.13	6.40	10.53
5270			#5 @ 32"		4.31	6.60	10.91
5280			#5 @ 16"		4.70	7.35	12.05
5300	8x8x16	125	#4 @ 48"		5.05	7.55	12.60
5320			#5 @ 32"		5.25	7.80	13.05
5330			#5 @ 16"		5.70	8.70	14.40
5350		105	#4 @ 48"		5.55	7.35	12.90
5370			#5 @ 32"		5.75	7.60	13.35
5380			#5 @ 16"		6.20	8.50	14.70
5500 Deep groove-75% solid	6x8x16	125	#4 @ 48"		4.90	6.85	11.75
5520			#5 @ 32"		5.05	7	12.05
5530			#5 @ 16"		5.35	7.55	12.90
5550		105	#4 @ 48"		5.15	6.65	11.80
5570			#5 @ 32"		5.30	6.90	12.20
5580			#5 @ 16"		5.60	7.35	12.95
5600	8x8x16	125	#4 @ 48"		6.30	7.40	13.70
5620			#5 @ 32"		6.45	7.65	14.10
5630			#5 @ 16"		6.75	8.20	14.95
5650		105	#4 @ 48"		6.95	7.20	14.15
5670			#5 @ 32"		7.10	7.45	14.55
5680			#5 @ 16"		7.40	8	15.40
5700 Deep groove-solid	2-4x8x16	125	#4 @ 48" E.W.		9.70	13.55	23.25
5730 Double wythe			#5 @ 16" E.W.		10.35	14.20	24.55
5750		105	#4 @ 48" E.W.		10.15	13.25	23.40
5780			#5 @ 16" E.W.		10.80	13.90	24.70
5800	2-6x8x16	125	#4 @ 48" E.W.		12.15	14.35	26.50
5830			#5 @ 16" E.W.		12.80	15	27.80
5850		105	#4 @ 48" E.W.		12.75	13.95	26.70
5880			#5 @ 16" E.W.		13.40	14.60	28
6100 Fluted	4x8x16	125	#4 @ 48"		3.06	6.55	9.61
6150		105	#4 @ 48"		3.15	6.35	9.50
6200	6x8x16	125	#4 @ 48"		3.80	6.95	10.75
6220			#5 @ 32"		3.98	7.15	11.13
6230			#5 @ 16"		4.37	7.90	12.27
6300	8x8x16	125	#4 @ 48"		4.55	7.55	12.10
6320			#5 @ 32"		4.76	7.80	12.56
6330			#5 @ 16"		5.20	8.70	13.90
6500 Hex-hollow	4x8x16	125	#4 @ 48"		2.25	6.55	8.80
6550		105	#4 @ 48"		2.39	6.35	8.74
6620			#5 @ 32"		3.36	7.45	10.81
6630			#5 @ 16"		3.75	8.20	11.95
6650	6x8x16	105	#4 @ 48"		3.42	7.05	10.47
6670			#5 @ 32"		3.60	7.25	10.85
6680			#5 @ 16"		3.99	8	11.99
6700	8x8x16	125	#4 @ 48"		4.19	7.80	11.99
6720			#5 @ 32"		4.40	8.05	12.45
6730			#5 @ 16"		4.85	8.95	13.80

Important: See the Reference Section for critical supporting data - Reference Numbers and City Cost Indexes

B20 Exterior Enclosure

B2010 Exterior Walls

| B2010 124 | Reinforced Misc. Block Walls | | | | | | |

	TYPE	SIZE (IN.)	WEIGHT (P.C.F.)	VERT. REINF. & GROUT SPACING		MAT.	INST.	TOTAL
6750	Hex-hollow		105	#4 @ 48"		4.48	7.55	12.03
6770				#5 @ 32"		4.69	7.80	12.49
6800	Hex-solid	2-4x8x16	125	#4 @ 48" E.W.		6.35	13.45	19.80
6830	Double wythes			#5 @ 16" E.W.		7.05	14.20	21.25
6850			105	#4 @ 48" E.W.		6.85	13.15	20
6880				#5 @ 16" E.W.		7.55	13.90	21.45
7100	Slump block	4x4x16	125	#4 @ 48"		4.03	5.45	9.48
7200		6x4x16	125	#4 @ 48"		5.65	5.65	11.30
7220				#5 @ 32"		5.85	6.10	11.95
7230				#5 @ 16"		6.25	6.60	12.85
7300	Slump block	8x4x16	125	#4 @ 48"		6.10	5.90	12
7320				#5 @ 32"		6.30	6.15	12.45
7330				#5 @ 16"		6.75	7.05	13.80
7400		12x4x16	125	#4 @ 48"		11.75	6.90	18.65
7420				#5 @ 32"		12	7.20	19.20
7430				#5 @ 16"		12.60	8.20	20.80
7600		6x6x16	125	#4 @ 48"		5.10	5.80	10.90
7620				#5 @ 32"		5.30	6	11.30
7630				#5 @ 16"		5.65	6.75	12.40
7700		8x6x16	125	#4 @ 48"		6.80	6.05	12.85
7720				#5 @ 32"		7	6.30	13.30
7730				#5 @ 16"		7.45	7.20	14.65
7800		12x6x16	125	#4 @ 48"		10.70	7.45	18.15
7820				#5 @ 32"		10.95	7.75	18.70
7830				#5 @ 16"		11.55	8.75	20.30

SHELL B

159

B2010 Exterior Walls

Exterior solid brick walls are defined in the following terms; structural reinforcement, type of brick, thickness, and bond. Sixteen different types of face bricks are presented, with single wythes shown in four different bonds. Six types of reinforced single wythe walls are also given, and twice that many for reinforced double wythe. Shelf angles are included in system components. These walls do not include ties and as such are not tied to a backup wall.

System Components	QUANTITY	UNIT	COST PER S.F.		
			MAT.	INST.	TOTAL
SYSTEM B2010 125 1000					
BRICK WALL, COMMON, SINGLE WYTHE, 4"THICK, RUNNING BOND					
Common brick wall, running bond	1.000	S.F.	2.84	8.75	11.59
Wash brick	1.000	S.F.	.03	.80	.83
Control joint, backer rod	.100	L.F.		.10	.10
Control joint, sealant	.100	L.F.	.02	.24	.26
Shelf angle	1.000	Lb.	.80	.81	1.61
Flashing	.100	S.F.	.08	.29	.37
TOTAL			3.77	10.99	14.76

B2010 125			Solid Brick Walls - Single Wythe					
	TYPE	THICKNESS (IN.)	BOND			COST PER S.F.		
						MAT.	INST.	TOTAL
1000	Common	4	running			3.77	11	14.77
1010			common			4.54	12.55	17.09
1050			Flemish			5.05	15.25	20.30
1100			English			5.65	16.10	21.75
1150	Standard	4	running			4.49	11.40	15.89
1160			common			5.30	13.10	18.40
1200			Flemish			5.90	15.65	21.55
1250			English			6.65	16.60	23.25
1300	Glazed	4	running			9.50	11.85	21.35
1310			common			11.35	13.75	25.10
1350			Flemish			12.60	16.60	29.20
1400			English			14.15	17.70	31.85
1450	Engineer	4	running			4.53	10	14.53
1460			common			5.35	11.40	16.75
1500			Flemish			5.90	13.75	19.65
1550			English			6.65	14.45	21.10
1600	Economy	4	running			4.39	8.75	13.14
1610			common			5.15	10	15.15
1650			Flemish			5.70	11.85	17.55
1700			English			13.85	12.55	26.40
1750	Double	4	running			8.75	7.15	15.90
1760			common			10.40	8.10	18.50
1800			Flemish			11.50	9.55	21.05
1850			English			12.95	10.15	23.10

Important: See the Reference Section for critical supporting data - Reference Numbers and City Cost Indexes

B2010 Exterior Walls

B2010 125			Solid Brick Walls - Single Wythe		COST PER S.F.		
	TYPE	THICKNESS (IN.)	BOND		MAT.	INST.	TOTAL
1900	Fire	4-1/2	running		8.70	10	18.70
1910			common		10.35	11.40	21.75
1950			Flemish		11.45	13.75	25.20
2000			English		12.90	14.45	27.35
2050	King	3-1/2	running		3.82	8.95	12.77
2060			common		4.49	10.30	14.79
2100			Flemish		5.45	12.30	17.75
2150			English		6.05	12.85	18.90
2200	Roman	4	running		6.30	10.30	16.60
2210			common		7.50	11.85	19.35
2250			Flemish		8.30	14.10	22.40
2300			English		9.40	15.25	24.65
2350	Norman	4	running		5.25	8.55	13.80
2360			common		6.25	9.70	15.95
2400			Flemish		14.15	11.60	25.75
2450			English		7.75	12.30	20.05
2500	Norwegian	4	running		4.42	7.60	12.02
2510			common		5.20	8.65	13.85
2550			Flemish		5.75	10.30	16.05
2600			English		6.50	10.80	17.30
2650	Utility	4	running		4.11	6.70	10.81
2660			common		4.83	7.55	12.38
2700			Flemish		5.30	8.95	14.25
2750			English		5.95	9.45	15.40
2800	Triple	4	running		3.23	6.05	9.28
2810			common		3.73	6.80	10.53
2850			Flemish		4.07	8	12.07
2900			English		4.51	8.35	12.86
2950	SCR	6	running		6.05	8.75	14.80
2960			common		7.25	10	17.25
3000			Flemish		8.10	11.85	19.95
3050			English		9.15	12.55	21.70
3100	Norwegian	6	running		5.20	7.75	12.95
3110			common		6.15	8.85	15
3150			Flemish		6.80	10.45	17.25
3200			English		7.65	11	18.65
3250	Jumbo	6	running		5.65	6.85	12.50
3260			common		6.75	7.75	14.50
3300			Flemish		7.50	9.20	16.70
3350			English		8.45	9.70	18.15

B2010 126			Solid Brick Walls - Double Wythe		COST PER S.F.		
	TYPE	THICKNESS (IN.)	COLLAR JOINT THICKNESS (IN.)		MAT.	INST.	TOTAL
4100	Common	8	3/4		7.25	18.40	25.65
4150	Standard	8	1/2		8.55	19.05	27.60
4200	Glazed	8	1/2		18.60	19.70	38.30
4250	Engineer	8	1/2		8.65	16.80	25.45
4300	Economy	8	3/4		8.35	14.45	22.80
4350	Double	8	3/4		10.65	11.70	22.35
4400	Fire	8	3/4		16.95	16.80	33.75
4450	King	7	3/4		7.25	14.80	22.05

SHELL B

B2010 Exterior Walls

B2010 126	Solid Brick Walls - Double Wythe					

	TYPE	THICKNESS (IN.)	COLLAR JOINT THICKNESS (IN.)		COST PER S.F.		
					MAT.	INST.	TOTAL
4500	Roman	8	1		12.15	17.30	29.45
4550	Norman	8	3/4		10.10	14.15	24.25
4600	Norwegian	8	3/4		8.45	12.55	21
4650	Utility	8	3/4		7.80	10.85	18.65
4700	Triple	8	3/4		6.05	9.75	15.80
4750	SCR	12	3/4		11.70	14.50	26.20
4800	Norwegian	12	3/4		10	12.80	22.80
4850	Jumbo	12	3/4		10.95	11.20	22.15

B2010 127	Solid Brick Walls, Reinforced					

	TYPE	THICKNESS (IN.)	WYTHE	REINF. & SPACING	COST PER S.F.		
					MAT.	INST.	TOTAL
7200	Common	4"	1	#4 @ 48"vert	3.94	11.40	15.34
7220				#5 @ 32"vert	4.07	11.55	15.62
7230				#5 @ 16"vert	4.30	12.10	16.40
7300	Utility	4"	1	#4 @ 48"vert	4.28	7.15	11.43
7320				#5 @ 32"vert	4.41	7.30	11.71
7330				#5 @ 16"vert	4.64	7.80	12.44
7400	SCR	6"	1	#4 @ 48"vert	6.25	9.15	15.40
7420				#5 @ 32"vert	6.40	9.30	15.70
7430				#5 @ 16"vert	6.70	9.85	16.55
7500	Jumbo	6"	1	#4 @ 48"vert	5.85	7.25	13.10
7520				#5 @ 32"vert	5.95	7.40	13.35
7530				#5 @ 16"vert	6.20	7.85	14.05
7600	Jumbo	8"	1	#4 @ 48"vert	6.95	7.45	14.40
7620				#5 @ 32"vert	7.10	7.65	14.75
7630				#5 @ 16"vert	7.45	8.25	15.70
7700	King	9"	2	#4 @ 48" E.W.	7.65	15.50	23.15
7720				#5 @ 32" E.W.	7.85	15.65	23.50
7730				#5 @ 16" E.W.	8.25	16.10	24.35
7800	Common	10"	2	#4 @ 48" E.W.	7.65	19.10	26.75
7820				#5 @ 32" E.W.	7.85	19.25	27.10
7830				#5 @ 16" E.W.	8.25	19.70	27.95
7900	Standard	10"	2	#4 @ 48" E.W.	8.95	19.75	28.70
7920				#5 @ 32" E.W.	9.15	19.90	29.05
7930				#5 @ 16" E.W.	9.55	20.50	30.05
8000	Engineer	10"	2	#4 @ 48" E.W.	9.05	17.50	26.55
8020				#5 @ 32" E.W.	9.25	17.65	26.90
8030				#5 @ 16" E.W.	9.65	18.10	27.75
8100	Economy	10"	2	#4 @ 48" E.W.	8.75	15.15	23.90
8120				#5 @ 32" E.W.	8.95	15.30	24.25
8130				#5 @ 16" E.W.	9.35	15.75	25.10
8200	Double	10"	2	#4 @ 48" E.W.	11.05	12.40	23.45
8220				#5 @ 32" E.W.	11.25	12.55	23.80
8230				#5 @ 16" E.W.	11.65	13	24.65
8300	Fire	10"	2	#4 @ 48" E.W.	17.35	17.50	34.85
8320				#5 @ 32" E.W.	17.55	17.65	35.20
8330				#5 @ 16" E.W.	17.95	18.10	36.05
8400	Roman	10"	2	#4 @ 48" E.W.	12.55	18	30.55
8420				#5 @ 32" E.W.	12.75	18.15	30.90
8430				#5 @ 16" E.W.	13.15	18.60	31.75

B SHELL

Important: See the Reference Section for critical supporting data - Reference Numbers and City Cost Indexes

B2010 Exterior Walls

B2010 127	Solid Brick Walls, Reinforced							

	TYPE	THICKNESS (IN.)	WYTHE	REINF. & SPACING		COST PER S.F.		
						MAT.	INST.	TOTAL
8500	Norman	10″	2	#4 @ 48″ E.W.		10.50	14.85	25.35
8520				#5 @ 32″ E.W.		10.70	15	25.70
8530				#5 @ 16″ E.W.		11.10	15.45	26.55
8600	Norwegian	10″	2	#4 @ 48″ E.W.		8.85	13.25	22.10
8620				#5 @ 32″ E.W.		9.05	13.40	22.45
8630				#5 @ 16″ E.W.		9.45	13.85	23.30
8700	Utility	10″	2	#4 @ 48″ E.W.		8.20	11.55	19.75
8720				#5 @ 32″ E.W.		8.40	11.70	20.10
8730				#5 @ 16″ E.W.		8.80	12.15	20.95
8800	Triple	10″	2	#4 @ 48″ E.W.		6.45	10.45	16.90
8820				#5 @ 32″ E.W.		6.70	10.60	17.30
8830				#5 @ 16″ E.W.		7.05	11.05	18.10
8900	SCR	14″	2	#4 @ 48″ E.W.		12.10	15.15	27.25
8920				#5 @ 32″ E.W.		12.30	15.30	27.60
8930				#5 @ 16″ E.W.		12.70	15.75	28.45
9000	Jumbo	14″	2	#4 @ 48″ E.W.		11.35	11.85	23.20
9030				#5 @ 16″ E.W.		11.95	12.45	24.40

B

SHELL

The table below lists costs per S.F. for stone veneer walls on various backup using different stone. Typical components for a system are shown in the component block below.

System Components	QUANTITY	UNIT	COST PER S.F.		
			MAT.	INST.	TOTAL
SYSTEM B2010 128 2000					
ASHLAR STONE VENEER 4″, 2″X4″ STUD 16″ O.C. BACK UP, 8′ HIGH					
Ashlar veneer, 4″ thick, sawn face, split joints, low priced stone	1.000	S.F.	11.85	12.10	23.95
Framing, 2x4 studs 8′ high 16″ O.C.	.920	B.F.	.46	1.11	1.57
Wall ties for stone veneer, galv. corrg 7/8″ x 7″, 22 gauge	.700	Ea.	.06	.30	.36
Asphalt felt sheathing paper, 15 lb.	1.000	S.F.	.04	.12	.16
Sheathing plywood on wall CDX 1/2″	1.000	S.F.	.68	.79	1.47
Fiberglass insulation, 3-1/2″, cr-11	1.000	S.F.	.52	.28	.80
Flashing, copper, paperback 1 side, 3 oz	.125	S.F.	.16	.16	.32
TOTAL			13.77	14.86	28.63

B2010 128	Stone Veneer	COST PER S.F.		
		MAT.	INST.	TOTAL
2000	Ashlar veneer,4″, 2″x4″ stud backup,16″ O.C., 8′ high, low priced stone	13.75	14.85	28.60
2100	Metal stud backup, 8′ high, 16″ O.C.	14.40	15.10	29.50
2150	24″ O.C.	14.10	14.70	28.80
2200	Conc. block backup, 4″ thick	14.65	18.10	32.75
2300	6″ thick	15.25	18.30	33.55
2350	8″ thick	15.45	18.80	34.25
2400	10″ thick	15.95	19.85	35.80
2500	12″ thick	16.35	21.50	37.85
3100	High priced stone, wood stud backup, 10′ high, 16″ O.C.	22.50	16.75	39.25
3200	Metal stud backup, 10′ high, 16″ O.C.	23	17	40
3250	24″ O.C.	23	16.60	39.60
3300	Conc. block backup, 10′ high, 4″ thick	23.50	20	43.50
3350	6″ thick	24	20	44
3400	8″ thick	24	20.50	44.50
3450	10″ thick	24.50	21.50	46
3500	12″ thick	25	23	48
4000	Indiana limestone 2″ thick,sawn finish,wood stud backup,10′ high,16″ O.C.	29.50	12.35	41.85
4100	Metal stud backup, 10′ high, 16″ O.C.	30	12.60	42.60
4150	24″ O.C.	30	12.20	42.20
4200	Conc. block backup, 4″ thick	30.50	15.60	46.10
4250	6″ thick	31	15.80	46.80
4300	8″ thick	31	16.30	47.30

B2010 Exterior Walls

B2010 128	Stone Veneer	COST PER S.F.		
		MAT.	INST.	TOTAL
4350	10" thick	31.50	17.35	48.85
4400	12" thick	32	18.85	50.85
4450	2" thick, smooth finish, wood stud backup, 8' high, 16" O.C.	29.50	12.35	41.85
4550	Metal stud backup, 8' high, 16" O.C.	30	12.40	42.40
4600	24" O.C.	30	12.05	42.05
4650	Conc. block backup, 4" thick	30.50	15.45	45.95
4700	6" thick	31	15.65	46.65
4750	8" thick	31	16.15	47.15
4800	10" thick	31.50	17.35	48.85
4850	12" thick	32	18.85	50.85
5350	4" thick, smooth finish, wood stud backup, 8' high, 16" O.C.	45.50	12.35	57.85
5450	Metal stud backup, 8' high, 16" O.C.	46	12.60	58.60
5500	24" O.C.	46	12.20	58.20
5550	Conc. block backup, 4" thick	46.50	15.60	62.10
5600	6" thick	47	15.80	62.80
5650	8" thick	47	16.30	63.30
5700	10" thick	47.50	17.35	64.85
5750	12" thick	48	18.85	66.85
6000	Granite, grey or pink, 2" thick, wood stud backup, 8' high, 16" O.C.	22.50	21	43.50
6100	Metal studs, 8' high, 16" O.C.	23	21.50	44.50
6150	24" O.C.	23	21	44
6200	Conc. block backup, 4" thick	23.50	24.50	48
6250	6" thick	24	24.50	48.50
6300	8" thick	24	25	49
6350	10" thick	24.50	26	50.50
6400	12" thick	25	27.50	52.50
6900	4" thick, wood stud backup, 8' high, 16" O.C.	36	24	60
7000	Metal studs, 8' high, 16" O.C.	36.50	24	60.50
7050	24" O.C.	36.50	23.50	60
7100	Conc. block backup, 4" thick	37	27	64
7150	6" thick	37.50	27	64.50
7200	8" thick	37.50	27.50	65
7250	10" thick	38	28.50	66.50
7300	12" thick	38.50	30	68.50

SHELL B

Exterior brick veneer/stud backup walls are defined in the following terms: type of brick and studs, stud spacing and bond. All systems include a back-up wall, a control joint every 20', a brick shelf every 12' of height, ties to the backup and the necessary dampproofing, flashing and insulation.

System Components	QUANTITY	UNIT	COST PER S.F.		
			MAT.	INST.	TOTAL
SYSTEM B2010 129 1100					
STANDARD BRICK VENEER, 2"X4" STUD BACKUP @ 16"O.C., RUNNING BOND					
Standard brick wall, 4" thick, running bond	1.000	S.F.	3.56	9.15	12.71
Wash smooth brick	1.000	S.F.	.03	.80	.83
Joint backer rod	.100	L.F.		.10	.10
Sealant	.100	L.F.	.02	.24	.26
Wall ties, corrugated, 7/8" x 7", 22 gauge	.003	Ea.	.03	.13	.16
Shelf angle	1.000	Lb.	.80	.81	1.61
Wood stud partition, backup, 2" x 4" @ 16" O.C.	1.000	S.F.	.43	.89	1.32
Sheathing, plywood, CDX, 1/2"	1.000	S.F.	.68	.63	1.31
Building paper, asphalt felt, 15 lb.	1.000	S.F.	.04	.12	.16
Fiberglass insulation, batts, 3-1/2" thick paper backing	1.000	S.F.	.52	.28	.80
Flashing, copper, paperbacked	.100	S.F.	.08	.29	.37
TOTAL			6.19	13.44	19.63

B2010 129		Brick Veneer/Wood Stud Backup				COST PER S.F.		
	FACE BRICK	STUD BACKUP	STUD SPACING (IN.)	BOND		MAT.	INST.	TOTAL
1100	Standard	2x4-wood	16	running		6.20	13.45	19.65
1120				common		7	15.15	22.15
1140				Flemish		7.60	17.70	25.30
1160				English		8.35	18.65	27
1400		2x6-wood	16	running		6.45	13.55	20
1420				common		7.25	15.25	22.50
1440				Flemish		7.85	17.80	25.65
1460				English		8.60	18.75	27.35
1500			24	running		6.30	13.30	19.60
1520				common		7.10	15	22.10
1540				Flemish		7.70	17.55	25.25
1560				English		8.40	18.50	26.90
1700	Glazed	2x4-wood	16	running		11.20	13.90	25.10
1720				common		13.05	15.80	28.85
1740				Flemish		14.30	18.65	32.95
1760				English		15.85	19.75	35.60

Important: See the Reference Section for critical supporting data - Reference Numbers and City Cost Indexes

B2010 Exterior Walls

| B2010 129 | | | Brick Veneer/Wood Stud Backup | | | | |

	FACE BRICK	STUD BACKUP	STUD SPACING (IN.)	BOND		COST PER S.F.		
						MAT.	INST.	TOTAL
2000	Glazed	2x6-wood	16	running		11.45	14	25.45
2020				common		13.30	15.90	29.20
2040				Flemish		14.55	18.75	33.30
2060				English		16.10	19.85	35.95
2100			24	running		11.25	13.75	25
2120				common		13.10	15.65	28.75
2140				Flemish		14.35	18.50	32.85
2160				English		15.90	19.60	35.50
2300	Engineer	2x4-wood	16	running		6.25	12.05	18.30
2320				common		7.05	13.45	20.50
2340				Flemish		7.60	15.80	23.40
2360				English		8.35	16.50	24.85
2600		2x6-wood	16	running		6.50	12.15	18.65
2620				common		7.30	13.55	20.85
2640				Flemish		7.85	15.90	23.75
2660				English		8.60	16.60	25.20
2700			24	running		6.30	11.90	18.20
2720				common		7.15	13.30	20.45
2740				Flemish		7.70	15.65	23.35
2760				English		8.40	16.35	24.75
2900	Roman	2x4-wood	16	running		8	12.35	20.35
2920				common		9.20	13.90	23.10
2940				Flemish		10	16.15	26.15
2960				English		11.10	17.30	28.40
3200		2x6-wood	16	running		8.25	12.45	20.70
3220				common		9.45	14	23.45
3240				Flemish		10.25	16.15	26.40
3260				English		11.35	17.40	28.75
3300			24	running		8.05	12.20	20.25
3320				common		9.25	13.75	23
3340				Flemish		10.05	16	26.05
3360				English		11.15	17.15	28.30
3500	Norman	2x4-wood	16	running		6.95	10.60	17.55
3520				common		7.95	11.75	19.70
3540				Flemish		15.85	13.65	29.50
3560				English		9.45	14.35	23.80
3800		2x6-wood	16	running		7.20	10.70	17.90
3820				common		8.20	11.85	20.05
3840				Flemish		15.45	13.15	28.60
3860				English		9.70	14.45	24.15
3900			24	running		7.05	10.45	17.50
3920				common		8	11.60	19.60
3940				Flemish		15.90	13.50	29.40
3960				English		9.50	14.20	23.70
4100	Norwegian	2x4-wood	16	running		6.10	9.65	15.75
4120				common		6.90	10.70	17.60
4140				Flemish		7.45	12.35	19.80
4160				English		8.20	12.85	21.05
4400		2x6-wood	16	running		6.35	9.75	16.10
4420				common		7.15	10.80	17.95
4440				Flemish		7.70	12.45	20.15
4460				English		8.45	12.95	21.40

B2010 Exterior Walls

B2010 129	Brick Veneer/Wood Stud Backup						

	FACE BRICK	STUD BACKUP	STUD SPACING (IN.)	BOND		COST PER S.F.	
						MAT.	INST.	TOTAL

	FACE BRICK	STUD BACKUP	STUD SPACING (IN.)	BOND		MAT.	INST.	TOTAL
4500	Norwegian	2x6-wood	24	running		6.20	9.50	15.70
4520				common		7	10.55	17.55
4540				Flemish		7.55	12.20	19.75
4560				English		8.25	12.70	20.95

B2010 130	Brick Veneer/Metal Stud Backup						

	FACE BRICK	STUD BACKUP	STUD SPACING (IN.)	BOND		COST PER S.F.		
						MAT.	INST.	TOTAL
5100	Standard	25ga.x6"NLB	24	running		5.75	13.15	18.90
5120				common		6.55	14.85	21.40
5140				Flemish		6.80	16.85	23.65
5160				English		7.90	18.35	26.25
5200		20ga.x3-5/8"NLB	16	running		5.90	13.70	19.60
5220				common		6.75	15.40	22.15
5240				Flemish		7.35	17.95	25.30
5260				English		8.05	18.90	26.95
5300			24	running		5.75	13.30	19.05
5320				common		6.60	15	21.60
5340				Flemish		7.20	17.55	24.75
5360				English		7.90	18.50	26.40
5400		16ga.x3-5/8"LB	16	running		6.45	13.90	20.35
5420				common		7.30	15.60	22.90
5440				Flemish		7.85	18.15	26
5460				English		8.60	19.10	27.70
5500			24	running		6.15	13.50	19.65
5520				common		7	15.20	22.20
5540				Flemish		7.55	17.75	25.30
5560				English		8.30	18.70	27
5700	Glazed	25ga.x6"NLB	24	running		10.75	13.60	24.35
5720				common		12.60	15.50	28.10
5740				Flemish		13.85	18.35	32.20
5760				English		15.40	19.45	34.85
5800		20ga.x3-5/8"NLB	24	running		10.75	13.75	24.50
5820				common		12.60	15.65	28.25
5840				Flemish		13.85	18.50	32.35
5860				English		15.40	19.60	35
6000		16ga.x3-5/8"LB	16	running		11.45	14.35	25.80
6020				common		13.30	16.25	29.55
6040				Flemish		14.55	19.10	33.65
6060				English		16.10	20	36.10
6100			24	running		11.15	13.95	25.10
6120				common		13	15.85	28.85
6140				Flemish		14.25	18.70	32.95
6160				English		15.80	19.80	35.60
6300	Engineer	25ga.x6"NLB	24	running		5.80	11.75	17.55
6320				common		6.60	13.15	19.75
6340				Flemish		7.15	15.50	22.65
6360				English		7.90	16.20	24.10
6400		20ga.x3-5/8"NLB	16	running		5.95	12.30	18.25
6420				common		6.80	13.70	20.50
6440				Flemish		7.35	16.05	23.40
6460				English		8.05	16.75	24.80

Important: See the Reference Section for critical supporting data - Reference Numbers and City Cost Indexes

B2010 Exterior Walls

| B2010 130 | | | | Brick Veneer/Metal Stud Backup | | | | |

	FACE BRICK	STUD BACKUP	STUD SPACING (IN.)	BOND		COST PER S.F.		
						MAT.	INST.	TOTAL
6500	Engineer	20ga.x3-5/8"NLB	24	running		5.45	11.20	16.65
6520				common		6.65	13.30	19.95
6540				Flemish		7.20	15.65	22.85
6560				English		7.90	16.35	24.25
6600		16ga.x3-5/8"LB	16	running		6.50	12.50	19
6620				common		7.30	13.90	21.20
6640				Flemish		7.90	16.25	24.15
6660				English		8.60	16.95	25.55
6700			24	running		6.20	12.10	18.30
6720				common		7	13.50	20.50
6740				Flemish		7.60	15.85	23.45
6760				English		8.30	16.55	24.85
6900	Roman	25ga.x6"NLB	24	running		7.55	12.05	19.60
6920				common		8.75	13.60	22.35
6940				Flemish		9.55	15.85	25.40
6960				English		10.65	17	27.65
7000		20ga.x3-5/8"NLB	16	running		7.70	12.60	20.30
7020				common		8.90	14.15	23.05
7040				Flemish		9.70	16.40	26.10
7060				English		10.80	17.55	28.35
7100			24	running		7.55	12.20	19.75
7120				common		8.75	13.75	22.50
7140				Flemish		9.55	16	25.55
7160				English		10.65	17.15	27.80
7200		16ga.x3-5/8"LB	16	running		8.25	12.80	21.05
7220				common		9.45	14.35	23.80
7240				Flemish		10.25	16.60	26.85
7260				English		11.35	17.75	29.10
7300			24	running		7.95	12.40	20.35
7320				common		9.15	13.95	23.10
7340				Flemish		9.95	16.20	26.15
7360				English		11.05	17.35	28.40
7500	Norman	25ga.x6"NLB	24	running		6.50	10.30	16.80
7520				common		7.50	11.45	18.95
7540				Flemish		15.40	13.35	28.75
7560				English		9	14.05	23.05
7600		20ga.x3-5/8"NLB	24	running		6.55	10.45	17
7620				common		7.50	11.60	19.10
7640				Flemish		15.40	13.50	28.90
7660				English		9	14.20	23.20
7800		16ga.x3-5/8"LB	16	running		7.20	11.05	18.25
7820				common		8.20	12.20	20.40
7840				Flemish		16.10	14.10	30.20
7860				English		9.70	14.80	24.50
7900			24	running		6.90	10.65	17.55
7920				common		7.90	11.80	19.70
7940				Flemish		15.80	13.70	29.50
7960				English		9.40	14.40	23.80
8100	Norwegian	25ga.x6"NLB	24	running		5.65	9.35	15
8120				common		6.45	10.40	16.85
8140				Flemish		7	12.05	19.05
8160				English		7.75	12.55	20.30

B2010 Exterior Walls

B2010 130			Brick Veneer/Metal Stud Backup					
	FACE BRICK	STUD BACKUP	STUD SPACING (IN.)	BOND		COST PER S.F.		
						MAT.	INST.	TOTAL
8200	Norwegian	20ga.x3-5/8"NLB	16	running		5.85	9.90	15.75
8220				common		6.65	10.95	17.60
8240				Flemish		7.20	12.60	19.80
8260				English		7.90	13.10	21
8300			24	running		5.70	9.50	15.20
8320				common		6.50	10.55	17.05
8340				Flemish		7.05	12.20	19.25
8360				English		7.75	12.70	20.45
8400		16ga.x3-5/8"LB	16	running		6.40	10.10	16.50
8420				common		7.20	11.15	18.35
8440				Flemish		7.70	12.80	20.50
8460				English		8.45	13.30	21.75
8500			24	running		6.10	9.70	15.80
8520				common		6.90	10.75	17.65
8540				Flemish		7.40	12.40	19.80
8560				English		8.15	12.90	21.05

Important: See the Reference Section for critical supporting data - Reference Numbers and City Cost Indexes

B2010 Exterior Walls

Exterior brick face composite walls are defined in the following terms: type of face brick and backup masonry, thickness of backup masonry and insulation. A special section is included on triple wythe construction at the back. Seven types of face brick are shown with various thicknesses of seven types of backup. All systems include a brick shelf, ties to the backup and necessary dampproofing, flashing, and control joints every 20'.

SHELL **B**

System Components	QUANTITY	UNIT	COST PER S.F.		
			MAT.	INST.	TOTAL
SYSTEM B2010 132 1120					
COMPOSITE WALL, STANDARD BRICK FACE, 6" C.M.U. BACKUP, PERLITE FILL					
Face brick veneer, standard, running bond	1.000	S.F.	3.56	9.15	12.71
Wash brick	1.000	S.F.	.03	.80	.83
Concrete block backup, 6" thick	1.000	S.F.	1.86	5.05	6.91
Wall ties	.300	Ea.	.06	.13	.19
Perlite insulation, poured	1.000	S.F.	.33	.26	.59
Flashing, aluminum	.100	S.F.	.08	.29	.37
Shelf angle	1.000	Lb.	.80	.81	1.61
Control joint	.050	L.F.	.06	.06	.12
Backer rod	.100	L.F.		.10	.10
Sealant	.100	L.F.	.02	.24	.26
Collar joint	1.000	S.F.	.34	.51	.85
TOTAL			7.14	17.40	24.54

B2010 132	Brick Face Composite Wall - Double Wythe						
	FACE BRICK	BACKUP MASONRY	BACKUP THICKNESS (IN.)	BACKUP CORE FILL		COST PER S.F.	
						MAT.	INST.	TOTAL

	FACE BRICK	BACKUP MASONRY	BACKUP THICKNESS (IN.)	BACKUP CORE FILL	MAT.	INST.	TOTAL
1000	Standard	common brick	4	none	7.80	21	28.80
1040		SCR brick	6	none	10.05	18.60	28.65
1080		conc. block	4	none	6.20	16.75	22.95
1120			6	perlite	7.15	17.40	24.55
1160				styrofoam	7.95	17.15	25.10
1200			8	perlite	7.45	17.75	25.20
1240				styrofoam	8.10	17.45	25.55
1280		L.W. block	4	none	6.40	16.65	23.05
1320			6	perlite	7.30	17.25	24.55
1360				styrofoam	8.10	17	25.10
1400			8	perlite	7.90	17.65	25.55
1440				styrofoam	8.55	17.35	25.90
1520		glazed block	4	none	12.20	17.95	30.15
1560			6	perlite	13.05	18.45	31.50
1600				styrofoam	13.80	18.20	32
1640			8	perlite	13.80	18.90	32.70
1680				styrofoam	14.40	18.60	33

B2010 132   Brick Face Composite Wall - Double Wythe

	FACE BRICK	BACKUP MASONRY	BACKUP THICKNESS (IN.)	BACKUP CORE FILL		MAT.	INST.	TOTAL
1720	Standard	clay tile	4	none		10.55	16.10	26.65
1760			6	none		11.30	16.55	27.85
1800			8	none		13.15	17.15	30.30
1840		glazed tile	4	none		12.70	21	33.70
1880								
2000	Glazed	common brick	4	none		12.80	21	33.80
2040		SCR brick	6	none		15.05	19.05	34.10
2080		conc. block	4	none		11.20	17.20	28.40
2120			6	perlite		12.15	17.85	30
2160				styrofoam		12.90	17.60	30.50
2200			8	perlite		12.45	18.20	30.65
2240				styrofoam		13.10	17.90	31
2280		L.W. block	4	none		11.40	17.10	28.50
2320			6	perlite		12.30	17.70	30
2360				styrofoam		13.10	17.45	30.55
2400			8	perlite		12.90	18.10	31
2440				styrofoam		13.55	17.80	31.35
2520		glazed block	4	none		17.20	18.40	35.60
2560			6	perlite		18	18.90	36.90
2600				styrofoam		18.80	18.65	37.45
2640			8	perlite		18.75	19.35	38.10
2680				styrofoam		19.40	19.05	38.45
2720		clay tile	4	none		15.55	16.55	32.10
2760			6	none		16.30	17	33.30
2800			8	none		18.15	17.60	35.75
2840		glazed tile	4	none		17.70	21.50	39.20
2880								
3000	Engineer	common brick	4	none		7.85	19.45	27.30
3040		SCR brick	6	none		10.10	17.20	27.30
3080		conc. block	4	none		6.25	15.35	21.60
3120			6	perlite		7.20	16	23.20
3160				styrofoam		7.95	15.75	23.70
3200			8	perlite		7.50	16.35	23.85
3240				styrofoam		8.15	16.05	24.20
3280		L.W. block	4	none		5.05	10.90	15.95
3320			6	perlite		7.35	15.85	23.20
3360				styrofoam		8.15	15.60	23.75
3400			8	perlite		7.95	16.25	24.20
3440				styrofoam		8.60	15.95	24.55
3520		glazed block	4	none		12.25	16.55	28.80
3560			6	perlite		13.05	17.05	30.10
3600				styrofoam		13.85	16.80	30.65
3640			8	perlite		13.80	17.50	31.30
3680				styrofoam		14.45	17.20	31.65
3720		clay tile	4	none		10.60	14.70	25.30
3760			6	none		11.35	15.15	26.50
3800			8	none		13.20	15.75	28.95
3840		glazed tile	4	none		12.75	19.85	32.60
4000	Roman	common brick	4	none		9.60	19.75	29.35
4040		SCR brick	6	none		11.85	17.50	29.35

Important: See the Reference Section for critical supporting data - Reference Numbers and City Cost Indexes

B2010   Exterior Walls

B2010 132	Brick Face Composite Wall - Double Wythe

	FACE BRICK	BACKUP MASONRY	BACKUP THICKNESS (IN.)	BACKUP CORE FILL		COST PER S.F.		
						MAT.	INST.	TOTAL
4080	Roman	conc. block	4	none		8	15.65	23.65
4120			6	perlite		8.95	16.30	25.25
4160				styrofoam		9.70	16.05	25.75
4200			8	perlite		9.25	16.65	25.90
4240				styrofoam		9.90	16.35	26.25
4280		L.W. block	4	none		8.20	15.55	23.75
4320			6	perlite		9.10	16.15	25.25
4360				styrofoam		9.90	15.90	25.80
4400			8	perlite		9.70	16.55	26.25
4440				styrofoam		10.35	16.25	26.60
4520		glazed block	4	none		14	16.85	30.85
4560			6	perlite		14.80	17.35	32.15
4600				styrofoam		15.60	17.10	32.70
4640			8	perlite		15.55	17.80	33.35
4680				styrofoam		16.20	17.50	33.70
4720		claytile	4	none		12.35	15	27.35
4760			6	none		13.10	15.45	28.55
4800			8	none		14.95	16.05	31
4840		glazed tile	4	none		14.50	20	34.50
5000	Norman	common brick	4	none		8.55	18	26.55
5040		SCR brick	6	none		10.80	15.75	26.55
5080		conc. block	4	none		6.95	13.90	20.85
5120			6	perlite		7.90	14.55	22.45
5160				styrofoam		8.70	14.30	23
5200			8	perlite		8.25	14.90	23.15
5240				styrofoam		8.85	14.60	23.45
5280		L.W. block	4	none		7.20	13.80	21
5320			6	perlite		8.10	14.55	22.65
5360				styrofoam		8.85	14.15	23
5400			8	perlite		8.70	14.80	23.50
5440				styrofoam		9.30	14.50	23.80
5520		glazed block	4	none		12.95	15.10	28.05
5560			6	perlite		13.80	15.60	29.40
5600				styrofoam		14.60	15.35	29.95
5640			8	perlite		14.55	16.05	30.60
5680				styrofoam		15.20	15.75	30.95
5720		clay tile	4	none		11.30	13.25	24.55
5760			6	none		12.05	13.70	25.75
5800			8	none		13.90	14.30	28.20
5840		glazed tile	4	none		13.45	18.40	31.85
6000	Norwegian	common brick	4	none		7.70	17.05	24.75
6040		SCR brick	6	none		10	14.80	24.80
6080		conc. block	4	none		6.15	12.95	19.10
6120			6	perlite		7.05	13.60	20.65
6160				styrofoam		7.85	13.35	21.20
6200			8	perlite		7.40	13.95	21.35
6240				styrofoam		8.05	13.65	21.70
6280		L.W. block	4	none		6.35	12.85	19.20
6320			6	perlite		7.25	13.45	20.70
6360				styrofoam		8	13.20	21.20
6400			8	perlite		7.85	13.85	21.70
6440				styrofoam		8.50	13.55	22.05

B2010 Exterior Walls

B2010 132	Brick Face Composite Wall - Double Wythe						

	FACE BRICK	BACKUP MASONRY	BACKUP THICKNESS (IN.)	BACKUP CORE FILL		COST PER S.F.		
						MAT.	INST.	TOTAL
6520	Norwegian	glazed block	4	none		12.15	14.15	26.30
6560			6	perlite		12.95	14.65	27.60
6600				styrofoam		13.75	14.40	28.15
6640			8	perlite		13.70	15.10	28.80
6680				styrofoam		14.35	14.80	29.15
6720		clay tile	4	none		10.50	12.30	22.80
6760			6	none		11.25	12.75	24
6800			8	none		13.10	13.35	26.45
6840		glazed tile	4	none		12.65	17.45	30.10
6880								
7000	Utility	common brick	4	none		7.40	16.15	23.55
7040		SCR brick	6	none		9.65	13.90	23.55
7080		conc. block	4	none		5.80	12.10	17.90
7120			6	perlite		6.75	12.70	19.45
7160				styrofoam		7.55	12.45	20
7200			8	perlite		7.10	13.10	20.20
7240				styrofoam		7.75	12.75	20.50
7280		L.W. block	4	none		6.05	12	18.05
7320			6	perlite		6.90	12.60	19.50
7360				styrofoam		7.70	12.30	20
7400			8	perlite		7.55	13	20.55
7440				styrofoam		8.20	12.65	20.85
7520		glazed block	4	none		11.80	13.25	25.05
7560			6	perlite		12.65	13.75	26.40
7600				styrofoam		13.45	13.50	26.95
7640			8	perlite		13.40	14.25	27.65
7680				styrofoam		14.05	13.90	27.95
7720		clay tile	4	none		10.15	11.45	21.60
7760			6	none		10.90	11.90	22.80
7800			8	none		12.75	12.45	25.20
7840		glazed tile	4	none		12.30	16.55	28.85
7880								

B2010 133	Brick Face Composite Wall - Triple Wythe						

	FACE BRICK	MIDDLE WYTHE	INSIDE MASONRY	TOTAL THICKNESS (IN.)		COST PER S.F.		
						MAT.	INST.	TOTAL
8000	Standard	common brick	standard brick	12		11.75	29.50	41.25
8100		4" conc. brick	standard brick	12		11.25	30.50	41.75
8120		4" conc. brick	common brick	12		10.55	30	40.55
8200	Glazed	common brick	standard brick	12		16.75	30	46.75
8300		4" conc. brick	standard brick	12		16.25	31	47.25
8320		4" conc. brick	glazed brick	12		15.55	30.50	46.05
8400	Engineer	common brick	standard brick	12		11.80	28.50	40.30
8500		4" conc. brick	standard brick	12		11.30	29	40.30
8520		4" conc. brick	engineer brick	12		10.60	29	39.60
8600	Roman	common brick	standard brick	12		13.55	28.50	42.05
8700		4" conc. brick	standard brick	12		13.05	29.50	42.55
8720		4" conc. brick	Roman brick	12		12.35	29	41.35
8800	Norman	common, brick	standard brick	12		12.55	27	39.55
8900		4" conc. brick	standard brick	12		12.05	27.50	39.55
8920		4" conc. brick	Norman brick	12		11.30	27.50	38.80

Important: See the Reference Section for critical supporting data - Reference Numbers and City Cost Indexes

B20 Exterior Enclosure

B2010 Exterior Walls

B2010 133			Brick Face Composite Wall - Triple Wythe					

	FACE BRICK	MIDDLE WYTHE	INSIDE MASONRY	TOTAL THICKNESS (IN.)		COST PER S.F.		
						MAT.	INST.	TOTAL
9000	Norwegian	common brick	standard brick	12		11.70	26	37.70
9100		4" conc. brick	standard brick	12		11.20	27	38.20
9120		4" conc. brick	Norwegian brick	12		10.50	26.50	37
9200	Utility	common brick	standard brick	12		11.40	25	36.40
9300		4" conc. brick	standard brick	12		10.80	26	36.80
9320		4" conc. brick	utility brick	12		10.15	25.50	35.65

SHELL B

175

B2010 Exterior Walls

Exterior brick face cavity walls are defined in the following terms: cavity treatment, type of face brick, backup masonry, total thickness and insulation. Seven types of face brick are shown with fourteen types of backup. All systems include a brick shelf, ties to the backups and necessary dampproofing, flashing, and control joints every 20'.

System Components	QUANTITY	UNIT	COST PER S.F.		
			MAT.	INST.	TOTAL
SYSTEM B2010 134 1000					
CAVITY WALL, STANDARD BRICK FACE, COMMON BRICK BACKUP, POLYSTYRENE					
Face brick veneer, standard, running bond	1.000	S.F.	3.56	9.15	12.71
Wash brick	1.000	S.F.	.03	.80	.83
Common brick wall backup, 4" thick	1.000	S.F.	2.84	8.75	11.59
Wall ties	.300	L.F.	.11	.13	.24
Polystyrene insulation board, 1" thick	1.000	S.F.	.24	.55	.79
Flashing, aluminum	.100	S.F.	.08	.29	.37
Shelf angle	1.000	Lb.	.80	.81	1.61
Control joint	.050	L.F.	.06	.06	.12
Backer rod	.100	L.F.		.10	.10
Sealant	.100	L.F.	.02	.24	.26
TOTAL			7.74	20.88	28.62

B2010 134		Brick Face Cavity Wall						
	FACE BRICK	BACKUP MASONRY	TOTAL THICKNESS (IN.)	CAVITY INSULATION		COST PER S.F.		
						MAT.	INST.	TOTAL
1000	Standard	4"common brick	10	polystyrene		7.75	21	28.75
1020				none		7.50	20.50	28
1040		6"SCR brick	12	polystyrene		10.15	18.65	28.80
1060				none		9.90	18.10	28
1080		4"conc. block	10	polystyrene		6.15	16.80	22.95
1100				none		5.90	16.25	22.15
1120		6"conc. block	12	polystyrene		6.90	17.20	24.10
1140				none		6.65	16.65	23.30
1160		4"L.W. block	10	polystyrene		6.35	16.70	23.05
1180				none		6.15	16.15	22.30
1200		6"L.W. block	12	polystyrene		7.05	17.05	24.10
1220				none		6.85	16.50	23.35
1240		4"glazed block	10	polystyrene		12.15	18	30.15
1260				none		11.90	17.45	29.35
1280		6"glazed block	12	polystyrene		12.30	18	30.30
1300				none		12.05	17.45	29.50

Important: See the Reference Section for critical supporting data - Reference Numbers and City Cost Indexes

B2010 Exterior Walls

| B2010 134 | | | | | Brick Face Cavity Wall | | | | |

	FACE BRICK	BACKUP MASONRY	TOTAL THICKNESS (IN.)	CAVITY INSULATION		COST PER S.F.		
						MAT.	INST.	TOTAL
1320	Standard	4"clay tile	10	polystyrene		10.50	16.15	26.65
1340				none		10.25	15.60	25.85
1360		4"glazed tile	10	polystyrene		12.65	21.50	34.15
1380				none		12.40	20.50	32.90
1500	Glazed	4" common brick	10	polystyrene		12.75	21.50	34.25
1520				none		12.50	21	33.50
1540		6" SCR brick	12	polystyrene		15.15	19.10	34.25
1560				none		14.90	18.55	33.45
1580		4" conc. block	10	polystyrene		11.15	17.25	28.40
1600				none		10.90	16.70	27.60
1620		6" conc. block	12	polystyrene		11.90	17.65	29.55
1640				none		11.65	17.10	28.75
1660		4" L.W. block	10	polystyrene		11.35	17.15	28.50
1680				none		11.10	16.60	27.70
1700		6"L.W. block	12	polystyrene		12.05	17.50	29.55
1720				none		11.80	16.95	28.75
1740		4" glazed block	10	polystyrene		17.15	18.45	35.60
1760				none		16.90	17.90	34.80
1780		6" glazed block	12	polystyrene		17.80	18.70	36.50
1800				none		17.55	18.15	35.70
1820		4" clay tile	10	polystyrene		15.50	16.60	32.10
1840				none		15.25	16.05	31.30
1860		4" glazed tile	10	polystyrene		16.25	17.05	33.30
1880				none		16	16.50	32.50
2000	Engineer	4" common brick	10	polystyrene		7.80	19.50	27.30
2020				none		7.55	18.95	26.50
2040		6" SCR brick	12	polystyrene		10.20	17.25	27.45
2060				none		9.95	16.70	26.65
2080		4" conc. block	10	polystyrene		6.20	15.40	21.60
2100				none		5.95	14.85	20.80
2120		6" conc. block	12	polystyrene		6.95	15.80	22.75
2140				none		6.70	15.25	21.95
2160		4" L.W. block	10	polystyrene		6.40	15.30	21.70
2180				none		6.15	14.75	20.90
2200		6"L.W. block	12	polystyrene		7.10	15.65	22.75
2220				none		6.85	15.10	21.95
2240		4" glazed block	10	polystyrene		12.20	16.60	28.80
2260				none		11.95	16.05	28
2280		6" glazed block	12	polystyrene		12.85	16.85	29.70
2300				none		12.60	16.30	28.90
2320		4" clay tile	10	polystyrene		10.55	14.75	25.30
2340				none		10.30	14.20	24.50
2360		4" glazed tile	10	polystyrene		12.70	19.90	32.60
2380				none		12.45	19.35	31.80
2500	Roman	4" common brick	10	polystyrene		9.55	19.80	29.35
2520				none		9.30	19.25	28.55
2540		6" SCR brick	12	polystyrene		11.95	17.55	29.50
2560				none		11.70	17	28.70

B2010 Exterior Walls

B2010 134					Brick Face Cavity Wall				

	FACE BRICK	BACKUP MASONRY	TOTAL THICKNESS (IN.)	CAVITY INSULATION		COST PER S.F.		
						MAT.	INST.	TOTAL
2580	Roman	4" conc. block	10	polystyrene		7.95	15.70	23.65
2600				none		7.70	15.15	22.85
2620		6" conc. block	12	polystyrene		8.70	16.10	24.80
2640				none		8.45	15.55	24
2660		4" L.W. block	10	polystyrene		8.15	15.60	23.75
2680				none		7.90	15.05	22.95
2700		6" L.W. block	12	polystyrene		8.85	15.95	24.80
2720				none		8.60	15.40	24
2740		4" glazed block	10	polystyrene		13.95	16.90	30.85
2760				none		13.70	16.35	30.05
2780		6" glazed block	12	polystyrene		14.60	17.15	31.75
2800				none		14.35	16.60	30.95
2820		4" clay tile	10	polystyrene		12.30	15.05	27.35
2840				none		12.05	14.50	26.55
2860		4" glazed tile	10	polystyrene		14.45	20	34.45
2880				none		14.20	19.65	33.85
3000	Norman	4" common brick	10	polystyrene		8.50	18.05	26.55
3020				none		8.25	17.50	25.75
3040		6" SCR brick	12	polystyrene		10.90	15.80	26.70
3060				none		10.65	15.25	25.90
3080		4" conc. block	10	polystyrene		6.90	13.95	20.85
3100				none		6.65	13.40	20.05
3120		6" conc. block	12	polystyrene		7.65	14.35	22
3140				none		7.45	13.80	21.25
3160		4" L.W. block	10	polystyrene		7.15	13.85	21
3180				none		6.90	13.30	20.20
3200		6" L.W. block	12	polystyrene		7.85	14.20	22.05
3220				none		7.60	13.65	21.25
3240		4" glazed block	10	polystyrene		12.90	15.15	28.05
3260				none		12.65	14.60	27.25
3280		6" glazed block	12	polystyrene		13.55	15.40	28.95
3300				none		13.30	14.85	28.15
3320		4" clay tile	10	polystyrene		11.25	13.30	24.55
3340				none		11	12.75	23.75
3360		4" glazed tile	10	polystyrene		13.40	18.45	31.85
3380				none		13.15	17.90	31.05
3500	Norwegian	4" common brick	10	polystyrene		7.65	17.10	24.75
3520				none		7.45	16.55	24
3540		6" SCR brick	12	polystyrene		10.10	14.85	24.95
3560				none		9.85	14.30	24.15
3580		4" conc. block	10	polystyrene		6.10	13	19.10
3600				none		5.85	12.45	18.30
3620		6" conc. block	12	polystyrene		6.85	13.40	20.25
3640				none		6.60	12.85	19.45

Important: See the Reference Section for critical supporting data - Reference Numbers and City Cost Indexes

B2010 Exterior Walls

B2010 134	Brick Face Cavity Wall							

	FACE BRICK	BACKUP MASONRY	TOTAL THICKNESS (IN.)	CAVITY INSULATION		COST PER S.F.		
						MAT.	INST.	TOTAL
3660	Norwegian	4" L.W. block	10	polystyrene		6.30	12.90	19.20
3680				none		6.05	12.35	18.40
3700		6"L.W. block	12	polystyrene		7	13.25	20.25
3720				none		6.75	12.70	19.45
3740		4" glazed block	10	polystyrene		12.10	14.20	26.30
3760				none		11.85	13.65	25.50
3780		6" glazed block	12	polystyrene		12.75	14.45	27.20
3800				none		12.50	13.90	26.40
3820		4" clay tile	10	polystyrene		10.45	12.35	22.80
3840				none		10.20	11.80	22
3860		4" glazed tile	10	polystyrene		12.60	17.50	30.10
3880				none		12.35	16.95	29.30
4000	Utility	4" common brick	10	polystyrene		7.35	16.20	23.55
4020				none		7.10	15.65	22.75
4040		6" SCR brick	12	polystyrene		9.75	13.95	23.70
4060				none		9.55	13.40	22.95
4080		4" conc. block	10	polystyrene		5.75	12.15	17.90
4100				none		5.55	11.60	17.15
4120		6" conc. block	12	polystyrene		6.55	12.50	19.05
4140				none		6.30	11.95	18.25
4160		4" L.W. block	10	polystyrene		6	12	18
4180				none		5.75	11.45	17.20
4200		6"L.W. block	12	polystyrene		6.70	12.35	19.05
4220				none		6.45	11.80	18.25
4240		4" glazed block	10	polystyrene		11.75	13.30	25.05
4260				none		11.55	12.75	24.30
4280		6" glazed block	12	polystyrene		12.40	13.55	25.95
4300				none		12.20	13	25.20
4320		4" clay tile	10	polystyrene		10.10	11.45	21.55
4340				none		9.90	10.90	20.80
4360		4" glazed tile	10	polystyrene		12.25	16.60	28.85
4380				none		12.05	16.05	28.10

B2010 135	Brick Face Cavity Wall - Insulated Backup							

	FACE BRICK	BACKUP MASONRY	TOTAL THICKNESS (IN.)	BACKUP CORE FILL		COST PER S.F.		
						MAT.	INST.	TOTAL
5100	Standard	6" conc. block	10	perlite		6.85	16.90	23.75
5120				styrofoam		7.65	16.65	24.30
5140		8" conc. block	12	perlite		7.35	17.25	24.60
5160				styrofoam		7.95	16.95	24.90
5180		6" L.W. block	10	perlite		7	16.75	23.75
5200				styrofoam		7.80	16.50	24.30
5220		8"L.W. block	12	perlite		7.80	17.15	24.95
5240				styrofoam		8.40	16.85	25.25
5260		6" glazed block	10	perlite		12.75	17.95	30.70
5280				styrofoam		13.55	17.70	31.25
5300		8" glazed block	12	perlite		13.65	18.40	32.05
5320				styrofoam		14.30	18.10	32.40
5340		6" clay tile	10	none		11	16.05	27.05
5360		8" clay tile	12	none		13	16.65	29.65
5600	Glazed	6" conc. block	10	perlite		11.85	17.35	29.20
5620				styrofoam		12.65	17.10	29.75

B2010 Exterior Walls

B2010 135 — Brick Face Cavity Wall - Insulated Backup

	FACE BRICK	BACKUP MASONRY	TOTAL THICKNESS (IN.)	BACKUP CORE FILL		COST PER S.F.		
						MAT.	INST.	TOTAL
5640	Glazed	8" conc. block	12	perlite		12.30	17.70	30
5660				styrofoam		12.95	17.40	30.35
5680		6" L.W. block	10	perlite		12	17.20	29.20
5700				styrofoam		12.80	16.95	29.75
5720		8" L.W. block	12	perlite		12.75	17.60	30.35
5740				styrofoam		13.40	17.30	30.70
5760		6" glazed block	10	perlite		17.75	18.40	36.15
5780				styrofoam		18.50	18.15	36.65
5800		8" glazed block	12	perlite		18.65	18.85	37.50
5820				styrofoam		19.25	18.55	37.80
5840		6" clay tile	10	none		16	16.50	32.50
5860		8"clay tile	8	none		18	17.10	35.10
6100	Engineer	6" conc. block	10	perlite		6.90	15.50	22.40
6120				styrofoam		7.70	15.25	22.95
6140		8" conc. block	12	perlite		7.35	15.85	23.20
6160				styrofoam		8	15.55	23.55
6180		6" L.W. block	10	perlite		7.05	15.35	22.40
6200				styrofoam		7.85	15.10	22.95
6220		8"L.W. block	12	perlite		7.80	15.75	23.55
6240				styrofoam		8.45	15.45	23.90
6260		6" glazed block	10	perlite		12.80	16.55	29.35
6280				styrofoam		13.55	16.30	29.85
6300		8" glazed block	12	perlite		13.70	17	30.70
6320				styrofoam		14.30	16.70	31
6340		6" clay tile	10	none		11.05	14.65	25.70
6360		8" clay tile	12	none		13.05	15.25	28.30
6600	Roman	6" conc. block	10	perlite		8.65	15.80	24.45
6620				styrofoam		9.45	15.55	25
6640		8" conc. block	12	perlite		9.10	16.15	25.25
6660				styrofoam		9.75	15.85	25.60
6680		6" L.W. block	10	perlite		8.80	15.65	24.45
6700				styrofoam		9.60	15.40	25
6720		8"L.W. block	12	perlite		9.55	16.05	25.60
6740				styrofoam		10.20	15.75	25.95
6760		6" glazed block	10	perlite		14.55	16.85	31.40
6780				styrofoam		15.30	16.60	31.90
6800		8" glazed block	12	perlite		15.45	17.30	32.75
6820				styrofoam		16.05	17	33.05
6840	Roman	6" clay tile	10	none		12.80	14.95	27.75
6860		8" clay tile	12	none		14.80	15.55	30.35
7100	Norman	6" conc. block	10	perlite		7.60	14.05	21.65
7120				styrofoam		8.40	13.80	22.20
7140		8" conc. block	12	perlite		8.10	14.40	22.50
7160				styrofoam		8.75	14.10	22.85
7180		6" L.W. block	10	perlite		7.75	13.90	21.65
7200				styrofoam		8.55	13.65	22.20
7220		8"L.W. block	12	perlite		8.55	14.30	22.85
7240				styrofoam		9.20	14	23.20
7260		6" glazed block	10	perlite		13.50	15.10	28.60
7280				styrofoam		14.30	14.85	29.15
7300		8" glazed block	12	perlite		14.40	15.55	29.95
7320				styrofoam		15.05	15.25	30.30

Important: See the Reference Section for critical supporting data - Reference Numbers and City Cost Indexes

B2010 Exterior Walls

B2010 135		Brick Face Cavity Wall - Insulated Backup							

	FACE BRICK	BACKUP MASONRY	TOTAL THICKNESS (IN.)	BACKUP CORE FILL		COST PER S.F.		
						MAT.	INST.	TOTAL
7340	Norman	6" clay tile	10	none		11.75	13.20	24.95
7360		8" clay tile	12	none		13.75	13.80	27.55
7600	Norwegian	6" conc. block	10	perlite		6.80	13.10	19.90
7620				styrofoam		7.55	12.85	20.40
7640		8" conc. block	12	perlite		7.25	13.45	20.70
7660				styrofoam		7.90	13.15	21.05
7680		6" L.W. block	10	perlite		6.95	12.95	19.90
7700				styrofoam		7.75	12.70	20.45
7720		8" L.W. block	12	perlite		7.70	13.35	21.05
7740				styrofoam		8.35	13.05	21.40
7760		6" glazed block	10	perlite		12.65	14.15	26.80
7780				styrofoam		13.45	13.90	27.35
7800		8" glazed block	12	perlite		13.55	14.60	28.15
7820				styrofoam		14.20	14.30	28.50
7840		6" clay tile	10	none		10.95	12.25	23.20
7860		8" clay tile	12	none		12.95	12.85	25.80
8100	Utility	6" conc. block	10	perlite		6.45	12.20	18.65
8120				styrofoam		7.25	11.95	19.20
8140		8" conc. block	12	perlite		6.95	12.60	19.55
8160				styrofoam		7.60	12.25	19.85
8180		6" L.W. block	10	perlite		6.65	12.05	18.70
8200				styrofoam		7.40	11.80	19.20
8220		8" L.W. block	12	perlite		7.40	12.50	19.90
8240				styrofoam		8.05	12.15	20.20
8260		6" glazed block	10	perlite		12.35	13.25	25.60
8280				styrofoam		13.15	13	26.15
8300		8" glazed block	12	perlite		13.25	13.75	27
8320				styrofoam		13.90	13.40	27.30
8340		6" clay tile	10	none		10.65	11.35	22
8360		8" clay tile	12	none		12.65	11.95	24.60

B2010 Exterior Walls

Exterior block face cavity walls are defined in the following terms: cavity treatment, type of face block, backup masonry, total thickness and insulation. Twelve types of face block are shown with four types of backup. All systems include a brick shelf and necessary dampproofing, flashing, and control joints every 20'.

System Components	QUANTITY	UNIT	COST PER S.F.		
			MAT.	INST.	TOTAL
SYSTEM B2010 137 1600					
CAVITY WALL, FLUTED BLOCK FACE, 4″ C.M.U. BACKUP, POLYSTYRENE BOARD					
Fluted block partition, 4″ thick	1.000	S.F.	4.37	6	10.37
Conc. block wall backup, 4″ thick	1.000	S.F.	1.25	4.68	5.93
Horizontal joint reinforcing	.800	L.F.	.10	.18	.28
Polystyrene insulation board, 1″ thick	1.000	S.F.	.24	.55	.79
Flashing, aluminum	.100	S.F.	.08	.29	.37
Shelf angle	1.000	Lb.	.80	.81	1.61
Control joint	.050	L.F.	.06	.06	.12
Backer rod	.100	L.F.		.10	.10
Sealant	.100	L.F.	.02	.24	.26
TOTAL			6.92	12.91	19.83

B2010 137		Block Face Cavity Wall						
	FACE BLOCK	BACKUP MASONRY	TOTAL THICKNESS (IN.)	CAVITY INSULATION		COST PER S.F.		
						MAT.	INST.	TOTAL
1000	Deep groove	4″ conc. block	10	polystyrene		6.90	12.90	19.80
1050	Reg. wt.			none		6.70	12.35	19.05
1060		6″ conc. block	12	polystyrene		7.55	13.30	20.85
1110				none		7.30	12.75	20.05
1120		4″ L.W. block	10	polystyrene		7.15	12.80	19.95
1170				none		6.90	12.25	19.15
1180		6″ L.W. block	12	polystyrene		7.70	13.15	20.85
1230				none		7.45	12.60	20.05
1300	Deep groove	4″ conc. block	10	polystyrene		7.15	12.75	19.90
1350	L.W.			none		6.90	12.20	19.10
1360		6″ conc. block	12	polystyrene		7.75	13.15	20.90
1410				none		7.55	12.60	20.15
1420		4″ L.W. block	10	polystyrene		7.35	12.65	20
1470				none		7.15	12.10	19.25
1480		6″ L.W. block	12	polystyrene		7.95	13	20.95
1530				none		7.70	12.45	20.15

Important: See the Reference Section for critical supporting data - Reference Numbers and City Cost Indexes

B20 Exterior Enclosure

B2010 Exterior Walls

| B2010 137 | | Block Face Cavity Wall | | | | | | |

	FACE BLOCK	BACKUP MASONRY	TOTAL THICKNESS (IN.)	CAVITY INSULATION		MAT.	INST.	TOTAL
1600	Fluted	4" conc. block	10	polystyrene		6.90	12.90	19.80
1650	Reg. wt.			none		5.10	12.20	17.30
1660		6" conc. block	12	polystyrene		5.95	13.15	19.10
1710				none		5.70	12.60	18.30
1720		4" L.W. block	10	polystyrene		5.55	12.65	18.20
1770				none		5.30	12.10	17.40
1780		6" L.W. block	12	polystyrene		6.10	13	19.10
1830				none		5.90	12.45	18.35
1900	Fluted	4" conc. block	10	polystyrene		5.45	12.55	18
1950	L.W.			none		5.20	12	17.20
1960		6" conc. block	12	polystyrene		6.05	12.95	19
2010				none		5.80	12.40	18.20
2020		4" L.W. block	10	polystyrene		5.65	12.45	18.10
2070				none		5.40	11.90	17.30
2080		6" L.W. block	12	polystyrene		6.20	12.80	19
2130				none		5.95	12.25	18.20
2200	Ground face	4" conc. block	10	polystyrene		9.35	13	22.35
2250	1 Score			none		9.10	12.45	21.55
2260		6" conc. block	12	polystyrene		9.95	13.40	23.35
2310				none		9.75	12.85	22.60
2320		4" L.W. block	10	polystyrene		9.55	12.90	22.45
2370				none		9.35	12.35	21.70
2380		6" L.W. block	12	polystyrene		10.15	13.25	23.40
2430				none		9.90	12.70	22.60
2500	Hexagonal	4" conc. block	10	polystyrene		4.53	12.75	17.28
2550	Reg. wt.			none		4.29	12.20	16.49
2560		6" conc. block	12	polystyrene		5.15	13.15	18.30
2610				none		4.91	12.60	17.51
2620		4" L.W. block	10	polystyrene		4.75	12.65	17.40
2670				none		4.51	12.10	16.61
2680		6" L.W. block	12	polystyrene		5.30	13	18.30
2730				none		5.05	12.45	17.50
2800	Hexagonal	4" conc. block	10	polystyrene		4.67	12.55	17.22
2850	L.W.			none		4.43	12	16.43
2860		6" conc. block	12	polystyrene		5.30	12.95	18.25
2910				none		5.05	12.40	17.45
2920		4" L.W. block	10	polystyrene		4.89	12.45	17.34
2970				none		4.65	11.90	16.55
2980		6" L.W. block	12	polystyrene		5.45	12.80	18.25
3030				none		5.20	12.25	17.45
3100	Slump block	4" conc. block	10	polystyrene		6.30	11.65	17.95
3150	4x16			none		6.05	11.10	17.15
3160		6" conc. block	12	polystyrene		6.95	12.05	19
3210				none		6.70	11.50	18.20
3220		4" L.W. block	10	polystyrene		6.55	11.55	18.10
3270				none		6.30	11	17.30
3280		6" L.W. block	12	polystyrene		7.10	11.90	19
3330				none		6.85	11.35	18.20

B2010 Exterior Walls

B2010 137	Block Face Cavity Wall

	FACE BLOCK	BACKUP MASONRY	TOTAL THICKNESS (IN.)	CAVITY INSULATION		COST PER S.F.		
						MAT.	INST.	TOTAL
3400	Split face	4" conc. block	10	polystyrene		6.40	12.90	19.30
3450	1 Score			none		6.15	12.35	18.50
3460	Reg. wt.	6" conc. block	12	polystyrene		7	13.30	20.30
3510				none		6.75	12.75	19.50
3520		4" L.W. block	10	polystyrene		6.60	12.80	19.40
3570				none		6.35	12.25	18.60
3580		6" L.W. block	12	polystyrene		7.15	13.15	20.30
3630				none		6.95	12.60	19.55
3700	Split face	4" conc. block	10	polystyrene		6.70	12.75	19.45
3750	1 Score			none		6.45	12.20	18.65
3760	L.W.	6" conc. block	12	polystyrene		7.35	13.15	20.50
3810				none		7.10	12.60	19.70
3820		4" L.W. block	10	polystyrene		6.95	12.65	19.60
3870				none		6.70	12.10	18.80
3880		6" L.W. block	12	polystyrene		7.50	13	20.50
3930				none		7.25	12.45	19.70
4000	Split rib	4" conc. block	10	polystyrene		6.75	12.90	19.65
4050	8 Rib			none		6.55	12.35	18.90
4060	Reg. wt.	6" conc. block	12	polystyrene		7.40	13.30	20.70
4110				none		7.15	12.75	19.90
4120		4" L.W. block	10	polystyrene		7	12.80	19.80
4170				none		6.75	12.25	19
4180		6" L.W. block	12	polystyrene		7.55	13.15	20.70
4230				none		7.30	12.60	19.90
4300	Split rib	4" conc. block	10	polystyrene		7.10	12.75	19.85
4350	8 Rib			none		6.90	12.20	19.10
4360	L.W.	6" conc. block	12	polystyrene		7.75	13.15	20.90
4410				none		7.50	12.60	20.10
4420		4" L.W. block	10	polystyrene		7.35	12.65	20
4470				none		7.10	12.10	19.20
4480		6" L.W. block	12	polystyrene		7.90	13	20.90
4530				none		7.65	12.45	20.10

B2010 138	Block Face Cavity Wall - Insulated Backup

	FACE BLOCK	BACKUP MASONRY	TOTAL THICKNESS (IN.)	BACKUP CORE INSULATION		COST PER S.F.		
						MAT.	INST.	TOTAL
5010	Deep groove	6" conc. block	10	perlite		7.60	13	20.60
5050	Reg. wt.			styrofoam		8.40	12.75	21.15
5060		8" conc. block	12	perlite		7.95	13.35	21.30
5110				styrofoam		8.60	13.05	21.65
5120		6" L.W. block	10	perlite		7.80	12.85	20.65
5170				styrofoam		8.55	12.60	21.15
5180		8" L.W. block	12	perlite		8.40	13.25	21.65
5230				styrofoam		9.05	12.95	22
5300	Deep groove	6" conc. block	10	perlite		7.85	12.85	20.70
5350	L.W.			styrofoam		8.65	12.60	21.25
5360		8" conc. block	12	perlite		8.20	13.20	21.40
5410				styrofoam		8.85	12.90	21.75
5420		6" L.W. block	10	perlite		8	12.70	20.70
5470				styrofoam		8.80	12.45	21.25
5480		8" L.W. block	12	perlite		8.65	13.10	21.75
5530				styrofoam		9.30	12.80	22.10

B2010 Exterior Walls

B2010 138		Block Face Cavity Wall - Insulated Backup					

	FACE BLOCK	BACKUP MASONRY	TOTAL THICKNESS (IN.)	BACKUP CORE INSULATION		COST PER S.F.		
						MAT.	INST.	TOTAL
5600	Fluted	6" conc. block	10	perlite		6.05	12.85	18.90
5650	Reg.			styrofoam		6.85	12.60	19.45
5660		8" conc. block	12	perlite		6.40	13.20	19.60
5710				styrofoam		7	12.90	19.90
5720		6" L.W. block	10	perlite		6.20	12.70	18.90
5770				styrofoam		7	12.45	19.45
5780		8" L.W. block	12	perlite		6.85	13.10	19.95
5830				styrofoam		7.45	12.80	20.25
5900	Fluted	6 conc. block	10	perlite		6.15	12.65	18.80
5950	L.W.			styrofoam		6.90	12.40	19.30
5960		8" conc. block	12	perlite		6.45	13	19.45
6010				styrofoam		7.10	12.70	19.80
6020		6" L.W. block	10	perlite		6.30	12.50	18.80
6070				styrofoam		7.10	12.25	19.35
6080		8" L.W. block	12	perlite		6.90	12.90	19.80
6130				styrofoam		7.55	12.60	20.15
6200	Ground face	6" conc. block	10	perlite		10.05	13.10	23.15
6250	1 score			styrofoam		10.85	12.85	23.70
6260		8" conc. block	12	perlite		10.40	13.45	23.85
6310				styrofoam		11.05	13.15	24.20
6320		6" L.W. block	10	perlite		10.20	12.95	23.15
6370				styrofoam		11	12.70	23.70
6380		8" L.W. block	12	perlite		10.85	13.35	24.20
6430				styrofoam		12.20	15.60	27.80
6500	Hexagonal	6" conc. block	10	perlite		5.25	12.85	18.10
6550	Reg. wt.			styrofoam		6	12.60	18.60
6560		8" conc. block	12	perlite		5.55	13.20	18.75
6610				styrofoam		6.20	12.90	19.10
6620		6" L.W. block	10	perlite		5.40	12.70	18.10
6670				styrofoam		6.20	12.45	18.65
6680		8" L.W. block	12	perlite		6	13.10	19.10
6730				styrofoam		6.65	12.80	19.45
6800	Hexagonal	6" conc. block	10	perlite		5.35	12.65	18
6850	L.W.			styrofoam		6.15	12.40	18.55
6860		8" conc. block	12	perlite		5.70	13	18.70
6910				styrofoam		6.35	12.70	19.05
6920		6" L.W. block	10	perlite		5.55	12.50	18.05
6970				styrofoam		6.30	12.25	18.55
6980		8" L.W. block	12	perlite		6.15	12.90	19.05
7030				styrofoam		6.80	12.60	19.40
7100	Slump block	6" conc. block	10	perlite		7	11.75	18.75
7150	4x16			styrofoam		7.80	11.50	19.30
7160		8" conc. block	12	perlite		7.35	12.10	19.45
7210				styrofoam		8	11.80	19.80
7220		6" L.W. block	10	perlite		7.15	11.60	18.75
7270				styrofoam		7.95	11.35	19.30
7280		8" L.W. block	12	perlite		7.80	12	19.80
7330				styrofoam		8.45	11.70	20.15

SHELL B

B2010 Exterior Walls

B2010 138	Block Face Cavity Wall - Insulated Backup

	FACE BLOCK	BACKUP MASONRY	TOTAL THICKNESS (IN.)	BACKUP CORE INSULATION		COST PER S.F.		
						MAT.	INST.	TOTAL
7400	Split face	6" conc. block	10	perlite		7.10	13	20.10
7450	1 Score			styrofoam		7.90	12.75	20.65
7460	Reg. wt.	8" conc. block	12	perlite		7.45	13.35	20.80
7510				styrofoam		8.05	13.05	21.10
7520		6" L.W. block	10	perlite		7.25	12.85	20.10
7570				styrofoam		8.05	12.60	20.65
7580		8" L.W. block	12	perlite		7.90	13.25	21.15
7630				styrofoam		8.50	12.95	21.45
7700	Split rib	6" conc. block	10	perlite		7.45	13	20.45
7750	8 Rib			styrofoam		8.25	12.75	21
7760	Reg. wt.	8" conc. block	12	perlite		7.80	13.35	21.15
7810				styrofoam		8.45	13.05	21.50
7820		6" L.W. block	10	perlite		7.65	12.85	20.50
7870				styrofoam		8.40	12.60	21
7880		8" L.W. block	12	perlite		8.25	13.25	21.50
7930				styrofoam		8.90	12.95	21.85
8000	Split rib	6" conc. block	10	perlite		7.80	12.85	20.65
8050	8 Rib			styrofoam		8.65	12.65	21.30
8060	L.W.	8" conc. block	12	perlite		8.15	13.20	21.35
8110				styrofoam		8.80	12.90	21.70
8120		6" L.W. block	10	perlite		8	12.70	20.70
8170				styrofoam		8.75	12.45	21.20
8180		8" L.W. block	12	perlite		8.60	13.10	21.70
8230				styrofoam		9.25	12.80	22.05

Important: See the Reference Section for critical supporting data - Reference Numbers and City Cost Indexes

B2010 Exterior Walls

Exterior block face composite walls are defined in the following terms: type of face block and backup masonry, total thickness and insulation. All systems include shelf angles and necessary dampproofing, flashing, and control joints every 20'.

SHELL **B**

System Components	QUANTITY	UNIT	COST PER S.F.		
			MAT.	INST.	TOTAL
SYSTEM B2010 139 1000					
COMPOSITE WALL, GROOVED BLOCK FACE, 4″ C.M.U. BACKUP, PERLITE FILL					
Deep groove block veneer, 4″ thick	1.000	S.F.	4.37	6	10.37
Concrete block wall, backup, 4″ thick	1.000	S.F.	1.25	4.68	5.93
Horizontal joint reinforcing	.800	L.F.	.11	.14	.25
Perlite insulation, poured	1.000	S.F.	.22	.16	.38
Flashing, aluminum	.100	S.F.	.08	.29	.37
Shelf angle	1.000	Lb.	.80	.81	1.61
Control joint	.050	L.F.	.06	.06	.12
Backer rod	.100	L.F.		.10	.10
Sealant	.100	L.F.	.02	.24	.26
Collar joint	1.000	S.F.	.34	.51	.85
TOTAL			7.25	12.99	20.24

B2010 139		Block Face Composite Wall						
	FACE BLOCK	BACKUP MASONRY	TOTAL THICKNESS (IN.)	BACKUP CORE INSULATION		COST PER S.F.		
						MAT.	INST.	TOTAL
1000	Deep groove	4″ conc. block	8	perlite		7.25	13	20.25
1050	Reg. wt.			none		7.05	12.85	19.90
1060		6″ conc. block	10	perlite		8	13.50	21.50
1110				styrofoam		8.75	13.25	22
1120		8″ conc. block	12	perlite		8.30	13.85	22.15
1170				styrofoam		8.95	13.55	22.50
1200		4″ L.W. block	8	perlite		7.45	12.90	20.35
1250				none		7.25	12.70	19.95
1260		6″ L.W. block	10	perlite		8.15	13.35	21.50
1310				styrofoam		8.95	13.10	22.05
1320		8″ L.W. block	12	perlite		8.75	13.75	22.50
1370				styrofoam		9.40	13.45	22.85

B2010 Exterior Walls

B2010 139	Block Face Composite Wall

B SHELL

	FACE BLOCK	BACKUP MASONRY	TOTAL THICKNESS (IN.)	BACKUP CORE INSULATION		COST PER S.F.		
						MAT.	INST.	TOTAL
1600	Deep groove	4" conc. block	8	perlite		7.50	12.85	20.35
1650	L.W.			none		7.25	12.70	19.95
1660		6" conc. block	10	perlite		8.20	13.35	21.55
1710				styrofoam		9	13.10	22.10
1720		8" conc. block	12	perlite		8.55	13.70	22.25
1770				styrofoam		9.20	13.40	22.60
1800		4" L.W. block	8	perlite		7.70	12.75	20.45
1850				none		7.50	12.55	20.05
1860		6" L.W. block	10	perlite		8.35	13.20	21.55
1910				styrofoam		9.15	12.95	22.10
1920		8" L.W. block	12	perlite		9	13.60	22.60
1970				styrofoam		9.65	13.30	22.95
2200	Fluted	4" conc. block	8	perlite		5.65	12.85	18.50
2250	Reg. wt.			none		5.45	12.70	18.15
2260		6"conc. block	10	perlite		6.40	13.35	19.75
2310				styrofoam		7.20	13.10	20.30
2320		8" conc. block	12	perlite		6.75	13.70	20.45
2370				styrofoam		7.35	13.40	20.75
2400		4" L.W. block	8	perlite		5.90	12.75	18.65
2450				styrofoam		5.65	12.55	18.20
2460		6" L.W. block	10	perlite		6.55	13.20	19.75
2510				styrofoam		7.35	12.95	20.30
2520		8" L.W. block	12	perlite		7.20	13.60	20.80
2570				styrofoam		7.80	13.30	21.10
2860	Fluted	6" conc. block	10	perlite		6.55	13.45	20
2910	L.W.			styrofoam		7.30	12.90	20.20
2920		8" conc. block	12	perlite		6.80	13.50	20.30
2970				styrofoam		7.45	13.20	20.65
3060		6" L.W. block	10	perlite		6.65	13	19.65
3110				styrofoam		7.45	12.75	20.20
3120		8" L.W. block	12	perlite		7.25	13.40	20.65
3170				styrofoam		7.90	13.10	21
3400	Ground face	4" conc. block	8	perlite		9.70	13.10	22.80
3450	1 Score			none		9.45	12.95	22.40
3460		6" conc. block	10	perlite		10.40	13.60	24
3510				styrofoam		11.20	13.35	24.55
3520		8" conc. block	12	perlite		10.75	14.05	24.80
3570				styrofoam		11.40	13.65	25.05
3600		4" L.W. block	8	perlite		9.90	13	22.90
3650				none		9.70	12.80	22.50
3660		6" L.W. block	10	perlite		10.55	13.45	24
3710				styrofoam		11.35	13.20	24.55
3720		8" L.W. block	12	perlite		11.20	13.85	25.05
3770				styrofoam		11.85	13.55	25.40

Important: See the Reference Section for critical supporting data - Reference Numbers and City Cost Indexes

B2010 Exterior Walls

B2010 139			Block Face Composite Wall					

	FACE BLOCK	BACKUP MASONRY	TOTAL THICKNESS (IN.)	BACKUP CORE INSULATION		COST PER S.F.		
						MAT.	INST.	TOTAL
4000	Hexagonal	4" conc. block	8	perlite		4.86	12.85	17.71
4050	Reg. wt.			none		4.64	12.70	17.34
4060		6" conc. block	10	perlite		5.60	13.35	18.95
4110				styrofoam		6.40	13.10	19.50
4120		8" conc. block	12	perlite		6	13.75	19.75
4170				styrofoam		6.55	13.40	19.95
4200		4" L.W. block	8	perlite		5.10	12.75	17.85
4250				none		4.86	12.55	17.41
4260		6" L.W. block	10	perlite		5.75	13.20	18.95
4310				styrofoam		6.55	12.95	19.50
4320		8" L.W. block	12	perlite		6.35	13.60	19.95
4370				styrofoam		7	13.30	20.30
4600	Hexagonal	4" conc. block	8	perlite		5	12.65	17.65
4650	L.W.			none		4.78	12.50	17.28
4660		6" conc. block	10	perlite		5.60	13.05	18.65
4710				styrofoam		6.50	12.90	19.40
4720		8" conc. block	12	perlite		6.05	13.50	19.55
4770				styrofoam		6.70	13.20	19.90
4800		4" L.W. block	8	perlite		5.20	12.55	17.75
4850				none		5	12.35	17.35
4860		6" L.W. block	10	perlite		5.90	13	18.90
4910				styrofoam		6.70	12.75	19.45
4920		8" L.W. block	12	perlite		6.50	13.40	19.90
4970				styrofoam		7.15	13.10	20.25
5200	Slump block	4" conc. block	8	perlite		6.65	11.75	18.40
5250	4x16			none		6.40	11.60	18
5260		6" conc. block	10	perlite		7.35	12.25	19.60
5310				styrofoam		8.15	12	20.15
5320		8" conc. block	12	perlite		7.70	12.60	20.30
5370				styrofoam		8.35	12.30	20.65
5400		4" L.W. block	8	perlite		6.85	11.65	18.50
5450				none		6.65	11.45	18.10
5460		6" L.W. block	10	perlite		7.55	12.10	19.65
5510				styrofoam		8.30	11.85	20.15
5520		8" L.W. block	12	perlite		8.15	12.50	20.65
5570				styrofoam		8.80	12.20	21
5800	Split face	4" conc. block	8	perlite		6.70	13	19.70
5850	1 Score			none		6.50	12.85	19.35
5860	Reg. wt.	6" conc. block	10	perlite		7.45	13.50	20.95
5910				styrofoam		8.25	13.25	21.50
5920		8" conc. block	12	perlite		7.80	13.85	21.65
5970				styrofoam		8.40	13.55	21.95
6000		4" L.W. block	8	perlite		6.95	12.90	19.85
6050				none		6.70	12.70	19.40
6060		6" L.W. block	10	perlite		7.60	13.35	20.95
6110				styrofoam		8.40	13.10	21.50
6120		8" L.W. block	12	perlite		8.25	13.75	22
6170				styrofoam		8.85	13.45	22.30

B2010 Exterior Walls

B2010 139	Block Face Composite Wall

	FACE BLOCK	BACKUP MASONRY	TOTAL THICKNESS (IN.)	BACKUP CORE INSULATION		COST PER S.F.		
						MAT.	INST.	TOTAL
6460	Split face	6" conc. block	10	perlite		7.75	13.35	21.10
6510	1 Score			styrofoam		8.55	13.10	21.65
6520	L.W.	8" conc. block	12	perlite		8.10	13.70	21.80
6570				styrofoam		8.75	13.40	22.15
6660		6" L.W. block	10	perlite		7.95	13.20	21.15
6710				styrofoam		8.80	13.10	21.90
6720		8" L.W. block	12	perlite		8.55	13.60	22.15
6770				styrofoam		9.25	13.45	22.70
7000	Split rib	4" conc. block	8	perlite		7.10	13	20.10
7050	8 Rib			none		6.90	12.85	19.75
7060	Reg. wt.	6" conc. block	10	perlite		7.85	13.50	21.35
7110				styrofoam		8.60	13.25	21.85
7120		8" conc. block	12	perlite		8.15	13.85	22
7170				styrofoam		8.80	13.55	22.35
7200		4" L.W. block	8	perlite		7.30	12.90	20.20
7250				none		7.10	12.70	19.80
7260		6" L.W. block	10	perlite		8	13.35	21.35
7310				styrofoam		8.80	13.10	21.90
7320		8" L.W. block	12	perlite		8.60	13.75	22.35
7370				styrofoam		9.25	13.45	22.70
7600	Split rib	4" conc. block	8	perlite		7.45	12.85	20.30
7650	8 Rib			none		7.25	12.70	19.95
7660	L.W.	6" conc. block	10	perlite		8.20	13.35	21.55
7710				styrofoam		8.95	13.10	22.05
7720		8" conc. block	12	perlite		8.50	13.70	22.20
7770				styrofoam		9.15	13.40	22.55
7800		4" conc. block	8	perlite		7.65	12.75	20.40
7850				none		7.45	12.55	20
7860		6" conc. block	10	perlite		8.35	13.20	21.55
7910				styrofoam		9.15	12.95	22.10
7920		8" conc. block	12	perlite		8.95	13.60	22.55
7970				styrofoam		9.60	13.30	22.90

Important: See the Reference Section for critical supporting data - Reference Numbers and City Cost Indexes

B2010 Exterior Walls

The table below lists costs per S.F. for glass block walls. Included in the costs are the following special accessories required for glass block walls.

Glass block accessories required for proper installation.

Wall ties: Galvanized double steel mesh full length of joint.

Fiberglass expansion joint at sides and top.

Silicone caulking: One gallon does 95 L.F.

Oakum: One lb. does 30 L.F.

Asphalt emulsion: One gallon does 600 L.F.

If block are not set in wall chase, use 2'-0" long wall anchors at 2'-0" O.C.

SHELL **B**

System Components	QUANTITY	UNIT	COST PER S.F. MAT.	COST PER S.F. INST.	COST PER S.F. TOTAL
SYSTEM B2010 140 2300					
GLASS BLOCK, 4" THICK, 6" X 6" PLAIN, UNDER 1,000 S.F.					
Glass block, 4" thick, 6" x 6" plain, under 1000 S.F.	4.100	Ea.	19.75	17.50	37.25
Glass block, cleaning blocks after installation, both sides add	2.000	S.F.	.11	2.01	2.12
TOTAL			19.86	19.51	39.37

B2010 140	Glass Block	COST PER S.F. MAT.	COST PER S.F. INST.	COST PER S.F. TOTAL
2300	Glass block 4" thick, 6"x6" plain, under 1,000 S.F.	19.85	19.50	39.35
2400	1,000 to 5,000 S.F.	19.45	16.90	36.35
2500	Over 5,000 S.F.	18.85	15.85	34.70
2600	Solar reflective, under 1,000 S.F.	28	26.50	54.50
2700	1,000 to 5,000 S.F.	27	23	50
2800	Over 5,000 S.F.	26.50	21.50	48
3500	8"x8" plain, under 1,000 S.F.	13.25	14.55	27.80
3600	1,000 to 5,000 S.F.	13	12.60	25.60
3700	Over 5,000 S.F.	12.60	11.35	23.95
3800	Solar reflective, under 1,000 S.F.	18.50	19.60	38.10
3900	1,000 to 5,000 S.F.	18.15	16.85	35
4000	Over 5,000 S.F.	17.60	15.10	32.70
5000	12"x12" plain, under 1,000 S.F.	15.80	13.50	29.30
5100	1,000 to 5,000 S.F.	15.50	11.35	26.85
5200	Over 5,000 S.F.	15	10.40	25.40
5300	Solar reflective, under 1,000 S.F.	22	18.10	40.10
5400	1,000 to 5,000 S.F.	21.50	15.10	36.60
5600	Over 5,000 S.F.	21	13.75	34.75
5800	3" thinline, 6"x6" plain, under 1,000 S.F.	21.50	19.50	41
5900	Over 5,000 S.F.	21.50	19.50	41
6000	Solar reflective, under 1,000 S.F.	30	26.50	56.50
6100	Over 5,000 S.F.	19.35	21.50	40.85
6200	8"x8" plain, under 1,000 S.F.	10.95	14.55	25.50
6300	Over 5,000 S.F.	8	11.35	19.35
6400	Solar reflective, under 1,000 S.F.	15.30	19.60	34.90
6500	Over 5,000 S.F.	11.15	15.10	26.25

Concrete Block Lintel

Concrete block specialties are divided into lintels, pilasters, and bond beams. Lintels are defined by span, thickness, height and wall loading. Span refers to the clear opening but the cost includes 8″ of bearing at both ends.

Bond beams and pilasters are defined by height thickness and weight of the masonry unit itself. Components for bond beams also include grout and reinforcing. For pilasters, components include four #5 reinforcing bars and type N mortar.

Bond Beam

Pilaster

System Components			COST PER LINTEL		
	QUANTITY	UNIT	MAT.	INST.	TOTAL
SYSTEM B2010 144 3100					
CONCRETE BLOCK LINTEL, 6″X8″, LOAD 300 LB/L.F. 3′-4″ SPAN					
Lintel blocks, 6″ x 8″ x 8″	7.000	Ea.	8.96	18.27	27.23
Joint reinforcing, #3 & #4 steel bars, horizontal	3.120	Lb.	1.47	3.09	4.56
Grouting, 8″ deep, 6″ thick, .15 C.F./L.F.	.700	C.F.	3.15	4.24	7.39
Temporary shoring, lintel forms	1.000	Set	.88	10.04	10.92
Temporary shoring, wood joists	1.000	Set		17.70	17.70
Temporary shoring, vertical members	1.000	Set		32.20	32.20
Mortar, masonry cement, 1:3 mix, type N	.270	C.F.	.90	.64	1.54
TOTAL			15.36	86.18	101.54

B2010 144		**Concrete Block Lintel**						
	SPAN	THICKNESS (IN.)	HEIGHT (IN.)	WALL LOADING P.L.F.		COST PER LINTEL		
						MAT.	INST.	TOTAL
3100	3′-4″	6	8	300		15.35	86.50	101.85
3150		6	16	1,000		24	93	117
3200		8	8	300		17.20	92.50	109.70
3300		8	8	1,000		18.70	96	114.70
3400		8	16	1,000		28	101	129
3450	4′-0″	6	8	300		18	96	114
3500		6	16	1,000		31.50	108	139.50
3600		8	8	300		20	100	120
3700		8	16	1,000		32	109	141

Important: See the Reference Section for critical supporting data - Reference Numbers and City Cost Indexes

B2010 Exterior Walls

B2010 144 — Concrete Block Lintel

	SPAN	THICKNESS (IN.)	HEIGHT (IN.)	WALL LOADING P.L.F.		COST PER LINTEL		
						MAT.	INST.	TOTAL
3800	4'-8"	6	8	300		19.85	97	116.85
3900		6	16	1,000		31.50	107	138.50
4000		8	8	300		22.50	107	129.50
4100		8	16	1,000		36	117	153
4200	5'-4"	6	8	300		24.50	108	132.50
4300		6	16	1,000		36	112	148
4400		8	8	300		27	118	145
4500		8	16	1,000		41.50	127	168.50
4600	6'-0"	6	8	300		29.50	131	160.50
4700		6	16	300		38.50	137	175.50
4800		6	16	1,000		40.50	142	182.50
4900		8	8	300		30	143	173
5000		8	16	1,000		46.50	156	202.50
5100	6'-8"	6	16	300		37.50	138	175.50
5200		6	16	1,000		40	144	184
5300		8	8	300		35	149	184
5400		8	16	1,000		51	165	216
5500	7'-4"	6	16	300		41	144	185
5600		6	16	1,000		52	155	207
5700		8	8	300		43	162	205
5800		8	16	300		60	173	233
5900		8	16	1,000		60	173	233
6000	8'-0"	6	16	300		44.50	150	194.50
6100		6	16	1,000		56.50	162	218.50
6200		8	16	300		66.50	183	249.50
6500		8	16	1,000		66.50	183	249.50

B2010 145 — Concrete Block Specialties

	TYPE	HEIGHT	THICKNESS	WEIGHT (P.C.F.)		COST PER L.F.		
						MAT.	INST.	TOTAL
7100	Bond beam	8	8	125		4.09	5.95	10.04
7200				105		3.95	5.90	9.85
7300			12	125		5.25	7.45	12.70
7400				105		5.60	7.35	12.95
7500	Pilaster	8	16	125		10.15	19.55	29.70
7600			20	125		12.20	23.50	35.70

The table below lists costs for metal siding of various descriptions, not including the steel frame, or the structural steel, of a building. Costs are per S.F. including all accessories and insulation.

For steel frame support see System B2010 154.

System Components	QUANTITY	UNIT	COST PER S.F.		
			MAT.	INST.	TOTAL
SYSTEM B2010 146 1400					
METAL SIDING ALUMINUM PANEL, CORRUGATED, .024" THICK, NATURAL					
Corrugated aluminum siding, industrial type, .024" thick	1.000	S.F.	1.20	2.22	3.42
Flashing alum mill finish .032 thick	.111	S.F.	.14	.32	.46
Tapes, sealant , p.v.c. foam adheasive, 1/16"x1"	.004	C.L.F.	.05		.05
Closure strips for aluminum siding, corrugated, .032" thick	.111	L.F.	.07	.24	.31
Pre-engineered steel building insulation, vinyl faced, 1-1/2" thick, R5	1.000	S.F.	.24	.39	.63
TOTAL			1.70	3.17	4.87

B2010 146	Metal Siding Panel	COST PER S.F.		
		MAT.	INST.	TOTAL
1400	Metal siding aluminum panel, corrugated, .024" thick, natural	1.71	3.17	4.88
1450	Painted	1.95	3.17	5.12
1500	.032" thick, natural	2.26	3.17	5.43
1550	Painted	2.73	3.17	5.90
1600	Ribbed 4" pitch, .032" thick, natural	2.22	3.17	5.39
1650	Painted	2.75	3.17	5.92
1700	.040" thick, natural	2.65	3.17	5.82
1750	Painted	3.04	3.17	6.21
1800	.050" thick, natural	3.03	3.17	6.20
1850	Painted	3.39	3.17	6.56
1900	8" pitch panel, .032" thick, natural	2.12	3.04	5.16
1950	Painted	2.63	3.05	5.68
2000	.040" thick, natural	2.54	3.06	5.60
2050	Painted	2.92	3.09	6.01
2100	.050" thick, natural	2.92	3.07	5.99
2150	Painted	3.30	3.10	6.40
3000	Steel, corrugated or ribbed, 29 Ga. .0135" thick, galvanized	1.65	2.89	4.54
3050	Colored	2.34	2.93	5.27
3100	26 Ga. .0179" thick, galvanized	1.71	2.91	4.62
3150	Colored	2.41	2.95	5.36
3200	24 Ga. .0239" thick, galvanized	2.17	2.92	5.09
3250	Colored	2.70	2.96	5.66
3300	22 Ga. .0299" thick, galvanized	2.18	2.93	5.11
3350	Colored	3.26	2.97	6.23
3400	20 Ga. .0359" thick, galvanized	2.18	2.93	5.11
3450	Colored	3.46	3.14	6.60

Important: See the Reference Section for critical supporting data - Reference Numbers and City Cost Indexes

B2010 Exterior Walls

B2010 146	Metal Siding Panel	COST PER S.F.		
		MAT.	INST.	TOTAL
4100	Sandwich panels, factory fab., 1" polystyrene, steel core, 26 Ga., galv.	4.39	4.53	8.92
4200	Colored, 1 side	5.35	4.53	9.88
4300	2 sides	6.90	4.53	11.43
4400	2" polystyrene, steel core, 26 Ga., galvanized	5.20	4.53	9.73
4500	Colored, 1 side	6.15	4.53	10.68
4600	2 sides	7.70	4.53	12.23
4700	22 Ga., baked enamel exterior	10.20	4.78	14.98
4800	Polyvinyl chloride exterior	10.75	4.78	15.53
5100	Textured aluminum, 4' x 8' x 5/16" plywood backing, single face	2.89	2.77	5.66
5200	Double face	4.25	2.77	7.02
5300	4' x 10' x 5/16" plywood backing, single face	3	2.77	5.77

SHELL B

B2010 Exterior Walls

The table below lists costs per S.F. for exterior walls with wood siding. A variety of systems are presented using both wood and metal studs at 16″ and 24″ O.C.

System Components	QUANTITY	UNIT	COST PER S.F.		
			MAT.	INST.	TOTAL
SYSTEM B2010 148 1400					
2″X4″ STUDS, 16″ O.C., INSUL. WALL, W/ 5/8″ TEXTURE 1-11 FIR PLYWOOD					
Partitions, 2″ x 4″ studs 8′ high 16″ O.C.	1.000	B.F.	.46	1.11	1.57
Sheathing plywood on wall CDX 1/2″	1.000	S.F.	.68	.79	1.47
Building paper, asphalt felt sheathing paper 15 lb	1.000	S.F.	.04	.12	.16
Fiberglass insulation batts, paper or foil back, 3-1/2″, R11	1.000	S.F.	.52	.28	.80
Siding plywood texture 1-11 fir 5/8″, natural	1.000	S.F.	1.11	1.31	2.42
Exterior wood stain on other than shingles, 2 coats & sealer	1.000	S.F.	.15	.90	1.05
TOTAL			2.96	4.51	7.47

B2010 148	Panel, Shingle & Lap Siding	COST PER S.F.		
		MAT.	INST.	TOTAL
1400	Wood siding w/2″x4″studs, 16″O.C., insul. wall, 5/8″text 1-11 fir plywood	2.96	4.51	7.47
1450	5/8″ text 1-11 cedar plywood	4.47	4.35	8.82
1500	1″ x 4″ vert T.&G. redwood	5.05	6.15	11.20
1600	1″ x 8″ vert T.&G. redwood	4.60	5.55	10.15
1650	1″ x 5″ rabbetted cedar bev. siding	5.35	4.97	10.32
1700	1″ x 6″ cedar drop siding	5.45	5	10.45
1750	1″ x 12″ rough sawn cedar	3.07	4.55	7.62
1800	1″ x 12″ sawn cedar, 1″ x 4″ battens	4.27	4.90	9.17
1850	1″ x 10″ redwood shiplap siding	4.52	4.75	9.27
1900	18″ no. 1 red cedar shingles, 5-1/2″ exposed	3.75	5.95	9.70
1950	6″ exposed	3.62	5.75	9.37
2000	6-1/2″ exposed	3.48	5.55	9.03
2100	7″ exposed	3.34	5.35	8.69
2150	7-1/2″ exposed	3.21	5.15	8.36
3000	8″ wide aluminum siding	3.03	4.02	7.05
3150	8″ plain vinyl siding	2.51	4.04	6.55
3250	8″ insulated vinyl siding	2.62	4.04	6.66
3400	2″ x 6″ studs, 16″ O.C., insul. wall, w/ 5/8″ text 1-11 fir plywood	3.27	4.68	7.95
3500	5/8″ text 1-11 cedar plywood	4.78	4.68	9.46
3600	1″ x 4″ vert T.&G. redwood	5.35	6.30	11.65
3700	1″ x 8″ vert T.&G. redwood	4.91	5.75	10.66
3800	1″ x 5″ rabbetted cedar bev siding	5.70	5.15	10.85
3900	1″ x 6″ cedar drop siding	5.75	5.20	10.95
4000	1″ x 12″ rough sawn cedar	3.38	4.72	8.10
4200	1″ x 12″ sawn cedar, 1″ x 4″ battens	4.58	5.05	9.63
4500	1″ x 10″ redwood shiplap siding	4.83	4.92	9.75

Important: See the Reference Section for critical supporting data - Reference Numbers and City Cost Indexes

B2010 Exterior Walls

B2010 148	Panel, Shingle & Lap Siding	COST PER S.F.		
		MAT.	INST.	TOTAL
4550	18" no. 1 red cedar shingles, 5-1/2" exposed	4.06	6.15	10.21
4600	6" exposed	3.93	5.95	9.88
4650	6-1/2" exposed	3.79	5.75	9.54
4700	7" exposed	3.65	5.55	9.20
4750	7-1/2" exposed	3.52	5.35	8.87
4800	8" wide aluminum siding	3.34	4.19	7.53
4850	8" plain vinyl siding	2.82	4.21	7.03
4900	8" insulated vinyl siding	2.93	4.21	7.14
5000	2" x 6" studs, 24" O.C., insul. wall, 5/8" text 1-11, fir plywood	3.11	4.41	7.52
5050	5/8" text 1-11 cedar plywood	4.62	4.41	9.03
5100	1" x 4" vert T.&G. redwood	5.20	6.05	11.25
5150	1" x 8" vert T.&G. redwood	4.75	5.45	10.20
5200	1" x 5" rabbetted cedar bev siding	5.50	4.87	10.37
5250	1" x 6" cedar drop siding	5.60	4.91	10.51
5300	1" x 12" rough sawn cedar	3.22	4.45	7.67
5400	1" x 12" sawn cedar, 1" x 4" battens	4.42	4.80	9.22
5450	1" x 10" redwood shiplap siding	4.67	4.65	9.32
5500	18" no. 1 red cedar shingles, 5-1/2" exposed	3.90	5.85	9.75
5550	6" exposed	3.77	5.65	9.42
5650	7" exposed	3.49	5.25	8.74
5700	7-1/2" exposed	3.36	5.05	8.41
5750	8" wide aluminum siding	3.18	3.92	7.10
5800	8" plain vinyl siding	2.66	3.94	6.60
5850	8" insulated vinyl siding	2.77	3.94	6.71
5900	3-5/8" metal studs, 16 Ga., 16" OC insul.wall, 5/8"text 1-11 fir plywood	3.60	4.74	8.34
5950	5/8" text 1-11 cedar plywood	5.10	4.74	9.84
6000	1" x 4" vert T.&G. redwood	5.70	6.40	12.10
6050	1" x 8" vert T.&G. redwood	5.25	5.80	11.05
6100	1" x 5" rabbetted cedar bev siding	6	5.20	11.20
6150	1" x 6" cedar drop siding	6.10	5.25	11.35
6200	1" x 12" rough sawn cedar	3.71	4.78	8.49
6250	1" x 12" sawn cedar, 1" x 4" battens	4.91	5.15	10.06
6300	1" x 10" redwood shiplap siding	5.15	4.98	10.13
6350	18" no. 1 red cedar shingles, 5-1/2" exposed	4.39	6.20	10.59
6500	6" exposed	4.26	6	10.26
6550	6-1/2" exposed	4.12	5.80	9.92
6600	7" exposed	3.98	5.60	9.58
6650	7-1/2" exposed	3.98	5.45	9.43
6700	8" wide aluminum siding	3.67	4.25	7.92
6750	8" plain vinyl siding	3.15	4.11	7.26
6800	8" insulated vinyl siding	3.26	4.11	7.37
7000	3-5/8" metal studs, 16 Ga. 24" OC insul wall, 5/8" text 1-11 fir plywood	3.30	4.20	7.50
7050	5/8" text 1-11 cedar plywood	4.81	4.20	9.01
7100	1" x 4" vert T.&G. redwood	5.40	6	11.40
7150	1" x 8" vert T.&G. redwood	4.94	5.40	10.34
7200	1" x 5" rabbetted cedar bev siding	5.70	4.82	10.52
7250	1" x 6" cedar drop siding	5.75	4.86	10.61
7300	1" x 12" rough sawn cedar	3.41	4.40	7.81
7350	1" x 12" sawn cedar 1" x 4" battens	4.61	4.75	9.36
7400	1" x 10" redwood shiplap siding	4.86	4.60	9.46
7450	18" no. 1 red cedar shingles, 5-1/2" exposed	4.23	6	10.23
7500	6" exposed	4.09	5.80	9.89
7550	6-1/2" exposed	3.96	5.60	9.56
7600	7" exposed	3.82	5.40	9.22
7650	7-1/2" exposed	3.55	5	8.55
7700	8" wide aluminum siding	3.37	3.87	7.24
7750	8" plain vinyl siding	2.85	3.89	6.74
7800	8" insul. vinyl siding	2.96	3.89	6.85

B2010 Exterior Walls

Exterior Stucco Wall

The table below lists costs for some typical stucco walls including all the components as demonstrated in the component block below. Prices are presented for backup walls using wood studs, metal studs and CMU.

System Components	QUANTITY	UNIT	COST PER S.F.		
			MAT.	INST.	TOTAL
SYSTEM B2010 151 2100					
7/8″ CEMENT STUCCO, PLYWOOD SHEATHING, STUD WALL, 2″ X 4″, 16″ O.C.					
Framing 2x4 studs 8′ high 16″ O.C.	1.000	S.F.	.46	1.11	1.57
Plywood sheathing on ext. stud wall, 5/8″ thick	1.000	S.F.	.73	.84	1.57
Building paper, asphalt felt sheathing paper 15 lb	1.000	S.F.	.04	.12	.16
Stucco, 3 coats 7/8″ thk, float finish, on frame construction	1.000	S.F.	.65	4.14	4.79
Fiberglass insulation batts, paper or foil back, 3-1/2″, R-11	1.000	S.F.	.52	.28	.80
Paint exterior stucco, brushwork, primer & 2 coats	1.000	S.F.	.21	.76	.97
TOTAL			2.61	7.25	9.86

B2010 151	Stucco Wall	COST PER S.F.		
		MAT.	INST.	TOTAL
2100	Cement stucco, 7/8″ th., plywood sheathing, stud wall, 2″ x 4″, 16″ O.C.	2.61	7.25	9.86
2200	24″ O.C.	2.51	7.05	9.56
2300	2″ x 6″, 16″ O.C.	2.92	7.40	10.32
2400	24″ O.C.	2.76	7.15	9.91
2500	No sheathing, metal lath on stud wall, 2″ x 4″, 16″ O.C.	1.84	6.30	8.14
2600	24″ O.C.	1.74	6.05	7.79
2700	2″ x 6″, 16″ O.C.	2.16	6.45	8.61
2800	24″ O.C.	1.99	6.20	8.19
2900	1/2″ gypsum sheathing, 3-5/8″ metal studs, 16″ O.C.	2.95	6.55	9.50
2950	24″ O.C.	2.65	6.20	8.85
3000	Cement stucco, 5/8″ th., 2 coats on std. CMU block, 8″x 16″, 8″ thick	2.28	9.05	11.33
3100	10″ Thick	3.06	9.35	12.41
3200	12″ Thick	3.13	10.80	13.93
3300	Std. light Wt. block 8″ x 16″, 8″ Thick	3	8.80	11.80
3400	10″ Thick	3.76	9	12.76
3500	12″ Thick	3.84	10.30	14.14
3600	3 coat stucco, self furring metal lath 3.4 Lb/SY, on 8″ x 16″, 8″ thick	2.32	9.55	11.87
3700	10″ Thick	3.10	9.85	12.95
3800	12″ Thick	3.17	11.30	14.47
3900	Lt. Wt. block, 8″ Thick	3.04	9.30	12.34
4000	10″ Thick	3.80	9.50	13.30
4100	12″ Thick	3.88	10.80	14.68

Important: See the Reference Section for critical supporting data - Reference Numbers and City Cost Indexes

B2010 Exterior Walls

B2010 152	E.I.F.S.	COST PER S.F.		
		MAT.	INST.	TOTAL
5100	E.I.F.S., plywood sheathing, stud wall, 2" x 4", 16" O.C., 1" EPS	3.99	8.95	12.94
5110	2" EPS	4.35	8.95	13.30
5120	3" EPS	4.50	8.95	13.45
5130	4" EPS	4.96	8.95	13.91
5140	2" x 6", 16" O.C., 1" EPS	4.30	9.10	13.40
5150	2" EPS	4.66	9.10	13.76
5160	3" EPS	4.81	9.10	13.91
5170	4" EPS	5.25	9.10	14.35
5180	Cement board sheathing, 3-5/8" metal studs, 16" O.C., 1" EPS	4.73	10.60	15.33
5190	2" EPS	5.10	10.60	15.70
5200	3" EPS	5.25	10.60	15.85
5210	4" EPS	6.55	10.60	17.15
5220	6" metal studs, 16" O.C., 1" EPS	5.15	10.70	15.85
5230	2" EPS	5.50	10.70	16.20
5240	3" EPS	5.65	10.70	16.35
5250	4" EPS	6.95	10.70	17.65
5260	CMU block, 8" x 8" x 16", 1" EPS	4.14	12.25	16.39
5270	2" EPS	4.50	12.25	16.75
5280	3" EPS	4.65	12.25	16.90
5290	4" EPS	5.10	12.25	17.35
5300	8" x 10" x 16", 1" EPS	4.92	12.55	17.47
5310	2" EPS	5.30	12.55	17.85
5320	3" EPS	5.45	12.55	18
5330	4" EPS	5.90	12.55	18.45
5340	8" x 12" x 16", 1" EPS	4.99	14	18.99
5350	2" EPS	5.35	14	19.35
5360	3" EPS	5.50	14	19.50
5370	4" EPS	5.95	14	19.95

SHELL B

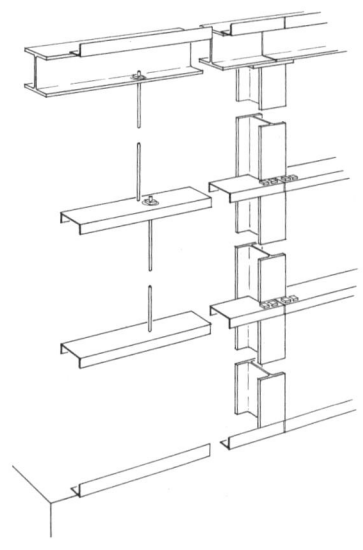

Description: The table below lists costs, $/S.F., for channel girts with sag rods and connector angles top and bottom for various column spacings, building heights and wind loads. Additive costs are shown for wind columns.

How to Use this Table: Add the cost of girts, sag rods, and angles to the framing costs of steel buildings clad in metal or composition siding. If the column spacing is in excess of the column spacing shown, use intermediate wind columns. Additive costs are shown under "Wind Columns".

Column Spacing 20'–0" Girts	$4.92
Wind Columns	1.13
Total Costs for Girts	$6.05/S.F.

Design and Pricing Assumptions: Structural steel is A36.

System Components	QUANTITY	UNIT	COST PER S.F.		
			MAT.	INST.	TOTAL
SYSTEM B2010 154 3000					
METAL SIDING SUPPORT, 18' HIGH, 20 PSF WIND LOAD, 20' SPACING					
Structural steel	1.417	Lb.	1.49	.39	1.88
Lightgage framing, angles less than 4"	.400	Lb.	.25	2.79	3.04
TOTAL			1.74	3.18	4.92
SYSTEM B2010 154 3100					
WIND COLUMNS					
Structural steel	.856	Lb.	.90	.23	1.13

B2010 154			Metal Siding Support					
	BLDG. HEIGHT (FT.)	WIND LOAD (P.S.F.)	COL. SPACING (FT.)		INTERMEDIATE COLUMNS	COST PER S.F.		
						MAT.	INST.	TOTAL
3000	18	20	20			1.74	3.17	4.91
3100					wind cols.	.90	.23	1.13
3200	RB2010 -430	20	25			1.90	3.21	5.11
3300					wind cols.	.72	.18	.90
3400		20	30			2.10	3.26	5.36
3500					wind cols.	.60	.15	.75
3600		20	35			2.31	3.32	5.63
3700					wind cols.	.51	.13	.64
3800		30	20			1.92	3.21	5.13
3900					wind cols.	.90	.23	1.13
4000		30	25			2.10	3.26	5.36
4100					wind cols.	.72	.18	.90
4200		30	30			2.32	3.32	5.64
4300					wind cols.	.81	.21	1.02
4400		30	35			3.09	3.52	6.61
4500					wind cols.	.81	.21	1.02
4600	30	20	20			1.69	2.23	3.92
4700					wind cols.	1.27	.32	1.59
4800		20	25			1.90	2.29	4.19
4900					wind cols.	1.02	.26	1.28

B20 Exterior Enclosure

B2010 Exterior Walls

B2010 154				Metal Siding Support				

	BLDG. HEIGHT (FT.)	WIND LOAD (P.S.F.)	COL. SPACING (FT.)		INTERMEDIATE COLUMNS	COST PER S.F.		
						MAT.	INST.	TOTAL
5000	30	20	30			2.14	2.35	4.49
5100					wind cols.	1	.26	1.26
5200		20	35			2.39	2.41	4.80
5300					wind cols.	.99	.26	1.25
5400		30	20			1.92	2.29	4.21
5500					wind cols.	1.50	.39	1.89
5600		30	25			2.13	2.35	4.48
5700					wind cols.	1.43	.37	1.80
5800		30	30			2.40	2.42	4.82
5900					wind cols.	1.35	.35	1.70
6000		30	35			3.32	2.65	5.97
6100					wind cols.	1.32	.34	1.66

SHELL B

201

The table below lists window systems by material, type and size. Prices between sizes listed can be interpolated with reasonable accuracy. Prices include frame, hardware, and casing as illustrated in the component block below.

System Components	QUANTITY	UNIT	COST PER UNIT		
			MAT.	INST.	TOTAL
SYSTEM B2020 102 3000					
WOOD, DOUBLE HUNG, STD. GLASS, 2'-8" X 4'-6"					
Framing rough opening, header w/jacks	14.680	B.F.	10.28	22.46	32.74
Casing, stock pine 11/16" x 2-1/2"	16.000	L.F.	22.56	29.44	52
Stool cap, pine, 11/16" x 3-1/2"	5.000	L.F.	12.10	11.05	23.15
Residential wood window, double hung, 2'-8" x 4'-6" standard glazed	1.000	Ea.	177	130	307
TOTAL			221.94	192.95	414.89

B2020 102				Wood Windows				
	MATERIAL	TYPE	GLAZING	SIZE	DETAIL	COST PER UNIT		
						MAT.	INST.	TOTAL
3000	Wood	double hung	std. glass	2'-8" x 4'-6"		222	193	415
3050				3'-0" x 5'-6"		292	222	514
3100			insul. glass	2'-8" x 4'-6"		237	193	430
3150				3'-0" x 5'-6"		315	222	537
3200		sliding	std. glass	3'-4" x 2'-7"		265	161	426
3250				4'-4" x 3'-3"		305	177	482
3300				5'-4" x 6'-0"		385	213	598
3350			insul. glass	3'-4" x 2'-7"		325	188	513
3400				4'-4" x 3'-3"		370	205	575
3450				5'-4" x 6'-0"		465	242	707
3500		awning	std. glass	2'-10" x 1'-9"		199	97	296
3600				4'-4" x 2'-8"		325	94	419
3700			insul. glass	2'-10" x 1'-9"		242	112	354
3800				4'-4" x 2'-8"		395	103	498
3900		casement	std. glass	1'-10" x 3'-2"	1 lite	253	127	380
3950				4'-2" x 4'-2"	2 lite	515	169	684
4000				5'-11" x 5'-2"	3 lite	795	218	1,013
4050				7'-11" x 6'-3"	4 lite	1,125	262	1,387
4100				9'-11" x 6'-3"	5 lite	1,475	305	1,780
4150			insul. glass	1'-10" x 3'-2"	1 lite	266	127	393
4200				4'-2" x 4'-2"	2 lite	545	169	714
4250				5'-11" x 5'-2"	3 lite	840	218	1,058
4300				7'-11" x 6'-3"	4 lite	1,175	262	1,437
4350				9'-11" x 6'-3"	5 lite	1,575	305	1,880

Important: See the Reference Section for critical supporting data - Reference Numbers and City Cost Indexes

B2020 Exterior Windows

B2020 102			Wood Windows					
	MATERIAL	TYPE	GLAZING	SIZE	DETAIL	COST PER UNIT		
						MAT.	INST.	TOTAL
4400	Wood	picture	std. glass	4'-6" x 4'-6"		420	218	638
4450				5'-8" x 4'-6"		470	243	713
4500		picture	insul. glass	4'-6" x 4'-6"		510	254	764
4550				5'-8" x 4'-6"		575	282	857
4600		fixed bay	std. glass	8' x 5'		1,450	400	1,850
4650				9'-9" x 5'-4"		1,025	565	1,590
4700			insul. glass	8' x 5'		2,075	400	2,475
4750				9'-9" x 5'-4"		1,125	565	1,690
4800		casement bay	std. glass	8' x 5'		1,450	460	1,910
4850			insul. glass	8' x 5'		1,650	460	2,110
4900		vert. bay	std. glass	8' x 5'		1,650	460	2,110
4950			insul. glass	8' x 5'		1,700	460	2,160

B2020 104			Steel Windows					
	MATERIAL	TYPE	GLAZING	SIZE	DETAIL	COST PER UNIT		
						MAT.	INST.	TOTAL
5000	Steel	double hung	1/4" tempered	2'-8" x 4'-6"		715	152	867
5050				3'-4" x 5'-6"		1,100	232	1,332
5100			insul. glass	2'-8" x 4'-6"		730	173	903
5150				3'-4" x 5'-6"		1,100	264	1,364
5200		horz. pivoted	std. glass	2' x 2'		230	50.50	280.50
5250				3' x 3'		515	114	629
5300				4' x 4'		920	202	1,122
5350				6' x 4'		1,375	305	1,680
5400			insul. glass	2' x 2'		235	58	293
5450				3' x 3'		530	130	660
5500				4' x 4'		940	231	1,171
5550				6' x 4'		1,400	345	1,745
5600		picture window	std. glass	3' x 3'		310	114	424
5650				6' x 4'		825	305	1,130
5700			insul. glass	3' x 3'		320	130	450
5750				6' x 4'		855	345	1,200
5800		industrial security	std. glass	2'-9" x 4'-1"		685	142	827
5850				4'-1" x 5'-5"		1,350	280	1,630
5900			insul. glass	2'-9" x 4'-1"		695	162	857
5950				4'-1" x 5'-5"		1,375	320	1,695
6000		comm. projected	std. glass	3'-9" x 5'-5"		1,125	257	1,382
6050				6'-9" x 4'-1"		1,525	350	1,875
6100			insul. glass	3'-9" x 5'-5"		1,150	294	1,444
6150				6'-9" x 4'-1"		1,550	400	1,950
6200		casement	std. glass	4'-2" x 4'-2"	2 lite	860	219	1,079
6250			insul. glass	4'-2" x 4'-2"		880	251	1,131
6300			std. glass	5'-11" x 5'-2"	3 lite	1,625	385	2,010
6350			insul. glass	5'-11" x 5'-2"		1,675	440	2,115

B2020 106			Aluminum Windows					
	MATERIAL	TYPE	GLAZING	SIZE	DETAIL	COST PER UNIT		
						MAT.	INST.	TOTAL
6400	Aluminum	projecting	std. glass	3'-1" x 3'-2"		244	115	359
6450				4'-5" x 5'-3"		345	144	489
6500			insul. glass	3'-1" x 3'-2"		293	138	431
6550				4'-5" x 5'-3"		415	173	588

B2020 Exterior Windows

B2020 106			Aluminum Windows				

	MATERIAL	TYPE	GLAZING	SIZE	DETAIL	COST PER UNIT		
						MAT.	INST.	TOTAL
6600	Aluminum	sliding	std. glass	3' x 2'		185	115	300
6650				5' x 3'		234	128	362
6700				8' x 4'		335	192	527
6750				9' x 5'		510	288	798
6800			insul. glass	3' x 2'		204	115	319
6850				5' x 3'		330	128	458
6900				8' x 4'		540	192	732
6950				9' x 5'		815	288	1,103
7000		single hung	std. glass	2' x 3'		164	115	279
7050				2'-8" x 6'-8"		350	144	494
7100				3'-4" x 5'0"		224	128	352
7150			insul. glass	2' x 3'		199	115	314
7200				2'-8" x 6'-8"		450	144	594
7250				3'-4" x 5'		315	128	443
7300		double hung	std. glass	2' x 3'		240	76	316
7350				2'-8" x 6'-8"		710	225	935
7400				3'-4" x 5'		665	211	876
7450			insul. glass	2' x 3'		247	86.50	333.50
7500				2'-8" x 6'-8"		730	257	987
7550				3'-4" x 5'-0"		685	241	926
7600		casements	std. glass	3'-1" x 3'-2"		198	123	321
7650				4'-5" x 5'-3"		480	299	779
7700			insul. glass	3'-1" x 3'-2"		210	141	351
7750				4'-5" x 5'-3"		510	340	850
7800		hinged swing	std. glass	3' x 4'		445	152	597
7850				4' x 5'		740	253	993
7900			insul. glass	3' x 4'		460	173	633
7950				4' x 5'		765	289	1,054
8000		folding type	std. glass	3'-0" x 4'-0"		228	124	352
8050				4'-0" x 5'-0"		320	124	444
8100			insul. glass	3'-0" x 4'-0"		286	124	410
8150				4'-0" x 5'-0"		415	124	539
8200		picture unit	std. glass	2'-0" x 3'-0"		128	76	204
8250				2'-8" x 6'-8"		380	225	605
8300				3'-4" x 5'-0"		355	211	566
8350			insul. glass	2'-0" x 3'-0"		135	86.50	221.50
8400				2'-8" x 6'-8"		400	257	657
8450				3'-4" x 5'-0"		375	241	616
8500		awning type	std. glass	3'-0" x 3'-0"	2 lite	405	82	487
8550				3'-0" x 4'-0"	3 lite	470	115	585
8600				3'-0" x 5'-4"	4 lite	570	115	685
8650				4'-0" x 5'-4"	4 lite	625	128	753
8700			insul. glass	3'-0" x 3'-0"	2 lite	430	82	512
8750				3'-0" x 4'-0"	3 lite	540	115	655
8800				3'-0" x 5'-4"	4 lite	670	115	785
8850				4'-0" x 5'-4"	4 lite	745	128	873
8900		jalousie type	std. glass	1'-7" x 3'-2"		150	115	265
8950				2'-3" x 4'-0"		214	115	329
9000				3'-1" x 2'-0"		169	115	284
9050				3'-1" x 5'-3"		305	115	420

Important: See the Reference Section for critical supporting data - Reference Numbers and City Cost Indexes

The table below lists costs per S.F of opening for framing with 1-3/4" x 4-1/2" clear anodized tubular aluminum framing. This is the type often used for 1/4" plate glass flush glazing.

For bronze finish, add 18% to material cost. For black finish, add 27% to material cost. For stainless steel, add 75% to material cost. For monumental grade, add 50% to material cost. This tube framing is usually installed by a glazing contractor.

Note: The costs below do not include the glass.

System Components	QUANTITY	UNIT	COST/S.F. OPNG.		
			MAT.	INST.	TOTAL
SYSTEM B2020 210 1250					
ALUM FLUSH TUBE, FOR 1/4"GLASS, 5'X20'OPENING, 3 INTER. HORIZONTALS					
Flush tube frame, alum mill fin, 1-3/4"x4" open headr for 1/4" glass	.450	L.F.	4.25	4.66	8.91
Flush tube frame alum mill fin 1-3/4"x4" open sill for 1/4" glass	.050	L.F.	.41	.51	.92
Flush tube frame alum mill fin,1-3/4"x4"closed back sill, 1/4" glass	.150	L.F.	1.88	1.46	3.34
Aluminum structural shapes, 1" to 10" members, under 1 ton	.040	Lb.	.10	.20	.30
Joints for tube frame, 90° clip type	.100	Ea.	2.20		2.20
Caulking/sealants, polysulfide, 1 or 2 part,1/2x1/4"bead 154 lf/gal	.500	L.F.	.16	1.65	1.81
TOTAL			9	8.48	17.48

B2020 210	Tubular Aluminum Framing	COST/S.F. OPNG.		
		MAT.	INST.	TOTAL
1100	Alum flush tube frame,for 1/4"glass,1-3/4"x4",5'x6'opng, no inter horizntls	10	10.25	20.25
1150	One intermediate horizontal	13.65	11.95	25.60
1200	Two intermediate horizontals	17.25	13.75	31
1250	5' x 20' opening, three intermediate horizontals	9	8.45	17.45
1400	1-3/4" x 4-1/2", 5' x 6' opening, no intermediate horizontals	11.40	10.25	21.65
1450	One intermediate horizontal	15.15	11.95	27.10
1500	Two intermediate horizontals	18.95	13.75	32.70
1550	5' x 20' opening, three intermediate horizontals	10.15	8.45	18.60
1700	For insulating glass, 2"x4-1/2", 5'x6' opening, no intermediate horizontals	12.25	10.75	23
1750	One intermediate horizontal	16.05	12.55	28.60
1800	Two intermediate horizontals	19.85	14.45	34.30
1850	5' x 20' opening, three intermediate horizontals	10.70	8.90	19.60
2000	Thermal break frame, 2-1/4"x4-1/2", 5'x6'opng, no intermediate horizontals	13.20	10.90	24.10
2050	One intermediate horizontal	17.70	13.05	30.75
2100	Two intermediate horizontals	22	15.25	37.25
2150	5' x 20' opening, three intermediate horizontals	12	9.30	21.30

B2020 Exterior Windows

The table below lists costs of curtain wall and spandrel panels per S.F. Costs do not include structural framing used to hang the panels from.

Spandrel Glass Panel

Sandwich Panel

B2020 220	Curtain Wall Panels	COST PER S.F.		
		MAT.	INST.	TOTAL
1000	Glazing panel, insulating, 1/2" thick, 2 lites 1/8" float, clear	8.65	8.70	17.35
1100	Tinted	12.70	8.70	21.40
1200	5/8" thick units, 2 lites 3/16" float, clear	10.50	9.20	19.70
1400	1" thick units, 2 lites, 1/4" float, clear	14.80	11.05	25.85
1700	Light and heat reflective glass, tinted	24	9.75	33.75
2000	Plate glass, 1/4" thick, clear	6.10	6.90	13
2050	Tempered	7.45	6.90	14.35
2100	Tinted	6	6.90	12.90
2200	3/8" thick, clear	9.70	11.05	20.75
2250	Tempered	15.05	11.05	26.10
2300	Tinted	12	11.05	23.05
2400	1/2" thick, clear	19.40	15.05	34.45
2450	Tempered	22.50	15.05	37.55
2500	Tinted	21	15.05	36.05
2600	3/4" thick, clear	27	23.50	50.50
2650	Tempered	31.50	23.50	55
3000	Spandrel glass, panels, 1/4" plate glass insul w/fiberglass, 1" thick	12.80	6.90	19.70
3100	2" thick	14.90	6.90	21.80
3200	Galvanized steel backing, add	4.43		4.43
3300	3/8" plate glass, 1" thick	21.50	6.90	28.40
3400	2" thick	23.50	6.90	30.40
4000	Polycarbonate, masked, clear or colored, 1/8" thick	6.60	4.88	11.48
4100	3/16" thick	7.95	5	12.95
4200	1/4" thick	8.80	5.35	14.15
4300	3/8" thick	16.20	5.55	21.75
5000	Facing panel, textured al, 4' x 8' x 5/16" plywood backing, sgl face	2.89	2.77	5.66
5100	Double face	4.25	2.77	7.02
5200	4' x 10' x 5/16" plywood backing, single face	3	2.77	5.77
5300	Double face	4.41	2.77	7.18
5400	4' x 12' x 5/16" plywood backing, single face	3.87	2.77	6.64
5500	Sandwich panel, 22 Ga. galv., both sides 2" insulation, enamel exterior	10.20	4.78	14.98
5600	Polyvinylidene floride exterior finish	10.75	4.78	15.53
5700	26 Ga., galv. both sides, 1" insulation, colored 1 side	5.35	4.53	9.88
5800	Colored 2 sides	6.90	4.53	11.43

Important: See the Reference Section for critical supporting data - Reference Numbers and City Cost Indexes

B2030 Exterior Doors

Costs are listed for exterior door systems by material, type and size. Prices between sizes listed can be interpolated with reasonable accuracy. Prices are per opening for a complete door system including frame.

B2030 110			**Glazed Doors, Steel or Aluminum**					
	MATERIAL	TYPE	DOORS	SPECIFICATION	OPENING	COST PER OPNG.		
						MAT.	INST.	TOTAL
5600	St. Stl. & glass	revolving	stock unit	manual oper.	6'-0" x 7'-0"	33,200	7,625	40,825
5650				auto Cntrls.	6'-10" x 7'-0"	46,800	8,125	54,925
5700	Bronze	revolving	stock unit	manual oper.	6'-10" x 7'-0"	38,800	15,300	54,100
5750				auto Cntrls.	6'-10" x 7'-0"	52,500	15,800	68,300
5800	St. Stl. & glass	balanced	standard	economy	3'-0" x 7'-0"	7,200	1,275	8,475
5850				premium	3'-0" x 7'-0"	12,500	1,675	14,175
6300	Alum. & glass	w/o transom	narrow stile	w/panic Hrdwre.	3'-0" x 7'-0"	1,275	835	2,110
6350				dbl. door, Hrdwre.	6'-0" x 7'-0"	2,200	1,375	3,575
6400			wide stile	hdwre.	3'-0" x 7'-0"	1,550	820	2,370
6450				dbl. door, Hdwre.	6'-0" x 7'-0"	3,050	1,625	4,675
6500			full vision	hdwre.	3'-0" x 7'-0"	1,900	1,300	3,200
6550				dbl. door, Hdwre.	6'-0" x 7'-0"	2,675	1,850	4,525
6600			non-standard	hdwre.	3'-0" x 7'-0"	1,100	820	1,920
6650				dbl. door, Hdwre.	6'-0" x 7'-0"	2,200	1,625	3,825
6700			bronze fin.	hdwre.	3'-0" x 7'-0"	1,175	820	1,995
6750				dbl. door, Hrdwre.	6'-0" x 7'-0"	2,350	1,625	3,975
6800			black fin.	hdwre.	3'-0" x 7'-0"	1,875	820	2,695
6850				dbl. door, Hdwre.	6'-0" x 7'-0"	3,750	1,625	5,375
6900		w/transom	narrow stile	hdwre.	3'-0" x 10'-0"	1,550	945	2,495
6950				dbl. door, Hdwre.	6'-0" x 10'-0"	2,175	1,650	3,825
7000			wide stile	hdwre.	3'-0" x 10'-0"	1,775	1,125	2,900
7050				dbl. door, Hdwre.	6'-0" x 10'-0"	2,350	1,975	4,325
7100			full vision	hdwre.	3'-0" x 10'-0"	1,950	1,275	3,225
7150				dbl. door, Hdwre.	6'-0" x 10'-0"	2,500	2,175	4,675
7200			non-standard	hdwre.	3'-0" x 10'-0"	1,150	880	2,030
7250				dbl. door, Hdwre.	6'-0" x 10'-0"	2,325	1,750	4,075
7300			bronze fin.	hdwre.	3'-0" x 10'-0"	1,225	880	2,105
7350				dbl. door, Hdwre.	6'-0" x 10'-0"	2,475	1,750	4,225
7400			black fin.	hdwre.	3'-0" x 10'-0"	1,950	880	2,830
7450				dbl. door, Hdwre.	6'-0" x 10'-0"	3,875	1,750	5,625
7500		revolving	stock design	minimum	6'-10" x 7'-0"	18,000	3,075	21,075
7550				average	6'-0" x 7'-0"	22,200	3,825	26,025
7600				maximum	6'-10" x 7'-0"	29,300	5,125	34,425
7650				min., automatic	6'-10" x 7'-0"	31,600	3,575	35,175

B2030 Exterior Doors

B2030 110	Glazed Doors, Steel or Aluminum

	MATERIAL	TYPE	DOORS	SPECIFICATION	OPENING	COST PER OPNG.		
						MAT.	INST.	TOTAL
7700	Alum. & glass	w/transom		avg., automatic	6'-10" x 7'-0"	35,800	4,325	40,125
7750				max., automatic	6'-10" x 7'-0"	42,900	5,625	48,525
7800		balanced	standard	economy	3'-0" x 7'-0"	5,825	1,275	7,100
7850				premium	3'-0" x 7'-0"	7,250	1,650	8,900
7900		mall front	sliding panels	alum. fin.	16'-0" x 9'-0"	2,675	640	3,315
7950					24'-0" x 9'-0"	3,875	1,175	5,050
8000				bronze fin.	16'-0" x 9'-0"	3,125	745	3,870
8050					24'-0" x 9'-0"	4,525	1,375	5,900
8100			fixed panels	alum. fin.	48'-0" x 9'-0"	7,225	920	8,145
8150				bronze fin.	48'-0" x 9'-0"	8,425	1,075	9,500
8200		sliding entrance	5' x 7' door	electric oper.	12'-0" x 7'-6"	6,825	1,175	8,000
8250		sliding patio	temp. glass	economy	6'-0" x 7'-0"	1,400	221	1,621
8300			temp. glass	economy	12'-0" x 7'-0"	2,425	295	2,720
8350				premium	6'-0" x 7'-0"	2,100	330	2,430
8400					12'-0" x 7'-0"	3,650	445	4,095

Important: See the Reference Section for critical supporting data - Reference Numbers and City Cost Indexes

B20 Exterior Enclosure

B2030   Exterior Doors

Costs are listed for exterior door systems by material, type and size. Prices between sizes listed can be interpolated with reasonable accuracy. Prices are per opening for a complete door system including frame as illustrated in the component block.

SHELL B

System Components	QUANTITY	UNIT	MAT.	INST.	TOTAL
SYSTEM B2030 210 2500					
WOOD DOOR, SOLID CORE, SINGLE, HINGED, 3'-0" X 7'-0"					
Commercial door, exterior flush solid core birch, 1-3/4" x 7' x 3' wide	1.000	Ea.	194	63.50	257.50
Wood door exterior frame, pine, w/trim, 5/4 x 5-3/16" deep	1.000	Set	148.75	40.12	188.87
Hinges,full mortise-high freq,brass base,4-1/2" x 4-1/2", US10	1.500	Pr.	114		114
Panic device for mortise locks, single door exit only	1.000	Ea.	555	111	666
Paint exterior door & frame one side 3' x 7', primer & 2 coats	1.000	Ea.	2.15	22.50	24.65
Sill, 8/4 x 8" deep, oak, 2" horns	1.000	Ea.	23	9.85	32.85
TOTAL			1,036.90	246.97	1,283.87

B2030 210   Wood Doors

	MATERIAL	TYPE	DOORS	SPECIFICATION	OPENING	MAT.	INST.	TOTAL
2350	Birch	solid core	single door	hinged	2'-6" x 6'-8"	1,025	241	1,266
2400					2'-6" x 7'-0"	1,025	242	1,267
2450					2'-8" x 7'-0"	1,025	242	1,267
2500					3'-0" x 7'-0"	1,025	247	1,272
2550			double door	hinged	2'-6" x 6'-8"	1,875	435	2,310
2600					2'-6" x 7'-0"	1,875	440	2,315
2650					2'-8" x 7'-0"	1,900	440	2,340
2700					3'-0" x 7'-0"	1,925	450	2,375
2750	Wood	combination	storm & screen	hinged	3'-0" x 6'-8"	320	59	379
2800					3'-0" x 7'-0"	350	65	415
2850		overhead	panels, H.D.	manual oper.	8'-0" x 8'-0"	665	445	1,110
2900					10'-0" x 10'-0"	1,000	490	1,490
2950					12'-0" x 12'-0"	1,450	590	2,040
3000					14'-0" x 14'-0"	2,125	680	2,805
3050					20'-0" x 16'-0"	4,350	1,350	5,700
3100				electric oper.	8'-0" x 8'-0"	1,425	665	2,090
3150					10'-0" x 10'-0"	1,775	710	2,485
3200					12'-0" x 12'-0"	2,225	810	3,035
3250					14'-0" x 14'-0"	2,900	900	3,800
3300					20'-0" x 16'-0"	5,200	1,800	7,000

B2030 Exterior Doors

| B2030 220 | | | | | Steel Doors | | |

	MATERIAL	TYPE	DOORS	SPECIFICATION	OPENING	COST PER OPNG.		
						MAT.	INST.	TOTAL
3350	Steel 18 Ga.	hollow metal	1 door w/frame	no label	2'-6" x 7'-0"	1,175	250	1,425
3400					2'-8" x 7'-0"	1,175	250	1,425
3450					3'-0" x 7'-0"	1,175	250	1,425
3500					3'-6" x 7'-0"	1,275	264	1,539
3550			1 door w/frame	no label	4'-0" x 8'-0"	1,475	265	1,740
3600			2 doors w/frame	no label	5'-0" x 7'-0"	2,250	465	2,715
3650					5'-4" x 7'-0"	2,250	465	2,715
3700					6'-0" x 7'-0"	2,250	465	2,715
3750					7'-0" x 7'-0"	2,400	490	2,890
3800					8'-0" x 8'-0"	2,850	490	3,340
3850			1 door w/frame	"A" label	2'-6" x 7'-0"	1,300	300	1,600
3900					2'-8" x 7'-0"	1,300	305	1,605
3950					3'-0" x 7'-0"	1,300	305	1,605
4000					3'-6" x 7'-0"	1,425	310	1,735
4050					4'-0" x 8'-0"	1,500	330	1,830
4100			2 doors w/frame	"A" label	5'-0" x 7'-0"	2,450	560	3,010
4150					5'-4" x 7'-0"	2,500	565	3,065
4200					6'-0" x 7'-0"	2,500	565	3,065
4250					7'-0" x 7'-0"	2,750	580	3,330
4300					8'-0" x 8'-0"	2,750	580	3,330
4350	Steel 24 Ga.	overhead	sectional	manual oper.	8'-0" x 8'-0"	585	445	1,030
4400					10'-0" x 10'-0"	805	490	1,295
4450					12'-0" x 12'-0"	1,100	590	1,690
4500					20'-0" x 14'-0"	3,250	1,275	4,525
4550				electric oper.	8'-0" x 8'-0"	1,350	665	2,015
4600					10'-0" x 10'-0"	1,575	710	2,285
4650					12'-0" x 12'-0"	1,875	810	2,685
4700					20'-0" x 14'-0"	4,100	1,725	5,825
4750	Steel	overhead	rolling	manual oper.	8'-0" x 8'-0"	890	720	1,610
4800					10'-0" x 10'-0"	1,175	820	1,995
4850					12'-0" x 12'-0"	1,525	960	2,485
4900					14'-0" x 14'-0"	2,000	1,450	3,450
4950					20'-0" x 12'-0"	2,550	1,275	3,825
5000					20'-0" x 16'-0"	2,975	1,925	4,900
5050				electric oper.	8'-0" x 8'-0"	1,750	950	2,700
5100					10'-0" x 10'-0"	2,025	1,050	3,075
5150					12'-0" x 12'-0"	2,375	1,200	3,575
5200					14'-0" x 14'-0"	2,850	1,675	4,525
5250					20'-0" x 12'-0"	3,400	1,500	4,900
5300					20'-0" x 16'-0"	3,825	2,150	5,975
5350				fire rated	10'-0" x 10'-0"	1,600	1,050	2,650
5400			rolling grill	manual oper.	10'-0" x 10'-0"	2,275	1,150	3,425
5450					15'-0" x 8'-0"	2,675	1,450	4,125
5500		vertical lift	1 door w/frame	motor operator	16'-0" x 16'-0"	19,400	4,675	24,075
5550					32'-0" x 24'-0"	38,600	3,100	41,700

Important: See the Reference Section for critical supporting data - Reference Numbers and City Cost Indexes

B2030 Exterior Doors

B2030 230	Aluminum Doors						

	MATERIAL	TYPE	DOORS	SPECIFICATION	OPENING	COST PER OPNG.		
						MAT.	INST.	TOTAL
6000	Aluminum	combination	storm & screen	hinged	3'-0" x 6'-8"	252	63.50	315.50
6050					3'-0" x 7'-0"	277	70	347
6100		overhead	rolling grill	manual oper.	12'-0" x 12'-0"	3,675	2,025	5,700
6150				motor oper.	12'-0" x 12'-0"	4,825	2,250	7,075
6200	Alum. & Fbrgls.	overhead	heavy duty	manual oper.	12'-0" x 12'-0"	2,100	590	2,690
6250				electric oper.	12'-0" x 12'-0"	2,875	810	3,685

SHELL B

B3010 Roof Coverings

Multiple ply roofing is the most popular covering for minimum pitch roofs. Lines 1200 through 6300 list the costs of the various types, plies and weights per S.F.

System Components	QUANTITY	UNIT	COST PER S.F.		
			MAT.	INST.	TOTAL
SYSTEM B3010 105 2500					
ASPHALT FLOOD COAT, W/GRAVEL, 4 PLY ORGANIC FELT					
Organic #30 base felt	1.000	S.F.	.08	.07	.15
Organic #15 felt, 3 plies	3.000	S.F.	.11	.21	.32
Asphalt mopping of felts	4.000	S.F.	.22	.65	.87
Asphalt flood coat	1.000	S.F.	.14	.52	.66
Gravel aggregate, washed river stone	4.000	Lb.	.06	.12	.18
TOTAL			.61	1.57	2.18

B3010 105	Built-Up	COST PER S.F.		
		MAT.	INST.	TOTAL
1200	Asphalt flood coat w/gravel; not incl. insul, flash., nailers			
1300				
1400	Asphalt base sheets & 3 plies #15 asphalt felt, mopped	.55	1.40	1.95
1500	On nailable deck	.61	1.47	2.08
1600	4 plies #15 asphalt felt, mopped	.79	1.55	2.34
1700	On nailable deck	.71	1.63	2.34
1800	Coated glass base sheet, 2 plies glass (type IV), mopped	.59	1.40	1.99
1900	For 3 plies	.70	1.55	2.25
2000	On nailable deck	.66	1.63	2.29
2300	4 plies glass fiber felt (type IV), mopped	.85	1.55	2.40
2400	On nailable deck	.77	1.63	2.40
2500	Organic base sheet & 4 plies #15 organic felt, mopped	.60	1.56	2.16
2600	On nailable deck	.59	1.63	2.22
2700	4 plies #15 organic felt, mopped	.73	1.40	2.13
2750				
2800	Asphalt flood coat, smooth surface			
2850				
2900	Asphalt base sheet & 3 plies #15 asphalt felt, mopped	.60	1.28	1.88
3000	On nailable deck	.57	1.34	1.91
3100	Coated glass fiber base sheet & 2 plies glass fiber felt, mopped	.54	1.24	1.78
3200	On nailable deck	.51	1.28	1.79
3300	For 3 plies, mopped	.65	1.34	1.99
3400	On nailable deck	.61	1.40	2.01
3700	4 plies glass fiber felt (type IV), mopped	.76	1.34	2.10
3800	On nailable deck	.72	1.40	2.12
3900	Organic base sheet & 3 plies #15 organic felt, mopped	.58	1.28	1.86
4000	On nailable decks	.54	1.34	1.88
4100	4 plies #15 organic felt, mopped	.68	1.40	2.08
4200	Coal tar pitch with gravel surfacing			
4300	4 plies #15 tarred felt, mopped	1.29	1.47	2.76
4400	3 plies glass fiber felt (type IV), mopped	1.06	1.63	2.69
4500	Coated glass fiber base sheets 2 plies glass fiber felt, mopped	1.05	1.63	2.68

Important: See the Reference Section for critical supporting data - Reference Numbers and City Cost Indexes

B3010 Roof Coverings

B3010 105	Built-Up	COST PER S.F.		
		MAT.	INST.	TOTAL
4600	On nailable decks	.94	1.72	2.66
4800	3 plies glass fiber felt (type IV), mopped	1.47	1.47	2.94
4900	On nailable decks	1.35	1.55	2.90
5300	Asphalt mineral surface, roll roofing			
5400	1 ply #15 organic felt, 1 ply mineral surfaced			
5500	Selvage roofing, lap 19", nailed and mopped	.46	1.15	1.61
5600	3 plies glass fiber felt (type IV), 1 ply mineral surfaced			
5700	Selvage roofing, lapped 19", mopped	.72	1.24	1.96
5800	Coated glass fiber base sheet			
5900	2 plies glass fiber, felt (type IV), 1 ply mineral surfaced			
6000	Selvage, roofing, lapped 19", mopped	.78	1.24	2.02
6100	On nailable deck	.73	1.28	2.01
6200	3 plies glass fiber felt (type IV), 1 ply mineral surfaced			
6300	Selvage roofing, lapped 19", mopped	.72	1.24	1.96

Fully Adhered

Ballasted

The systems listed below reflect only the cost for the single ply membrane.

For additional components see:

Insulation	B3010 320
Base Flashing	B3010 410
Roof Edge	B3010 420
Roof Openings	B3020 210

B3010 120	Single Ply Membrane	MAT.	INST.	TOTAL
		COST PER S.F.		
1000	CSPE (Chlorosulfonated polyethylene), 35 mils, fully adhered	1.36	.79	2.15
1100	Loosely laid with stone ballast	1.43	.40	1.83
2000	EPDM (Ethylene propylene diene monomer), 45 mils, fully adhered	.85	.79	1.64
2100	Loosely laid with stone ballast	.67	.40	1.07
2200	Mechanically fastened with batten strips	.59	.59	1.18
3300	60 mils, fully adhered	1	.79	1.79
3400	Loosely laid with stone ballast	.83	.40	1.23
3500	Mechanically fastened with batten strips	.74	.59	1.33
4000	Modified bit., SBS modified, granule surface cap sheet, mopped, 150 mils	.54	1.55	2.09
4100	Smooth surface cap sheet, mopped, 145 mils	.50	1.47	1.97
4500	APP modified, granule surface cap sheet, torched, 180 mils	.61	1.02	1.63
4600	Smooth surface cap sheet, torched, 170 mils	.55	.98	1.53
5000	PIB (Polyisobutylene), 100 mils, fully adhered with contact cement	1.82	.79	2.61
5100	Loosely laid with stone ballast	1.41	.40	1.81
5200	Partially adhered with adhesive	1.76	.59	2.35
5300	Hot asphalt attachment	1.68	.59	2.27
6000	Reinforced PVC, 48 mils, loose laid and ballasted with stone	1.07	.40	1.47
6100	Partially adhered with mechanical fasteners	.98	.59	1.57
6200	Fully adhered with adhesive	1.30	.79	2.09
6300	Reinforced PVC, 60 mils, loose laid and ballasted with stone	.99	.40	1.39
6400	Partially adhered with mechanical fasteners	.89	.59	1.48
6500	Fully adhered with adhesive	1.22	.79	2.01

B3010   Roof Coverings

Listed below are installed prices for preformed roofing materials with applicable specifications for each material.

B3010 130	Preformed Metal Roofing	COST PER S.F.		
		MAT.	INST.	TOTAL
0200	Corrugated roofing, aluminum, mill finish, .0175" thick, .272 P.S.F.	.76	1.43	2.19
0250	.0215 thick, .334 P.S.F.	.99	1.43	2.42
0300	.0240" thick, .412 P.S.F.	1.43	1.43	2.86
0350	.0320" thick, .552 P.S.F.	1.69	1.43	3.12
0400	Painted, .0175" thick, .280 P.S.F.	1.10	1.43	2.53
0450	.0215" thick, .344 P.S.F.	1.22	1.43	2.65
0500	.0240" thick, .426 P.S.F.	1.73	1.43	3.16
0550	.0320" thick, .569 P.S.F.	2.34	1.43	3.77
0700	Fiberglass, 6 oz., .375 P.S.F.	2.46	1.72	4.18
0750	8 oz., .5 P.S.F.	3.55	1.72	5.27
0900	Steel, galvanized, 29 ga., .72 P.S.F.	1.16	1.57	2.73
0930	26 ga., .91 P.S.F.	1.21	1.64	2.85
0950	24 ga., 1.26 P.S.F.	1.68	1.72	3.40
0970	22 ga., 1.45 P.S.F.	2.08	1.81	3.89
1000	Colored, 26 ga., 1.08 P.S.F.	1.19	1.64	2.83
1050	24 ga., 1.43 P.S.F.	1.82	1.72	3.54

B3010 Roof Coverings

Batten Seam

Flat Seam

Standing Seam

Formed metal roofing is practical on all sloped roofs from 1/4″ per foot of rise to vertical. Its use is more aesthetic than economical. Table below lists the various materials used and the weight per S.F.

System Components	QUANTITY	UNIT	COST PER S.F.		
			MAT.	INST.	TOTAL
SYSTEM B3010 135 1000					
BATTEN SEAM, FORMED COPPER ROOFING 3″ MIN. SLOPE, 16 OZ., 1.2 PSF.					
Copper roof, batten seam, 16 oz. over 10 squares	1.000	S.F.	4.85	4.70	9.55
Asphalt impregnated felt, 30 lb, 2 sq. per roll, not mopped	1.000	S.F.	.08	.07	.15
TOTAL			4.93	4.77	9.70

B3010 135	Formed Metal	COST PER S.F.		
		MAT.	INST.	TOTAL
1000	Batten seam, formed copper roofing, 3″min slope, 16 oz., 1.2 P.S.F.	4.93	4.77	9.70
1100	18 oz., 1.35 P.S.F.	5.50	5.25	10.75
1200	20 oz., 1.50 P.S.F.	6.05	5.25	11.30
1500	Lead, 3″ min slope, 3 Lb., 3.6 P.S.F.	4.43	4.42	8.85
2000	Zinc copper alloy, 3″ min slope, .020″ thick, .88 P.S.F.	6.30	4.42	10.72
2050	.027″ thick, 1.18 P.S.F.	7.55	4.57	12.12
2100	.032″ thick, 1.41 P.S.F.	8.50	4.77	13.27
2150	.040″ thick, 1.77 P.S.F.	10	5	15
3000	Flat seam, copper, 1/4″ min. slope, 16 oz., 1.2 P.S.F.	4.38	4.42	8.80
3100	18 oz., 1.35 P.S.F.	4.93	4.57	9.50
3200	20 oz., 1.50 P.S.F.	5.50	4.77	10.27
3500	Lead, 1/4″ min. slope, 3 Lb. 3.6 P.S.F.	4.43	4.07	8.50
4000	Zinc copper alloy, 1/4″ min. slope, .020″ thick, .84 P.S.F.	6.10	4.42	10.52
4050	.027″ thick, 1.13 P.S.F.	7.35	4.57	11.92
4100	.032″ thick, 1.33 P.S.F.	8.25	4.72	12.97
4150	.040″ thick, 1.67 P.S.F.	9.75	5	14.75
5000	Standing seam, copper, 2-1/2″ min. slope, 16 oz., 1.25 P.S.F.	4.73	4.07	8.80
5100	18 oz., 1.40 P.S.F.	5.30	4.42	9.72
5200	20 oz., 1.56 P.S.F.	5.70	4.77	10.47
5500	Lead, 2-1/2″ min. slope, 3 Lb., 3.6 P.S.F.	4.43	4.07	8.50
6000	Zinc copper alloy, 2-1/2″ min. slope, .020″ thick, .87 P.S.F.	6.15	4.42	10.57
6050	.027″ thick, 1.18 P.S.F.	7.40	4.57	11.97
6100	.032″ thick, 1.39 P.S.F.	8.35	4.77	13.12
6150	.040″ thick, 1.74 P.S.F.	9.85	5	14.85

Important: See the Reference Section for critical supporting data - Reference Numbers and City Cost Indexes

B3010 Roof Coverings

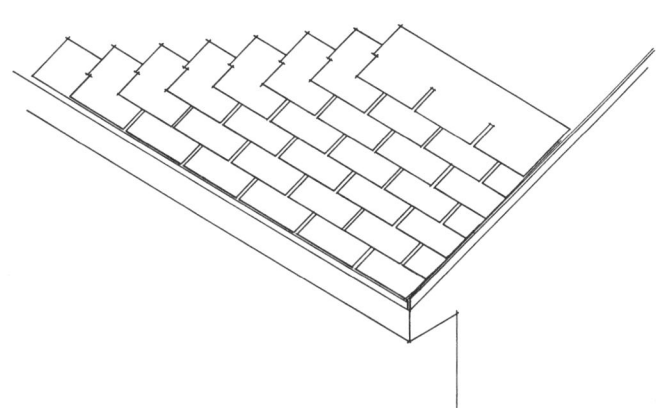

Shingles and tiles are practical in applications where the roof slope is more than 3-1/2″ per foot of rise. Table below lists the various materials and the weight per S.F.

System Components	QUANTITY	UNIT	COST PER S.F.		
			MAT.	INST.	TOTAL
SYSTEM B3010 140 1200					
STRIP SHINGLES, 4″ SLOPE, MULTI-LAYERED, CLASS A					
Class A, 260-300 lb./sq., 4 bundles/sq.	1.000	S.F.	.69	1.19	1.88
Building paper, asphalt felt sheathing paper 15 lb	1.000	S.F.	.04	.12	.16
TOTAL			.73	1.31	2.04

B3010 140	Shingle & Tile	COST PER S.F.		
		MAT.	INST.	TOTAL
1095	Asphalt roofing			
1100	Strip shingles,4″ slope,inorganic class A 210-235 lb/Sq.	.46	.87	1.33
1150	Organic, class C, 235-240 lb./sq.	.51	.95	1.46
1200	Premium laminated multi-layered, class A, 260-300 lb./Sq.	.73	1.31	2.04
1250	Class C, 300-385 lb/sq	.87	1.51	2.38
1545	Metal roofing			
1550	Alum., shingles, colors, 3″min slope, .019″thick, 0.4 PSF	1.99	.96	2.95
1600	.020″ thick, 0.6 PSF	2.19	.96	3.15
1850	Steel, colors, 3″ min slope, 26 gauge, 1.0 PSF	2.64	1.96	4.60
2000	24 gauge, 1.7 PSF	2.74	1.96	4.70
2295	Mineral fiber roofing			
2300	3″ min slope, class A, shingles, 14″ x 30″ 3.25 psf	3.28	1.96	5.24
2400	Shakes, 9″ x 16″, 5.0 PSF	2.99	1.96	4.95
2500	16″ x 16″, 5.4 PSF	2.03	1.46	3.49
2795	Slate roofing			
2800	4″ min. slope, shingles, 3/16″ thick, 8.0 PSF	6.50	2.44	8.94
2900	1/4″ thick, 9.0 PSF	6.85	2.44	9.29
3495	Wood roofing			
3500	4″ min slope, cedar shingles, 16″ x 5″, 5″ exposure 1.6 PSF	2.14	1.91	4.05
4000	Shakes, 18″, 8-1/2″ exposure, 2.8 PSF	1.06	2.35	3.41
4100	24″, 10″ exposure, 3.5 PSF	1.35	1.91	3.26
5095	Tile roofing			
5100	Aluminum, mission, 3″ min slope, .019″ thick, 0.65 PSF	6.35	1.84	8.19
5200	S type, 0.48 PSF	4.38	1.55	5.93
6000	Clay, Americana, 3″ minimum slope, 8 PSF	5.85	2.59	8.44
6500	Mission, 4″ minimum slope, 12.2 PSF	7.25	3.67	10.92

B3010 Roof Coverings

B3010 320	Roof Deck Rigid Insulation	COST PER S.F.		
		MAT.	INST.	TOTAL
0100	Fiberboard low density, 1/2" thick, R1.39			
0150	1" thick R2.78	.42	.52	.94
0300	1 1/2" thick R4.17	.63	.52	1.15
0350	2" thick R5.56	.84	.52	1.36
0370	Fiberboard high density, 1/2" thick R1.3	.22	.42	.64
0380	1" thick R2.5	.44	.52	.96
0390	1 1/2" thick R3.8	.66	.52	1.18
0410	Fiberglass, 3/4" thick R2.78	.55	.42	.97
0450	15/16" thick R3.70	.74	.42	1.16
0500	1-1/16" thick R4.17	.92	.42	1.34
0550	1-5/16" thick R5.26	1.27	.42	1.69
0600	2-1/16" thick R8.33	1.35	.52	1.87
0650	2 7/16" thick R10	1.54	.52	2.06
1000	Foamglass, 1 1/2" thick R4.55	1.43	.52	1.95
1100	3" thick R9.09	2.92	.59	3.51
1200	Tapered for drainage	1.24	.69	1.93
1260	Perlite, 1/2" thick R1.32	.31	.40	.71
1300	3/4" thick R2.08	.33	.52	.85
1350	1" thick R2.78	.35	.52	.87
1400	1 1/2" thick R4.17	.44	.52	.96
1450	2" thick R5.56	.70	.59	1.29
1510	Polyisocyanurate 2#/CF density, 1" thick R7.14	.72	.30	1.02
1550	1 1/2" thick R10.87	.76	.33	1.09
1600	2" thick R14.29	1	.38	1.38
1650	2 1/2" thick R16.67	1.10	.40	1.50
1700	3" thick R21.74	1.61	.42	2.03
1750	3 1/2" thick R25	1.63	.42	2.05
1800	Tapered for drainage	.84	.30	1.14
1810	Expanded polystyrene, 1#/CF density, 3/4" thick R2.89	.28	.28	.56
1820	2" thick R7.69	.58	.33	.91
1825	Extruded Polystyrene			
1830	Extruded polystyrene, 15 PSI compressive strength, 1" thick R5	.47	.28	.75
1835	2" thick R10	.73	.33	1.06
1840	3" thick R15	.95	.42	1.37
1900	25 PSI compressive strength, 1" thick R5	.69	.28	.97
1950	2" thick R10	1.33	.33	1.66
2000	3" thick R15	2.02	.42	2.44
2050	4" thick R20	2.70	.42	3.12
2150	Tapered for drainage	.55	.28	.83
2550	40 PSI compressive strength, 1" thick R5	.48	.28	.76
2600	2" thick R10	.95	.33	1.28
2650	3" thick R15	1.39	.42	1.81
2700	4" thick R20	1.86	.42	2.28
2750	Tapered for drainage	.69	.30	.99
2810	60 PSI compressive strength, 1" thick R5	.58	.29	.87
2850	2" thick R10	1.03	.35	1.38
2900	Tapered for drainage	.84	.30	1.14
4000	Composites with 1-1/2" polyisocyanurate			
4010	1" fiberboard	1.47	.52	1.99
4020	1" perlite	1.55	.49	2.04
4030	7/16" oriented strand board	1.78	.52	2.30

RB3010 -010

B
SHELL

Important: See the Reference Section for critical supporting data - Reference Numbers and City Cost Indexes

B30 Roofing

B3010   Roof Coverings

The table below lists the costs for various types of base flashing per L.F.

All the systems below are also based on using a 4″ x 4″ diagonally cut treated wood cant and a 2″ x 8″ treated wood nailer fastened at 4′-0″ O.C.

SHELL **B**

System Components		QUANTITY	UNIT	COST PER L.F.		
				MAT.	INST.	TOTAL
SYSTEM B3010 410 1000						
ALUMINUM BASE FLASHING MILL FINISH, .019″ THICK						
Treated 2″ x 8″ wood blocking		1.333	B.F.	1.07	1.37	2.44
Anchor bolts, 4′ O.C.		1.333	B.F.	.50	1.94	2.44
Treated wood cant, 4″ x 4″ split		1.000	L.F.	1.52	1.28	2.80
Roof covering, splice sheet		1.000	S.F.	1.43	.40	1.83
Base flashing, .019″ thick, aluminum		1.000	L.F.	.78	2.87	3.65
Reglet .025″ thick counter flashing .032″ thick		1.000	L.F.	1.12	1.97	3.09
Counter flashing, aluminum		1.000	L.F.	1.45	3.46	4.91
	TOTAL			7.87	13.29	21.16

B3010 410			Base Flashing					
	TYPE	DESCRIPTION	SPECIFICATION	REGLET	COUNTER FLASHING	COST PER L.F.		
						MAT.	INST.	TOTAL
1000	Aluminum	mill finish	.019″ thick	alum. .025	alum. .032	7.90	13.30	21.20
1100			.050″ thick	.025	.032	9.30	13.30	22.60
1200		fabric, 2 sides	.004″ thick	.025	.032	8.20	11.70	19.90
1300			.016″ thick	.025	.032	8.40	11.70	20.10
1400		mastic, 2 sides	.004″ thick	.025	.032	8.20	11.70	19.90
1500			.016″ thick	.025	.032	8.55	11.70	20.25
1700	Copper	sheets, plain	16 oz.	copper 10 oz.	copper 10 oz.	13.60	14.05	27.65
1800			20 oz.	10 oz.	10 oz.	15.05	14.20	29.25
1900			24 oz.	10 oz.	10 oz.	17.05	14.40	31.45
2000			32 oz.	10 oz.	10 oz.	18.85	14.60	33.45
2200	Sheets, Metal	galvanized	20 Ga.	galv. 24 Ga.	galv. 24 Ga.	9.05	30	39.05
2300			24 Ga.	24 Ga.	24 Ga.	8.20	24	32.20
2400			30 Ga.	24 Ga.	24 Ga.	7.30	18.25	25.55
2600	Rubber	butyl	1/32″ thick	galv. 24 Ga.	galv. 24 Ga.	6.95	11.90	18.85
2700			1/16″ thick	24 Ga.	24 Ga.	7.35	11.90	19.25
2800		neoprene	1/16″ thick	24 Ga.	24 Ga.	7.85	11.90	19.75
2900			1/8″ thick	24 Ga.	24 Ga.	9.65	11.90	21.55

B3010 Roof Coverings

The table below lists the costs for various types of roof perimeter edge treatments.

All systems below include the cost per L.F. for a 2″ x 8″ treated wood nailer fastened at 4′-0″ O.C. and a diagonally cut 4″ x 6″ treated wood cant.

Roof edge and base flashing are assumed to be made from the same material.

System Components	QUANTITY	UNIT	COST PER L.F.		
			MAT.	INST.	TOTAL
SYSTEM B3010 420 1000					
ALUMINUM ROOF EDGE, MILL FINISH, .050″ THICK, 4″ HIGH					
Treated 2″ x 8″ wood blocking	1.333	B.F.	1.07	1.37	2.44
Anchor bolts, 4′ O.C.	1.333	B.F.	.50	1.94	2.44
Treated wood cant	1.000	L.F.	1.52	1.28	2.80
Roof covering, splice sheet	1.000	S.F.	1.43	.40	1.83
Aluminum roof edge, .050″ thick, 4″ high	1.000	L.F.	4.71	3.58	8.29
TOTAL			9.23	8.57	17.80

B3010 420			Roof Edges			COST PER L.F.		
	EDGE TYPE	DESCRIPTION	SPECIFICATION	FACE HEIGHT		MAT.	INST.	TOTAL
1000	Aluminum	mill finish	.050″ thick	4″		9.25	8.60	17.85
1100				6″		9.60	8.85	18.45
1200				8″		10.85	9.15	20
1300		duranodic	.050″ thick	4″		9.05	8.60	17.65
1400				6″		9.90	8.85	18.75
1500				8″		10.70	9.15	19.85
1600		painted	.050″ thick	4″		9.80	8.60	18.40
1700				6″		10.80	8.85	19.65
1800				8″		10.80	9.15	19.95
2000	Copper	plain	16 oz.	4″		6.40	8.60	15
2100				6″		7.35	8.85	16.20
2200				8″		8.30	10.15	18.45
2300			20 oz.	4″		7	9.80	16.80
2400				6″		7.80	9.45	17.25
2500				8″		8.85	10.90	19.75
2700	Sheet Metal	galvanized	20 Ga.	4″		7.20	10.15	17.35
2800				6″		8.55	10.15	18.70
2900				8″		9.85	11.85	21.70
3000			24 Ga.	4″		6.40	8.60	15
3100				6″		7.30	8.60	15.90
3200				8″		8.25	9.80	18.05

Important: See the Reference Section for critical supporting data - Reference Numbers and City Cost Indexes

B3010 Roof Coverings

B3010 430						

Flashing

	MATERIAL	BACKING	SIDES	SPECIFICATION	QUANTITY	MAT.	INST.	TOTAL
						COST PER S.F.		
0040	Aluminum	none		.019"		.78	2.87	3.65
0050				.032"		1.29	2.87	4.16
0100				.040"		1.74	2.87	4.61
0150				.050"		2.21	2.87	5.08
0300		fabric	2	.004"		1.11	1.26	2.37
0350				.016"		1.31	1.26	2.57
0400		mastic		.004"		1.11	1.26	2.37
0450				.005"		1.31	1.26	2.57
0500				.016"		1.44	1.26	2.70
0510								
0700	Copper	none		16 oz.	<500 lbs.	3.04	3.61	6.65
0710					>2000 lbs.	3.37	2.68	6.05
0750				20 oz.	<500 lbs.	4.51	3.78	8.29
0760					>2000 lbs.	4.18	2.87	7.05
0800				24 oz.	<500 lbs.	5.40	3.96	9.36
0810					>2000 lbs.	5	3.08	8.08
0850				32 oz.	<500 lbs.	7.20	4.16	11.36
0860					>2000 lbs.	6.70	3.20	9.90
1000		paper backed	1	2 oz.		.99	1.26	2.25
1100				3 oz.		1.30	1.26	2.56
1200			2	2 oz.		1	1.26	2.26
1300				5 oz.		1.93	1.26	3.19
2000	Copper lead	fabric	1	2 oz.		1.69	1.26	2.95
2100	Coated			5 oz.		1.95	1.26	3.21
2200		mastic	2	2 oz.		1.32	1.26	2.58
2300				5 oz.		1.64	1.26	2.90
2400		paper	1	2 oz.		.99	1.26	2.25
2500				3 oz.		1.30	1.26	2.56
2600			2	2 oz.		1	1.26	2.26
2700				5 oz.		1.93	1.26	3.19
3000	Lead	none		2.5 lb.	to 12" wide	3.33	3.08	6.41
3100					over 12"	4.06	3.08	7.14
3500	PVC black	none		.010"		.19	1.46	1.65
3600				.020"		.26	1.46	1.72
3700				.030"		.35	1.46	1.81
3800				.056"		.84	1.46	2.30
4000	Rubber butyl	none		1/32"		.83	1.46	2.29
4100				1/16"		1.23	1.46	2.69
4200	Neoprene			1/16"		1.74	1.46	3.20
4300				1/8"		3.52	1.46	4.98
4500	Stainless steel	none		.015"	<500 lbs.	3.65	3.61	7.26
4600	Copper clad				>2000 lbs.	3.52	2.68	6.20
4700				.018"	<500 lbs.	4.83	4.16	8.99
4800					>2000 lbs.	3.53	2.87	6.40
5000	Plain			32 ga.		2.97	2.68	5.65
5100				28 ga.		3.69	2.68	6.37
5200				26 ga.		4.47	2.68	7.15
5300				24 ga.		5.80	2.68	8.48
5400	Terne coated			28 ga.		5.55	2.68	8.23
5500				26 ga.		6.30	2.68	8.98

SHELL B

B3010 Roof Coverings

B3010 610			Gutters					
	SECTION	MATERIAL	THICKNESS	SIZE	FINISH	COST PER L.F.		
						MAT.	INST.	TOTAL
0050	Box	aluminum	.027″	5″	enameled	1.38	4.33	5.71
0100					mill	1.46	4.33	5.79
0200			.032″	5″	enameled	2.60	4.33	6.93
0300					mill	2.59	4.33	6.92
0500		copper	16 Oz.	4″	lead coated	8.90	4.33	13.23
0600					mill	5.55	4.33	9.88
0700				5″	lead coated	9.75	4.33	14.08
0800					mill	6.05	4.33	10.38
1000		steel galv.	28 Ga.	5″	enameled	1.23	4.33	5.56
1100					mill	1.30	4.33	5.63
1200			26 Ga.	5″	mill	1.05	4.33	5.38
1300				6″	mill	1.76	4.33	6.09
1500		stainless steel		4″	mill	5.10	4.33	9.43
1600		stainless		5″	mill	5.50	4.33	9.83
1800		vinyl		4″	colors	.94	4.03	4.97
1900				5″	colors	1.10	4.03	5.13
2200		wood clear,		3″x4″	treated	6.90	4.43	11.33
2300		hemlock or fir		4″x5″	treated	8	4.43	12.43
3000	Half round	copper	16 Oz.	4″	lead coated	8.25	4.33	12.58
3100					mill	4.51	4.33	8.84
3200				5″	lead coated	9.50	4.98	14.48
3300					mill	5.75	4.33	10.08
3600		steel galv.	28 Ga.	5″	enameled	1.23	4.33	5.56
3700					mill	1.30	4.33	5.63
3800			26 Ga.	5″	mill	1.05	4.33	5.38
4000		stainless steel		4″	mill	5.10	4.33	9.43
4100				5″	mill	5.50	4.33	9.83
5000		vinyl		4″	white	.75	4.03	4.78

B3010 620			Downspouts					
	MATERIALS	SECTION	SIZE	FINISH	THICKNESS	COST PER V.L.F.		
						MAT.	INST.	TOTAL
0100	Aluminum	rectangular	2″x3″	embossed mill	.020″	.94	2.74	3.68
0150				enameled	.020″	1.39	2.74	4.13
0200				enameled	.024″	1.95	2.89	4.84
0250			3″x4″	enameled	.024″	2.75	3.71	6.46
0300		round corrugated	3″	enameled	.020″	1.50	2.74	4.24
0350			4″	enameled	.025″	2.23	3.71	5.94
0500	Copper	rectangular corr.	2″x3″	mill	16 Oz.	4.27	2.74	7.01
0550				lead coated	16 Oz.	8.60	2.74	11.34
0600		smooth		mill	16 Oz.	5.80	2.74	8.54
0650				lead coated	16 Oz.	8.10	2.74	10.84
0700		rectangular corr.	3″x4″	mill	16 Oz.	5.05	3.58	8.63
0750				lead coated	16 Oz.	7.60	3.10	10.70
0800		smooth		mill	16 Oz.	6.55	3.58	10.13
0850				lead coated	16 Oz.	8.75	3.58	12.33
0900		round corrugated	2″	mill	16 Oz.	5.15	2.82	7.97
0950		smooth		lead coated	16 Oz.	6.05	2.74	8.79
1000		corrugated	3″	mill	16 Oz.	5	2.84	7.84
1050		smooth		lead coated	16 Oz.	8.10	2.74	10.84

B3010 Roof Coverings

B3010 620	Downspouts							

	MATERIALS	SECTION	SIZE	FINISH	THICKNESS	COST PER V.L.F.		
						MAT.	INST.	TOTAL
1100	Copper	corrugated	4"	mill	16 Oz.	5.95	3.72	9.67
1150		smooth		lead coated	16 Oz.	10.70	3.58	14.28
1200		corrugated	5"	mill	16 Oz.	8.85	4.18	13.03
1250				lead coated	16 Oz.	15.15	4	19.15
1300	Steel	rectangular corr.	2"x3"	galvanized	28 Ga.	1.39	2.74	4.13
1350				epoxy coated	24 Ga.	1.71	2.74	4.45
1400		smooth		galvanized	28 Ga.	1.96	2.74	4.70
1450		rectangular corr.	3"x4"	galvanized	28 Ga.	1.39	3.58	4.97
1500				epoxy coated	24 Ga.	1.96	3.58	5.54
1550		smooth		galvanized	28 Ga.	1.74	3.58	5.32
1600		round corrugated	2"	galvanized	28 Ga.	1.47	2.74	4.21
1650			3"	galvanized	28 Ga.	1.47	2.74	4.21
1700			4"	galvanized	28 Ga.	1.54	3.58	5.12
1750			5"	galvanized	28 Ga.	2.57	4	6.57
1800				galvanized	26 Ga.	2.59	4	6.59
1850			6"	galvanized	28 Ga.	3.65	4.95	8.60
1900				galvanized	26 Ga.	3.58	4.95	8.53
2000	Steel pipe	round	4"	black	X.H.	19.95	26	45.95
2050			6"	black	X.H.	42.50	29	71.50
2500	Stainless steel	rectangular	2"x3"	mill		22.50	2.74	25.24
2550	Tubing sch.5		3"x4"	mill		28.50	3.58	32.08
2600			4"x5"	mill		59	3.85	62.85
2650		round	3"	mill		22.50	2.74	25.24
2700			4"	mill		28.50	3.58	32.08
2750			5"	mill		59	3.85	62.85

B3010 630	Gravel Stop							

	MATERIALS	SECTION	SIZE	FINISH	THICKNESS	COST PER L.F.		
						MAT.	INST.	TOTAL
5100	Aluminum	extruded	4"	mill	.050"	4.71	3.58	8.29
5200			4"	duranodic	.050"	4.54	3.58	8.12
5300			8"	mill	.050"	6.30	4.16	10.46
5400			8"	duranodic	.050"	6.15	4.16	10.31
5500			12"-2 pc.	mill	.050"	8.40	5.20	13.60
6000			12"-2 pc.	duranodic	.050"	7.75	5.20	12.95
6100	Stainless	formed	6"	mill	24 Ga.	8.80	3.85	12.65
6200			12"	mill	24 Ga.	18.50	5.20	23.70

Roof Hatch

Smoke Hatch

Skylight

B3020 110	Skylights	COST PER S.F.		
		MAT.	INST.	TOTAL
5100	Skylights, plastic domes, insul curbs, nom. size to 10 S.F., single glaze	15.35	10.75	26.10
5200	Double glazing	21	13.25	34.25
5300	10 S.F. to 20 S.F., single glazing	9.55	4.36	13.91
5400	Double glazing	18.55	5.45	24
5500	20 S.F. to 30 S.F., single glazing	12.05	3.70	15.75
5600	Double glazing	16.95	4.36	21.31
5700	30 S.F. to 65 S.F., single glazing	16.50	2.82	19.32
5800	Double glazing	12.90	3.70	16.60
6000	Sandwich panels fiberglass, 9-1/16"thick, 2 S.F. to 10 S.F.	17.10	8.60	25.70
6100	10 S.F. to 18 S.F.	15.30	6.50	21.80
6200	2-3/4" thick, 25 S.F. to 40 S.F.	24.50	5.85	30.35
6300	40 S.F. to 70 S.F.	20	5.20	25.20

B3020 210	Hatches	COST PER OPNG.		
		MAT.	INST.	TOTAL
0200	Roof hatches, with curb, and 1" fiberglass insulation, 2'-6"x3'-0",aluminum	605	172	777
0300	Galvanized steel 165 lbs.	420	172	592
0400	Primed steel 164 lbs.	490	172	662
0500	2'-6"x4'-6" aluminum curb and cover, 150 lbs.	715	191	906
0600	Galvanized steel 220 lbs.	595	191	786
0650	Primed steel 218 lbs.	750	191	941
0800	2'x6"x8'-0" aluminum curb and cover, 260 lbs.	1,500	261	1,761
0900	Galvanized steel, 360 lbs.	1,150	261	1,411
0950	Primed steel 358 lbs.	1,025	261	1,286
1200	For plexiglass panels, add to the above	410		410
2100	Smoke hatches, unlabeled not incl. hand winch operator, 2'-6"x3', galv	510	208	718
2200	Plain steel, 160 lbs.	595	208	803
2400	2'-6"x8'-0",galvanized steel, 360 lbs.	1,250	286	1,536
2500	Plain steel, 350 lbs.	1,125	286	1,411
3000	4'-0"x8'-0", double leaf low profile, aluminum cover, 359 lb.	1,700	215	1,915
3100	Galvanized steel 475 lbs.	1,500	215	1,715
3200	High profile, aluminum cover, galvanized curb, 361 lbs.	1,550	215	1,765

For information about Means Estimating Seminars, see yellow pages 12 and 13 in back of book

Important: See the Reference Section for critical supporting data - Reference Numbers and City Cost Indexes

C   INTERIORS

The Concrete Block Partition Systems are defined by weight and type of block, thickness, type of finish and number of sides finished. System components include joint reinforcing on alternate courses and vertical control joints.

System Components	QUANTITY	UNIT	COST PER S.F.		
			MAT.	INST.	TOTAL
SYSTEM C1010 102 1020					
CONC. BLOCK PARTITION, 8″ X 16″, 4″ TK., 2 CT. GYP. PLASTER 2 SIDES					
Conc. block partition, 4″ thick	1.000	S.F.	1.25	4.68	5.93
Control joint	.050	L.F.	.08	.12	.20
Horizontal joint reinforcing	.800	L.F.	.06	.06	.12
Gypsum plaster, 2 coat, on masonry	2.000	S.F.	.74	4.38	5.12
TOTAL			2.13	9.24	11.37

C1010 102		Concrete Block Partitions - Regular Weight				COST PER S.F.		
	TYPE	THICKNESS (IN.)	TYPE FINISH	SIDES FINISHED		MAT.	INST.	TOTAL
1000	Hollow	4	none	0		1.39	4.85	6.24
1010			gyp. plaster 2 coat	1		1.76	7.05	8.81
1020				2		2.13	9.20	11.33
1100			lime plaster - 2 coat	1		1.62	7.05	8.67
1150			lime portland - 2 coat	1		1.63	7.05	8.68
1200			portland - 3 coat	1		1.65	7.40	9.05
1400			5/8″ drywall	1		1.87	6.40	8.27
1410				2		2.35	7.95	10.30
1500		6	none	0		2.01	5.20	7.21
1510			gyp. plaster 2 coat	1		2.38	7.45	9.83
1520				2		2.75	9.60	12.35
1600			lime plaster - 2 coat	1		2.24	7.45	9.69
1650			lime portland - 2 coat	1		2.25	7.45	9.70
1700			portland - 3 coat	1		2.27	7.75	10.02
1900			5/8″ drywall	1		2.49	6.75	9.24
1910				2		2.97	8.30	11.27
2000		8	none	0		2.20	5.55	7.75
2010			gyp. plaster 2 coat	1		2.57	7.75	10.32
2020			gyp. plaster 2 coat	2		2.93	9.90	12.83
2100			lime plaster - 2 coat	1		2.42	7.75	10.17
2150			lime portland - 2 coat	1		2.44	7.75	10.19
2200			portland - 3 coat	1		2.46	8.10	10.56
2400			5/8″ drywall	1		2.68	7.10	9.78
2410				2		3.16	8.65	11.81

C10 Interior Construction

C1010 Partitions

C1010 102			Concrete Block Partitions - Regular Weight					
	TYPE	THICKNESS (IN.)	TYPE FINISH	SIDES FINISHED		COST PER S.F.		
						MAT.	INST.	TOTAL
2500	Hollow	10	none	0		2.74	5.85	8.59
2510			gyp. plaster 2 coat	1		3.11	8.05	11.16
2520				2		3.48	10.20	13.68
2600			lime plaster - 2 coat	1		2.97	8.05	11.02
2650			lime portland - 2 coat	1		2.98	8.05	11.03
2700			portland - 3 coat	1		3	8.35	11.35
2900			5/8" drywall	1		3.22	7.40	10.62
2910				2		3.70	8.95	12.65
3000	Solid	2	none	0		1.20	4.79	5.99
3010			gyp. plaster	1		1.57	7	8.57
3020				2		1.94	9.15	11.09
3100			lime plaster - 2 coat	1		1.43	7	8.43
3150			lime portland - 2 coat	1		1.44	7	8.44
3200			portland - 3 coat	1		1.46	7.35	8.81
3400			5/8" drywall	1		1.68	6.35	8.03
3410				2		2.16	7.90	10.06
3500		4	none	0		1.70	5	6.70
3510			gyp. plaster	1		2.14	7.25	9.39
3520				2		2.44	9.40	11.84
3600			lime plaster - 2 coat	1		1.93	7.25	9.18
3650			lime portland - 2 coat	1		1.94	7.25	9.19
3700			portland - 3 coat	1		1.96	7.55	9.51
3900			5/8" drywall	1		2.18	6.55	8.73
3910				2		2.66	8.10	10.76
4000		6	none	0		2.08	5.40	7.48
4010			gyp. plaster	1		2.45	7.65	10.10
4020				2		2.82	9.80	12.62
4100			lime plaster - 2 coat	1		2.31	7.65	9.96
4150			lime portland - 2 coat	1		2.32	7.65	9.97
4200			portland - 3 coat	1		2.34	7.95	10.29
4400			5/8" drywall	1		2.56	6.95	9.51
4410				2		3.04	8.50	11.54

C1010 104			Concrete Block Partitions - Lightweight					
	TYPE	THICKNESS (IN.)	TYPE FINISH	SIDES FINISHED		COST PER S.F.		
						MAT.	INST.	TOTAL
5000	Hollow	4	none	0		1.61	4.74	6.35
5010			gyp. plaster	1		1.98	6.95	8.93
5020				2		2.35	9.10	11.45
5100			lime plaster - 2 coat	1		1.84	6.95	8.79
5150			lime portland - 2 coat	1		1.85	6.95	8.80
5200			portland - 3 coat	1		1.87	7.30	9.17
5400			5/8" drywall	1		2.09	6.30	8.39
5410				2		2.57	7.85	10.42
5500		6	none	0		2.17	5.10	7.27
5520			gyp. plaster	2		2.91	9.45	12.36
5600			lime plaster - 2 coat	1		2.40	7.30	9.70
5650			lime portland - 2 coat	1		2.41	7.30	9.71
5700			portland - 3 coat	1		2.43	7.60	10.03
5900			5/8" drywall	1		2.65	6.65	9.30
5910				2		3.13	8.20	11.33
6000		8	none	0		2.65	5.45	8.10

C10 Interior Construction

C1010 Partitions

C1010 104						Concrete Block Partitions - Lightweight		
	TYPE	THICKNESS (IN.)	TYPE FINISH	SIDES FINISHED		COST PER S.F.		
						MAT.	INST.	TOTAL
6010	Hollow	8	gyp. plaster	1		3.02	7.65	10.67
6020				2		3.38	9.80	13.18
6100			lime plaster - 2 coat	1		2.87	7.65	10.52
6150			lime portland - 2 coat	1		2.89	7.65	10.54
6200			portland - 3 coat	1		2.91	8	10.91
6400			5/8" drywall	1		3.13	7	10.13
6410				2		3.61	8.55	12.16
6500		10	none	0		3.41	5.70	9.11
6700			portland - 3 coat	1		3.67	8.20	11.87
6900			5/8" drywall	1		3.89	7.25	11.14
6910				2		4.37	8.80	13.17
7000	Solid	4	none	0		2.22	4.96	7.18
7010			gyp. plaster	1		2.59	7.15	9.74
7020				2		2.96	9.35	12.31
7100			lime plaster - 2 coat	1		2.45	7.15	9.60
7150			lime portland - 2 coat	1		2.46	7.15	9.61
7200			portland - 3 coat	1		2.48	7.50	9.98
7400			5/8" drywall	1		2.70	6.50	9.20
7410				2		3.18	8.05	11.23
7500		6	none	0		2.56	5.30	7.86
7510			gyp. plaster	1		3	7.60	10.60
7520				2		3.30	9.70	13
7600			lime plaster - 2 coat	1		2.79	7.55	10.34
7650			lime portland - 2 coat	1		2.80	7.55	10.35
7700			portland - 3 coat	1		2.82	7.85	10.67
7900			5/8" drywall	1		3.04	6.85	9.89
7910				2		3.52	8.40	11.92
8000		8	none	0		3.36	5.70	9.06
8010			gyp. plaster	1		3.73	7.90	11.63
8020				2		4.09	10.05	14.14
8100			lime plaster - 2 coat	1		3.58	7.90	11.48
8150			lime portland - 2 coat	1		3.60	7.90	11.50
8200			portland - 3 coat	1		3.62	8.25	11.87
8400			5/8" drywall	1		3.84	7.25	11.09
8410				2		4.32	8.80	13.12

Important: See the Reference Section for critical supporting data - Reference Numbers and City Cost Indexes

C1010 Partitions

Structural Facing Tile
8W Series
8" x 16"

Structural Facing Tile
6T Series
5-1/3" x 12"

C1010 120	Tile Partitions	COST PER S.F.		
		MAT.	INST.	TOTAL
1000	8W series 8"x16", 4" thick wall, reinf every 2 courses, glazed 1 side	7.80	5.85	13.65
1100	Glazed 2 sides	11.65	6.20	17.85
1200	Glazed 2 sides, using 2 wythes of 2" thick tile	14.30	11.20	25.50
1300	6" thick wall, horizontal reinf every 2 courses, glazed 1 side	11.05	6.10	17.15
1400	Glazed 2 sides, each face different color, 2" and 4" tile	14.95	11.45	26.40
1500	8" thick wall, glazed 2 sides using 2 wythes of 4" thick tile	15.60	11.70	27.30
1600	10" thick wall, glazed 2 sides using 1 wythe of 4" tile & 1 wythe of 6" tile	18.85	11.95	30.80
1700	Glazed 2 sides cavity wall, using 2 wythes of 4" thick tile	15.60	11.70	27.30
1800	12" thick wall, glazed 2 sides using 2 wythes of 6" thick tile	22	12.20	34.20
1900	Glazed 2 sides cavity wall, using 2 wythes of 4" thick tile	15.60	11.70	27.30
2100	6T series 5-1/3"x12" tile, 4" thick, non load bearing glazed one side,	7.75	9.15	16.90
2200	Glazed two sides	12.30	10.30	22.60
2300	Glazed two sides, using two wythes of 2" thick tile	13.20	17.90	31.10
2400	6" thick, glazed one side	12	9.60	21.60
2500	Glazed two sides	14.10	10.85	24.95
2600	Glazed two sides using 2" thick tile and 4" thick tile	14.35	18.10	32.45
2700	8" thick, glazed one side	15.05	11.20	26.25
2800	Glazed two sides using two wythes of 4" thick tile	15.50	18.30	33.80
2900	Glazed two sides using 6" thick tile and 2" thick tile	18.60	18.55	37.15
3000	10" thick cavity wall, glazed two sides using two wythes of 4" tile	15.50	18.30	33.80
3100	12" thick, glazed two sides using 4" thick tile and 8" thick tile	23	20.50	43.50
3200	2" thick facing tile, glazed one side, on 6" concrete block	9	10.65	19.65
3300	On 8" concrete block	9.20	10.95	20.15
3400	On 10" concrete block	9.75	11.20	20.95

C1010 Partitions

Wood Stud Framing

Metal Stud Framing

The Drywall Partitions/Stud Framing Systems are defined by type of drywall and number of layers, type and spacing of stud framing, and treatment on the opposite face. Components include taping and finishing.

Cost differences between regular and fire resistant drywall are negligible, and terminology is interchangeable. In some cases fiberglass insulation is included for additional sound deadening.

System Components	QUANTITY	UNIT	COST PER S.F.		
			MAT.	INST.	TOTAL
SYSTEM C1010 124 1250					
DRYWALL PARTITION,5/8″ F.R.1 SIDE,5/8″ REG.1 SIDE,2″X4″STUDS,16″ O.C.					
Gypsum plasterboard, nailed/screwed to studs, 5/8″F.R. fire resistant	1.000	S.F.	.33	.44	.77
Gypsum plasterboard, nailed/screwed to studs, 5/8″ regular	1.000	S.F.	.33	.44	.77
Taping and finishing joints	2.000	S.F.	.08	.88	.96
Framing, 2 x 4 studs @ 16″ O.C., 10′ high	1.000	S.F.	.43	.89	1.32
TOTAL			1.17	2.65	3.82

C1010 124		Drywall Partitions/Wood Stud Framing					
	FACE LAYER	BASE LAYER	FRAMING	OPPOSITE FACE	INSULATION	COST PER S.F.		

	FACE LAYER	BASE LAYER	FRAMING	OPPOSITE FACE	INSULATION	MAT.	INST.	TOTAL
1200	5/8″ FR drywall	none	2 x 4, @ 16″ O.C.	same	0	1.17	2.65	3.82
1250				5/8″ reg. drywall	0	1.17	2.65	3.82
1300				nothing	0	.80	1.77	2.57
1400		1/4″ SD gypsum	2 x 4 @ 16″ O.C.	same	1-1/2″ fiberglass	2.31	4.07	6.38
1450				5/8″ FR drywall	1-1/2″ fiberglass	2.01	3.58	5.59
1500				nothing	1-1/2″ fiberglass	1.64	2.70	4.34
1600		resil. channels	2 x 4 @ 16″, O.C.	same	1-1/2″ fiberglass	2.02	5.15	7.17
1650				5/8″ FR drywall	1-1/2″ fiberglass	1.86	4.13	5.99
1700				nothing	1-1/2″ fiberglass	1.49	3.25	4.74
1800		5/8″ FR drywall	2 x 4 @ 24″ O.C.	same	0	1.73	3.35	5.08
1850				5/8″ FR drywall	0	1.40	2.91	4.31
1900				nothing	0	1.03	2.03	3.06
1950		5/8″ FR drywall	2 x 4, 16″ O.C.	same	0	1.83	3.53	5.36
1955				5/8″ FR drywall	0	1.50	3.09	4.59
2000				nothing	0	1.13	2.21	3.34
2010		5/8″ FR drywall	staggered, 6″ plate	same	0	2.27	4.43	6.70
2015				5/8″ FR drywall	0	1.94	3.99	5.93
2020				nothing	0	1.57	3.11	4.68
2200		5/8″ FR drywall	2 rows-2 x 4	same	2″ fiberglass	2.91	4.85	7.76
2250			16″O.C.	5/8″ FR drywall	2″ fiberglass	2.58	4.41	6.99
2300				nothing	2″ fiberglass	2.21	3.53	5.74
2400	5/8″ WR drywall	none	2 x 4, @ 16″ O.C.	same	0	1.11	2.65	3.76
2450				5/8″ FR drywall	0	1.14	2.65	3.79
2500				nothing	0	.77	1.77	2.54
2600		5/8″ FR drywall	2 x 4, @ 24″ O.C.	same	0	1.67	3.35	5.02
2650				5/8″ FR drywall	0	1.37	2.91	4.28
2700				nothing	0	1	2.03	3.03

Important: See the Reference Section for critical supporting data - Reference Numbers and City Cost Indexes

C1010 Partitions

C1010 124			Drywall Partitions/Wood Stud Framing					

	FACE LAYER	BASE LAYER	FRAMING	OPPOSITE FACE	INSULATION	COST PER S.F.		
						MAT.	INST.	TOTAL
2800	5/8 VF drywall	none	2 x 4, @ 16" O.C.	same	0	1.87	2.85	4.72
2850				5/8" FR drywall	0	1.52	2.75	4.27
2900				nothing	0	1.15	1.87	3.02
3000	5/8 VF drywall	5/8" FR drywall	2 x 4 , 24" O.C.	same	0	2.43	3.55	5.98
3050				5/8" FR drywall	0	1.75	3.01	4.76
3100				nothing	0	1.38	2.13	3.51
3200	1/2" reg drywall	3/8" reg drywall	2 x 4, @ 16" O.C.	same	0	1.63	3.53	5.16
3250				5/8" FR drywall	0	1.40	3.09	4.49
3300				nothing	0	1.03	2.21	3.24

C1010 126			Drywall Partitions/Metal Stud Framing					

	FACE LAYER	BASE LAYER	FRAMING	OPPOSITE FACE	INSULATION	COST PER S.F.		
						MAT.	INST.	TOTAL
5200	5/8" FR drywall	none	1-5/8" @ 24" O.C.	same	0	.92	2.34	3.26
5250				5/8" reg. drywall	0	.92	2.34	3.26
5300				nothing	0	.55	1.46	2.01
5400			3-5/8" @ 24" O.C.	same	0	.99	2.36	3.35
5450				5/8" reg. drywall	0	.99	2.36	3.35
5500				nothing	0	.62	1.48	2.10
5600		1/4" SD gypsum	1-5/8" @ 24" O.C.	same	0	1.52	3.32	4.84
5650				5/8" FR drywall	0	1.22	2.83	4.05
5700				nothing	0	.85	1.95	2.80
5800			2-1/2" @ 24" O.C.	same	0	1.56	3.33	4.89
5850				5/8" FR drywall	0	1.26	2.84	4.10
5900				nothing	0	.89	1.96	2.85
6000		5/8" FR drywall	2-1/2" @ 16" O.C.	same	0	1.88	3.78	5.66
6050				5/8" FR drywall	0	1.55	3.34	4.89
6100				nothing	0	1.18	2.46	3.64
6200			3-5/8" @ 24" O.C.	same	0	1.65	3.24	4.89
6250				5/8"FR drywall	3-1/2" fiberglass	1.84	3.08	4.92
6300				nothing	0	.95	1.92	2.87
6400	5/8" WR drywall	none	1-5/8" @ 24" O.C.	same	0	.86	2.34	3.20
6450				5/8" FR drywall	0	.89	2.34	3.23
6500				nothing	0	.52	1.46	1.98
6600			3-5/8" @ 24" O.C.	same	0	.93	2.36	3.29
6650				5/8" FR drywall	0	.96	2.36	3.32
6700				nothing	0	.59	1.48	2.07
6800		5/8" FR drywall	2-1/2" @ 16" O.C.	same	0	1.82	3.78	5.60
6850				5/8" FR drywall	0	1.52	3.34	4.86
6900				nothing	0	1.15	2.46	3.61
7000			3-5/8" @ 24" O.C.	same	0	1.59	3.24	4.83
7050				5/8"FR drywall	3-1/2" fiberglass	1.81	3.08	4.89
7100				nothing	0	.92	1.92	2.84
7200	5/8" VF drywall	none	1-5/8" @ 24" O.C.	same	0	1.62	2.54	4.16
7250				5/8" FR drywall	0	1.27	2.44	3.71
7300				nothing	0	.90	1.56	2.46
7400			3-5/8" @ 24" O.C.	same	0	1.69	2.56	4.25
7450				5/8" FR drywall	0	1.34	2.46	3.80
7500				nothing	0	.97	1.58	2.55

INTERIORS C

C1010 Partitions

| C1010 126 | Drywall Partitions/Metal Stud Framing |

	FACE LAYER	BASE LAYER	FRAMING	OPPOSITE FACE	INSULATION	COST PER S.F.		
						MAT.	INST.	TOTAL
7600	5/8" VF drywall	5/8" FR drywall	2-1/2" @ 16" O.C.	same	0	2.58	3.98	6.56
7650				5/8" FR drywall	0	1.90	3.44	5.34
7700				nothing	0	1.53	2.56	4.09
7800			3-5/8" @ 24" O.C.	same	0	2.35	3.44	5.79
7850				5/8"FR drywall	3-1/2" fiberglass	2.19	3.18	5.37
7900				nothing	0	1.30	2.02	3.32

Important: See the Reference Section for critical supporting data - Reference Numbers and City Cost Indexes

C1010 Partitions

C1010 128	Drywall Components	COST PER S.F.		
		MAT.	INST.	TOTAL
0140	Metal studs, 24" O.C. including track, load bearing, 18 gage, 2-1/2"	.58	.83	1.41
0160	3-5/8"	.69	.85	1.54
0180	4"	.62	.86	1.48
0200	6"	.92	.88	1.80
0220	16 gage, 2-1/2"	.67	.94	1.61
0240	3-5/8"	.80	.97	1.77
0260	4"	.85	.99	1.84
0280	6"	1.07	1.01	2.08
0300	Non load bearing, 25 gage, 1-5/8"	.18	.58	.76
0340	3-5/8"	.25	.60	.85
0360	4"	.27	.60	.87
0380	6"	.39	.61	1
0400	20 gage, 2-1/2"	.35	.74	1.09
0420	3-5/8"	.42	.75	1.17
0440	4"	.46	.75	1.21
0460	6"	.59	.76	1.35
0540	Wood studs including blocking, shoe and double top plate, 2"x4", 12"O.C.	.53	1.11	1.64
0560	16" O.C.	.43	.89	1.32
0580	24" O.C.	.33	.71	1.04
0600	2"x6", 12" O.C.	.85	1.27	2.12
0620	16" O.C.	.69	.99	1.68
0640	24" O.C.	.52	.77	1.29
0642	Furring one side only, steel channels, 3/4", 12" O.C.	.25	1.66	1.91
0644	16" O.C.	.23	1.47	1.70
0646	24" O.C.	.15	1.11	1.26
0647	1-1/2", 12" O.C.	.35	1.86	2.21
0648	16" O.C.	.32	1.62	1.94
0649	24" O.C.	.21	1.28	1.49
0650	Wood strips, 1" x 3", on wood, 12" O.C.	.31	.81	1.12
0651	16" O.C.	.23	.61	.84
0652	On masonry, 12" O.C.	.31	.89	1.20
0653	16" O.C.	.23	.67	.90
0654	On concrete, 12" O.C.	.31	1.70	2.01
0655	16" O.C.	.23	1.28	1.51
0665	Gypsum board, one face only, exterior sheathing, 1/2"	.68	.79	1.47
0680	Interior, fire resistant, 1/2"	.33	.44	.77
0700	5/8"	.33	.44	.77
0720	Sound deadening board 1/4"	.30	.49	.79
0740	Standard drywall 3/8"	.26	.44	.70
0760	1/2"	.30	.44	.74
0780	5/8"	.33	.44	.77
0800	Tongue & groove coreboard 1"	.53	1.84	2.37
0820	Water resistant, 1/2"	.28	.44	.72
0840	5/8"	.30	.44	.74
0860	Add for the following:, foil backing	.11		.11
0880	Fiberglass insulation, 3-1/2"	.52	.28	.80
0900	6"	.56	.33	.89
0920	Rigid insulation 1"	.47	.44	.91
0940	Resilient furring @ 16" O.C.	.20	1.39	1.59
0960	Taping and finishing	.04	.44	.48
0980	Texture spray	.06	.49	.55
1000	Thin coat plaster	.08	.54	.62
1040	2"x4" staggered studs 2"x6" plates & blocking	.62	.97	1.59

Plaster Partitions are defined as follows: type of plaster, type and spacing of framing, type of lath and treatment of opposite face.

Included in the system components are expansion joints. Metal studs are assumed to be non-loadbearing.

System Components	QUANTITY	UNIT	COST PER S.F.		
			MAT.	INST.	TOTAL
SYSTEM C1010 140 1000					
GYP PLASTER PART'N, 2 COATS,2-1/2″ MTL. STUD,16″O.C.,3/8″GYP. LATH					
Gypsum plaster, 2 coats, on walls	2.000	S.F.	1.01	4.16	5.17
Gypsum lath, 3/8″ thick, on metal studs	2.000	S.F.	1.32	1.15	2.47
Metal studs, 25 ga., 2-1/2″ @ 16″ O.C.	1.000	S.F.	.30	.90	1.20
Expansion joint	.100	L.F.	.11	.14	.25
TOTAL			2.74	6.35	9.09

C1010 140 — Plaster Partitions/Metal Stud Framing

	TYPE	FRAMING	LATH	OPPOSITE FACE		MAT.	INST.	TOTAL
1000	2 coat gypsum	2-1/2″ @ 16″O.C.	3/8″ gypsum	same		2.74	6.35	9.09
1010				nothing		1.57	3.70	5.27
1100		3-1/4″ @ 24″O.C.	1/2″ gypsum	same		2.40	6.15	8.55
1110				nothing		1.38	3.44	4.82
1500	2 coat vermiculite	2-1/2″ @ 16″O.C.	3/8″ gypsum	same		2.52	6.95	9.47
1510				nothing		1.46	3.98	5.44
1600		3-1/4″ @ 24″O.C.	1/2″ gypsum	same		2.18	6.70	8.88
1610				nothing		1.27	3.72	4.99
2000	3 coat gypsum	2-1/2″ @ 16″O.C.	3/8″ gypsum	same		2.62	7.25	9.87
2010				nothing		1.51	4.16	5.67
2020			3.4lb. diamond	same		2.10	7.25	9.35
2030				nothing		1.25	4.16	5.41
2040			2.75lb. ribbed	same		2.04	7.25	9.29
2050				nothing		1.22	4.16	5.38
2100		3-1/4″ @ 24″O.C.	1/2″ gypsum	same		2.28	7.05	9.33
2110				nothing		1.32	3.90	5.22
2120			3.4lb. ribbed	same		2.33	7.05	9.38
2130				nothing		1.34	3.90	5.24
2500	3 coat lime	2-1/2″ @ 16″O.C.	3.4lb. diamond	same		1.77	7.25	9.02
2510				nothing		1.09	4.16	5.25
2520			2.75lb. ribbed	same		1.71	7.25	8.96
2530				nothing		1.06	4.16	5.22
2600		3-1/4″ @ 24″O.C.	3.4lb. ribbed	same		2	7.05	9.05
2610				nothing		1.18	3.90	5.08

C1010 Partitions

C1010 140	Plaster Partitions/Metal Stud Framing

	TYPE	FRAMING	LATH	OPPOSITE FACE		COST PER S.F.		
						MAT.	INST.	TOTAL
3000	3 coat Portland	2-1/2" @ 16"O.C.	3.4lb. diamond	same		1.73	7.25	8.98
3010				nothing		1.07	4.16	5.23
3020			2.75lb. ribbed	same		1.67	7.25	8.92
3030				nothing		1.04	4.16	5.20
3100		3-1/4" @ 24"O.C.	3.4lb. ribbed	same		1.96	7.05	9.01
3110				nothing		1.16	3.90	5.06
3500	3 coat gypsum W/med. Keenes	2-1/2" @ 16"O.C.	3/8" gypsum	same		3.17	9.30	12.47
3510				nothing		1.79	5.15	6.94
3520			3.4lb. diamond	same		2.65	9.30	11.95
3530				nothing		1.53	5.15	6.68
3540			2.75lb. ribbed	same		2.59	9.30	11.89
3550				nothing		1.50	5.15	6.65
3600		3-1/4" @ 24"O.C.	1/2" gypsum	same		2.83	9.05	11.88
3610				nothing		1.59	4.91	6.50
3620			3.4lb. ribbed	same		2.87	9.05	11.92
3630				nothing		1.61	4.91	6.52
4000	3 coat gypsum W/hard Keenes	2-1/2" @ 16"O.C.	3/8" gypsum	same		3.17	10.30	13.47
4010				nothing		1.79	5.70	7.49
4020			3.4lb. diamond	same		2.65	10.30	12.95
4030				nothing		1.53	5.70	7.23
4040			2.75lb. ribbed	same		2.59	10.30	12.89
4050				nothing		1.50	5.70	7.20
4100		3-1/4" @ 24"O.C.	1/2" gypsum	same		2.83	10.10	12.93
4110				nothing		1.59	5.40	6.99
4120			3.4lb. ribbed	same		2.88	10.10	12.98
4130				nothing		1.62	5.40	7.02
4500	3 coat lime Portland	2-1/2" @ 16"O.C.	3.4lb. diamond	same		1.71	7.25	8.96
4510				nothing		1.06	4.16	5.22
4520			2.75lb. ribbed	same		1.65	7.25	8.90
4530				nothing		1.03	4.16	5.19
4600		3-1/4" @ 24"O.C.	3.4lb. ribbed	same		1.94	7.05	8.99
4610				nothing		1.15	3.90	5.05

C1010 142	Plaster Partitions/Wood Stud Framing

	TYPE	FRAMING	LATH	OPPOSITE FACE		COST PER S.F.		
						MAT.	INST.	TOTAL
5000	2 coat gypsum	2"x4" @ 16"O.C.	3/8" gypsum	same		2.87	6.20	9.07
5010				nothing		1.70	3.62	5.32
5100		2"x4" @ 24"O.C.	1/2" gypsum	same		2.47	6.10	8.57
5110				nothing		1.45	3.47	4.92
5500	2 coat vermiculite	2"x4" @ 16"O.C.	3/8" gypsum	same		2.65	6.80	9.45
5510				nothing		1.59	3.90	5.49
5600		2"x4" @ 24"O.C.	1/2" gypsum	same		2.25	6.70	8.95
5610				nothing		1.34	3.76	5.10

C1010 Partitions

C1010 142	Plaster Partitions/Wood Stud Framing

	TYPE	FRAMING	LATH	OPPOSITE FACE		COST PER S.F.		
						MAT.	INST.	TOTAL
6000	3 coat gypsum	2"x4" @ 16"O.C.	3/8" gypsum	same		2.75	7.10	9.85
6010				nothing		1.64	4.08	5.72
6020			3.4lb. diamond	same		2.23	7.20	9.43
6030				nothing		1.38	4.11	5.49
6040			2.75lb. ribbed	same		2.15	7.20	9.35
6050				nothing		1.34	4.13	5.47
6100		2"x4" @ 24"O.C.	1/2" gypsum	same		2.36	7	9.36
6110				nothing		1.40	3.93	5.33
6120			3.4lb. ribbed	same		2.07	7.10	9.17
6130				nothing		1.25	3.97	5.22
6500	3 coat lime	2"x4" @ 16"O.C.	3.4lb. diamond	same		1.90	7.20	9.10
6510				nothing		1.22	4.11	5.33
6520			2.75lb. ribbed	same		1.82	7.20	9.02
6530				nothing		1.18	4.13	5.31
6600		2"x4" @ 24"O.C.	3.4lb. ribbed	same		1.74	7.10	8.84
6610				nothing		1.09	3.97	5.06
7000	3 coat Portland	2"x4" @ 16"O.C.	3.4lb. diamond	same		1.86	7.20	9.06
7010				nothing		1.20	4.11	5.31
7020			2.75lb. ribbed	same		1.78	7.20	8.98
7030				nothing		1.16	4.13	5.29
7100		2"x4" @ 24"O.C.	3.4lb. ribbed	same		2.03	7.15	9.18
7110				nothing		1.23	4.01	5.24
7500	3 coat gypsum	2"x4" @ 16"O.C.	3/8" gypsum	same		3.30	9.15	12.45
7510	W/med Keenes			nothing		1.92	5.10	7.02
7520			3.4lb. diamond	same		2.78	9.20	11.98
7530				nothing		1.66	5.10	6.76
7540			2.75lb. ribbed	same		2.69	9.25	11.94
7550				nothing		1.62	5.15	6.77
7600		2"x4" @ 24"O.C.	1/2" gypsum	same		2.91	9	11.91
7610				nothing		1.67	4.95	6.62
7620			3.4lb. ribbed	same		2.95	9.15	12.10
7630				nothing		1.69	5	6.69
8000	3 coat gypsum	2"x4" @ 16"O.C.	3/8" gypsum	same		3.30	10.15	13.45
8010	W/hard Keenes			nothing		1.92	5.60	7.52
8020			3.4lb. diamond	same		2.78	10.25	13.03
8030				nothing		1.66	5.65	7.31
8040			2.75lb. ribbed	same		2.70	10.25	12.95
8050				nothing		1.62	5.65	7.27
8100		2"x4" @ 24"O.C.	1/2" gypsum	same		2.91	10.05	12.96
8110				nothing		1.67	5.45	7.12
8120			3.4lb. ribbed	same		2.95	10.20	13.15
8130				nothing		1.69	5.55	7.24
8500	3 coat lime	2"x4" @ 16"O.C.	3.4lb. diamond	same		1.84	7.20	9.04
8510	Portland			nothing		1.19	4.11	5.30
8520			2.75lb. ribbed	same		1.76	7.20	8.96
8530				nothing		1.15	4.13	5.28
8600		2"x4" @ 24"O.C.	3.4lb. ribbed	same		2.01	7.15	9.16
8610				nothing		1.22	4.01	5.23

Important: See the Reference Section for critical supporting data - Reference Numbers and City Cost Indexes

C1010 Partitions

C1010 144	Plaster Partition Components	COST PER S.F.		
		MAT.	INST.	TOTAL
0060	Metal studs, 16" O.C., including track, non load bearing, 25 gage, 1-5/8"	.30	.90	1.20
0080	2-1/2"	.30	.90	1.20
0100	3-1/4"	.34	.92	1.26
0120	3-5/8"	.34	.92	1.26
0140	4"	.36	.93	1.29
0160	6"	.53	.94	1.47
0180	Load bearing, 20 gage, 2-1/2"	.79	1.15	1.94
0200	3-5/8"	.94	1.17	2.11
0220	4"	.99	1.20	2.19
0240	6"	1.25	1.22	2.47
0260	16 gage 2-1/2"	.92	1.30	2.22
0280	3-5/8"	1.10	1.34	2.44
0300	4"	1.17	1.36	2.53
0320	6"	1.47	1.39	2.86
0340	Wood studs, including blocking, shoe and double plate, 2"x4", 12" O.C.	.53	1.11	1.64
0360	16" O.C.	.43	.89	1.32
0380	24" O.C.	.33	.71	1.04
0400	2"x6", 12" O.C.	.85	1.27	2.12
0420	16" O.C.	.69	.99	1.68
0440	24" O.C.	.52	.77	1.29
0460	Furring one face only, steel channels, 3/4", 12" O.C.	.25	1.66	1.91
0480	16" O.C.	.23	1.47	1.70
0500	24" O.C.	.15	1.11	1.26
0520	1-1/2", 12" O.C.	.35	1.86	2.21
0540	16" O.C.	.32	1.62	1.94
0560	24" O.C.	.21	1.28	1.49
0580	Wood strips 1"x3", on wood., 12" O.C.	.31	.81	1.12
0600	16" O.C.	.23	.61	.84
0620	On masonry, 12" O.C.	.31	.89	1.20
0640	16" O.C.	.23	.67	.90
0660	On concrete, 12" O.C.	.31	1.70	2.01
0680	16" O.C.	.23	1.28	1.51
0700	Gypsum lath. plain or perforated, nailed to studs, 3/8" thick	.66	.51	1.17
0720	1/2" thick	.52	.54	1.06
0740	Clipped to studs, 3/8" thick	.66	.58	1.24
0760	1/2" thick	.52	.62	1.14
0780	Metal lath, diamond painted, nailed to wood studs, 2.5 lb.	.35	.51	.86
0800	3.4 lb.	.40	.54	.94
0820	Screwed to steel studs, 2.5 lb.	.35	.54	.89
0840	3.4 lb.	.45	.58	1.03
0860	Rib painted, wired to steel, 2.75 lb	.37	.58	.95
0880	3.4 lb	.54	.62	1.16
0900	4.0 lb	.55	.67	1.22
0910				
0920	Gypsum plaster, 2 coats	.40	2.08	2.48
0940	3 coats	.57	2.54	3.11
0960	Perlite or vermiculite plaster, 2 coats	.40	2.36	2.76
0980	3 coats	.72	2.95	3.67
1000	Stucco, 3 coats, 1" thick, on wood framing	.65	4.14	4.79
1020	On masonry	.25	3.24	3.49
1100	Metal base galvanized and painted 2-1/2" high	.66	1.63	2.29

INTERIORS C

C1010 Partitions

Folding Accordion

Folding Leaf

Movable and Borrow Lites

C1010 205	Partitions	COST PER S.F.		
		MAT.	INST.	TOTAL
0360	Folding accordion, vinyl covered, acoustical, 3 lb. S.F., 17 ft max. hgt	25.50	8.85	34.35
0380	5 lb. per S.F. 27 ft max height	35	9.30	44.30
0400	5.5 lb. per S.F., 17 ft. max height	41	9.85	50.85
0420	Commercial, 1.75 lb per S.F., 8 ft. max height	22	3.94	25.94
0440	2.0 Lb per S.F., 17 ft. max height	22.50	5.90	28.40
0460	Industrial, 4.0. lb. per S.F. 27 ft. max height	32.50	11.80	44.30
0480	Vinyl clad wood or steel, electric operation 6 psf	47.50	5.55	53.05
0500	Wood, non acoustic, birch or mahogany	25.50	2.95	28.45
0560	Folding leaf, aluminum framed acoustical 12 ft.high.,5.5 lb per S.F.,min.	35	14.75	49.75
0580	Maximum	42	29.50	71.50
0600	6.5 lb. per S.F., minimum	36.50	14.75	51.25
0620	Maximum	45	29.50	74.50
0640	Steel acoustical, 7.5 per S.F., vinyl faced, minimum	51.50	14.75	66.25
0660	Maximum	63	29.50	92.50
0665	Steel with vinyl face, economy	51.50	14.75	66.25
0670	Deluxe	63	29.50	92.50
0680	Wood acoustic type, vinyl faced to 18' high 6 psf, minimum	48.50	14.75	63.25
0700	Average	58	19.70	77.70
0720	Maximum	75	29.50	104.50
0740	Formica or hardwood faced, minimum	50	14.75	64.75
0760	Maximum	53.50	29.50	83
0780	Wood, low acoustical type to 12 ft. high 4.5 psf	36.50	17.70	54.20
0840	Demountable, trackless wall, cork finish, semi acous, 1-5/8"th, min	35.50	2.72	38.22
0860	Maximum	33.50	4.66	38.16
0880	Acoustic, 2" thick, minimum	29	2.90	31.90
0900	Maximum	50	3.94	53.94
0920	In-plant modular office system, w/prehung steel door			
0940	3" thick honeycomb core panels			
0960	12' x 12', 2 wall	8.85	.54	9.39
0970	4 wall	8.20	.72	8.92
0980	16' x 16', 2 wall	9.20	.38	9.58
0990	4 wall	5.90	.38	6.28
1000	Gypsum, demountable, 3" to 3-3/4" thick x 9' high, vinyl clad	5.85	2.05	7.90
1020	Fabric clad	14.55	2.22	16.77
1040	1.75 system, vinyl clad hardboard, paper honeycomb core panel			
1060	1-3/4" to 2-1/2" thick x 9' high	9.80	2.05	11.85
1080	Unitized gypsum panel system, 2" to 2-1/2" thick x 9' high			
1100	Vinyl clad gypsum	12.65	2.05	14.70
1120	Fabric clad gypsum	21	2.22	23.22
1140	Movable steel walls, modular system			
1160	Unitized panels, 48" wide x 9' high			
1180	Baked enamel, pre-finished	14.20	1.64	15.84
1200	Fabric clad	20.50	1.75	22.25

Important: See the Reference Section for critical supporting data - Reference Numbers and City Cost Indexes

C10 Interior Construction

C1010 Partitions

C1010 210	Partitions	COST PER S.F.		
		MAT.	INST.	TOTAL
0100	Partitions, folding accordion, 1.25 lb. vinyl (residential), to 8' high	19.10	2.95	22.05
0110	4 lb. vinyl (industrial), to 27' high	32.50	11.80	44.30
0120	Acoustical, 3 lb. vinyl, to 17' high	25.50	8.85	34.35
0130	4.5 lb. vinyl, fire rated 40, 20' high	41	5.55	46.55
0485	Operable acoustic, no track, 1-5/8" thick, economy	29.50	2.36	31.86
0490	With track, 3" thick, deluxe	90	2.95	92.95

C1010 230	Demountable Partitions, L.F.	COST PER L.F.		
		MAT.	INST.	TOTAL
0100	Office-demountable gypsum, 3" to 3-3/4" thick x 9' high, vinyl clad	52.50	18.45	70.95
0110	Fabric clad	131	20	151
0120	1.75 system, vinyl clad hardboard, paper honeycomb			
0130	core panel, 1-3/4" to 2-1/2" thick x 9' high	88.50	18.45	106.95
0140	Steel, modular, unitized panels, 48" wide x 9' high			
0150	Baked enamel, pre-finished	128	14.75	142.75
0160	Portable, fiber core, 4' high, fabric face	88.50	5.05	93.55
0170	5' high, carpeted face, curved, 3'-0" long	91.50	9.85	101.35

C

INTERIORS

Alum. Tube Frame

Oakwood Frame

**Concealed Frame
Butt Glazed**

Interior Glazed Openings are defined as follows: framing material, glass, size and intermediate framing members. Components for each system include gasket setting or glazing bead and typical wood blocking.

System Components	QUANTITY	UNIT	COST PER OPNG.		
			MAT.	INST.	TOTAL
SYSTEM C1010 710 1000					
GLAZED OPENING, ALUMINUM TUBE FRAME, FLUSH, 1/4" FLOAT GLASS, 6' X 4'					
Aluminum tube frame, flush, anodized bronze, head & jamb	16.500	L.F.	155.93	170.78	326.71
Aluminum tube frame, flush, anodized bronze, open sill	7.000	L.F.	57.40	70.70	128.10
Joints for tube frame, clip type	4.000	Ea.	88		88
Gasket setting, add	20.000	L.F.	89.80		89.80
Wood blocking	8.000	B.F.	4.92	19.60	24.52
Float glass, 1/4" thick, clear, plain	24.000	S.F.	146.40	165.60	312
TOTAL			542.45	426.68	969.13

C1010 710 — Interior Glazed Opening

	FRAME	GLASS	OPENING-SIZE W X H	INTERMEDIATE MULLION	INTERMEDIATE HORIZONTAL	COST PER OPNG.		
						MAT.	INST.	TOTAL
1000	Aluminum flush	1/4" float	6'x4'	0	0	540	425	965
1040	Tube		12'x4'	3	0	1,300	915	2,215
1080			4'x5'	0	0	490	370	860
.1120			8'x5'	1	0	905	685	1,590
1160			12'x5'	2	0	1,325	995	2,320
1240		3/8" float	9'x6'	2	0	1,475	1,150	2,625
1280			4'x8'-6"	0	1	945	740	1,685
1320			16'x10'	3	1	3,825	3,025	6,850
1400		1/4" tempered	6'x4'	0	0	575	425	1,000
1440			12'x4'	3	0	1,375	915	2,290
1480			4'x5'	0	0	515	370	885
1520			8'x5'	1	0	960	685	1,645
1560			12'x5'	2	0	1,400	995	2,395
1640		3/8" tempered	9'x6'	2	0	1,750	1,150	2,900
1680			4'x8'-6"	0	0	990	690	1,680
1720			16'x10'	3	0	4,200	2,850	7,050
1800		1/4" one way mirror	6'x4'	0	0	795	420	1,215
1840			12'x4'	3	0	1,825	905	2,730

C1010 Partitions

| C1010 710 | Interior Glazed Opening |

	FRAME	GLASS	OPENING-SIZE W X H	INTERMEDIATE MULLION	INTERMEDIATE HORIZONTAL	COST PER OPNG.		
						MAT.	INST.	TOTAL
1880	Aluminum flush	1/4" one way mirror	4'x5'	0	0	700	365	1,065
1920	Tube		8'x5'	1	0	1,325	675	2,000
1960			12'x5'	2	0	1,950	980	2,930
2040		3/8" laminated safety	9'x6'	2	0	1,825	1,125	2,950
2080			4'x8'-6"	0	0	1,025	680	1,705
2120			16'x10'	3	0	4,400	2,775	7,175
2200		1-3/16" bullet proof	6'x4'	0	0	1,800	1,525	3,325
2240			12'x4'	3	0	3,800	3,100	6,900
2280			4'x5'	0	0	1,550	1,275	2,825
2320			8'x5'	1	0	2,975	2,500	5,475
2360			12'x5'	2	0	4,400	3,725	8,125
2440		1" acoustical	9'x6'	2	0	2,450	1,025	3,475
2480			4'x8'-6"	0	0	1,450	620	2,070
2520			16'x10'	3	0	6,125	2,450	8,575
3000	Oakwood	1/4" float	6'x4'	0	0	288	320	608
3040			12'x4'	3	0	695	775	1,470
3080			4'x5'	0	0	246	261	507
3120			8'x5'	1	0	500	550	1,050
3160			12'x5'	2	0	755	840	1,595
3240		3/8" float	9'x6'	2	0	910	1,000	1,910
3280			4'x8'-6"	0	1	555	590	1,145
3320			16'x10'	3	1	2,575	2,900	5,475
3400		1/4" tempered	6'x4'	0	0	320	320	640
3440			12'x4'	3	0	760	775	1,535
3480			4'x5'	0	0	273	261	534
3520			8'x5'	1	0	555	550	1,105
3560			12'x5'	2	0	835	840	1,675
3640		3/8" tempered	9'x6'	2	0	1,200	1,000	2,200
3680			4'x8'-6"	0	0	680	530	1,210
3720			16'x10'	3	0	3,200	2,600	5,800
3800		1/4" one way mirror	6'x4'	0	0	540	315	855
3840			12'x4'	3	0	1,200	765	1,965
3880			4'x5'	0	0	455	256	711
3920			8'x5'	1	0	920	540	1,460
3960			12'x5'	2	0	1,400	825	2,225
4040		3/8" laminated safety	9'x6'	2	0	1,275	980	2,255
4080			4'x8'-6"	0	0	725	515	1,240
4120			16'x10'	3	0	3,425	2,550	5,975
4200		1-3/16" bullet proof	6'x4'	0	0	1,475	1,400	2,875
4240			12'x4'	3	0	3,075	2,950	6,025
4280			4'x5'	0	0	1,225	1,175	2,400
4360			12'x5'	2	0	3,725	3,550	7,275
4440		1" acoustical	9'x6'	2	0	1,800	850	2,650
4520			16'x10'	3	0	4,950	2,175	7,125

INTERIORS C

C1010 Partitions

C1010 710	Interior Glazed Opening						

	FRAME	GLASS	OPENING-SIZE W X H	INTERMEDIATE MULLION	INTERMEDIATE HORIZONTAL	COST PER OPNG.		
						MAT.	INST.	TOTAL
5000	Concealed frame	1/2" float	6'x4'	1		715	780	1,495
5040	Butt glazed		9'x4'	2		1,025	1,125	2,150
5080			8'x5'	1		1,100	1,150	2,250
5120			16'x5'	3		2,075	2,150	4,225
5200		3/4" float	6'x8'	1		1,650	1,750	3,400
5240			9'x8'	2		2,375	2,500	4,875
5280			8'x10'	1		2,625	2,675	5,300
5320			16'x10'	3		5,000	5,025	10,025
5400		1/4" tempered	6'x4'	1		430	585	1,015
5440			9'x4'	2		595	820	1,415
5480			8'x5'	1		625	820	1,445
5520			16'x5'	3		1,125	1,500	2,625
5600		1/2" tempered	6'x8'	1		1,425	1,325	2,750
5640			9'x8'	2		2,050	1,875	3,925
5680			8'x10'	1		2,250	2,000	4,250
5720			16'x10'	3		4,275	3,675	7,950
5800		1/4" laminated safety	6'x4'	1		500	640	1,140
5840			9'x4'	2		705	905	1,610
5880			8'x5'	1		740	910	1,650
5920			16'x5'	3		1,375	1,675	3,050
6000		1/2" laminated safety	6'x8'	1		1,275	1,225	2,500
6040			9'x8'	2		1,825	1,725	3,550
6080			8'x10'	1		1,975	1,800	3,775
6120			16'x10'	3		3,725	3,300	7,025
6200		1-3/16" bullet proof	6'x4'	1		1,575	1,675	3,250
6240			9'x4'	2		2,325	2,450	4,775
6280			8'x5'	1		2,550	2,625	5,175
6320			16'x5'	3		4,975	5,100	10,075
6400		2" bullet proof	6'x8'	1		3,500	3,925	7,425
6440			9'x8'	2		5,150	5,775	10,925
6480			8'x10'	1		5,700	6,300	12,000
6520			16'x10'	3		11,100	12,300	23,400

Important: See the Reference Section for critical supporting data - Reference Numbers and City Cost Indexes

C1020 Interior Doors

Sliding Panel-Mall Front

Rolling Overhead Steel Door

INTERIORS C

C1020 102	Special Doors	COST PER OPNG.		
		MAT.	INST.	TOTAL
3800	Sliding entrance door and system mill finish	7,800	2,075	9,875
3900	Bronze finish	8,550	2,250	10,800
4000	Black finish	8,925	2,350	11,275
4100	Sliding panel mall front, 16'x9' opening, mill finish	2,675	640	3,315
4200	Bronze finish	3,475	830	4,305
4300	Black finish	4,275	1,025	5,300
4400	24'x9' opening mill finish	3,875	1,175	5,050
4500	Bronze finish	5,050	1,525	6,575
4600	Black finish	6,200	1,875	8,075
4700	48'x9' opening mill finish	7,225	920	8,145
4800	Bronze finish	9,400	1,200	10,600
4900	Black finish	11,600	1,475	13,075
5000	Rolling overhead steel door, manual, 8' x 8' high	890	720	1,610
5100	10' x 10' high	1,175	820	1,995
5200	20' x 10' high	2,475	1,150	3,625
5300	12' x 12' high	1,525	960	2,485
5400	Motor operated, 8' x 8' high	1,750	950	2,700
5500	10' x 10' high	2,025	1,050	3,075
5600	20' x 10' high	3,325	1,375	4,700
5700	12' x 12' high	2,375	1,200	3,575
5800	Roll up grille, aluminum, manual, 10' x 10' high, mill finish	2,550	1,400	3,950
5900	Bronze anodized	4,000	1,400	5,400
6000	Motor operated, 10' x 10' high, mill finish	3,700	1,625	5,325
6100	Bronze anodized	5,150	1,625	6,775
6200	Steel, manual, 10' x 10' high	2,275	1,150	3,425
6300	15' x 8' high	2,675	1,450	4,125
6400	Motor operated, 10' x 10' high	3,425	1,375	4,800
6500	15' x 8' high	3,825	1,675	5,500

C1020 Interior Doors

The Metal Door/Metal Frame Systems are defined as follows: door type, design and size, frame type and depth. Included in the components for each system is painting the door and frame. No hardware has been included in the systems.

Steel Door, Half Glass Steel Frame

Steel Door, Flush Steel Frame

Wood Door, Flush Wood Frame

System Components	QUANTITY	UNIT	COST EACH		
			MAT.	INST.	TOTAL
SYSTEM C1020 114 1200					
STEEL DOOR, HOLLOW, 20 GA., HALF GLASS, 2'-8"X6'-8", D.W.FRAME, 4-7/8" DP					
Steel door, flush, hollow core, 1-3/4" thick,half glass, 20 ga.,2'-8"x6'-8"	1.000	Ea.	420	44.50	464.50
Steel frame, KD, 16 ga., drywall, 4-7/8" deep, 2'-8"x6'-8", single	1.000	Ea.	114	59	173
Float glass, 3/16" thick, clear, tempered	5.000	S.F.	30	32	62
Paint door and frame each side, primer	1.000	Ea.	4.84	76	80.84
Paint door and frame each side, 2 coats	1.000	Ea.	7.56	127	134.56
TOTAL			576.40	338.50	914.90

C1020 114			Metal Door/Metal Frame					
	TYPE	DESIGN	SIZE	FRAME	DEPTH	COST EACH		
						MAT.	INST.	TOTAL
1000	Flush-hollow	20 ga. full panel	2'-8" x 6'-8"	drywall K.D.	4-7/8"	430	310	740
1020				butt welded	8-3/4"	535	350	885
1160			6'-0" x 7'-0"	drywall K.D.	4-7/8"	850	505	1,355
1180				butt welded	8-3/4"	970	545	1,515
1200		20 ga. half glass	2'-8" x 6'-8"	drywall K.D.	4-7/8"	575	340	915
1220				butt welded	8-3/4"	680	380	1,060
1360			6'-0" x 7'-0"	drywall K.D.	4-7/8"	1,100	560	1,660
1380				butt welded	8-3/4"	1,200	600	1,800
1800		18 ga. full panel	2'-8" x 6'-8"	drywall K.D.	4-7/8"	350	315	665
1820				butt welded	8-3/4"	455	355	810
1960			6'-0" x 7'-0"	drywall K.D.	4-7/8"	920	505	1,425
1980				butt welded	8-3/4"	1,050	545	1,595
2000		18 ga. half glass	2'-8" x 6'-8"	drywall K.D.	4-7/8"	625	345	970
2020				butt welded	8-3/4"	730	385	1,115
2160			6'-0" x 7'-0"	drywall K.D.	4-7/8"	1,175	565	1,740
2180				butt welded	8-3/4"	1,300	605	1,905

C10 Interior Construction

C1020 Interior Doors

C1020 116								

Labeled Metal Door/Metal Frames

	TYPE	DESIGN	SIZE	FRAME	DEPTH	COST EACH		
						MAT.	INST.	TOTAL
6000	Hollow-1/2 hour	20 ga. full panel	2'-8" x 6'-8"	drywall K.D.	4-7/8"	540	310	850
6020				butt welded	8-3/4"	600	350	950
6160			6'-0" x 7'-0"	drywall K.D.	4-7/8"	965	505	1,470
6180				butt welded	8-3/4"	1,050	545	1,595
6200		20 ga. vision	2'-8" x 6'-8"	drywall K.D.	4-7/8"	660	315	975
6220				butt welded	8-3/4"	720	360	1,080
6360			6'-0" x 7'-0"	drywall K.D.	4-7/8"	1,200	515	1,715
6380				butt welded	8-3/4"	1,300	555	1,855
6600		18 ga. full panel	2'-8" x 7'-0"	drywall K.D.	4-7/8"	610	315	925
6620				butt welded	8-3/4"	670	355	1,025
6760			6'-0" x 7'-0"	drywall K.D.	4-7/8"	1,025	510	1,535
6780				butt welded	8-3/4"	1,125	550	1,675
6800		18 ga. vision	2'-8" x 7'-0"	drywall K.D.	4-7/8"	730	320	1,050
6820				butt welded	8-3/4"	790	360	1,150
6960			6'-0" x 7'-0"	drywall K.D.	4-7/8"	1,275	525	1,800
6980				butt welded	8-3/4"	1,375	565	1,940
7200	Hollow-3 hour	18 ga. full panel	2'-8" x 7'-0"	butt K.D.	5-3/4"	610	315	925
7220				butt welded	8-3/4"	670	355	1,025
7360			6'-0" x 7'-0"	butt K.D.	5-3/4"	1,025	510	1,535
7380				butt welded	8-3/4"	1,125	550	1,675
8000	Composite-1-1/2 hour	20 ga. full panel	2'-8" x 6'-8"	drywall K.D.	4-3/8"	640	315	955
8020				butt welded	8-3/4"	700	355	1,055
8160			6'-0" x 7'-0"	drywall K.D.	4-7/8"	1,150	510	1,660
8180				butt welded	8-3/4"	1,250	550	1,800
8200		20 ga. vision	2'-8" x 6'-8"	drywall K.D.	4-7/8"	760	320	1,080
8220				butt welded	8-3/4"	820	360	1,180
8360			6'-0" x 7'-0"	drywall K.D.	4-7/8"	1,400	525	1,925
8400	Composite-3 hour	18 ga. full panel	2'-8" x 7'-0"	drywall K.D.	5-3/4"	590	320	910
8420				butt welded	8-3/4"	650	360	1,010
8560			6'-0" x 7'-0"	drywall K.D.	5-3/4"	1,025	515	1,540

C1020 120								

Wood Door/Wood Frame

	TYPE	FACE	SIZE	FRAME	DEPTH	COST EACH		
						MAT.	INST.	TOTAL
1600	Hollow core/flush	lauan	2'-8" x 6'-8"	pine	3-5/8"	142	229	371
1620					5-3/16"	145	245	390
1760			6'-0" x 6'-8"	pine	3-5/8"	220	360	580
1780					5-3/16"	224	380	604
1800		birch	2'-8" x 6'-8"	pine	3-5/8"	154	229	383
1820					5-3/16"	158	245	403
1960			6'-0" x 6'-8"	pine	3-5/8"	244	360	604
1980					5-3/16"	248	380	628
2000		oak	2'-8" x 6'-8"	oak	3-5/8"	121	195	316
2020					5-3/16"	204	248	452
2160			6'-0" x 6'-8"	oak	3-5/8"	310	365	675
2180					5-3/16"	330	385	715
3000	Particle core/flush	lauan	2'-8" x 6'-8"	pine	3-5/8"	186	236	422
3020					5-3/16"	189	252	441
3160			6'-0" x 7'-0"	pine	3-5/8"	320	390	710
3180					5-3/16"	325	410	735

INTERIORS C

245

C1020 Interior Doors

C1020 120	Wood Door/Wood Frame

	TYPE	FACE	SIZE	FRAME	DEPTH	COST EACH MAT.	COST EACH INST.	COST EACH TOTAL
3200	Particle core/flush	birch	2'-8" x 6'-8"	pine	3-5/8"	197	236	433
3220					5-3/16"	200	252	452
3360			6'-0" x 7'-0"	pine	3-5/8"	340	390	730
3380					5-3/16"	340	410	750
3398								
3400		oak	2'-8" x 6'-8"	oak	3-5/8"	201	239	440
3420					5-3/16"	220	255	475
3560			6'-0" x 7'-0"	oak	3-5/8"	355	395	750
3580					5-3/16"	380	415	795
5000	Solid core/flush	birch	2'-8" x 6'-8"	pine	3-5/8"	230	236	466
5020					5-3/16"	234	252	486
5160			6'-0" x 7'-0"	pine	3-5/8"	405	390	795
5180					5-3/16"	410	410	820
5200		oak	2'-8" x 6'-8"	oak	3-5/8"	234	239	473
5220					5-3/16"	253	255	508
5360			6'-0" x 7'-0"	oak	3-5/8"	420	395	815
5380					5-3/16"	445	415	860
5400		M.D. overlay	2'-8" x 6'-8"	pine	3-5/8"	248	380	628

C1020 122	Wood Door/Metal Frame

	TYPE	FACE	SIZE	FRAME	DEPTH	COST EACH MAT.	COST EACH INST.	COST EACH TOTAL
1600	Hollow core/flush	lauan	2'-8" x 6'-8"	drywall K.D.	4-7/8"	156	175	331
1620				butt welded	8-3/4"	259	183	442
1760			6'-0" x 7'-0"	drywall K.D.	4-7/8"	246	297	543
1780				butt welded	8-3/4"	365	305	670
1800		birch	2'-8" x 6'-8"	drywall K.D.	4-7/8"	168	175	343
1820				butt welded	8-3/4"	271	183	454
1960			6'-0" x 7'-0"	drywall K.D.	4-7/8"	270	297	567
1980				butt welded	8-3/4"	390	305	695
2000		oak	2'-8" x 6'-8"	drywall K.D.	4-7/8"	204	175	379
2020				butt welded	8-3/4"	305	183	488
2160			6'-0" x 7'-0"	drywall K.D.	4-7/8"	340	297	637
2180				butt welded	8-3/4"	460	305	765
3000	Particle core/flush	lauan	2'-8" x 6'-8"	drywall K.D.	4-7/8"	200	182	382
3020				butt welded	8-3/4"	305	190	495
3160			6'-0" x 7'-0"	drywall K.D.	4-7/8"	345	330	675
3180				butt welded	8-3/4"	465	340	805
3200		birch	2'-8" x 6'-8"	drywall K.D.	4-7/8"	211	182	393
3220				butt welded	8-3/4"	315	190	505
3360			6'-0" x 7'-0"	drywall K.D.	4-7/8"	365	330	695
3380				butt welded	8-3/4"	485	340	825
3400		oak	2'-8" x 6'-8"	drywall K.D.	4-7/8"	220	182	402
3420				butt welded	8-3/4"	325	190	515
3560			6'-0" x 7'-0"	drywall K.D.	4-7/8"	390	330	720
3580				butt welded	8-3/4"	510	340	850
5000	Solid core/flush	birch	2'-8" x 6'-8"	drywall K.D.	4-7/8"	244	182	426
5020				butt welded	8-3/4"	345	190	535
5160			6'-0" x 7'-0"	drywall K.D.	4-7/8"	430	330	760
5180				butt welded	8-3/4"	550	340	890

Important: See the Reference Section for critical supporting data - Reference Numbers and City Cost Indexes

C INTERIORS

C10 Interior Construction

C1020 Interior Doors

C1020 122 — Wood Door/Metal Frame

	TYPE	FACE	SIZE	FRAME	DEPTH	MAT.	INST.	TOTAL
5200	Solid core/flush	oak	2'-8" x 6'-8"	drywall K.D.	4-7/8"	254	182	436
5220				butt welded	8-3/4"	355	190	545
5360			6'-0" x 7'-0"	drywall K.D.	4-7/8"	455	330	785
5380				butt welded	8-3/4"	575	340	915
6000	1 hr/flush	birch	2'-8" x 6'-8"	drywall K.D.	4-7/8"	510	187	697
6020				butt welded	8-3/4"	570	195	765
6180				butt welded	8-3/4"	1,000	350	1,350
6200		oak	2'-8" x 6'-8"	drywall K.D.	4-7/8"	510	187	697
6220				butt welded	8-3/4"	570	195	765
6360			6'-0" x 7'-0"	drywall K.D.	4-7/8"	925	340	1,265
6380				butt welded	8-3/4"	1,000	350	1,350
6400	1 hr/flush	walnut	2'-8" x 6'-8"	drywall K.D.	4-7/8"	610	187	797
6420				butt welded	8-3/4"	670	195	865
6560			6'-0" x 7'-0"	drywall K.D.	4-7/8"	1,125	340	1,465
6580				butt welded	8-3/4"	1,225	350	1,575
7600	1-1/2 hr/flush	birch	2'-8" x 6'-8"	drywall K.D.	4-7/8"	460	187	647
7620				butt welded	8-3/4"	525	195	720
7760			6'-0" x 7'-0"	drywall K.D.	4-7/8"	915	340	1,255
7780				butt welded	8-3/4"	1,000	350	1,350
7800		oak	2'-8" x 6'-8"	drywall K.D.	4-7/8"	475	187	662
7820				butt welded	8-3/4"	540	195	735
7960			6'-0" x 7'-0"	drywall K.D.	4-7/8"	855	340	1,195
7980				butt welded	8-3/4"	940	350	1,290
8000		walnut	2'-8" x 6'-8"	drywall K.D.	4-7/8"	585	187	772
8020				butt welded	8-3/4"	645	195	840
8160			6'-0" x 7'-0"	drywall K. D.	4 7/8"	1,075	340	1,415
8180				butt welded	8-3/4"	1,175	350	1,525

C1020 124 — Wood Fire Doors/Metal Frames

	FIRE RATING (HOURS)	CORE MATERIAL	THICKNESS (IN.)	FACE MATERIAL	SIZE W(FT)XH(FT)	MAT.	INST.	TOTAL
0840	1	mineral	1-3/4"	birch	2/8 6/8	510	187	697
0860		3 ply stile			3/0 7/0	535	202	737
0880				oak	2/8 6/8	510	187	697
0900					3/0 7/0	535	202	737
0920				walnut	2/8 6/8	610	187	797
0940					3/0 7/0	640	202	842
0960				m.d. overlay	2/8 6/8	475	320	795
0980					3/0 7/0	505	330	835
1000	1-1/2	mineral	1-3/4"	birch	2/8 6/8	460	187	647
1020		3 ply stile			3/0 7/0	530	202	732
1040				oak	2/8 6/8	475	187	662
1060					3/0 7/0	500	202	702
1080				walnut	2/8 6/8	585	187	772
1100					3/0 7/0	615	202	817
1120				m.d. overlay	2/8 6/8	510	320	830
1140					3/0 7/0	535	330	865

INTERIORS C

247

C1020 Interior Doors

C1020 310	Hardware	COST EACH		
		MAT.	INST.	TOTAL
0060	**HINGES**			
0080				
0100	Full mortise, low frequency, steel base, 4-1/2″ x 4-1/2″, USP	5.20		5.20
0120	5″ x 5″ USP	14.25		14.25
0140	6″ x 6″ USP	28.50		28.50
0160	Average frequency, steel base, 4-1/2″ x 4-1/2″ USP	11.75		11.75
0180	5″ x 5″ USP	19.50		19.50
0200	6″ x 6″ USP	41.50		41.50
0220	High frequency, steel base, 4-1/2″ x 4-1/2″, USP	28.50	.	28.50
0240	5″ x 5″ USP	26.50		26.50
0260	6″ x 6″ USP	64.50		64.50
0280				
0300	**LOCKSETS**			
0320				
0340	Heavy duty cylindrical, passage door			
0360	Non-keyed, passage	47	37	84
0380	Privacy	58.50	37	95.50
0400	Keyed, single cylinder function	82	44.50	126.50
0420	Hotel	117	55.50	172.50
0440				
0460	For re-core cylinder, add	34		34
0480				
0500	**CLOSERS**			
0520				
0540	Rack & pinion			
0560	Adjustable backcheck, 3 way mount, all sizes, regular arm	151	74	225
0580	Hold open arm	151	74	225
0600	Fusible link	134	68	202
0620	Non-sized, regular arm	150	74	224
0640	4 way mount, non-sized, regular arm	206	74	280
0660	Hold open arm	221	74	295
0680				
0700	**PUSH, PULL**			
0720				
0740	Push plate, aluminum	7.55	37	44.55
0760	Bronze	19.10	37	56.10
0780	Pull handle, push bar, aluminum	133	40.50	173.50
0800	Bronze	172	44.50	216.50
0810				
0820	**PANIC DEVICES**			
0840				
0860	Narrow stile, rim mounted, bar, exit only	635	74	709
0880	Outside key & pull	690	88.50	778.50
0900	Bar and vertical rod, exit only	655	88.50	743.50
0920	Outside key & pull	655	111	766
0940	Bar and concealed rod, exit only	765	148	913
0960	Mortise, bar, exit only	555	111	666
0980	Touch bar, exit only	810	111	921
1000				
1020	**WEATHERSTRIPPING**			
1040				
1060	Interlocking, 3′ x 7′, zinc	15.15	148	163.15
1080	Spring type, 3′ x 7′, bronze	33	148	181

Important: See the Reference Section for critical supporting data - Reference Numbers and City Cost Indexes

Toilet Units

Entrance Screens

Urinal Screens

C1030 110	Toilet Partitions	COST PER UNIT		
		MAT.	INST.	TOTAL
0380	Toilet partitions, cubicles, ceiling hung, marble	1,700	425	2,125
0400	Painted metal	520	221	741
0420	Plastic laminate	630	221	851
0460	Stainless steel	1,375	221	1,596
0480	Handicap addition	455		455
0520	Floor and ceiling anchored, marble	1,850	340	2,190
0540	Painted metal	515	177	692
0560	Plastic laminate	730	177	907
0600	Stainless steel	1,600	177	1,777
0620	Handicap addition	315		315
0660	Floor mounted marble	1,100	283	1,383
0680	Painted metal	580	127	707
0700	Plastic laminate	625	127	752
0740	Stainless steel	1,700	127	1,827
0760	Handicap addition	315		315
0780	Juvenile deduction	43.50		43.50
0820	Floor mounted with handrail marble	1,050	283	1,333
0840	Painted metal	515	148	663
0860	Plastic laminate	850	148	998
0900	Stainless steel	1,575	148	1,723
0920	Handicap addition	335		335
0960	Wall hung, painted metal	655	127	782
1020	Stainless steel	1,550	127	1,677
1040	Handicap addition	335		335
1080	Entrance screens, floor mounted, 54" high, marble	685	94.50	779.50
1100	Painted metal	234	59	293
1140	Stainless steel	845	59	904
1300	Urinal screens, floor mounted, 24" wide, laminated plastic	390	111	501
1320	Marble	600	131	731
1340	Painted metal	217	111	328
1380	Stainless steel	705	111	816
1428	Wall mounted wedge type, painted metal	283	88.50	371.50
1430	Urinal screens, wall hung, plastic laminate/particle board	225	111	336
1460	Stainless steel	575	88.50	663.50
1500	Partitions, shower stall, single wall, painted steel, 2'-8" x 2'-8"	840	208	1,048
1510	Fiberglass, 2'-8" x 2'-8"	680	231	911
1520	Double wall, enameled steel, 2'-8" x 2'-8"	960	208	1,168
1530	Stainless steel, 2'-8" x 2'-8"	1,550	208	1,758
1540	Doors, plastic, 2' wide	120	57.50	177.50
1550	Tempered glass, chrome/brass frame-deluxe	805	520	1,325
1560	Tub enclosure, sliding panels, tempered glass, aluminum frame	355	260	615
1570	Chrome/brass frame-deluxe	1,025	345	1,370

C1030 Fittings

C1030 210	Fabricated Compact Units and Cubicles, EACH	COST EACH		
		MAT.	INST.	TOTAL
0500	Telephone enclosure, shelf type, wall hung, economy	1,125	177	1,302
0510	Deluxe	2,900	177	3,077
0520	Booth type, painted steel, economy	3,525	590	4,115
0530	Stainless steel, deluxe	11,800	590	12,390

C1030 310	Storage Specialties, EACH	COST EACH		
		MAT.	INST.	TOTAL
0200	Lockers, steel, single tier, 5' to 6' high, per opening, min.	124	37	161
0210	Maximum	210	43.50	253.50
0220	Two tier, minimum	89	20	109
0230	Maximum	115	26	141
0440	Parts bins, 3' wide x 7' high, 14 bins, 18" x 12" x 12"	281	68	349
0450	84 bins, 6" x 6" x 12"	625	85.50	710.50
0600	Shelving, metal industrial, braced, 3' wide, 1' deep	32	9.85	41.85
0610	2' deep	45.50	10.45	55.95
0620	Enclosed, 3' wide, 1' deep	49	11.50	60.50
0630	2' deep	60.50	13.55	74.05

C1030 510	Identifying/Visual Aid Specialties, EACH	COST EACH		
		MAT.	INST.	TOTAL
0100	Control boards, magnetic, porcelain finish, framed, 24" x 18"	186	111	297
0110	96" x 48"	820	177	997
0120	Directory boards, outdoor, black plastic, 36" x 24"	755	445	1,200
0130	36" x 36"	870	590	1,460
0140	Indoor, economy, open faced, 18" x 24"	125	127	252
0150	36" x 48"	257	148	405
0160	Building, aluminum, black felt panel, 24" x 18"	310	221	531
0170	48" x 72"	1,025	885	1,910
0500	Signs, interior electric exit sign, wall mounted, 6"	114	62.50	176.50
0510	Street, reflective alum., dbl. face, 4 way, w/bracket	122	29.50	151.50
0520	Letters, cast aluminum, 1/2" deep, 4" high	22.50	24.50	47
0530	1" deep, 10" high	37.50	24.50	62
0540	Plaques, cast aluminum, 20" x 30"	940	221	1,161
0550	Cast bronze, 36" x 48"	3,700	445	4,145

C1030 520	Identifying/Visual Aid Specialties, S.F.	COST PER S.F.		
		MAT.	INST.	TOTAL
0100	Bulletin board, cork sheets, no frame, 1/4" thick	3.96	3.05	7.01
0110	1/2" thick	6.75	3.05	9.80
0120	Aluminum frame, 1/4" thick, 3' x 5'	8.40	3.72	12.12
0130	4' x 8'	6.20	1.97	8.17
0200	Chalkboards, wall hung, alum, frame & chalktrough	11.30	1.97	13.27
0210	Wood frame & chalktrough	10.55	2.11	12.66
0220	Sliding board, one board with back panel	37.50	1.89	39.39
0230	Two boards with back panel	56.50	1.89	58.39
0240	Liquid chalk type, alum. frame & chalktrough	11.30	1.97	13.27
0250	Wood frame & chalktrough	21	1.97	22.97

C1030 710	Bath and Toilet Accessories, EACH	COST EACH		
		MAT.	INST.	TOTAL
0100	Specialties, bathroom accessories, st. steel, curtain rod, 5' long, 1" diam	33.50	34	67.50
0110	1-1/2" diam.	33.50	34	67.50
0120	Dispenser, towel, surface mounted	41.50	27.50	69
0130	Flush mounted with waste receptacle	296	44.50	340.50
0140	Grab bar, 1-1/4" diam., 12" long	22	18.45	40.45
0150	1-1/2" diam. 36" long	62.50	22	84.50

Important: See the Reference Section for critical supporting data - Reference Numbers and City Cost Indexes

C10 Interior Construction

C1030 Fittings

C1030 710	Bath and Toilet Accessories, EACH	COST EACH		
		MAT.	INST.	TOTAL
0160	Mirror, framed with shelf, 18" x 24"	172	22	194
0170	72" x 24"	375	74	449
0180	Toilet tissue dispenser, surface mounted, single roll	12.85	14.75	27.60
0190	Double roll	17.90	18.45	36.35
0200	Towel bar, 18" long	38.50	19.25	57.75
0210	30" long	78.50	21	99.50
0300	Medicine cabinets, sliding mirror doors, 20" x 16" x 4-3/4", unlighted	99.50	63.50	163
0310	24" x 19" x 8-1/2", lighted	157	88.50	245.50
0320	Triple door, 30" x 32", unlighted, plywood body	236	63.50	299.50
0330	Steel body	310	63.50	373.50
0340	Oak door, wood body, beveled mirror, single door	152	63.50	215.50
0350	Double door	370	74	444

C1030 730	Bath and Toilet Accessories, L.F.	COST PER L.F.		
		MAT.	INST.	TOTAL
0100	Partitions, hospital curtain, ceiling hung, polyester oxford cloth	16.70	4.32	21.02
0110	Designer oxford cloth	31.50	5.45	36.95

C1030 830	Fabricated Cabinets & Counters, L.F.	COST PER L.F.		
		MAT.	INST.	TOTAL
0110	Household, base, hardwood, one top drawer & one door below x 12" wide	152	35.50	187.50
0120	Four drawer x 24" wide	335	39.50	374.50
0130	Wall, hardwood, 30" high with one door x 12" wide	146	40.50	186.50
0140	Two doors x 48" wide	340	48	388
0150	Counter top-laminated plastic, stock, economy	9.75	14.75	24.50
0160	Custom-square edge, 7/8" thick	12.25	33	45.25
0170	School, counter, wood, 32" high	203	44.50	247.50
0180	Metal, 84" high	310	59	369

C1030 910	Other Fittings, EACH	COST EACH		
		MAT.	INST.	TOTAL
0500	Mail boxes, horizontal, rear loaded, aluminum, 5" x 6" x 15" deep	36	13	49
0510	Front loaded, aluminum, 10" x 12" x 15" deep	125	22	147
0520	Vertical, front loaded, aluminum, 15" x 5" x 6" deep	28.50	13	41.50
0530	Bronze, duranodic finish	48.50	13	61.50
0540	Letter slot, post office	242	55.50	297.50
0550	Mail counter, window, post office, with grille	565	221	786
0600	Partition, woven wire, wall panel, 4' x 7' high	126	35.50	161.50
0610	Wall panel with window & shelf, 5' x 8' high	345	59	404
0620	Doors, swinging w/ no transom, 3' x 7' high	244	148	392
0630	Sliding, 6' x 10' high	575	221	796
0700	Turnstiles, one way, 4' arm, 46" diam., manual	605	177	782
0710	Electric	990	740	1,730
0720	3 arm, 5'-5' diam. & 7' high, manual	2,575	885	3,460
0730	Electric	3,325	1,475	4,800

C2010 Stair Construction

General Design: See reference section for code requirements. Maximum height between landings is 12'; usual stair angle is 20° to 50° with 30° to 35° best. Usual relation of riser to treads is:

Riser + tread = 17.5.
2x (Riser) + tread = 25.
Riser x tread = 70 or 75.

Maximum riser height is 7" for commercial, 8-1/4" for residential.
Usual riser height is 6-1/2" to 7-1/4".

Minimum tread width is 11" for commercial and 9" for residential.

For additional information please see reference section.

Cost Per Flight: Table below lists the cost per flight for 4'-0" wide stairs.
Side walls are not included.
Railings are included.

System Components	QUANTITY	UNIT	COST PER FLIGHT		
			MAT.	INST.	TOTAL
SYSTEM C2010 110 0560					
STAIRS, C.I.P. CONCRETE WITH LANDING, 12 RISERS					
Concrete in place, free standing stairs not incl. safety treads	48.000	L.F.	280.80	1,525.44	1,806.24
Concrete in place, free standing stair landing	32.000	S.F.	153.28	423.04	576.32
Stair tread C.I. abrasive 4" wide	48.000	L.F.	566.40	266.40	832.80
Industrial railing, welded, 2 rail 3'-6" high 1-1/2" pipe	18.000	L.F.	423	171.72	594.72
Wall railing with returns, steel pipe	17.000	L.F.	188.70	162.18	350.88
TOTAL			1,612.18	2,548.78	4,160.96

C2010 110	Stairs	COST PER FLIGHT		
		MAT.	INST.	TOTAL
0470	Stairs, C.I.P. concrete, w/o landing, 12 risers, w/o nosing RC2010-100	895	1,850	2,745
0480	With nosing	1,450	2,125	3,575
0550	W/landing, 12 risers, w/o nosing	1,050	2,275	3,325
0560	With nosing	1,600	2,550	4,150
0570	16 risers, w/o nosing	1,275	2,875	4,150
0580	With nosing	2,025	3,225	5,250
0590	20 risers, w/o nosing	1,500	3,450	4,950
0600	With nosing	2,450	3,900	6,350
0610	24 risers, w/o nosing	1,750	4,025	5,775
0620	With nosing	2,875	4,575	7,450
0630	Steel, grate type w/nosing & rails, 12 risers, w/o landing	3,725	970	4,695
0640	With landing	4,875	1,325	6,200
0660	16 risers, with landing	6,100	1,675	7,775
0680	20 risers, with landing	7,350	1,975	9,325
0700	24 risers, with landing	8,600	2,300	10,900
0710	Cement fill metal pan & picket rail, 12 risers, w/o landing	4,850	970	5,820
0720	With landing	6,400	1,450	7,850
0740	16 risers, with landing	8,025	1,775	9,800
0760	20 risers, with landing	9,625	2,100	11,725
0780	24 risers, with landing	11,300	2,425	13,725
0790	Cast iron tread & pipe rail, 12 risers, w/o landing	4,850	970	5,820
0800	With landing	6,400	1,450	7,850
0820	16 risers, with landing	8,025	1,775	9,800
0840	20 risers, with landing	9,625	2,100	11,725
0860	24 risers, with landing	11,300	2,425	13,725
0870	Pan tread & flat bar rail, pre-assembled, 12 risers, w/o landing	4,025	730	4,755
0880	With landing	7,300	1,150	8,450
0900	16 risers, with landing	8,050	1,275	9,325
0920	20 risers, with landing	9,400	1,525	10,925
0940	24 risers, with landing	10,700	1,775	12,475

C20 Stairs

C2010 Stair Construction

C2010 110	Stairs	COST PER FLIGHT		
		MAT.	INST.	TOTAL
0950	Spiral steel, industrial checkered plate 4'-6' dia., 12 risers	4,675	650	5,325
0960	16 risers	6,250	865	7,115
0970	20 risers	7,800	1,100	8,900
0980	24 risers	9,350	1,300	10,650
0990	Spiral steel, industrial checkered plate 6'-0" dia., 12 risers	5,650	725	6,375
1000	16 risers	7,525	970	8,495
1010	20 risers	9,400	1,200	10,600
1020	24 risers	11,300	1,450	12,750
1030	Aluminum, spiral, stock units, 5'-0" dia., 12 risers	5,225	650	5,875
1040	16 risers	6,950	865	7,815
1050	20 risers	8,700	1,100	9,800
1060	24 risers	10,400	1,300	11,700
1070	Custom 5'-0" dia., 12 risers	9,850	650	10,500
1080	16 risers	13,100	865	13,965
1090	20 risers	16,400	1,100	17,500
1100	24 risers	19,700	1,300	21,000
1120	Wood, prefab box type, oak treads, wood rails 3'-6" wide, 14 risers	1,800	370	2,170
1150	Prefab basement type, oak treads, wood rails 3'-0" wide, 14 risers	890	91	981

C3010 Wall Finishes

C3010 230	Paint & Covering	COST PER S.F.		
		MAT.	INST.	TOTAL
0060	Painting, interior on plaster and drywall, brushwork, primer & 1 coat	.11	.59	.70
0080	Primer & 2 coats	.16	.78	.94
0100	Primer & 3 coats	.21	.96	1.17
0120	Walls & ceilings, roller work, primer & 1 coat	.11	.39	.50
0140	Primer & 2 coats	.16	.50	.66
0160	Woodwork incl. puttying, brushwork, primer & 1 coat	.11	.85	.96
0180	Primer & 2 coats	.16	1.12	1.28
0200	Primer & 3 coats	.21	1.53	1.74
0260	Cabinets and casework, enamel, primer & 1 coat	.10	.96	1.06
0280	Primer & 2 coats	.15	1.18	1.33
0300	Masonry or concrete, latex, brushwork, primer & 1 coat	.19	.80	.99
0320	Primer & 2 coats	.25	1.14	1.39
0340	Addition for block filler	.12	.98	1.10
0380	Fireproof paints, intumescent, 1/8" thick 3/4 hour	1.87	.78	2.65
0420	7/16" thick 2 hour	5.40	2.73	8.13
0440	1-1/16" thick 3 hour	8.80	5.45	14.25
0500	Gratings, primer & 1 coat	.20	1.19	1.39
0600	Pipes over 12" diameter	.51	3.82	4.33
0720	Heavy framing 50-100 S.F./Ton	.06	.76	.82
0740	Spraywork, light framing 300-500 S.F./Ton	.07	.34	.41
0800	Varnish, interior wood trim, no sanding sealer & 1 coat	.08	.96	1.04
0820	Hardwood floor, no sanding 2 coats	.14	.20	.34
0860	Maximum	.59	1.25	1.84
0880	Epoxy coatings, minimum	.36	.73	1.09
1080	Maximum	.96	2.63	3.59
1100	High build epoxy 50 mil, minimum	.62	.98	1.60
1120	Maximum	1.06	4.02	5.08
1140	Laminated epoxy with fiberglass minimum	.67	1.30	1.97
1160	Maximum	1.19	2.63	3.82
1180	Sprayed perlite or vermiculite 1/16" thick, minimum	.24	.13	.37
1200	Maximum	.68	.60	1.28
1260	Wall coatings, vinyl plastic, minimum	.31	.52	.83
1280	Maximum	.75	1.59	2.34
1300	Urethane on smooth surface, 2 coats, minimum	.23	.34	.57
1320	Maximum	.51	.57	1.08
1340	3 coats, minimum	.31	.45	.76
1360	Maximum	.68	.81	1.49
1380	Ceramic-like glazed coating, cementitious, minimum	.44	.87	1.31
1420	Resin base, minimum	.31	.60	.91
1440	Maximum	.50	1.16	1.66
1460	Wall coverings, aluminum foil	.96	1.39	2.35
1500	Vinyl backing	5.15	1.60	6.75
1520	Cork tiles, 12"x12", light or dark, 3/16" thick	4.11	1.60	5.71
1540	5/16" thick	3.51	1.63	5.14
1580	Natural, non-directional, 1/2" thick	6.90	1.60	8.50
1600	12"x36", granular, 3/16" thick	1.17	1	2.17
1620	1" thick	1.51	1.04	2.55
1660	5/16" thick	5.20	1.63	6.83
1661	Paneling, prefinished plywood, birch	1.41	2.11	3.52
1662	Mahogany, African	2.63	2.21	4.84
1664	Oak or cherry	3.38	2.21	5.59
1665	Rosewood	4.80	2.77	7.57
1666	Teak	3.38	2.21	5.59
1720	Gypsum based, fabric backed, minimum	.75	.48	1.23
1740	Average	1.13	.53	1.66
1760	Maximum	1.25	.60	1.85
1780	Vinyl wall covering, fabric back, light weight	.65	.60	1.25
1800	Medium weight	.80	.80	1.60

Important: See the Reference Section for critical supporting data - Reference Numbers and City Cost Indexes

C30 Interior Finishes

C3010 Wall Finishes

C3010 230	Paint & Covering	COST PER S.F.		
		MAT.	INST.	TOTAL
1820	Heavy weight	1.29	.88	2.17
1840	Wall paper, double roll, solid pattern, avg. workmanship	.33	.60	.93
1860	Basic pattern, avg. workmanship	.72	.72	1.44
1880	Basic pattern, quality workmanship	1.68	.88	2.56
1900	Grass cloths with lining paper, minimum	.72	.96	1.68
1920	Maximum	2.30	1.10	3.40
1940	Ceramic tile, thin set, 4-1/4" x 4-1/4"	2.19	3.76	5.95

C3010 235	Paint Trim	COST PER L.F.		
		MAT.	INST.	TOTAL
2040	Painting, wood trim, to 6" wide, enamel, primer & 1 coat	.10	.48	.58
2060	Primer & 2 coats	.15	.61	.76
2080	Misc. metal brushwork, ladders	.40	4.78	5.18
2120	6" to 8" dia.	.14	2.01	2.15
2140	10" to 12" dia.	.40	3.01	3.41
2160	Railings, 2" pipe	.14	2.39	2.53
2180	Handrail, single	.10	.96	1.06

INTERIORS C

255

C3020 Floor Finishes

C3020 410	Tile & Covering	COST PER S.F.		
		MAT.	INST.	TOTAL
0060	Carpet tile, nylon, fusion bonded, 18" x 18" or 24" x 24", 24 oz.	3.72	.56	4.28
0080	35 oz.	3.94	.56	4.50
0100	42 oz.	5.15	.56	5.71
0140	Carpet, tufted, nylon, roll goods, 12' wide, 26 oz.	4.05	.64	4.69
0160	36 oz.	6.45	.64	7.09
0180	Woven, wool, 36 oz.	10.20	.64	10.84
0200	42 oz.	10.40	.64	11.04
0220	Padding, add to above, minimum	.48	.30	.78
0240	Maximum	.83	.30	1.13
0260	Composition flooring, acrylic, 1/4" thick	1.43	4.20	5.63
0280	3/8" thick	1.88	4.85	6.73
0300	Epoxy, minimum	2.53	3.24	5.77
0320	Maximum	3.07	4.46	7.53
0340	Epoxy terrazzo, minimum	6	4.92	10.92
0360	Maximum	9.20	6.55	15.75
0380	Mastic, hot laid, 1-1/2" thick, minimum	3.71	3.17	6.88
0400	Maximum	4.76	4.20	8.96
0420	Neoprene 1/4" thick, minimum	3.65	4.01	7.66
0440	Maximum	5.05	5.10	10.15
0460	Polyacrylate with ground granite 1/4", minimum	3.34	2.97	6.31
0480	Maximum	6.45	4.55	11
0500	Polyester with colored quart 2 chips 1/16", minimum	3.25	2.05	5.30
0520	Maximum	5.05	3.24	8.29
0540	Polyurethane with vinyl chips, minimum	7.10	2.05	9.15
0560	Maximum	10.25	2.54	12.79
0620	1" thick	.43	3.53	3.96
0640	2" thick	.85	4.06	4.91
0680	Maximum	.29	6.35	6.64
0700	For colors, add to above, minimum	.66	1.23	1.89
0740	Exposed aggregate finish, minimum	.18	.65	.83
0760	Maximum	.26	.87	1.13
0800	Maximum	.49	.48	.97
0820	Dust on coloring, add, minimum	.66	.31	.97
0860	1/2" integral, minimum	5.25	3.44	8.69
0880	Maximum	5.70	3.44	9.14
1040	1/2" thick	.19	2.13	2.32
1060	3/4" thick	.29	2.39	2.68
1080	1" thick	.38	2.71	3.09
1100	Terrazzo, minimum	2.81	7.55	10.36
1120	Maximum	5.25	16.35	21.60
1280	Resilient, asphalt tile, 1/8" thick on concrete, minimum	1.17	1.01	2.18
1300	Maximum	1.28	1.01	2.29
1320	On wood, add for felt underlay	.21		.21
1340	Cork tile, minimum	4.66	1.28	5.94
1360	Maximum	10.85	1.28	12.13
1380	Polyethylene, in rolls, minimum	2.83	1.46	4.29
1400	Maximum	5.50	1.46	6.96
1420	Polyurethane, thermoset, minimum	4.35	4.02	8.37
1440	Maximum	5.15	8.05	13.20
1460	Rubber, sheet goods, minimum	4.68	3.35	8.03
1480	Maximum	7.70	4.47	12.17
1500	Tile, minimum	5.35	1.01	6.36
1520	Maximum	8.70	1.46	10.16
1540	Synthetic turf, minimum	3.29	1.92	5.21
1560	Maximum	8.45	2.12	10.57
1580	Vinyl, composition tile, minimum	.95	.80	1.75
1600	Maximum	2.26	.80	3.06
1620	Vinyl tile, minimum	2.31	.80	3.11

Important: See the Reference Section for critical supporting data - Reference Numbers and City Cost Indexes

C3020 Floor Finishes

C3020 410	Tile & Covering	COST PER S.F.		
		MAT.	INST.	TOTAL
1660	Sheet goods, minimum	2.43	1.61	4.04
1680	Maximum	4.68	2.01	6.69
1720	Tile, ceramic natural clay	4.38	3.90	8.28
1730	Marble, synthetic 12"x12"x5/8"	6.95	11.90	18.85
1740	Porcelain type, minimum	4.70	3.90	8.60
1800	Quarry tile, mud set, minimum	3.45	5.10	8.55
1820	Maximum	5.75	6.50	12.25
1840	Thin set, deduct		1.02	1.02
1860	Terrazzo precast, minimum	4.18	5.35	9.53
1880	Maximum	9.10	5.35	14.45
1900	Non-slip, minimum	17.80	27	44.80
1920	Maximum	19.80	37.50	57.30
2020	Wood block, end grain factory type, natural finish, 2" thick	3.91	3.54	7.45
2040	Fir, vertical grain, 1"x4", no finish, minimum	2.60	1.74	4.34
2060	Maximum	2.76	1.74	4.50
2080	Prefinished white oak, prime grade, 2-1/4" wide	6.80	2.60	9.40
2100	3-1/4" wide	8.75	2.39	11.14
2120	Maple strip, sanded and finished, minimum	4.03	3.76	7.79
2140	Maximum	4.88	3.76	8.64
2160	Oak strip, sanded and finished, minimum	4.08	3.76	7.84
2180	Maximum	4.87	3.76	8.63
2200	Parquetry, sanded and finished, minimum	4.55	3.93	8.48
2220	Maximum	6.65	5.60	12.25
2260	Add for sleepers on concrete, treated, 24" O.C., 1"x2"	.66	2.28	2.94
2280	1"x3"	.83	1.78	2.61
2300	2"x4"	.80	.90	1.70
2320	2"x6"	.83	.68	1.51
2340	Underlayment, plywood, 3/8" thick	.86	.59	1.45
2350	1/2" thick	.96	.61	1.57
2360	5/8" thick	1.34	.63	1.97
2370	3/4" thick	1.45	.68	2.13
2380	Particle board, 3/8" thick	.36	.59	.95
2390	1/2" thick	.41	.61	1.02
2400	5/8" thick	.45	.63	1.08
2410	3/4" thick	.62	.68	1.30
2420	Hardboard, 4' x 4', .215" thick	.43	.59	1.02

C3030 Ceiling Finishes

2 Coats of Plaster on Gypsum Lath on Wood Furring

Fiberglass Board on Exposed Suspended Grid System

Plaster and Metal Lath on Metal Furring

System Components	QUANTITY	UNIT	COST PER S.F.		
			MAT.	INST.	TOTAL
SYSTEM C3030 105 2400					
GYPSUM PLASTER, 2 COATS, 3/8"GYP. LATH, WOOD FURRING, FRAMING					
Gypsum plaster 2 coats no lath, on ceilings	.110	S.Y.	.40	2.36	2.76
Gypsum lath plain/perforated nailed, 3/8" thick	.110	S.Y.	.66	.51	1.17
Add for ceiling installation	.110	S.Y.		.20	.20
Furring, 1" x 3" wood strips on wood joists	.750	L.F.	.23	.95	1.18
Paint, primer and one coat	1.000	S.F.	.11	.39	.50
TOTAL			1.40	4.41	5.81

C3030 105				Plaster Ceilings				
	TYPE	LATH	FURRING	SUPPORT		COST PER S.F.		
						MAT.	INST.	TOTAL
2400	2 coat gypsum	3/8" gypsum	1"x3" wood, 16" O.C.	wood		1.40	4.41	5.81
2500	Painted			masonry		1.40	4.50	5.90
2600				concrete		1.40	5.05	6.45
2700	3 coat gypsum	3.4# metal	1"x3" wood, 16" O.C.	wood		1.31	4.79	6.10
2800	Painted			masonry		1.31	4.88	6.19
2900				concrete		1.31	5.40	6.71
3000	2 coat perlite	3/8" gypsum	1"x3" wood, 16" O.C.	wood		1.40	4.62	6.02
3100	Painted			masonry		1.40	4.70	6.10
3200				concrete		1.40	5.25	6.65
3300	3 coat perlite	3.4# metal	1"x3" wood, 16" O.C.	wood		1.14	4.73	5.87
3400	Painted			masonry		1.14	4.81	5.95
3500				concrete		1.14	5.35	6.49
3600	2 coat gypsum	3/8" gypsum	3/4" CRC, 12" O.C.	1-1/2" CRC, 48"O.C.		1.42	5.35	6.77
3700	Painted		3/4" CRC, 16" O.C.	1-1/2" CRC, 48"O.C.		1.40	4.81	6.21
3800			3/4" CRC, 24" O.C.	1-1/2" CRC, 48"O.C.		1.32	4.39	5.71
3900	2 coat perlite	3/8" gypsum	3/4" CRC, 12" O.C.	1-1/2" CRC, 48"O.C		1.42	5.70	7.12
4000	Painted		3/4" CRC, 16" O.C.	1-1/2" CRC, 48"O.C.		1.40	5.20	6.60
4100			3/4" CRC, 24" O.C.	1-1/2" CRC, 48"O.C.		1.32	4.80	6.12
4200	3 coat gypsum	3.4# metal	3/4" CRC, 12" O.C.	1-1/2" CRC, 36" O.C.		1.54	7	8.54
4300	Painted		3/4" CRC, 16" O.C.	1-1/2" CRC, 36" O.C.		1.52	6.50	8.02
4400			3/4" CRC, 24" O.C.	1-1/2" CRC, 36" O.C.		1.44	6.05	7.49
4500	3 coat perlite	3.4# metal	3/4" CRC, 12" O.C.	1-1/2" CRC,36" O.C.		1.69	7.65	9.34
4600	Painted		3/4" CRC, 16" O.C.	1-1/2" CRC, 36" O.C.		1.67	7.15	8.82
4700			3/4" CRC, 24" O.C.	1-1/2" CRC, 36" O.C.		1.59	6.75	8.34

Important: See the Reference Section for critical supporting data - Reference Numbers and City Cost Indexes

C3030 Ceiling Finishes

C3030 110			Drywall Ceilings					

	TYPE	FINISH	FURRING	SUPPORT		COST PER S.F.		
						MAT.	INST.	TOTAL
4800	1/2" F.R. drywall	painted and textured	1"x3" wood, 16" O.C.	wood		.77	2.75	3.52
4900				masonry		.77	2.84	3.61
5000				concrete		.77	3.38	4.15
5100	5/8" F.R. drywall	painted and textured	1"x3" wood, 16" O.C.	wood		.77	2.75	3.52
5200				masonry		.77	2.84	3.61
5300				concrete		.77	3.38	4.15
5400	1/2" F.R. drywall	painted and textured	7/8"resil. channels	24" O.C.		.67	2.67	3.34
5500			1"x2" wood	stud clips		.81	2.57	3.38
5600			1-5/8"metal studs	24" O.C.		.72	2.38	3.10
5700	5/8" F.R. drywall	painted and textured	1-5/8"metal studs	24" O.C.		.72	2.38	3.10
5702								

INTERIORS

C

259

C3030 Ceiling Finishes

C3030 140	Plaster Ceiling Components	COST PER S.F.		
		MAT.	INST.	TOTAL
0060	Plaster, gypsum incl. finish, 2 coats			
0080	3 coats	.57	2.83	3.40
0100	Perlite, incl. finish, 2 coats	.40	2.77	3.17
0120	3 coats	.72	3.47	4.19
0140	Thin coat on drywall	.08	.54	.62
0200	Lath, gypsum, 3/8" thick	.66	.71	1.37
0220	1/2" thick	.52	.74	1.26
0240	5/8" thick	.55	.87	1.42
0260	Metal, diamond, 2.5 lb.	.35	.58	.93
0280	3.4 lb.	.40	.62	1.02
0300	Flat rib, 2.75 lb.	.37	.58	.95
0320	3.4 lb.	.54	.62	1.16
0440	Furring, steel channels, 3/4" galvanized , 12" O.C.	.25	1.86	2.11
0460	16" O.C.	.23	1.35	1.58
0480	24" O.C.	.15	.93	1.08
0500	1-1/2" galvanized , 12" O.C.	.35	2.05	2.40
0520	16" O.C.	.32	1.50	1.82
0540	24" O.C.	.21	1	1.21
0560	Wood strips, 1"x3", on wood, 12" O.C.	.31	1.27	1.58
0580	16" O.C.	.23	.95	1.18
0600	24" O.C.	.16	.64	.80
0620	On masonry, 12" O.C.	.31	1.38	1.69
0640	16" O.C.	.23	1.04	1.27
0660	24" O.C.	.16	.69	.85
0680	On concrete, 12" O.C.	.31	2.11	2.42
0700	16" O.C.	.23	1.58	1.81
0720	24" O.C.	.16	1.06	1.22

C3030 210		Acoustical Ceilings						
	TYPE	TILE	GRID	SUPPORT		COST PER S.F.		
						MAT.	INST.	TOTAL
5800	5/8" fiberglass board	24" x 48"	tee	suspended		1.16	1.27	2.43
5900		24" x 24"	tee	suspended		1.30	1.40	2.70
6000	3/4" fiberglass board	24" x 48"	tee	suspended		1.90	1.30	3.20
6100		24" x 24"	tee	suspended		2.04	1.43	3.47
6500	5/8" mineral fiber	12" x 12"	1"x3" wood, 12" O.C.	wood		1.93	1.71	3.64
6600				masonry		1.93	1.82	3.75
6700				concrete		1.93	2.55	4.48
6800	3/4" mineral fiber	12" x 12"	1"x3" wood, 12" O.C.	wood		1.86	1.71	3.57
6900				masonry		1.86	1.71	3.57
7000				concrete		1.86	1.71	3.57
7100	3/4"mineral fiber on	12" x 12"	25 ga. channels	runners		2.08	2.30	4.38
7102	5/8" F.R. drywall							
7200	5/8" plastic coated	12" x 12"		adhesive backed		1.99	.44	2.43
7201	Mineral fiber							
7202								
7300	3/4" plastic coated	12" x 12"		adhesive backed		1.92	.44	2.36
7301	Mineral fiber							
7302								
7400	3/4" mineral fiber	12" x 12"	conceal 2" bar & channels	suspended		2.04	2.24	4.28
7401								
7402								

Important: See the Reference Section for critical supporting data - Reference Numbers and City Cost Indexes

C3030 Ceiling Finishes

C3030 240	Acoustical Ceiling Components	COST PER S.F.		
		MAT.	INST.	TOTAL
2480	Ceiling boards, eggcrate, acrylic, 1/2" x 1/2" x 1/2" cubes	1.63	.89	2.52
2500	Polystyrene, 3/8" x 3/8" x 1/2" cubes	1.36	.87	2.23
2520	1/2" x 1/2" x 1/2" cubes	1.83	.89	2.72
2540	Fiberglass boards, plain, 5/8" thick	.57	.71	1.28
2560	3/4" thick	1.31	.74	2.05
2580	Grass cloth faced, 3/4" thick	1.89	.89	2.78
2600	1" thick	2.09	.91	3
2620	Luminous panels, prismatic, acrylic	1.98	1.11	3.09
2640	Polystyrene	1.01	1.11	2.12
2660	Flat or ribbed, acrylic	3.44	1.11	4.55
2680	Polystyrene	2.35	1.11	3.46
2700	Drop pan, white, acrylic	5.05	1.11	6.16
2720	Polystyrene	4.22	1.11	5.33
2740	Mineral fiber boards, 5/8" thick, standard	.70	.66	1.36
2760	Plastic faced	1.10	1.11	2.21
2780	2 hour rating	.96	.66	1.62
2800	Perforated aluminum sheets, .024 thick, corrugated painted	2.01	.90	2.91
2820	Plain	3.41	.89	4.30
3080	Mineral fiber, plastic coated, 12" x 12" or 12" x 24", 5/8" thick	1.62	.44	2.06
3100	3/4" thick	1.55	.44	1.99
3120	Fire rated, 3/4" thick, plain faced	1.40	.44	1.84
3140	Mylar faced	1.50	.44	1.94
3160	Add for ceiling primer	.13		.13
3180	Add for ceiling cement	.37		.37
3240	Suspension system, furring, 1" x 3" wood 12" O.C.	.31	1.27	1.58
3260	T bar suspension system, 2' x 4' grid	.58	.55	1.13
3280	2' x 2' grid	.72	.68	1.40
3300	Concealed Z bar suspension system 12" module	.52	.85	1.37
3320	Add to above for 1-1/2" carrier channels 4' O.C.	.11	.94	1.05
3340	Add to above for carrier channels for recessed lighting	.20	.96	1.16

For information about Means Estimating Seminars, see yellow pages 12 and 13 in back of book

D SERVICES

D1010 Elevators and Lifts

The hydraulic elevator obtains its motion from the movement of liquid under pressure in the piston connected to the car bottom. These pistons can provide travel to a maximum rise of 70′ and are sized for the intended load. As the rise reaches the upper limits the cost tends to exceed that of a geared electric unit.

System Components	QUANTITY	UNIT	COST EACH		
			MAT.	INST.	TOTAL
SYSTEM D1010 110 2000					
PASS. ELEV. HYDRAULIC 2500 LB. 5 FLOORS, 100 FPM					
Passenger elevator, hydraulic, 1500 lb capacity, 2 stop, 100 FPM	1.000	Ea.	33,700	11,900	45,600
Over 10′ travel height, passenger elevator, hydraulic, add	40.000	V.L.F.	15,400	6,600	22,000
Passenger elevator, hydraulic, 2500lb capacity over standard, add	1.000	Ea.	2,200		2,200
Over 2 stops, passenger elevator, hydraulic, add	3.000	Stop	10,050	13,275	23,325
Hall lantern	5.000	Ea.	1,900	745	2,645
Maintenance agreement for pass. elev. 9 months	1.000	Ea.	3,425		3,425
Position indicator at lobby	1.000	Ea.	72.50	37.50	110
TOTAL			66,747.50	32,557.50	99,305

D1010 110	Hydraulic		COST EACH		
			MAT.	INST.	TOTAL
1300	Pass. elev., 1500 lb., 2 Floors, 100 FPM	RD1010 -010	37,900	12,200	50,100
1400	5 Floors, 100 FPM		64,500	32,600	97,100
1600	2000 lb., 2 Floors, 100 FPM		39,000	12,200	51,200
1700	5 floors, 100 FPM		65,500	32,600	98,100
1900	2500 lb., 2 Floors, 100 FPM		40,100	12,200	52,300
2000	5 floors, 100 FPM		66,500	32,600	99,100
2200	3000 lb., 2 Floors, 100 FPM		41,900	12,200	54,100
2300	5 floors, 100 FPM		68,500	32,600	101,100
2500	3500 lb., 2 Floors, 100 FPM		43,800	12,200	56,000
2600	5 floors, 100 FPM		70,500	32,600	103,100
2800	4000 lb., 2 Floors, 100 FPM		45,000	12,200	57,200
2900	5 floors, 100 FPM		71,500	32,600	104,100
3100	4500 lb., 2 Floors, 100 FPM		46,500	12,200	58,700
3200	5 floors, 100 FPM		73,000	32,600	105,600
4000	Hospital elevators, 3500 lb., 2 Floors, 100 FPM		56,500	12,200	68,700
4100	5 floors, 100 FPM		84,000	32,600	116,600
4300	4000 lb., 2 Floors, 100 FPM		56,500	12,200	68,700
4400	5 floors, 100 FPM		84,000	32,600	116,600
4600	4500 lb., 2 Floors, 100 FPM		63,000	12,200	75,200
4800	5 floors, 100 FPM		90,500	32,600	123,100
4900	5000 lb., 2 Floors, 100 FPM		66,000	12,200	78,200
5000	5 floors, 100 FPM		93,500	32,600	126,100
6700	Freight elevators (Class "B"), 3000 lb., 2 Floors, 50 FPM		58,000	15,500	73,500
6800	5 floors, 100 FPM		89,500	40,800	130,300
7000	4000 lb., 2 Floors, 50 FPM		60,500	15,500	76,000
7100	5 floors, 100 FPM		92,500	40,800	133,300
7500	10,000 lb., 2 Floors, 50 FPM		75,500	15,500	91,000
7600	5 floors, 100 FPM		107,500	40,800	148,300
8100	20,000 lb., 2 Floors, 50 FPM		92,000	15,500	107,500
8200	5 Floors, 100 FPM		124,000	40,800	164,800

D1010   Elevators and Lifts

Geared traction elevators are the intermediate group both in cost and in operating areas. These are in buildings of four to fifteen floors and speed ranges from 200' to 500' per minute.

System Components	QUANTITY	UNIT	COST EACH		
			MAT.	INST.	TOTAL
SYSTEM D1010 140 1600					
PASSENGER 2500 LB., 5 FLOORS, 200 FPM					
Passenger elevator, geared, 2000 lb capacity, 4 stop, 200 FPM	1.000	Ea.	75,500	23,900	99,400
Over 40' travel height, passenger elevator electric, add	10.000	V.L.F.	1,180	1,650	2,830
Passenger elevator, electric, 2500 lb. cap., add	1.000	Ea.	3,075		3,075
Over 4 stops, passenger elevator, electric, add	1.000	Stop	2,625	4,425	7,050
Hall lantern	5.000	Ea.	1,900	745	2,645
Maintenance agreement for passenger elevator, 9 months	1.000	Ea.	3,425		3,425
Position indicator at lobby	1.000	Ea.	72.50	37.50	110
TOTAL			87,777.50	30,757.50	118,535

D1010 140	Traction Geared Elevators		COST EACH		
			MAT.	INST.	TOTAL
1300	Passenger, 2000 Lb., 5 floors, 200 FPM		84,500	30,800	115,300
1500	15 floors, 350 FPM		133,000	93,000	226,000
1600	2500 Lb., 5 floors, 200 FPM	RD1010 -010	88,000	30,800	118,800
1800	15 floors, 350 FPM		136,000	93,000	229,000
1900	3000 Lb., 5 floors, 200 FPM		89,500	30,800	120,300
2100	15 floors, 350 FPM		137,500	93,000	230,500
2200	3500 Lb., 5 floors, 200 FPM		91,000	30,800	121,800
2400	15 floors, 350 FPM		139,500	93,000	232,500
2500	4000 Lb., 5 floors, 200 FPM		91,500	30,800	122,300
2700	15 floors, 350 FPM		140,000	93,000	233,000
2800	4500 Lb., 5 floors, 200 FPM		94,000	30,800	124,800
3000	15 floors, 350 FPM		142,000	93,000	235,000
3100	5000 Lb., 5 floors, 200 FPM		96,500	30,800	127,300
3300	15 floors, 350 FPM		144,500	93,000	237,500
4000	Hospital, 3500 Lb., 5 floors, 200 FPM		104,000	30,800	134,800
4200	15 floors, 350 FPM		179,500	93,000	272,500
4300	4000 Lb., 5 floors, 200 FPM		104,000	30,800	134,800
4500	15 floors, 350 FPM		179,500	93,000	272,500
4600	4500 Lb., 5 floors, 200 FPM		110,000	30,800	140,800
4800	15 floors, 350 FPM		185,000	93,000	278,000
4900	5000 Lb., 5 floors, 200 FPM		111,500	30,800	142,300
5100	15 floors, 350 FPM		187,000	93,000	280,000
6000	Freight, 4000 Lb., 5 floors, 50 FPM class 'B'		93,000	31,800	124,800
6200	15 floors, 200 FPM class 'B'		171,000	110,500	281,500
6300	8000 Lb., 5 floors, 50 FPM class 'B'		111,500	31,800	143,300
6500	15 floors, 200 FPM class 'B'		189,500	110,500	300,000
7000	10,000 Lb., 5 floors, 50 FPM class 'B'		129,000	31,800	160,800
7200	15 floors, 200 FPM class 'B'		347,000	110,500	457,500
8000	20,000 Lb., 5 floors, 50 FPM class 'B'		143,000	31,800	174,800
8200	15 floors, 200 FPM class 'B'		361,000	110,500	471,500

D1010 Elevators and Lifts

Gearless traction elevators are used in high rise situations where speeds to over 1000 FPM are required to move passengers efficiently. This type of installation is also the most costly.

System Components	QUANTITY	UNIT	COST EACH		
			MAT.	INST.	TOTAL
SYSTEM D1010 150 1700					
PASSENGER, 2500 LB., 10 FLOORS, 200 FPM					
Passenger elevator, geared, 2000 lb capacity, 4 stop 100 FPM	1.000	Ea.	75,500	23,900	99,400
Over 10' travel height, passenger elevator electric, add	60.000	V.L.F.	7,080	9,900	16,980
Passenger elevator, electric, 2500 lb. cap., add	1.000	Ea.	3,075		3,075
Over 2 stops, passenger elevator, electric, add	6.000	Stop	15,750	26,550	42,300
Gearless variable voltage machinery, overhead	1.000	Ea.	60,000	17,000	77,000
Hall lantern	9.000	Ea.	3,800	1,490	5,290
Position indicator at lobby	1.000	Ea.	72.50	37.50	110
Fireman services control	1.000	Ea.	1,650	995	2,645
Emergency power manual switching, add	1.000	Ea.	370	149	519
Maintenance agreement for passenger elevator, 9 months	1.000	Ea.	3,425		3,425
TOTAL			170,722.50	80,021.50	250,744

D1010 150	Traction Gearless Elevators	COST EACH		
		MAT.	INST.	TOTAL
1700	Passenger, 2500 Lb., 10 floors, 200 FPM	170,500	80,000	250,500
1900	30 floors, 600 FPM	299,000	204,500	503,500
2000	3000 Lb., 10 floors, 200 FPM	172,500	80,000	252,500
2200	30 floors, 600 FPM	300,500	204,500	505,000
2300	3500 Lb., 10 floors, 200 FPM	174,000	80,000	254,000
2500	30 floors, 600 FPM	302,500	204,500	507,000
2700	50 floors, 800 FPM	395,000	329,000	724,000
2800	4000 lb., 10 floors, 200 FPM	174,500	80,000	254,500
3000	30 floors, 600 FPM	303,000	204,500	507,500
3200	50 floors, 800 FPM	395,500	329,000	724,500
3300	4500 lb., 10 floors, 200 FPM	177,000	80,000	257,000
3500	30 floors, 600 FPM	305,000	204,500	509,500
3700	50 floors, 800 FPM	397,500	329,000	726,500
3800	5000 lb., 10 floors, 200 FPM	179,500	80,000	259,500
4000	30 floors, 600 FPM	307,500	204,500	512,000
4200	50 floors, 800 FPM	400,000	329,000	729,000
6000	Hospital, 3500 Lb., 10 floors, 200 FPM	200,500	80,000	280,500
6200	30 floors, 600 FPM	383,000	204,500	587,500
6400	4000 Lb., 10 floors, 200 FPM	200,500	80,000	280,500
6600	30 floors, 600 FPM	383,000	204,500	587,500
6800	4500 Lb., 10 floors, 200 FPM	206,500	80,000	286,500
7000	30 floors, 600 FPM	388,500	204,500	593,000
7200	5000 Lb., 10 floors, 200 FPM	208,000	80,000	288,000
7400	30 floors, 600 FPM	390,000	204,500	594,500

RD1010 -010

Important: See the Reference Section for critical supporting data - Reference Numbers and City Cost Indexes

D1020 Escalators and Moving Walks

PLAN

Floor Opening Enclosure

Upper Level

Ballustrade

Lower Level

ELEVATION

Moving stairs can be used for buildings where 600 or more people are to be carried to the second floor or beyond. Freight cannot be carried on escalators and at least one elevator must be available for this function.

Carrying capacity is 5000 to 8000 people per hour. Power requirement is 2 KW to 3 KW per hour and incline angle is 30°.

D1020 110				Moving Stairs		COST EACH		
	TYPE	HEIGHT	WIDTH	BALUSTRADE MATERIAL		MAT.	INST.	TOTAL
0100	Escalator	10ft	32"	glass		70,500	33,300	103,800
0150				metal		72,000	33,300	105,300
0200			48"	glass		77,500	33,300	110,800
0250				metal		77,500	33,300	110,800
0300		15ft	32"	glass		74,000	38,800	112,800
0350				metal		74,000	38,800	112,800
0400			48"	glass		78,000	38,800	116,800
0450				metal		78,000	38,800	116,800
0500		20ft	32"	glass		76,000	47,400	123,400
0550				metal		76,000	47,400	123,400
0600			48"	glass		81,500	47,400	128,900
0650				metal		81,500	47,400	128,900
0700		25ft	32"	glass		82,000	58,000	140,000
0750				metal		82,000	58,000	140,000
0800			48"	glass		88,000	58,000	146,000
0850				metal		88,000	58,000	146,000

D1020 210				Moving Walks		COST PER L.F.		
	TYPE	STORY HEIGHT	DEGREE SLOPE	WIDTH	COST RANGE	MAT.	INST.	TOTAL
1500	Ramp	10'-23'	12	3'-4"	minimum	2,125	650	2,775
1600					maximum	2,675	800	3,475
2000	Walk	0'	0	2'-0"	minimum	790	360	1,150
2500					maximum	1,100	525	1,625
3000	Walk	0'	0	3'-4"	minimum	1,775	525	2,300
3500					maximum	2,075	610	2,685

D1090 Other Conveying Systems

D1090 410	Miscellaneous Types	COST EACH		
		MAT.	INST.	TOTAL
1050	Conveyors, Horiz. belt, ctr drive, 60FPM, cap 40 lb/LF, belt x 26.5'L	2,925	1,750	4,675
1100	24" belt, 42' length	4,525	2,200	6,725
1200	24" belt,62' length	6,050	2,925	8,975
1500	Inclined belt, horiz. loading and end idler assembly, 16" belt x 27.5'L	8,525	2,925	11,450
1600	24" belt, 27.5' length	9,575	5,850	15,425
2000	Motorized car distribution system, single type, capacity, 20 lb./car			
2010	Automatic selective central control, base price 2 stop	56,000	11,700	67,700
2100	Automatic service station to station, add per station	4,375	8,775	13,150
3500	Pneumatic tube system, single, 2 station-100', stock economy model, 3" dia.	5,600	8,650	14,250
3600	4" diameter	6,300	11,500	17,800
4000	Twin tube system base cost, minimum	5,400	1,375	6,775
4100	Maximum	10,700	4,150	14,850
4200	Add per LF, of system, minimum	6.95	11.10	18.05
4300	Maximum	20.50	27.50	48
5000	Completely automatic system, 15 to 50 stations, 4" dia., cost/station	17,000	3,575	20,575
5100	50 to 144 stations, 4" diameter, cost per station	13,200	3,250	16,450
6000	Vertical conveyor, automatic selective control, cart 300 lb. cap., 2 stop	25,500	6,600	32,100
6100	Automatic service additional floors		2,000	2,000
7100	Corresp. lift, Self cont., 45 lb. cap., 2 stop, elec. enam. steel doors	7,725	5,000	12,725

D1090 510	Chutes	COST EACH		
		MAT.	INST.	TOTAL
0100	Chutes, linen or refuse incl. sprinklers, galv. steel, 18" diam., per floor	955	297	1,252
0110	Aluminized steel, 36" diam., per floor	1,300	370	1,670
0120	Mail, 8-3/4" x 3-1/2", aluminum & glass, per floor	660	208	868
0130	Bronze or stainless, per floor	1,025	231	1,256

Important: See the Reference Section for critical supporting data - Reference Numbers and City Cost Indexes

D2010 Plumbing Fixtures

Systems are complete with trim seat and rough-in (supply, waste and vent) for connection to supply branches and waste mains.

| One Piece Wall Hung | Supply | Waste/Vent | Floor Mount |

SERVICES D

System Components	QUANTITY	UNIT	COST EACH		
			MAT.	INST.	TOTAL
SYSTEM D2010 110 1840					
WATER CLOSET, VITREOUS CHINA, ELONGATED					
TANK TYPE, WALL HUNG, ONE PIECE					
Wtr closet tank type vit china wall hung 1 pc w/seat supply & stop	1.000	Ea.	430	175	605
Pipe steel galvanized, schedule 40, threaded, 2" diam	4.000	L.F.	41.60	57.80	99.40
Pipe, CI soil, no hub, cplg 10' OC, hanger 5' OC, 4" diam	2.000	L.F.	20.10	31.90	52
Pipe, coupling, standard coupling, CI soil, no hub, 4" diam	2.000	Ea.	18.60	56	74.60
Copper tubing type L, solder joint, hanger 10'OC, 1/2" diam	6.000	L.F.	11.70	38.10	49.80
Wrought copper 90° elbow for solder joints 1/2" diam	2.000	Ea.	1.12	51	52.12
Wrought copper Tee for solder joints 1/2" diam	1.000	Ea.	.96	39.50	40.46
Support/carrier, for water closet, siphon jet, horiz, single, 4" waste	1.000	Ea.	395	96.50	491.50
TOTAL			919.08	545.80	1,464.88

D2010 110	Water Closet Systems		COST EACH		
			MAT.	INST.	TOTAL
1800	Water closet, vitreous china, elongated				
1840	Tank type, wall hung, one piece		920	545	1,465
1880	Close coupled two piece	RD2010 -400	1,025	545	1,570
1920	Floor mount, one piece		775	580	1,355
1960	One piece low profile		665	580	1,245
2000	Two piece close coupled		415	580	995
2040	Bowl only with flush valve				
2080	Wall hung		910	620	1,530
2120	Floor mount		570	590	1,160

D2010 Plumbing Fixtures

Systems are complete with trim, seat, flush valve and rough-in (supply, waste and vent) for connection to supply branches and waste mains.

Side by Side

Back to Back

Supply

Waste/Vent

Supply

Waste/Vent

System Components	QUANTITY	UNIT	COST EACH		
			MAT.	INST.	TOTAL
SYSTEM D2010 120 1760					
WATER CLOSETS, BATTERY MOUNT, WALL HUNG, SIDE BY SIDE, FIRST CLOSET					
Water closet, bowl only w/flush valve, seat, wall hung	1.000	Ea.	385	160	545
Pipe, CI soil, no hub, cplg 10' OC, hanger 5' OC, 4" diam	3.000	L.F.	30.15	47.85	78
Coupling, standard, CI, soil, no hub, 4" diam	2.000	Ea.	18.60	56	74.60
Copper tubing type L, solder joints, hangers 10' OC, 1" diam	6.000	L.F.	24.12	45.30	69.42
Copper tubing, type DWV, solder joints, hangers 10'OC, 2" diam	6.000	L.F.	48.90	70.20	119.10
Wrought copper 90° elbow for solder joints 1" diam	1.000	Ea.	3.10	32	35.10
Wrought copper Tee for solder joints, 1" diam	1.000	Ea.	7.15	51.50	58.65
Support/carrier, siphon jet, horiz, adjustable single, 4" pipe	1.000	Ea.	395	96.50	491.50
Valve, gate, bronze, 125 lb, NRS, soldered 1" diam	1.000	Ea.	35	27	62
Wrought copper, DWV, 90° elbow, 2" diam	1.000	Ea.	11.20	51.50	62.70
TOTAL			958.22	637.85	1,596.07

D2010 120	Water Closets, Group		COST EACH		
			MAT.	INST.	TOTAL
1760	Water closets, battery mount, wall hung, side by side, first closet		960	640	1,600
1800	Each additional water closet, add		910	605	1,515
3000	Back to back, first pair of closets	RD2010 -400	1,700	845	2,545
3100	Each additional pair of closets, back to back		1,675	835	2,510

Important: See the Reference Section for critical supporting data - Reference Numbers and City Cost Indexes

D2010 Plumbing Fixtures

Systems are complete with trim, flush valve and rough-in (supply, waste and vent) for connection to supply branches and waste mains.

Stall Type   **Supply**   **Waste/Vent**   **Wall Hung**

SERVICES **D**

System Components	QUANTITY	UNIT	COST EACH MAT.	COST EACH INST.	COST EACH TOTAL
SYSTEM D2010 210 2000					
URINAL, VITREOUS CHINA, WALL HUNG					
Urinal, wall hung, vitreous china, incl. hanger	1.000	Ea.	345	310	655
Pipe, steel, galvanized, schedule 40, threaded, 1-1/2" diam	5.000	L.F.	39	57.75	96.75
Copper tubing type DWV, solder joint, hangers 10'OC, 2" diam	3.000	L.F.	24.45	35.10	59.55
Combination Y & 1/8 bend for CI soil pipe, no hub, 3" diam	1.000	Ea.	11.70		11.70
Pipe, CI, no hub, cplg 10' OC, hanger 5' OC, 3" diam	4.000	L.F.	31.80	57.80	89.60
Pipe coupling standard, CI soil, no hub, 3" diam	3.000	Ea.	15.70	49	64.70
Copper tubing type L, solder joint, hanger 10' OC 3/4" diam	5.000	L.F.	14.30	33.75	48.05
Wrought copper 90° elbow for solder joints 3/4" diam	1.000	Ea.	1.25	27	28.25
Wrought copper Tee for solder joints, 3/4" diam	1.000	Ea.	2.32	43	45.32
TOTAL			485.52	613.40	1,098.92

D2010 210	Urinal Systems	COST EACH MAT.	COST EACH INST.	COST EACH TOTAL
2000	Urinal, vitreous china, wall hung	485	615	1,100
2040	Stall type	780	730	1,510

D2010 Plumbing Fixtures

Systems are complete with trim and rough-in (supply, waste and vent) to connect to supply branches and waste mains.

| Vanity Top | Supply | Waste/Vent | Wall Hung |

System Components			COST EACH		
	QUANTITY	UNIT	MAT.	INST.	TOTAL
SYSTEM D2010 310 1560					
LAVATORY W/TRIM, VANITY TOP, P.E. ON C.I., 20″ X 18″					
Lavatory w/trim, PE on Cl, white, vanity top, 20″ x 18″ oval	1.000	Ea.	234	145	379
Pipe, steel, galvanized, schedule 40, threaded, 1-1/4″ diam	4.000	L.F.	26.40	41.60	68
Copper tubing type DWV, solder joint, hanger 10′OC 1-1/4″ diam	4.000	L.F.	19.60	34.20	53.80
Wrought copper DWV, Tee, sanitary, 1-1/4″ diam	1.000	Ea.	10.45	57	67.45
P trap w/cleanout, 20 ga, 1-1/4″ diam	1.000	Ea.	35	28.50	63.50
Copper tubing type L, solder joint, hanger 10′ OC 1/2″ diam	10.000	L.F.	19.50	63.50	83
Wrought copper 90° elbow for solder joints 1/2″ diam	2.000	Ea.	1.12	51	52.12
Wrought copper Tee for solder joints, 1/2″ diam	2.000	Ea.	1.92	79	80.92
Stop, chrome, angle supply, 1/2″ diam	2.000	Ea.	10.40	47	57.40
TOTAL			358.39	546.80	905.19

D2010 310	Lavatory Systems		COST EACH		
			MAT.	INST.	TOTAL
1560	Lavatory w/trim, vanity top, PE on Cl, 20″ x 18″, Vanity top by others.		360	545	905
1600	19″ x 16″ oval		276	545	821
1640	18″ round	RD2010 -400	345	545	890
1680	Cultured marble, 19″ x 17″		305	545	850
1720	25″ x 19″		315	545	860
1760	Stainless, self-rimming, 25″ x 22″		395	545	940
1800	17″ x 22″		390	545	935
1840	Steel enameled, 20″ x 17″		276	560	836
1880	19″ round		273	560	833
1920	Vitreous china, 20″ x 16″		370	575	945
1960	19″ x 16″		370	575	945
2000	22″ x 13″		380	575	955
2040	Wall hung, PE on Cl, 18″ x 15″		660	605	1,265
2080	19″ x 17″		640	605	1,245
2120	20″ x 18″		610	605	1,215
2160	Vitreous china, 18″ x 15″		565	620	1,185
2200	19″ x 17″		530	620	1,150
2240	24″ x 20″		695	620	1,315

Important: See the Reference Section for critical supporting data - Reference Numbers and City Cost Indexes

D2010 Plumbing Fixtures

Systems are complete with trim and rough-in (supply, waste and vent) to connect to supply branches and waste mains.

Countertop Single Bowl

Supply

Waste/Vent

Countertop Double Bowl

System Components	QUANTITY	UNIT	COST EACH		
			MAT.	INST.	TOTAL
SYSTEM D2010 410 1720					
KITCHEN SINK W/TRIM, COUNTERTOP, P.E. ON C.I., 24″ X 21″, SINGLE BOWL					
Kitchen sink, counter top, PE on CI, 1 bowl, 24″ x 21″ OD	1.000	Ea.	255	165	420
Pipe, steel, galvanized, schedule 40, threaded, 1-1/4″ diam	4.000	L.F.	26.40	41.60	68
Copper tubing, type DWV, solder, hangers 10′ OC 1-1/2″ diam	6.000	L.F.	36.60	57	93.60
Wrought copper, DWV, Tee, sanitary, 1-1/2″ diam	1.000	Ea.	14.10	64.50	78.60
P trap, standard, copper, 1-1/2″ diam	1.000	Ea.	34	30	64
Copper tubing, type L, solder joints, hangers 10′ OC 1/2″ diam	10.000	L.F.	19.50	63.50	83
Wrought copper 90° elbow for solder joints 1/2″ diam	2.000	Ea.	1.12	51	52.12
Wrought copper Tee for solder joints, 1/2″ diam	2.000	Ea.	1.92	79	80.92
Stop, angle supply, chrome, 1/2″ CTS	2.000	Ea.	10.40	47	57.40
TOTAL			399.04	598.60	997.64

D2010 410	Kitchen Sink Systems	COST EACH		
		MAT.	INST.	TOTAL
1720	Kitchen sink w/trim, countertop, PE on CI, 24″x21″, single bowl	400	600	1,000
1760	30″ x 21″ single bowl	385	600	985
1800	32″ x 21″ double bowl	500	645	1,145
1840	42″ x 21″ double bowl	690	655	1,345
1880	Stainless steel, 19″ x 18″ single bowl	560	600	1,160
1920	25″ x 22″ single bowl	610	600	1,210
1960	33″ x 22″ double bowl	825	645	1,470
2000	43″ x 22″ double bowl	935	655	1,590
2040	44″ x 22″ triple bowl	1,050	680	1,730
2080	44″ x 24″ corner double bowl	620	655	1,275
2120	Steel, enameled, 24″ x 21″ single bowl	269	600	869
2160	32″ x 21″ double bowl	320	645	965
2240	Raised deck, PE on CI, 32″ x 21″, dual level, double bowl	455	815	1,270
2280	42″ x 21″ dual level, triple bowl	895	890	1,785

D20 Plumbing

D2010 Plumbing Fixtures

Systems are complete with trim and rough-in (supply, waste and vent) to connect to supply branches and waste mains.

Single Compartment Sink

Supply   Waste/Vent

Double Compartment Sink

System Components	QUANTITY	UNIT	COST EACH MAT.	INST.	TOTAL
SYSTEM D2010 420 1760					
LAUNDRY SINK W/TRIM, PE ON CI, BLACK IRON FRAME					
24" X 20" OD, SINGLE COMPARTMENT					
Laundry sink PE on CI w/trim & frame, 24" x 21" OD, 1 compartment	1.000	Ea.	380	154	534
Pipe, steel, galvanized, schedule 40, threaded, 1-1/4" diam	4.000	L.F.	26.40	41.60	68
Copper tubing, type DWV, solder joint, hanger 10' OC 1-1/2"diam	6.000	L.F.	36.60	57	93.60
Wrought copper, DWV, Tee, sanitary, 1-1/2" diam	1.000	Ea.	14.10	64.50	78.60
P trap, standard, copper, 1-1/2" diam	1.000	Ea.	34	30	64
Copper tubing type L, solder joints, hangers 10' OC, 1/2" diam	10.000	L.F.	19.50	63.50	83
Wrought copper 90° elbow for solder joints 1/2" diam	2.000	Ea.	1.12	51	52.12
Wrought copper Tee for solder joints, 1/2" diam	2.000	Ea.	1.92	79	80.92
Stop, angle supply, 1/2" diam	2.000	Ea.	10.40	47	57.40
TOTAL			524.04	587.60	1,111.64

D2010 420	Laundry Sink Systems		COST EACH MAT.	INST.	TOTAL
1740	Laundry sink w/trim, PE on CI, black iron frame				
1760	24" x 20", single compartment		525	590	1,115
1800	24" x 23" single compartment	RD2010 -400	520	590	1,110
1840	48" x 21" double compartment		445	640	1,085
1920	Molded stone, on wall, 22" x 21" single compartment		280	590	870
1960	45"x 21" double compartment		440	640	1,080
2040	Plastic, on wall or legs, 18" x 23" single compartment		248	575	823
2080	20" x 24" single compartment		282	575	857
2120	36" x 23" double compartment		325	620	945
2160	40" x 24" double compartment		405	620	1,025

SERVICES

D

Important: See the Reference Section for critical supporting data - Reference Numbers and City Cost Indexes

D2010 Plumbing Fixtures

Corrosion resistant laboratory sink systems are complete with trim and rough-in (supply, waste and vent) to connect to supply branches and waste mains.

Laboratory Sink   **Supply**   **Waste/Vent**   **Polypropylene Cup Sink**

SERVICES **D**

System Components	QUANTITY	UNIT	COST EACH		
			MAT.	INST.	TOTAL
SYSTEM D2010 430 1600					
LABORATORY SINK W/TRIM, POLYETHYLENE, SINGLE BOWL					
DOUBLE DRAINBOARD, 54" X 24" OD					
Sink w/trim, polyethylene, 1 bowl, 2 drainboards 54" x 24" OD	1.000	Ea.	940	310	1,250
Pipe, polypropylene, schedule 40, acid resistant 1-1/2" diam	10.000	L.F.	12.40	136	148.40
Tee, sanitary, polypropylene, acid resistant, 1-1/2" diam	1.000	Ea.	23	51.50	74.50
P trap, polypropylene, acid resistant, 1-1/2" diam	1.000	Ea.	76.50	30	106.50
Copper tubing type L, solder joint, hanger 10' OC 1/2" diam	10.000	L.F.	19.50	63.50	83
Wrought copper 90° elbow for solder joints 1/2" diam	2.000	Ea.	1.12	51	52.12
Wrought copper Tee for solder joints, 1/2" diam	2.000	Ea.	1.92	79	80.92
Stop, angle supply, chrome, 1/2" diam	2.000	Ea.	10.40	47	57.40
TOTAL			1,084.84	768	1,852.84

D2010 430	Laboratory Sink Systems	COST EACH		
		MAT.	INST.	TOTAL
1580	Laboratory sink w/trim, polyethylene, single bowl,			
1600	Double drainboard, 54" x 24" O.D.	1,075	770	1,845
1640	Single drainboard, 47" x 24"O.D.	1,225	770	1,995
1680	70" x 24" O.D.	1,250	770	2,020
1760	Flanged, 14-1/2" x 14-1/2" O.D.	325	690	1,015
1800	18-1/2" x 18-1/2" O.D.	410	690	1,100
1840	23-1/2" x 20-1/2" O.D.	465	690	1,155
1920	Polypropylene, cup sink, oval, 7" x 4" O.D.	265	610	875
1960	10" x 4-1/2" O.D.	259	610	869

RD2010 -400

D20 Plumbing

D2010 Plumbing Fixtures

Corrosion resistant laboratory sink systems are complete with trim and rough–in (supply, waste and vent) to connect to supply branches and waste mains.

Wall Hung

Supply

Waste/Vent

Corner, Floor

System Components	QUANTITY	UNIT	COST EACH MAT.	COST EACH INST.	COST EACH TOTAL
SYSTEM D2010 440 4260					
SERVICE SINK, PE ON CI, CORNER FLOOR, 28"X28", W/RIM GUARD & TRIM					
Service sink, corner floor, PE on CI, 28" x 28", w/rim guard & trim	1.000	Ea.	645	210	855
Copper tubing type DWV, solder joint, hanger 10'OC 3" diam	6.000	L.F.	87	95.70	182.70
Copper tubing type DWV, solder joint, hanger 10'OC 2" diam	4.000	L.F.	32.60	46.80	79.40
Wrought copper DWV, Tee, sanitary, 3" diam	1.000	Ea.	59.50	132	191.50
P trap with cleanout & slip joint, copper 3" diam	1.000	Ea.	126	46.50	172.50
Copper tubing, type L, solder joints, hangers 10' OC, 1/2" diam	10.000	L.F.	19.50	63.50	83
Wrought copper 90° elbow for solder joints 1/2" diam	2.000	Ea.	1.12	51	52.12
Wrought copper Tee for solder joints, 1/2" diam	2.000	Ea.	1.92	79	80.92
Stop, angle supply, chrome, 1/2" diam	2.000	Ea.	10.40	47	57.40
TOTAL			983.04	771.50	1,754.54

D2010 440	Service Sink Systems	MAT.	INST.	TOTAL
4260	Service sink w/trim, PE on CI, corner floor, 28" x 28", w/rim guard	985	770	1,755
4300	Wall hung w/rim guard, 22" x 18"	1,100	895	1,995
4340	24" x 20"	1,175	895	2,070
4380	Vitreous china, wall hung 22" x 20"	1,050	895	1,945

Important: See the Reference Section for critical supporting data - Reference Numbers and City Cost Indexes

D2010 Plumbing Fixtures

Systems are complete with trim and rough-in (supply, waste and vent) to connect to supply branches and waste mains.

Recessed Bathtub | Supply | Waste/Vent | Corner Bathtub

SERVICES **D**

System Components	QUANTITY	UNIT	COST EACH		
			MAT.	INST.	TOTAL
SYSTEM D2010 510 2000					
BATHTUB, RECESSED, PORCELAIN ENAMEL ON CAST IRON,, 48″ x 42″					
Bath tub, porcelain enamel on cast iron, w/fittings, 48″ x 42″	1.000	Ea.	1,775	231	2,006
Pipe, steel, galvanized, schedule 40, threaded, 1-1/4″ diam	4.000	L.F.	26.40	41.60	68
Pipe, CI no hub soil w/couplings 10′ OC, hangers 5′ OC, 4″ diam	3.000	L.F.	30.15	47.85	78
Combination Y and 1/8 bend for C.I. soil pipe, no hub, 4″ pipe size	1.000	Ea.	26.50		26.50
Drum trap, 3″ x 5″, copper, 1-1/2″ diam	1.000	Ea.	47.50	32	79.50
Copper tubing type L, solder joints, hangers 10′ OC 1/2″ diam	10.000	L.F.	19.50	63.50	83
Wrought copper 90° elbow, solder joints, 1/2″ diam	2.000	Ea.	1.12	51	52.12
Wrought copper Tee, solder joints, 1/2″ diam	2.000	Ea.	1.92	79	80.92
Stop, angle supply, 1/2″ diameter	2.000	Ea.	10.40	47	57.40
Copper tubing type DWV, solder joints, hanger 10′ OC 1-1/2″ diam	3.000	L.F.	18.30	28.50	46.80
Pipe coupling, standard, C.I. soil no hub, 4″ pipe size	2.000	Ea.	18.60	56	74.60
TOTAL			1,975.39	677.45	2,652.84

D2010 510	Bathtub Systems		COST EACH		
			MAT.	INST.	TOTAL
2000	Bathtub, recessed, P.E. on CI., 48″ x 42″	RD2010 -400	1,975	675	2,650
2040	72″ x 36″		2,025	755	2,780
2080	Mat bottom, 5′ long		1,000	655	1,655
2120	5′-6″ long		1,450	675	2,125
2160	Corner, 48″ x 42″		1,975	655	2,630
2200	Formed steel, enameled, 4′-6″ long		565	605	1,170

D2010 Plumbing Fixtures

Systems are complete with trim, flush valve and rough-in (supply, waste and vent) for connection to supply branches and waste mains.

Circular Fountain

Supply

Waste/Vent

Semi-Circular Fountain

System Components	QUANTITY	UNIT	COST EACH		
			MAT.	INST.	TOTAL
SYSTEM D2010 610 1760					
GROUP WASH FOUNTAIN, PRECAST TERRAZZO					
CIRCULAR, 36" DIAMETER					
Wash fountain, group, precast terrazzo, foot control 36" diam	1.000	Ea.	3,250	480	3,730
Copper tubing type DWV, solder joint, hanger 10'OC, 2" diam	10.000	L.F.	81.50	117	198.50
P trap, standard, copper, 2" diam	1.000	Ea.	52.50	34.50	87
Wrought copper, Tee, sanitary, 2" diam	1.000	Ea.	16.45	73.50	89.95
Copper tubing type L, solder joint, hanger 10' OC 1/2" diam	20.000	L.F.	39	127	166
Wrought copper 90° elbow for solder joints 1/2" diam	3.000	Ea.	1.68	76.50	78.18
Wrought copper Tee for solder joints, 1/2" diam	2.000	Ea.	1.92	79	80.92
TOTAL			3,443.05	987.50	4,430.55

D2010 610	Group Wash Fountain Systems	COST EACH		
		MAT.	INST.	TOTAL
1740	Group wash fountain, precast terrazzo			
1760	Circular, 36" diameter	3,450	990	4,440
1800	54" diameter	4,150	1,075	5,225
1840	Semi-circular, 36" diameter	3,125	990	4,115
1880	54" diameter	3,800	1,075	4,875
1960	Stainless steel, circular, 36" diameter	3,925	920	4,845
2000	54" diameter	5,125	1,025	6,150
2040	Semi-circular, 36" diameter	3,425	920	4,345
2080	54" diameter	4,450	1,025	5,475
2160	Thermoplastic, circular, 36" diameter	3,025	750	3,775
2200	54" diameter	3,525	870	4,395
2240	Semi-circular, 36" diameter	2,875	750	3,625
2280	54" diameter	3,375	870	4,245

D2010 Plumbing Fixtures

Systems are complete with trim and rough-in (supply, waste and vent) for connection to supply branches and waste mains.

Three Wall   **Supply**   **Waste/Vent**   **Corner Angle**

SERVICES D

System Components	QUANTITY	UNIT	COST EACH		
			MAT.	INST.	TOTAL
SYSTEM D2010 710 1560					
SHOWER, STALL, BAKED ENAMEL, MOLDED STONE RECEPTOR, 30" SQUARE					
Shower stall, enameled steel, molded stone receptor, 30" square	1.000	Ea.	400	178	578
Copper tubing type DWV, solder joints, hangers 10'OC, 2" diam	6.000	L.F.	36.60	57	93.60
Wrought copper DWV, Tee, sanitary, 2" diam	1.000	Ea.	14.10	64.50	78.60
Trap, standard, copper, 2" diam	1.000	Ea.	34	30	64
Copper tubing type L, solder joint, hanger 10' OC 1/2" diam	16.000	L.F.	31.20	101.60	132.80
Wrought copper 90° elbow for solder joints 1/2" diam	3.000	Ea.	1.68	76.50	78.18
Wrought copper Tee for solder joints, 1/2" diam	2.000	Ea.	1.92	79	80.92
Stop and waste, straightway, bronze, solder joint 1/2" diam	2.000	Ea.	5.96	43	48.96
TOTAL			525.46	629.60	1,155.06

D2010 710	Shower Systems		COST EACH		
			MAT.	INST.	TOTAL
1560	Shower, stall, baked enamel, molded stone receptor, 30" square		525	630	1,155
1600	32" square		505	635	1,140
1640	Terrazzo receptor, 32" square	RD2010 -400	930	635	1,565
1680	36" square		1,075	645	1,720
1720	36" corner angle		1,000	645	1,645
1800	Fiberglass one piece, three walls, 32" square		585	620	1,205
1840	36" square		635	620	1,255
1880	Polypropylene, molded stone receptor, 30" square		485	915	1,400
1920	32" square		585	915	1,500
1960	Built-in head, arm, bypass, stops and handles		147	237	384

D2010 Plumbing Fixtures

Systems are complete with trim and rough-in (supply, waste and vent) to connect to supply branches and waste mains.

Supply   **Waste/Vent**

Wall Mounted, No Back   **Wall Mounted, Low Back**

System Components	QUANTITY	UNIT	COST EACH		
			MAT.	INST.	TOTAL
SYSTEM D2010 810 1800					
DRINKING FOUNTAIN, ONE BUBBLER, WALL MOUNTED					
NON RECESSED, BRONZE, NO BACK					
Drinking fountain, wall mount, bronze, 1 bubbler	1.000	Ea.	1,175	129	1,304
Copper tubing, type L, solder joint, hanger 10' OC 3/8" diam	5.000	L.F.	8.40	30.50	38.90
Stop, supply, straight, chrome, 3/8" diam	1.000	Ea.	4.65	21.50	26.15
Wrought copper 90° elbow for solder joints 3/8" diam	1.000	Ea.	1.77	23.50	25.27
Wrought copper Tee for solder joints, 3/8" diam	1.000	Ea.	3	36.50	39.50
Copper tubing, type DWV, solder joint, hanger 10' OC 1-1/4" diam	4.000	L.F.	19.60	34.20	53.80
P trap, standard, copper drainage, 1-1/4" diam	1.000	Ea.	35	28.50	63.50
Wrought copper, DWV, Tee, sanitary, 1-1/4" diam	1.000	Ea.	10.45	57	67.45
TOTAL			1,257.87	360.70	1,618.57

D2010 810	Drinking Fountain Systems		COST EACH		
			MAT.	INST.	TOTAL
1740	Drinking fountain, one bubbler, wall mounted				
1760	Non recessed				
1800	Bronze, no back	RD2010 -400	1,250	360	1,610
1840	Cast iron, enameled, low back		605	360	965
1880	Fiberglass, 12" back		735	360	1,095
1920	Stainless steel, no back		1,025	360	1,385
1960	Semi-recessed, poly marble		830	360	1,190
2040	Stainless steel		1,050	360	1,410
2080	Vitreous china		705	360	1,065
2120	Full recessed, poly marble		925	360	1,285
2200	Stainless steel		850	360	1,210
2240	Floor mounted, pedestal type, aluminum		590	490	1,080
2320	Bronze		1,375	490	1,865
2360	Stainless steel		1,375	490	1,865

Important: See the Reference Section for critical supporting data - Reference Numbers and City Cost Indexes

D20 Plumbing

D2010 Plumbing Fixtures

Systems are complete with trim and rough-in (supply, waste and vent) for connection to supply branches and waste mains.

Wall Hung

Supply

Waste/Vent

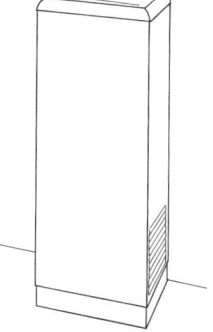

Floor Mounted

SERVICES **D**

System Components	QUANTITY	UNIT	COST EACH MAT.	INST.	TOTAL
SYSTEM D2010 820 1840					
WATER COOLER, ELECTRIC, SELF CONTAINED, WALL HUNG, 8.2 GPH					
Water cooler, wall mounted, 8.2 GPH	1.000	Ea.	615	231	846
Copper tubing type DWV, solder joint, hanger 10'OC 1-1/4" diam	4.000	L.F.	19.60	34.20	53.80
Wrought copper DWV, Tee, sanitary 1-1/4" diam	1.000	Ea.	10.45	57	67.45
P trap, copper drainage, 1-1/4" diam	1.000	Ea.	35	28.50	63.50
Copper tubing type L, solder joint, hanger 10' OC 3/8" diam	5.000	L.F.	8.40	30.50	38.90
Wrought copper 90° elbow for solder joints 3/8" diam	1.000	Ea.	1.77	23.50	25.27
Wrought copper Tee for solder joints, 3/8" diam	1.000	Ea.	3	36.50	39.50
Stop and waste, straightway, bronze, solder, 3/8" diam	1.000	Ea.	4.65	21.50	26.15
TOTAL			697.87	462.70	1,160.57

D2010 820	Water Cooler Systems	MAT.	INST.	TOTAL
1840	Water cooler, electric, wall hung, 8.2 GPH	700	465	1,165
1880	Dual height, 14.3 GPH	1,025	475	1,500
1920	Wheelchair type, 7.5 G.P.H.	1,650	465	2,115
1960	Semi recessed, 8.1 G.P.H.	1,000	465	1,465
2000	Full recessed, 8 G.P.H.	1,525	495	2,020
2040	Floor mounted, 14.3 G.P.H.	770	405	1,175
2080	Dual height, 14.3 G.P.H.	1,075	490	1,565
2120	Refrigerated compartment type, 1.5 G.P.H.	1,300	405	1,705

RD2010 -400

D2010 Plumbing Fixtures

Two Fixture Bathroom Systems consisting of a lavatory, water closet, and rough-in service piping.

• Prices for plumbing and fixtures only.

*Common wall is with an adjacent bathroom

System Components	QUANTITY	UNIT	COST EACH		
			MAT.	INST.	TOTAL
SYSTEM D2010 920 1180					
BATHROOM, LAVATORY & WATER CLOSET, 2 WALL PLUMBING, STAND ALONE					
Water closet, 2 Pc close cpld vit china flr mntd w/seat, supply & stop	1.000	Ea.	186	175	361
Water closet, rough-in waste & vent	1.000	Set	210	305	515
Lavatory w/ftngs, wall hung, white, PE on CI, 20" x 18"	1.000	Ea.	265	116	381
Lavatory, rough-in waste & vent	1.000	Set	340	555	895
Copper tubing type L, solder joint, hanger 10' OC 1/2" diam	10.000	L.F.	19.50	63.50	83
Pipe, steel, galvanized, schedule 40, threaded, 2" diam	12.000	L.F.	124.80	173.40	298.20
Pipe, CI soil, no hub, coupling 10' OC, hanger 5' OC, 4" diam	7.000	L.F.	68.60	117.60	186.20
TOTAL			1,213.90	1,505.50	2,719.40

D2010 920	Two Fixture Bathroom, Two Wall Plumbing	COST EACH		
		MAT.	INST.	TOTAL
1180	Bathroom, lavatory & water closet, 2 wall plumbing, stand alone	1,225	1,500	2,725
1200	Share common plumbing wall*	1,100	1,300	2,400

D2010 922	Two Fixture Bathroom, One Wall Plumbing	COST EACH		
		MAT.	INST.	TOTAL
2220	Bathroom, lavatory & water closet, one wall plumbing, stand alone	1,125	1,350	2,475
2240	Share common plumbing wall*	1,000	1,150	2,150

Important: See the Reference Section for critical supporting data - Reference Numbers and City Cost Indexes

D2010 Plumbing Fixtures

Three Fixture Bathroom Systems consisting of a lavatory, water closet, bathtub or shower and rough-in service piping.

- Prices for plumbing and fixtures only.

SERVICES D

*Common wall is with an adjacent bathroom

System Components	QUANTITY	UNIT	COST EACH		
			MAT.	INST.	TOTAL
SYSTEM D2010 924 1170					
BATHROOM, LAVATORY, WATER CLOSET & BATHTUB					
ONE WALL PLUMBING, STAND ALONE					
Wtr closet, 2 pc close cpld vit china flr mntd w/seat supply & stop	1.000	Ea.	186	175	361
Water closet, rough-in waste & vent	1.000	Set	210	305	515
Lavatory w/ftngs, wall hung, white, PE on CI, 20" x 18"	1.000	Ea.	265	116	381
Lavatory, rough-in waste & vent	1.000	Set	340	555	895
Bathtub, white PE on CI, w/ftgs, mat bottom, recessed, 5' long	1.000	Ea.	805	210	1,015
Baths, rough-in waste and vent	1.000	Set	163.80	400.50	564.30
TOTAL			1,969.80	1,761.50	3,731.30

D2010 924	Three Fixture Bathroom, One Wall Plumbing	COST EACH		
		MAT.	INST.	TOTAL
1150	Bathroom, three fixture, one wall plumbing			
1160	Lavatory, water closet & bathtub			
1170	Stand alone	1,975	1,750	3,725
1180	Share common plumbing wall *	1,675	1,275	2,950

D2010 926	Three Fixture Bathroom, Two Wall Plumbing	COST EACH		
		MAT.	INST.	TOTAL
2130	Bathroom, three fixture, two wall plumbing			
2140	Lavatory, water closet & bathtub			
2160	Stand alone	1,975	1,775	3,750
2180	Long plumbing wall common *	1,750	1,425	3,175
3610	Lavatory, bathtub & water closet			
3620	Stand alone	2,150	2,025	4,175
3640	Long plumbing wall common *	2,000	1,825	3,825
4660	Water closet, corner bathtub & lavatory			
4680	Stand alone	2,950	1,800	4,750
4700	Long plumbing wall common *	2,775	1,350	4,125
6100	Water closet, stall shower & lavatory			
6120	Stand alone	1,800	2,025	3,825
6140	Long plumbing wall common *	1,700	1,875	3,575
7060	Lavatory, corner stall shower & water closet			
7080	Stand alone	2,150	1,800	3,950
7100	Short plumbing wall common *	1,800	1,200	3,000

D2010 Plumbing Fixtures

Four Fixture Bathroom Systems consisting of a lavatory, water closet, bathtub, shower and rough-in service piping.

• Prices for plumbing and fixtures only.

*Common wall is with an adjacent bathroom

System Components	QUANTITY	UNIT	COST EACH		
			MAT.	INST.	TOTAL
SYSTEM D2010 928 1160					
BATHROOM, BATHTUB, WATER CLOSET, STALL SHOWER & LAVATORY					
TWO WALL PLUMBING, STAND ALONE					
Wtr closet, 2 pc close cpld vit china flr mntd w/seat supply & stop	1.000	Ea.	186	175	361
Water closet, rough-in waste & vent	1.000	Set	210	305	515
Lavatory w/ftngs, wall hung, white PE on CI, 20" x 18"	1.000	Ea.	265	116	381
Lavatory, rough-in waste & vent	1.000	Set	34	55.50	89.50
Bathtub, white PE on CI, w/ftgs, mat bottom, recessed, 5' long	1.000	Ea.	805	210	1,015
Baths, rough-in waste and vent	1.000	Set	182	445	627
Shower stall, bkd enam, molded stone receptor, door & trim 32" sq.	1.000	Ea.	380	185	565
Shower stall, rough-in supply, waste & vent	1.000	Set	268	450	718
TOTAL			2,330	1,941.50	4,271.50

D2010 928	Four Fixture Bathroom, Two Wall Plumbing	COST EACH		
		MAT.	INST.	TOTAL
1140	Bathroom, four fixture, two wall plumbing			
1150	Bathtub, water closet, stall shower & lavatory			
1160	Stand alone	2,325	1,950	4,275
1180	Long plumbing wall common *	2,150	1,500	3,650
2260	Bathtub, lavatory, corner stall shower & water closet			
2280	Stand alone	2,825	1,950	4,775
2320	Long plumbing wall common *	2,650	1,500	4,150
3620	Bathtub, stall shower, lavatory & water closet			
3640	Stand alone	2,625	2,450	5,075
3660	Long plumbing wall (opp. door) common *	2,450	2,000	4,450

D2010 930	Four Fixture Bathroom, Three Wall Plumbing	COST EACH		
		MAT.	INST.	TOTAL
4680	Bathroom, four fixture, three wall plumbing			
4700	Bathtub, stall shower, lavatory & water closet			
4720	Stand alone	3,275	2,675	5,950
4760	Long plumbing wall (opposite door) common *	3,150	2,475	5,625

Important: See the Reference Section for critical supporting data - Reference Numbers and City Cost Indexes

D2010 Plumbing Fixtures

Five Fixture Bathroom Systems consisting of two lavatories, a water closet, bathtub, shower and rough-in service piping.

- Prices for plumbing and fixtures only.

*Common wall is with an adjacent bathroom.

System Components	QUANTITY	UNIT	COST EACH		
			MAT.	INST.	TOTAL
SYSTEM D2010 932 1360					
BATHROOM, BATHTUB, WATER CLOSET, STALL SHOWER & TWO LAVATORIES					
TWO WALL PLUMBING, STAND ALONE					
Wtr closet, 2 pc close cpld vit china flr mntd incl seat,supply & stop	1.000	Ea.	186	175	361
Water closet, rough-in waste & vent	1.000	Set	210	305	515
Lavatory w/ftngs, wall hung, white PE on CI, 20" x 18"	2.000	Ea.	530	232	762
Lavatory, rough-in waste & vent	2.000	Set	680	1,110	1,790
Bathtub, white PE on CI, w/ftgs, mat bottom, recessed, 5' long	1.000	Ea.	805	210	1,015
Baths, rough-in waste and vent	1.000	Set	182	445	627
Shower stall, bkd enam molded stone receptor, door & ftng, 32" sq.	1.000	Ea.	380	185	565
Shower stall, rough-in supply, waste & vent,	1.000	Set	268	450	718
TOTAL			3,241	3,112	6,353

D2010 932	Five Fixture Bathroom, Two Wall Plumbing	COST EACH		
		MAT.	INST.	TOTAL
1320	Bathroom, five fixture, two wall plumbing			
1340	Bathtub, water closet, stall shower & two lavatories			
1360	Stand alone	3,250	3,100	6,350
1400	One short plumbing wall common *	3,050	2,675	5,725
1500	Bathtub, two lavatories, corner stall shower & water closet			
1520	Stand alone	3,750	3,125	6,875
1540	Long plumbing wall common*	3,325	2,425	5,750

D2010 934	Five Fixture Bathroom, Three Wall Plumbing	COST EACH		
		MAT.	INST.	TOTAL
2360	Bathroom, five fixture, three wall plumbing			
2380	Water closet, bathtub, two lavatories & stall shower			
2400	Stand alone	3,750	3,125	6,875
2440	One short plumbing wall common *	3,550	2,675	6,225

D2010 936	Five Fixture Bathroom, One Wall Plumbing	COST EACH		
		MAT.	INST.	TOTAL
4080	Bathroom, five fixture, one wall plumbing			
4100	Bathtub, two lavatories, corner stall shower & water closet			
4120	Stand alone	3,525	2,775	6,300
4160	Share common wall *	2,925	1,800	4,725

SERVICES

D

D2010 Plumbing Fixtures

Example of Plumbing Cost Calculations: The bathroom system includes the individual fixtures such as bathtub, lavatory, shower and water closet. These fixtures are listed below as separate items merely as a checklist.

D

SERVICES

D2010 951	Plumbing Systems 20 Unit, 2 Story Apartment Building							
	FIXTURE	SYSTEM	LINE	QUANTITY	UNIT	COST EACH		
						MAT.	INST.	TOTAL
0440	Bathroom	D2010 926	3640	20	Ea.	40,200	36,700	76,900
0480	Bathtub							
0520	Booster pump[1]	not req'd.						
0560	Drinking fountain							
0600	Garbage disposal[1]	not incl.						
0660								
0680	Grease interceptor							
0720	Water heater	D2020 250	2140	1	Ea.	7,475	2,500	9,975
0760	Kitchen sink	D2010 410	1960	20	Ea.	16,500	12,900	29,400
0800	Laundry sink	D2010 420	1840	4	Ea.	1,775	2,550	4,325
0840	Lavatory							
0900								
0920	Roof drain, 1 floor	D2040 210	4200	2	Ea.	1,000	1,325	2,325
0960	Roof drain, add'l floor	D2040 210	4240	20	L.F.	196	335	531
1000	Service sink	D2010 420	4300	1	Ea.	1,100	895	1,995
1040	Sewage ejector[1]	not req'd.						
1080	Shower							
1100								
1160	Sump pump							
1200	Urinal							
1240	Water closet							
1320								
1360	SUB TOTAL					68,500	57,000	125,500
1480	Water controls	RD2010-031		10%[2]		6,825	5,725	12,550
1520	Pipe & fittings[3]	RD2010-031		30%[2]		20,500	17,200	37,700
1560	Other							
1600	Quality/complexity	RD2010-031		15%[2]		10,200	8,575	18,775
1680								
1720	TOTAL					106,000	89,000	195,000
1740								

[1]**Note:** Cost for items such as booster pumps, backflow preventers, sewage ejectors, water meters, etc., may be obtained from RSMeans "Plumbing Cost Data" book, or be added as an increase in the Water Controls percentage. Water Controls, Pipe and Fittings, and the Quality/Complexity factors come from Table RD2010-031.

[2]Percentage of subtotal.

[3]Long easily discernable runs of pipe would be more accurately priced from Table D2090 810. If this is done, reduce the miscellaneous percentage in proportion.

D20 Plumbing

D2020 Domestic Water Distribution

Systems below include piping and fittings within 10' of heater. Electric water heaters do not require venting.

System Components	QUANTITY	UNIT	COST EACH		
			MAT.	INST.	TOTAL
SYSTEM D2020 240 1820					
ELECTRIC WATER HEATER, COMMERCIAL, 100° F RISE					
50 GALLON TANK, 9 KW, 37 GPH					
Water heater, commercial, electric, 50 Gal, 9 KW, 37 GPH	1.000	Ea.	3,000	286	3,286
Copper tubing, type L, solder joint, hanger 10' OC, 3/4" diam	34.000	L.F.	97.24	229.50	326.74
Wrought copper 90° elbow for solder joints 3/4" diam	5.000	Ea.	6.25	135	141.25
Wrought copper Tee for solder joints, 3/4" diam	2.000	Ea.	4.64	86	90.64
Wrought copper union for soldered joints, 3/4" diam	2.000	Ea.	15.20	57	72.20
Valve, gate, bronze, 125 lb, NRS, soldered 3/4" diam	2.000	Ea.	49	51	100
Relief valve, bronze, press & temp, self-close, 3/4" IPS	1.000	Ea.	94	18.35	112.35
Wrought copper adapter, copper tubing to male, 3/4" IPS	1.000	Ea.	1.97	30	31.97
TOTAL			3,268.30	892.85	4,161.15

D2020 240	Electric Water Heaters - Commercial Systems		COST EACH		
			MAT.	INST.	TOTAL
1800	Electric water heater, commercial, 100° F rise				
1820	50 gallon tank, 9 KW 37 GPH	RD2020 -100	3,275	895	4,170
1860	80 gal, 12 KW 49 GPH		4,175	1,100	5,275
1900	36 KW 147 GPH		5,775	1,200	6,975
1940	120 gal, 36 KW 147 GPH		6,675	1,300	7,975
1980	150 gal, 120 KW 490 GPH		19,100	1,375	20,475
2020	200 gal, 120 KW 490 GPH		20,300	1,425	21,725
2060	250 gal, 150 KW 615 GPH		22,500	1,650	24,150
2100	300 gal, 180 KW 738 GPH		24,100	1,750	25,850
2140	350 gal, 30 KW 123 GPH		18,500	1,875	20,375
2180	180 KW 738 GPH		24,800	1,875	26,675
2220	500 gal, 30 KW 123 GPH		23,700	2,200	25,900
2260	240 KW 984 GPH		33,800	2,200	36,000
2300	700 gal, 30 KW 123 GPH		28,400	2,525	30,925
2340	300 KW 1230 GPH		42,200	2,525	44,725
2380	1000 gal, 60 KW 245 GPH		34,500	3,500	38,000
2420	480 KW 1970 GPH		42,900	3,525	46,425
2460	1500 gal, 60 KW 245 GPH		50,500	4,350	54,850
2500	480 KW 1970 GPH		69,000	4,350	73,350

D2020 Domestic Water Distribution

Units may be installed in multiples for increased capacity.

Included below is the heater with self-energizing gas controls, safety pilots, insulated jacket, hi-limit aquastat and pressure relief valve.

Installation includes piping and fittings within 10' of heater. Gas heaters require vent piping (not included in these prices).

System Components	QUANTITY	UNIT	COST EACH		
			MAT.	INST.	TOTAL
SYSTEM D2020 250 1780					
GAS FIRED WATER HEATER, COMMERCIAL, 100° F RISE					
75.5 MBH INPUT, 63 GPH					
Water heater, commercial, gas, 75.5 MBH, 63 GPH	1.000	Ea.	1,725	365	2,090
Copper tubing, type L, solder joint, hanger 10' OC, 1-1/4" diam	30.000	L.F.	168	265.50	433.50
Wrought copper 90° elbow for solder joints 1-1/4" diam	4.000	Ea.	18.76	138	156.76
Wrought copper Tee for solder joints, 1-1/4" diam	2.000	Ea.	20.60	114	134.60
Wrought copper union for soldered joints, 1-1/4" diam	2.000	Ea.	45	73	118
Valve, gate, bronze, 125 lb, NRS, soldered 1-1/4" diam	2.000	Ea.	110	69	179
Relief valve, bronze, press & temp, self-close, 3/4" IPS	1.000	Ea.	94	18.35	112.35
Copper tubing, type L, solder joints, 3/4" diam	8.000	L.F.	22.88	54	76.88
Wrought copper 90° elbow for solder joints 3/4" diam	1.000	Ea.	1.25	27	28.25
Wrought copper, adapter, CTS to MPT, 3/4" IPS	1.000	Ea.	1.97	30	31.97
Pipe steel black, schedule 40, threaded, 3/4" diam	10.000	L.F.	26.40	84.50	110.90
Pipe, 90° elbow, malleable iron black, 150 lb threaded, 3/4" diam	2.000	Ea.	4	73	77
Pipe, union with brass seat, malleable iron black, 3/4" diam	1.000	Ea.	8.40	39.50	47.90
Valve, gas stop w/o check, brass, 3/4" IPS	1.000	Ea.	15.65	23.50	39.15
TOTAL			2,261.91	1,374.35	3,636.26

D2020 250	Gas Fired Water Heaters - Commercial Systems	COST EACH		
		MAT.	INST.	TOTAL
1760	Gas fired water heater, commercial, 100° F rise			
1780	75.5 MBH input, 63 GPH	2,250	1,375	3,625
1860	100 MBH input, 91 GPH	3,825	1,450	5,275
1980	155 MBH input, 150 GPH	5,075	1,675	6,750
2060	200 MBH input, 192 GPH	5,475	2,025	7,500
2140	300 MBH input, 278 GPH	7,475	2,500	9,975
2180	390 MBH input, 374 GPH	8,350	2,525	10,875
2220	500 MBH input, 480 GPH	10,200	2,725	12,925
2260	600 MBH input, 576 GPH	11,600	2,950	14,550

RD2020 -100

Important: See the Reference Section for critical supporting data - Reference Numbers and City Cost Indexes

D2020 Domestic Water Distribution

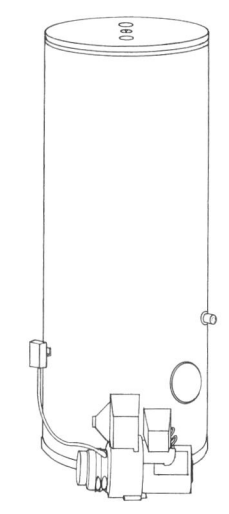

Units may be installed in multiples for increased capacity.

Included below is the heater, wired-in flame retention burners, cadmium cell primary controls, hi-limit controls, ASME pressure relief valves, draft controls, and insulated jacket.

Oil fired water heater systems include piping and fittings within 10′ of heater. Oil fired heaters require vent piping (not included in these systems).

SERVICES **D**

System Components	QUANTITY	UNIT	COST EACH		
			MAT.	INST.	TOTAL
SYSTEM D2020 260 1820					
OIL FIRED WATER HEATER, COMMERCIAL, 100° F RISE					
103 MBH OUTPUT, 116 GPH					
Water heater, commercial, oil, 140 gal., 140 MBH input, 134 GPH	1.000	Ea.	1,250	435	1,685
Copper tubing, type L, solder joint, hanger 10′ OC, 3/4″ diam	34.000	L.F.	97.24	229.50	326.74
Wrought copper 90° elbow for solder joints 3/4″ diam	5.000	Ea.	6.25	135	141.25
Wrought copper Tee for solder joints, 3/4″ diam	2.000	Ea.	4.64	86	90.64
Wrought copper union for soldered joints, 3/4″ diam	2.000	Ea.	15.20	57	72.20
Valve, bronze, 125 lb, NRS, soldered 3/4″ diam	2.000	Ea.	49	51	100
Relief valve, bronze, press & temp, self-close, 3/4″ IPS	1.000	Ea.	94	18.35	112.35
Wrought copper adapter, copper tubing to male, 3/4″ IPS	1.000	Ea.	1.97	30	31.97
Copper tubing, type L, solder joint, hanger 10′ OC, 3/8″ diam	10.000	L.F.	16.80	61	77.80
Wrought copper 90° elbow for solder joints 3/8″ diam	2.000	Ea.	3.54	47	50.54
Valve, globe, fusible, 3/8″ IPS	1.000	Ea.	8.30	21.50	29.80
TOTAL			1,546.94	1,171.35	2,718.29

D2020 260	Oil Fired Water Heaters - Commercial Systems	COST EACH		
		MAT.	INST.	TOTAL
1800	Oil fired water heater, commercial, 100° F rise			
1820	140 gal., 140 MBH input, 134 GPH	1,550	1,175	2,725
1900	140 gal., 255 MBH input, 247 GPH	13,800	1,475	15,275
1940	140 gal., 270 MBH input, 259 GPH	17,000	1,675	18,675
1980	140 gal., 400 MBH input, 384 GPH	17,600	1,925	19,525
2060	140 gal., 720 MBH input, 691 GPH	18,700	2,000	20,700
2100	221 gal., 300 MBH input, 288 GPH	24,700	2,225	26,925
2140	221 gal., 600 MBH input, 576 GPH	27,500	2,250	29,750
2180	221 gal., 800 MBH input, 768 GPH	27,900	2,325	30,225
2220	201 gal., 1000 MBH input, 960 GPH	28,500	2,350	30,850
2260	201 gal., 1250 MBH input, 1200 GPH	29,100	2,400	31,500
2300	201 gal., 1500 MBH input, 1441 GPH	31,500	2,475	33,975
2340	411 gal., 600 MBH input, 576 GPH	31,800	2,500	34,300
2380	411 gal., 800 MBH input, 768 GPH	31,900	2,575	34,475
2420	411 gal., 1000 MBH input, 960 GPH	34,200	2,975	37,175
2460	411 gal., 1250 MBH input, 1200 GPH	35,000	3,075	38,075
2500	397 gal., 1500 MBH input, 1441 GPH	37,200	3,175	40,375
2540	397 gal., 1750 MBH input, 1681 GPH	38,000	3,275	41,275

Note (by rows 1900–1940): RD2020 -100

D SERVICES

In this closed-loop indirect collection system, fluid with a low freezing temperature, such as propylene glycol, transports heat from the collectors to water storage. The transfer fluid is contained in a closed-loop consisting of collectors, supply and return piping, and a remote heat exchanger. The heat exchanger transfers heat energy from the

fluid in the collector loop to potable water circulated in a storage loop. A typical two-or-three panel system contains 5 to 6 gallons of heat transfer fluid.

When the collectors become approximately 20° F warmer than the storage temperature, a controller activates the circulator on the collector and storage

loops. The circulators will move the fluid and potable water through the heat exchanger until heat collection no longer occurs. At that point, the system shuts down. Since the heat transfer medium is a fluid with a very low freezing temperature, there is no need for it to be drained from the system between periods of collection.

D20 Plumbing

D2020 Domestic Water Distribution

System Components		QUANTITY	UNIT	COST EACH		
				MAT.	INST.	TOTAL
	SYSTEM D2020 265 2760					
	SOLAR, CLOSED LOOP, ADD-ON HOT WATER SYS., EXTERNAL HEAT EXCHANGER					
	3/4" TUBING, TWO 3'X7' BLACK CHROME COLLECTORS					
A,B,G,L,K,M	Heat exchanger fluid-fluid pkg incl 2 circulators, expansion tank,					
	Check valve, relief valve, controller, hi temp cutoff, & 2 sensors	1.000	Ea.	765	370	1,135
C	Thermometer, 2" dial	3.000	Ea.	66	96	162
D, T	Fill & drain valve, brass, 3/4" connection	1.000	Ea.	6.05	21.50	27.55
E	Air vent, manual, 1/8" fitting	2.000	Ea.	5.46	32.10	37.56
F	Air purger	1.000	Ea.	47.50	43	90.50
H	Strainer, Y type, bronze body, 3/4" IPS	1.000	Ea.	20.50	27.50	48
I	Valve, gate, bronze, NRS, soldered 3/4" diam	6.000	Ea.	147	153	300
J	Neoprene vent flashing	2.000	Ea.	19.60	51	70.60
N-1, N	Relief valve temp & press, 150 psi 210°F self-closing 3/4" IPS	1.000	Ea.	10	17.15	27.15
O	Pipe covering, urethane, ultraviolet cover, 1" wall 3/4" diam	20.000	L.F.	41.80	96.80	138.60
P	Pipe covering, fiberglass, all service jacket, 1" wall, 3/4" diam	50.000	L.F.	53.50	193.50	247
Q	Collector panel solar energy blk chrome on copper, 1/8" temp glass 3'x7'	2.000	Ea.	1,380	195	1,575
	Roof clamps for solar energy collector panels	2.000	Set	5.18	26.40	31.58
R	Valve, swing check, bronze, regrinding disc, 3/4" diam	2.000	Ea.	89	51	140
S	Pressure gauge, 60 psi, 2" dial	1.000	Ea.	24.50	16.05	40.55
U	Valve, water tempering, bronze, sweat connections, 3/4" diam	1.000	Ea.	69.50	25.50	95
W-2, V	Tank water storage w/heating element, drain, relief valve, existing	1.000	Ea.			
	Copper tubing type L, solder joint, hanger 10' OC 3/4" diam	20.000	L.F.	57.20	135	192.20
	Copper tubing, type M, solder joint, hanger 10' OC 3/4" diam	70.000	L.F.	165.20	462	627.20
	Sensor wire, #22-2 conductor multistranded	.500	C.L.F.	5.68	25	30.68
	Solar energy heat transfer fluid, propylene glycol anti-freeze	6.000	Gal.	58.20	110.10	168.30
	Wrought copper fittings & solder, 3/4" diam	76.000	Ea.	95	2,052	2,147
	TOTAL			3,131.87	4,199.60	7,331.47

D2020 265	Solar, Closed Loop, Add-On Hot Water Systems		COST EACH		
			MAT.	INST.	TOTAL
2550	Solar, closed loop, add-on hot water system, external heat exchanger				
2700	3/4" tubing, 3 ea 3'x7' black chrome collectors		3,825	4,300	8,125
2720	3 ea 3'x7' flat black absorber plate collectors	RD3010 -600	3,525	4,325	7,850
2740	2 ea 4'x9' flat black w/plastic glazing collectors		3,075	4,350	7,425
2760	2 ea 3'x7' black chrome collectors		3,125	4,200	7,325
2780	1" tubing, 4 ea 2'x9' plastic absorber & glazing collectors		4,275	4,875	9,150
2800	4 ea 3'x7' black chrome absorber collectors		4,850	4,900	9,750
2820	4 ea 3'x7' flat black absorber collectors		4,450	4,925	9,375

In the drainback indirect-collection system, the heat transfer fluid is distilled water contained in a loop consisting of collectors, supply and return piping, and an unpressurized holding tank. A large heat exchanger containing incoming potable water is immersed in the holding tank. When a controller activates solar collection, the distilled water is pumped through the collectors and heated and pumped back down to the holding tank. When the temperature differential between the water in the collectors and water in storage is such that collection no longer occurs, the pump turns off and gravity causes the distilled water in the collector loop to drain back to the holding tank. All the loop piping is pitched so that the water can drain out of the collectors and piping and not freeze there. As hot water is needed in the home, incoming water first flows through the holding tank with the immersed heat exchanger and is warmed and then flows through a conventional heater for any supplemental heating that is necessary.

Important: See the Reference Section for critical supporting data - Reference Numbers and City Cost Indexes

D2020 Domestic Water Distribution

System Components	QUANTITY	UNIT	COST EACH		
			MAT.	INST.	TOTAL
SYSTEM D2020 270 2760					
SOLAR, DRAINBACK, ADD ON, HOT WATER, IMMERSED HEAT EXCHANGER					
3/4″ TUBING, THREE EA 3′X7′ BLACK CHROME COLLECTOR					
A, B Differential controller 2 sensors, thermostat, solar energy system	1.000	Ea.	113	43	156
C Thermometer 2″ dial	3.000	Ea.	66	96	162
D, T Fill & drain valve, brass, 3/4″ connection	1.000	Ea.	6.05	21.50	27.55
E-1 Automatic air vent 1/8″ fitting	1.000	Ea.	12.40	16.05	28.45
H Strainer, Y type, bronze body, 3/4″ IPS	1.000	Ea.	20.50	27.50	48
I Valve, gate, bronze, NRS, soldered 3/4″ diam	2.000	Ea.	49	51	100
J Neoprene vent flashing	2.000	Ea.	19.60	51	70.60
L Circulator, solar heated liquid, 1/20 HP	1.000	Ea.	170	77	247
N Relief valve temp. & press. 150 psi 210°F self-closing 3/4″ IPS	1.000	Ea.	10	17.15	27.15
O Pipe covering, urethane, ultraviolet cover, 1″ wall, 3/4″ diam	20.000	L.F.	41.80	96.80	138.60
P Pipe covering, fiberglass, all service jacket, 1″ wall, 3/4″ diam	50.000	L.F.	53.50	193.50	247
Q Collector panel solar energy blk chrome on copper, 1/8″ temp glas 3′x7′	3.000	Ea.	2,070	292.50	2,362.50
Roof clamps for solar energy collector panels	3.000	Set	7.77	39.60	47.37
R Valve, swing check, bronze, regrinding disc, 3/4″ diam	1.000	Ea.	44.50	25.50	70
U Valve, water tempering, bronze sweat connections, 3/4″ diam	1.000	Ea.	69.50	25.50	95
V Tank, water storage w/heating element, drain, relief valve, existing	1.000	Ea.			
W Tank, water storage immersed heat exchr elec 2″x1/2# insul 120 gal	1.000	Ea.	1,025	365	1,390
X Valve, globe, bronze, rising stem, 3/4″ diam, soldered	3.000	Ea.	187.50	76.50	264
Y Flow control valve	1.000	Ea.	64.50	23.50	88
Z Valve, ball, bronze, solder 3/4″ diam, solar loop flow control	1.000	Ea.	14.70	25.50	40.20
Copper tubing, type L, solder joint, hanger 10′ OC 3/4″ diam	20.000	L.F.	57.20	135	192.20
Copper tubing, type M, solder joint, hanger 10′ OC 3/4″ diam	70.000	L.F.	165.20	462	627.20
Sensor wire, #22-2 conductor, multistranded	.500	C.L.F.	5.68	25	30.68
Wrought copper fittings & solder, 3/4″ diam	76.000	Ea.	95	2,052	2,147
TOTAL			**4,368.40**	**4,238.10**	**8,606.50**

D2020 270	Solar, Drainback, Hot Water Systems		COST EACH		
			MAT.	INST.	TOTAL
2550	Solar, drainback, hot water, immersed heat exchanger				
2560	3/8″ tubing, 3 ea. 4′ x 4′-4″ vacuum tube collectors		4,950	3,825	8,775
2760	3/4″ tubing, 3 ea 3′x7′ black chrome collectors, 120 gal tank	RD3010 -600	4,375	4,250	8,625
2780	3 ea. 3′x7′ flat black absorber collectors, 120 gal tank		4,075	4,250	8,325
2800	2 ea. 4′x9′ flat blk w/plastic glazing collectors 120 gal tank		3,625	4,275	7,900
2840	1″ tubing, 4 ea. 2′x9′ plastic absorber & glazing collectors, 120 gal tank		4,850	4,800	9,650
2860	4 ea. 3′x7′ black chrome absorber collectors, 120 gal tank		5,450	4,825	10,275
2880	4 ea. 3′x7′ flat black absorber collectors, 120 gal tank		5,050	4,850	9,900

SERVICES D

D

SERVICES

In the draindown direct-collection system, incoming domestic water is heated in the collectors. When the controller activates solar collection, domestic water is first heated as it flows through the collectors and is then pumped to storage. When conditions are no longer suitable for heat collection, the pump shuts off and the water in the loop drains down and out of the system by means of solenoid valves and properly pitched piping.

D20 Plumbing

D2020 Domestic Water Distribution

System Components	QUANTITY	UNIT	COST EACH		
			MAT.	INST.	TOTAL
SYSTEM D2020 275 2760					
SOLAR, DRAINDOWN, HOT WATER, DIRECT COLLECTION					
3/4″ TUBING, THREE 3′X7′ BLACK CHROME COLLECTORS					
A, B Differential controller, 2 sensors, thermostat, solar energy system	1.000	Ea.	113	43	156
A-1 Solenoid valve, solar heating loop, brass, 3/4″ diam, 24 volts	3.000	Ea.	699	171	870
B-1 Solar energy sensor, freeze prevention	1.000	Ea.	24	16.05	40.05
C Thermometer, 2″ dial	3.000	Ea.	66	96	162
E-1 Vacuum relief valve, 3/4″ diam	1.000	Ea.	29	16.05	45.05
F-1 Air vent, automatic, 1/8″ fitting	1.000	Ea.	12.40	16.05	28.45
H Strainer, Y type, bronze body, 3/4″ IPS	1.000	Ea.	20.50	27.50	48
I Valve, gate, bronze, NRS, soldered, 3/4″ diam	2.000	Ea.	49	51	100
J Vent flashing neoprene	2.000	Ea.	19.60	51	70.60
K Circulator, solar heated liquid, 1/25 HP	1.000	Ea.	162	66	228
N Relief valve temp & press 150 psi 210°F self-closing 3/4″ IPS	1.000	Ea.	10	17.15	27.15
O Pipe covering, urethane, ultraviolet cover, 1″ wall, 3/4″ diam	20.000	L.F.	41.80	96.80	138.60
P Pipe covering, fiberglass, all service jacket, 1″ wall, 3/4″ diam	50.000	L.F.	53.50	193.50	247
Roof clamps for solar energy collector panels	3.000	Set	7.77	39.60	47.37
Q Collector panel solar energy blk chrome on copper, 1/8″ temp glass 3′x7′	3.000	Ea.	2,070	292.50	2,362.50
R Valve, swing check, bronze, regrinding disc, 3/4″ diam, soldered	2.000	Ea.	89	51	140
T Drain valve, brass, 3/4″ connection	2.000	Ea.	12.10	43	55.10
U Valve, water tempering, bronze, sweat connections, 3/4″ diam	1.000	Ea.	69.50	25.50	95
W-2, W Tank, water storage elec elem 2″x1/2# insul 120 gal	1.000	Ea.	1,025	365	1,390
X Valve, globe, bronze, rising stem, 3/4″ diam, soldered	1.000	Ea.	62.50	25.50	88
Copper tubing, type L, solder joints, hangers 10′ OC 3/4″ diam	20.000	L.F.	57.20	135	192.20
Copper tubing, type M, solder joints, hangers 10′ OC 3/4″ diam	70.000	L.F.	165.20	462	627.20
Sensor wire, #22-2 conductor, multistranded	.500	C.L.F.	5.68	25	30.68
Wrought copper fittings & solder, 3/4″ diam	76.000	Ea.	95	2,052	2,147
TOTAL			4,958.75	4,377.20	9,335.95

D2020 275	Solar, Draindown, Hot Water Systems	COST EACH		
		MAT.	INST.	TOTAL
2550	Solar, draindown, hot water			
2580	1/2″ tubing, 4 ea. 4′ x 4′-4″ vacuum tube collectors, 80 gal tank	6,450	4,275	10,725
2760	3/4″ tubing, 3 ea 3′x7′ black chrome collectors, 120 gal tank	4,950	4,375	9,325
2780	3 ea 3′x7′ flat collectors, 120 gal tank	4,650	4,400	9,050
2800	2 ea. 4′x9′ flat black & plastic glazing collectors, 120 gal tank	4,225	4,400	8,625
2840	1″ tubing, 4 ea. 2′x9′ plastic absorber & glazing collectors, 120 gal tank	5,400	4,925	10,325
2860	4 ea 3′x7′ black chrome absorber collectors, 120 gal tank	5,975	4,950	10,925
2880	4 ea 3′x7′ flat black absorber collectors, 120 gal tank	5,575	4,975	10,550

(Note for 2760 row: RD3010 -600)

SERVICES

D

In this closed-loop indirect collection system, fluid with a low freezing temperature, such as propylene glycol, transports heat from the collectors to water storage. The transfer fluid is contained in a closed-loop consisting of collectors, supply and return piping, and a heat exchanger immersed in the storage tank. A typical two-or-three panel system contains 5 to 6 gallons of heat transfer fluid.

When the collectors become approximately 20° F warmer than the storage temperature, a controller activates the circulator. The circulator moves the fluid continuously through the collectors until the temperature difference between the collectors and storage is such that heat collection no longer occurs; at that point, the circulator shuts off. Since the heat transfer fluid has a very low freezing temperature, there is no need for it to be drained from the collectors between periods of collection.

Important: See the Reference Section for critical supporting data - Reference Numbers and City Cost Indexes

D2020 Domestic Water Distribution

System Components	QUANTITY	UNIT	COST EACH		
			MAT.	INST.	TOTAL
SYSTEM D2020 295 2760					
SOLAR, CLOSED LOOP, HOT WATER SYSTEM, IMMERSED HEAT EXCHANGER					
3/4″ TUBING, THREE 3′ X 7′ BLACK CHROME COLLECTORS					
A, B Differential controller, 2 sensors, thermostat, solar energy system	1.000	Ea.	113	43	156
C Thermometer 2″ dial	3.000	Ea.	66	96	162
D, T Fill & drain valves, brass, 3/4″ connection	3.000	Ea.	18.15	64.50	82.65
E Air vent, manual, 1/8″ fitting	1.000	Ea.	2.73	16.05	18.78
F Air purger	1.000	Ea.	47.50	43	90.50
G Expansion tank	1.000	Ea.	67.50	16.05	83.55
I Valve, gate, bronze, NRS, soldered 3/4″ diam	3.000	Ea.	73.50	76.50	150
J Neoprene vent flashing	2.000	Ea.	19.60	51	70.60
K Circulator, solar heated liquid, 1/25 HP	1.000	Ea.	162	66	228
N-1, N Relief valve, temp & press 150 psi 210°F self-closing 3/4″ IPS	2.000	Ea.	20	34.30	54.30
O Pipe covering, urethane ultraviolet cover, 1″ wall, 3/4″ diam	20.000	L.F.	41.80	96.80	138.60
P Pipe covering, fiberglass, all service jacket, 1″ wall, 3/4″ diam	50.000	L.F.	53.50	193.50	247
Roof clamps for solar energy collector panel	3.000	Set	7.77	39.60	47.37
Q Collector panel solar blk chrome on copper, 1/8″ temp glass, 3′x7′	3.000	Ea.	2,070	292.50	2,362.50
R-1 Valve, swing check, bronze, regrinding disc, 3/4″ diam, soldered	1.000	Ea.	44.50	25.50	70
S Pressure gauge, 60 psi, 2-1/2″ dial	1.000	Ea.	24.50	16.05	40.55
U Valve, water tempering, bronze, sweat connections, 3/4″ diam	1.000	Ea.	69.50	25.50	95
W-2, W Tank, water storage immersed heat exchr elec elem 2″x2# insul 120 Gal	1.000	Ea.	1,025	365	1,390
X Valve, globe, bronze, rising stem, 3/4″ diam, soldered	1.000	Ea.	62.50	25.50	88
Copper tubing type L, solder joint, hanger 10′ OC 3/4″ diam	20.000	L.F.	57.20	135	192.20
Copper tubing, type M, solder joint, hanger 10′ OC 3/4″ diam	70.000	L.F.	165.20	462	627.20
Sensor wire, #22-2 conductor multistranded	.500	C.L.F.	5.68	25	30.68
Solar energy heat transfer fluid, propylene glycol, anti-freeze	6.000	Gal.	58.20	110.10	168.30
Wrought copper fittings & solder, 3/4″ diam	76.000	Ea.	95	2,052	2,147
TOTAL			4,370.33	4,370.45	8,740.78

D2020 295	Solar, Closed Loop, Hot Water Systems		COST EACH		
			MAT.	INST.	TOTAL
2550	Solar, closed loop, hot water system, immersed heat exchanger				
2560	3/8″ tubing, 3 ea. 4′ x 4′-4″ vacuum tube collectors, 80 gal. tank		4,975	3,975	8,950
2580	1/2″ tubing, 4 ea. 4′ x 4′-4″ vacuum tube collectors, 80 gal. tank	RD3010 -600	5,875	4,275	10,150
2600	120 gal. tank		5,975	4,300	10,275
2640	2 ea. 3′x7′ black chrome collectors, 80 gal. tank		3,400	4,075	7,475
2660	120 gal. tank		3,475	4,075	7,550
2760	3/4″ tubing, 3 ea. 3′x7′ black chrome collectors, 120 gal. tank		4,375	4,375	8,750
2780	3 ea. 3′x7′ flat black collectors, 120 gal. tank		4,075	4,375	8,450
2840	1″ tubing, 4 ea. 2′x9′ plastic absorber & glazing collectors 120 gal. tank		4,775	4,900	9,675
2860	4 ea. 3′x7′ black chrome collectors, 120 gal. tank		5,350	4,925	10,275

SERVICES | D

D2040 Rain Water Drainage

Design Assumptions: Vertical conductor size is based on a maximum rate of rainfall of 4″ per hour. To convert roof area to other rates multiply "Max. S.F. Roof Area" shown by four and divide the result by desired local rate. The answer is the local roof area that may be handled by the indicated pipe diameter.

Basic cost is for roof drain, 10′ of vertical leader and 10′ of horizontal, plus connection to the main.

Pipe Dia.	Max. S.F. Roof Area	Gallons per Min.
2″	544	23
3″	1610	67
4″	3460	144
5″	6280	261
6″	10,200	424
8″	22,000	913

System Components	QUANTITY	UNIT	COST EACH MAT.	COST EACH INST.	COST EACH TOTAL
SYSTEM D2040 210 1880					
ROOF DRAIN, DWV PVC PIPE, 2″ DIAM., 10′ HIGH					
Drain, roof, main, PVC, dome type 2″ pipe size	1.000	Ea.	97.50	66	163.50
Clamp, roof drain, underdeck	1.000	Ea.	21	38.50	59.50
Pipe, Tee, PVC DWV, schedule 40, 2″ pipe size	1.000	Ea.	4.55	46.50	51.05
Pipe, PVC, DWV, schedule 40, 2″ diam.	20.000	L.F.	34.60	314	348.60
Pipe, elbow, PVC schedule 40, 2″ diam.	2.000	Ea.	4.34	51	55.34
TOTAL			161.99	516	677.99

D2040 210	Roof Drain Systems	COST EACH MAT.	COST EACH INST.	COST EACH TOTAL
1880	Roof drain, DWV PVC, 2″ diam., piping, 10′ high	162	515	677
1920	For each additional foot add	1.73	15.70	17.43
1960	3″ diam., 10′ high	210	600	810
2000	For each additional foot add	3.26	17.45	20.71
2040	4″ diam., 10′ high	250	675	925
2080	For each additional foot add	4.22	19.30	23.52
2120	5″ diam., 10′ high	650	780	1,430
2160	For each additional foot add	6.30	21.50	27.80
2200	6″ diam., 10′ high	705	860	1,565
2240	For each additional foot add	7.80	23.50	31.30
2280	8″ diam., 10′ high	1,525	1,475	3,000
2320	For each additional foot add	12.55	30	42.55
3940	C.I., soil, single hub, service wt., 2″ diam. piping 10′ high	310	565	875
3980	For each additional foot add	5.50	14.70	20.20
4120	3″ diam., 10′ high	435	610	1,045
4160	For each additional foot add	7.65	15.40	23.05
4200	4″ diam., 10′ high	505	665	1,170
4240	For each additional foot add	9.80	16.80	26.60
4280	5″ diam., 10′ high	785	735	1,520
4320	For each additional foot add	13.50	18.95	32.45
4360	6″ diam., 10′ high	885	785	1,670
4400	For each additional foot add	16.55	19.70	36.25
4440	8″ diam., 10′ high	1,850	1,600	3,450
4480	For each additional foot add	25.50	33	58.50
6040	Steel galv. sch 40 threaded, 2″ diam. piping 10′ high	480	545	1,025
6080	For each additional foot add	10.40	14.45	24.85
6120	3″ diam., 10′ high	985	785	1,770
6160	For each additional foot add	22	21.50	43.50

Important: See the Reference Section for critical supporting data - Reference Numbers and City Cost Indexes

D2040 Rain Water Drainage

D2040 210	Roof Drain Systems	COST EACH		
		MAT.	INST.	TOTAL
6200	4" diam., 10' high	1,425	1,000	2,425
6240	For each additional foot add	31.50	25.50	57
6280	5" diam., 10' high	2,400	1,275	3,675
6320	For each additional foot add	43.50	35.50	79
6360	6" diam, 10' high	3,025	1,575	4,600
6400	For each additional foot add	58	46.50	104.50
6440	8" diam., 10' high	6,325	2,150	8,475
6480	For each additional foot add	89.50	53.50	143

SERVICES

D

Pipe Material Considerations:

1. Malleable iron fittings should be used for gas service.

2. Malleable fittings are used where there are stresses/strains due to expansion and vibration.

3. Cast iron fittings may be broken as an aid to disassembling of piping joints frozen by long use, temperature and mineral deposits.

4. Cast iron pipe is extensively used for underground and submerged service.

5. Type M (light wall) copper tubing is available in hard temper only and is used for nonpressure and less severe applications than K and L.

6. Type L (medium wall) copper tubing, available hard or soft for interior service.

7. Type K (heavy wall) copper tubing, available in hard or soft temper for use where conditions are severe. For underground and interior service.

8. Hard drawn tubing requires fewer hangers or supports but should not be bent. Silver brazed fittings should be used.

Piping costs include hangers spaced for the material type and size, and the couplings required.

D2090 810	Piping - Installed - Unit Costs		COST PER L.F.		
			MAT.	INST.	TOTAL
0840	Cast iron, soil, B & S, service weight, 2" diameter	RD2010 -030	5.50	14.70	20.20
0860	3" diameter		7.65	15.40	23.05
0880	4" diameter		9.80	16.80	26.60
0900	5" diameter		13.50	18.95	32.45
0920	6" diameter		16.55	19.70	36.25
0940	8" diameter		25.50	33	58.50
0960	10" diameter		46.50	36.50	83
0980	12" diameter		65	41	106
1040	No hub, 1-1/2" diameter		5.75	13.05	18.80
1060	2" diameter		5.95	13.80	19.75
1080	3" diameter		7.95	14.45	22.40
1100	4" diameter		10.05	15.95	26
1120	5" diameter		14.30	17.35	31.65
1140	6" diameter		17.90	18.20	36.10
1160	8" diameter		28	28.50	56.50
1180	10" diameter		50	32	82
1220	Copper tubing, hard temper, solder, type K, 1/2" diameter		2.50	6.60	9.10
1260	3/4" diameter		3.95	6.95	10.90
1280	1" diameter		5.55	7.80	13.35
1300	1-1/4" diameter		6.85	9.20	16.05
1320	1-1/2" diameter		8.55	10.30	18.85
1340	2" diameter		13.10	12.85	25.95
1360	2-1/2" diameter		19.90	15.40	35.30
1380	3" diameter		28	17.15	45.15
1400	4" diameter		47	24.50	71.50
1480	5" diameter		122	29	151
1500	6" diameter		177	38	215
1520	8" diameter		315	42.50	357.50
1560	Type L, 1/2" diameter		1.95	6.35	8.30
1600	3/4" diameter		2.86	6.75	9.61
1620	1" diameter		4.02	7.55	11.57
1640	1-1/4" diameter		5.60	8.85	14.45
1660	1-1/2" diameter		7.10	9.90	17
1680	2" diameter		11.05	12.25	23.30
1700	2-1/2" diameter		17.10	14.90	32
1720	3" diameter		23	16.50	39.50
1740	4" diameter		39	23.50	62.50
1760	5" diameter		110	27	137
1780	6" diameter		136	36	172
1800	8" diameter		241	40	281
1840	Type M, 1/2" diameter		1.53	6.10	7.63
1880	3/4" diameter		2.36	6.60	8.96
1900	1" diameter		3.08	7.35	10.43
1920	1-1/4" diameter		4.57	8.55	13.12

Important: See the Reference Section for critical supporting data - Reference Numbers and City Cost Indexes

D2090 Other Plumbing Systems

D2090 810	Piping - Installed - Unit Costs	COST PER L.F.		
		MAT.	INST.	TOTAL
1940	1-1/2" diameter	6.15	9.50	15.65
1960	2" diameter	9.60	11.70	21.30
1980	2-1/2" diameter	14.65	14.45	29.10
2000	3" diameter	19.50	15.95	35.45
2020	4" diameter	35	23	58
2040	5" diameter	106	25.50	131.50
2060	6" diameter	136	34.50	170.50
2080	8" diameter	226	38	264
2120	Type DWV, 1-1/4" diameter	4.90	8.55	13.45
2160	1-1/2" diameter	6.10	9.50	15.60
2180	2" diameter	8.15	11.70	19.85
2200	3" diameter	14.50	15.95	30.45
2220	4" diameter	25.50	23	48.50
2240	5" diameter	75.50	25.50	101
2260	6" diameter	108	34.50	142.50
2280	8" diameter	259	38	297
2800	Plastic,PVC, DWV, schedule 40, 1-1/4" diameter	1.51	12.25	13.76
2820	1-1/2" diameter	1.54	14.30	15.84
2830	2" diameter	1.73	15.70	17.43
2840	3" diameter	3.26	17.45	20.71
2850	4" diameter	4.22	19.30	23.52
2890	6" diameter	7.80	23.50	31.30
3010	Pressure pipe 200 PSI, 1/2" diameter	1.21	9.50	10.71
3030	3/4" diameter	1.26	10.10	11.36
3040	1" diameter	1.39	11.15	12.54
3050	1-1/4" diameter	1.57	12.25	13.82
3060	1-1/2" diameter	1.69	14.30	15.99
3070	2" diameter	1.92	15.70	17.62
3080	2-1/2" diameter	3.14	16.50	19.64
3090	3" diameter	3.55	17.45	21
3100	4" diameter	4.68	19.30	23.98
3110	6" diameter	8.75	23.50	32.25
3120	8" diameter	13.80	30	43.80
4000	Steel, schedule 40, threaded, black, 1/2" diameter	2.25	8.15	10.40
4020	3/4" diameter	2.64	8.45	11.09
4030	1" diameter	3.82	9.70	13.52
4040	1-1/4" diameter	5	10.40	15.40
4050	1-1/2" diameter	5.90	11.55	17.45
4060	2" diameter	7.80	14.45	22.25
4070	2-1/2" diameter	12.10	18.50	30.60
4080	3" diameter	15.70	21.50	37.20
4090	4" diameter	23	25.50	48.50
4100	5" diameter	34.50	35.50	70
4110	6" diameter	51	46.50	97.50
4120	8" diameter	78	53.50	131.50
4130	10" diameter	109	62.50	171.50
4140	12" diameter	122	80	202
4200	Galvanized, 1/2" diameter	3.06	8.15	11.21
4220	3/4" diameter	3.66	8.45	12.11
4230	1" diameter	4.97	9.70	14.67
4240	1-1/4" diameter	6.60	10.40	17
4250	1-1/2" diameter	7.80	11.55	19.35
4260	2" diameter	10.40	14.45	24.85
4270	2-1/2" diameter	17.15	18.50	35.65
4280	3" diameter	22	21.50	43.50
4290	4" diameter	31.50	25.50	57
4300	5" diameter	43.50	35.50	79
4310	6" diameter	58	46.50	104.50

SERVICES D

D2090 Other Plumbing Systems

D2090 810	Piping - Installed - Unit Costs	COST PER L.F.		
		MAT.	INST.	TOTAL
4320	8" diameter	89.50	53.50	143
4330	10" diameter	100	62.50	162.50
4340	12" diameter	139	80	219
5010	Flanged, black, 1" diameter	9.40	14.30	23.70
5020	1-1/4" diameter	11.20	15.65	26.85
5030	1-1/2" diameter	11.90	17.25	29.15
5040	2" diameter	14	22	36
5050	2-1/2" diameter	18.15	27.50	45.65
5060	3" diameter	21	31.50	52.50
5070	4" diameter	28.50	38.50	67
5080	5" diameter	38	47.50	85.50
5090	6" diameter	49	60.50	109.50
5100	8" diameter	74.50	80	154.50
5110	10" diameter	130	95	225
5120	12" diameter	168	108	276
5720	Grooved joints, black, 3/4" diameter	3.28	7.25	10.53
5730	1" diameter	3.98	8.15	12.13
5740	1-1/4" diameter	5.25	8.85	14.10
5750	1-1/2" diameter	6.10	10.10	16.20
5760	2" diameter	7.85	12.85	20.70
5770	2-1/2" diameter	11.20	16.25	27.45
5900	3" diameter	14	18.50	32.50
5910	4" diameter	19	20.50	39.50
5920	5" diameter	24.50	25	49.50
5930	6" diameter	34.50	34.50	69
5940	8" diameter	52	39	91
5950	10" diameter	83.50	46.50	130
5960	12" diameter	103	53.50	156.50

D2090 820	Standard Pipe Fittings - Installed - Unit Costs	COST EACH		
		MAT.	INST.	TOTAL
0100	Cast iron, soil, B&S service weight, 1/8 Bend, 2" diameter	8.10	58	66.10
0110	3" diameter	12.70	66	78.70
0120	4" diameter	18.55	71	89.55
0130	5" diameter	25.50	80	105.50
0140	6" diameter	31.50	84.50	116
0150	8" diameter	94	178	272
0160	10" diameter	135	196	331
0170	12" diameter	256	218	474
0220	1/4 Bend, 2" diameter	11.35	58	69.35
0230	3" diameter	15.20	66	81.20
0240	4" diameter	23.50	71	94.50
0250	5" diameter	33	80	113
0260	6" diameter	41.50	84.50	126
0270	8" diameter	124	178	302
0280	10" diameter	182	196	378
0290	12" diameter	247	218	465
0360	Tee, 2" diameter	21	92.50	113.50
0370	3" diameter	34	103	137
0380	4" diameter	44	116	160
0390	5" diameter	67.50	120	187.50
0400	6" diameter	91.50	131	222.50
0410	8" diameter	201	280	481
0810	No hub, 1/8 Bend, 1-1/2" diameter	5.35		5.35
0820	2" diameter	5.95		5.95
0830	3" diameter	8.05		8.05
0840	4" diameter	10.20		10.20

Important: See the Reference Section for critical supporting data - Reference Numbers and City Cost Indexes

D2090 Other Plumbing Systems

D2090 820	Standard Pipe Fittings - Installed - Unit Costs	COST EACH		
		MAT.	INST.	TOTAL
0850	5" diameter	21.50		21.50
0860	6" diameter	23.50		23.50
0870	8" diameter	68		68
0880	10" diameter	127		127
0960	1/4 Bend, 1-1/2" diameter	6.40		6.40
0970	2" diameter	7		7
0980	3" diameter	9.65		9.65
0990	4" diameter	13.95		13.95
1000	5" diameter	32		32
1010	6" diameter	35		35
1020	8" diameter	96.50		96.50
1080	Sanitary tee, 1-1/2" diameter	8.85		8.85
1090	2" diameter	8.85		8.85
1110	3" diameter	11.70		11.70
1120	4" diameter	18.15		18.15
1130	5" diameter	52		52
1140	6" diameter	53		53
1150	8" diameter	214		214
1300	Coupling clamp & gasket, 1-1/2" diameter	8.50	19.30	27.80
1310	2" diameter	8.80	21	29.80
1320	3" diameter	9.20	24.50	33.70
1330	4" diameter	10.80	28	38.80
1340	5" diameter	20.50	32.50	53
1350	6" diameter	28	36	64
1360	8" diameter	48.50	59.50	108
1370	10" diameter	75	75.50	150.50
1510	Steel pipe fitting, Cl, threaded, 125 Lb. black, 45° Elbow, 1/2" diameter	4.76	34.50	39.26
1520	3/4" diameter	4.80	36.50	41.30
1530	1" diameter	5.60	39.50	45.10
1540	1-1/4" diameter	7.55	42	49.55
1550	1-1/2" diameter	12.55	46.50	59.05
1560	2" diameter	14.45	51.50	65.95
1570	2-1/2" diameter	37.50	66	103.50
1580	3" diameter	59.50	92.50	152
1590	4" diameter	124	154	278
1600	5" diameter	272	185	457
1610	6" diameter	350	206	556
1620	8" diameter	695	240	935
1700	90° Elbow, 1/2" diameter	3.12	34.50	37.62
1710	3/4" diameter	3.26	36.50	39.76
1720	1" diameter	3.85	39.50	43.35
1730	1-1/4" diameter	5.45	42	47.45
1740	1-1/2" diameter	7.55	46.50	54.05
1750	2" diameter	11.80	51.50	63.30
1760	2-1/2" diameter	28.50	66	94.50
1770	3" diameter	46.50	92.50	139
1780	4" diameter	86	154	240
1790	5" diameter	183	185	368
1800	6" diameter	237	206	443
1810	8" diameter	580	240	820
1900	Tee, 1/2" diameter	4.85	57	61.85
1910	3/4" diameter	5.65	57	62.65
1920	1" diameter	5.05	64.50	69.55
1930	1-1/4" diameter	9.20	66	75.20
1940	1-1/2" diameter	11.95	71	82.95
1950	2" diameter	16.65	84	100.65
1960	2-1/2" diameter	43.50	103	146.50
1970	3" diameter	66.50	154	220.50

D2090   Other Plumbing Systems

D2090 820	Standard Pipe Fittings - Installed - Unit Costs	COST EACH		
		MAT.	INST.	TOTAL
1980	4" diameter	129	231	360
1990	5" diameter	345	310	655
2000	6" diameter	390	360	750
2010	8" diameter	885	480	1,365
2300	Copper, wrought, solder joints, 45° Elbow, 1/2" diameter	1.03	25.50	26.53
2310	3/4" diameter	1.80	27	28.80
2320	1" diameter	4.54	32	36.54
2330	1-1/4" diameter	6.45	34.50	40.95
2340	1-1/2" diameter	7.80	39.50	47.30
2350	2" diameter	13	46.50	59.50
2360	2-1/2" diameter	27.50	71	98.50
2370	3" diameter	41	71	112
2380	4" diameter	87.50	103	190.50
2390	5" diameter	340	154	494
2400	6" diameter	535	160	695
2500	90° Elbow, 1/2" diameter	.56	25.50	26.06
2510	3/4" diameter	1.25	27	28.25
2520	1" diameter	3.10	32	35.10
2530	1-1/4" diameter	4.69	34.50	39.19
2540	1-1/2" diameter	7.30	39.50	46.80
2550	2" diameter	13.30	46.50	59.80
2560	2-1/2" diameter	28	71	99
2610	3" diameter	37.50	84	121.50
2620	4" diameter	96	103	199
2630	5" diameter	400	154	554
2640	6" diameter	535	160	695
2700	Tee, 1/2" diameter	.96	39.50	40.46
2710	3/4" diameter	2.32	43	45.32
2720	1" diameter	7.15	51.50	58.65
2730	1-1/4" diameter	10.30	57	67.30
2740	1-1/2" diameter	15.90	64.50	80.40
2750	2" diameter	25	73.50	98.50
2760	2-1/2" diameter	50	116	166
2770	3" diameter	76.50	132	208.50
2780	4" diameter	185	185	370
2790	5" diameter	605	231	836
2800	6" diameter	825	240	1,065
2880	Coupling, 1/2" diameter	.44	23.50	23.94
2890	3/4" diameter	.84	24.50	25.34
2900	1" diameter	1.72	28.50	30.22
2910	1-1/4" diameter	3.19	30	33.19
2920	1-1/2" diameter	4.22	34.50	38.72
2930	2" diameter	7.05	39.50	46.55
2940	2-1/2" diameter	14.95	61.50	76.45
2950	3" diameter	22.50	71	93.50
2960	4" diameter	47.50	132	179.50
2970	5" diameter	116	154	270
2980	6" diameter	192	180	372
3200	Malleable iron, 150 Lb. threaded, black, 45° elbow, 1/2" diameter	2.73	34.50	37.23
3210	3/4" diameter	3.38	36.50	39.88
3220	1" diameter	4.26	39.50	43.76
3230	1-1/4" diameter	7.55	42	49.55
3240	1-1/2" diameter	9.35	46.50	55.85
3250	2" diameter	14.10	51.50	65.60
3260	2-1/2" diameter	41	66	107
3270	3" diameter	53	92.50	145.50
3280	4" diameter	104	154	258
3290	5" diameter	410	185	595

Important: See the Reference Section for critical supporting data - Reference Numbers and City Cost Indexes

D20 Plumbing

D2090 Other Plumbing Systems

D2090 820	Standard Pipe Fittings - Installed - Unit Costs	COST EACH		
		MAT.	INST.	TOTAL
3300	6" diameter	410	206	616
3400	90° Elbow, 1/2" diameter	1.67	34.50	36.17
3500	3/4" diameter	2	36.50	38.50
3510	1" diameter	3.48	39.50	42.98
3520	1-1/4" diameter	5.70	42	47.70
3530	1-1/2" diameter	7.50	46.50	54
3540	2" diameter	13	51.50	64.50
3550	2-1/2" diameter	29	66	95
3560	3" diameter	42.50	92.50	135
3570	4" diameter	91	154	245
3580	5" diameter	258	185	443
3590	6" diameter	305	206	511
3700	Tee, 1/2" diameter	2.21	57	59.21
3800	3/4" diameter	3.19	57	60.19
3810	1" diameter	5.45	64.50	69.95
3820	1-1/4" diameter	8.80	66	74.80
3830	1-1/2" diameter	10.95	71	81.95
3910	2" diameter	18.70	84	102.70
3920	2-1/2" diameter	40.50	103	143.50
3930	3" diameter	59.50	154	213.50
3940	4" diameter	143	231	374
3950	5" diameter	410	310	720
3960	6" diameter	455	360	815
4000	Coupling, 1/2" diameter	2.28	27	29.28
4010	3/4" diameter	2.68	28.50	31.18
4020	1" diameter	4.06	34.50	38.56
4030	1-1/4" diameter	5.20	35.50	40.70
4040	1-1/2" diameter	7.05	38.50	45.55
4050	2" diameter	10.40	44	54.40
4060	2-1/2" diameter	29	51.50	80.50
4070	3" diameter	39	66	105
4080	4" diameter	78.50	92.50	171
4090	5" diameter	225	154	379
4100	6" diameter	225	180	405
5100	Plastic, PVC, high impact/pressure sch 40, 45° Elbow, 1/2" diameter	.59	15.45	16.04
5110	3/4" diameter	.92	17.95	18.87
5120	1" diameter	1.11	20.50	21.61
5130	1-1/4" diameter	1.54	23	24.54
5140	1-1/2" diameter	1.94	25.50	27.44
5150	2" diameter	2.53	25.50	28.03
5160	3" diameter	9.55	40.50	50.05
5170	4" diameter	18.35	51	69.35
5180	6" diameter	46.50	83.50	130
5190	8" diameter	145	140	285
5260	90° Elbow, 1/2" diameter	.36	15.45	15.81
5270	3/4" diameter	.41	17.95	18.36
5280	1" diameter	.73	20.50	21.23
5290	1-1/4" diameter	1.29	23	24.29
5300	1-1/2" diameter	1.38	25.50	26.88
5400	2" diameter	2.17	25.50	27.67
5410	3" diameter	7.90	40.50	48.40
5420	4" diameter	14.10	51	65.10
5430	6" diameter	45	83.50	128.50
5440	8" diameter	98	140	238
5500	Tee, 1/2" diameter	.45	23	23.45
5510	3/4" diameter	.52	27	27.52
5520	1" diameter	.97	31	31.97
5530	1-1/4" diameter	1.51	34.50	36.01

D2090 Other Plumbing Systems

D2090 820	Standard Pipe Fittings - Installed - Unit Costs	COST EACH		
		MAT.	INST.	TOTAL
5540	1-1/2" diameter	1.84	38.50	40.34
5550	2" diameter	2.66	38	40.66
5560	3" diameter	11.55	61	72.55
5570	4" diameter	21	76.50	97.50
5580	6" diameter	70.50	125	195.50
5590	8" diameter	163	212	375
5680	Coupling, 1/2" diameter	.24	15.45	15.69
5690	3/4" diameter	.33	17.95	18.28
5700	1" diameter	.57	20.50	21.07
5710	1-1/4" diameter	.78	23	23.78
5720	1-1/2" diameter	.78	23	23.78
5730	2" diameter	1.30	25.50	26.80
5740	3" diameter	4.47	40.50	44.97
5750	4" diameter	6.45	51	57.45
6010	6" diameter	20.50	83.50	104
6020	8" diameter	51	140	191
7810	DWW, socket joints, sch. 40, 1/8 Bend, 1-1/4" diameter	2.96	25.50	28.46
7820	1-1/2" diameter	1.73	28	29.73
7830	2" diameter	2.67	28	30.67
7840	3" diameter	7.30	44.50	51.80
7850	4" diameter	12.25	56	68.25
7860	6" diameter	50.50	91.50	142
7960	1/4 bend, 1-1/4" diameter	4.95	25.50	30.45
8000	1-1/2" diameter	1.88	28	29.88
8010	2" diameter	2.77	28	30.77
8020	3" diameter	7.95	44.50	52.45
8030	4" diameter	13.75	56	69.75
8210	6" diameter	52.50	91.50	144
8300	Sanitary tee, 1-1/4" diameter	4.53	38	42.53
8310	1-1/2" diameter	3.25	42.50	45.75
8320	2" diameter	4.55	46.50	51.05
8330	3" diameter	10.75	66.50	77.25
8340	4" diameter	18.65	84	102.65
8400	Coupling, 1-1/4" diameter	18.50	25.50	44
8410	1-1/2" diameter	19.90	28	47.90
8420	2" diameter	23.50	28	51.50
8430	3" diameter	49	44.50	93.50
8440	4" diameter	100	56	156

Important: See the Reference Section for critical supporting data - Reference Numbers and City Cost Indexes

D30 HVAC

D3010　Energy Supply

Basis for Heat Loss Estimate, Apartment Type Structures:

1. Masonry walls and flat roof are insulated. U factor is assumed at .08.
2. Window glass area taken as BOCA minimum, 1/10th of floor area. Double insulating glass with 1/4″ air space, U = .65.
3. Infiltration = 0.3 C.F. per hour per S.F. of net wall.
4. Concrete floor loss is 2 BTUH per S.F.
5. Temperature difference taken as 70° F.
6. Ventilating or makeup air has not been included and must be added if desired. Air shafts are not used.

SERVICES D

System Components	QUANTITY	UNIT	MAT.	INST.	TOTAL
SYSTEM D3010 510 1760					
HEATING SYSTEM, FIN TUBE RADIATION, FORCED HOT WATER					
1,000 S.F. AREA, 10,000 C.F. VOLUME					
Boiler, oil fired, CI, burner, ctrls/insul/breech/pipe/ftng/valves, 109 MBH	1.000	Ea.	4,331.25	2,887.50	7,218.75
Circulating pump, CI flange connection, 1/12 HP	1.000	Ea.	263	154	417
Expansion tank, painted steel, ASME 18 Gal capacity	1.000	Ea.	415	66.50	481.50
Storage tank, steel, above ground, 275 Gal capacity w/supports	1.000	Ea.	355	187	542
Copper tubing type L, solder joint, hanger 10' OC, 3/4″ diam	100.000	L.F.	286	675	961
Radiation, 3/4″ copper tube w/alum fin baseboard pkg, 7″ high	30.000	L.F.	220.50	483	703.50
Pipe covering, calcium silicate w/cover, 1″ wall, 3/4″ diam	100.000	L.F.	211	525	736
TOTAL			6,081.75	4,978	11,059.75
COST PER S.F.			6.08	4.98	11.06

D3010 510	Apartment Building Heating - Fin Tube Radiation		MAT.	INST.	TOTAL
1740	Heating systems, fin tube radiation, forced hot water				
1760	1,000 S.F. area, 10,000 C.F. volume	RD3020 -010	6.08	4.98	11.06
1800	10,000 S.F. area, 100,000 C.F. volume	RD3020 -020	1.85	3	4.85
1840	20,000 S.F. area, 200,000 C.F. volume		1.89	3.38	5.27
1880	30,000 S.F. area, 300,000 C.F. volume		1.76	3.27	5.03
1890					

307

D30 HVAC

D3010 Energy Supply

Fin Tube Radiator

Basis for Heat Loss Estimate, Factory or Commercial Type Structures:

1. Walls and flat roof are of lightly insulated concrete block or metal. U factor is assumed at .17.
2. Windows and doors are figured at 1-1/2 S.F. per linear foot of building perimeter. The U factor for flat single glass is 1.13.
3. Infiltration is approximately 5 C.F. per hour per S.F. of wall.
4. Concrete floor loss is 2 BTUH per S.F. of floor.
5. Temperature difference is assumed at 70° F.
6. Ventilation or makeup air has not been included and must be added if desired.

System Components	QUANTITY	UNIT	MAT.	INST.	TOTAL
SYSTEM D3010 520 1960					
HEATING SYSTEM, FIN TUBE RADIATION, FORCED HOT WATER					
1,000 S.F. BLDG., ONE FLOOR					
Boiler, oil fired, CI, burner/ctrls/insul/breech/pipe/ftngs/valves, 109 MBH	1.000	Ea.	4,331.25	2,887.50	7,218.75
Expansion tank, painted steel, ASME 18 Gal capacity	1.000	Ea.	1,875	78	1,953
Storage tank, steel, above ground, 550 Gal capacity w/supports	1.000	Ea.	1,750	345	2,095
Circulating pump, CI flanged, 1/8 HP	1.000	Ea.	440	154	594
Pipe, steel, black, sch. 40, threaded, cplg & hngr 10'OC, 1-1/2" diam.	260.000	L.F.	1,534	3,003	4,537
Pipe covering, calcium silicate w/cover , 1" wall, 1-1/2" diam	260.000	L.F.	559	1,404	1,963
Radiation, steel 1-1/4" tube & 4-1/4" fin w/cover & damper, wall hung	42.000	L.F.	1,155	1,092	2,247
Rough in, steel fin tube radiation w/supply & balance valves	4.000	Set	720	2,280	3,000
TOTAL			12,364.25	11,243.50	23,607.75
COST PER S.F.			12.36	11.24	23.60

D3010 520	Commercial Building Heating - Fin Tube Radiation		MAT.	INST.	TOTAL
1940	Heating systems, fin tube radiation, forced hot water				
1960	1,000 S.F. bldg, one floor		12.40	11.20	23.60
2000	10,000 S.F., 100,000 C.F., total two floors	RD3020 -010	3.05	4.29	7.34
2040	100,000 S.F., 1,000,000 C.F., total three floors		1.45	1.94	3.39
2080	1,000,000 S.F., 10,000,000 C.F., total five floors	RD3020 -020	.76	1.03	1.79
2090					

Important: See the Reference Section for critical supporting data - Reference Numbers and City Cost Indexes

D3010 Energy Supply

Unit Heater

Basis for Heat Loss Estimate, Factory or Commercial Type Structures:

1. Walls and flat roof are of lightly insulated concrete block or metal. U factor is assumed at .17.
2. Windows and doors are figured at 1-1/2 S.F. per linear foot of building perimeter. The U factor for flat single glass is 1.13.
3. Infiltration is approximately 5 C.F. per hour per S.F. of wall.
4. Concrete floor loss is 2 BTUH per S.F. of floor.
5. Temperature difference is assumed at 70° F.
6. Ventilation or makeup air has not been included and must be added if desired.

SERVICES **D**

System Components	QUANTITY	UNIT	COST EACH		
			MAT.	INST.	TOTAL
SYSTEM D3010 530 1880					
HEATING SYSTEM, TERMINAL UNIT HEATERS, FORCED HOT WATER					
1,000 S.F. BLDG., ONE FLOOR					
Boiler oil fired, CI, burner/ctrls/insul/breech/pipe/ftngs/valves, 109 MBH	1.000	Ea.	4,331.25	2,887.50	7,218.75
Expansion tank, painted steel, ASME 18 Gal capacity	1.000	Ea.	1,875	78	1,953
Storage tank, steel, above ground, 550 Gal capacity w/supports	1.000	Ea.	1,750	345	2,095
Circulating pump, CI, flanged, 1/8 HP	1.000	Ea.	440	154	594
Pipe, steel, black, schedule 40, threaded, cplg & hngr 10'OC 1-1/2" diam	260.000	L.F.	1,534	3,003	4,537
Pipe covering, calcium silicate w/cover, 1" wall, 1-1/2" diam	260.000	L.F.	559	1,404	1,963
Unit heater, 1 speed propeller, horizontal, 200° EWT, 26.9 MBH	2.000	Ea.	940	234	1,174
Unit heater piping hookup with controls	2.000	Set	840	2,200	3,040
TOTAL			12,269.25	10,305.50	22,574.75
COST PER S.F.			12.27	10.31	22.58

D3010 530	Commercial Bldg. Heating - Terminal Unit Heaters		COST PER S.F.		
			MAT.	INST.	TOTAL
1860	Heating systems, terminal unit heaters, forced hot water				
1880	1,000 S.F. bldg., one floor		12.30	10.30	22.60
1920	10,000 S.F. bldg., 100,000 C.F. total two floors	RD3020 -010	2.81	3.55	6.36
1960	100,000 S.F. bldg., 1,000,000 C.F. total three floors		1.51	1.78	3.29
2000	1,000,000 S.F. bldg., 10,000,000 C.F. total five floors	RD3020 -020	.97	1.15	2.12
2010					

D | **SERVICES**

CLOCK TIMER

PUMP

FILTER

AUXILIARY HEATER

POOL

This draindown pool system uses a differential thermostat similar to those used in solar domestic hot water and space heating applications. To heat the pool, the pool water passes through the conventional pump-filter loop and then flows through the collectors. When collection is not possible, or when the pool temperature is reached, all water drains from the solar loop back to the pool through the existing piping. The modes are controlled by solenoid valves or other automatic valves in conjunction with a vacuum breaker relief valve, which facilitates draindown.

Important: See the Reference Section for critical supporting data - Reference Numbers and City Cost Indexes

D3010 Energy Supply

System Components	QUANTITY	UNIT	COST EACH		
			MAT.	INST.	TOTAL
SYSTEM D3010 660 2640					
SOLAR SWIMMING POOL HEATER, ROOF MOUNTED COLLECTORS					
TEN 4' X 10' FULLY WETTED UNGLAZED PLASTIC ABSORBERS					
A Differential thermostat/controller, 110V, adj pool pump system	1.000	Ea.	305	257	562
A-1 Solenoid valve, PVC, normally 1 open 1 closed (included)	2.000	Ea.			
B Sensor, thermistor type (included)	2.000	Ea.			
E-1 Valve, vacuum relief	1.000	Ea.	29	16.05	45.05
Q Collector panel, solar energy, plastic, liquid full wetted, 4' x 10'	10.000	Ea.	2,150	1,850	4,000
R Valve, ball check, PVC, socket, 1-1/2" diam	1.000	Ea.	79	25.50	104.50
Z Valve, ball, PVC, socket, 1-1/2" diam	3.000	Ea.	106.50	76.50	183
Pipe, PVC, sch 40, 1-1/2" diam	80.000	L.F.	148.80	1,144	1,292.80
Pipe fittings, PVC sch 40, socket joint, 1-1/2" diam	10.000	Ea.	13.80	255	268.80
Sensor wire, #22-2 conductor, multistranded	.500	C.L.F.	5.68	25	30.68
Roof clamps for solar energy collector panels	10.000	Set	25.90	132	157.90
Roof strap, teflon for solar energy collector panels	26.000	L.F.	271.70	65.26	336.96
TOTAL			3,135.38	3,846.31	6,981.69

D3010 660	Solar Swimming Pool Heater Systems	COST EACH		
		MAT.	INST.	TOTAL
2530	Solar swimming pool heater systems, roof mounted collectors			
2540	10 ea. 3'x7' black chrome absorber, 1/8" temp. glass	7,875	2,975	10,850
2560	10 ea. 4'x8' black chrome absorber, 3/16" temp. glass	9,175	3,525	12,700
2580	10 ea. 3'8"x6' flat black absorber, 3/16" temp. glass	6,875	3,025	9,900
2600	10 ea. 4'x9' flat black absorber, plastic glazing	7,625	3,675	11,300
2620	10 ea. 2'x9' rubber absorber, plastic glazing	11,900	3,975	15,875
2640	10 ea. 4'x10' fully wetted unglazed plastic absorber	3,125	3,850	6,975
2660	Ground mounted collectors			
2680	10 ea. 3'x7' black chrome absorber, 1/8" temp. glass	8,100	3,400	11,500
2700	10 ea. 4'x8' black chrome absorber, 3/16" temp glass	9,400	3,975	13,375
2720	10 ea. 3'8"x6' flat blk absorber, 3/16" temp. glass	7,100	3,475	10,575
2740	10 ea. 4'x9' flat blk absorber, plastic glazing	7,850	4,125	11,975
2760	10 ea. 2'x9' rubber absorber, plastic glazing	12,100	4,275	16,375
2780	10 ea. 4'x10' fully wetted unglazed plastic absorber	3,350	4,275	7,625

(Note: between rows 2560 and 2580 a box labeled **RD3010-600** appears.)

SERVICES D

D3010 Energy Supply

Solar Direct Gain Glazing — Door Panels: In a Direct Gain System the collection, absorption, and storage of solar energy occur directly within the building's living space. In this system, a section of the building's south wall is removed and replaced with glazing. The glazing in the System Component example consists of three standard insulating glass door panels preassembled in wood frames. A direct gain system also includes a thermal storage mass and night insulation.

During the heating season, solar radiation enters the living space directly through the glazing and is converted to heat, which either warms the air or is stored in the mass. The storage mass is typically masonry or water. Masonry often serves as a basic structural component of the building (e.g., a wall or floor). In this system the cost of adding mass to the building is not included. To reduce heat loss through the south facing glass, thereby increasing overall thermal performance, night insulation is included in this system.

System Components	QUANTITY	UNIT	COST EACH MAT.	COST EACH INST.	COST EACH TOTAL
SYSTEM D3010 682 2600					
SOLAR PASSIVE HEATING, DIRECT GAIN, 3'X6'8" DOUBLE GLAZED DOOR PANEL					
3 PANELS WIDE					
A Framing headers over openings, fir 2" x 12"	.025	M.B.F.	26.40	38.25	64.65
A-1 Treated wood sleeper, 1" x 2"	.002	M.B.F.	1.32	4.55	5.87
B Aluminum flashing, mill finish, .013 thick	5.000	S.F.	3.36	22.96	26.32
Studs, 8' high wall, 2" x 4"	.020	M.B.F.	12.30	19.30	31.60
L-2 Interior casing, stock pine, 11/16" x 2-1/2"	25.000	L.F.	35.25	46	81.25
L-4 Valance 1" x 8"	12.000	L.F.	18.96	26.52	45.48
M Wood frame sliding door, 5/8" tempered insulating glass, 9' wide, econ	1.000	Ea.	1,792	282.88	2,074.88
N Caulking/sealants, silicone rubber, cartridges	.250	Gal.	10.50		10.50
O Drywall, gypsum plasterboard, nailed to studs, 5/8" standard	39.000	S.F.	12.87	17.16	30.03
O For taping and finishing, add	39.000	S.F.	1.56	17.16	18.72
P Mylar shade, dbl. layer heat reflective	49.640	S.F.	431.87	32.27	464.14
Q Stool cap, pine, 11/16" x 3-1/2"	12.000	L.F.	29.04	26.52	55.56
Paint wood trim	40.000	S.F.	4	19.20	23.20
Demolition, framing	84.000	S.F.		14.28	14.28
Demolition, shingles	84.000	S.F.		82.32	82.32
Demolition, drywall	84.000	S.F.		28.56	28.56
TOTAL			2,379.43	677.93	3,057.36

D3010 682	Solar Direct Gain Glazing, Door Panels	COST EACH MAT.	COST EACH INST.	COST EACH TOTAL
2550	Solar passive heating, direct gain, 3'x6'8", double glazed door			
2560	One panel wide	1,175	365	1,540
2580	Two panels wide	1,800	525	2,325
2600	Three panels wide	2,375	680	3,055

D3010 Energy Supply

Solar Direct Gain Glazing — Window Units: In a Direct Gain System, collection, absorption, and storage of solar energy occur directly within the building's living space. In this system, a section of the building's south wall is removed and replaced with glazing. The glazing consists of standard insulating replacement glass units in wood site-built frames. A direct gain system also includes a thermal storage mass and night insulation.

During the heating season, solar radiation enters the living space directly through the glazing and is converted to heat, which either warms the air or is stored in the mass. The thermal storage mass is typically masonry or water. Masonry often serves as a basic structural component of the building (e.g., a wall or floor). In this system the cost of adding mass to the building is not included. To reduce heat loss through the south facing glass, thereby increasing overall thermal performance, night insulation is included in this system.

SERVICES **D**

System Components	QUANTITY	UNIT	COST EACH MAT.	COST EACH INST.	COST EACH TOTAL
SYSTEM D3010 684 2600					
SOLAR PASSIVE HEATING, DIRECT GAIN, 2'-6"X5' DBL. GLAZED WINDOW PANEL					
3 PANELS WIDE					
A Framing headers over openings, fir 2" x 8"	.022	M.B.F.	15.51	43.45	58.96
A-1 Wood exterior sill, oak, no horns, 8/4 x 8" deep	8.000	L.F.	154.80	70.80	225.60
B Aluminum flashing, mill finish, .013 thick	4.160	S.F.	1.75	11.94	13.69
C Glazing, insul glass 2 lite 3/16" float 5/8" thick unit, 15-30 S.F.	38.000	S.F.	399	349.60	748.60
F 2"x4" sill blocking	.013	M.B.F.	8	33.80	41.80
Mullion	.027	M.B.F.	16.61	26.06	42.67
J-1 Exterior window stop 1" x 2"	45.000	L.F.	18	60.30	78.30
Exterior sill stock water repellent 1" x 3"	8.000	L.F.	19.36	17.68	37.04
K Interior window stop 1" x 3"	45.000	L.F.	34.20	82.80	117
L-1 Exterior casing	28.000	L.F.	39.60	53.04	92.64
L-2 Interior casing 1" x 4"	28.000	L.F.	56.28	57.68	113.96
L-3 Apron	8.000	M.B.F.	11.12	16.08	27.20
L-4 Valance 1" x 8"	10.000	L.F.	15.80	22.10	37.90
M-1 Window frame 2" x 6"	52.000	L.F.	100.36	139.36	239.72
O Drywall, gypsum plasterboard, nailed/screw to studs, 5/8" standard	39.000	S.F.	12.87	17.16	30.03
O For taping and finishing, add	39.000	S.F.	1.56	17.16	18.72
Demolition, wall framing	60.000	S.F.		11.56	11.56
Demolition, shingles	60.000	S.F.		58.80	58.80
P Mylar triple layer heat reflective shade (see desc. sheet)	37.500	S.F.	380.63	24.38	405.01
Demolition, drywall	60.000	S.F.		20.40	20.40
Paint interior & exterior trim	200.000	L.F.	20	96	116
TOTAL			1,305.45	1,230.15	2,535.60

D3010 684	Solar Direct Gain Glazing-Window Panels	COST EACH MAT.	COST EACH INST.	COST EACH TOTAL
2550	Solar passive heating, direct gain, 2'-6"x5', double glazed window			
2560	One panel wide	485	445	930
2580	Two panels wide	1,000	1,175	2,175
2600	Three panels wide	1,300	1,225	2,525

D
SERVICES

Solar Indirect Gain — Thermal Wall: The thermal storage wall is an Indirect Gain System because collection, absorption, and storage of solar energy occur outside the living space. This system consists of an unvented existing masonry wall located between south facing glazing and the space to be heated. The outside surface of the masonry wall is painted black to enable it to absorb maximum thermal energy. Standard double glazed units in wooden site-built frames are added to the south side of the building.

Thermal energy conducts through the masonry wall and radiates into the living space. A masonry thermal storage wall can also contain vents that allow warm air to flow into the living space through convection.

Indirect gain solar systems can be retrofitted to masonry walls constructed from a wide variety of building materials, including concrete, concrete block (solid and filled), brick, stone and adobe.

System Components	QUANTITY	UNIT	COST EACH MAT.	COST EACH INST.	COST EACH TOTAL
SYSTEM D3010 686 2600					
SOLAR PASSIVE HEATING, INDIRECT GAIN, THERMAL STORAGE WALL					
3' X 6'-8" DOUBLE GLAZED PANEL, THREE PANELS WIDE					
B Aluminum flashing, mill finish, .013 thick	5.500	S.F.	2.31	15.79	18.10
C Glazing, insul glass 2 lite 3/16" float 5/8" thick unit, 15-30 S.F.	54.000	S.F.	567	496.80	1,063.80
F Framing 2" x 4" lumber	.030	M.B.F.	18.45	53.25	71.70
K Molding, miscellaneous, 1" x 3"	44.000	L.F.	33.44	80.96	114.40
L-2 Wood batten, 1" x 4", custom pine or cedar	44.000	L.F.	33.44	97.24	130.68
M Drill holes, 4 per panel 1/2" diameter	30.000	Inch		29.40	29.40
N Caulking/sealants, silicone rubber, cartridges	.250	Gal.	10.50		10.50
X Lag screws 3/8" x 3" long	20.000	Ea.	5.80	59	64.80
X-1 Reglet	8.000	L.F.		25.28	25.28
Trim, primer & 1 coat	44.000	L.F.	6.60	26.84	33.44
Layout and drill holes for lag screws	20.000	Ea.	1.80	141	142.80
Expansion shields for lag screws	20.000	Ea.	7.20	55.40	62.60
TOTAL			686.54	1,080.96	1,767.50

D3010 686	Passive Solar Indirect Gain Wall	COST EACH MAT.	COST EACH INST.	COST EACH TOTAL
2550	Solar passive heating, indirect gain			
2560	3'x6'8", double glazed thermal storage wall, one panel wide	245	410	655
2580	Two panels wide	465	730	1,195
2600	Three panels wide	685	1,075	1,760

D3010 Energy Supply

MULLION DETAIL

SECTION

JAMB DETAIL

Solar Indirect Gain — Thermosyphon Panel: The Thermosyphon Air Panel (TAP) System is an indirect gain system used primarily for daytime heating because it does not include thermal mass. This site-built TAP System makes use of double glazing and a ribbed aluminum absorber plate. The panel is framed in wood and attached to the south side of a building. Solar radiation passes through the glass and is absorbed by the absorber plate. As the absorber plate heats, the air between it and the glass also heats. Vents cut through the building's south wall and the absorber plate near the top and bottom of the TAP allow hot air to circulate into the building by means of natural convection.

SERVICES **D**

System Components	QUANTITY	UNIT	COST EACH MAT.	COST EACH INST.	COST EACH TOTAL
SYSTEM D3010 688 2600					
SOLAR PASSIVE HEATING, INDIRECT GAIN, THERMOSYPHON PANEL					
3'X6'8" DOUBLE GLAZED PANEL, THREE PANELS WIDE					
B Aluminum flashing and sleeves, mill finish, .013 thick	8.000	S.F.	9.66	66.01	75.67
B-1 Aluminum ribbed, 4" pitch, on steel frame .032" mill finish	45.000	S.F.	76.95	99.90	176.85
B-2 Angle brackets	9.000	Ea.	6.93	22.77	29.70
C Glazing, insul glass 2 lite 3/16" float 5/8" thick unit, 15-30 S.F.	45.000	S.F.	472.50	414	886.50
D Support brackets, fir 2"x6"	.003	M.B.F.	1.85	7.80	9.65
D-1 Framing fir 2" x 6"	.040	M.B.F.	33	89	122
E Corrugated end closure	15.000	L.F.	9.75	32.25	42
E-1 Register, multilouvre operable 6" x 30"	3.000	Ea.	184.50	87	271.50
E-3 Grille 6" x 30"	3.000	Ea.	85.50	67.50	153
E-4 Back draft damper	.038	C.S.F.	.23	.45	.68
F House wall blocking 2" x 4"	.012	M.B.F.	24.60	43	67.60
J Absorber supports & glass stop	54.000	L.F.	19.98	88.56	108.54
K Exterior & bottom trim & interior glazing stop	40.000	L.F.	73.32	166.38	239.70
L Support ledger & side trim	24.000	L.F.	32.64	42.48	75.12
L-2 Trim cap	8.000	L.F.	12.64	17.68	30.32
N Caulking/sealants, silicone rubber, cartridges	.250	Gal.	10.50		10.50
O Drywall, nailed & taped, 5/8" thick	60.000	S.F.	19.80	26.40	46.20
O For taping and finishing, add	60.000	S.F.	2.40	26.40	28.80
O-1 Insulation foil faced exterior wall	45.000	S.F.	20.70	14.85	35.55
Demolition, framing	60.000	S.F.		14.28	14.28
Demolition, drywall	60.000	S.F.		20.40	20.40
Paint absorber plate black	45.000	S.F.	3.60	43.20	46.80
Finish paint	125.000	S.F.	12.50	60	72.50
Paint walls & ceiling with roller, primer & 1 coat	60.000	S.F.	6.60	23.40	30
TOTAL			1,120.15	1,473.71	2,593.86

D3010 688	Passive Solar Indirect Gain Panel	COST EACH MAT.	COST EACH INST.	COST EACH TOTAL
2550	Solar passive heating, indirect gain			
2560	3'x6'8", double glazed thermosyphon panel, one panel wide	390	520	910
2580	Two panels wide	755	995	1,750
2600	Three panels wide	1,125	1,475	2,600

D3010 Energy Supply

END ELEVATIONS

FRONT ELEVATION

CROSS SECTION

CONTINUOUS FROST WALL (as req'd)

Solar Attached Sunspace: The attached sunspace, a room adjacent to the south side of a building, is designed solely for heat collection. During the heating season, solar radiation enters the sunspace through south facing glass. Radiation is absorbed by elements in the space and converted to heat. This sunspace is thermally isolated from the building. A set of sliding glass doors connecting the sunspace to the building space is closed to prevent night heat losses from the living space. This attached sunspace uses standard size, double glazed units and has a concrete slab floor. Since the required depth of footings (if required) is dependent on local frost conditions, costs of footings at depths of 2′, 3′ and 4′ are provided with the base cost of system as shown.

System Components			COST EACH		
	QUANTITY	UNIT	MAT.	INST.	TOTAL
SYSTEM D3010 690 2560					
SOLAR PASSIVE HEATING, DIRECT GAIN, ATTACHED SUNSPACE					
WOOD FRAMED, SLAB ON GRADE					
B Aluminum, flashing, mill finish	21.000	S.F.	8.82	60.27	69.09
C Glazing, double, tempered glass	72.000	S.F.	756	662.40	1,418.40
D-2-4 Knee & end wall framing	.188	M.B.F.	122.20	166.38	288.58
F Ledgers, headers and rafters	.028	M.B.F.	111.32	194.58	305.90
Building paper, asphalt felt sheathing paper 15 lb	240.000	S.F.	9	28.68	37.68
G Sheathing plywood on roof CDX 1/2"	240.000	S.F.	163.20	151.20	314.40
H Asphalt shingles inorganic class A 210-235 lb/sq	1.000	Sq.	42	75.50	117.50
I Wood siding, match existing	164.000	S.F.	577.28	290.28	867.56
Window stop, interior or external	96.000	L.F.	76.80	257.28	334.08
L-1 Exterior trim	44.000	L.F.	72.60	77.88	150.48
L-2, L-3 Exterior trim 1" x 4"	206.000	L.F.	156.56	455.26	611.82
O Drywall, taping & finishing joints	200.000	S.F.	8	88	96
O Drywall, 1/2" on walls, standard, no finish	200.000	S.F.	60	88	148
S, S-1 Windows, wood, awning type, double glazed, with screens	4.000	Ea.	2,100	222	2,322
T Insulation batts, fiberglass, faced	210.000	S.F.	117.60	69.30	186.90
T-1 Polyethylene vapor barrier, standard, .004" thick	2.500	C.S.F.	6.02	23.90	29.92
Y Button vents	6.000	Ea.	13.08	44.40	57.48
Z Concrete floor, reinforced, 10' x 10' x 4"	1.234	C.Y.	144.38	86.94	231.32
Floor finishing, steel trowel	100.000	S.F.		74	74
Z-2 Mesh, welded wire 6 x 6, #10/10	1.000	C.S.F.	13.20	29.50	42.70
Paint interior, exterior and trim	330.000	S.F.	42.90	211.20	254.10
TOTAL			4,600.96	3,356.95	7,957.91

D3010 690	Passive Solar Sunspace	COST EACH		
		MAT.	INST.	TOTAL
2550	Solar passive heating, direct gain, attached sunspace,100 S.F.			
2560	Wood framed, slab on grade	4,600	3,350	7,950
2580	2' deep frost wall	5,000	4,125	9,125
2600	3' deep frost wall	5,125	4,400	9,525
2620	4' deep frost wall	5,250	4,700	9,950

D3020 Heat Generating Systems

Boiler   **Baseboard Radiation**

Small Electric Boiler
System Considerations:
1. Terminal units are fin tube baseboard radiation rated at 720 BTU/hr with 200° water temperature or 820 BTU/hr steam.
2. Primary use being for residential or smaller supplementary areas, the floor levels are based on 7-1/2′ ceiling heights.
3. All distribution piping is copper for boilers through 205 MBH. All piping for larger systems is steel pipe.

SERVICES D

System Components	QUANTITY	UNIT	COST EACH		
			MAT.	INST.	TOTAL
SYSTEM D3020 102 1120					
SMALL HEATING SYSTEM, HYDRONIC, ELECTRIC BOILER					
1,480 S.F., 61 MBH, STEAM, 1 FLOOR					
Boiler, electric steam, std cntrls, trim, ftngs and valves, 18 KW, 61.4 MBH	1.000	Ea.	3,712.50	1,320	5,032.50
Copper tubing type L, solder joint, hanger 10′OC, 1-1/4″ diam	160.000	L.F.	896	1,416	2,312
Radiation, 3/4″ copper tube w/alum fin baseboard pkg 7″ high	60.000	L.F.	441	966	1,407
Rough in baseboard panel or fin tube with valves & traps	10.000	Set	1,260	4,900	6,160
Pipe covering, calcium silicate w/cover 1″ wall 1-1/4″ diam	160.000	L.F.	337.60	864	1,201.60
Low water cut-off, quick hookup, in gage glass tappings	1.000	Ea.	191	32.50	223.50
TOTAL			6,838.10	9,498.50	16,336.60
COST PER S.F.			4.62	6.42	11.04

D3020 102	Small Heating Systems, Hydronic, Electric Boilers		COST PER S.F.		
			MAT.	INST.	TOTAL
1100	Small heating systems, hydronic, electric boilers				
1120	Steam, 1 floor, 1480 S.F., 61 M.B.H.		4.63	6.42	11.05
1160	3,000 S.F., 123 M.B.H.	RD3020 -010	3.44	5.60	9.04
1200	5,000 S.F., 205 M.B.H.		2.94	5.20	8.14
1240	2 floors, 12,400 S.F., 512 M.B.H.	RD3020 -020	2.43	5.15	7.58
1280	3 floors, 24,800 S.F., 1023 M.B.H.		2.56	5.10	7.66
1320	34,750 S.F., 1,433 M.B.H.		2.30	4.94	7.24
1360	Hot water, 1 floor, 1,000 S.F., 41 M.B.H.		7.95	3.56	11.51
1400	2,500 S.F., 103 M.B.H.		5.40	6.40	11.80
1440	2 floors, 4,850 S.F., 205 M.B.H.		5.05	7.70	12.75
1480	3 floors, 9,700 S.F., 410 M.B.H.		4.99	7.95	12.94

D3020 Heat Generating Systems

Boiler

Unit Heater

Large Electric Boiler System Considerations:

1. Terminal units are all unit heaters of the same size. Quantities are varied to accommodate total requirements.
2. All air is circulated through the heaters a minimum of three times per hour.
3. As the capacities are adequate for commercial use, floor levels are based on 10' ceiling heights.
4. All distribution piping is black steel pipe.

System Components	QUANTITY	UNIT	COST EACH		
			MAT.	INST.	TOTAL
SYSTEM D3020 104 1240					
LARGE HEATING SYSTEM, HYDRONIC, ELECTRIC BOILER					
9,280 S.F., 135 KW, 461 MBH, 1 FLOOR					
Boiler, electric hot water, std ctrls, trim, ftngs, valves, 135 KW, 461 MBH	1.000	Ea.	8,962.50	2,850	11,812.50
Expansion tank, painted steel, 60 Gal capacity ASME	1.000	Ea.	2,175	156	2,331
Circulating pump, CI, close cpld, 50 GPM, 2 HP, 2" pipe conn	1.000	Ea.	1,800	310	2,110
Unit heater, 1 speed propeller, horizontal, 200° EWT, 72.7 MBH	7.000	Ea.	4,760	1,190	5,950
Unit heater piping hookup with controls	7.000	Set	2,940	7,700	10,640
Pipe, steel, black, schedule 40, welded, 2-1/2" diam	380.000	L.F.	4,864	8,101.60	12,965.60
Pipe covering, calcium silicate w/cover, 1" wall, 2-1/2" diam	380.000	L.F.	995.60	2,109	3,104.60
TOTAL			26,497.10	22,416.60	48,913.70
COST PER S.F.			2.86	2.42	5.28

D3020 104	Large Heating Systems, Hydronic, Electric Boilers		COST PER S.F.		
			MAT.	INST.	TOTAL
1230	Large heating systems, hydronic, electric boilers				
1240	9,280 S.F., 135 K.W., 461 M.B.H., 1 floor		2.86	2.41	5.27
1280	14,900 S.F., 240 K.W., 820 M.B.H., 2 floors	RD3020 -010	3.82	3.94	7.76
1320	18,600 S.F., 296 K.W., 1,010 M.B.H., 3 floors		3.82	4.28	8.10
1360	26,100 S.F., 420 K.W., 1,432 M.B.H., 4 floors	RD3020 -020	3.78	4.20	7.98
1400	39,100 S.F., 666 K.W., 2,273 M.B.H., 4 floors		3.21	3.51	6.72
1440	57,700 S.F., 900 K.W., 3,071 M.B.H., 5 floors		3.10	3.45	6.55
1480	111,700 S.F., 1,800 K.W., 6,148 M.B.H., 6 floors		2.79	2.96	5.75
1520	149,000 S.F., 2,400 K.W., 8,191 M.B.H., 8 floors		2.78	2.95	5.73
1560	223,300 S.F., 3,600 K.W., 12,283 M.B.H., 14 floors		2.99	3.36	6.35

D3020 Heat Generating Systems

Boiler Selection: The maximum allowable working pressures are limited by ASME "Code for Heating Boilers" to 15 PSI for steam and 160 PSI for hot water heating boilers, with a maximum temperature limitation of 250°F. Hot water boilers are generally rated for a working pressure of 30 PSI. High pressure boilers are governed by the ASME "Code for Power Boilers" which is used almost universally for boilers operating over 15 PSIG. High pressure boilers used for a combination of heating/process loads are usually designed for 150 PSIG.

Boiler ratings are usually indicated as either Gross or Net Output. The Gross Load is equal to the Net Load plus a piping and pickup allowance. When this allowance cannot be determined, divide the gross output rating by 1.25 for a value equal to or greater than the next heat loss requirement of the building.

Table below lists installed cost per boiler and includes insulating jacket, standard controls, burner and safety controls. Costs do not include piping or boiler base pad. Outputs are Gross.

SERVICES **D**

D3020 106	Boilers, Hot Water & Steam		COST EACH		
			MAT.	INST.	TOTAL
0600	Boiler, electric, steel, hot water, 12 K.W., 41 M.B.H.		3,375	1,100	4,475
0620	30 K.W., 103 M.B.H.		4,025	1,200	5,225
0640	60 K.W., 205 M.B.H.	RD3020 -010	4,925	1,300	6,225
0660	120 K.W., 410 M.B.H.		5,525	1,600	7,125
0680	210 K.W., 716 M.B.H.	RD3020 -020	6,450	2,400	8,850
0700	510 K.W., 1,739 M.B.H.		16,100	4,425	20,525
0720	720 K.W., 2,452 M.B.H.		19,400	5,000	24,400
0740	1,200 K.W., 4,095 M.B.H.		25,200	5,750	30,950
0760	2,100 K.W., 7,167 M.B.H.		48,100	7,225	55,325
0780	3,600 K.W., 12,283 M.B.H.		74,000	12,200	86,200
0820	Steam, 6 K.W., 20.5 M.B.H.		3,200	1,200	4,400
0840	24 K.W., 81.8 M.B.H.		3,900	1,300	5,200
0860	60 K.W., 205 M.B.H.		5,400	1,425	6,825
0880	150 K.W., 512 M.B.H.		7,800	2,200	10,000
0900	510 K.W., 1,740 M.B.H.		20,700	5,425	26,125
0920	1,080 K.W., 3,685 M.B.H.		29,000	7,800	36,800
0940	2,340 K.W., 7,984 M.B.H.		59,500	12,200	71,700
0980	Gas, cast iron, hot water, 80 M.B.H.		3,575	1,350	4,925
1000	100 M.B.H.		4,300	1,475	5,775
1020	163 M.B.H.		2,450	1,975	4,425
1040	280 M.B.H.		2,900	2,200	5,100
1060	544 M.B.H.		7,525	3,900	11,425
1080	1,088 M.B.H.		11,300	4,950	16,250
1100	2,000 M.B.H.		14,800	7,725	22,525
1120	2,856 M.B.H.		20,100	9,875	29,975
1140	4,720 M.B.H.		59,500	13,600	73,100
1160	6,970 M.B.H.		76,000	22,200	98,200
1180	For steam systems under 2,856 M.B.H., add 8%				
1520	Oil, cast iron, hot water, 109 M.B.H.		2,475	1,650	4,125
1540	173 M.B.H.		3,150	1,975	5,125
1560	236 M.B.H.		3,450	2,325	5,775
1580	1,084 M.B.H.		8,500	5,250	13,750
1600	1,600 M.B.H.		11,200	7,550	18,750
1620	2,480 M.B.H.		16,600	9,650	26,250
1640	3,550 M.B.H.		21,900	11,600	33,500
1660	Steam systems same price as hot water				

Unit Heater

Fossil Fuel Boiler System Considerations:

1. Terminal units are horizontal unit heaters. Quantities are varied to accommodate total heat loss per building.
2. Unit heater selection was determined by their capacity to circulate the building volume a minimum of three times per hour in addition to the BTU output.

3. Systems shown are forced hot water. Steam boilers cost slightly more than hot water boilers. However, this is compensated for by the smaller size or fewer terminal units required with steam.
4. Floor levels are based on 10' story heights.
5. MBH requirements are gross boiler output.

SERVICES

D

System Components	QUANTITY	UNIT	COST EACH		
			MAT.	INST.	TOTAL
SYSTEM D3020 108 1280					
HEATING SYSTEM, HYDRONIC, FOSSIL FUEL, TERMINAL UNIT HEATERS					
CAST IRON BOILER, GAS, 80 MBH, 1,070 S.F. BUILDING					
Boiler, gas, hot water, CI, burner, controls, insulation, breeching, 80 MBH	1.000	Ea.	1,732.50	1,417.50	3,150
Pipe, steel, black, schedule 40, threaded, cplg & hngr 10'OC, 2" diam	200.000	L.F.	1,560	2,890	4,450
Unit heater, 1 speed propeller, horizontal, 200° EWT, 72.7 MBH	2.000	Ea.	1,360	340	1,700
Unit heater piping hookup with controls	2.000	Set	840	2,200	3,040
Expansion tank, painted steel, ASME, 18 Gal capacity	1.000	Ea.	1,875	78	1,953
Circulating pump, CI, flange connection, 1/12 HP	1.000	Ea.	263	154	417
Pipe covering, calcium silicate w/cover, 1" wall, 2" diam	200.000	L.F.	492	1,110	1,602
TOTAL			8,122.50	8,189.50	16,312
COST PER S.F.			7.59	7.65	15.24

D3020 108	Heating Systems, Unit Heaters		COST PER S.F.		
			MAT.	INST.	TOTAL
1260	Heating systems, hydronic, fossil fuel, terminal unit heaters,				
1280	Cast iron boiler, gas, 80 M.B.H., 1,070 S.F. bldg.		7.60	7.65	15.25
1320	163 M.B.H., 2,140 S.F. bldg.		5	5.20	10.20
1360	544 M.B.H., 7,250 S.F. bldg.	RD3020 -010	3.76	3.71	7.47
1400	1,088 M.B.H., 14,500 S.F. bldg.		3.31	3.51	6.82
1440	3,264 M.B.H., 43,500 S.F. bldg.	RD3020 -020	2.55	2.66	5.21
1480	5,032 M.B.H., 67,100 S.F. bldg.		3.11	2.73	5.84
1520	Oil, 109 M.B.H., 1,420 S.F. bldg.		8.10	6.80	14.90
1560	235 M.B.H., 3,150 S.F. bldg.		5.30	4.90	10.20
1600	940 M.B.H., 12,500 S.F. bldg.		3.75	3.21	6.96
1640	1,600 M.B.H., 21,300 S.F. bldg.		3.58	3.08	6.66
1680	2,480 M.B.H., 33,100 S.F. bldg.		3.56	2.78	6.34
1720	3,350 M.B.H., 44,500 S.F. bldg.		3.15	2.85	6
1760	Coal, 148 M.B.H., 1,975 S.F. bldg.		6.15	4.45	10.60
1800	300 M.B.H., 4,000 S.F. bldg.		4.48	3.49	7.97
1840	2,360 M.B.H., 31,500 S.F. bldg.		3.46	2.83	6.29

D3020 Heat Generating Systems

Fin Tube Radiator

Fossil Fuel Boiler System Considerations:

1. Terminal units are commercial steel fin tube radiation. Quantities are varied to accommodate total heat loss per building.
2. Systems shown are forced hot water. Steam boilers cost slightly more than hot water boilers. However, this is compensated for by the smaller size or fewer terminal units required with steam.
3. Floor levels are based on 10′ story heights.
4. MBH requirements are gross boiler output.

SERVICES D

System Components	QUANTITY	UNIT	COST EACH MAT.	COST EACH INST.	COST EACH TOTAL
SYSTEM D3020 110 3240					
HEATING SYSTEM, HYDRONIC, FOSSIL FUEL, FIN TUBE RADIATION					
CAST IRON BOILER, GAS, 80 MBH, 1,070 S.F. BLDG.					
Boiler, gas, hot water, CI, burner, controls, insulation, breeching, 80 MBH	1.000	Ea.	1,732.50	1,417.50	3,150
Pipe, steel, black, schedule 40, threaded, 2″ diam	200.000	L.F.	1,560	2,890	4,450
Radiation,steel 1-1/4″ tube & 4-1/4″ fin w/cover & damper, wall hung	80.000	L.F.	2,200	2,080	4,280
Rough-in, wall hung steel fin radiation with supply & balance valves	8.000	Set	1,440	4,560	6,000
Expansion tank, painted steel, ASME, 18 Gal capacity	1.000	Ea.	1,875	78	1,953
Circulating pump, CI flanged, 1/12 HP	1.000	Ea.	263	154	417
Pipe covering, calcium silicate w/cover, 1″ wall, 2″ diam	200.000	L.F.	492	1,110	1,602
TOTAL			9,562.50	12,289.50	21,852
COST PER S.F.			8.94	11.49	20.43

D3020 110	Heating System, Fin Tube Radiation		COST PER S.F. MAT.	COST PER S.F. INST.	COST PER S.F. TOTAL
3230	Heating systems, hydronic, fossil fuel, fin tube radiation				
3240	Cast iron boiler, gas, 80 MBH, 1,070 S.F. bldg.		10.84	11.50	22.34
3280	169 M.B.H., 2,140 S.F. bldg.	RD3020 -010	5.60	7.30	12.90
3320	544 M.B.H., 7,250 S.F. bldg.		4.77	6.25	11.02
3360	1,088 M.B.H., 14,500 S.F. bldg.	RD3020 -020	4.39	6.15	10.54
3400	3,264 M.B.H., 43,500 S.F. bldg.		3.74	5.45	9.19
3440	5,032 M.B.H., 67,100 S.F. bldg.		4.31	5.50	9.81
3480	Oil, 109 M.B.H., 1,420 S.F. bldg.		10.60	12.55	23.15
3520	235 M.B.H., 3,150 S.F. bldg.		6.30	7.40	13.70
3560	940 M.B.H., 12,500 S.F. bldg.		4.85	5.90	10.75
3600	1,600 M.B.H., 21,300 S.F. bldg.		4.77	5.85	10.62
3640	2,480 M.B.H., 33,100 S.F. bldg.		4.76	5.55	10.31
3680	3,350 M.B.H., 44,500 S.F. bldg.		4.33	5.60	9.93
3720	Coal, 148 M.B.H., 1,975 S.F. bldg.		7.10	6.95	14.05
3760	300 M.B.H., 4,000 S.F. bldg.		5.45	5.95	11.40
3800	2,360 M.B.H., 31,500 S.F. bldg.		4.61	5.55	10.16
4080	Steel boiler, oil, 97 M.B.H., 1,300 S.F. bldg.		7.70	10.30	18
4120	315 M.B.H., 4,550 S.F. bldg.		4.95	5.65	10.60
4160	525 M.B.H., 7,000 S.F. bldg.		5.75	6.05	11.80
4200	1,050 M.B.H., 14,000 S.F. bldg.		5.15	6.05	11.20
4240	2,310 M.B.H., 30,800 S.F. bldg.		4.70	5.60	10.30
4280	3,150 M.B.H., 42,000 S.F. bldg.		4.41	5.65	10.06

D3020 Heat Generating Systems

D3020 310	Gas Vent, Galvanized Steel, Double Wall	COST PER V.L.F.		
		MAT.	INST.	TOTAL
0030	Gas vent, galvanized steel, double wall, round pipe diameter, 3"	4.36	13	17.36
0040	4"	5.45	13.75	19.20
0050	5"	6.35	14.60	20.95
0060	6"	7.35	15.60	22.95
0070	7"	10.85	16.70	27.55
0080	8"	12.05	18	30.05
0090	10"	25.50	19.50	45
0100	12"	34	21.50	55.50
0110	14"	57	22.50	79.50
0120	16	77.50	23.50	101
0130	18"	100	24.50	124.50
0140	20"	118	40.50	158.50
0150	22"	149	43	192
0160	24"	184	45.50	229.50
0180	30"	246	52	298
0220	38"	375	60.50	435.50
0300	48"	590	76.50	666.50
0400				

D3020 310	Gas Vent Fittings, Galvanized Steel, Double Wall	COST EACH		
		MAT.	INST.	TOTAL
1020	Gas vent, fittings, 45° Elbow, adjustable thru 8" diameter			
1030	Pipe diameter, 3"	10.80	26	36.80
1040	4"	12.75	27.50	40.25
1050	5"	15	29	44
1060	6"	17.95	31	48.95
1070	7"	26	33.50	59.50
1080	8"	30	36	66
1090	10"	74	39	113
1100	12"	93.50	42.50	136
1110	14"	131	44.50	175.50
1120	16"	164	47	211
1130	18"	221	49	270
1140	20"	249	81	330
1150	22"	375	85.50	460.50
1170	24"	475	91	566
1200	30"	1,025	104	1,129
1250	38"	1,400	121	1,521
1300	48"	1,825	145	1,970
1410	90° Elbow, adjustable, pipe diameter, 3"	18	26	44
1420	4"	22	27.50	49.50
1430	5"	25.50	29	54.50
1440	6"	30.50	31	61.50
1450	7"	48	33.50	81.50
1460	8"	51.50	36	87.50
1510	Tees, standard, pipe diameter, 3"	26.50	34.50	61
1520	4"	27.50	36	63.50
1530	5"	30	37.50	67.50
1540	6"	33.50	39	72.50
1550	7"	35.50	40.50	76
1560	8"	49	42.50	91.50
1570	10"	133	44.50	177.50
1580	12"	159	47	206
1590	14"	260	52	312
1600	16"	370	58.50	428.50
1610	18"	460	67	527
1620	20"	625	85.50	710.50

Important: See the Reference Section for critical supporting data - Reference Numbers and City Cost Indexes

D3020 Heat Generating Systems

D3020 310	Gas Vent Fittings, Galvanized Steel, Double Wall	COST EACH		
		MAT.	INST.	TOTAL
1630	22"	775	112	887
1640	24"	910	121	1,031
1660	30"	1,575	145	1,720
1700	38"	2,200	182	2,382
1750	48"	3,350	291	3,641
1810	Tee, access caps, pipe diameter, 3"	1.88	21	22.88
1820	4"	2.13	22.50	24.63
1830	5"	2.85	23.50	26.35
1840	6"	3.30	25.50	28.80
1850	7"	5.35	26.50	31.85
1860	8"	6.25	27.50	33.75
1870	10"	11.95	29	40.95
1880	12"	16	31	47
1890	14"	33	33.50	66.50
1900	16"	38	37.50	75.50
1910	18"	43	39	82
1920	20"	46	54	100
1930	22"	53	66	119
1940	24"	59.50	69.50	129
1950	30"	141	81	222
1970	38"	221	104	325
1990	48"	330	132	462
2110	Tops, bird proof, pipe diameter, 3"	10	20.50	30.50
2120	4"	10.45	21.50	31.95
2130	5"	14.30	22.50	36.80
2140	6"	18.30	23.50	41.80
2150	7"	27.50	24.50	52
2160	8"	35	26	61
2170	10"	82	27.50	109.50
2180	12"	124	29	153
2190	14"	142	31	173
2200	16"	255	33.50	288.50
2210	18"	450	36	486
2220	20"	500	52	552
2230	22"	680	66	746
2240	24"	845	72.50	917.50
2410	Roof flashing, tall cone/adjustable, pipe diameter, 3"	6.50	26	32.50
2420	4"	7.35	27.50	34.85
2430	5"	11.35	29	40.35
2440	6"	15.95	31	46.95
2450	7"	20	33.50	53.50
2460	8"	22	36	58
2470	10"	59.50	39	98.50
2480	12"	76	42.50	118.50
2490	14"	93.50	47	140.50
2500	16"	116	52	168
2510	18"	157	58.50	215.50
2520	20"	203	81	284
2530	22"	257	104	361
2540	24"	310	121	431

Diagram labels: Insulated Pipe, Roof Thimble, Full Angle Ring, Wall Support, 15° Adjustable Elbow, Adjustable Length, Plate Support, Wall Guide, Standard Tee, Half Angle Ring, Drain

D3020 Heat Generating Systems

D3020 315	Chimney, Stainless Steel, Insulated, Wood & Oil	COST PER V.L.F.		
		MAT.	INST.	TOTAL
0530	Chimney, stainless steel, insulated, diameter, 6"	33.50	15.60	49.10
0540	7"	43.50	16.70	60.20
0550	8"	51	18	69
0560	10"	74.50	19.50	94
0570	12"	99.50	21.50	121
0580	14"	131	22.50	153.50

Important: See the Reference Section for critical supporting data - Reference Numbers and City Cost Indexes

D3030 Cooling Generating Systems

Chilled Water Supply & Return Piping

Air Cooled Water Chiller Unit

Roof

Insulate
Return
Fan Coil Unit
Supply — Finish Ceiling

Design Assumptions: The chilled water, air cooled systems priced, utilize reciprocating hermetic compressors and propeller-type condenser fans. Piping with pumps and expansion tanks is included based on a two pipe system. No ducting is included and the fan-coil units are cooling only. Water treatment and balancing are not included. Chilled water piping is insulated. Area distribution is through the use of multiple fan coil units. Fewer but larger fan coil units with duct distribution would be approximately the same S.F. cost.

System Components	QUANTITY	UNIT	COST EACH		
			MAT.	INST.	TOTAL
SYSTEM D3030 110 1200					
PACKAGED CHILLER, AIR COOLED, WITH FAN COIL UNIT					
APARTMENT CORRIDORS, 3,000 S.F., 5.50 TON					
Fan coil air conditioning unit, cabinet mounted & filters chilled water	1.000	Ea.	3,436.88	427.09	3,863.97
Water chiller, air conditioning unit, reciprocating, air cooled,	1.000	Ea.	6,132.50	1,615.63	7,748.13
Chilled water unit coil connections	1.000	Ea.	790	1,175	1,965
Chilled water distribution piping	440.000	L.F.	9,240	15,840	25,080
TOTAL			19,599.38	19,057.72	38,657.10
COST PER S.F.			6.53	6.35	12.88

D3030 110	Chilled Water, Air Cooled Condenser Systems	COST PER S.F.		
		MAT.	INST.	TOTAL
1180	Packaged chiller, air cooled, with fan coil unit			
1200	Apartment corridors, 3,000 S.F., 5.50 ton	6.53	6.37	12.90
1360	40,000 S.F., 73.33 ton	3.89	2.82	6.71
1440	Banks and libraries, 3,000 S.F., 12.50 ton	9.65	7.50	17.15
1560	20,000 S.F., 83.33 ton	6.70	3.85	10.55
1680	Bars and taverns, 3,000 S.F., 33.25 ton	15.85	8.85	24.70
1760	10,000 S.F., 110.83 ton	11.30	2.20	13.50
1920	Bowling alleys, 3,000 S.F., 17.00 ton	12	8.40	20.40
2040	20,000 S.F., 113.33 ton	7.70	3.89	11.59
2160	Department stores, 3,000 S.F., 8.75 ton	8.85	7.05	15.90
2320	40,000 S.F., 116.66 ton	4.91	2.91	7.82
2400	Drug stores, 3,000 S.F., 20.00 ton	13.60	8.70	22.30
2520	20,000 S.F., 133.33 ton	9.15	4.10	13.25
2640	Factories, 2,000 S.F., 10.00 ton	8.45	7.20	15.65
2800	40,000 S.F., 133.33 ton	5.35	2.99	8.34
2880	Food supermarkets, 3,000 S.F., 8.50 ton	8.70	7.05	15.75
3040	40,000 S.F., 113.33 ton	4.80	2.91	7.71
3120	Medical centers, 3,000 S.F., 7.00 ton	7.80	6.85	14.65
3280	40,000 S.F., 93.33 ton	4.34	2.86	7.20
3360	Offices, 3,000 S.F., 9.50 ton	7.90	7	14.90
3520	40,000 S.F., 126.66 ton	5.25	3.01	8.26
3600	Restaurants, 3,000 S.F., 15.00 ton	10.80	7.80	18.60
3720	20,000 S.F., 100.00 ton	7.70	4.07	11.77
3840	Schools and colleges, 3,000 S.F., 11.50 ton	9.15	7.35	16.50
3960	20,000 S.F., 76.66 ton	6.50	3.88	10.38

RD3030 -010

D3030 Cooling Generating Systems

General: Water cooled chillers are available in the same sizes as air cooled units. They are also available in larger capacities.

Design Assumptions: The chilled water systems with water cooled condenser include reciprocating hermetic compressors, water cooling tower, pumps, piping and expansion tanks and are based on a two pipe system. Chilled water piping is insulated. No ducts are included and fan-coil units are cooling only. Area distribution is through use of multiple fan coil units. Fewer but larger fan coil units with duct distribution would be approximately the same S.F. cost. Water treatment and balancing are not included.

System Components	QUANTITY	UNIT	COST EACH		
			MAT.	INST.	TOTAL
SYSTEM D3030 115 1320					
PACKAGED CHILLER, WATER COOLED, WITH FAN COIL UNIT					
APARTMENT CORRIDORS, 4,000 S.F., 7.33 TON					
Fan coil air conditioner unit, cabinet mounted & filters, chilled water	2.000	Ea.	4,580.63	569.22	5,149.85
Water chiller, water cooled, 1 compressor, hermetic scroll,	1.000	Ea.	4,694.50	2,793.80	7,488.30
Cooling tower, draw thru single flow, belt drive	1.000	Ea.	700.02	118.38	818.40
Cooling tower pumps & piping	1.000	System	351.84	278.54	630.38
Chilled water unit coil connections	2.000	Ea.	1,580	2,350	3,930
Chilled water distribution piping	520.000	L.F.	10,920	18,720	29,640
TOTAL			22,826.99	24,829.94	47,656.93
COST PER S.F.			5.71	6.21	11.92
*Cooling requirements would lead to choosing a water cooled unit					

D3030 115	Chilled Water, Cooling Tower Systems		COST PER S.F.		
			MAT.	INST.	TOTAL
1300	Packaged chiller, water cooled, with fan coil unit				
1320	Apartment corridors, 4,000 S.F., 7.33 ton		5.70	6.20	11.90
1600	Banks and libraries, 4,000 S.F., 16.66 ton	RD3030 -010	8.95	6.80	15.75
1800	60,000 S.F., 250.00 ton		6.65	5.50	12.15
1880	Bars and taverns, 4,000 S.F., 44.33 ton		14.20	8.60	22.80
2000	20,000 S.F., 221.66 ton		12.85	7.55	20.40
2160	Bowling alleys, 4,000 S.F., 22.66 ton		10.30	7.50	17.80
2320	40,000 S.F., 226.66 ton		7.55	5.40	12.95
2440	Department stores, 4,000 S.F., 11.66 ton		6.55	6.75	13.30
2640	60,000 S.F., 175.00 ton		5.55	5.15	10.70
2720	Drug stores, 4,000 S.F., 26.66 ton		11.05	7.70	18.75
2880	40,000 S.F., 266.67 ton		8.70	5.90	14.60
3000	Factories, 4,000 S.F., 13.33 ton		7.80	6.50	14.30
3200	60,000 S.F., 200.00 ton		6	5.30	11.30
3280	Food supermarkets, 4,000 S.F., 11.33 ton		6.45	6.70	13.15
3480	60,000 S.F., 170.00 ton		5.50	5.15	10.65
3560	Medical centers, 4.000 S.F., 9.33 ton		5.70	6.15	11.85
3760	60,000 S.F., 140.00 ton		5.10	5.10	10.20
3840	Offices, 4,000 S.F., 12.66 ton		7.55	6.45	14
4040	60,000 S.F., 190.00 ton		5.80	5.20	11
4120	Restaurants, 4,000 S.F., 20.00 ton		9.35	6.95	16.30
4320	60,000 S.F., 300.00 ton		7.35	5.70	13.05
4400	Schools and colleges, 4,000 S.F., 15.33 ton		8.50	6.70	15.20
4600	60,000 S.F., 230.00 ton		6.20	5.30	11.50

D3050 Terminal & Package Units

System Description: Rooftop single zone units are electric cooling and gas heat. Duct systems are low velocity, galvanized steel supply and return. Price variations between sizes are due to several factors. Jumps in the cost of the rooftop unit occur when the manufacturer shifts from the largest capacity unit on a small frame to the smallest capacity on the next larger frame, or changes from one compressor to two. As the unit capacity increases for larger areas the duct distribution grows in proportion. For most applications there is a tradeoff point where it is less expensive and more efficient to utilize smaller units with short simple distribution systems. Larger units also require larger initial supply and return ducts which can create a space problem. Supplemental heat may be desired in colder locations. The table below is based on one unit supplying the area listed. The 10,000 S.F. unit for bars and taverns is not listed because a nominal 110 ton unit would be required and this is above the normal single zone rooftop capacity.

System Components	QUANTITY	UNIT	COST EACH		
			MAT.	INST.	TOTAL
SYSTEM D3050 150 1280					
ROOFTOP, SINGLE ZONE, AIR CONDITIONER					
APARTMENT CORRIDORS, 500 S.F., .92 TON					
Rooftop air-conditioner, 1 zone, electric cool, standard controls, curb	1.000	Ea.	1,483.50	609.50	2,093
Ductwork package for rooftop single zone units	1.000	System	299	897	1,196
TOTAL			1,782.50	1,506.50	3,289
COST PER S.F.			3.57	3.01	6.58

D3050 150	Rooftop Single Zone Unit Systems	COST PER S.F.		
		MAT.	INST.	TOTAL
1260	Rooftop, single zone, air conditioner			
1280	Apartment corridors, 500 S.F., .92 ton	3.55	3	6.55
1480	10,000 S.F., 18.33 ton	2.43	2.06	4.49
1560	Banks or libraries, 500 S.F., 2.08 ton	8.05	6.80	14.85
1760	10,000 S.F., 41.67 ton	5.60	4.65	10.25
1840	Bars and taverns, 500 S.F. 5.54 ton	13.15	9.10	22.25
2000	5,000 S.F., 55.42 ton	11.75	6.95	18.70
2080	Bowling alleys, 500 S.F., 2.83 ton	8.45	7.70	16.15
2280	10,000 S.F., 56.67 ton	6.95	6.30	13.25
2360	Department stores, 500 S.F., 1.46 ton	5.65	4.78	10.43
2560	10,000 S.F., 29.17 ton	3.97	3.25	7.22
2640	Drug stores, 500 S.F., 3.33 ton	9.95	9.10	19.05
2840	10,000 S.F., 66.67 ton	8.15	7.45	15.60
2920	Factories, 500 S.F., 1.67 ton	6.45	5.45	11.90
3120	10,000 S.F., 33.33 ton	4.54	3.72	8.26
3200	Food supermarkets, 500 S.F., 1.42 ton	5.50	4.65	10.15
3400	10,000 S.F., 28.33 ton	3.86	3.16	7.02
3480	Medical centers, 500 S.F., 1.17 ton	4.52	3.83	8.35
3680	10,000 S.F., 23.33 ton	3.09	2.62	5.71
3760	Offices, 500 S.F., 1.58 ton	6.15	5.20	11.35
3960	10,000 S.F., 31.67 ton	4.31	3.53	7.84
4000	Restaurants, 500 S.F., 2.50 ton	9.70	8.20	17.90
4200	10,000 S.F., 50.00 ton	6.10	5.60	11.70
4240	Schools and colleges, 500 S.F., 1.92 ton	7.45	6.30	13.75
4440	10,000 S.F., 38.33 ton	5.15	4.28	9.43

Note: Row 1560 includes a reference box: RD3030 -010

D3050 Terminal & Package Units

System Description: Rooftop units are multizone with up to 12 zones, and include electric cooling, gas heat, thermostats, filters, supply and return fans complete. Duct systems are low velocity, galvanized steel supply and return with insulated supplies.

Multizone units cost more per ton of cooling than single zone. However, they offer flexibility where load conditions are varied due to heat generating areas or exposure to radiational heating. For example, perimeter offices on the "sunny side" may require cooling at the same

time "shady side" or central offices may require heating. It is possible to accomplish similar results using duct heaters in branches of the single zone unit. However, heater location could be a problem and total system operating energy efficiency could be lower.

System Components	QUANTITY	UNIT	COST EACH		
			MAT.	INST.	TOTAL
SYSTEM D3050 155 1280					
ROOFTOP, MULTIZONE, AIR CONDITIONER					
APARTMENT CORRIDORS, 3,000 S.F., 5.50 TON					
Rooftop multizone unit, standard controls, curb	1.000	Ea.	36,960	1,430	38,390
Ductwork package for rooftop multizone units	1.000	System	2,502.50	10,312.50	12,815
TOTAL			39,462.50	11,742.50	51,205
COST PER S.F.			13.15	3.91	17.06

Note A: Small single zone unit recommended

D3050 155	Rooftop Multizone Unit Systems		COST PER S.F.		
			MAT.	INST.	TOTAL
1240	Rooftop, multizone, air conditioner				
1260	Apartment corridors, 1,500 S.F., 2.75 ton. See Note A.				
1280	3,000 S.F., 5.50 ton	RD3030 -010	13.17	3.90	17.07
1440	25,000 S.F., 45.80 ton		7.30	3.78	11.08
1520	Banks or libraries, 1,500 S.F., 6.25 ton		30	8.90	38.90
1640	15,000 S.F., 62.50 ton		11.05	8.55	19.60
1720	25,000 S.F., 104.00 ton		9.95	8.50	18.45
1800	Bars and taverns, 1,500 S.F., 16.62 ton		64.50	12.80	77.30
1840	3,000 S.F., 33.24 ton		48.50	12.35	60.85
1880	10,000 S.F., 110.83 ton		24	12.30	36.30
2080	Bowling alleys, 1,500 S.F., 8.50 ton		40.50	12.10	52.60
2160	10,000 S.F., 56.70 ton		22.50	11.70	34.20
2240	20,000 S.F., 113.00 ton		13.55	11.55	25.10
2640	Drug stores, 1,500 S.F., 10.00 ton		48	14.25	62.25
2680	3,000 S.F., 20.00 ton		35	13.75	48.75
2760	15,000 S.F., 100.00 ton		16	13.65	29.65
3760	Offices, 1,500 S.F., 4.75 ton, See Note A				
3880	15,000 S.F., 47.50 ton		12.55	6.55	19.10
3960	25,000 S.F., 79.16 ton		8.40	6.50	14.90
4000	Restaurants, 1,500 S.F., 7.50 ton		36	10.70	46.70
4080	10,000 S.F., 50.00 ton		19.85	10.30	30.15
4160	20,000 S.F., 100.00 ton		12	10.25	22.25
4240	Schools and colleges, 1,500 S.F., 5.75 ton		27.50	8.20	35.70
4360	15,000 S.F., 57.50 ton		10.15	7.85	18
4441	25,000 S.F., 95.83 ton		9.20	7.85	17.05

D3050 Terminal & Package Units

System Description: Self-contained, single package water cooled units include cooling tower, pump, piping allowance. Systems for 1000 S.F. and up include duct and diffusers to provide for even distribution of air. Smaller units distribute air through a supply air plenum, which is integral with the unit.

Returns are not ducted and supplies are not insulated.

Hot water or steam heating coils are included but piping to boiler and the boiler itself is not included.

Where local codes or conditions permit single pass cooling for the smaller units, deduct 10%.

System Components			COST EACH		
	QUANTITY	UNIT	MAT.	INST.	TOTAL
SYSTEM D3050 160 1300					
SELF-CONTAINED, WATER COOLED UNIT					
APARTMENT CORRIDORS, 500 S.F., .92 TON					
Self contained, water cooled, single package air conditioner unit	1.000	Ea.	1,041.60	487.20	1,528.80
Ductwork package for water or air cooled packaged units	1.000	System	88.32	680.80	769.12
Cooling tower, draw thru single flow, belt drive	1.000	Ea.	87.86	14.86	102.72
Cooling tower pumps & piping	1.000	System	44.16	34.96	79.12
TOTAL			1,261.94	1,217.82	2,479.76
COST PER S.F.			2.52	2.44	4.96

D3050 160	Self-contained, Water Cooled Unit Systems		COST PER S.F.		
			MAT.	INST.	TOTAL
1280	Self-contained, water cooled unit		2.52	2.44	4.96
1300	Apartment corridors, 500 S.F., .92 ton		2.50	2.45	4.95
1440	10,000 S.F., 18.33 ton	RD3030 -010	3.13	1.71	4.84
1520	Banks or libraries, 500 S.F., 2.08 ton		5.30	2.44	7.74
1680	10,000 S.F., 41.66 ton		6.10	3.87	9.97
1760	Bars and taverns, 500 S.F., 5.54 ton		11.35	4.11	15.46
1920	10,000 S.F., 110.00 ton		14.95	6	20.95
2000	Bowling alleys, 500 S.F., 2.83 ton		7.25	3.32	10.57
2160	10,000 S.F., 56.66 ton		8.10	5.25	13.35
2200	Department stores, 500 S.F., 1.46 ton		3.74	1.71	5.45
2360	10,000 S.F., 29.17 ton		4.48	2.62	7.10
2440	Drug stores, 500 S.F., 3.33 ton		8.55	3.90	12.45
2600	10,000 S.F., 66.66 ton		9.55	6.15	15.70
2680	Factories, 500 S.F., 1.66 ton		4.26	1.95	6.21
2840	10,000 S.F., 33.33 ton		5.10	2.99	8.09
2920	Food supermarkets, 500 S.F., 1.42 ton		3.62	1.66	5.28
3080	10,000 S.F., 28.33 ton		4.36	2.54	6.90
3160	Medical centers, 500 S.F., 1.17 ton		2.98	1.36	4.34
3320	10,000 S.F., 23.33 ton		3.98	2.17	6.15
3400	Offices, 500 S.F., 1.58 ton		4.05	1.85	5.90
3560	10,000 S.F., 31.67 ton		4.86	2.84	7.70
3640	Restaurants, 500 S.F., 2.50 ton		6.40	2.93	9.33
3800	10,000 S.F., 50.00 ton		5.10	4.28	9.38
3880	Schools and colleges, 500 S.F., 1.92 ton		4.90	2.24	7.14
4040	10,000 S.F., 38.33 ton		5.65	3.57	9.22

D3050 Terminal & Package Units

System Description: Self-contained air cooled units with remote air cooled condenser and interconnecting tubing. Systems for 1000 S.F. and up include duct and diffusers. Smaller units distribute air directly.

Returns are not ducted and supplies are not insulated.

Potential savings may be realized by using a single zone rooftop system or through-the-wall unit, especially in the smaller capacities, if the application permits.

Hot water or steam heating coils are included but piping to boiler and the boiler itself is not included.

Condenserless models are available for 15% less where remote refrigerant source is available.

System Components	QUANTITY	UNIT	COST EACH MAT.	COST EACH INST.	COST EACH TOTAL
SYSTEM D3050 165 1320					
SELF-CONTAINED, AIR COOLED UNIT					
APARTMENT CORRIDORS, 500 S.F., .92 TON					
Air cooled, package unit	1.000	Ea.	1,113.50	317.90	1,431.40
Ductwork package for water or air cooled packaged units	1.000	System	88.32	680.80	769.12
Refrigerant piping	1.000	System	203.32	427.80	631.12
Air cooled condenser, direct drive, propeller fan	1.000	Ea.	279	120.90	399.90
TOTAL			1,684.14	1,547.40	3,231.54
COST PER S.F.			3.37	3.09	6.46

D3050 165	Self-contained, Air Cooled Unit Systems		COST PER S.F. MAT.	COST PER S.F. INST.	COST PER S.F. TOTAL
1300	Self-contained, air cooled unit				
1320	Apartment corridors, 500 S.F., .92 ton		3.35	3.10	6.45
1480	10,000 S.F., 18.33 ton		2.51	2.51	5.02
1560	Banks or libraries, 500 S.F., 2.08 ton	RD3030 -010	7.10	3.89	10.99
1720	10,000 S.F., 41.66 ton		4.76	5.60	10.36
1800	Bars and taverns, 500 S.F., 5.54 ton		14.10	8.45	22.55
1960	10,000 S.F., 110.00 ton		12.05	10.70	22.75
2040	Bowling alleys, 500 S.F., 2.83 ton		9.75	5.30	15.05
2200	10,000 S.F., 56.66 ton		6.35	7.60	13.95
2240	Department stores, 500 S.F., 1.46 ton		5	2.73	7.73
2400	10,000 S.F., 29.17 ton		3.40	3.91	7.31
2480	Drug stores, 500 S.F., 3.33 ton		11.45	6.25	17.70
2640	10,000 S.F., 66.66 ton		7.55	8.95	16.50
2720	Factories, 500 S.F., 1.66 ton		5.80	3.15	8.95
2880	10,000 S.F., 33.33 ton		3.89	4.47	8.36
3200	Medical centers, 500 S.F., 1.17 ton		4.03	2.19	6.22
3360	10,000 S.F., 23.33 ton		3.20	3.20	6.40
3440	Offices, 500 S.F., 1.58 ton		5.45	2.97	8.42
3600	10,000 S.F., 31.66 ton		3.70	4.24	7.94
3680	Restaurants, 500 S.F., 2.50 ton		8.55	4.67	13.22
3840	10,000 S.F., 50.00 ton		5.80	6.65	12.45
3920	Schools and colleges, 500 S.F., 1.92 ton		6.60	3.59	10.19
4080	10,000 S.F., 38.33 ton		4.39	5.15	9.54

Important: See the Reference Section for critical supporting data - Reference Numbers and City Cost Indexes

D3050 Terminal & Package Units

Air Cooled Condensing Unit
Refrigerant Piping
Roof
Supply Duct
Fin. Ceiling
Return Grille
DX Air Handling Unit
Supply Diffuser

General: Split systems offer several important advantages which should be evaluated when a selection is to be made. They provide a greater degree of flexibility in component selection which permits an accurate match-up of the proper equipment size and type with the particular needs of the building. This allows for maximum use of modern energy saving concepts in heating and cooling. Outdoor installation of the air cooled condensing unit allows space savings in the building and also isolates the equipment operating sounds from building occupants.

Design Assumptions: The systems below are comprised of a direct expansion air handling unit and air cooled condensing unit with interconnecting copper tubing. Ducts and diffusers are also included for distribution of air. Systems are priced for cooling only. Heat can be added as desired either by putting hot water/steam coils into the air unit or into the duct supplying the particular area of need. Gas fired duct furnaces are also available. Refrigerant liquid line is insulated.

System Components	QUANTITY	UNIT	COST EACH		
			MAT.	INST.	TOTAL
SYSTEM D3050 170 1280					
SPLIT SYSTEM, AIR COOLED CONDENSING UNIT					
APARTMENT CORRIDORS, 1,000 S.F., 1.80 TON					
Fan coil AC unit, cabinet mntd & filters direct expansion air cool	1.000	Ea.	353.19	113.46	466.65
Ductwork package, for split system, remote condensing unit	1.000	System	82.81	649.65	732.46
Refrigeration piping	1.000	System	174.77	713.70	888.47
Condensing unit, air cooled, incls compressor & standard controls	1.000	Ea.	854	457.50	1,311.50
TOTAL			1,464.77	1,934.31	3,399.08
COST PER S.F.			1.46	1.93	3.39

D3050 170	Split Systems With Air Cooled Condensing Units		COST PER S.F.		
			MAT.	INST.	TOTAL
1260	Split system, air cooled condensing unit				
1280	Apartment corridors, 1,000 S.F., 1.83 ton		1.48	1.93	3.41
1440	20,000 S.F., 36.66 ton		1.87	2.72	4.59
1520	Banks and libraries, 1,000 S.F., 4.17 ton	RD3030 -010	2.74	4.43	7.17
1680	20,000 S.F., 83.32 ton		4.94	6.45	11.39
1760	Bars and taverns, 1,000 S.F., 11.08 ton		8.75	8.80	17.55
1880	10,000 S.F., 110.84 ton		11.95	10.20	22.15
2000	Bowling alleys, 1,000 S.F., 5.66 ton		3.93	8.35	12.28
2160	20,000 S.F., 113.32 ton		7.40	9.10	16.50
2320	Department stores, 1,000 S.F., 2.92 ton		1.95	3.05	5
2480	20,000 S.F., 58.33 ton		2.97	4.33	7.30
2560	Drug stores, 1,000 S.F., 6.66 ton		4.62	9.80	14.42
2720	20,000 S.F., 133.32 ton*				
2800	Factories, 1,000 S.F., 3.33 ton		2.22	3.48	5.70
2960	20,000 S.F., 66.66 ton		3.44	5.20	8.64
3040	Food supermarkets, 1,000 S.F., 2.83 ton		1.89	2.97	4.86
3200	20,000 S.F., 56.66 ton		2.89	4.21	7.10
3280	Medical centers, 1,000 S.F., 2.33 ton		1.57	2.40	3.97
3440	20,000 S.F., 46.66 ton		2.38	3.47	5.85
3520	Offices, 1,000 S.F., 3.17 ton		2.11	3.31	5.42
3680	20,000 S.F., 63.32 ton		3.26	4.92	8.18
3760	Restaurants, 1,000 S.F., 5.00 ton		3.47	7.35	10.82
3920	20,000 S.F., 100.00 ton		6.55	8.05	14.60
4000	Schools and colleges, 1,000 S.F., 3.83 ton		2.53	4.08	6.61
4160	20,000 S.F., 76.66 ton		3.97	5.95	9.92

SERVICES

D

D3050 Terminal & Package Units

Computer rooms impose special requirements on air conditioning systems. A prime requirement is reliability, due to the potential monetary loss that could be incurred by a system failure. A second basic requirement is the tolerance of control with which temperature and humidity are regulated, and dust eliminated. As the air conditioning system reliability is so vital, the additional cost of reserve capacity and redundant components is often justified.

System Descriptions: Computer areas may be environmentally controlled by one of three methods as follows:

1. Self-contained Units
These are units built to higher standards of performance and

reliability. They usuallycontain alarms and controls to indicate component operation failure, filter change, etc. It should be remembered that these units in the room will occupy space that is relatively expensive to build and that all alterations and service of the equipment will also have to be accomplished within the computer area.

2. Decentralized Air Handling Units In operation these are similar to the self-contained units except that their cooling capability comes from remotely located refrigeration equipment as refrigerant or chilled water. As no compressors or refrigerating equipment are required in the air units, they are smaller and require less service than

self-contained units. An added plus for this type of system occurs if some of the computer components themselves also require chilled water for cooling.

3. Central System Supply Cooling is obtained from a central source which, since it is not located within the computer room, may have excess capacity and permit greater flexibility without interfering with the computer components. System performance criteria must still be met.

Note: The costs shown below do not include an allowance for ductwork or piping.

D3050 185	Computer Room Cooling Units		COST EACH		
			MAT.	INST.	TOTAL
0560	Computer room unit, air cooled, includes remote condenser				
0580	3 ton		9,900	1,875	11,775
0600	5 ton		13,200	2,075	15,275
0620	8 ton	RD3030 -010	27,600	3,450	31,050
0640	10 ton		28,700	3,725	32,425
0660	15 ton		32,100	4,250	36,350
0680	20 ton		38,300	5,000	43,300
0700	23 ton		40,200	5,175	45,375
0800	Chilled water, for connection to existing chiller system				
0820	5 ton		11,200	1,250	12,450
0840	8 ton		16,000	1,875	17,875
0860	10 ton		17,700	1,900	19,600
0880	15 ton		19,000	1,950	20,950
0900	20 ton		19,400	2,025	21,425
0920	23 ton		20,400	2,225	22,625
1000	Glycol system, complete except for interconnecting tubing				
1020	3 ton		12,100	2,325	14,425
1040	5 ton		15,800	2,450	18,250
1060	8 ton		34,100	4,050	38,150
1080	10 ton		37,000	4,450	41,450
1100	15 ton		44,800	5,575	50,375
1120	20 ton		48,000	6,050	54,050
1140	23 ton		52,500	6,600	59,100
1240	Water cooled, not including condenser water supply or cooling tower				
1260	3 ton		8,925	1,500	10,425
1280	5 ton		12,600	1,725	14,325
1300	8 ton		26,000	2,825	28,825
1320	15 ton		31,800	3,450	35,250
1340	20 ton		37,500	3,825	41,325
1360	23 ton		39,700	4,275	43,975

Important: See the Reference Section for critical supporting data - Reference Numbers and City Cost Indexes

D4010 Sprinklers

Dry Pipe System: A system employing automatic sprinklers attached to a piping system containing air under pressure, the release of which as from the opening of sprinklers permits the water pressure to open a valve known as a "dry pipe valve". The water then flows into the piping system and out the opened sprinklers.

All areas are assumed to be open.

System Components	QUANTITY	UNIT	COST EACH		
			MAT.	INST.	TOTAL
SYSTEM D4010 310 0580					
DRY PIPE SPRINKLER, STEEL BLACK, SCH. 40 PIPE					
LIGHT HAZARD, ONE FLOOR, 2000 S.F.					
Valve, gate, iron body 125 lb, OS&Y, flanged, 4" pipe size	1.000	Ea.	341.25	232.50	573.75
Valve, swing check, bronze, 125 lb, regrinding disc, 2-1/2" pipe size	1.000	Ea.	199.50	46.13	245.63
Valve, angle, bronze, 150 lb, rising stem, threaded, 2" pipe size	1.000	Ea.	285	34.88	319.88
*Alarm valve, 2-1/2" pipe size	1.000	Ea.	701.25	228.75	930
Alarm, water motor, complete with gong,	1.000	Ea.	155.25	94.50	249.75
Fire alarm horn, electric	1.000	Ea.	30	56.25	86.25
Valve swing check w/balldrip CI with brass trim, 4" pipe size	1.000	Ea.	135	228.75	363.75
Pipe, steel, black, schedule 40, 4" diam	10.000	L.F.	157.50	202.88	360.38
Dry pipe valve, trim & gauges, 4" pipe size	1.000	Ea.	1,068.75	682.50	1,751.25
Pipe, steel, black, schedule 40, threaded, cplg & hngr 10'OC 2-1/2" diam	20.000	L.F.	181.50	277.50	459
Pipe, steel, black, schedule 40, threaded, cplg & hngr 10'OC 2" diam	12.500	L.F.	73.13	135.47	208.60
Pipe, steel, black, schedule 40, threaded, cplg & hngr 10'OC 1-1/4" diam	37.500	L.F.	140.63	292.50	433.13
Pipe, steel, black, schedule 40, threaded, cplg & hngr 10'OC 1" diam	112.000	L.F.	320.88	814.80	1,135.68
Pipe Tee, malleable iron black, 150 lb threaded, 4" pipe size	2.000	Ea.	214.50	346.50	561
Pipe Tee, malleable iron black, 150 lb threaded, 2-1/2" pipe size	2.000	Ea.	60.75	154.50	215.25
Pipe Tee, malleable iron black, 150 lb threaded, 2" pipe size	1.000	Ea.	14.03	63	77.03
Pipe Tee, malleable iron black, 150 lb threaded, 1-1/4" pipe size	5.000	Ea.	33	247.50	280.50
Pipe Tee, malleable iron black, 150 lb threaded, 1" pipe size	4.000	Ea.	16.35	193.50	209.85
Pipe 90° elbow malleable iron black, 150 lb threaded, 1" pipe size	6.000	Ea.	15.66	177.75	193.41
Sprinkler head dry 1/2" orifice 1" NPT, 3" to 4-3/4" length	12.000	Ea.	582	432	1,014
Air compressor, 200 Gal sprinkler system capacity, 1/3 HP	1.000	Ea.	615	292.50	907.50
*Standpipe connection, wall, flush, brs w/plug & chain 2-1/2"x2-1/2"	1.000	Ea.	83.25	136.50	219.75
Valve gate bronze, 300 psi, NRS, class 150, threaded, 1" pipe size	1.000	Ea.	25.13	20.25	45.38
TOTAL			5,449.31	5,391.41	10,840.72
COST PER S.F.			2.72	2.70	5.42

D4010 310	Dry Pipe Sprinkler Systems		COST PER S.F.		
			MAT.	INST.	TOTAL
0520	Dry pipe sprinkler systems, steel, black, sch. 40 pipe				
0530	Light hazard, one floor, 500 S.F.		5.30	4.57	9.87
0560	1000 S.F.	RD4010 -100	3.01	2.66	5.67
0580	2000 S.F.		2.73	2.70	5.43
0600	5000 S.F.	RD4020 -300	1.45	1.83	3.28
0620	10,000 S.F.		1.05	1.51	2.56
0640	50,000 S.F.		.87	1.35	2.22
0660	Each additional floor, 500 S.F.		1.21	2.22	3.43

D4010 310	Dry Pipe Sprinkler Systems	COST PER S.F.		
		MAT.	INST.	TOTAL
0680	1000 S.F.	.97	1.83	2.80
0700	2000 S.F.	.95	1.70	2.65
0720	5000 S.F.	.77	1.45	2.22
0740	10,000 S.F.	.73	1.34	2.07
0760	50,000 S.F.	.66	1.20	1.86
1000	Ordinary hazard, one floor, 500 S.F.	5.35	4.62	9.97
1020	1000 S.F.	3.05	2.69	5.74
1040	2000 S.F.	2.81	2.82	5.63
1060	5000 S.F.	1.65	1.96	3.61
1080	10,000 S.F.	1.36	1.99	3.35
1100	50,000 S.F.	1.34	1.98	3.32
1140	Each additional floor, 500 S.F.	1.29	2.27	3.56
1160	1000 S.F.	1.11	2.05	3.16
1180	2000 S.F.	1.11	1.87	2.98
1200	5000 S.F.	1.01	1.61	2.62
1220	10,000 S.F.	.92	1.57	2.49
1240	50,000 S.F.	.94	1.44	2.38
1500	Extra hazard, one floor, 500 S.F.	7.20	5.75	12.95
1520	1000 S.F.	4.28	4.15	8.43
1540	2000 S.F.	3.11	3.62	6.73
1560	5000 S.F.	1.94	2.73	4.67
1580	10,000 S.F.	2	2.60	4.60
1600	50,000 S.F.	2.22	2.53	4.75
1660	Each additional floor, 500 S.F.	1.64	2.82	4.46
1680	1000 S.F.	1.61	2.68	4.29
1700	2000 S.F.	1.51	2.69	4.20
1720	5000 S.F.	1.29	2.35	3.64
1740	10,000 S.F.	1.46	2.15	3.61
1760	50,000 S.F.	1.48	2.06	3.54
2020	Grooved steel, black, sch. 40 pipe, light hazard, one floor, 2000 S.F.	2.63	2.31	4.94
2060	10,000 S.F.	1.04	1.31	2.35
2100	Each additional floor, 2000 S.F.	.93	1.37	2.30
2150	10,000 S.F.	.73	1.14	1.87
2200	Ordinary hazard, one floor, 2000 S.F.	2.74	2.45	5.19
2250	10,000 S.F.	1.28	1.68	2.96
2300	Each additional floor, 2000 S.F.	1.04	1.51	2.55
2350	10,000 S.F.	.97	1.51	2.48
2400	Extra hazard, one floor, 2000 S.F.	3.09	3.09	6.18
2450	10,000 S.F.	1.75	2.17	3.92
2500	Each additional floor, 2000 S.F.	1.49	2.21	3.70
2550	10,000 S.F.	1.33	1.89	3.22
3050	Grooved steel black sch. 10 pipe, light hazard, one floor, 2000 S.F.	2.60	2.29	4.89
3100	10,000 S.F.	1.03	1.29	2.32
3150	Each additional floor, 2000 S.F.	.90	1.35	2.25
3200	10,000 S.F.	.71	1.12	1.83
3250	Ordinary hazard, one floor, 2000 S.F.	2.71	2.44	5.15
3300	10,000 S.F.	1.26	1.65	2.91
3350	Each additional floor, 2000 S.F.	1.01	1.49	2.50
3400	10,000 S.F.	.94	1.48	2.42
3450	Extra hazard, one floor, 2000 S.F.	3.06	3.07	6.13
3500	10,000 S.F.	1.68	2.14	3.82
3550	Each additional floor, 2000 S.F.	1.46	2.20	3.66
3600	10,000 S.F.	1.29	1.87	3.16
4050	Copper tubing, type M, light hazard, one floor, 2000 S.F.	2.67	2.29	4.96
4100	10,000 S.F.	1.11	1.31	2.42
4150	Each additional floor, 2000 S.F.	.98	1.37	2.35
4200	10,000 S.F.	.80	1.14	1.94
4250	Ordinary hazard, one floor, 2000 S.F.	2.83	2.55	5.38

Important: See the Reference Section for critical supporting data - Reference Numbers and City Cost Indexes

D4010 Sprinklers

D4010 310	Dry Pipe Sprinkler Systems	COST PER S.F.		
		MAT.	INST.	TOTAL
4300	10,000 S.F.	1.35	1.55	2.90
4350	Each additional floor, 2000 S.F.	1.31	1.58	2.89
4400	10,000 S.F.	1.01	1.35	2.36
4450	Extra hazard, one floor, 2000 S.F.	3.21	3.11	6.32
4500	10,000 S.F.	2.61	2.34	4.95
4550	Each additional floor, 2000 S.F.	1.62	2.23	3.85
4600	10,000 S.F.	1.79	2.04	3.83
5050	Copper tubing, type M, T-drill system, light hazard, one floor			
5060	2000 S.F.	2.67	2.14	4.81
5100	10,000 S.F.	1.06	1.09	2.15
5150	Each additional floor, 2000 S.F.	.99	1.22	2.21
5200	10,000 S.F.	.74	.92	1.66
5250	Ordinary hazard, one floor, 2000 S.F.	2.74	2.19	4.93
5300	10,000 S.F.	1.31	1.38	2.69
5350	Each additional floor, 2000 S.F.	1.04	1.24	2.28
5400	10,000 S.F.	.94	1.14	2.08
5450	Extra hazard, one floor, 2000 S.F.	2.97	2.58	5.55
5500	10,000 S.F.	2.20	1.72	3.92
5550	Each additional floor, 2000 S.F.	1.37	1.70	3.07
5600	10,000 S.F.	1.38	1.42	2.80

SERVICES

D

Pre-Action System: A system employing automatic sprinklers attached to a piping system containing air that may or may not be under pressure, with a supplemental heat responsive system of generally more sensitive characteristics than the automatic sprinklers themselves, installed in the same areas as the sprinklers. Actuation of the heat responsive system, as from a fire, opens a valve which permits water to flow into the sprinkler piping system and to be discharged from those sprinklers which were opened by heat from the fire.

All areas are assumed to be open.

System Components	QUANTITY	UNIT	COST EACH MAT.	COST EACH INST.	COST EACH TOTAL
SYSTEM D4010 350 0580					
PREACTION SPRINKLER SYSTEM, STEEL BLACK SCH. 40 PIPE					
LIGHT HAZARD, 1 FLOOR, 2000 S.F.					
Valve, gate, iron body 125 lb, OS&Y, flanged, 4″ pipe size	1.000	Ea.	341.25	232.50	573.75
*Valve, swing check w/ball drip CI with brass trim 4″ pipe size	1.000	Ea.	135	228.75	363.75
Valve, swing check, bronze, 125 lb, regrinding disc, 2-1/2″ pipe size	1.000	Ea.	199.50	46.13	245.63
Valve, angle, bronze, 150 lb, rising stem, threaded, 2″ pipe size	1.000	Ea.	285	34.88	319.88
*Alarm valve, 2-1/2″ pipe size	1.000	Ea.	701.25	228.75	930
Alarm, water motor, complete with gong	1.000	Ea.	155.25	94.50	249.75
Fire alarm horn, electric	1.000	Ea.	30	56.25	86.25
Thermostatic release for release line	2.000	Ea.	495	37.50	532.50
Pipe, steel, black, schedule 40, 4″ diam	10.000	L.F.	157.50	202.88	360.38
Dry pipe valve, trim & gauges, 4″ pipe size	1.000	Ea.	1,068.75	682.50	1,751.25
Pipe, steel, black, schedule 40, threaded, cplg & hngr 10′OC 2-1/2″ diam	20.000	L.F.	181.50	277.50	459
Pipe steel black, schedule 40, threaded, cplg & hngr 10′OC 2″ diam	12.500	L.F.	73.13	135.47	208.60
Pipe, steel, black, schedule 40, threaded, cplg & hngr 10′OC 1-1/4″ diam	37.500	L.F.	140.63	292.50	433.13
Pipe, steel, black, schedule 40, threaded, cplg & hngr 10′OC 1″ diam	112.000	L.F.	320.88	814.80	1,135.68
Pipe, Tee, malleable iron, black, 150 lb threaded, 4″ diam	2.000	Ea.	214.50	346.50	561
Pipe, Tee, malleable iron, black, 150 lb threaded, 2-1/2″ pipe size	2.000	Ea.	60.75	154.50	215.25
Pipe, Tee, malleable iron, black, 150 lb threaded, 2″ pipe size	1.000	Ea.	14.03	63	77.03
Pipe, Tee, malleable iron, black, 150 lb threaded, 1-1/4″ pipe size	5.000	Ea.	33	247.50	280.50
Pipe, Tee, malleable iron, black, 150 lb threaded, 1″ pipe size	4.000	Ea.	16.35	193.50	209.85
Pipe, 90° elbow, malleable iron, blk, 150 lb threaded, 1″ pipe size	6.000	Ea.	15.66	177.75	193.41
Sprinkler head, std spray, brass 135°-286°F 1/2″ NPT, 3/8″ orifice	12.000	Ea.	88.80	378	466.80
Air compressor auto complete 200 Gal sprinkler sys cap, 1/3 HP	1.000	Ea.	615	292.50	907.50
*Standpipe conn.,wall,flush,brass w/plug & chain 2-1/2″ x 2-1/2″	1.000	Ea.	83.25	136.50	219.75
Valve, gate, bronze, 300 psi, NRS, class 150, threaded, 1″ pipe size	1.000	Ea.	25.13	20.25	45.38
TOTAL			5,451.11	5,374.91	10,826.02
COST PER S.F.			2.73	2.69	5.42

*Not included in systems under 2000 S.F.

D4010 350	Preaction Sprinkler Systems		COST PER S.F. MAT.	COST PER S.F. INST.	COST PER S.F. TOTAL
0520	Preaction sprinkler systems, steel, black, sch. 40 pipe	RD4010 -100			
0530	Light hazard, one floor, 500 S.F.		5.30	3.65	8.95
0560	1000 S.F.	RD4020 -300	3.08	2.73	5.81
0580	2000 S.F.		2.73	2.69	5.42

Important: See the Reference Section for critical supporting data - Reference Numbers and City Cost Indexes

D4010 350	Preaction Sprinkler Systems	COST PER S.F.		
		MAT.	INST.	TOTAL
0600	5000 S.F.	1.46	1.82	3.28
0620	10,000 S.F.	1.05	1.51	2.56
0640	50,000 S.F.	.86	1.34	2.20
0660	Each additional floor, 500 S.F.	1.29	1.98	3.27
0680	1000 S.F.	1.03	1.82	2.85
0700	2000 S.F.	1.01	1.70	2.71
0720	5000 S.F.	.77	1.44	2.21
0740	10,000 S.F.	.74	1.33	2.07
0760	50,000 S.F.	.74	1.23	1.97
1000	Ordinary hazard, one floor, 500 S.F.	5.55	3.95	9.50
1020	1000 S.F.	3.05	2.68	5.73
1040	2000 S.F.	2.94	2.82	5.76
1060	5000 S.F.	1.60	1.95	3.55
1080	10,000 S.F.	1.29	1.97	3.26
1100	50,000 S.F.	1.25	1.95	3.20
1140	Each additional floor, 500 S.F.	1.53	2.29	3.82
1160	1000 S.F.	1.01	1.84	2.85
1180	2000 S.F.	.99	1.85	2.84
1200	5000 S.F.	1.07	1.73	2.80
1220	10,000 S.F.	.98	1.80	2.78
1240	50,000 S.F.	.98	1.68	2.66
1500	Extra hazard, one floor, 500 S.F.	7.25	5.05	12.30
1520	1000 S.F.	4.12	3.79	7.91
1540	2000 S.F.	2.99	3.59	6.58
1560	5000 S.F.	2.02	2.94	4.96
1580	10,000 S.F.	1.96	2.88	4.84
1600	50,000 S.F.	2.14	2.81	4.95
1660	Each additional floor, 500 S.F.	1.76	2.81	4.57
1680	1000 S.F.	1.49	2.66	4.15
1700	2000 S.F.	1.39	2.67	4.06
1720	5000 S.F.	1.19	2.36	3.55
1740	10,000 S.F.	1.29	2.17	3.46
1760	50,000 S.F.	1.25	2.03	3.28
2020	Grooved steel, black, sch. 40 pipe, light hazard, one floor, 2000 S.F.	2.69	2.31	5
2060	10,000 S.F.	1.04	1.31	2.35
2100	Each additional floor of 2000 S.F.	.99	1.36	2.35
2150	10,000 S.F.	.73	1.13	1.86
2200	Ordinary hazard, one floor, 2000 S.F.	2.74	2.45	5.19
2250	10,000 S.F.	1.21	1.67	2.88
2300	Each additional floor, 2000 S.F.	1.04	1.50	2.54
2350	10,000 S.F.	.90	1.50	2.40
2400	Extra hazard, one floor, 2000 S.F.	2.96	3.07	6.03
2450	10,000 S.F.	1.57	2.15	3.72
2500	Each additional floor, 2000 S.F.	1.37	2.19	3.56
2550	10,000 S.F.	1.13	1.87	3
3050	Grooved steel, black, sch. 10 pipe light hazard, one floor, 2000 S.F.	2.66	2.29	4.95
3100	10,000 S.F.	1.03	1.29	2.32
3150	Each additional floor, 2000 S.F.	.96	1.34	2.30
3200	10,000 S.F.	.71	1.11	1.82
3250	Ordinary hazard, one floor, 2000 S.F.	2.67	2.28	4.95
3300	10,000 S.F.	1.06	1.63	2.69
3350	Each additional floor, 2000 S.F.	1.01	1.48	2.49
3400	10,000 S.F.	.87	1.46	2.33
3450	Extra hazard, one floor, 2000 S.F.	2.94	3.05	5.99
3500	10,000 S.F.	1.49	2.11	3.60
3550	Each additional floor, 2000 S.F.	1.34	2.17	3.51
3600	10,000 S.F.	1.09	1.84	2.93
4050	Copper tubing, type M, light hazard, one floor, 2000 S.F.	2.73	2.29	5.02

D4010 Sprinklers

D4010 350	Preaction Sprinkler Systems	COST PER S.F.		
		MAT.	INST.	TOTAL
4100	10,000 S.F.	1.11	1.31	2.42
4150	Each additional floor, 2000 S.F.	1.04	1.37	2.41
4200	10,000 S.F.	.67	1.13	1.80
4250	Ordinary hazard, one floor, 2000 S.F.	2.83	2.54	5.37
4300	10,000 S.F.	1.28	1.53	2.81
4350	Each additional floor, 2000 S.F.	1.05	1.40	2.45
4400	10,000 S.F.	.85	1.23	2.08
4450	Extra hazard, one floor, 2000 S.F.	3.09	3.09	6.18
4500	10,000 S.F.	2.39	2.30	4.69
4550	Each additional floor, 2000 S.F.	1.49	2.21	3.70
4600	10,000 S.F.	1.59	2.01	3.60
5050	Copper tubing, type M, T-drill system, light hazard, one floor			
5060	2000 S.F.	2.74	2.14	4.88
5100	10,000 S.F.	1.06	1.09	2.15
5150	Each additional floor, 2000 S.F.	1.05	1.22	2.27
5200	10,000 S.F.	.74	.92	1.66
5250	Ordinary hazard, one floor, 2000 S.F.	2.74	2.18	4.92
5300	10,000 S.F.	1.23	1.36	2.59
5350	Each additional floor, 2000 S.F.	1.05	1.24	2.29
5400	10,000 S.F.	.92	1.19	2.11
5450	Extra hazard, one floor, 2000 S.F.	2.84	2.56	5.40
5500	10,000 S.F.	1.98	1.69	3.67
5550	Each additional floor, 2000 S.F.	1.25	1.68	2.93
5600	10,000 S.F.	1.18	1.40	2.58

Important: See the Reference Section for critical supporting data - Reference Numbers and City Cost Indexes

D4010 Sprinklers

Deluge System: A system employing open sprinklers attached to a piping system connected to a water supply through a valve which is opened by the operation of a heat responsive system installed in the same areas as the sprinklers. When this valve opens, water flows into the piping system and discharges from all sprinklers attached thereto.

All areas are assumed to be open.

SERVICES **D**

System Components	QUANTITY	UNIT	COST EACH		
			MAT.	INST.	TOTAL
SYSTEM D4010 370 0580					
DELUGE SPRINKLER SYSTEM, STEEL BLACK SCH. 40 PIPE					
LIGHT HAZARD, 1 FLOOR, 2000 S.F.					
Valve, gate, iron body 125 lb, OS&Y, flanged, 4" pipe size	1.000	Ea.	341.25	232.50	573.75
Valve, swing check w/ball drip, CI w/brass ftngs, 4" pipe size	1.000	Ea.	135	228.75	363.75
Valve, swing check, bronze, 125 lb, regrinding disc, 2-1/2" pipe size	1.000	Ea.	199.50	46.13	245.63
Valve, angle, bronze, 150 lb, rising stem, threaded, 2" pipe size	1.000	Ea.	285	34.88	319.88
*Alarm valve, 2-1/2" pipe size	1.000	Ea.	701.25	228.75	930
Alarm, water motor, complete with gong	1.000	Ea.	155.25	94.50	249.75
Fire alarm horn, electric	1.000	Ea.	30	56.25	86.25
Thermostatic release for release line	2.000	Ea.	495	37.50	532.50
Pipe, steel, black, schedule 40, 4" diam	10.000	L.F.	157.50	202.88	360.38
Deluge valve trim, pressure relief, emergency release, gauge, 4" pipe size	1.000	Ea.	1,218.75	682.50	1,901.25
Deluge system pressured monitoring panel, 120V	1.000	Ea.	727.50	21	748.50
Pipe, steel, black, schedule 40, threaded, cplg & hngr 10'OC 2-1/2" diam	20.000	L.F.	181.50	277.50	459
Pipe, steel, black, schedule 40, threaded, cplg & hngr 10'OC 2" diam	12.500	L.F.	73.13	135.47	208.60
Pipe, steel, black, schedule 40, threaded, cplg & hngr 10'OC 1-1/4" diam	37.500	L.F.	140.63	292.50	433.13
Pipe, steel, black, schedule 40, threaded, cplg & hngr 10'OC 1" diam	112.000	L.F.	320.88	814.80	1,135.68
Pipe, Tee, malleable iron, black, 150 lb threaded, 4" pipe size	2.000	Ea.	214.50	346.50	561
Pipe, Tee, malleable iron, black, 150 lb threaded, 2-1/2" pipe size	2.000	Ea.	60.75	154.50	215.25
Pipe, Tee, malleable iron, black, 150 lb threaded, 2" pipe size	1.000	Ea.	14.03	63	77.03
Pipe, Tee, malleable iron, black, 150 lb threaded, 1-1/4" pipe size	5.000	Ea.	33	247.50	280.50
Pipe, Tee, malleable iron, black, 150 lb threaded, 1" pipe size	4.000	Ea.	16.35	193.50	209.85
Pipe, 90° elbow, malleable iron, black, 150 lb threaded 1" pipe size	6.000	Ea.	15.66	177.75	193.41
Sprinkler head, std spray, brass 135°-286°F 1/2" NPT, 3/8" orifice	9.720	Ea.	88.80	378	466.80
Air compressor, auto, complete, 200 Gal sprinkler sys cap, 1/3 HP	1.000	Ea.	615	292.50	907.50
*Standpipe connection, wall, flush w/plug & chain 2-1/2"x2-1/2"	1.000	Ea.	83.25	136.50	219.75
Valve, gate, bronze, 300 psi, NRS, class 150, threaded, 1" pipe size	1.000	Ea.	25.13	20.25	45.38
TOTAL			6,328.61	5,395.91	11,724.52
COST PER S.F.			3.16	2.70	5.86
*Not included in systems under 2000 S.F.					

D4010 370	Deluge Sprinkler Systems		COST PER S.F.		
			MAT.	INST.	TOTAL
0520	Deluge sprinkler systems, steel, black, sch. 40 pipe				
0530	Light hazard, one floor, 500 S.F.	RD4010 -100	6.40	3.69	10.09

D4010 Sprinklers

D4010 370	Deluge Sprinkler Systems	MAT.	INST.	TOTAL
		COST PER S.F.		
0560	1000 S.F.	3.60	2.62	6.22
0580	2000 S.F.	3.16	2.70	5.86
0600	5000 S.F.	1.63	1.83	3.46
0620	10,000 S.F.	1.14	1.51	2.65
0640	50,000 S.F.	.88	1.34	2.22
0660	Each additional floor, 500 S.F.	1.29	1.98	3.27
0680	1000 S.F.	1.03	1.82	2.85
0700	2000 S.F.	1.01	1.70	2.71
0720	5000 S.F.	.77	1.44	2.21
0740	10,000 S.F.	.74	1.33	2.07
0760	50,000 S.F.	.74	1.23	1.97
1000	Ordinary hazard, one floor, 500 S.F.	6.95	4.22	11.17
1020	1000 S.F.	3.59	2.70	6.29
1040	2000 S.F.	3.37	2.83	6.20
1060	5000 S.F.	1.78	1.95	3.73
1080	10,000 S.F.	1.38	1.97	3.35
1100	50,000 S.F.	1.31	1.98	3.29
1140	Each additional floor, 500 S.F.	1.53	2.29	3.82
1160	1000 S.F.	1.01	1.84	2.85
1180	2000 S.F.	.99	1.85	2.84
1200	5000 S.F.	.96	1.60	2.56
1220	10,000 S.F.	.91	1.59	2.50
1240	50,000 S.F.	.92	1.55	2.47
1500	Extra hazard, one floor, 500 S.F.	8.35	5.10	13.45
1520	1000 S.F.	4.87	3.93	8.80
1540	2000 S.F.	3.43	3.60	7.03
1560	5000 S.F.	1.97	2.71	4.68
1580	10,000 S.F.	1.93	2.64	4.57
1600	50,000 S.F.	2.10	2.58	4.68
1660	Each additional floor, 500 S.F.	1.76	2.81	4.57
1680	1000 S.F.	1.49	2.66	4.15
1700	2000 S.F.	1.39	2.67	4.06
1720	5000 S.F.	1.19	2.36	3.55
1740	10,000 S.F.	1.33	2.25	3.58
1760	50,000 S.F.	1.33	2.17	3.50
2000	Grooved steel, black, sch. 40 pipe, light hazard, one floor			
2020	2000 S.F.	3.13	2.32	5.45
2060	10,000 S.F.	1.14	1.32	2.46
2100	Each additional floor, 2,000 S.F.	.99	1.36	2.35
2150	10,000 S.F.	.73	1.13	1.86
2200	Ordinary hazard, one floor, 2000 S.F.	2.74	2.45	5.19
2250	10,000 S.F.	1.30	1.67	2.97
2300	Each additional floor, 2000 S.F.	1.04	1.50	2.54
2350	10,000 S.F.	.90	1.50	2.40
2400	Extra hazard, one floor, 2000 S.F.	3.40	3.08	6.48
2450	10,000 S.F.	1.67	2.16	3.83
2500	Each additional floor, 2000 S.F.	1.37	2.19	3.56
2550	10,000 S.F.	1.13	1.87	3
3000	Grooved steel, black, sch. 10 pipe, light hazard, one floor			
3050	2000 S.F.	2.78	2.20	4.98
3100	10,000 S.F.	1.12	1.29	2.41
3150	Each additional floor, 2000 S.F.	.96	1.34	2.30
3200	10,000 S.F.	.71	1.11	1.82
3250	Ordinary hazard, one floor, 2000 S.F.	3.14	2.44	5.58
3300	10,000 S.F.	1.15	1.63	2.78
3350	Each additional floor, 2000 S.F.	1.01	1.48	2.49
3400	10,000 S.F.	.87	1.46	2.33
3450	Extra hazard, one floor, 2000 S.F.	3.37	3.06	6.43

RD4020 -300

Important: See the Reference Section for critical supporting data - Reference Numbers and City Cost Indexes

D SERVICES

D4010 Sprinklers

D4010 370	Deluge Sprinkler Systems	COST PER S.F.		
		MAT.	INST.	TOTAL
3500	10,000 S.F.	1.57	2.11	3.68
3550	Each additional floor, 2000 S.F.	1.34	2.17	3.51
3600	10,000 S.F.	1.09	1.84	2.93
4000	Copper tubing, type M, light hazard, one floor			
4050	2000 S.F.	3.17	2.30	5.47
4100	10,000 S.F.	1.20	1.31	2.51
4150	Each additional floor, 2000 S.F.	1.04	1.37	2.41
4200	10,000 S.F.	.67	1.13	1.80
4250	Ordinary hazard, one floor, 2000 S.F.	3.27	2.55	5.82
4300	10,000 S.F.	1.37	1.53	2.90
4350	Each additional floor, 2000 S.F.	1.05	1.40	2.45
4400	10,000 S.F.	.85	1.23	2.08
4450	Extra hazard, one floor, 2000 S.F.	3.53	3.10	6.63
4500	10,000 S.F.	2.51	2.32	4.83
4550	Each additional floor, 2000 S.F.	1.49	2.21	3.70
4600	10,000 S.F.	1.59	2.01	3.60
5000	Copper tubing, type M, T-drill system, light hazard, one floor			
5050	2000 S.F.	3.18	2.15	5.33
5100	10,000 S.F.	1.15	1.09	2.24
5150	Each additional floor, 2000 S.F.	1.06	1.24	2.30
5200	10,000 S.F.	.74	.92	1.66
5250	Ordinary hazard, one floor, 2000 S.F.	3.18	2.19	5.37
5300	10,000 S.F.	1.32	1.37	2.69
5350	Each additional floor, 2000 S.F.	1.04	1.23	2.27
5400	10,000 S.F.	.92	1.19	2.11
5450	Extra hazard, one floor, 2000 S.F.	3.28	2.57	5.85
5500	10,000 S.F.	2.07	1.69	3.76
5550	Each additional floor, 2000 S.F.	1.25	1.68	2.93
5600	10,000 S.F.	1.18	1.40	2.58

Firecycle is a fixed fire protection sprinkler system utilizing water as its extinguishing agent. It is a time delayed, recycling, preaction type which automatically shuts the water off when heat is reduced below the detector operating temperature and turns the water back on when that temperature is exceeded.

The system senses a fire condition through a closed circuit electrical detector system which controls water flow to the fire automatically. Batteries supply up to 90 hour emergency power supply for system operation. The piping system is dry (until water is required) and is monitored with pressurized air. Should

any leak in the system piping occur, an alarm will sound, but water will not enter the system until heat is sensed by a Firecycle detector.

All areas are assumed to be open.

System Components

System Components	QUANTITY	UNIT	MAT.	INST.	TOTAL
SYSTEM D4010 390 0580					
FIRECYCLE SPRINKLER SYSTEM, STEEL BLACK SCH. 40 PIPE					
LIGHT HAZARD, ONE FLOOR, 2000 S.F.					
Valve, gate, iron body 125 lb, OS&Y, flanged, 4" pipe size	1.000	Ea.	341.25	232.50	573.75
Valve, angle, bronze, 150 lb, rising stem, threaded, 2" pipe size	1.000	Ea.	285	34.88	319.88
Valve, swing check, bronze, 125 lb, regrinding disc, 2-1/2" pipe size	1.000	Ea.	199.50	46.13	245.63
*Alarm valve, 2-1/2" pipe size	1.000	Ea.	701.25	228.75	930
Alarm, water motor, complete with gong	1.000	Ea.	155.25	94.50	249.75
Pipe, steel, black, schedule 40, 4" diam	10.000	L.F.	157.50	202.88	360.38
Fire alarm, horn, electric	1.000	Ea.	30	56.25	86.25
Pipe, steel, black, schedule 40, threaded, cplg & hngr 10'OC 2-1/2" diam	20.000	L.F.	181.50	277.50	459
Pipe, steel, black, schedule 40, threaded, cplg & hngr 10'OC 2" diam	12.500	L.F.	73.13	135.47	208.60
Pipe, steel, black, schedule 40, threaded, cplg & hngr 10'OC 1-1/4" diam	37.500	L.F.	140.63	292.50	433.13
Pipe, steel, black, schedule 40, threaded, cplg & hngr 10'OC 1" diam	112.000	L.F.	320.88	814.80	1,135.68
Pipe, Tee, malleable iron, black, 150 lb threaded, 4" pipe size	2.000	Ea.	214.50	346.50	561
Pipe, Tee, malleable iron, black, 150 lb threaded, 2-1/2" pipe size	2.000	Ea.	60.75	154.50	215.25
Pipe, Tee, malleable iron, black, 150 lb threaded, 2" pipe size	1.000	Ea.	14.03	63	77.03
Pipe, Tee, malleable iron, black, 150 lb threaded, 1-1/4" pipe size	5.000	Ea.	33	247.50	280.50
Pipe, Tee, malleable iron, black, 150 lb threaded, 1" pipe size	4.000	Ea.	16.35	193.50	209.85
Pipe, 90° elbow, malleable iron, black, 150 lb threaded, 1" pipe size	6.000	Ea.	15.66	177.75	193.41
Sprinkler head std spray, brass 135°-286°F 1/2" NPT, 3/8" orifice	12.000	Ea.	88.80	378	466.80
Firecycle controls, incls panel, battery, solenoid valves, press switches	1.000	Ea.	7,256.25	1,443.75	8,700
Detector, firecycle system	2.000	Ea.	532.50	47.25	579.75
Firecycle pkg, swing check & flow control valves w/trim 4" pipe size	1.000	Ea.	1,893.75	682.50	2,576.25
Air compressor, auto, complete, 200 Gal sprinkler sys cap, 1/3 HP	1.000	Ea.	615	292.50	907.50
*Standpipe connection, wall, flush, brass w/plug & chain 2-1/2"x2-1/2"	1.000	Ea.	83.25	136.50	219.75
Valve, gate, bronze, 300 psi, NRS, class 150, threaded, 1" diam	1.000	Ea.	25.13	20.25	45.38
TOTAL			13,434.86	6,599.66	20,034.52
COST PER S.F.			6.72	3.30	10.02

*Not included in systems under 2000 S.F.

Important: See the Reference Section for critical supporting data - Reference Numbers and City Cost Indexes

D4010 Sprinklers

D4010 390	Firecycle Sprinkler Systems	COST PER S.F.		
		MAT.	INST.	TOTAL
0520	Firecycle sprinkler systems, steel black sch. 40 pipe			
0530	Light hazard, one floor, 500 S.F. RD4010 -100	21.50	7.15	28.65
0560	1000 S.F. RD4020 -300	11.10	4.48	15.58
0580	2000 S.F.	6.70	3.30	10
0600	5000 S.F.	3.06	2.07	5.13
0620	10,000 S.F.	1.89	1.63	3.52
0640	50,000 S.F.	1.05	1.37	2.42
0660	Each additional floor of 500 S.F.	1.33	1.99	3.32
0680	1000 S.F.	1.05	1.83	2.88
0700	2000 S.F.	.90	1.69	2.59
0720	5000 S.F.	.79	1.45	2.24
0740	10,000 S.F.	.78	1.34	2.12
0760	50,000 S.F.	.76	1.23	1.99
1000	Ordinary hazard, one floor, 500 S.F.	21.50	7.45	28.95
1020	1000 S.F.	11.05	4.44	15.49
1040	2000 S.F.	6.80	3.42	10.22
1060	5000 S.F.	3.20	2.19	5.39
1080	10,000 S.F.	2.12	2.10	4.22
1100	50,000 S.F.	1.63	2.20	3.83
1140	Each additional floor, 500 S.F.	1.57	2.29	3.86
1160	1000 S.F.	1.03	1.85	2.88
1180	2000 S.F.	1.07	1.70	2.77
1200	5000 S.F.	.98	1.60	2.58
1220	10,000 S.F.	.88	1.56	2.44
1240	50,000 S.F.	.92	1.48	2.40
1500	Extra hazard, one floor, 500 S.F.	23.50	8.60	32.10
1520	1000 S.F.	12.05	5.55	17.60
1540	2000 S.F.	7	4.21	11.21
1560	5000 S.F.	3.40	2.95	6.35
1580	10,000 S.F.	2.76	2.99	5.75
1600	50,000 S.F.	2.61	3.32	5.93
1660	Each additional floor, 500 S.F.	1.80	2.82	4.62
1680	1000 S.F.	1.50	2.66	4.16
1700	2000 S.F.	1.41	2.67	4.08
1720	5000 S.F.	1.20	2.36	3.56
1740	10,000 S.F.	1.33	2.17	3.50
1760	50,000 S.F.	1.33	2.10	3.43
2020	Grooved steel, black, sch. 40 pipe, light hazard, one floor			
2030	2000 S.F.	6.70	2.92	9.62
2060	10,000 S.F.	2.02	1.99	4.01
2100	Each additional floor, 2000 S.F.	1.01	1.37	2.38
2150	10,000 S.F.	.77	1.14	1.91
2200	Ordinary hazard, one floor, 2000 S.F.	6.75	3.06	9.81
2250	10,000 S.F.	2.17	1.90	4.07
2300	Each additional floor, 2000 S.F.	1.06	1.50	2.56
2350	10,000 S.F.	.94	1.50	2.44
2400	Extra hazard, one floor, 2000 S.F.	6.95	3.68	10.63
2450	10,000 S.F.	2.37	2.26	4.63
2500	Each additional floor, 2000 S.F.	1.39	2.19	3.58
2550	10,000 S.F.	1.17	1.87	3.04
3050	Grooved steel, black, sch. 10 pipe light hazard, one floor,			
3060	2000 S.F.	6.65	2.91	9.56
3100	10,000 S.F.	1.86	1.41	3.27
3150	Each additional floor, 2000 S.F.	.98	1.35	2.33
3200	10,000 S.F.	.75	1.12	1.87
3250	Ordinary hazard, one floor, 2000 S.F.	6.70	3.04	9.74

SERVICES D

D4010 Sprinklers

D4010 390	Firecycle Sprinkler Systems	COST PER S.F.		
		MAT.	INST.	TOTAL
3300	10,000 S.F.	2.02	1.76	3.78
3350	Each additional floor, 2000 S.F.	1.02	1.49	2.51
3400	10,000 S.F.	.91	1.47	2.38
3450	Extra hazard, one floor, 2000 S.F.	6.95	3.67	10.62
3500	10,000 S.F.	2.31	2.22	4.53
3550	Each additional floor, 2000 S.F.	1.36	2.18	3.54
3600	10,000 S.F.	1.13	1.85	2.98
4060	Copper tubing, type M, light hazard, one floor, 2000 S.F.	6.70	2.90	9.60
4100	10,000 S.F.	1.95	1.43	3.38
4150	Each additional floor, 2000 S.F.	1.06	1.37	2.43
4200	10,000 S.F.	.84	1.14	1.98
4250	Ordinary hazard, one floor, 2000 S.F.	6.80	3.16	9.96
4300	10,000 S.F.	2.11	1.66	3.77
4350	Each additional floor, 2000 S.F.	1.07	1.40	2.47
4400	10,000 S.F.	.88	1.21	2.09
4450	Extra hazard, one floor, 2000 S.F.	7.10	3.70	10.80
4500	10,000 S.F.	3.28	2.45	5.73
4550	Each additional floor, 2000 S.F.	1.51	2.22	3.73
4600	10,000 S.F.	1.63	2.02	3.65
5060	Copper tubing, type M, T-drill system, light hazard, one floor 2000 S.F.	6.75	2.75	9.50
5100	10,000 S.F.	1.89	1.21	3.10
5150	Each additional floor, 2000 S.F.	1.18	1.30	2.48
5200	10,000 S.F.	.78	.92	1.70
5250	Ordinary hazard, one floor, 2000 S.F.	6.75	2.79	9.54
5300	10,000 S.F.	2.07	1.49	3.56
5350	Each additional floor, 2000 S.F.	1.06	1.24	2.30
5400	10,000 S.F.	.96	1.20	2.16
5450	Extra hazard, one floor, 2000 S.F.	6.85	3.17	10.02
5500	10,000 S.F.	2.80	1.80	4.60
5550	Each additional floor, 2000 S.F.	1.27	1.69	2.96
5600	10,000 S.F.	1.22	1.40	2.62

Important: See the Reference Section for critical supporting data - Reference Numbers and City Cost Indexes

Wet Pipe System. A system employing automatic sprinklers attached to a piping system containing water and connected to a water supply so that water discharges immediately from sprinklers opened by heat from a fire.

All areas are assumed to be open.

System Components	QUANTITY	UNIT	COST EACH		
			MAT.	INST.	TOTAL
SYSTEM D4010 410 0580					
WET PIPE SPRINKLER, STEEL, BLACK, SCH. 40 PIPE					
LIGHT HAZARD, ONE FLOOR, 2000 S.F.					
Valve, gate, iron body, 125 lb, OS&Y, flanged, 4" diam	1.000	Ea.	341.25	232.50	573.75
Valve, swing check, bronze, 125 lb, regrinding disc, 2-1/2" pipe size	1.000	Ea.	199.50	46.13	245.63
Valve, angle, bronze, 150 lb, rising stem, threaded, 2" diam	1.000	Ea.	285	34.88	319.88
*Alarm valve, 2-1/2" pipe size	1.000	Ea.	701.25	228.75	930
Alarm, water motor, complete with gong	1.000	Ea.	155.25	94.50	249.75
Valve, swing check, w/balldrip CI with brass trim 4" pipe size	1.000	Ea.	135	228.75	363.75
Pipe, steel, black, schedule 40, 4" diam	10.000	L.F.	157.50	202.88	360.38
*Flow control valve, trim & gauges, 4" pipe size	1.000	Set	1,350	513.75	1,863.75
Fire alarm horn, electric	1.000	Ea.	30	56.25	86.25
Pipe, steel, black, schedule 40, threaded, cplg & hngr 10'OC, 2-1/2" diam	20.000	L.F.	181.50	277.50	459
Pipe, steel, black, schedule 40, threaded, cplg & hngr 10'OC, 2" diam	12.500	L.F.	73.13	135.47	208.60
Pipe, steel, black, schedule 40, threaded, cplg & hngr 10'OC, 1-1/4" diam	37.500	L.F.	140.63	292.50	433.13
Pipe steel, black, schedule 40, threaded cplg & hngr 10'OC, 1" diam	112.000	L.F.	320.88	814.80	1,135.68
Pipe Tee, malleable iron black, 150 lb threaded, 4" pipe size	2.000	Ea.	214.50	346.50	561
Pipe Tee, malleable iron black, 150 lb threaded, 2-1/2" pipe size	2.000	Ea.	60.75	154.50	215.25
Pipe Tee, malleable iron black, 150 lb threaded, 2" pipe size	1.000	Ea.	14.03	63	77.03
Pipe Tee, malleable iron black, 150 lb threaded, 1-1/4" pipe size	5.000	Ea.	33	247.50	280.50
Pipe Tee, malleable iron black, 150 lb threaded, 1" pipe size	4.000	Ea.	16.35	193.50	209.85
Pipe 90° elbow, malleable iron black, 150 lb threaded, 1" pipe size	6.000	Ea.	15.66	177.75	193.41
Sprinkler head, standard spray, brass 135°-286°F 1/2" NPT, 3/8" orifice	12.000	Ea.	88.80	378	466.80
Valve, gate, bronze, NRS, class 150, threaded, 1" pipe size	1.000	Ea.	25.13	20.25	45.38
*Standpipe connection, wall, single, flush w/plug & chain 2-1/2"x2-1/2"	1.000	Ea.	83.25	136.50	219.75
TOTAL			4,622.36	4,876.16	9,498.52
COST PER S.F.			2.31	2.44	4.75

*Not included in systems under 2000 S.F.

D4010 410	Wet Pipe Sprinkler Systems		COST PER S.F.		
			MAT.	INST.	TOTAL
0520	Wet pipe sprinkler systems, steel, black, sch. 40 pipe				
0530	Light hazard, one floor, 500 S.F.		1.51	2.34	3.85
0560	1000 S.F.	RD4010 -100	2.15	2.42	4.57
0580	2000 S.F.		2.31	2.44	4.75
0600	5000 S.F.	RD4020 -300	1.19	1.72	2.91
0620	10,000 S.F.		.85	1.45	2.30

D4010 Sprinklers

D4010 410	Wet Pipe Sprinkler Systems	COST PER S.F.		
		MAT.	INST.	TOTAL
0640	50,000 S.F.	.70	1.32	2.02
0660	Each additional floor, 500 S.F.	.81	1.99	2.80
0680	1000 S.F.	.79	1.85	2.64
0700	2000 S.F.	.77	1.68	2.45
0720	5000 S.F.	.58	1.43	2.01
0740	10,000 S.F.	.56	1.32	1.88
0760	50,000 S.F.	.47	1.04	1.51
1000	Ordinary hazard, one floor, 500 S.F.	1.67	2.51	4.18
1020	1000 S.F.	2.11	2.37	4.48
1040	2000 S.F.	2.40	2.56	4.96
1060	5000 S.F.	1.34	1.84	3.18
1080	10,000 S.F.	1.08	1.91	2.99
1100	50,000 S.F.	1.07	1.90	2.97
1140	Each additional floor, 500 S.F.	1.04	2.25	3.29
1160	1000 S.F.	.76	1.82	2.58
1180	2000 S.F.	.86	1.84	2.70
1200	5000 S.F.	.89	1.75	2.64
1220	10,000 S.F.	.80	1.79	2.59
1240	50,000 S.F.	.82	1.67	2.49
1500	Extra hazard, one floor, 500 S.F.	3.97	3.87	7.84
1520	1000 S.F.	2.65	3.36	6.01
1540	2000 S.F.	2.59	3.44	6.03
1560	5000 S.F.	1.91	3	4.91
1580	10,000 S.F.	1.75	2.85	4.60
1600	50,000 S.F.	1.94	2.77	4.71
1660	Each additional floor, 500 S.F.	1.27	2.78	4.05
1680	1000 S.F.	1.24	2.64	3.88
1700	2000 S.F.	1.14	2.65	3.79
1720	5000 S.F.	.99	2.34	3.33
1740	10,000 S.F.	1.12	2.16	3.28
1760	50,000 S.F.	1.10	2.06	3.16
2020	Grooved steel, black sch. 40 pipe, light hazard, one floor, 2000 S.F.	2.28	2.06	4.34
2060	10,000 S.F.	.97	1.30	2.27
2100	Each additional floor, 2000 S.F.	.75	1.35	2.10
2150	10,000 S.F.	.55	1.12	1.67
2200	Ordinary hazard, one floor, 2000 S.F.	2.33	2.20	4.53
2250	10,000 S.F.	1	1.61	2.61
2300	Each additional floor, 2000 S.F.	.79	1.48	2.27
2350	10,000 S.F.	.72	1.49	2.21
2400	Extra hazard, one floor, 2000 S.F.	2.55	2.82	5.37
2450	10,000 S.F.	1.34	2.09	3.43
2500	Each additional floor, 2000 S.F.	1.12	2.17	3.29
2550	10,000 S.F.	.95	1.85	2.80
3050	Grooved steel black sch. 10 pipe, light hazard, one floor, 2000 S.F.	2.25	2.04	4.29
3100	10,000 S.F.	.82	1.23	2.05
3150	Each additional floor, 2000 S.F.	.71	1.33	2.04
3200	10,000 S.F.	.54	1.10	1.64
3250	Ordinary hazard, one floor, 2000 S.F.	2.29	2.18	4.47
3300	10,000 S.F.	.98	1.58	2.56
3350	Each additional floor, 2000 S.F.	.76	1.46	2.22
3400	10,000 S.F.	.70	1.45	2.15
3450	Extra hazard, one floor, 2000 S.F.	2.52	2.80	5.32
3500	10,000 S.F.	1.28	2.05	3.33
3550	Each additional floor, 2000 S.F.	1.09	2.16	3.25
3600	10,000 S.F.	.91	1.83	2.74
4050	Copper tubing, type M, light hazard, one floor, 2000 S.F.	2.31	2.04	4.35
4100	10,000 S.F.	.91	1.25	2.16
4150	Each additional floor, 2000 S.F.	.80	1.35	2.15

Important: See the Reference Section for critical supporting data - Reference Numbers and City Cost Indexes

SERVICES

D

D4010 Sprinklers

D4010 410	Wet Pipe Sprinkler Systems	COST PER S.F.		
		MAT.	INST.	TOTAL
4200	10,000 S.F.	.63	1.12	1.75
4250	Ordinary hazard, one floor, 2000 S.F.	2.42	2.30	4.72
4300	10,000 S.F.	1.07	1.47	2.54
4350	Each additional floor, 2000 S.F.	.92	1.50	2.42
4400	10,000 S.F.	.76	1.32	2.08
4450	Extra hazard, one floor, 2000 S.F.	2.67	2.84	5.51
4500	10,000 S.F.	2.18	2.24	4.42
4550	Each additional floor, 2000 S.F.	1.25	2.19	3.44
4600	10,000 S.F.	1.41	2	3.41
5050	Copper tubing, type M, T-drill system, light hazard, one floor			
5060	2000 S.F.	2.32	1.89	4.21
5100	10,000 S.F.	.85	1.03	1.88
5150	Each additional floor, 2000 S.F.	.80	1.20	2
5200	10,000 S.F.	.57	.90	1.47
5250	Ordinary hazard, one floor, 2000 S.F.	2.33	1.93	4.26
5300	10,000 S.F.	1.03	1.30	2.33
5350	Each additional floor, 2000 S.F.	.80	1.21	2.01
5400	10,000 S.F.	.75	1.18	1.93
5450	Extra hazard, one floor, 2000 S.F.	2.43	2.31	4.74
5500	10,000 S.F.	1.77	1.63	3.40
5550	Each additional floor, 2000 S.F.	1.06	1.70	2.76
5600	10,000 S.F.	1.01	1.38	2.39

SERVICES **D**

D4020 Standpipes

Roof → Roof connections with hose gate valves (for combustible roof)

Hose connections on each floor (size based on class of service)

Check Valve →

Siamese inlet connections (for fire department use)

System Components	QUANTITY	UNIT	COST PER FLOOR		
			MAT.	INST.	TOTAL
SYSTEM D4020 310 0560					
WET STANDPIPE RISER, CLASS I, STEEL, BLACK, SCH. 40 PIPE, 10' HEIGHT					
4" DIAMETER PIPE, ONE FLOOR					
Pipe, steel, black, schedule 40, threaded, 4" diam	20.000	L.F.	460	510	970
Pipe, Tee, malleable iron, black, 150 lb threaded, 4" pipe size	2.000	Ea.	286	462	748
Pipe, 90° elbow, malleable iron, black, 150 lb threaded 4" pipe size	1.000	Ea.	91	154	245
Pipe, nipple, steel, black, schedule 40, 2-1/2" pipe size x 3" long	2.000	Ea.	19.50	116	135.50
Fire valve, gate, 300 lb, brass w/handwheel, 2-1/2" pipe size	1.000	Ea.	163	72	235
Fire valve, pressure restricting, adj, rgh brs, 2-1/2" pipe size	1.000	Ea.	218	144	362
Valve, swing check, w/ball drip, CI w/brs ftngs, 4" pipe size	1.000	Ea.	180	305	485
Standpipe conn wall dble flush brs w/plugs & chains 2-1/2"x2-1/2"x4"	1.000	Ea.	360	182	542
Valve, swing check, bronze, 125 lb, regrinding disc, 2-1/2" pipe size	1.000	Ea.	266	61.50	327.50
Roof manifold, fire, w/valves & caps, horiz/vert brs 2-1/2"x2-1/2"x4"	1.000	Ea.	113	189	302
Fire, hydrolator, vent & drain, 2-1/2" pipe size	1.000	Ea.	46	42	88
Valve, gate, iron body 125 lb, OS&Y, threaded, 4" pipe size	1.000	Ea.	455	310	765
TOTAL			2,657.50	2,547.50	5,205

D4020 310	Wet Standpipe Risers, Class I		COST PER FLOOR		
			MAT.	INST.	TOTAL
0550	Wet standpipe risers, Class I, steel black sch. 40, 10' height				
0560	4" diameter pipe, one floor		2,650	2,550	5,200
0580	Additional floors		710	790	1,500
0600	6" diameter pipe, one floor		4,625	4,475	9,100
0620	Additional floors	RD4020 -300	1,250	1,250	2,500
0640	8" diameter pipe, one floor		7,100	5,425	12,525
0660	Additional floors		1,800	1,550	3,350
0680					

D4020 310	Wet Standpipe Risers, Class II		COST PER FLOOR		
			MAT.	INST.	TOTAL
1030	Wet standpipe risers, Class II, steel black sch. 40, 10' height				
1040	2" diameter pipe, one floor		1,150	920	2,070
1060	Additional floors		390	355	745
1080	2-1/2" diameter pipe, one floor		1,525	1,350	2,875
1100	Additional floors		440	415	855
1120					

Important: See the Reference Section for critical supporting data - Reference Numbers and City Cost Indexes

D40 Fire Protection

D4020 Standpipes

D4020 310	Wet Standpipe Risers, Class III	COST PER FLOOR		
		MAT.	INST.	TOTAL
1530	Wet standpipe risers, Class III, steel black sch. 40, 10' height			
1540	4" diameter pipe, one floor	2,725	2,550	5,275
1560	Additional floors	570	660	1,230
1580	6" diameter pipe, one floor	4,700	4,475	9,175
1600	Additional floors	1,275	1,250	2,525
1620	8" diameter pipe, one floor	7,150	5,425	12,575
1640	Additional floors	1,825	1,550	3,375

SERVICES

D

349

Roof

Roof connections with hose gate valves (for combustible roof)

Hose connections on each floor (size based on class of service)

Check Valve

Siamese inlet connections (for fire department use)

System Components			COST PER FLOOR		
	QUANTITY	UNIT	MAT.	INST.	TOTAL
SYSTEM D4020 330 0540					
DRY STANDPIPE RISER, CLASS I, PIPE, STEEL, BLACK, SCH 40, 10' HEIGHT					
4" DIAMETER PIPE, ONE FLOOR					
Pipe, steel, black, schedule 40, threaded, 4" diam	20.000	L.F.	460	510	970
Pipe, Tee, malleable iron, black, 150 lb threaded, 4" pipe size	2.000	Ea.	286	462	748
Pipe, 90° elbow, malleable iron, black, 150 lb threaded 4" pipe size	1.000	Ea.	91	154	245
Pipe, nipple, steel, black, schedule 40, 2-1/2" pipe size x 3" long	2.000	Ea.	19.50	116	135.50
Fire valve gate NRS 300 lb, brass w/handwheel, 2-1/2" pipe size	1.000	Ea.	163	72	235
Fire valve, pressure restricting, adj, rgh brs, 2-1/2" pipe size	1.000	Ea.	109	72	181
Standpipe conn wall dble flush brs w/plugs & chains 2-1/2"x2-1/2"x4"	1.000	Ea.	360	182	542
Valve swing check w/ball drip Cl w/brs ftngs, 4"pipe size	1.000	Ea.	180	305	485
Roof manifold, fire, w/valves & caps, horiz/vert brs 2-1/2"x2-1/2"x4"	1.000	Ea.	113	189	302
TOTAL			1,781.50	2,062	3,843.50

D4020 330	Dry Standpipe Risers, Class I	COST PER FLOOR		
		MAT.	INST.	TOTAL
0530	Dry standpipe riser, Class I, steel black sch. 40, 10' height			
0540	4" diameter pipe, one floor	1,775	2,050	3,825
0560	Additional floors	665	745	1,410
0580	6" diameter pipe, one floor	3,525	3,550	7,075
0600	Additional floors	1,200	1,225	2,425
0620	8" diameter pipe, one floor	5,400	4,325	9,725
0640	Additional floors	1,750	1,500	3,250
0660				

D4020 330	Dry Standpipe Risers, Class II	COST PER FLOOR		
		MAT.	INST.	TOTAL
1030	Dry standpipe risers, Class II, steel black sch. 40, 10' height			
1040	2" diameter pipe, one floor	985	955	1,940
1060	Additional floor	345	315	660
1080	2-1/2" diameter pipe, one floor	1,200	1,125	2,325
1100	Additional floors	395	370	765
1120				

Important: See the Reference Section for critical supporting data - Reference Numbers and City Cost Indexes

D4020 Standpipes

D4020 330	Dry Standpipe Risers, Class III	COST PER FLOOR		
		MAT.	INST.	TOTAL
1530	Dry standpipe risers, Class III, steel black sch. 40, 10' height			
1540	4" diameter pipe, one floor	1,800	2,025	3,825
1560	Additional floors	535	675	1,210
1580	6" diameter pipe, one floor	3,550	3,550	7,100
1600	Additional floors	1,250	1,225	2,475
1620	8" diameter pipe, one floor	5,425	4,325	9,750
1640	Additional floor	1,775	1,500	3,275

SERVICES D

D4020 410	Fire Hose Equipment	COST EACH		
		MAT.	INST.	TOTAL
0100	Adapters, reducing, 1 piece, FxM, hexagon, cast brass, 2-1/2" x 1-1/2"	39		39
0200	Pin lug, 1-1/2" x 1"	33		33
0250	3" x 2-1/2"	87		87
0300	For polished chrome, add 75% mat.			
0400	Cabinets, D.S. glass in door, recessed, steel box, not equipped			
0500	Single extinguisher, steel door & frame	77	114	191
0550	Stainless steel door & frame	164	114	278
0600	Valve, 2-1/2" angle, steel door & frame	94.50	75.50	170
0650	Aluminum door & frame	115	75.50	190.50
0700	Stainless steel door & frame	154	75.50	229.50
0750	Hose rack assy, 2-1/2" x 1-1/2" valve & 100' hose, steel door & frame	176	151	327
0800	Aluminum door & frame	260	151	411
0850	Stainless steel door & frame	345	151	496
0900	Hose rack assy,& extinguisher,2-1/2"x1-1/2" valve & hose,steel door & frame	193	182	375
0950	Aluminum	330	182	512
1000	Stainless steel	370	182	552
1550	Compressor, air, dry pipe system, automatic, 200 gal., 1/3 H.P.	820	390	1,210
1600	520 gal., 1 H.P.	970	390	1,360
1650	Alarm, electric pressure switch (circuit closer)	163	19.40	182.40
2500	Couplings, hose, rocker lug, cast brass, 1-1/2"	37		37
2550	2-1/2"	48		48
3000	Escutcheon plate, for angle valves, polished brass, 1-1/2"	12.15		12.15
3050	2-1/2"	29		29
3500	Fire pump, electric, w/controller, fittings, relief valve			
3550	4" pump, 30 H.P., 500 G.P.M.	17,100	2,825	19,925
3600	5" pump, 40 H.P., 1000 G.P.M.	25,100	3,200	28,300
3650	5" pump, 100 H.P., 1000 G.P.M.	28,100	3,575	31,675
3700	For jockey pump system, add	3,075	455	3,530
5000	Hose, per linear foot, synthetic jacket, lined,			
5100	300 lb. test, 1-1/2" diameter	2.04	.35	2.39
5150	2-1/2" diameter	3.38	.41	3.79
5200	500 lb. test, 1-1/2" diameter	2.09	.35	2.44
5250	2-1/2" diameter	3.63	.41	4.04
5500	Nozzle, plain stream, polished brass, 1-1/2" x 10"	33		33
5550	2-1/2" x 15" x 13/16" or 1-1/2"	60.50		60.50
5600	Heavy duty combination adjustable fog and straight stream w/handle 1-1/2"	279		279
5650	2-1/2" direct connection	420		420
6000	Rack, for 1-1/2" diameter hose 100 ft. long, steel	36.50	45.50	82
6050	Brass	56	45.50	101.50
6500	Reel, steel, for 50 ft. long 1-1/2" diameter hose	91.50	65	156.50
6550	For 75 ft. long 2-1/2" diameter hose	148	65	213
7050	Siamese, w/plugs & chains, polished brass, sidewalk, 4" x 2-1/2" x 2-1/2"	345	365	710
7100	6" x 2-1/2" x 2-1/2"	580	455	1,035
7200	Wall type, flush, 4" x 2-1/2" x 2-1/2"	360	182	542
7250	6" x 2-1/2" x 2-1/2"	505	198	703
7300	Projecting, 4" x 2-1/2" x 2-1/2"	300	182	482
7350	6" x 2-1/2" x 2-1/2"	495	198	693
7400	For chrome plate, add 15% mat.			
8000	Valves, angle, wheel handle, 300 Lb., rough brass, 1-1/2"	33	42	75
8050	2-1/2"	56.50	72	128.50
8100	Combination pressure restricting, 1-1/2"	59.50	42	101.50
8150	2-1/2"	121	72	193
8200	Pressure restricting, adjustable, satin brass, 1-1/2"	77.50	42	119.50
8250	2-1/2"	109	72	181
8300	Hydrolator, vent and drain, rough brass, 1-1/2"	46	42	88
8350	2-1/2"	46	42	88
8400	Cabinet assy, incls. adapter, rack, hose, and nozzle	510	274	784

Important: See the Reference Section for critical supporting data - Reference Numbers and City Cost Indexes

D

SERVICES

D40 Fire Protection

D4090 Other Fire Protection Systems

General: Automatic fire protection (suppression) systems other than water sprinklers may be desired for special environments, high risk areas, isolated locations or unusual hazards. Some typical applications would include:

Paint dip tanks
Securities vaults
Electronic data processing
Tape and data storage
Transformer rooms
Spray booths
Petroleum storage
High rack storage

Piping and wiring costs are dependent on the individual application and must be added to the component costs shown below.

All areas are assumed to be open.

SERVICES D

D4090 910	Unit Components	COST EACH		
		MAT.	INST.	TOTAL
0020	Detectors with brackets			
0040	Fixed temperature heat detector	31	62.50	93.50
0060	Rate of temperature rise detector	37	62.50	99.50
0080	Ion detector (smoke) detector	82.50	81	163.50
0200	Extinguisher agent			
0240	200 lb FM200, container	6,375	231	6,606
0280	75 lb carbon dioxide cylinder	1,050	154	1,204
0320	Dispersion nozzle			
0340	FM200 1-1/2" dispersion nozzle	55	36.50	91.50
0380	Carbon dioxide 3" x 5" dispersion nozzle	55	28.50	83.50
0420	Control station			
0440	Single zone control station with batteries	1,400	500	1,900
0470	Multizone (4) control station with batteries	2,700	1,000	3,700
0490				
0500	Electric mechanical release	138	254	392
0520				
0550	Manual pull station	49.50	85.50	135
0570				
0640	Battery standby power 10" x 10" x 17"	760	125	885
0700				
0740	Bell signalling device	54.50	62.50	117

D4090 920	FM200 Systems	COST PER C.F.		
		MAT.	INST.	TOTAL
0820	Average FM200 system, minimum			1.38
0840	Maximum			2.75

System Components		QUANTITY	UNIT	COST PER L.F.		
				MAT.	INST.	TOTAL
SYSTEM D5010 110 0200						
HIGH VOLTAGE CABLE, NEUTRAL AND CONDUIT INCLUDED, COPPER #2, 5 kV						
Shielded cable, no splice/termn, copper, XLP shielding, 5 kV, #2		.030	C.L.F.	5.76	7.50	13.26
Wire 600 volt, type THW, copper, stranded, #4		.010	C.L.F.	.55	.95	1.50
Rigid galv steel conduit to 15' H, 2" diam, w/ term, ftng & support		1.000	L.F.	9.10	11.15	20.25
	TOTAL			15.41	19.60	35.01

D5010 110	High Voltage Shielded Conductors	COST PER L.F.		
		MAT.	INST.	TOTAL
0200	High voltage cable, neutral & conduit included, copper #2, 5 kV	15.40	19.60	35
0240	Copper #1, 5 kV	16.55	19.75	36.30
0280	15 kV	29.50	28.50	58
0320	Copper 1/0, 5 kV	17.25	20	37.25
0360	15 kV	31	29	60
0400	25 kV	34.50	29.50	64
0440	35 kV	40.50	33	73.50
0480	Copper 2/0, 5 kV	27.50	24	51.50
0520	15 kV	32.50	29.50	62
0560	25 kV	41	33	74
0600	35 kV	46.50	35.50	82
0640	Copper 4/0, 5 kV	34	31	65
0680	15 kV	41.50	34	75.50
0720	25 kV	45	35	80
0760	35 kV	51	37.50	88.50
0800	Copper 250 kcmil, 5 kV	37	32	69
0840	15 kV	43.50	35	78.50
0880	25 kV	53.50	38	91.50
0920	35 kV	87	47	134
0960	Copper 350 kcmil, 5 kV	42	34	76
1000	15 kV	52	39	91
1040	25 kV	58.50	40	98.50
1080	35 kV	93	49.50	142.50
1120	Copper 500 kcmil, 5 kV	52.50	38	90.50
1160	15 kV	89.50	48.50	138
1200	25 kV	96.50	50	146.50
1240	35 kV	99.50	51.50	151

Important: See the Reference Section for critical supporting data - Reference Numbers and City Cost Indexes

D5010 Electrical Service/Distribution

Service Entrance Cap

Conduit

Circuit Breaker or Safety Switch

Meter Socket

Ground Rod

SERVICES D

System Components	QUANTITY	UNIT	COST EACH		
			MAT.	INST.	TOTAL
SYSTEM D5010 120 0220					
SERVICE INSTALLATION, INCLUDES BREAKERS, METERING, 20' CONDUIT & WIRE					
3 PHASE, 4 WIRE, 60 A					
Circuit breaker, enclosed (NEMA 1), 600 volt, 3 pole, 60 A	1.000	Ea.	550	179	729
Meter socket, single position, 4 terminal, 100 A	1.000	Ea.	39.50	157	196.50
Rigid galvanized steel conduit, 3/4", including fittings	20.000	L.F.	61.20	125	186.20
Wire, 600V type XHHW, copper stranded #6	.900	C.L.F.	38.70	69.30	108
Service entrance cap 3/4" diameter	1.000	Ea.	8.65	38.50	47.15
Conduit LB fitting with cover, 3/4" diameter	1.000	Ea.	12.85	38.50	51.35
Ground rod, copper clad, 8' long, 3/4" diameter	1.000	Ea.	31.50	94.50	126
Ground rod clamp, bronze, 3/4" diameter	1.000	Ea.	6.35	15.65	22
Ground wire, bare armored, #6-1 conductor	.200	C.L.F.	22.40	55.60	78
TOTAL			771.15	773.05	1,544.20

D5010 120	Electric Service, 3 Phase - 4 Wire		COST EACH		
			MAT.	INST.	TOTAL
0200	Service installation, includes breakers, metering, 20' conduit & wire				
0220	3 phase, 4 wire, 120/208 volts, 60 A		770	775	1,545
0240	100 A		965	930	1,895
0280	200 A		1,300	1,425	2,725
0320	400 A	RD5010 -110	3,050	2,625	5,675
0360	600 A		5,900	3,550	9,450
0400	800 A		7,825	4,275	12,100
0440	1000 A		9,875	4,925	14,800
0480	1200 A		12,000	5,025	17,025
0520	1600 A		23,600	7,225	30,825
0560	2000 A		26,400	8,250	34,650
0570	Add 25% for 277/480 volt				
0580					
0610	1 phase, 3 wire, 120/240 volts, 100 A		460	840	1,300
0620	200 A		960	1,225	2,185

D5010 Electrical Service/Distribution

System Components	QUANTITY	UNIT	COST PER L.F.		
			MAT.	INST.	TOTAL
SYSTEM D5010 230 0200					
FEEDERS, INCLUDING STEEL CONDUIT & WIRE, 60 A					
Rigid galvanized steel conduit, 3/4", including fittings	1.000	L.F.	3.06	6.25	9.31
Wire 600 volt, type XHHW copper stranded #6	.040	C.L.F.	1.72	3.08	4.80
TOTAL			4.78	9.33	14.11

D5010 230	Feeder Installation		COST PER L.F.		
			MAT.	INST.	TOTAL
0200	Feeder installation 600 V, including RGS conduit and XHHW wire, 60 A		4.78	9.35	14.13
0240	100 A		8.90	12.35	21.25
0280	200 A	RD5010	16.40	19.15	35.55
0320	400 A	-140	33	38.50	71.50
0360	600 A		71	62	133
0400	800 A		95	74.50	169.50
0440	1000 A		113	95.50	208.50
0480	1200 A		135	98	233
0520	1600 A		190	149	339
0560	2000 A		226	191	417
1200	Branch installation 600 V, including EMT conduit and THW wire, 15 A		.98	4.77	5.75
1240	20 A		.98	4.77	5.75
1280	30 A		1.56	5.85	7.41
1320	50 A		2.68	6.75	9.43
1360	65 A		3.11	7.15	10.26
1400	85 A		4.80	8.45	13.25
1440	100 A		5.75	8.95	14.70
1480	130 A		7.35	10.05	17.40
1520	150 A		8.70	11.60	20.30
1560	200 A		10.50	13	23.50

Important: See the Reference Section for critical supporting data - Reference Numbers and City Cost Indexes

D5010 Electrical Service/Distribution

System Components	QUANTITY	UNIT	COST EACH		
			MAT.	INST.	TOTAL
SYSTEM D5010 240 0240					
SWITCHGEAR INSTALLATION, INCL SWBD, PANELS & CIRC BREAKERS, 600 A					
Panelboard, NQOD 225A 4W 120/208V main CB, w/20A bkrs 42 circ	1.000	Ea.	2,000	1,800	3,800
Switchboard, alum. bus bars, 120/208V, 4 wire, 600V	1.000	Ea.	3,800	1,000	4,800
Distribution sect., alum. bus bar, 120/208 or 277/480 V, 4 wire, 600A	1.000	Ea.	1,625	1,000	2,625
Feeder section circuit breakers, KA frame, 70 to 225 A	3.000	Ea.	3,000	471	3,471
TOTAL			10,425	4,271	14,696

D5010 240	Switchgear		COST EACH		
			MAT.	INST.	TOTAL
0200	Switchgear inst., incl. swbd., panels & circ bkr, 400 A, 120/208volt		3,675	3,150	6,825
0240	600 A		10,400	4,275	14,675
0280	800 A	RD5010 -110	13,100	6,100	19,200
0320	1200 A		16,200	9,325	25,525
0360	1600 A		21,900	13,100	35,000
0400	2000 A		27,700	16,700	44,400
0410	Add 20% for 277/480 volt				

Duplex Receptacle

System Components		QUANTITY	UNIT	COST PER S.F.		
				MAT.	INST.	TOTAL
SYSTEM D5020 110 0200						
RECEPTACLES INCL. PLATE, BOX, CONDUIT, WIRE & TRANS. WHEN REQUIRED						
2.5 PER 1000 S.F., .3 WATTS PER S.F.						
Steel intermediate conduit, (IMC) 1/2″ diam		167.000	L.F.	.39	.84	1.23
Wire 600V type THWN-THHN, copper solid #12		3.382	C.L.F.	.02	.15	.17
Wiring device, receptacle, duplex, 120V grounded, 15 amp		2.500	Ea.		.03	.03
Wall plate, 1 gang, brown plastic		2.500	Ea.		.02	.02
Steel outlet box 4″ square		2.500	Ea.	.01	.06	.07
Steel outlet box 4″ plaster rings		2.500	Ea.	.01	.02	.03
	TOTAL			.43	1.12	1.55

D5020 110	Receptacle (by Wattage)		COST PER S.F.		
			MAT.	INST.	TOTAL
0190	Receptacles include plate, box, conduit, wire & transformer when required				
0200	2.5 per 1000 S.F., .3 watts per S.F.		.43	1.12	1.55
0240	With transformer	RD5010	.49	1.17	1.66
0280	4 per 1000 S.F., .5 watts per S.F.	-110	.49	1.30	1.79
0320	With transformer		.57	1.38	1.95
0360	5 per 1000 S.F., .6 watts per S.F.		.57	1.54	2.11
0400	With transformer		.68	1.65	2.33
0440	8 per 1000 S.F., .9 watts per S.F.		.60	1.70	2.30
0480	With transformer		.74	1.85	2.59
0520	10 per 1000 S.F., 1.2 watts per S.F.		.63	1.85	2.48
0560	With transformer		.87	2.10	2.97
0600	16.5 per 1000 S.F., 2.0 watts per S.F.		.72	2.31	3.03
0640	With transformer		1.14	2.74	3.88
0680	20 per 1000 S.F.,2.4 watts per S.F.		.75	2.51	3.26
0720	With transformer		1.23	3.01	4.24

Important: See the Reference Section for critical supporting data - Reference Numbers and City Cost Indexes

D5020 Lighting and Branch Wiring

Underfloor Receptacle System

Description: Table D5020 115 includes installed costs of raceways and copper wire from panel to and including receptacle.

National Electrical Code prohibits use of undercarpet system in residential, school or hospital buildings. Can only be used with carpet squares.

Low density = (1) Outlet per 259 S.F. of floor area.

High density = (1) Outlet per 127 S.F. of floor area.

SERVICES D

System Components	QUANTITY	UNIT	COST PER S.F.		
			MAT.	INST.	TOTAL
SYSTEM D5020 115 0200					
RECEPTACLE SYSTEMS, UNDERFLOOR DUCT, 5' ON CENTER, LOW DENSITY					
Underfloor duct 3-1/8" x 7/8" w/insert 24" on center	.190	L.F.	2.74	1.36	4.10
Vertical elbow for underfloor duct, 3-1/8", included					
Underfloor duct conduit adapter, 2" x 1-1/4", included					
Underfloor duct junction box, single duct, 3-1/8"	.003	Ea.	.98	.38	1.36
Underfloor junction box carpet pan	.003	Ea.	.80	.02	.82
Underfloor duct outlet, high tension receptacle	.004	Ea.	.30	.25	.55
Wire 600V type THWN-THHN copper solid #12	.010	C.L.F.	.07	.46	.53
TOTAL			4.89	2.47	7.36

D5020 115	Receptacles, Floor		COST PER S.F.		
			MAT.	INST.	TOTAL
0200	Receptacle systems, underfloor duct, 5' on center, low density		4.88	2.46	7.34
0240	High density		5.25	3.16	8.41
0280	7' on center, low density	RD5010 -110	3.86	2.11	5.97
0320	High density		4.23	2.82	7.05
0400	Poke thru fittings, low density		.98	1.13	2.11
0440	High density		1.95	2.23	4.18
0520	Telepoles, using Romex, low density		.81	.74	1.55
0560	High density		1.61	1.48	3.09
0600	Using EMT, low density		.85	.98	1.83
0640	High density		1.71	1.97	3.68
0720	Conduit system with floor boxes, low density		.87	.84	1.71
0760	High density		1.74	1.70	3.44
0840	Undercarpet power system, 3 conductor with 5 conductor feeder, low density		1.30	.31	1.61
0880	High density		2.54	.62	3.16

D5020 Lighting and Branch Wiring

Duplex Receptacle

Wall Switch

System Components			COST PER EACH		
	QUANTITY	UNIT	MAT.	INST.	TOTAL
SYSTEM D5020 125 0520					
RECEPTACLES AND WALL SWITCHES, RECEPTACLE DUPLEX 120 V GROUNDED, 15 A					
Electric metallic tubing conduit, (EMT), 3/4" diam	22.000	L.F.	23.76	84.70	108.46
Wire, 600 volt, type THWN-THHN, copper, solid #12	.630	C.L.F.	4.44	28.67	33.11
Steel outlet box 4" square	1.000	Ea.	2.46	25	27.46
Steel outlet box, 4" square, plaster rings	1.000	Ea.	2.23	7.85	10.08
Receptacle, duplex, 120 volt grounded, 15 amp	1.000	Ea.	1.29	12.50	13.79
Wall plate, 1 gang, brown plastic	1.000	Ea.	.34	6.25	6.59
TOTAL			34.52	164.97	199.49

D5020 125	Receptacles & Switches by Each	COST PER EACH		
		MAT.	INST.	TOTAL
0460	Receptacles & Switches, with box, plate, 3/4" EMT conduit & wire			
0520	Receptacle duplex 120 V grounded, 15 A	34.50	165	199.50
0560	20 A	43	171	214
0600	Receptacle duplex ground fault interrupting, 15 A	65.50	171	236.50
0640	20 A	67.50	171	238.50
0680	Toggle switch single, 15 A	38.50	165	203.50
0720	20 A	41	171	212
0760	3 way switch, 15 A	41	174	215
0800	20 A	42	180	222
0840	4 way switch, 15 A	56	186	242
0880	20 A	71	198	269

D5020 Lighting and Branch Wiring

Description: Table D5020 130 includes the cost for switch, plate, box, conduit in slab or EMT exposed and copper wire. Add 20% for exposed conduit.

No power required for switches.

Federal energy guidelines recommend the maximum lighting area controlled per switch shall not exceed 1000 S.F. and that areas over 500 S.F. shall be so controlled that total illumination can be reduced by at least 50%.

SERVICES **D**

System Components	QUANTITY	UNIT	COST PER S.F.		
			MAT.	INST.	TOTAL
SYSTEM D5020 130 0360					
WALL SWITCHES, 5.0 PER 1000 S.F.					
Steel, intermediate conduit (IMC), 1/2" diameter	88.000	L.F.	.21	.44	.65
Wire, 600V type THWN-THHN, copper solid #12	1.710	C.L.F.	.01	.08	.09
Toggle switch, single pole, 15 amp	5.000	Ea.	.03	.06	.09
Wall plate, 1 gang, brown plastic	5.000	Ea.		.03	.03
Steel outlet box 4" plaster rings	5.000	Ea.	.01	.13	.14
Plaster rings	5.000	Ea.	.01	.04	.05
TOTAL			.27	.78	1.05

D5020 130	Wall Switch by Sq. Ft.	COST PER S.F.		
		MAT.	INST.	TOTAL
0200	Wall switches, 1.0 per 1000 S.F.	.07	.18	.25
0240	1.2 per 1000 S.F.	.07	.20	.27
0280	2.0 per 1000 S.F.	.10	.29	.39
0320	2.5 per 1000 S.F.	.12	.36	.48
0360	5.0 per 1000 S.F.	.27	.78	1.05
0400	10.0 per 1000 S.F.	.55	1.57	2.12

D5020 Lighting and Branch Wiring

System D5020 135 includes all wiring and connections.

System Components

System Components	QUANTITY	UNIT	COST PER S.F.		
			MAT.	INST.	TOTAL
SYSTEM D5020 135 0200					
MISCELLANEOUS POWER, TO .5 WATTS					
Steel intermediate conduit, (IMC) 1/2″ diam	15.000	L.F.	.04	.08	.12
Wire 600V type THWN-THHN, copper solid #12	.325	C.L.F.		.01	.01
TOTAL			.04	.09	.13

D5020 135	Miscellaneous Power	COST PER S.F.		
		MAT.	INST.	TOTAL
0200	Miscellaneous power, to .5 watts	.04	.09	.13
0240	.8 watts	.05	.13	.18
0280	1 watt	.07	.16	.23
0320	1.2 watts	.08	.20	.28
0360	1.5 watts	.10	.23	.33
0400	1.8 watts	.11	.27	.38
0440	2 watts	.14	.32	.46
0480	2.5 watts	.17	.40	.57
0520	3 watts	.20	.47	.67

Important: See the Reference Section for critical supporting data - Reference Numbers and City Cost Indexes

D5020 Lighting and Branch Wiring

System D5020 140 includes all wiring and connections for central air conditioning units.

System Components	QUANTITY	UNIT	COST PER S.F.		
			MAT.	INST.	TOTAL
SYSTEM D5020 140 0200					
CENTRAL AIR CONDITIONING POWER, 1 WATT					
Steel intermediate conduit, 1/2" diam.	.030	L.F.	.07	.15	.22
Wire 600V type THWN-THHN, copper solid #12	.001	C.L.F.	.01	.05	.06
TOTAL			.08	.20	.28

D5020 140	Central A. C. Power (by Wattage)	COST PER S.F.		
		MAT.	INST.	TOTAL
0200	Central air conditioning power, 1 watt	.08	.20	.28
0220	2 watts	.09	.23	.32
0240	3 watts	.11	.25	.36
0280	4 watts	.17	.34	.51
0320	6 watts	.30	.47	.77
0360	8 watts	.36	.49	.85
0400	10 watts	.43	.57	1

D5020 Lighting and Branch Wiring

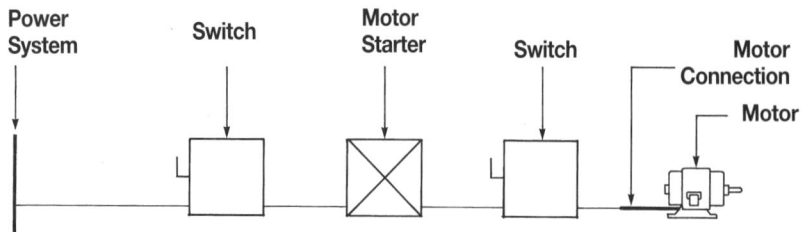

System D5020 145 installed cost of motor wiring as per Reference Table D5010 17D using 50' of rigid conduit and copper wire. **Cost and setting of motor not included.**

System Components	QUANTITY	UNIT	COST EACH		
			MAT.	INST.	TOTAL
SYSTEM D5020 145 0200					
MOTOR INSTALLATION, SINGLE PHASE, 115V, TO AND INCLUDING 1/3 HP MOTOR SIZE					
Wire 600V type THWN-THHN, copper solid #12	1.250	C.L.F.	8.81	56.88	65.69
Steel intermediate conduit, (IMC) 1/2" diam	50.000	L.F.	118	250	368
Magnetic FVNR, 115V, 1/3 HP, size 00 starter	1.000	Ea.	185	125	310
Safety switch, fused, heavy duty, 240V 2P 30 amp	1.000	Ea.	109	143	252
Safety switch, non fused, heavy duty, 600V, 3 phase, 30 A	1.000	Ea.	128	157	285
Flexible metallic conduit, Greenfield 1/2" diam	1.500	L.F.	.62	3.75	4.37
Connectors for flexible metallic conduit Greenfield 1/2" diam	1.000	Ea.	1.82	6.25	8.07
Coupling for Greenfield to conduit 1/2" diam flexible metalic conduit	1.000	Ea.	.89	10	10.89
Fuse cartridge nonrenewable, 250V 30 amp	1.000	Ea.	1.55	10	11.55
TOTAL			553.69	761.88	1,315.57

D5020 145	Motor Installation		COST EACH		
			MAT.	INST.	TOTAL
0200	Motor installation, single phase, 115V, to and including 1/3 HP motor size		555	760	1,315
0240	To and incl. 1 HP motor size		575	760	1,335
0280	To and incl. 2 HP motor size	RD5010 -170	615	810	1,425
0320	To and incl. 3 HP motor size		685	825	1,510
0360	230V, to and including 1 HP motor size		555	770	1,325
0400	To and incl. 2 HP motor size		580	770	1,350
0440	To and incl. 3 HP motor size		650	830	1,480
0520	Three phase, 200V, to and including 1-1/2 HP motor size		645	850	1,495
0560	To and incl. 3 HP motor size		690	925	1,615
0600	To and incl. 5 HP motor size		735	1,025	1,760
0640	To and incl. 7-1/2 HP motor size		755	1,050	1,805
0680	To and incl. 10 HP motor size		1,200	1,325	2,525
0720	To and incl. 15 HP motor size		1,550	1,450	3,000
0760	To and incl. 20 HP motor size		1,950	1,675	3,625
0800	To and incl. 25 HP motor size		1,975	1,700	3,675
0840	To and incl. 30 HP motor size		3,075	2,000	5,075
0880	To and incl. 40 HP motor size		3,775	2,350	6,125
0920	To and incl. 50 HP motor size		6,450	2,750	9,200
0960	To and incl. 60 HP motor size		6,650	2,925	9,575
1000	To and incl. 75 HP motor size		8,650	3,325	11,975
1040	To and incl. 100 HP motor size		17,800	3,900	21,700
1080	To and incl. 125 HP motor size		18,200	4,300	22,500
1120	To and incl. 150 HP motor size		22,100	5,050	27,150
1160	To and incl. 200 HP motor size		26,500	6,150	32,650
1240	230V, to and including 1-1/2 HP motor size		615	840	1,455
1280	To and incl. 3 HP motor size		660	915	1,575
1320	To and incl. 5 HP motor size		700	1,025	1,725
1360	To and incl. 7-1/2 HP motor size		700	1,025	1,725
1400	To and incl. 10 HP motor size		1,075	1,250	2,325
1440	To and incl. 15 HP motor size		1,225	1,350	2,575
1480	To and incl. 20 HP motor size		1,825	1,625	3,450
1520	To and incl. 25 HP motor size		1,950	1,675	3,625

Important: See the Reference Section for critical supporting data - Reference Numbers and City Cost Indexes

D5020 Lighting and Branch Wiring

D5020 145	Motor Installation	COST EACH		
		MAT.	INST.	TOTAL
1560	To and incl. 30 HP motor size	1,975	1,700	3,675
1600	To and incl. 40 HP motor size	3,700	2,300	6,000
1640	To and incl. 50 HP motor size	3,800	2,425	6,225
1680	To and incl. 60 HP motor size	6,450	2,775	9,225
1720	To and incl. 75 HP motor size	7,875	3,125	11,000
1760	To and incl. 100 HP motor size	9,150	3,500	12,650
1800	To and incl. 125 HP motor size	18,000	4,025	22,025
1840	To and incl. 150 HP motor size	19,500	4,575	24,075
1880	To and incl. 200 HP motor size	21,100	5,075	26,175
1960	460V, to and including 2 HP motor size	735	845	1,580
2000	To and incl. 5 HP motor size	780	920	1,700
2040	To and incl. 10 HP motor size	810	1,025	1,835
2080	To and incl. 15 HP motor size	1,050	1,175	2,225
2120	To and incl. 20 HP motor size	1,100	1,250	2,350
2160	To and incl. 25 HP motor size	1,200	1,300	2,500
2200	To and incl. 30 HP motor size	1,525	1,400	2,925
2240	To and incl. 40 HP motor size	1,925	1,525	3,450
2280	To and incl. 50 HP motor size	2,150	1,675	3,825
2320	To and incl. 60 HP motor size	3,225	1,975	5,200
2360	To and incl. 75 HP motor size	3,700	2,175	5,875
2400	To and incl. 100 HP motor size	4,000	2,425	6,425
2440	To and incl. 125 HP motor size	6,650	2,775	9,425
2480	To and incl. 150 HP motor size	8,225	3,100	11,325
2520	To and incl. 200 HP motor size	9,675	3,500	13,175
2600	575V, to and including 2 HP motor size	735	845	1,580
2640	To and incl. 5 HP motor size	780	920	1,700
2680	To and incl. 10 HP motor size	810	1,025	1,835
2720	To and incl. 20 HP motor size	1,050	1,175	2,225
2760	To and incl. 25 HP motor size	1,100	1,250	2,350
2800	To and incl. 30 HP motor size	1,525	1,400	2,925
2840	To and incl. 50 HP motor size	1,625	1,475	3,100
2880	To and incl. 60 HP motor size	3,200	1,975	5,175
2920	To and incl. 75 HP motor size	3,225	1,975	5,200
2960	To and incl. 100 HP motor size	3,700	2,175	5,875
3000	To and incl. 125 HP motor size	6,525	2,725	9,250
3040	To and incl. 150 HP motor size	6,650	2,775	9,425
3080	To and incl. 200 HP motor size	8,400	3,150	11,550

SERVICES

D

D5020 Lighting and Branch Wiring

System Components	QUANTITY	UNIT	COST PER L.F.		
			MAT.	INST.	TOTAL
SYSTEM D5020 155 0200					
MOTOR FEEDER SYSTEMS, SINGLE PHASE, UP TO 115V, 1HP OR 230V, 2HP					
Steel intermediate conduit, (IMC) 1/2" diam	1.000	L.F.	2.36	5	7.36
Wire 600V type THWN-THHN, copper solid #12	.020	C.L.F.	.14	.91	1.05
TOTAL			2.50	5.91	8.41

D5020 155	Motor Feeder		COST PER L.F.		
			MAT.	INST.	TOTAL
0200	Motor feeder systems, single phase, feed up to 115V 1HP or 230V 2 HP		2.50	5.90	8.40
0240	115V 2HP, 230V 3HP		2.57	6	8.57
0280	115V 3HP	RD5010 -170	2.79	6.25	9.04
0360	Three phase, feed to 200V 3HP, 230V 5HP, 460V 10HP, 575V 10HP		2.57	6.35	8.92
0440	200V 5HP, 230V 7.5HP, 460V 15HP, 575V 20HP		2.68	6.50	9.18
0520	200V 10HP, 230V 10HP, 460V 30HP, 575V 30HP		3.01	6.90	9.91
0600	200V 15HP, 230V 15HP, 460V 40HP, 575V 50HP		3.86	7.85	11.71
0680	200V 20HP, 230V 25HP, 460V 50HP, 575V 60HP		5.65	10	15.65
0760	200V 25HP, 230V 30HP, 460V 60HP, 575V 75HP		6.10	10.15	16.25
0840	200V 30HP		7.10	10.50	17.60
0920	230V 40HP, 460V 75HP, 575V 100HP		8.60	11.45	20.05
1000	200V 40HP		9.85	12.25	22.10
1080	230V 50HP, 460V 100HP, 575V 125HP		10.45	13.55	24
1160	200V 50HP, 230V 60HP, 460V 125HP, 575V 150HP		12.50	14.35	26.85
1240	200V 60HP, 460V 150HP		14.65	16.85	31.50
1320	230V 75HP, 575V 200HP		17.30	17.50	34.80
1400	200V 75HP		18.80	17.90	36.70
1480	230V 100HP, 460V 200HP		27.50	21	48.50
1560	200V 100HP		37	26	63
1640	230V 125HP		40	28.50	68.50
1720	200V 125HP, 230V 150HP		43	32.50	75.50
1800	200V 150HP		48.50	35	83.50
1880	200V 200HP		74	52.50	126.50
1960	230V 200HP		61.50	39	100.50

Important: See the Reference Section for critical supporting data - Reference Numbers and City Cost Indexes

D5020 Lighting and Branch Wiring

Starters are full voltage, type NEMA 1 for general purpose indoor application with motor overload protection and include mounting and wire connections.

SERVICES **D**

System Components	QUANTITY	UNIT	COST EACH		
			MAT.	INST.	TOTAL
SYSTEM D5020 160 0200					
MAGNETIC STARTER,SIZE 00 TO 1/3 HP, 1 PHASE 115V OR 1 HP 230V	1.000	Ea.	185	125	310
TOTAL			185	125	310

D5020 160	Magnetic Starter	COST EACH		
		MAT.	INST.	TOTAL
0200	Magnetic starter, size 00, to 1/3 HP, 1 phase, 115V or 1 HP 230V	185	125	310
0280	Size 00, to 1-1/2 HP, 3 phase, 200-230V or 2 HP 460-575V	203	143	346
0360	Size 0, to 1 HP, 1 phase, 115V or 2 HP 230V	206	125	331
0440	Size 0, to 3 HP, 3 phase, 200-230V or 5 HP 460-575V	246	218	464
0520	Size 1, to 2 HP, 1 phase, 115V or 3 HP 230V	237	167	404
0600	Size 1, to 7-1/2 HP, 3 phase, 200-230V or 10 HP 460-575V	276	315	591
0680	Size 2, to 10 HP, 3 phase, 200V, 15 HP-230V or 25 HP 460-575V	520	455	975
0760	Size 3, to 25 HP, 3 phase, 200V, 30 HP-230V or 50 HP 460-575V	845	555	1,400
0840	Size 4, to 40 HP, 3 phase, 200V, 50 HP-230V or 100 HP 460-575V	1,875	835	2,710
0920	Size 5, to 75 HP, 3 phase, 200V, 100 HP-230V or 200 HP 460-575V	4,400	1,125	5,525
1000	Size 6, to 150 HP, 3 phase, 200V, 200 HP-230V or 400 HP 460-575V	12,500	1,250	13,750

D5020 Lighting and Branch Wiring

Safety switches are type NEMA 1 for general purpose indoor application, and include time delay fuses, insulation and wire terminations.

System Components	QUANTITY	UNIT	COST EACH		
			MAT.	INST.	TOTAL
SYSTEM D5020 165 0200					
SAFETY SWITCH, 30A FUSED, 1 PHASE, 115V OR 230V.					
Safety switch fused, hvy duty, 240V 2p 30 amp	1.000	Ea.	109	143	252
Fuse, dual element time delay 250V, 30 amp	2.000	Ea.	7.92	20	27.92
TOTAL			116.92	163	279.92

D5020 165	Safety Switches	COST EACH		
		MAT.	INST.	TOTAL
0200	Safety switch, 30A fused, 1 phase, 2HP 115V or 3HP, 230V.	117	163	280
0280	3 phase, 5HP, 200V or 7 1/2HP, 230V	157	187	344
0360	15HP, 460V or 20HP, 575V	272	195	467
0440	60A fused, 3 phase, 15HP 200V or 15HP 230V	268	248	516
0520	30HP 460V or 40HP 575V	345	256	601
0600	100A fused, 3 phase, 20HP 200V or 25HP 230V	440	300	740
0680	50HP 460V or 60HP 575V	640	305	945
0760	200A fused, 3 phase, 50HP 200V or 60HP 230V	775	425	1,200
0840	125HP 460V or 150HP 575V	975	435	1,410
0920	400A fused, 3 phase, 100HP 200V or 125HP 230V	1,900	605	2,505
1000	250HP 460V or 350HP 575V	2,425	620	3,045

Important: See the Reference Section for critical supporting data - Reference Numbers and City Cost Indexes

D5020   Lighting and Branch Wiring

Straight
Connector

Angle
Connector

Flexible Conduit

Table below includes costs for the
flexible conduit. Not included are wire
terminations and testing motor for
correct rotation.

System Components			QUANTITY	UNIT	COST EACH		
					MAT.	INST.	TOTAL
SYSTEM D5020 170 0200							
MOTOR CONNECTIONS, SINGLE PHASE, 115V/230V UP TO 1 HP							
Motor connection, flexible conduit & fittings, 1 HP motor 115V			1.000	Ea.	5.18	59.38	64.56
		TOTAL			5.18	59.38	64.56

D5020 170	Motor Connections		COST EACH		
			MAT.	INST.	TOTAL
0200	Motor connections, single phase, 115/230V, up to 1 HP		5.20	59.50	64.70
0240	Up to 3 HP		5.05	69.50	74.55
0280	Three phase, 200/230/460/575V, up to 3 HP	RD5010 -170	5.60	77	82.60
0320	Up to 5 HP		5.60	77	82.60
0360	Up to 7-1/2 HP		9.35	91	100.35
0400	Up to 10 HP		10.10	119	129.10
0440	Up to 15 HP		19.30	152	171.30
0480	Up to 25 HP		22	185	207
0520	Up to 50 HP		60.50	228	288.50
0560	Up to 100 HP		155	335	490

Manual
Starter

Magnetic
Starter

Induction Motor

For 230/460 Volt A.C., 3 phase, 60 cycle ball bearing squirrel cage induction motors, NEMA Class B standard line. Installation included.

No conduit, wire, or terminations included.

System Components	QUANTITY	UNIT	COST EACH		
			MAT.	INST.	TOTAL
SYSTEM D5020 175 0220					
MOTOR, DRIPPROOF CLASS B INSULATION, 1.15 SERVICE FACTOR, WITH STARTER					
1 H.P., 1200 RPM WITH MANUAL STARTER					
Motor, dripproof, class B insul, 1.15 serv fact, 1200 RPM, 1 HP	1.000	Ea.	259	111	370
Motor starter, manual, 3 phase, 1 HP motor	1.000	Ea.	168	143	311
TOTAL			427	254	681

D5020 175	Motor & Starter	COST EACH		
		MAT.	INST.	TOTAL
0190	Motor, dripproof, premium efficient, 1.15 service factor			
0200	1 HP, 1200 RPM, motor only	259	111	370
0220	With manual starter	425	254	679
0240	With magnetic starter	460	254	714
0260	1800 RPM, motor only	221	111	332
0280	With manual starter	390	254	644
0300	With magnetic starter	425	254	679
0320	2 HP, 1200 RPM, motor only	365	111	476
0340	With manual starter	535	254	789
0360	With magnetic starter	610	330	940
0380	1800 RPM, motor only	261	111	372
0400	With manual starter	430	254	684
0420	With magnetic starter	505	330	835
0440	3600 RPM, motor only	274	111	385
0460	With manual starter	440	254	694
0480	With magnetic starter	520	330	850
0500	3 HP, 1200 RPM, motor only	485	111	596
0520	With manual starter	655	254	909
0540	With magnetic starter	730	330	1,060
0560	1800 RPM, motor only	262	111	373
0580	With manual starter	430	254	684
0600	With magnetic starter	510	330	840
0620	3600 RPM, motor only	315	111	426
0640	With manual starter	485	254	739
0660	With magnetic starter	560	330	890
0680	5 HP, 1200 RPM, motor only	585	111	696
0700	With manual starter	785	360	1,145
0720	With magnetic starter	860	425	1,285
0740	1800 RPM, motor only	340	111	451
0760	With manual starter	540	360	900
0780	With magnetic starter	615	425	1,040
0800	3600 RPM, motor only	340	111	451

Note: RD5010-170 reference appears next to lines 0220/0240.

Important: See the Reference Section for critical supporting data - Reference Numbers and City Cost Indexes

D5020 Lighting and Branch Wiring

D5020 175	Motor & Starter	COST EACH		
		MAT.	INST.	TOTAL
0820	With manual starter	540	360	900
0840	With magnetic starter	615	425	1,040
0860	7.5 HP, 1800 RPM, motor only	490	119	609
0880	With manual starter	690	370	1,060
0900	With magnetic starter	1,000	575	1,575
0920	10 HP, 1800 RPM, motor only	605	125	730
0940	With manual starter	805	375	1,180
0960	With magnetic starter	1,125	580	1,705
0980	15 HP, 1800 RPM, motor only	770	157	927
1000	With magnetic starter	1,300	610	1,910
1040	20 HP, 1800 RPM, motor only	960	193	1,153
1060	With magnetic starter	1,800	750	2,550
1100	25 HP, 1800 RPM, motor only	1,125	200	1,325
1120	With magnetic starter	1,975	755	2,730
1160	30 HP, 1800 RPM, motor only	1,300	209	1,509
1180	With magnetic starter	2,150	765	2,915
1220	40 HP, 1800 RPM, motor only	1,875	250	2,125
1240	With magnetic starter	3,750	1,075	4,825
1280	50 HP, 1800 RPM, motor only	2,125	315	2,440
1300	With magnetic starter	4,000	1,150	5,150
1340	60 HP, 1800 RPM, motor only	2,775	360	3,135
1360	With magnetic starter	7,175	1,475	8,650
1400	75 HP, 1800 RPM, motor only	3,150	415	3,565
1420	With magnetic starter	7,550	1,550	9,100
1460	100 HP, 1800 RPM, motor only	4,150	555	4,705
1480	With magnetic starter	8,550	1,675	10,225
1520	125 HP, 1800 RPM, motor only	4,825	715	5,540
1540	With magnetic starter	17,300	1,975	19,275
1580	150 HP, 1800 RPM, motor only	6,275	835	7,110
1600	With magnetic starter	18,800	2,075	20,875
1640	200 HP, 1800 RPM, motor only	8,100	1,000	9,100
1660	With magnetic starter	20,600	2,250	22,850
1680	Totally encl, premium efficient, 1.0 ser. fac., 1HP, 1200RPM, motor only	299	111	410
1700	With manual starter	465	254	719
1720	With magnetic starter	500	254	754
1740	1800 RPM, motor only	256	111	367
1760	With manual starter	425	254	679
1780	With magnetic starter	460	254	714
1800	2 HP, 1200 RPM, motor only	365	111	476
1820	With manual starter	535	254	789
1840	With magnetic starter	610	330	940
1860	1800 RPM, motor only	310	111	421
1880	With manual starter	480	254	734
1900	With magnetic starter	555	330	885
1920	3600 RPM, motor only	296	111	407
1940	With manual starter	465	254	719
1960	With magnetic starter	540	330	870
1980	3 HP, 1200 RPM, motor only	485	111	596
2000	With manual starter	655	254	909
2020	With magnetic starter	730	330	1,060
2040	1800 RPM, motor only	340	111	451
2060	With manual starter	510	254	764
2080	With magnetic starter	585	330	915
2100	3600 RPM, motor only	345	111	456
2120	With manual starter	515	254	769
2140	With magnetic starter	590	330	920
2160	5 HP, 1200 RPM, motor only	715	111	826
2180	With manual starter	915	360	1,275

D5020 Lighting and Branch Wiring

D5020 175	Motor & Starter	COST EACH		
		MAT.	INST.	TOTAL
2200	With magnetic starter	990	425	1,415
2220	1800 RPM, motor only	390	111	501
2240	With manual starter	590	360	950
2260	With magnetic starter	665	425	1,090
2280	3600 RPM, motor only	430	111	541
2300	With manual starter	630	360	990
2320	With magnetic starter	705	425	1,130
2340	7.5 HP, 1800 RPM, motor only	560	119	679
2360	With manual starter	760	370	1,130
2380	With magnetic starter	1,075	575	1,650
2400	10 HP, 1800 RPM, motor only	675	125	800
2420	With manual starter	875	375	1,250
2440	With magnetic starter	1,200	580	1,780
2460	15 HP, 1800 RPM, motor only	905	157	1,062
2480	With magnetic starter	1,425	610	2,035
2500	20 HP, 1800 RPM, motor only	1,100	193	1,293
2520	With magnetic starter	1,950	750	2,700
2540	25 HP, 1800 RPM, motor only	1,350	200	1,550
2560	With magnetic starter	2,200	755	2,955
2580	30 HP, 1800 RPM, motor only	1,575	209	1,784
2600	With magnetic starter	2,425	765	3,190
2620	40 HP, 1800 RPM, motor only	2,025	250	2,275
2640	With magnetic starter	3,900	1,075	4,975
2660	50 HP, 1800 RPM, motor only	2,500	315	2,815
2680	With magnetic starter	4,375	1,150	5,525
2700	60 HP, 1800 RPM, motor only	3,750	360	4,110
2720	With magnetic starter	8,150	1,475	9,625
2740	75 HP, 1800 RPM, motor only	4,775	415	5,190
2760	With magnetic starter	9,175	1,550	10,725
2780	100 HP, 1800 RPM, motor only	5,675	555	6,230
2800	With magnetic starter	10,100	1,675	11,775
2820	125 HP, 1800 RPM, motor only	8,250	715	8,965
2840	With magnetic starter	20,800	1,975	22,775
2860	150 HP, 1800 RPM, motor only	9,350	835	10,185
2880	With magnetic starter	21,900	2,075	23,975
2900	200 HP, 1800 RPM, motor only	11,200	1,000	12,200
2920	With magnetic starter	23,700	2,250	25,950

Important: See the Reference Section for critical supporting data - Reference Numbers and City Cost Indexes

D5020 Lighting and Branch Wiring

A. Strip Fixture

B. Surface Mounted

C. Recessed

D. Pendent Mounted

Design Assumptions:

1. A 100 footcandle average maintained level of illumination.
2. Ceiling heights range from 9' to 11'.
3. Average reflectance values are assumed for ceilings, walls and floors.
4. Cool white (CW) fluorescent lamps with 3150 lumens for 40 watt lamps and 6300 lumens for 8' slimline lamps.
5. Four 40 watt lamps per 4' fixture and two 8' lamps per 8' fixture.
6. Average fixture efficiency values and spacing to mounting height ratios.
7. Installation labor is average U.S. rate as of January 1.

SERVICES | **D**

System Components	QUANTITY	UNIT	COST PER S.F. MAT.	COST PER S.F. INST.	COST PER S.F. TOTAL
SYSTEM D5020 208 0520					
FLUORESCENT FIXTURES MOUNTED 9'-11" ABOVE FLOOR, 100 FC					
TYPE A, 8 FIXTURES PER 400 S.F.					
Steel intermediate conduit, (IMC) 1/2" diam	.404	L.F.	.95	2.02	2.97
Wire, 600V, type THWN-THHN, copper, solid, #12	.008	C.L.F.	.06	.36	.42
Fluorescent strip fixture 8' long, surface mounted, two 75W SL	.020	Ea.	1.13	1.62	2.75
Steel outlet box 4" concrete	.020	Ea.	.24	.50	.74
Steel outlet box plate with stud, 4" concrete	.020	Ea.	.10	.13	.23
TOTAL			2.48	4.63	7.11

D5020 208	Fluorescent Fixtures (by Type)		COST PER S.F. MAT.	COST PER S.F. INST.	COST PER S.F. TOTAL
0520	Fluorescent fixtures, type A, 8 fixtures per 400 S.F.		2.48	4.63	7.11
0560	11 fixtures per 600 S.F.		2.37	4.48	6.85
0600	17 fixtures per 1000 S.F.	RD5020 -200	2.30	4.42	6.72
0640	23 fixtures per 1600 S.F.		2.13	4.18	6.31
0680	28 fixtures per 2000 S.F.		2.13	4.18	6.31
0720	41 fixtures per 3000 S.F.		2.10	4.17	6.27
0800	53 fixtures per 4000 S.F.		2.06	4.07	6.13
0840	64 fixtures per 5000 S.F.		2.06	4.07	6.13
0880	Type B, 11 fixtures per 400 S.F.		4.73	6.75	11.48
0920	15 fixtures per 600 S.F.		4.42	6.50	10.92
0960	24 fixtures per 1000 S.F.		4.33	6.45	10.78
1000	35 fixtures per 1600 S.F.		4.07	6.15	10.22
1040	42 fixtures per 2000 S.F.		4.02	6.20	10.22
1080	61 fixtures per 3000 S.F.		3.97	5.95	9.92
1160	80 fixtures per 4000 S.F.		3.91	6.10	10.01
1200	98 fixtures per 5000 S.F.		3.90	6.05	9.95
1240	Type C, 11 fixtures per 400 S.F.		3.68	7.15	10.83
1280	14 fixtures per 600 S.F.		3.36	6.70	10.06
1320	23 fixtures per 1000 S.F.		3.34	6.65	9.99
1360	34 fixtures per 1600 S.F.		3.27	6.55	9.82
1400	43 fixtures per 2000 S.F.		3.26	6.50	9.76
1440	63 fixtures per 3000 S.F.		3.20	6.40	9.60
1520	81 fixtures per 4000 S.F.		3.14	6.30	9.44
1560	101 fixtures per 5000 S.F.		3.14	6.30	9.44
1600	Type D, 8 fixtures per 400 S.F.		3.40	5.55	8.95
1640	12 fixtures per 600 S.F.		3.39	5.50	8.89
1680	19 fixtures per 1000 S.F.		3.28	5.35	8.63
1720	27 fixtures per 1600 S.F.		3.12	5.25	8.37
1760	34 fixtures per 2000 S.F.		3.09	5.20	8.29
1800	48 fixtures per 3000 S.F.		2.98	5.05	8.03
1880	64 fixtures per 4000 S.F.		2.98	5.05	8.03
1920	79 fixtures per 5000 S.F.		2.98	5.05	8.03

D5020 Lighting and Branch Wiring

Type C. Recessed, mounted on grid ceiling suspension system, 2' x 4', four 40 watt lamps, acrylic prismatic diffusers.

5.3 watts per S.F. for 100 footcandles.

3 watts per S.F. for 57 footcandles.

System Components			COST PER S.F.		
	QUANTITY	UNIT	MAT.	INST.	TOTAL
SYSTEM D5020 210 0200					
FLUORESCENT FIXTURES RECESS MOUNTED IN CEILING					
1 WATT PER S.F., 20 FC, 5 FIXTURES PER 1000 S.F.					
Steel intermediate conduit, (IMC) 1/2" diam	.128	L.F.	.30	.64	.94
Wire, 600 volt, type THW, copper, solid, #12	.003	C.L.F.	.02	.14	.16
Fluorescent fixture, recessed, 2'x 4', four 40W, w/ lens, for grid ceiling	.005	Ea.	.33	.54	.87
Steel outlet box 4" square	.005	Ea.	.06	.13	.19
Fixture whip, Greenfield w/#12 THHN wire	.005	Ea.	.02	.03	.05
TOTAL			.73	1.48	2.21

D5020 210	Fluorescent Fixtures (by Wattage)	COST PER S.F.		
		MAT.	INST.	TOTAL
0190	Fluorescent fixtures recess mounted in ceiling			
0200	1 watt per S.F., 20 FC, 5 fixtures per 1000 S.F.	.73	1.47	2.20
0240	2 watts per S.F., 40 FC, 10 fixtures per 1000 S.F.	1.46	2.89	4.35
0280	3 watts per S.F., 60 FC, 15 fixtures per 1000 S.F	2.19	4.36	6.55
0320	4 watts per S.F., 80 FC, 20 fixtures per 1000 S.F.	2.92	5.80	8.72
0400	5 watts per S.F., 100 FC, 25 fixtures per 1000 S.F.	3.65	7.25	10.90

RD5020
-200

D50 Electrical

D5020 Lighting and Branch Wiring

Type A

Type B

Type C

Type D

Type A. Recessed wide distribution reflector with flat glass lens 150 W.

Maximum spacing = 1.2 x mounting height.

13 watts per S.F. for 100 footcandles.

Type C. Recessed PAR–38 flood lamp with concentric louver 150 W.

Maximum spacing = 0.5 x mounting height.

19 watts per S.F. for 100 footcandles.

Type B. Recessed reflector down light with baffles 150 W.

Maximum spacing = 0.8 x mounting height.

18 watts per S.F. for 100 footcandles.

Type D. Recessed R–40 flood lamp with reflector skirt.

Maximum spacing = 0.7 x mounting height.

15 watts per S.F. for 100 footcandles.

SERVICES **D**

System Components	QUANTITY	UNIT	COST PER S.F.		
			MAT.	INST.	TOTAL
SYSTEM D5020 214 0400					
INCANDESCENT FIXTURE RECESS MOUNTED, 100 FC					
TYPE A, 34 FIXTURES PER 400 S.F.					
Steel intermediate conduit, (IMC) 1/2″ diam	1.060	L.F.	2.50	5.30	7.80
Wire, 600V, type THWN-THHN, copper, solid, #12	.033	C.L.F.	.23	1.50	1.73
Steel outlet box 4″ square	.085	Ea.	1.02	2.13	3.15
Fixture whip, Greenfield w/#12 THHN wire	.085	Ea.	.42	.53	.95
Incandescent fixture, recessed, w/lens, prewired, square trim, 200W	.085	Ea.	7.69	6.38	14.07
TOTAL			11.86	15.84	27.70

D5020 214	Incandescent Fixture (by Type)	COST PER S.F.		
		MAT.	INST.	TOTAL
0380	Incandescent fixture recess mounted, 100 FC			
0400	Type A, 34 fixtures per 400 S.F.	11.85	15.85	27.70
0440	49 fixtures per 600 S.F.	11.60	15.55	27.15
0480	63 fixtures per 800 S.F.	11.30	15.40	26.70
0520	90 fixtures per 1200 S.F.	10.95	15.05	26
0560	116 fixtures per 1600 S.F.	10.75	14.95	25.70
0600	143 fixtures per 2000 S.F.	10.65	14.90	25.55
0640	Type B, 47 fixtures per 400 S.F.	11.10	19.95	31.05
0680	66 fixtures per 600 S.F.	10.70	19.45	30.15
0720	88 fixtures per 800 S.F.	10.70	19.45	30.15
0760	127 fixtures per 1200 S.F.	10.50	19.30	29.80
0800	160 fixtures per 1600 S.F.	10.35	19.15	29.50
0840	206 fixtures per 2000 S.F.	10.30	19.05	29.35
0880	Type C, 51 fixtures per 400 S.F.	15	21	36
0920	74 fixtures per 600 S.F.	14.60	20.50	35.10
0960	97 fixtures per 800 S.F.	14.45	20.50	34.95
1000	142 fixtures per 1200 S.F.	14.25	20	34.25
1040	186 fixtures per 1600 S.F.	14.05	20	34.05
1080	230 fixtures per 2000 S.F.	14	20	34
1120	Type D, 39 fixtures per 400 S.F.	14	16.45	30.45
1160	57 fixtures per 600 S.F.	13.65	16.20	29.85
1200	75 fixtures per 800 S.F.	13.60	16.20	29.80
1240	109 fixtures per 1200 S.F.	13.30	16	29.30
1280	143 fixtures per 1600 S.F.	13.05	15.85	28.90
1320	176 fixtures per 2000 S.F.	12.95	15.80	28.75

Note: In the row for 0440/0480 there is a reference box: RD5020 -200

Type A. Recessed, wide distribution reflector with flat glass lens.

150 watt inside frost—2500 lumens per lamp.

PS–25 extended service lamp.

Maximum spacing = 1.2 x mounting height.

13 watts per S.F. for 100 footcandles.

System Components			COST PER S.F.		
	QUANTITY	UNIT	MAT.	INST.	TOTAL
SYSTEM D5020 216 0200					
INCANDESCENT FIXTURE RECESS MOUNTED, TYPE A					
1 WATT PER S.F., 8 FC, 6 FIXT PER 1000 S.F.					
Steel intermediate conduit, (IMC) 1/2" diam	.091	L.F.	.21	.46	.67
Wire, 600V, type THWN-THHN, copper, solid, #12	.002	C.L.F.	.01	.09	.10
Incandescent fixture, recessed, w/lens, prewired, square trim, 200W	.006	Ea.	.54	.45	.99
Steel outlet box 4" square	.006	Ea.	.07	.15	.22
Fixture whip, Greenfield w/#12 THHN wire	.006	Ea.	.03	.04	.07
TOTAL			.86	1.19	2.05

D5020 216	Incandescent Fixture (by Wattage)	COST PER S.F.		
		MAT.	INST.	TOTAL
0190	Incandescent fixture recess mounted, type A			
0200	1 watt per S.F., 8 FC, 6 fixtures per 1000 S.F.	.87	1.18	2.05
0240	2 watt per S.F., 16 FC, 12 fixtures per 1000 S.F. RD5020-200	1.75	2.37	4.12
0280	3 watt per S.F., 24 FC, 18 fixtures, per 1000 S.F.	2.61	3.51	6.12
0320	4 watt per S.F., 32 FC, 24 fixtures per 1000 S.F.	3.49	4.69	8.18
0400	5 watt per S.F., 40 FC, 30 fixtures per 1000 S.F.	4.36	5.85	10.21

D5020 Lighting and Branch Wiring

HIGH BAY FIXTURES

B. Metal halide 400 watt

C. High Pressure sodium 400 watt

E. Metal halide 1000 watt

F. High pressure sodium 1000 watt

G. Metal halide 1000 watt
   125,000 lumen lamp

System Components	QUANTITY	UNIT	COST PER S.F.		
			MAT.	INST.	TOTAL
SYSTEM D5020 220 0880					
HIGH INTENSITY DISCHARGE FIXTURE, 8'-10' ABOVE WORK PLANE, 100 FC					
TYPE B, 12 FIXTURES PER 900 S.F.					
Steel intermediate conduit, (IMC) 1/2" diam	.460	L.F.	1.09	2.30	3.39
Wire, 600V, type THWN-THHN, copper, solid, #10	.009	C.L.F.	.10	.45	.55
Steel outlet box 4" concrete	.009	Ea.	.11	.23	.34
Steel outlet box plate with stud 4" concrete	.009	Ea.	.04	.06	.10
Metal halide, hi bay, aluminum reflector, 400 W lamp	.009	Ea.	3.42	1.96	5.38
TOTAL			4.76	5	9.76

D5020 220	H.I.D. Fixture, High Bay, 8'-10' (by Type)		COST PER S.F.		
			MAT.	INST.	TOTAL
0500	High intensity discharge fixture, 8'-10' above work plane, 100 FC				
0880	Type B, 8 fixtures per 900 S.F.		4.75	4.99	9.74
0920	15 fixtures per 1800 S.F.	RD5020 -200	4.37	4.79	9.16
0960	24 fixtures per 3000 S.F.		4.37	4.79	9.16
1000	31 fixtures per 4000 S.F.		4.38	4.79	9.17
1040	38 fixtures per 5000 S.F.		4.38	4.79	9.17
1080	60 fixtures per 8000 S.F.		4.38	4.79	9.17
1120	72 fixtures per 10000 S.F.		3.94	4.44	8.38
1160	115 fixtures per 16000 S.F.		3.94	4.44	8.38
1200	230 fixtures per 32000 S.F.		3.94	4.44	8.38
1240	Type C, 4 fixtures per 900 S.F.		2.32	3.05	5.37
1280	8 fixtures per 1800 S.F.		2.32	3.05	5.37
1320	13 fixtures per 3000 S.F.		2.32	3.05	5.37
1360	17 fixtures per 4000 S.F.		2.32	3.05	5.37
1400	21 fixtures per 5000 S.F.		2.24	2.80	5.04
1440	33 fixtures per 8000 S.F.		2.23	2.75	4.98
1480	40 fixtures per 10000 S.F.		2.16	2.60	4.76
1520	63 fixtures per 16000 S.F.		2.16	2.60	4.76
1560	126 fixtures per 32000 S.F.		2.16	2.60	4.76

HIGH BAY FIXTURES

B. Metal halide 400 watt

C. High Pressure sodium 400 watt

E. Metal halide 1000 watt

F. High pressure sodium 1000 watt

G. Metal halide 1000 watt
   125,000 lumen lamp

System Components	QUANTITY	UNIT	COST PER S.F.		
			MAT.	INST.	TOTAL
SYSTEM D5020 222 0240					
HIGH INTENSITY DISCHARGE FIXTURE, 8'-10' ABOVE WORK PLANE					
1 WATT/S.F., TYPE B, 29 FC, 2 FIXTURES/1000 S.F.					
Steel intermediate conduit, (IMC) 1/2" diam	.100	L.F.	.24	.50	.74
Wire, 600V, type THWN-THHN, copper, solid, #10	.002	C.L.F.	.02	.10	.12
Steel outlet box 4" concrete	.002	Ea.	.02	.05	.07
Steel outlet box plate with stud, 4" concrete	.002	Ea.	.01	.01	.02
Metal halide, hi bay, aluminum reflector, 400 W lamp	.002	Ea.	.76	.44	1.20
TOTAL			1.05	1.10	2.15

D5020 222	H.I.D. Fixture, High Bay, 8'-10' (by Wattage)		COST PER S.F.		
			MAT.	INST.	TOTAL
0190	High intensity discharge fixture, 8'-10' above work plane				
0240	1 watt/S.F., type B, 29 FC, 2 fixtures/1000 S.F.		1.05	1.10	2.15
0280	Type C, 54 FC, 2 fixtures/1000 S.F.	RD5020 -200	.88	.83	1.71
0400	2 watt/S.F., type B, 59 FC, 4 fixtures/1000 S.F.		2.07	2.13	4.20
0440	Type C, 108 FC, 4 fixtures/1000 S.F.		1.74	1.62	3.36
0560	3 watt/S.F., type B, 103 FC, 7 fixtures/1000 S.F.		3.48	3.39	6.87
0600	Type C, 189 FC, 6 fixtures/1000 S.F.		2.69	2.09	4.78
0720	4 watt/S.F., type B, 133 FC, 9 fixtures/1000 S.F.		4.52	4.45	8.97
0760	Type C, 243 FC, 9 fixtures/1000 S.F.		3.88	3.61	7.49
0880	5 watt/S.F., type B, 162 FC, 11 fixtures/1000 S.F.		5.55	5.50	11.05
0920	Type C, 297 FC, 11 fixtures/1000 S.F.		4.78	4.48	9.26

Important: See the Reference Section for critical supporting data - Reference Numbers and City Cost Indexes

D50 Electrical

D5020 Lighting and Branch Wiring

HIGH BAY FIXTURES

B. Metal halide 400 watt

C. High Pressure sodium 400 watt

E. Metal halide 1000 watt

F. High pressure sodium 1000 watt

G. Metal halide 1000 watt
125,000 lumen lamp

System Components	QUANTITY	UNIT	COST PER S.F.		
			MAT.	INST.	TOTAL
SYSTEM D5020 224 1240					
HIGH INTENSITY DISCHARGE FIXTURE, 16' ABOVE WORK PLANE, 100 FC					
TYPE C, 5 FIXTURES PER 900 S.F.					
Steel intermediate conduit, (IMC) 1/2" diam	.260	L.F.	.61	1.30	1.91
Wire, 600V, type THWN-THHN, copper, solid, #10	.007	C.L.F.	.07	.35	.42
Steel outlet box 4" concrete	.006	Ea.	.07	.15	.22
Steel outlet box plate with stud, 4" concrete	.006	Ea.	.03	.04	.07
High pressure sodium, hi bay, aluminum reflector, 400 W lamp	.006	Ea.	2.10	1.31	3.41
TOTAL			2.88	3.15	6.03

D5020 224	H.I.D. Fixture, High Bay, 16' (by Type)	COST PER S.F.		
		MAT.	INST.	TOTAL
0510	High intensity discharge fixture, 16' above work plane, 100 FC			
1240	Type C, 5 fixtures per 900 S.F.	2.89	3.15	6.04
1280	9 fixtures per 1800 S.F.	2.70	3.30	6
1320	15 fixtures per 3000 S.F.	2.70	3.30	6
1360	18 fixtures per 4000 S.F.	2.53	2.95	5.48
1400	22 fixtures per 5000 S.F.	2.53	2.95	5.48
1440	36 fixtures per 8000 S.F.	2.53	2.95	5.48
1480	42 fixtures per 10,000 S.F.	2.33	3.05	5.38
1520	65 fixtures per 16,000 S.F.	2.33	3.05	5.38
1600	Type G, 4 fixtures per 900 S.F.	3.87	4.98	8.85
1640	6 fixtures per 1800 S.F.	3.27	4.64	7.91
1720	9 fixtures per 4000 S.F.	2.79	4.51	7.30
1760	11 fixtures per 5000 S.F.	2.79	4.51	7.30
1840	21 fixtures per 10,000 S.F.	2.58	4.06	6.64
1880	33 fixtures per 16,000 S.F.	2.58	4.06	6.64

RD5020 -200 (at rows 1280/1320)

SERVICES D

D5020 Lighting and Branch Wiring

HIGH BAY FIXTURES

B. Metal halide 400 watt

C. High Pressure sodium 400 watt

E. Metal halide 1000 watt

F. High pressure sodium 1000 watt

G. Metal halide 1000 watt
125,000 lumen lamp

System Components	QUANTITY	UNIT	COST PER S.F.		
			MAT.	INST.	TOTAL
SYSTEM D5020 226 0240					
HIGH INTENSITY DISCHARGE FIXTURE, 16' ABOVE WORK PLANE					
1 WATT/S.F., TYPE E, 42 FC, 1 FIXTURE/1000 S.F.					
Steel intermediate conduit, (IMC) 1/2" diam	.160	L.F.	.38	.80	1.18
Wire, 600V, type THWN-THHN, copper, solid, #10	.003	C.L.F.	.03	.15	.18
Steel outlet box 4" concrete	.001	Ea.	.01	.03	.04
Steel outlet box plate with stud, 4" concrete	.001	Ea.		.01	.01
Metal halide, hi bay, aluminum reflector, 1000 W lamp	.001	Ea.	.55	.25	.80
TOTAL			.97	1.24	2.21

D5020 226	H.I.D. Fixture, High Bay, 16' (by Wattage)	COST PER S.F.		
		MAT.	INST.	TOTAL
0190	High intensity discharge fixture, 16' above work plane			
0240	1 watt/S.F., type E, 42 FC, 1 fixture/1000 S.F.	.97	1.23	2.20
0280	Type G, 52 FC, 1 fixture/1000 S.F.	.97	1.23	2.20
0320	Type C, 54 FC, 2 fixture/1000 S.F.	1.11	1.38	2.49
0440	2 watt/S.F., type E, 84 FC, 2 fixture/1000 S.F.	1.95	2.51	4.46
0480	Type G, 105 FC, 2 fixture/1000 S.F.	1.95	2.51	4.46
0520	Type C, 108 FC, 4 fixture/1000 S.F.	2.22	2.76	4.98
0640	3 watt/S.F., type E, 126 FC, 3 fixture/1000 S.F.	2.93	3.74	6.67
0680	Type G, 157 FC, 3 fixture/1000 S.F.	2.93	3.74	6.67
0720	Type C, 162 FC, 6 fixture/1000 S.F.	3.33	4.14	7.47
0840	4 watt/S.F., type E, 168 FC, 4 fixture/1000 S.F.	3.91	5.05	8.96
0880	Type G, 210 FC, 4 fixture/1000 S.F.	3.91	5.05	8.96
0920	Type C, 243 FC, 9 fixture/1000 S.F.	4.81	5.75	10.56
1040	5 watt/S.F., type E, 210 FC, 5 fixture/1000 S.F.	4.88	6.25	11.13
1080	Type G, 262 FC, 5 fixture/1000 S.F.	4.88	6.25	11.13
1120	Type C, 297 FC, 11 fixture/1000 S.F.	5.90	7.15	13.05

RD5020 -200 (between rows 0280 and 0320)

Important: See the Reference Section for critical supporting data - Reference Numbers and City Cost Indexes

D5020 Lighting and Branch Wiring

HIGH BAY FIXTURES

B. Metal halide 400 watt

C. High Pressure sodium 400 watt

E. Metal halide 1000 watt

F. High pressure sodium 1000 watt

G. Metal halide 1000 watt
125,000 lumen lamp

SERVICES D

System Components			COST PER S.F.		
	QUANTITY	UNIT	MAT.	INST.	TOTAL
SYSTEM D5020 228 1240 **HIGH INTENSITY DISCHARGE FIXTURE, 20' ABOVE WORK PLANE, 100 FC** **TYPE C, 6 FIXTURES PER 900 S.F.**					
Steel intermediate conduit, (IMC) 1/2" diam	.350	L.F.	.83	1.75	2.58
Wire, 600V, type THWN-THHN, copper, solid, #10.	.011	C.L.F.	.12	.55	.67
Steel outlet box 4" concrete	.007	Ea.	.08	.18	.26
Steel outlet box plate with stud, 4" concrete	.007	Ea.	.03	.04	.07
High pressure sodium, hi bay, aluminum reflector, 400 W lamp	.007	Ea.	2.45	1.53	3.98
TOTAL			3.51	4.05	7.56

D5020 228	H.I.D. Fixture, High Bay, 20' (by Type)	COST PER S.F.		
		MAT.	INST.	TOTAL
0510	High intensity discharge fixture 20' above work plane, 100 FC			
1240	Type C, 6 fixtures per 900 S.F.	3.51	4.04	7.55
1280	10 fixtures per 1800 S.F.	3.14	3.80	6.94
1320	16 fixtures per 3000 S.F.	2.81	3.65	6.46
1360	20 fixtures per 4000 S.F.	2.81	3.65	6.46
1400	24 fixtures per 5000 S.F.	2.81	3.65	6.46
1440	38 fixtures per 8000 S.F.	2.74	3.50	6.24
1520	68 fixtures per 16000 S.F.	2.67	3.85	6.52
1560	132 fixtures per 32000 S.F.	2.67	3.85	6.52
1600	Type G, 4 fixtures per 900 S.F.	3.80	4.88	8.68
1640	6 fixtures per 1800 S.F.	3.25	4.64	7.89
1680	7 fixtures per 3000 S.F.	3.18	4.49	7.67
1720	10 fixtures per 4000 S.F.	3.16	4.44	7.60
1760	11 fixtures per 5000 S.F.	2.87	4.76	7.63
1800	18 fixtures per 8000 S.F.	2.87	4.76	7.63
1840	22 fixtures per 10000 S.F.	2.87	4.76	7.63
1880	34 fixtures per 16000 S.F.	2.80	4.61	7.41
1920	66 fixtures per 32000 S.F.	2.61	4.21	6.82

RD5020
-200

D5020 Lighting and Branch Wiring

HIGH BAY FIXTURES

B. Metal halide 400 watt

C. High Pressure sodium 400 watt

E. Metal halide 1000 watt

F. High pressure sodium 1000 watt

G. Metal halide 1000 watt
   125,000 lumen lamp

System Components	QUANTITY	UNIT	COST PER S.F.		
			MAT.	INST.	TOTAL
SYSTEM D5020 230 0240					
HIGH INTENSITY DISCHARGE FIXTURE, 20' ABOVE WORK PLANE					
1 WATT/S.F., TYPE E, 40 FC, 1 FIXTURE 1000 S.F.					
Steel intermediate conduit, (IMC) 1/2" diam	.160	L.F.	.38	.80	1.18
Wire, 600V, type THWN-THHN, copper, solid, #10	.005	C.L.F.	.05	.25	.30
Steel outlet box 4" concrete	.001	Ea.	.01	.03	.04
Steel outlet box plate with stud, 4" concrete	.001	Ea.		.01	.01
Metal halide, hi bay, aluminum reflector, 1000 W lamp	.001	Ea.	.55	.25	.80
TOTAL			.99	1.34	2.33

D5020 230	H.I.D. Fixture, High Bay, 20' (by Wattage)		COST PER S.F.		
			MAT.	INST.	TOTAL
0190	High intensity discharge fixture, 20' above work plane				
0240	1 watt/S.F., type E, 40 FC, 1 fixture/1000 S.F.		.99	1.33	2.32
0280	Type G, 50 FC, 1 fixture/1000 S.F.	RD5020 -200	.99	1.33	2.32
0320	Type C, 52 FC, 2 fixtures/1000 S.F.		1.16	1.54	2.70
0440	2 watt/S.F., type E, 81 FC, 2 fixtures/1000 S.F.		1.99	2.66	4.65
0480	Type G, 101 FC, 2 fixtures/1000 S.F.		1.99	2.66	4.65
0520	Type C, 104 FC, 4 fixtures/1000 S.F.		2.31	3.03	5.34
0640	3 watt/S.F., type E, 121 FC, 3 fixtures/1000 S.F.		2.98	3.99	6.97
0680	Type G, 151 FC, 3 fixtures/1000 S.F.		2.98	3.99	6.97
0720	Type C, 155 FC, 6 fixtures/1000 S.F.		3.47	4.57	8.04
0840	4 watt/S.F., type E, 161 FC, 4 fixtures/1000 S.F.		3.97	5.35	9.32
0880	Type G, 202 FC, 4 fixtures/1000 S.F.		3.97	5.35	9.32
0920	Type C, 233 FC, 9 fixtures/1000 S.F.		4.99	6.30	11.29
1040	5 watt/S.F., type E, 202 FC, 5 fixtures/1000 S.F.		4.96	6.65	11.61
1080	Type G, 252 FC, 5 fixtures/1000 S.F.		4.96	6.65	11.61
1120	Type C, 285 FC, 11 fixtures/1000 S.F.		6.15	7.85	14

Important: See the Reference Section for critical supporting data - Reference Numbers and City Cost Indexes

D5020 Lighting and Branch Wiring

HIGH BAY FIXTURES

B. Metal halide 400 watt

C. High Pressure sodium 400 watt

E. Metal halide 1000 watt

F. High pressure sodium 1000 watt

G. Metal halide 1000 watt
   125,000 lumen lamp

SERVICES **D**

System Components	QUANTITY	UNIT	COST PER S.F.		
			MAT.	INST.	TOTAL
SYSTEM D5020 232 1240					
HIGH INTENSITY DISCHARGE FIXTURE, 30' ABOVE WORK PLANE, 100 FC					
TYPE F, 4 FIXTURES PER 900 S.F.					
Steel intermediate conduit, (IMC) 1/2" diam	.580	L.F.	1.37	2.90	4.27
Wire, 600V, type THWN-THHN, copper, solid, #10	.018	C.L.F.	.19	.90	1.09
Steel outlet box 4" concrete	.004	Ea.	.05	.10	.15
Steel outlet box plate with stud, 4" concrete	.004	Ea.	.02	.03	.05
High pressure sodium, hi bay, aluminum, 1000 W lamp	.004	Ea.	2.02	1	3.02
TOTAL			3.65	4.93	8.58

D5020 232	H.I.D. Fixture, High Bay, 30' (by Type)		COST PER S.F.		
			MAT.	INST.	TOTAL
0510	High intensity discharge fixture, 30' above work plane, 100 FC				
1240	Type F, 4 fixtures per 900 S.F.		3.65	4.93	8.58
1280	6 fixtures per 1800 S.F.	R16510	3.15	4.69	7.84
1320	8 fixtures per 3000 S.F.	-105	3.15	4.69	7.84
1360	9 fixtures per 4000 S.F.		2.65	4.46	7.11
1400	10 fixtures per 5000 S.F.		2.65	4.46	7.11
1440	17 fixtures per 8000 S.F.		2.65	4.46	7.11
1480	18 fixtures per 10,000 S.F.		2.49	4.11	6.60
1520	27 fixtures per 16,000 S.F.		2.49	4.11	6.60
1560	52 fixtures per 32000 S.F.		2.48	4.06	6.54
1600	Type G, 4 fixtures per 900 S.F.		3.96	5.25	9.21
1640	6 fixtures per 1800 S.F.		3.31	4.74	8.05
1680	9 fixtures per 3000 S.F.		3.24	4.59	7.83
1720	11 fixtures per 4000 S.F.		3.16	4.44	7.60
1760	13 fixtures per 5000 S.F.		3.16	4.44	7.60
1800	21 fixtures per 8000 S.F.		3.16	4.44	7.60
1840	23 fixtures per 10,000 S.F.		2.72	4.71	7.43
1880	36 fixtures per 16,000 S.F.		2.72	4.71	7.43
1920	70 fixtures per 32,000 S.F.		2.72	4.71	7.43

D5020 Lighting and Branch Wiring

HIGH BAY FIXTURES

B. Metal halide 400 watt

C. High Pressure sodium 400 watt

E. Metal halide 1000 watt

F. High pressure sodium 1000 watt

G. Metal halide 1000 watt
   125,000 lumen lamp

System Components	QUANTITY	UNIT	COST PER S.F.		
			MAT.	INST.	TOTAL
SYSTEM D5020 234 0240					
HIGH INTENSITY DISCHARGE FIXTURE, 30' ABOVE WORK PLANE					
1 WATT/S.F., TYPE E, 37 FC, 1 FIXTURE/1000 S.F.					
Steel intermediate conduit, (IMC) 1/2" diam	.196	L.F.	.46	.98	1.44
Wire, 600V type THWN-THHN, copper, solid, #10	.006	C.L.F.	.06	.30	.36
Steel outlet box 4" concrete	.001	Ea.	.01	.03	.04
Steel outlet box plate with stud, 4" concrete	.001	Ea.		.01	.01
Metal halide, hi bay, aluminum reflector, 1000 W lamp	.001	Ea.	.55	.25	.80
TOTAL			1.08	1.57	2.65

D5020 234	H.I.D. Fixture, High Bay, 30' (by Wattage)	COST PER S.F.		
		MAT.	INST.	TOTAL
0190	High intensity discharge fixture, 30' above work plane			
0240	1 watt/S.F., type E, 37 FC, 1 fixture/1000 S.F.	1.09	1.56	2.65
0280	Type G, 45 FC., 1 fixture/1000 S.F.	1.09	1.56	2.65
0320	Type F, 50 FC, 1 fixture/1000 S.F.	.91	1.22	2.13
0440	2 watt/S.F., type E, 74 FC, 2 fixtures/1000 S.F.	2.18	3.12	5.30
0480	Type G, 92 FC, 2 fixtures/1000 S.F.	2.18	3.12	5.30
0520	Type F, 100 FC, 2 fixtures/1000 S.F.	1.84	2.49	4.33
0640	3 watt/S.F., type E, 110 FC, 3 fixtures/1000 S.F.	3.28	4.73	8.01
0680	Type G, 138FC, 3 fixtures/1000 S.F.	3.28	4.73	8.01
0720	Type F, 150 FC, 3 fixtures/1000 S.F.	2.75	3.71	6.46
0840	4 watt/S.F., type E, 148 FC, 4 fixtures/1000 S.F.	4.36	6.30	10.66
0880	Type G, 185 FC, 4 fixtures/1000 S.F.	4.36	6.30	10.66
0920	Type F, 200 FC, 4 fixtures/1000 S.F.	3.68	4.99	8.67
1040	5 watt/S.F., type E, 185 FC, 5 fixtures/1000 S.F.	5.45	7.85	13.30
1080	Type G, 230 FC, 5 fixtures/1000 S.F.	5.45	7.85	13.30
1120	Type F, 250 FC, 5 fixtures/1000 S.F.	4.59	6.20	10.79

(Reference: RD5020 -200, at row 0320)

D5020 Lighting and Branch Wiring

LOW BAY FIXTURES

J. Metal halide 250 watt

K. High pressure sodium 150 watt

System Components	QUANTITY	UNIT	COST PER S.F.		
			MAT.	INST.	TOTAL
SYSTEM D5020 236 0920					
HIGH INTENSITY DISCHARGE FIXTURE, 8'-10' ABOVE WORK PLANE, 50 FC					
TYPE J, 13 FIXTURES PER 1800 S.F.					
Steel intermediate conduit, (IMC) 1/2" diam	.550	L.F.	1.30	2.75	4.05
Wire, 600V, type THWN-THHN, copper, solid, #10	.012	C.L.F.	.13	.60	.73
Steel outlet box 4" concrete	.007	Ea.	.08	.18	.26
Steel outlet box plate with stud, 4" concrete	.007	Ea.	.03	.04	.07
Metal halide, lo bay, aluminum reflector, 250 W DX lamp	.007	Ea.	2.59	1.10	3.69
TOTAL			4.13	4.67	8.80

D5020 236	H.I.D. Fixture, Low Bay, 8'-10' (by Type)	COST PER S.F.		
		MAT.	INST.	TOTAL
0510	High intensity discharge fixture, 8'-10' above work plane, 50 FC			
0880	Type J, 7 fixtures per 900 S.F.	4.46	4.71	9.17
0920	13 fixtures per 1800 S.F.	4.13	4.67	8.80
0960	21 fixtures per 3000 S.F.	4.13	4.67	8.80
1000	28 fixtures per 4000 S.F.	4.13	4.67	8.80
1040	35 fixtures per 5000 S.F.	4.13	4.67	8.80
1120	62 fixtures per 10,000 S.F.	3.75	4.48	8.23
1160	99 fixtures per 16,000 S.F.	3.75	4.48	8.23
1200	199 fixtures per 32,000 S.F.	3.75	4.48	8.23
1240	Type K, 9 fixtures per 900 S.F.	4.13	4.03	8.16
1280	16 fixtures per 1800 S.F.	3.83	3.89	7.72
1320	26 fixtures per 3000 S.F.	3.81	3.84	7.65
1360	31 fixtures per 4000 S.F.	3.53	3.76	7.29
1400	39 fixtures per 5000 S.F.	3.53	3.76	7.29
1440	62 fixtures per 8000 S.F.	3.53	3.76	7.29
1480	78 fixtures per 10,000 S.F.	3.53	3.76	7.29
1520	124 fixtures per 16,000 S.F.	3.50	3.66	7.16
1560	248 fixtures per 32,000 S.F.	3.43	4.07	7.50

RD5020 -200

D50 Electrical

D5020 Lighting and Branch Wiring

LOW BAY FIXTURES
J. Metal halide 250 watt
K. High pressure sodium 150 watt

System Components	QUANTITY	UNIT	COST PER S.F.		
			MAT.	INST.	TOTAL
SYSTEM D5020 238 0240					
HIGH INTENSITY DISCHARGE FIXTURE, 8'-10' ABOVE WORK PLANE					
1 WATT/S.F., TYPE J, 30 FC, 4 FIXTURES/1000 S.F.					
Steel intermediate conduit, (IMC) 1/2" diam	.280	L.F.	.66	1.40	2.06
Wire, 600V, type THWN-THHN, copper, solid, #10	.008	C.L.F.	.09	.40	.49
Steel outlet box 4" concrete	.004	Ea.	.05	.10	.15
Steel outlet box plate with stud, 4" concrete	.004	Ea.	.02	.03	.05
Metal halide, lo bay, aluminum reflector, 250 W DX lamp	.004	Ea.	1.48	.63	2.11
TOTAL			2.30	2.56	4.86

D5020 238	H.I.D. Fixture, Low Bay, 8'-10' (by Wattage)	COST PER S.F.		
		MAT.	INST.	TOTAL
0190	High intensity discharge fixture, 8'-10' above work plane			
0240	1 watt/S.F., type J, 30 FC, 4 fixtures/1000 S.F.	2.29	2.55	4.84
0280	Type K, 29 FC, 5 fixtures/1000 S.F.	2.17	2.29	4.46
0400	2 watt/S.F., type J, 52 FC, 7 fixtures/1000 S.F.	4.09	4.68	8.77
0440	Type K, 63 FC, 11 fixtures/1000 S.F.	4.67	4.82	9.49
0560	3 watt/S.F., type J, 81 FC, 11 fixtures/1000 S.F.	6.30	7.05	13.35
0600	Type K, 92 FC, 16 fixtures/1000 S.F.	6.85	7.10	13.95
0720	4 watt/S.F., type J, 103 FC, 14 fixtures/1000 S.F.	8.15	9.30	17.45
0760	Type K, 127 FC, 22 fixtures/1000 S.F.	9.35	9.65	19
0880	5 watt/S.F., type J, 133 FC, 18 fixtures/1000 S.F.	10.40	11.75	22.15
0920	Type K, 155 FC, 27 fixtures/1000 S.F.	11.50	11.95	23.45

(Note at 0280/0400 row: RD5020 -200)

Important: See the Reference Section for critical supporting data - Reference Numbers and City Cost Indexes

D50 Electrical

D5020 Lighting and Branch Wiring

LOW BAY FIXTURES
J. Metal halide 250 watt
K. High pressure sodium 150 watt

System Components	QUANTITY	UNIT	COST PER S.F. MAT.	INST.	TOTAL
SYSTEM D5020 240 0880					
HIGH INTENSITY DISCHARGE FIXTURE, 16' ABOVE WORK PLANE, 50 FC					
TYPE J, 9 FIXTURES PER 900 S.F.					
Steel intermediate conduit, (IMC) 1/2" diam	.630	L.F.	1.49	3.15	4.64
Wire, 600V type, THWN-THHN, copper, solid, #10	.012	C.L.F.	.13	.60	.73
Steel outlet box 4" concrete	.010	Ea.	.12	.25	.37
Steel outlet box plate with stud, 4" concrete	.010	Ea.	.05	.06	.11
Metal halide, lo bay, aluminum reflector, 250 W DX lamp	.010	Ea.	3.70	1.57	5.27
TOTAL			5.49	5.63	11.12

D5020 240	H.I.D. Fixture, Low Bay, 16' (by Type)	MAT.	INST.	TOTAL
0510	High intensity discharge fixture, 16' above work plane, 50 FC			
0880	Type J, 9 fixtures per 900 S.F.	5.50	5.65	11.15
0920	14 fixtures per 1800 S.F.	4.72	5.30	10.02
0960	24 fixtures per 3000 S.F.	4.75	5.40	10.15
1000	32 fixtures per 4000 S.F.	4.75	5.40	10.15
1040	35 fixtures per 5000 S.F.	4.40	5.25	9.65
1080	56 fixtures per 8000 S.F.	4.40	5.25	9.65
1120	70 fixtures per 10,000 S.F.	4.38	5.25	9.63
1160	111 fixtures per 16,000 S.F.	4.39	5.25	9.64
1200	222 fixtures per 32,000 S.F.	4.39	5.25	9.64
1240	Type K, 11 fixtures per 900 S.F.	5.15	5.25	10.40
1280	20 fixtures per 1800 S.F.	4.75	4.87	9.62
1320	29 fixtures per 3000 S.F.	4.46	4.78	9.24
1360	39 fixtures per 4000 S.F.	4.43	4.73	9.16
1400	44 fixtures per 5000 S.F.	4.15	4.64	8.79
1440	62 fixtures per 8000 S.F.	4.15	4.64	8.79
1480	87 fixtures per 10,000 S.F.	4.15	4.64	8.79
1520	138 fixtures per 16,000 S.F.	4.15	4.64	8.79

SERVICES D

D5020 Lighting and Branch Wiring

LOW BAY FIXTURES
J. Metal halide 250 watt
K. High pressure sodium 150 watt

System Components	QUANTITY	UNIT	COST PER S.F.		
			MAT.	INST.	TOTAL
SYSTEM D5020 242 0240					
HIGH INTENSITY DISCHARGE FIXTURE, 16' ABOVE WORK PLANE					
1 WATT/S.F., TYPE J, 28 FC, 4 FIXTURES/1000 S.F.					
Steel intermediate conduit, (IMC) 1/2" diam	.328	L.F.	.77	1.64	2.41
Wire, 600V, type THWN-THHN, copper, solid, #10	.010	C.L.F.	.11	.50	.61
Steel outlet box 4" concrete	.004	Ea.	.05	.10	.15
Steel outlet box plate with stud, 4" concrete	.004	Ea.	.02	.03	.05
Metal halide, lo bay, aluminum reflector, 250 W DX lamp	.004	Ea.	1.48	.63	2.11
TOTAL			2.43	2.90	5.33

D5020 242	H.I.D. Fixture, Low Bay, 16' (by Wattage)		COST PER S.F.		
			MAT.	INST.	TOTAL
0190	High intensity discharge fixture, mounted 16' above work plane				
0240	1 watt/S.F., type J, 28 FC, 4 fixt./1000 S.F.		2.43	2.89	5.32
0280	Type K, 27 FC, 5 fixt./1000 S.F.	RD5020 -200	2.55	3.24	5.79
0400	2 watt/S.F., type J, 48 FC, 7 fixt/1000 S.F.		4.47	5.60	10.07
0440	Type K, 58 FC, 11 fixt/1000 S.F.		5.40	6.60	12
0560	3 watt/S.F., type J, 75 FC, 11 fixt/1000 S.F.		6.90	8.50	15.40
0600	Type K, 85 FC, 16 fixt/1000 S.F.		7.95	9.85	17.80
0720	4 watt/S.F., type J, 95 FC, 14 fixt/1000 S.F.		8.95	11.20	20.15
0760	Type K, 117 FC, 22 fixt/1000 S.F.		10.85	13.25	24.10
0880	5 watt/S.F., type J, 122 FC, 18 fixt/1000 S.F.		11.35	14.10	25.45
0920	Type K, 143 FC, 27 fixt/1000 S.F.		13.40	16.50	29.90

Important: See the Reference Section for critical supporting data - Reference Numbers and City Cost Indexes

D50 Electrical

D5030 Communications and Security

Description: System below includes telephone fitting installed. Does not include cable.

When poke thru fittings and telepoles are used for power, they can also be used for telephones at a negligible additional cost.

SERVICES **D**

System Components	QUANTITY	UNIT	COST PER S.F.		
			MAT.	INST.	TOTAL
SYSTEM D5030 310 0200					
TELEPHONE SYSTEMS, UNDERFLOOR DUCT, 5' ON CENTER, LOW DENSITY					
Underfloor duct 7-1/4" w/insert 2' O.C. 1-3/8" x 7-1/4" super duct	.190	L.F.	4.75	1.90	6.65
Vertical elbow for underfloor superduct, 7-1/4", included					
Underfloor duct conduit adapter, 2" x 1-1/4", included					
Underfloor duct junction box, single duct, 7-1/4" x 3 1/8"	.003	Ea.	1.13	.38	1.51
Underfloor junction box carpet pan	.003	Ea.	.80	.02	.82
Underfloor duct outlet, low tension	.004	Ea.	.30	.25	.55
TOTAL			6.98	2.55	9.53

D5030 310	Telephone Systems	COST PER S.F.		
		MAT.	INST.	TOTAL
0200	Telephone systems, underfloor duct, 5' on center, low density	7	2.54	9.54
0240	5' on center, high density	7.30	2.79	10.09
0280	7' on center, low density	5.60	2.11	7.71
0320	7' on center, high density	5.90	2.36	8.26
0400	Poke thru fittings, low density	.92	.83	1.75
0440	High density	1.84	1.67	3.51
0520	Telepoles, low density	.78	.54	1.32
0560	High density	1.55	1.08	2.63
0640	Conduit system with floor boxes, low density	.99	.89	1.88
0680	High density	1.98	1.76	3.74

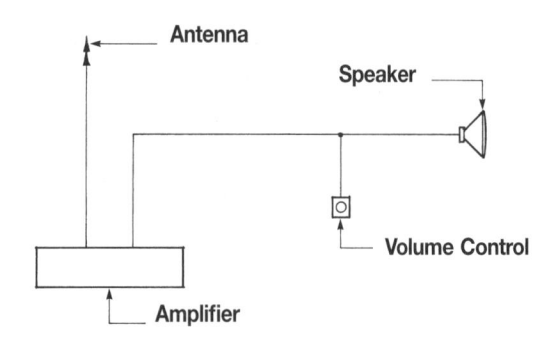

Sound System Includes AM–FM antenna, outlets, rigid conduit, and copper wire.
Fire Detection System Includes pull stations, signals, smoke and heat detectors, rigid conduit, and copper wire.
Intercom System Includes master and remote stations, rigid conduit, and copper wire.
Master Clock System Includes clocks, bells, rigid conduit, and copper wire.
Master TV Antenna Includes antenna, VHF–UHF reception and distribution, rigid conduit, and copper wire.

System Components	QUANTITY	UNIT	COST EACH		
			MAT.	INST.	TOTAL
SYSTEM D5030 910 0220					
SOUND SYSTEM, INCLUDES OUTLETS, BOXES, CONDUIT & WIRE					
Steel intermediate conduit, (IMC) 1/2" diam	1200.000	L.F.	2,832	6,000	8,832
Wire sound shielded w/drain, #22-2 conductor	15.500	C.L.F.	294.50	968.75	1,263.25
Sound system speakers ceiling or wall	12.000	Ea.	1,236	750	1,986
Sound system volume control	12.000	Ea.	912	750	1,662
Sound system amplifier, 250 Watts	1.000	Ea.	1,525	500	2,025
Sound system antenna, AM FM	1.000	Ea.	190	125	315
Sound system monitor panel	1.000	Ea.	340	125	465
Sound system cabinet	1.000	Ea.	740	500	1,240
Steel outlet box 4" square	12.000	Ea.	29.52	300	329.52
Steel outlet box 4" plaster rings	12.000	Ea.	26.76	94.20	120.96
TOTAL			8,125.78	10,112.95	18,238.73

D5030 910	Communication & Alarm Systems		COST EACH		
			MAT.	INST.	TOTAL
0200	Communication & alarm systems, includes outlets, boxes, conduit & wire				
0210	Sound system, 6 outlets		5,650	6,325	11,975
0220	12 outlets	RD5010 -110	8,125	10,100	18,225
0240	30 outlets		14,200	19,100	33,300
0280	100 outlets		43,700	64,000	107,700
0320	Fire detection systems, 12 detectors		2,850	5,250	8,100
0360	25 detectors		4,925	8,875	13,800
0400	50 detectors		9,550	17,400	26,950
0440	100 detectors		17,200	31,400	48,600
0480	Intercom systems, 6 stations		3,300	4,325	7,625
0520	12 stations		6,675	8,625	15,300
0560	25 stations		11,700	16,600	28,300
0600	50 stations		22,400	31,100	53,500
0640	100 stations		44,100	61,000	105,100
0680	Master clock systems, 6 rooms		4,550	7,075	11,625
0720	12 rooms		7,050	12,000	19,050
0760	20 rooms		9,400	17,000	26,400
0800	30 rooms		16,100	31,400	47,500
0840	50 rooms		26,200	53,000	79,200
0880	100 rooms		50,500	105,000	155,500
0920	Master TV antenna systems, 6 outlets		2,925	4,475	7,400
0960	12 outlets		5,475	8,325	13,800
1000	30 outlets		11,300	19,300	30,600
1040	100 outlets		37,700	63,000	100,700

D5090 Other Electrical Systems

Engine · Muffler · Sleeve · Transfer Switch · Charger · Battery

Description: System below tabulates the installed cost for generators by kW. Included in costs are battery, charger, muffler, and transfer switch.

No conduit, wire, or terminations included.

SERVICES **D**

System Components	QUANTITY	UNIT	COST PER kW		
			MAT.	INST.	TOTAL
SYSTEM D5090 210 0200 **GENERATOR SET, INCL. BATTERY, CHARGER, MUFFLER & TRANSFER SWITCH** **GAS/GASOLINE OPER., 3 PHASE, 4 WIRE, 277/480V, 7.5 kW**					
Generator set, gas or gasoline operated, 3 ph 4 W, 277/480 V, 7.5 kW	.133	Ea.	880	226.93	1,106.93
TOTAL			880	226.93	1,106.93

D5090 210	Generators (by kW)	COST PER kW		
		MAT.	INST.	TOTAL
0190	Generator sets, include battery, charger, muffler & transfer switch			
0200	Gas/gasoline operated, 3 phase, 4 wire, 277/480 volt, 7.5 kW	880	227	1,107
0240	11.5 kW RD5010-110	815	173	988
0280	20 kW	550	112	662
0320	35 kW	375	73	448
0360	80 kW	283	44.50	327.50
0400	100 kW	248	42.50	290.50
0440	125 kW	405	40.50	445.50
0480	185 kW	360	30.50	390.50
0560	Diesel engine with fuel tank, 30 kW	585	85	670
0600	50 kW	435	67	502
0720	125 kW	264	38.50	302.50
0760	150 kW	252	36	288
0800	175 kW	236	32.50	268.50
0840	200 kW	213	29.50	242.50
0880	250 kW	200	24.50	224.50
0920	300 kW	180	21.50	201.50
0960	350 kW	176	20	196
1000	400 kW	189	18.50	207.50
1040	500 kW	190	15.60	205.60

D5090 Other Electrical Systems

Low Density and Medium Density Baseboard Radiation

The costs shown in Table below are based on the following system considerations:

1. The heat loss per square foot is based on approximately 34 BTU/hr. per S.F. of floor or 10 watts per S.F. of floor.
2. Baseboard radiation is based on the low watt density type rated 187 watts per L.F. and the medium density type rated 250 watts per L.F.
3. Thermostat is not included.
4. Wiring costs include branch circuit wiring.

System Components	QUANTITY	UNIT	COST PER S.F.		
			MAT.	INST.	TOTAL
SYSTEM D5090 510 1000					
ELECTRIC BASEBOARD RADIATION, LOW DENSITY, 900 S.F., 31 MBH, 9 kW					
Electric baseboard radiator, 5' long 935 watt	.011	Ea.	.67	.97	1.64
Steel intermediate conduit, (IMC) 1/2" diam	.170	L.F.	.40	.85	1.25
Wire 600 volt, type THW, copper, solid, #12	.005	C.L.F.	.04	.23	.27
TOTAL			1.11	2.05	3.16

D5090 510	Electric Baseboard Radiation (Low Density)	COST PER S.F.		
		MAT.	INST.	TOTAL
1000	Electric baseboard radiation, low density, 900 S.F., 31 MBH, 9 kW	1.10	2.05	3.15
1200	1500 S.F., 51 MBH, 15 kW	1.08	2	3.08
1400	2100 S.F., 72 MBH, 21 kW	.98	1.79	2.77
1600	3000 S.F., 102 MBH, 30 kW	.88	1.63	2.51
2000	Medium density, 900 S.F., 31 MBH, 9 kW	.98	1.87	2.85
2200	1500 S.F., 51 MBH, 15 kW	.96	1.82	2.78
2400	2100 S.F., 72 MBH, 21 kW	.91	1.70	2.61
2600	3000 S.F., 102 MBH, 30 kW	.82	1.54	2.36

D50 Electrical

D5090 Other Electrical Systems

Commercial Duty Baseboard Radiation
The costs shown in Table below are based on the following system considerations:

1. The heat loss per square foot is based on approximately 41 BTU/hr. per S.F. of floor or 12 watts per S.F.
2. The baseboard radiation is of the commercial duty type rated 250 watts per L.F. served by 277 volt, single phase power.
3. Thermostat is not included.
4. Wiring costs include branch circuit wiring.

System Components	QUANTITY	UNIT	COST PER S.F.		
			MAT.	INST.	TOTAL
SYSTEM D5090 520 1000					
ELECTRIC BASEBOARD RADIATION, MEDIUM DENSITY, 1230 S.F., 51 MBH, 15 kW					
Electric baseboard radiator, 5' long	.013	Ea.	.79	1.14	1.93
Steel intermediate conduit, (IMC) 1/2" diam	.154	L.F.	.36	.77	1.13
Wire 600 volt, type THW, copper, solid, #12	.004	C.L.F.	.03	.18	.21
TOTAL			1.18	2.09	3.27

D5090 520	Electric Baseboard Radiation (Medium Density)	COST PER S.F.		
		MAT.	INST.	TOTAL
1000	Electric baseboard radiation, medium density, 1230 SF, 51 MBH, 15 kW	1.18	2.10	3.28
1200	2500 S.F. floor area, 106 MBH, 31 kW	1.10	1.96	3.06
1400	3700 S.F. floor area, 157 MBH, 46 kW	1.07	1.91	2.98
1600	4800 S.F. floor area, 201 MBH, 59 kW	.98	1.77	2.75
1800	11,300 S.F. floor area, 464 MBH, 136 kW	.93	1.62	2.55
2000	30,000 S.F. floor area, 1229 MBH, 360 kW	.91	1.58	2.49

For information about Means Estimating Seminars, see yellow pages 12 and 13 in back of book

SERVICES **D**

E EQUIPMENT & FURNISHINGS

E1010 Commercial Equipment

E1010 110	Security/Vault, EACH	COST EACH		
		MAT.	INST.	TOTAL
0100	Bank equipment, drive up window, drawer & mike, no glazing, economy	5,175	1,150	6,325
0110	Deluxe	8,725	2,300	11,025
0120	Night depository, economy	7,150	1,150	8,300
0130	Deluxe	10,100	2,300	12,400
0140	Pneumatic tube systems, 2 station, standard	24,500	3,800	28,300
0150	Teller, automated, 24 hour, single unit	44,000	3,800	47,800
0160	Teller window, bullet proof glazing, 44" x 60"	3,100	690	3,790
0170	Pass through, painted steel, 72" x 40"	3,150	1,450	4,600
0300	Safe, office type, 1 hr. rating, 34" x 20" x 20"	1,900		1,900
0310	4 hr. rating, 62" x 33" x 20"	7,825		7,825
0320	Data storage, 4 hr. rating, 63" x 44" x 16"	10,500		10,500
0330	Jewelers, 63" x 44" x 16"	16,800		16,800
0340	Money, "B" label, 9" x 14" x 14"	455		455
0350	Tool and torch resistive, 24" x 24" x 20"	7,550	171	7,721
0500	Security gates-scissors type, painted steel, single, 6' high, 5-1/2' wide	156	288	444
0510	Double gate, 7-1/2' high, 14' wide	430	575	1,005

E1010 510	Mercantile Equipment, EACH	COST EACH		
		MAT.	INST.	TOTAL
0015	Barber equipment, chair, hydraulic, economy	475	18.45	493.45
0020	Deluxe	3,025	27.50	3,052.50
0030	Console, including mirrors, deluxe	455	96.50	551.50
0040	Sink, hair washing basin	310	64.50	374.50
0100	Checkout counter, single belt	2,525	68	2,593
0110	Double belt, power take-away	4,250	76	4,326
0200	Display cases, freestanding, glass and aluminum, 3'-6" x 3' x 1'-0" deep	1,175	111	1,286
0210	5'-10" x 4' x 1'-6" deep	1,800	148	1,948
0220	Wall mounted, glass and aluminum, 3' x 4' x 1'-4" deep	1,250	177	1,427
0230	16' x 4' x 1'-4" deep	3,600	590	4,190
0240	Table exhibit, flat, 3' x 4' x 2'-0" wide	815	177	992
0250	Sloping, 3' x 4' x 3'-0" wide	1,250	295	1,545
0300	Refrigerated food cases, dairy, multi-deck, rear sliding doors, 6 ft. long	10,800	310	11,110
0310	Delicatessen case, service type, 12 ft. long	6,375	239	6,614
0320	Frozen food, chest type, 12 ft. long	6,400	283	6,683
0330	Glass door, reach-in, 5 door	12,200	310	12,510
0340	Meat case, 12 ft. long, single deck	5,250	283	5,533
0350	Multi-deck	9,050	300	9,350
0360	Produce case, 12 ft. long, single deck	6,950	283	7,233
0370	Multi-deck	7,750	300	8,050

E1010 610	Laundry/Dry Cleaning, EACH	COST EACH		
		MAT.	INST.	TOTAL
0100	Laundry equipment, dryers, gas fired, residential, 16 lb. capacity	715	171	886
0110	Commercial, 30 lb. capacity, single	3,025	171	3,196
0120	Dry cleaners, electric, 20 lb. capacity	36,100	5,075	41,175
0130	30 lb. capacity	49,500	6,775	56,275
0140	Ironers, commercial, 120" with canopy, 8 roll	173,500	14,500	188,000
0150	Institutional, 110", single roll	30,300	2,500	32,800
0160	Washers, residential, 4 cycle	765	171	936
0170	Commercial, coin operated, deluxe	3,225	171	3,396

E

EQUIPMENT & FURNISHINGS

E10 Equipment

E1020 Institutional Equipment

E1020 110	Ecclesiastical Equipment, EACH	COST EACH		
		MAT.	INST.	TOTAL
0090	Church equipment, altar, wood, custom, plain	2,100	315	2,415
0100	Granite, custom, deluxe	26,700	4,250	30,950
0110	Baptistry, fiberglass, economy	3,200	1,150	4,350
0120	Bells & carillons, keyboard operation	17,200	7,100	24,300
0130	Confessional, wood, single, economy	2,675	740	3,415
0140	Double, deluxe	16,700	2,225	18,925
0150	Steeples, translucent fiberglas, 30" square, 15' high	4,150	1,475	5,625
0160	Porcelain enamel, custom, 60' high	21,800	9,750	31,550

E1020 130	Ecclesiastical Equipment, L.F.	COST PER L.F.		
		MAT.	INST.	TOTAL
0100	Arch. equip., church equip. pews, bench type, hardwood, economy	74	22	96
0110	Deluxe	147	29.50	176.50

E1020 210	Library Equipment, EACH	COST EACH		
		MAT.	INST.	TOTAL
0110	Library equipment, carrels, metal, economy	258	88.50	346.50
0120	Hardwood, deluxe	940	111	1,051

E1020 230	Library Equipment, L.F.	COST PER L.F.		
		MAT.	INST.	TOTAL
0100	Library equipment, book shelf, metal, single face, 90" high x 10" shelf	176	37	213
0110	Double face, 90" high x 10" shelf	330	80	410
0120	Charging desk, built-in, with counter, plastic laminate	460	63.50	523.50

E1020 310	Theater and Stage Equipment, EACH	COST EACH		
		MAT.	INST.	TOTAL
0200	Movie equipment, changeover, economy	450		450
0210	Film transport, incl. platters and autowind, economy	4,875		4,875
0220	Lamphouses, incl. rectifiers, xenon, 1000W	6,450	250	6,700
0230	4000W	9,125	335	9,460
0240	Projector mechanisms, 35 mm, economy	10,700		10,700
0250	Deluxe	14,700		14,700
0260	Sound systems, incl. amplifier, single, economy	3,200	555	3,755
0270	Dual, Dolby/super sound	16,000	1,250	17,250
0280	Projection screens, wall hung, manual operation, 50 S.F., economy	283	88.50	371.50
0290	Electric operation, 100 S.F., deluxe	1,925	445	2,370
0400	Stage equipment, control boards, incl. dimmers and breakers, economy	12,700	500	13,200
0410	Deluxe	116,500	2,500	119,000
0420	Spotlight, stationary, quartz, 6" lens	104	125	229
0430	Follow, incl. transformer, 2100W	1,300	125	1,425

E1020 320	Theater and Stage Equipment, S.F.	COST PER S.F.		
		MAT.	INST.	TOTAL
0090	Movie equipment, projection screens, rigid in wall, acrylic, 1/4" thick	40.50	4.25	44.75
0100	1/2" thick	47	6.40	53.40
0110	Stage equipment, curtains, velour, medium weight	7.50	1.48	8.98
0120	Silica based yarn, fireproof	14.45	17.70	32.15
0130	Stages, portable with steps, folding legs, 8" high	25		25
0140	Telescoping platforms, aluminum, deluxe	36.50	23	59.50

E1020 330	Theater and Stage Equipment, L.F.	COST PER L.F.		
		MAT.	INST.	TOTAL
0100	Stage equipment, curtain track, heavy duty	50	49	99
0110	Lights, border, quartz, colored	164	25	189

EQUIPMENT & FURNISHINGS

E

E1020 Institutional Equipment

E1020 610	Detention Equipment, EACH	COST EACH		
		MAT.	INST.	TOTAL
0110	Detention equipment, cell front rolling door, 7/8" bars, 5' x 7' high	5,325	1,225	6,550
0120	Cells, prefab., including front, 5' x 7' x 7' deep	8,800	1,625	10,425
0130	Doors and frames, 3' x 7', single plate	4,000	605	4,605
0140	Double plate	4,925	605	5,530
0150	Toilet apparatus, incl wash basin	3,300	760	4,060
0160	Visitor cubicle, vision panel, no intercom	2,900	1,225	4,125

E1020 710	Laboratory Equipment, EACH	COST EACH		
		MAT.	INST.	TOTAL
0110	Laboratory equipment, glassware washer, distilled water, economy	6,950	565	7,515
0120	Deluxe	9,475	1,025	10,500
0130	Glove box, fiberglass, bacteriological	15,300		15,300
0140	Radio isotope	13,100		13,100
0150	Titration unit, 4-2000 ml. reservoirs	9,950		9,950

E1020 720	Laboratory Equipment, S.F.	COST PER S.F.		
		MAT.	INST.	TOTAL
0100	Arch. equip., lab equip., counter tops, acid proof, economy	29	10.80	39.80
0110	Stainless steel	85.50	10.80	96.30

E1020 730	Laboratory Equipment, L.F.	COST PER L.F.		
		MAT.	INST.	TOTAL
0110	Laboratory equipment, cabinets, wall, open	120	44.50	164.50
0120	Base, drawer units	395	49	444
0130	Fume hoods, not incl. HVAC, economy	600	164	764
0140	Deluxe incl. fixtures	1,425	370	1,795

E1020 810	Medical Equipment, EACH	COST EACH		
		MAT.	INST.	TOTAL
0100	Dental equipment, central suction system, economy	3,325	430	3,755
0110	Compressor-air, deluxe	25,400	910	26,310
0120	Chair, hydraulic, economy	4,175	910	5,085
0130	Deluxe	8,875	1,825	10,700
0140	Drill console with accessories, economy	3,775	284	4,059
0150	Deluxe	10,800	284	11,084
0160	X-ray unit, portable	5,025	114	5,139
0170	Panoramic unit	22,800	755	23,555
0300	Medical equipment, autopsy table, standard	8,425	515	8,940
0310	Deluxe	15,200	855	16,055
0320	Incubators, economy	3,425		3,425
0330	Deluxe	19,100		19,100
0700	Station, scrub-surgical, single, economy	2,250	171	2,421
0710	Dietary, medium, with ice	15,200		15,200
0720	Sterilizers, general purpose, single door, 20" x 20" x 28"	14,500		14,500
0730	Floor loading, double door, 28" x 67" x 52"	193,500		193,500
0740	Surgery tables, standard	13,200	820	14,020
0750	Deluxe	29,900	1,150	31,050
0770	Tables, standard, with base cabinets, economy	1,225	295	1,520
0780	Deluxe	3,650	445	4,095
0790	X-ray, mobile, economy	12,700		12,700
0800	Stationary, deluxe	206,500		206,500

Important: See the Reference Section for critical supporting data - Reference Numbers and City Cost Indexes

E10 Equipment

E1030 Vehicular Equipment

E1030 110	Vehicular Service Equipment, EACH	COST EACH		
		MAT.	INST.	TOTAL
0110	Automotive equipment, compressors, electric, 1-1/2 H.P., std. controls	310	825	1,135
0120	5 H.P., dual controls	1,950	1,225	3,175
0130	Hoists, single post, 4 ton capacity, swivel arms	4,975	3,100	8,075
0140	Dual post, 12 ton capacity, adjustable frame	8,525	8,250	16,775
0150	Lube equipment, 3 reel type, with pumps	7,875	2,475	10,350
0160	Product dispenser, 6 nozzles, w/vapor recovery, not incl. piping, installed	20,800		20,800
0800	Scales, dial type, built in floor, 5 ton capacity, 8' x 6' platform	7,175	2,650	9,825
0810	10 ton capacity, 9' x 7' platform	10,600	3,800	14,400
0820	Truck (including weigh bridge), 20 ton capacity, 24' x 10'	11,000	4,425	15,425
0830	Digital type, truck, 60 ton capacity, 75' x 10'	33,600	11,100	44,700
0840	Concrete foundations, 8' x 6'	915	3,350	4,265
0850	70' x 10'	3,950	11,100	15,050

E1030 210	Parking Control Equipment, EACH	COST EACH		
		MAT.	INST.	TOTAL
0110	Parking equipment, automatic gates, 8 ft. arm, one way	3,825	910	4,735
0120	Traffic detectors, single treadle	2,425	415	2,840
0130	Booth for attendant, economy	6,025		6,025
0140	Deluxe	25,600		25,600
0150	Ticket printer/dispenser, rate computing	8,200	715	8,915
0160	Key station on pedestal	740	244	984

E1030 310	Loading Dock Equipment, EACH	COST EACH		
		MAT.	INST.	TOTAL
0110	Dock bumpers, rubber blocks, 4-1/2" thick, 10" high, 14" long	49	17.05	66.05
0120	6" thick, 20" high, 11" long	142	34	176
0130	Dock boards, H.D., 5' x 5', aluminum, 5000 lb. capacity	1,375		1,375
0140	16,000 lb. capacity	1,425		1,425
0150	Dock levelers, hydraulic, 7' x 8', 10 ton capacity	9,050	1,200	10,250
0160	Dock lifters, platform, 6' x 6', portable, 3000 lb. capacity	8,875		8,875
0170	Dock shelters, truck, scissor arms, economy	1,425	445	1,870
0180	Deluxe	2,075	885	2,960

EQUIPMENT & FURNISHINGS E

399

E1090 Other Equipment

E1090 110	Maintenance Equipment, EACH	MAT.	INST.	TOTAL
0110	Vacuum cleaning, central, residential, 3 valves	705	505	1,210
0120	7 valves	1,200	1,125	2,325

E1090 210	Solid Waste Handling Equipment, EACH	COST EACH		
		MAT.	INST.	TOTAL
0110	Waste handling, compactors, single bag, 250 lbs./hr., hand fed	7,875	515	8,390
0120	Heavy duty industrial, 5 C.Y. capacity	18,200	2,475	20,675
0130	Incinerator, electric, 100 lbs./hr., economy	10,400	2,475	12,875
0140	Gas, 2000 lbs./hr., deluxe	220,000	19,600	239,600
0150	Shredder, no baling, 35 tons/hr.	272,000		272,000
0160	Incl. baling, 50 tons/day	543,000		543,000

E1090 350	Food Service Equipment, EACH	COST EACH		
		MAT.	INST.	TOTAL
0110	Kitchen equipment, bake oven, single deck	5,150	116	5,266
0120	Broiler, without oven	3,250	116	3,366
0130	Commercial dish washer, semiautomatic, 50 racks/hr.	7,625	710	8,335
0140	Automatic, 275 racks/hr.	21,900	3,050	24,950
0150	Cooler, beverage, reach-in, 6 ft. long	4,150	154	4,304
0160	Food warmer, counter, 1.65 kw	580		580
0170	Fryers, with submerger, single	1,850	132	1,982
0180	Double	1,750	185	1,935
0185	Ice maker, 1000 lb. per day, with bin	7,275	925	8,200
0190	Kettles, steam jacketed, 20 gallons	5,625	211	5,836
0200	Range, restaurant type, burners, 2 ovens and 24" griddle	5,100	154	5,254
0210	Range hood, incl. carbon dioxide system, economy	3,775	310	4,085
0220	Deluxe	27,900	925	28,825

E1090 360	Food Service Equipment, S.F.	COST PER S.F.		
		MAT.	INST.	TOTAL
0110	Refrigerators, prefab, walk-in, 7'-6" high, 6' x 6'	124	16.15	140.15
0120	12' x 20'	78	8.10	86.10

E1090 410	Residential Equipment, EACH	COST EACH		
		MAT.	INST.	TOTAL
0110	Arch. equip., appliances, range, cook top, 4 burner, economy	211	83.50	294.50
0120	Built in, single oven 30" wide, economy	505	83.50	588.50
0130	Standing, single oven-21" wide, economy	281	68	349
0140	Double oven-30" wide, deluxe	1,775	68	1,843
0150	Compactor, residential, economy	470	88.50	558.50
0160	Deluxe	530	148	678
0170	Dish washer, built-in, 2 cycles, economy	282	254	536
0180	4 or more cycles, deluxe	615	505	1,120
0190	Garbage disposer, sink type, economy	48	101	149
0200	Deluxe	165	101	266
0210	Refrigerator, no frost, 10 to 12 C.F., economy	500	68	568
0220	21 to 29 C.F., deluxe	2,350	227	2,577
0300	Washing machine, automatic	1,150	515	1,665

E1090 550	Darkroom Equipment, EACH	COST EACH		
		MAT.	INST.	TOTAL
0110	Darkroom equipment, developing tanks, 2' x 4', 5" deep	4,875	465	5,340
0120	2' x 9', 10" deep	2,200	615	2,815
0130	Combination, tray & tank sinks, washers & dry tables	9,375	2,050	11,425
0140	Dryers, dehumidified filtered air, 3' x 2' x 5'-8" high	4,775	246	5,021

Important: See the Reference Section for critical supporting data - Reference Numbers and City Cost Indexes

E1090 Other Equipment

E1090 550	Darkroom Equipment, EACH	COST EACH		
		MAT.	INST.	TOTAL
0150	Processors, manual, 16" x 20" print size	9,425	740	10,165
0160	Automatic, color print, deluxe	13,000	2,450	15,450
0170	Washers, round, 11" x 14" sheet size	3,000	465	3,465
0180	Square, 50" x 56" sheet size	4,925	1,150	6,075

E1090 610	School Equipment, EACH	COST EACH		
		MAT.	INST.	TOTAL
0110	School equipment, basketball backstops, wall mounted, wood, fixed	1,175	770	1,945
0120	Suspended type, electrically operated	4,125	1,500	5,625
0130	Bleachers-telescoping, manual operation, 15 tier, economy (per seat)	78	27.50	105.50
0140	Power operation, 30 tier, deluxe (per seat)	297	49	346
0150	Weight lifting gym, universal, economy	2,825	680	3,505
0160	Deluxe	15,600	1,375	16,975
0170	Scoreboards, basketball, 1 side, economy	2,250	680	2,930
0180	4 sides, deluxe	26,000	9,375	35,375
0800	Vocational shop equipment, benches, metal	425	177	602
0810	Wood	495	177	672
0820	Dust collector, not incl. ductwork, 6' diam.	3,000	470	3,470
0830	Planer, 13" x 6"	1,900	221	2,121

E1090 620	School Equipment, S.F.	COST PER S.F.		
		MAT.	INST.	TOTAL
0110	School equipment, gym mats, naugahyde cover, 2" thick	3.88		3.88
0120	Wrestling, 1" thick, heavy duty	5.20		5.20

E1090 810	Athletic, Recreational, and Therapeutic Equipment, EACH	COST EACH		
		MAT.	INST.	TOTAL
0110	Sauna, prefabricated, incl. heater and controls, 7' high, 6' x 4'	4,050	670	4,720
0120	10' x 12'	9,925	1,475	11,400
0130	Heaters, wall mounted, to 200 C.F.	520		520
0140	Floor standing, to 1000 C.F., 12500 W	1,825	167	1,992
0610	Shooting range incl. bullet traps, controls, separators, ceilings, economy	24,200	5,800	30,000
0620	Deluxe	44,000	9,775	53,775
0650	Sport court, squash, regulation, in existing building, economy			16,800
0660	Deluxe			31,400
0670	Racketball, regulation, in existing building, economy	2,225	8,350	10,575
0680	Deluxe	29,800	16,700	46,500
0700	Swimming pool equipment, diving stand, stainless steel, 1 meter	5,675	330	6,005
0710	3 meter	6,375	2,225	8,600
0720	Diving boards, 16 ft. long, aluminum	2,775	330	3,105
0730	Fiberglass	2,225	330	2,555
0740	Filter system, sand, incl. pump, 6000 gal./hr.	1,225	570	1,795
0750	Lights, underwater, 12 volt with transformer, 300W	159	1,250	1,409
0760	Slides, fiberglass with aluminum handrails & ladder, 6' high, straight	2,550	555	3,105
0780	12' high, straight with platform	1,900	740	2,640

E1090 820	Athletic, Recreational, and Therapeutic Equipment, S.F.	COST PER S.F.		
		MAT.	INST.	TOTAL
0110	Swimming pools, residential, vinyl liner, metal sides	10.75	5.85	16.60
0120	Concrete sides	12.85	10.85	23.70
0130	Gunite shell, plaster finish, 350 S.F.	21.50	22.50	44
0140	800 S.F.	17.10	13.05	30.15
0150	Motel, gunite shell, plaster finish	26	28.50	54.50
0160	Municipal, gunite shell, tile finish, formed gutters	110	48.50	158.50

E1090 Other Equipment

E1090 910	Other Equipment, EACH	COST EACH		
		MAT.	INST.	TOTAL
0200	Steam bath heater, incl. timer and head, single, to 140 C.F.	1,175	430	1,605
0210	Commercial size, to 2500 C.F.	7,225	645	7,870
0900	Wine Cellar, redwood, refrigerated, 6'-8" high, 6'W x 6'D	3,425	590	4,015
0910	6'W x 12'D	7,125	885	8,010

E20 Furnishings

E2010 Fixed Furnishings

E2010 310	Window Treatment, EACH	COST EACH		
		MAT.	INST.	TOTAL
0110	Furnishings, blinds, exterior, aluminum, louvered, 1'-4" wide x 3'-0" long	50	44.50	94.50
0120	1'-4" wide x 6'-8" long	101	49	150
0130	Hemlock, solid raised, 1'-4" wide x 3'-0" long	42	44.50	86.50
0140	1'-4" wide x 6'-9" long	68	49	117
0150	Polystyrene, louvered, 1'-3" wide x 3'-3" long	28.50	44.50	73
0160	1'-3" wide x 6'-8" long	48	49	97
0200	Interior, wood folding panels, louvered, 7" x 20" (per pair)	46.50	26	72.50
0210	18" x 40" (per pair)	144	26	170

E2010 320	Window Treatment, S.F.	COST PER S.F.		
		MAT.	INST.	TOTAL
0110	Furnishings, blinds-interior, venetian-aluminum, stock, 2" slats, economy	1.76	.75	2.51
0120	Custom, 1" slats, deluxe	7.15	1.01	8.16
0130	Vertical, PVC or cloth, T&B track, economy	8.65	.96	9.61
0140	Deluxe	10.50	1.11	11.61
0150	Draperies, unlined, economy	2.57		2.57
0160	Lightproof, deluxe	7.70		7.70
0510	Shades, mylar, wood roller, single layer, non-reflective	6.35	.65	7
0520	Metal roller, triple layer, heat reflective	10.40	.65	11.05
0530	Vinyl, light weight, 4 ga.	.47	.65	1.12
0540	Heavyweight, 6 ga.	1.49	.65	2.14
0550	Vinyl coated cotton, lightproof decorator shades	1.93	.65	2.58
0560	Woven aluminum, 3/8" thick, light and fireproof	4.79	1.27	6.06

E2010 420	Fixed Floor Grilles and Mats, S.F.	COST PER S.F.		
		MAT.	INST.	TOTAL
0110	Floor mats, recessed, inlaid black rubber, 3/8" thick, solid	23	2.20	25.20
0120	Colors, 1/2" thick, perforated	32.50	2.20	34.70
0130	Link-including nosings, steel-galvanized, 3/8" thick	8.80	2.20	11
0140	Vinyl, in colors	20.50	2.20	22.70

E2010 510	Fixed Multiple Seating, EACH	COST EACH		
		MAT.	INST.	TOTAL
0110	Seating, painted steel, upholstered, economy	128	25.50	153.50
0120	Deluxe	410	31.50	441.50
0400	Seating, lecture hall, pedestal type, economy	140	40.50	180.50
0410	Deluxe	485	61	546
0500	Auditorium chair, veneer construction	154	40.50	194.50
0510	Fully upholstered, spring seat	195	40.50	235.50

EQUIPMENT & FURNISHINGS   E

E2020 Moveable Furnishings

E2020 210	Furnishings/EACH	COST EACH		
		MAT.	INST.	TOTAL
0200	Hospital furniture, beds, manual, economy	855		855
0210	Deluxe	2,600		2,600
0220	All electric, economy	1,375		1,375
0230	Deluxe	4,225		4,225
0240	Patient wall systems, no utilities, economy, per room	1,050		1,050
0250	Deluxe, per room	1,950		1,950
0300	Hotel furnishings, standard room set, economy, per room	1,650		1,650
0310	Deluxe, per room	8,675		8,675
0500	Office furniture, standard employee set, economy, per person	360		360
0510	Deluxe, per person	2,350		2,350
0550	Posts, portable, pedestrian traffic control, economy	97		97
0560	Deluxe	315		315
0700	Restaurant furniture, booth, molded plastic, stub wall and 2 seats, economy	330	221	551
0710	Deluxe	1,125	295	1,420
0720	Upholstered seats, foursome, single-economy	940	88.50	1,028.50
0730	Foursome, double-deluxe	1,700	148	1,848

E2020 220	Furniture and Accessories, L.F.	COST PER L.F.		
		MAT.	INST.	TOTAL
0210	Dormitory furniture, desk top (built-in),laminated plastc, 24"deep, economy	28	17.70	45.70
0220	30" deep, deluxe	156	22	178
0230	Dressing unit, built-in, economy	198	74	272
0240	Deluxe	595	111	706
0310	Furnishings, cabinets, hospital, base, laminated plastic	243	88.50	331.50
0320	Stainless steel	335	88.50	423.50
0330	Counter top, laminated plastic, no backsplash	42	22	64
0340	Stainless steel	116	22	138
0350	Nurses station, door type, laminated plastic	282	88.50	370.50
0360	Stainless steel	355	88.50	443.50
0710	Restaurant furniture, bars, built-in, back bar	176	88.50	264.50
0720	Front bar	242	88.50	330.50
0910	Wardrobes & coatracks, standing, steel, single pedestal, 30" x 18" x 63"	75		75
0920	Double face rack, 39" x 26" x 70"	68		68
0930	Wall mounted rack, steel frame & shelves, 12" x 15" x 26"	63	6.30	69.30
0940	12" x 15" x 50"	34	3.30	37.30

For information about Means Estimating Seminars, see yellow pages 12 and 13 in back of book

Important: See the Reference Section for critical supporting data - Reference Numbers and City Cost Indexes

F SPECIAL CONSTRUCTION

F1010 120	Air-Supported Structures, S.F.	COST PER S.F.		
		MAT.	INST.	TOTAL
0110	Air supported struc., polyester vinyl fabric, 24oz., warehouse, 5000 S.F.	18.45	.27	18.72
0120	50,000 S.F.	8.45	.22	8.67
0130	Tennis, 7,200 S.F.	17.20	.23	17.43
0140	24,000 S.F.	11.70	.23	11.93
0150	Woven polyethylene, 6 oz., shelter, 3,000 S.F.	7.25	.46	7.71
0160	24,000 S.F.	6.15	.23	6.38
0170	Teflon coated fiberglass, stadium cover, economy	44	.12	44.12
0180	Deluxe	51.50	.16	51.66
0190	Air supported storage tank covers, reinf. vinyl fabric, 12 oz., 400 S.F.	16.60	.38	16.98
0200	18,000 S.F.	5.10	.35	5.45

F1010 210	Pre-Engineered Structures, EACH	COST EACH		
		MAT.	INST.	TOTAL
0600	Radio towers, guyed, 40 lb. section, 50' high, 70 MPH basic wind speed	2,650	1,150	3,800
0610	90 lb. section, 400' high, wind load 70 MPH basic wind speed	34,300	12,200	46,500
0620	Self supporting, 60' high, 70 MPH basic wind speed	5,500	2,150	7,650
0630	190' high, wind load 90 MPH basic wind speed	32,800	8,550	41,350
0700	Shelters, aluminum frame, acrylic glazing, 8' high, 3' x 9'	2,800	1,000	3,800
0710	9' x 12'	4,950	1,575	6,525

F1010 320	Other Special Structures, S.F.	COST PER S.F.		
		MAT.	INST.	TOTAL
0110	Swimming pool enclosure, transluscent, freestanding, economy	13.10	4.43	17.53
0120	Deluxe	202	12.65	214.65
0510	Tension structures, steel frame, polyester vinyl fabric, 12,000 S.F.	10.50	2	12.50
0520	20,800 S.F.	10.75	1.80	12.55

F1010 330	Special Structures, EACH	COST EACH		
		MAT.	INST.	TOTAL
0110	Kiosks, round, 5' diam., 1/4" fiberglass wall, 8' high	6,175		6,175
0120	Rectangular, 5' x 9', 1" insulated dbl. wall fiberglass, 7'-6" high	10,700		10,700
0200	Silos, concrete stave, industrial, 12' diam., 35' high	21,400	18,300	39,700
0210	25' diam., 75' high	70,000	40,200	110,200
0220	Steel prefab, 30,000 gal., painted, economy	19,000	4,675	23,675
0230	Epoxy-lined, deluxe	39,300	9,300	48,600

F1010 340	Special Structures, S.F.	COST PER S.F.		
		MAT.	INST.	TOTAL
0110	Comfort stations, prefab, mobile on steel frame, economy	40		40
0120	Permanent on concrete slab, deluxe	227	40	267
0210	Domes, bulk storage, wood framing, wood decking, 50' diam.	42.50	1.47	43.97
0220	116' diam.	25.50	1.69	27.19
0230	Steel framing, metal decking, 150' diam.	30.50	9.70	40.20
0240	400' diam.	25	7.40	32.40
0250	Geodesic, wood framing, wood panels, 30' diam.	30.50	1.61	32.11
0260	60' diam.	23	1.04	24.04
0270	Aluminum framing, acrylic panels, 40' diam.			103
0280	Aluminum panels, 400' diam.			22
0310	Garden house, prefab, wood, shell only, 48 S.F.	21	4.43	25.43
0320	200 S.F.	38	18.45	56.45
0410	Greenhouse, shell-stock, residential, lean-to, 8'-6" long x 3'-10" wide	42.50	26	68.50
0420	Freestanding, 8'-6" long x 13'-6" wide	33	8.20	41.20
0430	Commercial-truss frame, under 2000 S.F., deluxe			47.50
0440	Over 5,000 S.F., economy			28.50
0450	Institutional-rigid frame, under 500 S.F., deluxe			124
0460	Over 2,000 S.F., economy			48.50

Important: See the Reference Section for critical supporting data - Reference Numbers and City Cost Indexes

F1010   Special Structures

F1010 340	Special Structures, S.F.	COST PER S.F.		
		MAT.	INST.	TOTAL
0510	Hangar, prefab, galv. roof and walls, bottom rolling doors, economy	10.75	3.84	14.59
0520	Electric bifolding doors, deluxe	11.70	5	16.70

F1020 Integrated Construction

F1020 110	Integrated Construction, EACH	COST EACH		
		MAT.	INST.	TOTAL
0110	Integrated ceilings, radiant electric, 2' x 4' panel, manila finish	96.50	20	116.50
0120	ABS plastic finish	112	38.50	150.50

F1020 120	Integrated Construction, S.F.	COST PER S.F.		
		MAT.	INST.	TOTAL
0110	Integrated ceilings, Luminaire, suspended, 5' x 5' modules, 50% lighted	4.32	10.60	14.92
0120	100% lighted	6.45	19.05	25.50
0130	Dimensionaire, 2' x 4' module tile system, no air bar	2.61	1.77	4.38
0140	With air bar, deluxe	3.83	1.77	5.60
0220	Pedestal access floors, stl pnls, no stringers, vinyl covering, >6000 S.F.	15.95	1.97	17.92
0230	With stringers, high pressure laminate covering, under 6000 S.F.	20.50	2.21	22.71
0240	Carpet covering, under 6000 S.F.	22.50	2.21	24.71
0250	Aluminum panels, no stringers, no covering	34	1.77	35.77

F1020 250	Special Purpose Room, EACH	COST EACH		
		MAT.	INST.	TOTAL
0110	Portable booth, acoustical, 27 db 1000 hz., 15 S.F. floor	3,325		3,325
0120	55 S.F. flr.	6,825		6,825

F1020 260	Special Purpose Room, S.F.	COST PER S.F.		
		MAT.	INST.	TOTAL
0110	Anechoic chambers, 7' high, 100 cps cutoff, 25 S.F.			2,275
0120	200 cps cutoff, 100 S.F.			1,200
0130	Audiometric rooms, under 500 S.F.	51	18.05	69.05
0140	Over 500 S.F.	48.50	14.75	63.25
0300	Darkrooms, shell, not including door, 240 S.F., 8' high	40.50	7.40	47.90
0310	64 S.F., 12' high	93.50	13.85	107.35
0510	Music practice room, modular, perforated steel, under 500 S.F.	31	12.65	43.65
0520	Over 500 S.F.	26.50	11.05	37.55

F1020 330	Special Construction, L.F.	COST PER L.F.		
		MAT.	INST.	TOTAL
0110	Spec. const., air curtains, shipping & receiving, 8'high x 5'wide, economy	345	104	449
0120	20' high x 8' wide, heated, deluxe	1,400	260	1,660
0130	Customer entrance, 10' high x 5' wide, economy	335	104	439
0140	12' high x 4' wide, heated, deluxe	560	130	690

F1030 Special Construction Systems

F1030 120	Sound, Vibration, and Seismic Construction, S.F.	COST PER S.F.		
		MAT.	INST.	TOTAL
0020	Special construction, acoustical, enclosure, 4" thick, 8 psf panels	32	18.45	50.45
0030	Reverb chamber, 4" thick, parallel walls	45	22	67
0110	Sound absorbing panels, 2'-6" x 8', painted metal	11.95	6.20	18.15
0120	Vinyl faced	9.30	5.55	14.85
0130	Flexible transparent curtain, clear	7.20	7.25	14.45
0140	With absorbing foam, 75% coverage	10.05	7.25	17.30
0150	Strip entrance, 2/3 overlap	5.70	11.55	17.25
0160	Full overlap	6	13.55	19.55
0200	Audio masking system, plenum mounted, over 10,000 S.F.	.70	.23	.93
0210	Ceiling mounted, under 5,000 S.F.	1.31	.42	1.73

F1030 210	Radiation Protection, EACH	COST EACH		
		MAT.	INST.	TOTAL
0110	Shielding, lead x-ray protection, radiography room, 1/16" lead, economy	6,600	3,125	9,725
0120	Deluxe	8,250	5,200	13,450

F1030 220	Radiation Protection, S.F.	COST PER S.F.		
		MAT.	INST.	TOTAL
0110	Shielding, lead, gypsum board, 5/8" thick, 1/16" lead	6.20	5.20	11.40
0120	1/8" lead	12.20	5.90	18.10
0130	Lath, 1/16" thick	5.95	5.80	11.75
0140	1/8" thick	12.15	6.50	18.65
0150	Radio frequency, copper, prefab screen type, economy	26.50	4.92	31.42
0160	Deluxe	34	6.10	40.10

F1030 910	Other Special Construction Systems, EACH	COST EACH		
		MAT.	INST.	TOTAL
0110	Disappearing stairways, folding, pine, 8'-6" ceiling	117	111	228
0120	9'-6" ceiling	128	111	239
0200	Fire escape, galvanized steel, 8'-0" to 10'-6" ceiling	1,500	885	2,385
0210	10'-6" to 13'-6" ceiling	1,875	885	2,760
0220	Automatic electric, wood, 8' to 9' ceiling	7,700	885	8,585
0230	Aluminum, 14' to 15' ceiling	8,850	1,275	10,125
0300	Fireplace prefabricated, freestanding or wall hung, economy	1,075	340	1,415
0310	Deluxe	3,700	490	4,190
0320	Woodburning stoves, cast iron, economy	1,000	680	1,680
0330	Deluxe	2,575	1,100	3,675

F1040 Special Facilities

F1040 210	Ice Rinks, EACH	COAT EACH		
		MAT.	INST.	TOTAL
0100	Ice skating rink, 85' x 200', 55° system, 5 mos., 100 ton			545,500
0110	90° system, 12 mos., 135 ton			605,500
0120	Dash boards, acrylic screens, polyethylene coated plywood	126,500	29,900	156,400
0130	Fiberglass and aluminum construction	176,000	29,900	205,900

F1040 510	Liquid & Gas Storage Tanks, EACH	COST EACH		
		MAT.	INST.	TOTAL
0100	Tanks, steel, ground level, 100,000 gal.			150,500
0110	10,000,000 gal.			3,020,000
0120	Elevated water, 50,000 gal.		155,000	237,500
0130	1,000,000 gal.		723,000	1,109,000
0150	Cypress wood, ground level, 3,000 gal.	8,225	8,800	17,025
0160	Redwood, ground level, 45,000 gal.	62,500	23,900	86,400

F1040 910	Special Construction, EACH	COST EACH		
		MAT.	INST.	TOTAL
0110	Special construction, bowling alley incl. pinsetter, scorer etc., economy	38,600	8,850	47,450
0120	Deluxe	50,500	9,850	60,350
0130	For automatic scorer, economy, add	7,000		7,000
0140	Deluxe, add	7,925		7,925
0200	Chimney, metal, high temp. steel jacket, factory lining, 24" diameter	10,500	4,100	14,600
0210	60" diameter	76,500	17,800	94,300
0220	Poured concrete, brick lining, 10' diam., 200' high			1,345,000
0230	20' diam., 500' high			5,900,000
0240	Radial brick, 3'-6" I.D., 75' high			225,000
0250	7' I.D., 175' high			669,500
0300	Control tower, modular, 12' x 10', incl. instrumentation, economy			540,500
0310	Deluxe			753,500
0400	Garage costs, residential, prefab, wood, single car economy	3,500	885	4,385
0410	Two car deluxe	10,500	1,775	12,275
0500	Hangars, prefab, galv. steel, bottom rolling doors, economy (per plane)	11,100	4,250	15,350
0510	Electrical bi-folding doors, deluxe (per plane)	15,500	5,850	21,350

For information about Means Estimating Seminars, see yellow pages 12 and 13 in back of book

Important: See the Reference Section for critical supporting data - Reference Numbers and City Cost Indexes

G BUILDING SITEWORK

G1030 Site Earthwork

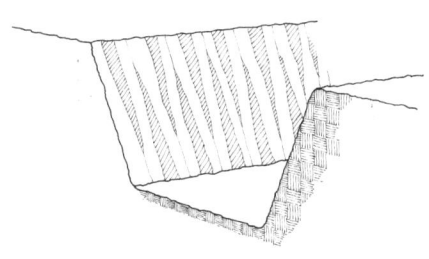

Trenching Systems are shown on a cost per linear foot basis. The systems include: excavation; backfill and removal of spoil; and compaction for various depths and trench bottom widths. The backfill has been reduced to accommodate a pipe of suitable diameter and bedding.

The slope for trench sides varies from 0:1 to 2:1.

The Expanded System Listing shows Trenching Systems that range from 2' to 12' in width. Depths range from 2' to 25'.

System Components			COST PER L.F.		
	QUANTITY	UNIT	EQUIP.	LABOR	TOTAL
SYSTEM G1030 805 1310					
TRENCHING, BACKHOE, 0 TO 1 SLOPE, 2' WIDE, 2' DP, 3/8 C.Y. BUCKET					
Excavation, trench, hyd. backhoe, track mtd., 3/8 C.Y. bucket	.174	C.Y.	.29	.91	1.20
Backfill and load spoil, from stockpile	.174	C.Y.	.09	.27	.36
Compaction by rammer tamper, 8" lifts, 4 passes	.014	C.Y.		.04	.04
Remove excess spoil, 6 C.Y. dump truck, 2 mile roundtrip	.160	C.Y.	.57	.55	1.12
TOTAL			.95	1.77	2.72

G1030 805	Trenching	COST PER L.F.		
		EQUIP.	LABOR	TOTAL
1310	Trenching, backhoe, 0 to 1 slope, 2' wide, 2' deep, 3/8 C.Y. bucket	.96	1.77	2.73
1320	3' deep, 3/8 C.Y. bucket	1.18	2.59	3.77
1330	4' deep, 3/8 C.Y. bucket	1.40	3.41	4.81
1340	6' deep, 3/8 C.Y. bucket	1.77	4.34	6.11
1350	8' deep, 1/2 C.Y. bucket	2.08	5.50	7.58
1360	10' deep, 1 C.Y. bucket	3.05	6.55	9.60
1400	4' wide, 2' deep, 3/8 C.Y. bucket	2.01	3.56	5.57
1410	3' deep, 3/8 C.Y. bucket	2.85	5.30	8.15
1420	4' deep, 1/2 C.Y. bucket	2.88	5.30	8.18
1430	6' deep, 1/2 C.Y. bucket	3.70	8.10	11.80
1440	8' deep, 1/2 C.Y. bucket	5.65	10.40	16.05
1450	10' deep, 1 C.Y. bucket	6.35	12.90	19.25
1460	12' deep, 1 C.Y. bucket	8.05	16.55	24.60
1470	15' deep, 1-1/2 C.Y. bucket	6.85	14.20	21.05
1480	18' deep, 2-1/2 C.Y. bucket	9.40	20.50	29.90
1520	6' wide, 6' deep, 5/8 C.Y. bucket	7.30	11.20	18.50
1530	8' deep, 3/4 C.Y. bucket	8.90	13.45	22.35
1540	10' deep, 1 C.Y. bucket	9.55	16.20	25.75
1550	12' deep, 1-1/4 C.Y. bucket	8.40	12.80	21.20
1560	16' deep, 2 C.Y. bucket	11.05	14.60	25.65
1570	20' deep, 3-1/2 C.Y. bucket	15	26	41
1580	24' deep, 3-1/2 C.Y. bucket	24	43.50	67.50
1640	8' wide, 12' deep, 1-1/4 C.Y. bucket	12.40	17.50	29.90
1650	15' deep, 1-1/2 C.Y. bucket	14.65	21.50	36.15
1660	18' deep, 2-1/2 C.Y. bucket	17.10	22	39.10
1680	24' deep, 3-1/2 C.Y. bucket	33.50	58	91.50
1730	10' wide, 20' deep, 3-1/2 C.Y. bucket	27	44	71
1740	24' deep, 3-1/2 C.Y. bucket	42	73	115
1780	12' wide, 20' deep, 3-1/2 C.Y. bucket	33	53	86
1790	25' deep, bucket	56.50	97.50	154
1800	1/2 to 1 slope, 2' wide, 2' deep, 3/8 C.Y. bucket	1.29	2.27	3.56
1810	3' deep, 3/8 C.Y. bucket	1.78	3.75	5.53
1820	4' deep, 3/8 C.Y. bucket	2.27	5.50	7.77
1840	6' deep, 3/8 C.Y. bucket	3.38	8.45	11.83

Note at rows 1330/1340: RG1030 -020

G BUILDING SITEWORK

G1030 Site Earthwork

G1030 805	Trenching	COST PER L.F.		
		EQUIP.	LABOR	TOTAL
1860	8' deep, 1/2 C.Y. bucket	4.73	12.85	17.58
1880	10' deep, 1 C.Y. bucket	8.45	17.70	26.15
2300	4' wide, 2' deep, 3/8 C.Y. bucket	2.14	3.79	5.93
2310	3' deep, 3/8 C.Y. bucket	3.53	6.25	9.78
2320	4' deep, 1/2 C.Y. bucket	3.78	6.50	10.28
2340	6' deep, 1/2 C.Y. bucket	5.30	11.30	16.60
2360	8' deep, 1/2 C.Y. bucket	8.75	15.35	24.10
2380	10' deep, 1 C.Y. bucket	11.45	22.50	33.95
2400	12' deep, 1 C.Y. bucket	12.10	24.50	36.60
2430	15' deep, 1-1/2 C.Y. bucket	15.35	30.50	45.85
2460	18' deep, 2-1/2 C.Y. bucket	21	36.50	57.50
2840	6' wide, 6' deep, 5/8 C.Y. bucket	8.65	12.35	21
2860	8' deep, 3/4 C.Y. bucket	12.35	18.50	30.85
2880	10' deep, 1 C.Y. bucket	14.45	24.50	38.95
2900	12' deep, 1-1/4 C.Y. bucket	13.50	20.50	34
2940	16' deep, 2 C.Y. bucket	21.50	32.50	54
2980	20' deep, 3-1/2 C.Y. bucket	32.50	56	88.50
3020	24' deep, 3-1/2 C.Y. bucket	59	104	163
3100	8' wide, 12' deep, 1-1/4 C.Y. bucket	18.10	25	43.10
3120	15' deep, 1-1/2 C.Y. bucket	23	33.50	56.50
3140	18' deep, 2-1/2 C.Y. bucket	29.50	36.50	66
3180	24' deep, 3-1/2 C.Y. bucket	68	117	185
3270	10' wide, 20' deep, 3-1/2 C.Y. bucket	42	69.50	111.50
3280	24' deep, 3-1/2 C.Y. bucket	77.50	132	209.50
3370	12' wide, 20' deep, 3-1/2 C.Y. bucket	48	77.50	125.50
3380	25' deep, 3-1/2 C.Y. bucket	90.50	153	243.50
3500	1 to 1 slope, 2' wide, 2' deep, 3/8 C.Y. bucket	1.50	2.66	4.16
3520	3' deep, 3/8 C.Y. bucket	2.31	4.65	6.96
3540	4' deep, 3/8 C.Y. bucket	2.99	7	9.99
3560	6' deep, 3/8 C.Y. bucket	4.61	11.15	15.76
3580	8' deep, 1/2 C.Y. bucket	6.65	17.55	24.20
3600	10' deep, 1 C.Y. bucket	14.05	28	42.05
3800	4' wide, 2' deep, 3/8 C.Y. bucket	2.56	4.53	7.09
3820	3' deep, 3/8 C.Y. bucket	4.51	7.95	12.46
3840	4' deep, 1/2 C.Y. bucket	4.59	7.30	11.89
3860	6' deep, 1/2 C.Y. bucket	6.60	13.55	20.15
3880	8' deep, 1/2 C.Y. bucket	11.70	20	31.70
3900	10' deep, 1 C.Y. bucket	15.25	29	44.25
3920	12' deep, 1 C.Y. bucket	22	42.50	64.50
3940	15' deep, 1-1/2 C.Y. bucket	21.50	39.50	61
3960	18' deep, 2-1/2 C.Y. bucket	30.50	47.50	78
4030	6' wide, 6' deep, 5/8 C.Y. bucket	10.50	15	25.50
4040	8' deep, 3/4 C.Y. bucket	15.40	24	39.40
4050	10' deep, 1 C.Y. bucket	18.60	33.50	52.10
4060	12' deep, 1-1/4 C.Y. bucket	18	30	48
4070	16' deep, 2 C.Y. bucket	28.50	42.50	71
4080	20' deep, 3-1/2 C.Y. bucket	46.50	88.50	135
4090	24' deep, 3-1/2 C.Y. bucket	88	172	260
4500	8' wide, 12' deep, 1-1/4 C.Y. bucket	23	34	57
4550	15' deep, 1-1/2 C.Y. bucket	30.50	48.50	79
4600	18' deep, 2-1/2 C.Y. bucket	40	56.50	96.50
4650	24' deep, 3-1/2 C.Y. bucket	96.50	183	279.50
4800	10' wide, 20' deep, 3-1/2 C.Y. bucket	58.50	103	161.50
4850	24' deep, 3-1/2 C.Y. bucket	105	195	300
4950	12' wide, 20' deep, 3-1/2 C.Y. bucket	64	110	174
4980	25' deep, 3-1/2 C.Y. bucket	127	237	364
5000	1-1/2 to 1 slope, 2' wide, 2' deep, 3/8 C.Y. bucket	1.74	3.07	4.81
5020	3' deep, 3/8 C.Y. bucket	2.83	5.50	8.33

G1030 Site Earthwork

G1030 805	Trenching	COST PER L.F.		
		EQUIP.	LABOR	TOTAL
5040	4' deep, 3/8 C.Y. bucket	3.67	8.25	11.92
5060	6' deep, 3/8 C.Y. bucket	5.70	13.10	18.80
5080	8' deep, 1/2 C.Y. bucket	8.35	20.50	28.85
5100	10' deep, 1 C.Y. bucket	15.70	29	44.70
5300	4' wide, 2' deep, 3/8 C.Y. bucket	2.42	4.27	6.69
5320	3' deep, 3/8 C.Y. bucket	4.39	7.75	12.14
5340	4' deep, 1/2 C.Y. bucket	5.45	8.55	14
5360	6' deep, 1/2 C.Y. bucket	7.85	15.35	23.20
5380	8' deep, 1/2 C.Y. bucket	14.10	23	37.10
5400	10' deep, 1 C.Y. bucket	18.60	33	51.60
5420	12' deep, 1 C.Y. bucket	27.50	49	76.50
5450	15' deep, 1-1/2 C.Y. bucket	26.50	43.50	70
5480	18' deep, 2-1/2 C.Y. bucket	38	51.50	89.50
5660	6' wide, 6' deep, 5/8 C.Y. bucket	12.20	16.80	29
5680	8' deep, 3/4 C.Y. bucket	18.05	27	45.05
5700	10' deep, 1 C.Y. bucket	22	37.50	59.50
5720	12' deep, 1-1/4 C.Y. bucket	21	32	53
5760	16' deep, 2 C.Y. bucket	35	45.50	80.50
5800	20' deep, 3-1/2 C.Y. bucket	57.50	100	157.50
5840	24' deep, 3-1/2 C.Y. bucket	109	195	304
6020	8' wide, 12' deep, 1-1/4 C.Y. bucket	27.50	37.50	65
6050	15' deep, 1-1/2 C.Y. bucket	36.50	53	89.50
6080	18' deep, 2-1/2 C.Y. bucket	48	60.50	108.50
6140	24' deep, 3-1/2 C.Y. bucket	119	206	325
6300	10' wide, 20' deep, 3-1/2 C.Y. bucket	70	114	184
6350	24' deep, 3-1/2 C.Y. bucket	126	217	343
6450	12' wide, 20' deep, 3-1/2 C.Y. bucket	76.50	121	197.50
6480	25' deep, 3-1/2 C.Y. bucket	153	261	414
6600	2 to 1 slope, 2' wide, 2' deep, 3/8 C.Y. bucket	2.72	2.80	5.52
6620	3' deep, 3/8 C.Y. bucket	3.24	6.10	9.34
6640	4' deep, 3/8 C.Y. bucket	4.19	9.15	13.34
6660	6' deep, 3/8 C.Y. bucket	6.15	13.85	20
6680	8' deep, 1/2 C.Y. bucket	9.45	23.50	32.95
6700	10' deep, 1 C.Y. bucket	17.80	33	50.80
6900	4' wide, 2' deep, 3/8 C.Y. bucket	2.42	4.26	6.68
6920	3' deep, 3/8 C.Y. bucket	4.51	7.95	12.46
6940	4' deep, 1/2 C.Y. bucket	6.10	9.10	15.20
6960	6' deep, 1/2 C.Y. bucket	8.75	16.65	25.40
6980	8' deep, 1/2 C.Y. bucket	15.65	25	40.65
7000	10' deep, 1 C.Y. bucket	20.50	36.50	57
7020	12' deep, 1 C.Y. bucket	30.50	54.50	85
7050	15' deep, 1-1/2 C.Y. bucket	30	49	79
7080	18' deep, 2-1/2 C.Y. bucket	43	58.50	101.50
7260	6' wide, 6' deep, 5/8 C.Y. bucket	13.45	17.90	31.35
7280	8' deep, 3/4 C.Y. bucket	19.80	29	48.80
7300	10' deep, 3/4 C.Y. bucket	26	41.50	67.50
7320	12' deep, 1-1/4 C.Y. bucket	23.50	36	59.50
7360	16' deep, 2 C.Y. bucket	39	50.50	89.50
7400	20' deep, 3-1/2 C.Y. bucket	64.50	112	176.50
7440	24' deep, 3-1/2 C.Y. bucket	123	220	343
7620	8' wide, 12' deep, 1-1/4 C.Y. bucket	30.50	41	71.50
7650	15' deep, 1-1/2 C.Y. bucket	40.50	58	98.50
7680	18' deep, 2-1/2 C.Y. bucket	53.50	67	120.50
7740	24' deep, 3-1/2 C.Y. bucket	133	230	363
7920	10' wide, 20' deep, 3-1/2 C.Y. bucket	77.50	125	202.50
7940	24' deep, 3-1/2 C.Y. bucket	140	239	379
8060	12' wide, 20' deep, 3-1/2 C.Y. bucket	84.50	132	216.50
8080	25' deep, 3-1/2 C.Y. bucket	169	286	455

Important: See the Reference Section for critical supporting data - Reference Numbers and City Cost Indexes

G1030 Site Earthwork

The Pipe Bedding System is shown for various pipe diameters. Compacted bank sand is used for pipe bedding and to fill 12″ over the pipe. No backfill is included. Various side slopes are shown to accommodate different soil conditions. Pipe sizes vary from 6″ to 84″ diameter.

System Components	QUANTITY	UNIT	COST PER L.F. MAT.	INST.	TOTAL
SYSTEM G1030 815 1440 PIPE BEDDING, SIDE SLOPE 0 TO 1, 1′ WIDE, PIPE SIZE 6″ DIAMETER					
Borrow, bank sand, 2 mile haul, machine spread	.067	C.Y.	.31	.38	.69
Compaction, vibrating plate	.067	C.Y.		.12	.12
TOTAL			.31	.50	.81

G1030 815	Pipe Bedding		COST PER L.F. MAT.	INST.	TOTAL
1440	Pipe bedding, side slope 0 to 1, 1′ wide, pipe size 6″ diameter		.31	.50	.81
1460	2′ wide, pipe size 8″ diameter		.69	1.11	1.80
1480	Pipe size 10″ diameter	RG1030 -020	.70	1.14	1.84
1500	Pipe size 12″ diameter		.72	1.17	1.89
1520	3′ wide, pipe size 14″ diameter		1.19	1.92	3.11
1540	Pipe size 15″ diameter		1.20	1.94	3.14
1560	Pipe size 16″ diameter		1.22	1.96	3.18
1580	Pipe size 18″ diameter		1.25	2.02	3.27
1600	4′ wide, pipe size 20″ diameter		1.80	2.90	4.70
1620	Pipe size 21″ diameter		1.82	2.93	4.75
1640	Pipe size 24″ diameter		1.86	3	4.86
1660	Pipe size 30″ diameter		1.90	3.08	4.98
1680	6′ wide, pipe size 32″ diameter		3.33	5.40	8.73
1700	Pipe size 36″ diameter		3.42	5.50	8.92
1720	7′ wide, pipe size 48″ diameter		4.45	7.20	11.65
1740	8′ wide, pipe size 60″ diameter		5.55	8.95	14.50
1760	10′ wide, pipe size 72″ diameter		8	12.95	20.95
1780	12′ wide, pipe size 84″ diameter		10.90	17.60	28.50
2140	Side slope 1/2 to 1, 1′ wide, pipe size 6″ diameter		.65	1.06	1.71
2160	2′ wide, pipe size 8″ diameter		1.08	1.75	2.83
2180	Pipe size 10″ diameter		1.18	1.90	3.08
2200	Pipe size 12″ diameter		1.25	2.02	3.27
2220	3′ wide, pipe size 14″ diameter		1.80	2.91	4.71
2240	Pipe size 15″ diameter		1.85	2.98	4.83
2260	Pipe size 16″ diameter		1.90	3.08	4.98
2280	Pipe size 18″ diameter		2.01	3.25	5.26
2300	4′ wide, pipe size 20″ diameter		2.66	4.29	6.95
2320	Pipe size 21″ diameter		2.72	4.40	7.12
2340	Pipe size 24″ diameter		2.92	4.71	7.63
2360	Pipe size 30″ diameter		3.27	5.30	8.57
2380	6′ wide, pipe size 32″ diameter		4.82	7.80	12.62
2400	Pipe size 36″ diameter		5.15	8.35	13.50
2420	7′ wide, pipe size 48″ diameter		7.05	11.35	18.40
2440	8′ wide, pipe size 60″ diameter		9.15	14.80	23.95
2460	10′ wide, pipe size 72″ diameter		12.85	21	33.85
2480	12′ wide, pipe size 84″ diameter		17.10	27.50	44.60
2620	Side slope 1 to 1, 1′ wide, pipe size 6″ diameter		1	1.61	2.61
2640	2′ wide, pipe size 8″ diameter		1.50	2.42	3.92

G1030   Site Earthwork

G1030 815	Pipe Bedding	COST PER L.F.		
		MAT.	INST.	TOTAL
2660	Pipe size 10" diameter	1.64	2.65	4.29
2680	Pipe size 12" diameter	1.80	2.91	4.71
2700	3' wide, pipe size 14" diameter	2.41	3.90	6.31
2720	Pipe size 15" diameter	2.50	4.04	6.54
2740	Pipe size 16" diameter	2.59	4.18	6.77
2760	Pipe size 18" diameter	2.78	4.50	7.28
2780	4' wide, pipe size 20" diameter	3.52	5.70	9.22
2800	Pipe size 21" diameter	3.63	5.85	9.48
2820	Pipe size 24" diameter	3.96	6.40	10.36
2840	Pipe size 30" diameter	4.63	7.50	12.13
2860	6' wide, pipe size 32" diameter	6.30	10.20	16.50
2880	Pipe size 36" diameter	6.90	11.15	18.05
2900	7' wide, pipe size 48" diameter	9.65	15.60	25.25
2920	8' wide, pipe size 60" diameter	12.80	20.50	33.30
2940	10' wide, pipe size 72" diameter	17.65	28.50	46.15
2960	12' wide, pipe size 84" diameter	23.50	37.50	61
3000	Side slope 1-1/2 to 1, 1' wide, pipe size 6" diameter	1.34	2.16	3.50
3020	2' wide, pipe size 8" diameter	1.90	3.06	4.96
3040	Pipe size 10" diameter	2.10	3.40	5.50
3060	Pipe size 12" diameter	2.33	3.77	6.10
3080	3' wide, pipe size 14" diameter	3.02	4.88	7.90
3100	Pipe size 15" diameter	3.14	5.10	8.24
3120	Pipe size 16" diameter	3.27	5.30	8.57
3140	Pipe size 18" diameter	3.56	5.75	9.31
3160	4' wide, pipe size 20" diameter	4.39	7.10	11.49
3180	Pipe size 21" diameter	4.53	7.30	11.83
3200	Pipe size 24" diameter	5	8.10	13.10
3220	Pipe size 30" diameter	6	9.70	15.70
3240	6' wide, pipe size 32" diameter	7.80	12.60	20.40
3260	Pipe size 36" diameter	8.60	13.95	22.55
3280	7' wide, pipe size 48" diameter	12.25	19.80	32.05
3300	8' wide, pipe size 60" diameter	16.40	26.50	42.90
3320	10' wide, pipe size 72" diameter	22.50	36.50	59
3340	12' wide, pipe size 84" diameter	29.50	48	77.50
3400	Side slope 2 to 1, 1' wide, pipe size 6" diameter	1.71	2.77	4.48
3420	2' wide, pipe size 8" diameter	2.30	3.71	6.01
3440	Pipe size 10" diameter	2.57	4.15	6.72
3460	Pipe size 12" diameter	2.87	4.63	7.50
3480	3' wide, pipe size 14" diameter	3.63	5.85	9.48
3500	Pipe size 15" diameter	3.77	6.10	9.87
3520	Pipe size 16" diameter	3.96	6.40	10.36
3540	Pipe size 18" diameter	4.33	7	11.33
3560	4' wide, pipe size 20" diameter	5.25	8.50	13.75
3580	Pipe size 21" diameter	5.45	8.75	14.20
3600	Pipe size 24" diameter	6.05	9.80	15.85
3620	Pipe size 30" diameter	7.40	11.95	19.35
3640	6' wide, pipe size 32" diameter	9.30	15	24.30
3660	Pipe size 36" diameter	10.35	16.75	27.10
3680	7' wide, pipe size 48" diameter	14.85	24	38.85
3700	8' wide, pipe size 60" diameter	20	32.50	52.50
3720	10' wide, pipe size 72" diameter	27.50	44	71.50
3740	12' wide, pipe size 84" diameter	35.50	57.50	93

Important: See the Reference Section for critical supporting data - Reference Numbers and City Cost Indexes

G2010 Roadways

The Roadway System includes surveying; 8″ of compacted bank-run gravel; fine grading with a grader and roller; and bituminous wearing course. Roadway Systems are shown on a cost per linear foot basis.

The Expanded System Listing shows Roadway Systems with widths that vary from 20′ to 34′; for various pavement thicknesses.

System Components	QUANTITY	UNIT	COST PER L.F.		
			MAT.	INST.	TOTAL
SYSTEM G2010 210 1500					
ROADWAYS, BITUMINOUS CONC. PAVING, 2-1/2″ THICK, 20′ WIDE					
4 man surveying crew	.010	Ea.		19.14	19.14
Bank gravel, 2 mi. haul, place and spread by dozer	.494	C.Y.	9.86	2.78	12.64
Compaction, granular material to 98%	.494	C.Y.		2.44	2.44
Grading, fine grade, 3 passes with grader plus rolling	2.220	S.Y.		8.69	8.69
Bituminous, paving, wearing course 2″ - 2-1/2″	2.220	S.Y.	12.54	2.86	15.40
Curbs, granite, split face, straight, 5″ x 16″	2.000	L.F.	22.40	12.34	34.74
Painting lines, reflectorized, 4″ wide	1.000	L.F.	.19	.12	.31
TOTAL			44.99	48.37	93.36

G2010 210	Roadway Pavement		COST PER L.F.		
			MAT.	INST.	TOTAL
1500	Roadways, bituminous conc. paving, 2-1/2″ thick, 20′ wide		45	48.50	93.50
1520	24′ wide		49.50	50	99.50
1540	26′ wide	RG2010 -100	51.50	51.50	103
1560	28′ wide		54	53	107
1580	30′ wide		56	54.50	110.50
1600	32′ wide		58.50	56	114.50
1620	34′ wide		60.50	58	118.50
1800	3″ thick, 20′ wide		47.50	47.50	95
1820	24′ wide		52.50	50.50	103
1840	26′ wide		55	51.50	106.50
1860	28′ wide		57.50	53	110.50
1880	30′ wide		60	55	115
1900	32′ wide		62.50	56.50	119
1920	34′ wide		65	58	123
2100	4″ thick, 20′ wide		53	47.50	100.50
2120	24′ wide		59.50	50	109.50
2140	26′ wide		62.50	51.50	114
2160	28′ wide		65.50	53	118.50
2180	30′ wide		68.50	55	123.50
2200	32′ wide		71.50	56.50	128

G2010 Roadways

G2010 210	Roadway Pavement	COST PER L.F.		
		MAT.	INST.	TOTAL
2220	34' wide	74.50	58	132.50
2390				
2400	5" thick, 20' wide	63	47.50	110.50
2420	24' wide	71	51	122
2440	26' wide	75	52.50	127.50
2460	28' wide	79	54.50	133.50
2480	30' wide	83	56	139
2500	32' wide	87.50	58	145.50
2520	34' wide	91.50	59.50	151
3000	6" thick, 20' wide	84.50	49.50	134
3020	24' wide	97	53	150
3040	26' wide	103	54.50	157.50
3060	28' wide	109	56	165
3080	30' wide	116	58.50	174.50
3100	32' wide	122	60	182
3120	34' wide	128	62	190
3300	8" thick 20' wide	105	49.50	154.50
3320	24' wide	121	53.50	174.50
3340	26' wide	129	55.50	184.50
3360	28' wide	137	57	194
3380	30' wide	146	58.50	204.50
3400	32' wide	154	61	215
3420	34' wide	162	62.50	224.50
3600	12" thick 20' wide	130	53	183
3620	24' wide	152	57	209
3640	26' wide	163	59.50	222.50
3660	28' wide	173	61	234
3700	32' wide	195	65.50	260.50
3720	34' wide	206	68	274

G2010 305	Curbs & Berms	COST PER L.F.		
		MAT.	INST.	TOTAL
1000	Bituminous, curbs, 8" wide, 6" high	.90	1.65	2.55
1100	8" high	1.04	1.82	2.86
1500	Berm, 12" wide, 3" to 6" high	1.18	2.35	3.53
1600	1-1/2" to 4" high	.72	1.65	2.37
2000	Concrete, curb, 6" wide, 18" high, cast-in-place	3.17	5.10	8.27
2100	Precast	8.75	5	13.75
2500	Curb and gutter, monolithic, 6" high, 24" wide	15.70	6.80	22.50
2600	30" wide	17.15	7.50	24.65
3000	Granite, curb, 4-1/2" wide, 12" high precast	5.60	11.70	17.30
3100	5" wide, 16" high	11.20	6.15	17.35
3200	6" wide, 18" high	14.75	6.85	21.60
3290				

Important: See the Reference Section for critical supporting data - Reference Numbers and City Cost Indexes

G2020 Parking Lots

The Parking Lot System includes: compacted bank-run gravel; fine grading with a grader and roller; and bituminous concrete wearing course. All Parking Lot systems are on a cost per car basis. There are three basic types of systems: 90° angle, 60° angle, and 45° angle. The gravel base is compacted to 98%. Final stall design and lay-out of the parking lot with precast bumpers, sealcoating and white paint is also included.

The Expanded System Listing shows the three basic parking lot types with various depths of both gravel base and wearing course. The gravel base depths range from 6″ to 10″. The bituminous paving wearing course varies from a depth of 3″ to 6″.

BUILDING SITEWORK G

System Components			COST PER CAR		
	QUANTITY	UNIT	MAT.	INST.	TOTAL
SYSTEM G2020 210 1500					
PARKING LOT, 90° ANGLE PARKING, 3″ BITUMINOUS PAVING, 6″ GRAVEL BASE					
Surveying crew for layout, 4 man crew	.020	Day		38.28	38.28
Borrow, bank run gravel, haul 2 mi., spread w/dozer, no compaction	5.556	C.Y.	110.84	31.22	142.06
Grading, fine grade 3 passes with motor grader	33.333	S.Y.		54.74	54.74
Compact w/ vibrating plate, 8″ lifts, granular mat'l. to 98%	5.556	C.Y.		27.45	27.45
Bituminous paving, 3″ thick	33.333	S.Y.	225	48	273
Seal coating, petroleum resistant under 1,000 S.Y.	33.333	S.Y.	24.33	33.33	57.66
Painting lines on pavement, parking stall, white	1.000	Ea.	3.77	5.67	9.44
Precast concrete parking bar, 6″ x 10″ x 6′-0″	1.000	Ea.	44	14.45	58.45
TOTAL			407.94	253.14	661.08

G2020 210	Parking Lots		COST PER CAR		
			MAT.	INST.	TOTAL
1500	Parking lot, 90° angle parking, 3″ bituminous paving, 6″ gravel base		410	254	664
1520	8″ gravel base		445	273	718
1540	10″ gravel base		480	293	773
1560	4″ bituminous paving, 6″ gravel base	RG2020 -500	495	252	747
1580	8″ gravel base		530	272	802
1600	10″ gravel base		570	291	861
1620	6″ bituminous paving, 6″ gravel base		640	263	903
1640	8″ gravel base		680	282	962
1661	10″ gravel base		715	300	1,015
1800	60° angle parking, 3″ bituminous paving, 6″ gravel base		470	277	747
1820	8″ gravel base		510	299	809
1840	10″ gravel base		555	325	880
1860	4″ bituminous paving, 6″ gravel base		570	276	846
1880	8″ gravel base		610	298	908
1900	10″ gravel base		655	320	975
1920	6″ bituminous paving, 6″ gravel base		740	288	1,028
1940	8″ gravel base		785	310	1,095
1961	10″ gravel base		825	335	1,160
2200	45° angle parking, 3″ bituminous paving, 6″ gravel base		530	300	830
2220	8″ gravel base		575	325	900
2240	10″ gravel base		625	350	975
2260	4″ bituminous paving, 6″ gravel base		645	299	944
2280	8″ gravel base		695	325	1,020
2300	10″ gravel base		740	350	1,090
2320	6″ bituminous paving, 6″ gravel base		840	315	1,155
2340	8″ gravel base		890	340	1,230
2361	10″ gravel base		940	520	1,460

G2030 Pedestrian Paving

The Bituminous and Concrete Sidewalk Systems include excavation, hand grading and compacted gravel base. Pavements are shown for two conditions of each of the following variables; pavement thickness, gravel base thickness and pavement width. Costs are given on a linear foot basis.

The Plaza Systems listed include several brick and tile paving surfaces on two different bases: gravel and slab on grade. The type of bedding for the pavers depends on the base being used, and alternate bedding may be desirable. Also included in the paving costs are edging and precast grating costs and where concrete bases are involved, expansion joints. Costs are given on a square foot basis.

G2030 110	Bituminous Sidewalks	MAT.	INST.	TOTAL
		COST PER L.F.		
1580	Bituminous sidewalk, 1" thick paving, 4" gravel base, 3' width	1.47	3.24	4.71
1600	4' width	1.96	3.54	5.50
1640	6" gravel base, 3' width	1.85	3.42	5.27
1660	4' width	2.46	3.77	6.23
2120	2" thick paving, 4" gravel base, 3' width	2.21	3.94	6.15
2140	4' width	2.94	4.40	7.34
2180	6" gravel base, 3' width	2.59	4.11	6.70
2201	4' width	3.43	4.63	8.06

G2030 120	Concrete Sidewalks	MAT.	INST.	TOTAL
		COST PER L.F.		
1580	Concrete sidewalk, 4" thick, 4" gravel base, 3' wide	5.55	8.90	14.45
1600	4' wide	7.40	11	18.40
1640	6" gravel base, 3' wide	5.90	9.05	14.95
1660	4' wide	7.90	11.20	19.10
2120	6" thick concrete, 4" gravel base, 3' wide	8.20	10.15	18.35
2140	4' wide	10.90	12.60	23.50
2180	6" gravel base, 3' wide	8.55	10.35	18.90
2201	4' wide	11.40	13.35	24.75

G2030 150	Brick & Tile Plazas	MAT.	INST.	TOTAL
		COST PER S.F.		
2050	Brick pavers, 4" x 8" x 2-1/4", aggregate base, course washed sand bedding	4.98	4.50	9.48
2100	Slab on grade, asphalt bedding	6.70	6.50	13.20
3550	Concrete paving stone, 4" x 8" x 2-1/2", gravel base, sand bedding	3.70	3.13	6.83
3600	Slab on grade, asphalt bedding	5.10	4.76	9.86
4050	Concrete patio blocks, 8" x 16" x 2", gravel base, sand bedding	3.46	4.11	7.57
4100	Slab on grade, asphalt bedding	5.05	6.05	11.10
6050	Granite pavers, 3-1/2" x 3-1/2" x 3-1/2", gravel base, sand bedding	9.25	10.05	19.30
6101	Slab on grade, mortar bedding	11	13.45	24.45

G — BUILDING SITEWORK

Important: See the Reference Section for critical supporting data - Reference Numbers and City Cost Indexes

G20 Site Improvements

G2040 Site Development

G2040 105	Fence & Guardrails	COST PER L.F.		
		MAT.	INST.	TOTAL
1000	Fence, chain link, 2" post, 1-5/8" rail, 9 ga. galv. wire, 5' high	12.15	3.89	16.04
1050	6' high w/barb. wire	14	4.87	18.87
1100	6 ga. galv. wire, 6' high	21.50	4.67	26.17
1150	6' high w/ barb. wire	22	4.87	26.87
1200	1-3/8" rail, 11 ga., galv. wire, 10' high	13.85	10.85	24.70
1250	12' high	16.20	12.10	28.30
1300	9 ga. vinyl covered wire, 10' high	16.20	10.85	27.05
1350	12' high	27	12.10	39.10
1500	1-5/8" post, 1-3/8" rail, 11 ga. galv. wire, 3' high	4.25	2.34	6.59
1550	4' high	4.82	2.92	7.74
1600	9 ga. vinyl covered wire, 3' high	4.59	2.34	6.93
1650	4' high	5.35	2.92	8.27
2000	Guardrail, corrugated steel, galvanized steel posts	18.05	2.42	20.47
2100	Timber, 6" x 8" posts, 4" x 8" rails	24	2.14	26.14
2200	Steel box beam, 6" x 6" rails	27.50	17.15	44.65
2250	6" x 8" rails	34	9.55	43.55
3000	Barrier, median, precast concrete, single face	43.50	9.25	52.75
3050	Double face	50	10.30	60.30

G2040 810	Flagpoles	COST EACH		
		MAT.	INST.	TOTAL
0110	Flagpoles, on grade, aluminum, tapered, 20' high	870	485	1,355
0120	70' high	8,575	1,225	9,800
0130	Fiberglass, tapered, 23' high	1,175	485	1,660
0140	59' high	6,900	1,075	7,975
0150	Concrete, internal halyard, 20' high	1,475	390	1,865
0160	100' high	14,800	975	15,775

G2040 950	Other Site Development, EACH	COST EACH		
		MAT.	INST.	TOTAL
0110	Grandstands, permanent, closed deck, steel, economy (per seat)			33
0120	Deluxe (per seat)			88
0130	Composite design, economy (per seat)			82.50
0140	Deluxe (per seat)			154

G3010 110	Water Distribution Piping	COST PER L.F.		
		MAT.	INST.	TOTAL
2000	Piping, excav. & backfill excl., ductile iron class 250, mech. joint			
2130	4" diameter	14.85	13.40	28.25
2150	6" diameter	17.25	16.80	34.05
2160	8" diameter	19.05	20	39.05
2170	10" diameter	26	23.50	49.50
2180	12" diameter	31.50	25.50	57
2210	16" diameter	44	37	81
2220	18" diameter	55.50	39	94.50
3000	Tyton joint			
3130	4" diameter	8.35	6.75	15.10
3150	6" diameter	9.80	8.05	17.85
3160	8" diameter	13.25	13.40	26.65
3170	10" diameter	20.50	14.75	35.25
3180	12" diameter	21.50	16.80	38.30
3210	16" diameter	33.50	23.50	57
3220	18" diameter	37	27	64
3230	20" diameter	41	30	71
3250	24" diameter	54.50	34.50	89
4000	Copper tubing, type K			
4050	3/4" diameter	3.54	2.31	5.85
4060	1" diameter	5.05	2.89	7.94
4080	1-1/2" diameter	7.75	3.49	11.24
4090	2" diameter	12	4.02	16.02
4110	3" diameter	25	6.90	31.90
4130	4" diameter	41.50	9.75	51.25
4150	6" diameter	125	15.15	140.15
5000	Polyvinyl chloride class 160, S.D.R. 26			
5130	1-1/2" diameter	.51	1.55	2.06
5150	2" diameter	1.22	1.69	2.91
5160	4" diameter	4.24	3.10	7.34
5170	6" diameter	9.15	3.68	12.83
5180	8" diameter	15.50	4.47	19.97
6000	Polyethylene 160 psi, S.D.R. 7			
6050	3/4" diameter	.34	2.21	2.55
6060	1" diameter	.78	2.40	3.18
6080	1-1/2" diameter	1.09	2.58	3.67
6090	2" diameter	1.71	3.18	4.89

G30 Site Mechanical Utilities

G3020 Sanitary Sewer

G3020 110	Drainage & Sewage Piping	COST PER L.F.		
		MAT.	INST.	TOTAL
2000	Piping, excavation & backfill excluded, PVC, plain			
2130	4" diameter	2.78	3.10	5.88
2150	6" diameter	5.85	3.32	9.17
2160	8" diameter	8.85	3.47	12.32
2900	Box culvert, precast, 8' long			
3000	6' x 3'	185	24.50	209.50
3020	6' x 7'	280	27.50	307.50
3040	8' x 3'	255	25.50	280.50
3060	8' x 8'	345	34	379
3080	10' x 3'	380	31	411
3100	10' x 8'	430	42.50	472.50
3120	12' x 3'	365	34	399
3140	12' x 8'	630	51	681
4000	Concrete, nonreinforced			
4150	6" diameter	4.91	9.10	14.01
4160	8" diameter	5.40	10.75	16.15
4170	10" diameter	6	11.15	17.15
4180	12" diameter	7.35	12.05	19.40
4200	15" diameter	8.60	13.40	22
4220	18" diameter	10.55	16.70	27.25
4250	24" diameter	15.95	24	39.95
4400	Reinforced, no gasket			
4580	12" diameter	13.35	16	29.35
4600	15" diameter	17.05	16	33.05
4620	18" diameter	17.95	18.25	36.20
4650	24" diameter	26.50	24	50.50
4670	30" diameter	36.50	37.50	74
4680	36" diameter	51.50	46	97.50
4690	42" diameter	70.50	50.50	121
4700	48" diameter	87.50	57	144.50
4720	60" diameter	140	76	216
4730	72" diameter	198	90.50	288.50
4740	84" diameter	335	113	448
4800	With gasket			
4980	12" diameter	14.60	9.30	23.90
5000	15" diameter	17.50	9.75	27.25
5020	18" diameter	22	10.25	32.25
5050	24" diameter	33	11.50	44.50
5070	30" diameter	43.50	37.50	81
5080	36" diameter	65.50	46	111.50
5090	42" diameter	85.50	45.50	131
5100	48" diameter	107	57	164
5120	60" diameter	158	72.50	230.50
5130	72" diameter	277	90.50	367.50
5140	84" diameter	400	152	552
5700	Corrugated metal, alum. or galv. bit. coated			
5760	8" diameter	10.25	7.30	17.55
5770	10" diameter	12.30	9.25	21.55
5780	12" diameter	14.70	11.45	26.15
5800	15" diameter	17.90	12.05	29.95
5820	18" diameter	23	12.65	35.65
5850	24" diameter	28	15.05	43.05
5870	30" diameter	37	27.50	64.50
5880	36" diameter	54.50	27.50	82
5900	48" diameter	82.50	33	115.50
5920	60" diameter	107	48.50	155.50
5930	72" diameter	160	81	241
6000	Plain			

G3020 Sanitary Sewer

G3020 110	Drainage & Sewage Piping	COST PER L.F.		
		MAT.	INST.	TOTAL
6060	8" diameter	8.35	6.75	15.10
6070	10" diameter	9.20	8.60	17.80
6080	12" diameter	10.50	10.95	21.45
6100	15" diameter	13.35	10.95	24.30
6120	18" diameter	17.65	11.70	29.35
6140	24" diameter	25.50	13.70	39.20
6170	30" diameter	32.50	25.50	58
6180	36" diameter	53.50	25.50	79
6200	48" diameter	71.50	30	101.50
6220	60" diameter	112	46.50	158.50
6230	72" diameter	157	113	270
6300	Steel or alum. oval arch, coated & paved invert			
6400	15" equivalent diameter	32	12.05	44.05
6420	18" equivalent diameter	41	16	57
6450	24" equivalent diameter	59	19.25	78.25
6470	30" equivalent diameter	72	24	96
6480	36" equivalent diameter	107	33	140
6490	42" equivalent diameter	129	36.50	165.50
6500	48" equivalent diameter	144	44	188
6600	Plain			
6700	15" equivalent diameter	17.30	10.70	28
6720	18" equivalent diameter	20.50	13.70	34.20
6750	24" equivalent diameter	33	16	49
6770	30" equivalent diameter	41.50	30.50	72
6780	36" equivalent diameter	68	30.50	98.50
6790	42" equivalent diameter	80.50	36	116.50
6800	48" equivalent diameter	71	44	115
8000	Polyvinyl chloride SDR 35			
8130	4" diameter	2.78	3.10	5.88
8150	6" diameter	5.85	3.32	9.17
8160	8" diameter	8.85	3.47	12.32
8170	10" diameter	12.80	4.73	17.53
8180	12" diameter	12.70	4.88	17.58
8200	15" diameter	18.65	8.25	26.90

Important: See the Reference Section for critical supporting data - Reference Numbers and City Cost Indexes

G3030 Storm Sewer

Manhole   Catch Basin

The Manhole and Catch Basin System includes: excavation with a backhoe; a formed concrete footing; frame and cover; cast iron steps and compacted backfill.

The Expanded System Listing shows manholes that have a 4', 5' and 6' inside diameter riser. Depths range from 4' to 14'. Construction material shown is either concrete, concrete block, precast concrete, or brick.

BUILDING SITEWORK   G

System Components	QUANTITY	UNIT	COST PER EACH		
			MAT.	INST.	TOTAL
SYSTEM G3030 210 1920					
MANHOLE/CATCH BASIN, BRICK, 4' I.D. RISER, 4' DEEP					
Excavation, hydraulic backhoe, 3/8 C.Y. bucket	14.815	C.Y.		87.41	87.41
Trim sides and bottom of excavation	64.000	S.F.		46.72	46.72
Forms in place, manhole base, 4 uses	20.000	SFCA	13	80.60	93.60
Reinforcing in place footings, #4 to #7	.019	Ton	17.01	18.81	35.82
Concrete, 3000 psi	.925	C.Y.	88.34		88.34
Place and vibrate concrete, footing, direct chute	.925	C.Y.		36.88	36.88
Catch basin or MH, brick, 4' ID, 4' deep	1.000	Ea.	395	785	1,180
Catch basin or MH steps; heavy galvanized cast iron	1.000	Ea.	15.35	11.15	26.50
Catch basin or MH frame and cover	1.000	Ea.	246	174	420
Fill, granular	12.954	C.Y.	114.64		114.64
Backfill, spread with wheeled front end loader	12.954	C.Y.		23.45	23.45
Backfill compaction, 12" lifts, air tamp	12.954	C.Y.		85.75	85.75
TOTAL			889.34	1,349.77	2,239.11

G3030 210	Manholes & Catch Basins	COST PER EACH		
		MAT.	INST.	TOTAL
1920	Manhole/catch basin, brick, 4' I.D. riser, 4' deep	890	1,350	2,240
1940	6' deep	1,175	1,900	3,075
1960	8' deep	1,500	2,575	4,075
1980	10' deep	2,000	3,200	5,200
3000	12' deep	2,625	3,450	6,075
3020	14' deep	3,300	4,800	8,100
3200	Block, 4' I.D. riser, 4' deep	825	1,075	1,900
3220	6' deep	1,075	1,550	2,625
3240	8' deep	1,325	2,125	3,450
3260	10' deep	1,550	2,625	4,175
3280	12' deep	1,900	3,300	5,200
3300	14' deep	2,275	4,025	6,300
4620	Concrete, cast-in-place, 4' I.D. riser, 4' deep	1,025	1,875	2,900
4640	6' deep	1,375	2,525	3,900
4660	8' deep	1,850	3,650	5,500
4680	10' deep	2,225	4,525	6,750
4700	12' deep	2,700	5,600	8,300
4720	14' deep	3,200	6,725	9,925
5820	Concrete, precast, 4' I.D. riser, 4' deep	1,275	1,000	2,275
5840	6' deep	1,625	1,350	2,975

G3030 Storm Sewer

G3030 210	Manholes & Catch Basins	COST PER EACH		
		MAT.	INST.	TOTAL
5860	8' deep	1,950	1,875	3,825
5880	10' deep	2,400	2,325	4,725
5900	12' deep	2,950	2,850	5,800
5920	14' deep	3,550	3,600	7,150
6000	5' I.D. riser, 4' deep	1,375	1,100	2,475
6020	6' deep	1,800	1,525	3,325
6040	8' deep	2,250	2,025	4,275
6060	10' deep	2,825	2,575	5,400
6080	12' deep	3,450	3,275	6,725
6100	14' deep	4,125	3,975	8,100
6200	6' I.D. riser, 4' deep	1,975	1,425	3,400
6220	6' deep	2,575	1,900	4,475
6240	8' deep	3,150	2,625	5,775
6260	10' deep	3,950	3,350	7,300
6280	12' deep	4,800	4,200	9,000
6300	14' deep	5,700	5,100	10,800

Important: See the Reference Section for critical supporting data - Reference Numbers and City Cost Indexes

G3060 Fuel Distribution

G3060 110	Gas Service Piping	COST PER L.F.		
		MAT.	INST.	TOTAL
2000	Piping, excavation & backfill excluded, polyethylene			
2070	1-1/4" diam, SDR 10	1.04	3	4.04
2090	2" diam, SDR 11	1.29	3.35	4.64
2110	3" diam, SDR 11.5	2.67	4	6.67
2130	4" diam, SDR 11	6.15	7.65	13.80
2150	6" diam, SDR 21	19.10	8.20	27.30
2160	8" diam, SDR 21	26	9.85	35.85
3000	Steel, schedule 40, plain end, tarred & wrapped			
3060	1" diameter	3.14	6.80	9.94
3090	2" diameter	4.92	7.25	12.17
3110	3" diameter	8.15	7.85	16
3130	4" diameter	10.60	12.25	22.85
3140	5" diameter	15.35	14.25	29.60
3150	6" diameter	18.80	17.40	36.20
3160	8" diameter	30	22.50	52.50
3170	10" diameter	103	31.50	134.50
3180	12" diameter	129	39	168
3190	14" diameter	137	41.50	178.50
3200	16" diameter	150	44.50	194.50
3210	18" diameter	193	48.50	241.50
3220	20" diameter	300	52	352
3230	24" diameter	345	62.50	407.50

G

BUILDING SITEWORK

G4020   Site Lighting

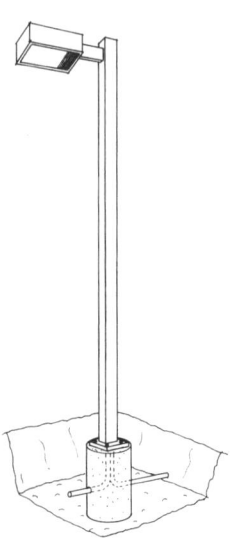

The Site Lighting System includes the complete unit from foundation to electrical fixtures. Each system includes: excavation; concrete base; backfill by hand; compaction with a plate compacter; pole of specified material; all fixtures; and lamps.

The Expanded System Listing shows Site Lighting Systems that use one of three types of lamps: high pressure sodium; mercury vapor; and metal halide. Systems are listed for 400-watt and 1000-watt lamps. Pole height varies from 20′ to 40′. There are four types of poles listed: aluminum, fiberglass, steel and wood.

G4020 110	Site Lighting	COST EACH		
		MAT.	INST.	TOTAL
2320	Site lighting, high pressure sodium, 400 watt, aluminum pole, 20′ high	1,425	1,050	2,475
2360	40′ high	2,650	1,675	4,325
2920	Wood pole, 20′ high	870	1,050	1,920
2960	40′ high	1,150	1,600	2,750
3120	1000 watt, aluminum pole, 20′ high	1,500	1,075	2,575
3160	40′ high	2,725	1,700	4,425
3520	Wood pole, 20′ high	945	1,050	1,995
3560	40′ high	1,225	1,625	2,850
5820	Metal halide, 400 watt, aluminum pole, 20′ high	1,425	1,050	2,475
5860	40′ high	2,625	1,675	4,300
7620	1000 watt, aluminum pole, 20′ high	1,475	1,075	2,550
7661	40′ high	2,700	1,700	4,400

Table G4020 210 Procedure for Calculating Floodlights Required for Various Footcandles
Poles should not be spaced more than 4 times the fixture mounting height for good light distribution. To maintain 1 footcandle

Estimating Chart
Select Lamp type.

Determine total square feet.

Chart will show quantity of fixtures to provide 1 footcandle initial, at intersection of lines. Multiply fixture quantity by desired footcandle level.

Chart based on use of wide beam luminaires in an area whose dimensions are large compared to mounting height and is approximate only.

Over a large area use these watts per square foot:

Incandescent	0.15
Metal Halide	0.032
Mercury Vapor	0.05
High Pressure Sodium	0.024

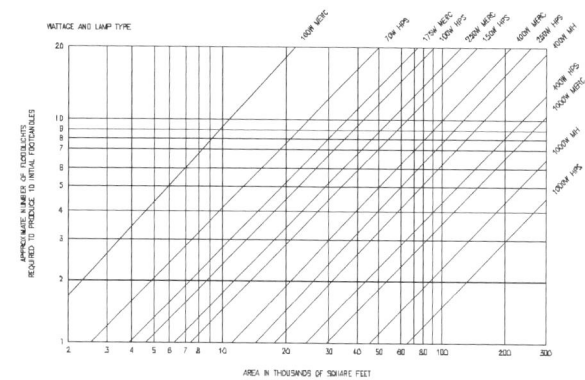

BUILDING SITEWORK G

System Components			COST EACH		
	QUANTITY	UNIT	MAT.	INST.	TOTAL
SYSTEM G4020 210 0200					
LIGHT POLES, ALUMINUM, 20' HIGH, 1 ARM BRACKET					
Aluminum light pole, 20', no concrete base	1.000	Ea.	840	483	1,323
Bracket arm for Aluminum light pole	1.000	Ea.	110	62.50	172.50
Excavation by hand, pits to 6' deep, heavy soil or clay	2.368	C.Y.		202.46	202.46
Footing, concrete incl forms, reinforcing, spread, under 1 C.Y.	.465	C.Y.	63.71	72.03	135.74
Backfill by hand	1.903	C.Y.		58.99	58.99
Compaction vibrating plate	1.903	C.Y.		8.01	8.01
TOTAL			1,013.71	886.99	1,900.70

G4020 210	Light Pole (Installed)		COST EACH		
			MAT.	INST.	TOTAL
0200	Light pole, aluminum, 20' high, 1 arm bracket		1,025	890	1,915
0240	2 arm brackets		1,125	890	2,015
0280	3 arm brackets		1,225	920	2,145
0320	4 arm brackets	RD5020 -250	1,350	920	2,270
0360	30' high, 1 arm bracket		1,825	1,125	2,950
0400	2 arm brackets		1,925	1,125	3,050
0440	3 arm brackets		2,025	1,150	3,175
0480	4 arm brackets		2,150	1,150	3,300
0680	40' high, 1 arm bracket		2,225	1,475	3,700
0720	2 arm brackets		2,350	1,475	3,825
0760	3 arm brackets		2,450	1,525	3,975
0800	4 arm brackets		2,550	1,525	4,075
0840	Steel, 20' high, 1 arm bracket		1,225	945	2,170
0880	2 arm brackets		1,325	945	2,270
0920	3 arm brackets		1,350	975	2,325
0960	4 arm brackets		1,450	975	2,425
1000	30' high, 1 arm bracket		1,425	1,200	2,625
1040	2 arm brackets		1,500	1,200	2,700
1080	3 arm brackets		1,525	1,225	2,750
1120	4 arm brackets		1,625	1,225	2,850
1320	40' high, 1 arm bracket		1,850	1,625	3,475
1360	2 arm brackets		1,950	1,625	3,575

G4020 Site Lighting

G4020 210	Light Pole (Installed)	COST EACH		
		MAT.	INST.	TOTAL
1400	3 arm brackets	1,975	1,650	3,625
1440	4 arm brackets	2,075	1,650	3,725

For information about Means Estimating Seminars, see yellow pages 12 and 13 in back of book

Important: See the Reference Section for critical supporting data - Reference Numbers and City Cost Indexes

Reference Section

All the reference information is in one section, making it easy to find what you need to know . . . and easy to use the book on a daily basis. This section is visually identified by a vertical gray bar on the page edges.

In the reference number information that follows, you'll see the background that relates to the reference numbers that appeared in the Assembly Cost Sections. You'll find reference tables, explanations, and estimating information that support how we derive the systems data. Also included are alternate pricing methods, technical data, and estimating procedures along with information on design and economy in construction.

Also in this Reference Section, we've included Historical Cost Indexes for cost comparison over time; City Cost Indexes and Location Factors for adjusting costs to the region you are in; Square Foot Costs; Reference Aids; Estimating Forms; and an explanation of all Abbreviations used in the book.

Table of Contents

General: A spread footing is used to convert a concentrated load (from one superstructure column, or substructure grade beams) into an allowable area load on supporting soil.

Because of punching action from the column load, a spread footing is usually thicker than strip footings which support wall loads. One or two story commercial or residential buildings should have no less than 1' thick spread footings. Heavier loads require no less than 2' thick. Spread footings may be square, rectangular or octagonal in plan.

Spread footings tend to minimize excavation and foundation materials, as well as labor and equipment. Another advantage is that footings and soil conditions can be readily examined. They are the most widely used type of footing, especially in mild climates and for buildings of four stories or under. This is because they are usually more economical than other types, if suitable soil and site conditions exist.

They are used when suitable supporting soil is located within several feet of the surface or line of subsurface excavation. Suitable soil types include sands and gravels, gravels with a small amount of clay or silt, hardpan, chalk, and rock. Pedestals may be used to bring the column base load down to the top of footing. Alternately, undesirable soil between underside of footing and top of bearing level can be removed and replaced with lean concrete mix or compacted granular material.

Depth of footing should be below topsoil, uncompacted fill, muck, etc. It must be lower than frost penetration (see local code or Table L1030-502 in Section L) but should be above the water table. It must not be at the ground surface because of potential surface erosion. If the ground slopes, approximately three horizontal feet of edge protection must remain. Differential footing elevations may overlap soil stresses or cause excavation problems if clear spacing between footings is less than the difference in depth.

Other footing types are usually used for the following reasons:

A. Bearing capacity of soil is low.

B. Very large footings are required, at a cost disadvantage.

C. Soil under footing (shallow or deep) is very compressible, with probability of causing excessive or differential settlement.

D. Good bearing soil is deep.

E. Potential for scour action exists.

F. Varying subsoil conditions within building perimeter.

Cost of spread footings for a building is determined by:

1. The soil bearing capacity.

2. Typical bay size.

3. Total load (live plus dead) per S.F. for roof and elevated floor levels.

4. The size and shape of the building.

5. Footing configuration. Does the building utilize outer spread footings or are there continuous perimeter footings only or a combination of spread footings plus continuous footings?

COST DETERMINATION

1. Determine Soil Bearing Capacity by a known value or using Table A1010-121 as a guide.

Table A1010-121   Soil Bearing Capacity in Kips per S.F.

Bearing Material	Typical Allowable Bearing Capacity
Hard sound rock	120 KSF
Medium hard rock	80
Hardpan overlaying rock	24
Compact gravel and boulder-gravel; very compact sandy gravel	20
Soft rock	16
Loose gravel; sandy gravel; compact sand; very compact sand-inorganic silt	12
Hard dry consolidated clay	10
Loose coarse to medium sand; medium compact fine sand	8
Compact sand-clay	6
Loose fine sand; medium compact sand-inorganic silts	4
Firm or stiff clay	3
Loose saturated sand-clay; medium soft clay	2

Table A1010-122 Working Load Determination

2. Determine Bay Size in S.F.

3. Determine Load to the Footing

Type Load	Reference	Working Loads			
		Roof	Floor	Column	Total
1. Total Load	Floor System	PSF	PSF		
2. Whole Bay Load/Level	(Line 1) x Bay Area ÷ 1000	Kips	Kips		
3. Load to Column	[(Roof + (Floor x No. Floors)]	Kips	Kips		Kips
4. Column Weight	Systems B1010 201 to B1010 208			Kips	Kips
5. Fireproofing	System B1010 720			Kips	Kips
6. Total Load to Footing	Roof + Floor + Column				Kips

Working loads represent the highest actual load a structural member is designed to support.

4. Determine Size and Shape of Building

Figure A1010-123

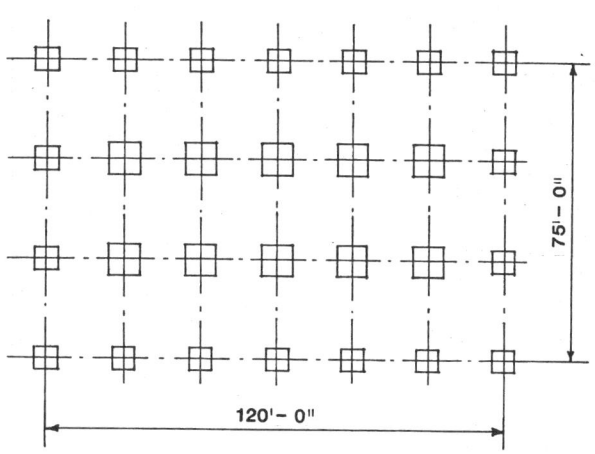

Figure A1010-124

5. Footing Configuration Possibilities

a. Building has spread footings throughout (including exterior and corners).

b. Building has continuous perimeter footings to support exterior columns, corners and exterior walls (as well as interior spread footings).

c. Building has both perimeter continuous footings for walls plus exterior spread footings for column loads. If there are interior spread footings also, use "b".
  1. Figure continuous footings from tables in Section A1010-110.
  2. Add total dollars for spread footings and continuous footings to determine total foundation dollars.

Table A1010-126 Working Load Determination

Tabulate interior footing costs and add continuous footing costs from Systems A1010 210 and A1010 110.

Example: Spread Footing Cost Determination

Office Building . . . . 5 Story
Story Height . . . . 12'
Dimensions . . . . 75' x 120'
Bay Size . . . . 20' x 25'
Soil Bearing Capacity . . . . 6 KSF
Foundation . . . . Spread Footings
Slab on Grade . . . . 5" Reinforced
Code . . . . BOCA
Construction . . . . From Example RB1010-112
Using concrete flat plate

Figure A1010-125

Determine load to footings Table A1010-122 or Example RB1010-112.

Type Load	Reference	Working Loads			
		Roof	Floor	Column	Total
Determine floor system	System in Div. B1010 (Example: Flat Plate)				
1. Total Load	System # B1010 223	146PSF	188PSF		
2. Whole Bay Load/Level	Bay Area x Line 1/1000	73K	94K		
3. Load To Column	Roof + (Floor x No. Floors)	73K	276K		449K
4. Column Weight	Systems B1010 201 to B1010 208			25K	25K
5. Fireproofing	N/A			—	
6. Load to Footing	Roof + Floor + Column				474K

1. Total load working from System B1010 223: 40# superimposed load for roof and 75# superimposed load for floors.
2. Bay Area 20' x 25' = 500 S.F. x Line 1
   500 x 146/1000 = 73K roof
   500 x 188/1000 = 94K floor
3. Roof + (floor x no. floors)
   73K + (4 x 94K) = 449K

4. Enter System B1010 203 with a total working load of 474K and 12' story height. A 16" square tied column with 6 KSI concrete is most economical. Minimum column size from System B1010 223 is 20". Use the larger. Column weight = 394 #/V.L.F. x 12' = approx. 5K.
5. No fireproofing is required for concrete column.
6. Add roof, floors and column loads for a total load to each interior footing.

Footing Cost: Enter System A1010 210 with total working load and allowable soil pressure. Determine the cost per footing using the closest higher load on the table. Add the total costs for interior, exterior, and corner footings.

A10 Foundations — RA1010-120 — Spread Footings

Table A1010-127   Typical Spread Footing, Square

EXTERIOR FOOTINGS

Exterior vs. Interior Footings

System A1010 210 contains the cost for individual spread interior footings for various loading and soil bearing conditions. To determine the loads for exterior footings:

1. Multiply whole bay working load by 60% for half bay footings and by 45% for quarter bay (or corner) footings. This will give the approximate column load for half and quarter bay footings.
2. Enter System A1010 210 and find cost of footing with loads determined in step 1.
3. Determine the number of each type of footing on the job and multiply the cost per footing by the appropriate quantity. This will give the total dollar cost for exterior and interior spread footings.

Example: Using soil capacity, 6KSF
Total working load from Table A1010-126 . . . . = 474K
Half Bay Footings . . . . 474K x 60% = 284K
Corner Footings . . . . 474K x 45% = 213K

Table A1010-128   Spread Footing Cost

Footing Type	Working Load (Kips)	Total Quantity	System A1010 210 Unit Cost	Total
Interior	474	10	$2,100	$21,000
Exterior	284	14	1,170	16,380
Corner	213	4	1,170	4,680
Total				$42,060

SUBSTRUCTURE

REFERENCE TABLES

Table A1010-141 Section Thru Typical Strip Footing

General: Strip (continuous) footings are a type of spread footing, but are used to convert the lineal type wall into an allowable area load on supporting soil.

Many of the general comments on isolated spread footings (Section A1010 210) apply here as well.

Strip footings may be used under walls of concrete, brick, block or stone. They are constructed of continuously placed concrete, 2000 psi or greater. Normally they are not narrower than twice the wall thickness, nor are they thinner than the wall thickness. Plain concrete footings should be at least 8" thick, and reinforced footings require at least 6" of thickness above the bottom reinforcement. Where there is no basement, the bottom of footing is usually 3' to 5' below finished exterior grade, and 12" below the average frost penetration (see Figure A1010-141), resting on undisturbed soil. In some cases strip footings serve as leveling pads (stone walls, etc.).

Method of construction is affected by reinforcement and soil type. If steel mats and/or dowels are present and require accurate placement (frequently for reinforced concrete or reinforced masonry walls), then side forms are required, as they are in wet or sandy soil. If there is no reinforcement or only longitudinal reinforcement, and the soil will, "stand," forms might be eliminated and economically should.

Preferably, strip footings are used on sand and gravel. The presence of a small amount of clay or dense silty sand in gravel is acceptable, as is bearing on rock or chalk. They are sometimes used in uniform, firm or stiff clays with little nearby ground vegetation, and placed at least 3-1/2' into the clay. If the clay is sloped, there is potential for downhill creep. Footings up to

3-1/2' wide are sometimes used in soft or silty clays, but settlement must be expected and provision made.

Sloping terrain requires stepping the footing to maintain depth, since a sloped footing will create horizontal thrust which may distress the structure. Steps should be at least 2' long horizontally and each vertical step no greater than three-fourths the horizontal distance between steps. Vertical risers must be at least 6" thick and of footing width. Potential for erosion of surrounding soil must be considered.

Strip footings should be used under walls constructed of unreinforced concrete, or of units bonded with mortar. They are also used when bearing capacity of soil is inadequate to support wall thickness alone.

Footings require horizontal reinforcement at right angles to the wall line when side projection of footings beyond face of wall is greater than 1/2 to 2/3 footing thickness. Horizontal reinforcement parallel to wall line should be used if soil bearing capacity is low, or soil compressibility is variable, or the footing spans pipe trenches, etc. Vertical reinforcing dowels depend upon wall design.

Alternate foundation types include bored piles, grade beams, and mat. Short, bored compacted concrete piles are sometimes competitive with strip footings, especially in shrinkable clay soils where they may provide an added factor of safety. When soil bearing capacity is inadequate, grade beams can be used to transfer wall loads across the inadequate ground to column support foundations. Mat footings replace both strip and isolated spread footings. They may eliminate differential settlement problems.

**Figure A1010-141
Section Thru Typical Strip Footing**

Example: Strip Footing Cost Determination for Bearing Wall

Office Building . . . . .5 Story
Story Height . . . . .12 VLF
Dimensions . . . . .50′ x 80′ = 4000 S.F.G.
Exterior Wall . . . . .12″ Brick and Block
Interior Wall . . . . .8″ Block
Floor . . . . Prestressed Concrete Slabs + 2″ Topping
Roof . . . . .Prestressed Concrete Slabs, No Topping
Soil Bearing Capacity . . . . .6 KSF
Superimposed Load:
    floor . . . . .100 PSF
    roof . . . . .40 PSF

Figure A1010-142 Footing Plan

A. Total load to each footing

Table A1010-143   Strip Footing, Load Determination

Type Load	Reference	Working Loads			
		Roof	Floor	Wall	Total
1. Total Load	System B1010 229	90 PSF	175 PSF	95 PSF	
2. Load Span or Height	Building Design	10 LF	10 LF	12 VLF	
3. Whole Bay Load/Level	(Line 1 x Line 2) ÷ 1000	.9 KIPS	1.75 KIPS	1.1 KIPS	
4. Load to Footing/L.F.	Roof + (Floor + Wall)	.9 KLF	7.0 KLF	5.9 KLF	13.8 KLF

1. Total load from System B1010-229.

2. Roof and floor load spans for exterior strip footings are 1/2 total span of 20′ or 10′ (wall height for our example is 12′).

3. Multiply Line 1 by Line 2 in each column and divide by 1,000Lb/Kip.

4. Add 1 roof + 4 elevated floors + 5.33 stories of wall (includes .33 added for a frost wall) for the total load to the footing.

B. Enter System A1010-110 with a load of 13.8 KLF and interpolate between line items 2700 and 3100.

$$\text{Interpolating: } \$30.95 + \left[\frac{13.8 \text{-} 11.1}{14.8 \text{-} 11.1} \times (\$35.55 - \$30.95) \right] = \$34.40$$

C. If the building design incorporates other strip footing loads, repeat steps A and B for each type of footing load.

Table A1010-144   Cost Determination, Strip Footings

Description	Strip Footing Type or Size	Total Length Each Type (Ft.)	Cost From System A1010 110 ($/L.F.)	Cost x Length ($/Each Type)	Total Cost All Types
End Bearing Wall	1	100	$34.40	$ 3,440	
Non Bearing Wall	2	160	30.95	$ 4,952	
Interior Bearing Wall	3	150	72.00	$10,800	$19,192

Note: Type 2 wall: use "wall" loads only in Step A above.
      Type 3 wall: same as Step A above with load span increased to 20′.
      Interpolated values from System A1010 110 are rounded to $.05.

General: The function of a reinforced concrete pile cap is to transfer superstructure load from isolated column or pier to each pile in its supporting cluster. To do this, the cap must be thick and rigid, with all piles securely embedded into and bonded to it.

SEE "SLAB-ON-GRADE"
SEE "COLUMNS"
EXCAV. & FILL INCLUDED IN COST
f'c = 3000 psi
fy = 60,000 psi

Figure A1010-331 Section Through Pile Cap

Table A1010-332   Concrete Quantities for Pile Caps

Load	Number of Piles @ 3'-0" O.C. Per Footing Cluster									
Working (K)	2 (CY)	4 (CY)	6 (CY)	8 (CY)	10 (CY)	12 (CY)	14 (CY)	16 (CY)	18 (CY)	20 (CY)
50	(.9)	(1.9)	(3.3)	(4.9)	(5.7)	(7.8)	(9.9)	(11.1)	(14.4)	(16.5)
100	(1.0)	(2.2)	(3.3)	(4.9)	(5.7)	(7.8)	(9.9)	(11.1)	(14.4)	(16.5)
200	(1.0)	(2.2)	(4.0)	(4.9)	(5.7)	(7.8)	(9.9)	(11.1)	(14.4)	(16.5)
400	(1.1)	(2.6)	(5.2)	(6.3)	(7.4)	(8.2)	(13.7)	(11.1)	(14.4)	(16.5)
800		(2.9)	(5.8)	(7.5)	(9.2)	(13.6)	(17.6)	(15.9)	(19.7)	(22.1)
1200			(5.8)	(8.3)	(9.7)	(14.2)	(18.3)	(20.4)	(21.2)	(22.7)
1600				(9.8)	(11.4)	(14.5)	(19.5)	(20.4)	(24.6)	(27.2)
2000				(9.8)	(11.4)	(16.6)	(24.1)	(21.7)	(26.0)	(28.8)
3000						(17.5)		(26.5)	(30.3)	(32.9)
4000								(30.2)	(30.7)	(36.5)

Table A1010-333   Concrete Quantities for Pile Caps

Load	Number of Piles @ 4'-6" O.C. Per Footing					
Working (K)	2 (CY)	3 (CY)	4 (CY)	5 (CY)	6 (CY)	7 (CY)
50	(2.3)	(3.6)	(5.6)	(11.0)	(13.7)	(12.9)
100	(2.3)	(3.6)	(5.6)	(11.0)	(13.7)	(12.9)
200	(2.3)	(3.6)	(5.6)	(11.0)	(13.7)	(12.9)
400	(3.0)	(3.6)	(5.6)	(11.0)	(13.7)	(12.9)
800			(6.2)	(11.5)	(14.0)	(12.9)
1200				(13.0)	(13.7)	(13.4)
1600						(14.0)

General: Piles are column-like shafts which receive superstructure loads, overturning forces, or uplift forces. They receive these loads from isolated column or pier foundations (pile caps), foundation walls, grade beams, or foundation mats. The piles then transfer these loads through shallower poor soil strata to deeper soil of adequate support strength and acceptable settlement with load.

Be sure that other foundation types aren't better suited to the job. Consider ground and settlement, as well as loading, when reviewing. Piles usually are associated with difficult foundation problems and substructure condition. Ground conditions determine type of pile (different pile types have been developed to suit ground conditions). Decide each case by technical study, experience, and sound engineering judgment—not rules of thumb. A full investigation of ground conditions, early, is essential to provide maximum information for professional foundation engineering and an acceptable structure.

Piles support loads by end bearing and friction. Both are generally present; however, piles are designated by their principal method of load transfer to soil.

Boring should be taken at expected pile locations. Ground strata (to bedrock or depth of 1-1/2 building width) must be located and identified with appropriate strengths and compressibilities. The sequence of strata determines if end-bearing or friction piles are best suited. See **Table A1020-105** for site investigation costs.

End-bearing piles have shafts which pass through soft strata or thin hard strata and tip bear on bedrock or penetrate some distance into a dense, adequate soil (sand or gravel.)

Friction piles have shafts which may be entirely embedded in cohesive soil (moist clay), and develop required support mainly by adhesion or "skin-friction" between soil and shaft area.

Piles pass through soil by either one of two ways:

1. Displacement piles force soil out of the way. This may cause compaction, ground heaving, remolding of sensitive soils, damage to adjacent structures, or hard driving.
2. Non-displacement piles have either a hole bored and the pile cast or placed in the hole, or open-ended pipe (casing) driven and the soil core removed. They tend to eliminate heaving or lateral pressure damage to adjacent structures of piles. Steel "HP" piles are considered of small displacement.

Placement of piles (attitude) is most often vertical; however, they are sometimes battered (placed at a small angle from vertical) to advantageously resist lateral loads. Seldom are piles installed singly but rather in clusters (or groups). Codes require a minimum of three piles per major column load or two per foundation wall or grade beam. Single pile capacity is limited by pile structural strength or support strength of soil. Support capacity of a pile cluster is almost always less than the sum of its individual pile capacities due to overlapping of bearing the friction stresses. See **Table A1020-101** for minimum pile spacing requirements.

Large rigs for heavy, long piles create large soil surface loads and additional expense on weak ground. See **Table A1020-103** for percent of cost increases.

Fewer piles create higher costs per pile. See **Table A1020-102** for effect of mobilization.

Pile load tests are frequently required by code, ground situation, or pile type. See **Table A1020-104** for costs. Test load is twice the design load.

Table A1020-101   Min. Pile Spacing by Pile Type

Type Support		Min. Pile Spacing	
		X's Butt Diameter	Foot
End Bearing	Bedrock	2	2'-0"
	Hard Strata	2.5	2'-6"
Friction		3 to 5	3'-6"

Table A1020-102   Add cost for Mobilization (Setup and Removal)

Job Size	Cost	
	$/Job	$/L.F. Pile
Small (12,000 L.F.)	$20,300	$1.69
Large (25,000 L.F.)	28,800	1.15

Table A1020-103   Add Costs for Special Soil Conditions

Special Conditions	Add % of Total
Soft, damp ground	40%
Swampy, wet ground	40%
Barge-mounted drilling rig	30%

Table A1020-104   Testing Costs, if Required, Any Pile Type

Test Weight (Tons)	$/Test
50-100T	$16,050
100-200T	20,300
150-300T	25,700
200-400T	28,800

Table A1020-105   Cost/L.F. of Boring Types (4") and Per Job Cost of Other Items

Type Soil	Type Boring	Sample	$/L.F. Boring	Total $
Earth	Auger	None	$24.50	
	Cased	Yes	51.00	
Rock, "BX"	Core	None	55.50	
	Cased	Yes	78.00	
Filed survey, mobilization, demobilization & engineering report				$2,275

SUBSTRUCTURE

General: Caissons, as covered in this section, are drilled cylindrical foundation shafts which function primarily as short column-like compression members. They transfer superstructure loads through inadequate soils to bedrock or hard stratum. They may be either reinforced or unreinforced and either straight or belled out at the bearing level.

Shaft diameters range in size from 20″ to 84″ with the most usual sizes beginning at 34″. If inspection of bottom is required, the minimum diameter practical is 30″. If handwork is required (in addition to mechanical belling, etc.) the minimum diameter is 32″. The most frequently used shaft diameter is probably 36″ with a 5′ or 6′ bell diameter. The maximum bell diameter practical is three times the shaft diameter.

Plain concrete is commonly used, poured directly against the excavated face of soil. Permanent casings add to cost and economically should be avoided. Wet or loose strata are undesirable. The associated installation sometimes involves a mudding operation with bentonite clay slurry to keep walls of excavation stable (costs not included here).

Reinforcement is sometimes used, especially for heavy loads. It is required if uplift, bending moment, or lateral loads exist. A small amount of reinforcement is desirable at the top portion of each caisson, even if the above conditions theoretically are not present. This will provide for construction eccentricities and other possibilities. Reinforcement, if present, should extend below the soft strata. Horizontal reinforcement is not required for belled bottoms.

There are three basic types of caisson bearing details:

1. Belled, which are generally recommended to provide reduced bearing pressure on soil. These are not for shallow depths or poor soils. Good soils for belling include most clays, hardpan, soft shale, and decomposed rock.

Soils requiring handwork include hard shale, limestone, and sandstone.

Soils not recommended include sand, gravel, silt, and igneous rock. Compact sand and gravel above water table may stand. Water in the bearing strata is undesirable.

2. Straight shafted, which have no bell but the entire length is enlarged to permit safe bearing pressures. They are most economical for light loads on high bearing capacity soil.

3. Socketed (or keyed), which are used for extremely heavy loads. They involve sinking the shaft into rock for combined friction and bearing support action. Reinforcement of shaft is usually necessary. Wide flange cores are frequently used here.

Advantages include:

A. Shafts can pass through soils that piles cannot

B. No soil heaving or displacement during installation

C. No vibration during installation

D. Less noise than pile driving

E. Bearing strata can be visually inspected & tested

Uses include:

A. Situations where unsuitable soil exists to moderate depth

B. Tall structures

C. Heavy structures

D. Underpinning (extensive use)

See Table A1010-121 for Soil Bearing Capacities.

Figure A1020-201 Design Assumptions

Figure A1020-202 Size Range

REFERENCE TABLES

Table A1020-231   Grade Beam Detail

General: Grade beams are stiff self-supporting structural members, partly exposed above grade. They support wall loads and carry them across unacceptable soil to column footings, support piles or caissons.

They should be deep enough to be below frost depth. Therefore, they are more frequently used in mild climates.

They must be stiff enough to prevent cracking of the supported wall. Usually, they are designed as simply supported beams, so as to minimize effects of unequal footing settlement.

Heavy column loads and light wall loads tend to make grade beams an economical consideration. Light column loads but heavy wall loads tend to make strip footings more economical.

Conditions which favor other foundation types include:

  A. Deep frost penetration requiring deep members

  B. Good soil bearing near surface

  C. Large live loads to grade beam

  D. Ground floor 2′ or 3′ above finished grade requiring wall depth beams (not included here).

  E. Varying ground elevations

  F. Basement, requiring walls

Design Assumption: See Figure A1020-231
   Concrete placed by chute
   Forms, four uses
   Simply supported
   Max. deflection = L/480
   Design span = Bay width less 2′

Figure A1020-231 Grade Beam Detail

Table A1030-202 Thickness and Loading Assumptions by Type of Use

General: Grade slabs are classified on the basis of use. Thickness is generally controlled by the heaviest concentrated load supported. If load area is greater than 80 sq. in., soil bearing may be important. The base granular fill must be a uniformly compacted material of limited capillarity, such as gravel or crushed rock. Concrete is placed on this surface or the vapor barrier on top of base.

Grade slabs are either single or two course floors. Single course are widely used. Two course floors have a subsequent wear resistant topping.

Reinforcement is provided to maintain tightly closed cracks.

Control joints limit crack locations and provide for differential horizontal movement only. Isolation joints allow both horizontal and vertical differential movement.

Use of Table: Determine appropriate type of slab (A, B, C, or D) by considering type of use or amount of abrasive wear of traffic type.

Determine thickness by maximum allowable wheel load or uniform load, opposite 1st column thickness. Increase the controlling thickness if details require, and select either plain or reinforced slab thickness and type.

Figure A1030-201 Section, Slab-on-Grade

Table A1030-202 Thickness and Loading Assumptions by Type of Use

SLAB THICKNESS (IN.)	TYPE	A Non Little Foot Only Load* (K)	B Light Light Pneumatic Wheels Load* (K)	C Normal Moderate Solid Rubber Wheels Load* (K)	D Heavy Severe Steel Tires Load* (K)	◄ Slab I.D. ◄ Industrial ◄ Abrasion ◄ Type of Traffic Max. Uniform Load to Slab ▼ (PSF)
4″	Reinf. Plain	4K				100
5″	Reinf. Plain	6K	4K			200
6″	Reinf. Plain		8K	6K	6K	500 to 800
7″	Reinf. Plain			9K	8K	1,500
8″	Reinf. Plain				11K	
10″	Reinf. Plain				14K	* Max. Wheel Load in Kips (incl. impact)
12″	Reinf. Plain					
D E S I G N A S S U M P T I O N S	Concrete, Chuted	f'c = 3.5 KSI	4 KSI	4.5 KSI	Slab @ 3.5 KSI	ASSUMPTIONS BY SLAB TYPE
	Topping			1″ Integral	1″ Bonded	
	Finish	Steel Trowel	Steel Trowel	Steel Trowel	Screed & Steel Trowel	
	Compacted Granular Base	4″ deep for 4″ slab thickness 6″ deep for 5″ slab thickness & greater				ASSUMPTIONS FOR ALL SLAB TYPES
	Vapor Barrier	6 mil polyethylene				
	Forms & Joints	Allowances included				
	Reinforcement	WWF As required ≥ 60,000 psi				

Table A2020-212 Minimum Wall Reinforcement Weight (PSF)

General: Foundation wall heights depend upon depth of frost penetration, basement configuration or footing requirements. For certain light load conditions, good soil bearing and adequate wall thickness, strip footings may not be required.

Thickness requirements are based upon code requirements (usually 8" min. for foundation, exposed basement, fire and party walls), unsupported height or length, configuration of wall above, and structural load considerations.

Shrinkage and temperature forces make longitudinal reinforcing desirable, while high vertical and lateral loads make vertical reinforcing necessary. Earthquake design also necessitates reinforcing.

Figure A2020-211 shows the range in wall sizes for which costs are provided in System A2020 110. Other pertinent information is also given.

Design Assumptions: Reinforced concrete wall designs are based upon ACI 318-71-14, Empirical Design of Walls. Plain concrete wall designs are based upon nonreinforced load bearing masonry criteria (fm = 1700 psi). Earthquake, lateral or shear load requirements may add to reinforcement costs, and are not included.

Minimum Wall Reinforcement: Table below lists the weight of wall for minimum quantities of reinforcing steel for various wall thicknesses.

Design Assumptions:
Reinforcing steel is A 615, grade 60 (fy = 60 k).
Total percent of steel area is 0.20% for horizontal steel, 0.12% for vertical steel.
For steel at both faces, mats are identical.
For single layer, mat is in center of wall.
For grade 40 steel, add 25% to weights and costs.
If other than the minimum amount of steel area is required, factor weights directly with increased percentage.

**Figure A2020-211
Concrete Foundation Walls
Size Range for System A2020 110**

Note: Excavation and fill costs are not included in System A2020 110, but are provided in STRIP FOOTING COST, System A1010 110

Table A2020-212 Minimum Wall Reinforcement Weight (PSF)

Location	Wall Thickness	Horizontal Steel				Vertical Steel				Horizontal & Vertical Steel
		Bar Size	Spacing C/C	Sq. In. per L.F.	Total Wt. per S.F.	Bar Size	Spacing C/C	Sq. In. per L.F.	Total Wt. per S.F.	Total Wt. per S.F.
Both Faces	10"	#4	18"	.13	.891#	#3	18"	.07	.501#	1.392#
	12"	#4	16"	.15	1.002	#3	15"	.09	.602	1.604
	14"	#4	14"	.17	1.145	#3	13"	.10	.694	1.839
	16"	#4	12"	.20	1.336	#3	11"	.12	.820	2.156
	18"	#5	17"	.22	1.472	#4	18"	.13	.891	2.363
One Face	6"	#3	9"	.15	.501	#3	18"	.07	.251	.752
	8"	#4	12"	.20	.668	#3	11"	.12	.410	1.078
	10"	#5	15"	.25	.834	#4	16"	.15	.501	1.335

Table A2020-213 Weight of Steel Reinforcing per S.F. in Walls (PSF)

Reinforced Weights: Table below lists the weight per S.F. for reinforcing steel in walls.

Design Assumptions: Reinforcing Steel is A615 grade 60 (for costs).

Weights will be correct for any grade steel reinforcing. For bars in two directions, add weights for each size and spacing.

Table A2020-213 Weight of Steel Reinforcing per S.F. in Walls (PSF)

C/C Spacing in Inches	Bar size								
	#3 Wt. (PSF)	#4 Wt. (PSF)	#5 Wt. (PSF)	#6 Wt. (PSF)	#7 Wt. (PSF)	#8 Wt. (PSF)	#9 Wt. (PSF)	#10 Wt. (PSF)	#11 Wt. (PSF)
2"	2.26	4.01	6.26	9.01	12.27				
3"	1.50	2.67	4.17	6.01	8.18	0.68	13.60	17.21	21.25
4"	1.13	2.01	3.13	4.51	6.13	8.10	10.20	12.91	15.94
5"	.90	1.60	2.50	3.60	4.91	6.41	8.16	10.33	12.75
6"	.752	1.34	2.09	3.00	4.09	5.34	6.80	8.61	10.63
8"	.564	1.00	1.57	2.25	3.07	4.01	5.10	6.46	7.97
10"	.451	.802	1.25	1.80	2.45	3.20	4.08	5.16	6.38
12"	.376	.668	1.04	1.50	2.04	2.67	3.40	4.30	5.31
18"	.251	.445	.695	1.00	1.32	1.78	2.27	2.86	3.54
24"	.188	.334	.522	.751	1.02	1.34	1.70	2.15	2.66
30"	.150	.267	.417	.600	.817	1.07	1.36	1.72	2.13
36"	.125	.223	.348	.501	.681	.890	1.13	1.43	1.77
42"	.107	.191	.298	.429	.584	.763	.97	1.17	1.52
48"	.094	.167	.261	.376	.511	.668	.85	1.08	1.33

Selecting a Floor System:

1. Determine size of building — total S.F. and occupancy
   (i.e., office building, dormitory, etc.).
2. Determine number of floors and S.F./floor.
3. Select a possible bay size and layout.
4. From Table L1010-101 determine the minimum design live load.
   Add partition load, ceiling load and miscellaneous
   loads such as mechanical, light fixtures and flooring.

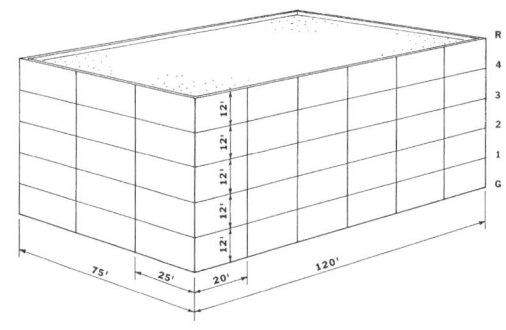

Example:

Live Load Office Building	50	PSF	BOCA Code Table L1010-101
Partitions	20	⎤	Assume using Tables L1010-201
Ceiling	5	⎦	through L1010-226
	75	PSF	Total Superimposed Load

Suspended Floors					
Building Dimension (Ft.)	Suspended Floor (S.F.)	Number of Suspended Floors	Total Suspended Floors (S.F.)	Bay Size (Ft.)	Superimposed Load PSF
75 x 120	9,000	4	36,000	20 x 25	75

5. Enter Tables with 75 PSF superimposed load and 20 x 25 bay size.

Table B1010-223	C.I.P. Flat Plate	$12.85 /S.F.
Table B1010-256	Composite Beam, Deck & Slab	14.47 /S.F.
Table B1010-250	Steel Joists & Beams on Col.	12.72 /S.F.
	(col. not included)	

It appears, from the costs listed above, that the most economical system for framing the suspended floors with 20' x 25' bays, 75 PSF superimposed load, is the open web joists, beams, slab form and 2-1/2" concrete slabs. However, the system will require a fireproof ceiling for fireproofing the structure, which will ultimately raise the cost.

The concrete flat plate may be more economical if no hung ceiling is required and the soil capacity is sufficient to justify the additional load to the foundations.

Some of the factors affecting the selection of a floor system are:

A. location
B. owner's preference
C. fire code
D. availability of materials
E. subsoil conditions
F. economy
G. clear span
H. acoustical characteristics
I. contractor capability

Table B1010-101 Comparative Costs ($/S.F.) of Floor Systems/Type (Table Number), Bay Size, & Load

Bay Size	Cast-In-Place Concrete						Precast Concrete			Structural Steel					Wood	
	1 Way BM & Slab B1010-219	2 Way BM & Slab B1010-220	Flat Slab B1010-222	Flat Plate B1010-223	Joist Slab B1010-226	Waffle Slab B1010-227	Beams & Hollow Core Slabs B1010-236	Beams & Hollow Core Slabs Topped B1010-238	Beams & Double Tees Topped B1010-239	Bar Joists on Cols. & Brg. Walls B1010-248	Bar Joists & Beams on Cols. B1010-250	Composite Beams & C.I.P. Slab B1010-252	Composite Deck & Slab, W Shapes B1010-254	Composite Beam & DK., Lt. Wt. Slab B1010-256	Wood Beams & Joists B1010-264	Laminated Wood Beams & Joists B1010-265
Superimposed Load = 40 PSF																
15 x 15	13.46	12.67	11.30	10.86	14.05	—	—	—	—	—	—	—	—	—	9.24	9.73
15 x 20	13.40	13.11	11.73	11.57	14.15	—	—	—	—	8.45	9.92	—	14.05	—	12.51	9.88
20 x 20	13.55	13.75	12.07	11.57	14.10	18.05	17.50	19.90	—	8.66	10.58	—	15.05	—	12.36	10.09
20 x 25	13.71	14.55	12.90	12.60	14.25	18.25	16.61	19.00	—	9.92	12.15	15.95	15.15	14.21	—	—
25 x 25	13.80	14.85	13.30	12.85	14.10	18.55	17.41	20.35	—	10.49	12.88	16.45	16.85	13.83	—	—
25 x 30	14.05	15.50	14.00	—	14.70	18.70	16.75	19.05	18.51	10.80	13.44	17.75	16.95	14.00	—	—
30 x 30	15.55	16.65	14.65	—	15.15	19.40	17.21	20.25	19.68	10.63	13.35	17.75	18.50	14.19	—	—
30 x 35	16.15	17.90	15.60	—	15.75	19.70	17.45	19.89		11.81	14.91	18.45	20.00	14.75	—	—
35 x 35	17.20	18.65	15.85	—	15.80	20.35	18.40	20.39	—	12.14	15.25	19.10	20.30	15.43	—	—
35 x 40	17.55	19.65	—	—	16.35	20.75	18.54	21.50	17.44	—	—	20.65	21.25	16.85	—	—
40 x 40	—	—	—	—	16.80	21.30	19.44	22.30	19.60	—	—	—	—	—	—	—
40 x 45	—	—	—	—	17.45	21.80	—	—	—	—	—	—	—	—	—	—
40 x 50	—	—	—	—	—	—	—	—	18.47	—	—	—	—	—	—	—
Superimposed Load = 75 PSF																
15 x 15	13.48	12.93	11.44	10.89	14.15	—	—	—	—	—	—	—	—	—	11.76	12.02
15 x 20	13.83	14.15	12.09	11.96	14.50	—	—	—	—	9.17	11.22	—	15.25	—	15.18	12.90
20 x 20	14.65	14.90	12.50	12.05	14.90	18.25	18.30	20.70	—	9.76	12.14	—	16.75	—	15.14	12.76
20 x 25	14.95	16.15	13.55	12.85	14.95	18.55	16.61	19.50	—	11.27	12.72	17.75	18.40	14.57	—	—
25 x 25	14.90	16.00	13.80	13.35	14.90	18.90	18.06	20.35	—	11.14	13.92	18.80	19.10	14.90	—	—
25 x 30	15.05	16.75	14.70	—	15.25	19.10	17.23	20.12	18.51	11.29	14.25	19.90	19.95	14.75	—	—
30 x 30	16.70	18.05	15.45	—	15.65	19.75	17.91	21.30	19.68	12.14	15.26	19.65	21.25	15.10	—	—
30 x 35	16.95	18.70	16.45	—	15.90	19.70	18.35	20.36	—	13.95	17.30	21.10	22.85	15.75	—	—
35 x 35	18.65	19.45	16.85	—	16.45	20.70	19.15	22.25	—	14.90	18.55	21.75	23.30	17.20	—	—
35 x 40	18.90	20.35	—	—	17.05	21.30	—	—	18.47	—	—	23.20	24.35	17.95	—	—
40 x 40	—	—	—	—	17.25	21.90	—	—	20.15	—	—	—	—	—	—	—
40 x 45	—	—	—	—	17.60	22.40	—	—	—	—	—	—	—	—	—	—
40 x 50	—	—	—	—	—	—	—	—	18.48	—	—	—	—	—	—	—
Superimposed Load = 125 PSF																
15 x 15	13.68	13.43	11.80	11.15	14.35	—	—	—	—	—	—	—	—	—	17.59	17.45
15 x 20	14.42	15.30	12.55	12.65	15.15	—	—	—	—	10.76	13.27	—	17.40	—	21.00	19.10
20 x 20	15.30	15.45	13.30	12.65	15.10	18.50	—	—	—	12.08	13.78	—	18.70	—	25.65	18.85
20 x 25	15.85	16.50	14.45	13.50	15.75	18.80	—	—	—	12.01	14.85	19.80	21.35	17.50	—	—
25 x 25	17.20	17.40	14.55	13.85	16.50	19.35	—	—	—	12.94	16.20	21.35	22.65	15.75	—	—
25 x 30	17.25	18.15	15.20	—	16.30	19.45	—	—	—	13.85	17.20	23.30	22.45	16.90	—	—
30 x 30	17.70	19.00	16.00	—	16.50	20.00	—	—	—	15.30	19.15	23.10	24.75	17.80	—	—
30 x 35	19.85	20.45	16.95	—	16.60	20.40	—	—	—	15.65	19.50	25.80	26.40	18.30	—	—
35 x 35	19.90	21.20	17.30	—	16.65	21.05	—	—	—	17.45	20.30	25.10	27.50	20.00	—	—
35 x 40	20.05	21.30	—	—	16.95	21.95	—	—	—	—	—	27.05	28.35	20.45	—	—
40 x 40	—	—	—	—	17.95	22.25	—	—	—	—	—	—	—	—	—	—
40 x 45	—	—	—	—	18.20	22.75	—	—	—	—	—	—	—	—	—	—
40 x 50	—	—	—	—	—	—	—	—	—	—	—	—	—	—	—	—
Superimposed Load = 200 PSF																
15 x 15	14.30	14.35	12.19	—	14.85	—	—	—	—	—	—	—	—	—	31.50	27.40
15 x 20	15.55	16.45	12.75	—	15.55	—	—	—	—	—	—	—	20.75	—	28.30	24.75
20 x 20	16.70	16.70	13.55	—	15.80	19.25	—	—	—	—	—	—	21.95	—	—	26.05
20 x 25	17.15	18.00	14.85	—	16.50	19.40	—	—	—	—	—	24.20	23.85	19.45	—	—
25 x 25	18.60	19.50	15.05	—	17.00	19.70	—	—	—	—	—	25.20	25.85	20.30	—	—
25 x 30	18.75	19.70	15.90	—	17.20	20.70	—	—	—	—	—	26.75	27.30	20.35	—	—
30 x 30	19.70	20.30	16.85	—	17.35	21.25	—	—	—	—	—	28.05	32.60	21.10	—	—
30 x 35	19.85	21.50	—	—	17.80	21.90	—	—	—	—	—	30.40	31.70	21.25	—	—
35 x 35	21.45	22.25	—	—	17.80	22.25	—	—	—	—	—	31.05	34.70	22.65	—	—
35 x 40	21.65	22.80	—	—	18.15	23.45	—	—	—	—	—	33.60	35.25	24.30	—	—
40 x 40	—	—	—	—	—	—	—	—	—	—	—	—	—	—	—	—
40 x 45	—	—	—	—	—	—	—	—	—	—	—	—	—	—	—	—
40 x 50	—	—	—	—	—	—	—	—	—	—	—	—	—	—	—	—

Design Assumptions:
Bay Size 20' x 25' = 500 S.F.
Use Concrete Flat Plate

Roof Load: Superimposed Load   40 PSF
                            Total Load   146 PSF

Floor Load: Superimposed Load   75 PSF
                            Total Load   188 PSF

Minimum Column size 20"

ROOF   Total Roof Load .146 KSF x 500 S.F. =   73 K

     73 K

Column Load 1.67' x 1.67' x .15 KCF x 12' =   5 K

     78 K

FOURTH   Total Floor Load .188 K x 500 S.F. =   94 K

     172 K

Column   5 K

     177 K

THIRD   Total Floor Load   94 K

     271 K

Column   5 K

     276 K

SECOND   Total Floor Load   94 K

     370 K

Column   5 K

     375 K

FIRST   Total Floor Load   94 K

     469 K

Column   5 K

     474 K

GROUND

     474 K

Total Load

Description:

1. Multiply roof load x bay area.
2. Show total load at top of column.
3. Multiply est. column weight x floor to floor height.
4. Add to roof load.
5. Multiply floor load x bay area.
6. Add to roof and column load.
7. Multiply est. column weight x floor to floor height.
8. Add to total loads above.
9. Repeat steps above for remainder of floors & columns.
10. Enter the minimum reinforced portion of the table with total load on the column and the minimum allowable column size for the selected cast-in-place floor system.

    If the total load on the column does not exceed the allowable load shown, use the cost per L.F. multiplied by the length of columns required to obtain the column cost.

11. If the total load on the column exceeds the allowable working load shown in the minimum reinforced portion of the table enter the first portion of the table with the total load on the column and the minimum allowable column size from the selected cast-in-place floor system.

    Select a cost per L.F. for bottom level columns by total load or minimum allowable column size.

    Select a cost per L.F. for top level columns using the column size required for bottom level columns from the minimum reinforced portion of the table.

$$\frac{\text{Bottom \& Top Col. Costs/L.F.}}{2} = \text{Average Column Cost/L.F.}$$

Column Cost = Average Col. Cost/L.F. x Length of Cols. Req'd.

Description: Below is an example of steel column determination when roof and floor loads are known.

Design Assumptions:

Bay Size	35' x 35' = 1,225 S.F.		
Roof Load:	Superimposed Load	40	PSF
	Dead Load	44	PSF
	Total Load	84	PSF
Floor Load:	Superimposed Load	125	PSF
	Dead Load	45	PSF
	Total Load	170	PSF

Calculation of Total Loads

ROOF Total Roof Load .084 KSF x 1225 S.F. =   102.9 K

102.9 K

Column Estimated .087 K x 12' =   + 1.0 K

103.9 K

W12 x 79

FOURTH Total Floor Load = .17 KSF x 1225 S.F. =   208.3 K

312.2 K

Column Estimated .087 K x 12'   + 1.0 K

313.2 K

— Assumed Splice

+ 3'

THIRD Total Floor Load = .17 KSF x 1225 S.F. =   208.3 K

521.5 K

Column Estimated .145 K x 12'   + 1.7 K

523.2 K

W14 x 145

SECOND Total Floor Load = .17 KSF x 1225 S.F. =   208.3 K

731.5 K

Column Estimated .145 K x 12'   + 1.7 K

733.2 K

— Assumed Splice

+ 3'

FIRST Total Floor Load = .17 KSF x 1225 S.F. =   208.3 K

941.5 K

Column Estimated .176 K x 12'   + 2.1 K

943.6 K

W14 x 176

Slab on grade
GROUND

Total Load to Foundation   944 K

5 Floors at 12'-0" = 60'-0"

Description:

1. Multiply roof load x bay area.
2. Show total load at top of column.
3. Multiply estimated column weight* x floor to floor height.
4. Add to roof load.
5. Multiply floor load x bay area.
6. Add to roof and column load.
7. Multiply estimated column weight x floor to floor height.
8. Add to total loads above.
9. Choose column from **System B1010 208** using unsupported height.
10. Interpolate or use higher loading to obtain cost/L.F.
11. Repeat steps above for remainder of floors and columns.
12. Multiply number of columns by the height of the column times the cost per foot to obtain the cost of each type of column.

* To Estimate Column Weight

Roof Load	.084 KSF
Floor Load x No Floors above Splice	
170 x 1	.170 KSF
Total	.254 KSF

Total Load (KSF) x Bay Area (S.F.) = Load to Col.
.254 KSF x 1,225 KSF = 311 K

From **System B1010 208**, choose a column by:

Load,	Height,	Weight
400 K	10'	79 lb.

Table B1010-241 One, Two and Three Member Beams, Maximum Load for Various Spans

Description: Table below lists the maximum uniform load (W) or concentrated load (P) allowable for various size beams at various spans.

Design Assumptions: Fiber strength (f) is 1,000 psi.

Maximum deflection does not exceed 1/360 the span of the beam.

Modulus of elasticity (E) is 1,100,000 psi. Anything less will result in excessive deflection in the longer members so that spans must be reduced or member size increased.

The uniform loads (W) are in pounds per foot; concentrated loads (P), at the midpoint, are in pounds.

The span is in feet and is the unsupported clear span.

The member sizes are from 2″ x 6″ to 4″ x 12″ and include one, two and three pieces of each member size to arrive at spans and loading.

Individual Member		Span and Load														
		6′		8′		10′		12′		14′		16′		18′	
		W	P	W	P	W	P	W	P	W	P	W	P	W	P
Size	No.	#/L.F.	#	#/L.F.	#	#/L.F.	#	#/L.F.	#	#/L.F.	#	#/L.F.	#	#/L.F.	#
2″ x 6″	2	238	840	—	—										
	3	357	1260	—	—										
2″ x 8″	2	314	1460	235	1095										
	3	471	2190	353	1642	232	1164								
2″ x 10″	2	400	2376	300	1782	240	1426	—	—						
	3	600	3565	450	2673	360	2139	279	1679						
2″ x 12″	2	487	2925	365	2636	292	2109	243	1757	208	1479	—	—		
	3	731	4387	548	3955	548	3164	365	2636	313	2219	212	1699		
2″ x 14″	2	574	3445	430	3445	344	2926	287	2438	246	2090	215	1828	—	—
	3	861	5167	645	5167	516	4389	430	3657	369	3135	322	2743	243	2193
3″ x 6″	1	—	—	—	—										
	2	397	1400	200	882										
	3	595	2100	300	1323										
3″ x 8″	1	261	1216	—	—	—	—	—	—						
	2	523	2433	392	1825	258	1293	—	—						
	3	785	3650	589	2737	388	1940	224	1347						
3″ x 10″	1	334	1980	250	1485	200	1188	—	—	—	—				
	2	668	3961	500	2970	400	2376	310	1865	—	—	—	—		
	3	1002	5941	750	4456	600	3565	466	2798	293	2056				
3″ x 12″	1	406	2437	304	2197	243	1757	203	1464	—	—	—	—	—	—
	2	812	4875	608	4394	487	3515	406	2929	348	2466	236	1888	—	—
	3	1218	7312	912	6591	731	5273	609	4394	522	3699	354	2832	248	2237
4″ x 6″	1	278	980	—	—	—	—	—	—						
	2	556	1960	308	1235	—	—	—	—						
4″ x 8″	1	366	1703	274	1277	—	—	—	—						
	2	733	3406	549	2555	362	1811	209	1257						
4″ x 10″	1	467	2772	350	2079	280	1663	217	1306	—	—				
	2	935	5545	700	4159	560	3327	435	2612	274	1919				
4″ x 12″	1	568	3412	426	3076	341	2460	284	2050	243	1726	—	—		
	2	1137	6825	853	6152	682	4921	568	4100	487	3452	330	2643	232	2088

Table B1010-711 Maximum Floor Joist Spans

Description: Table below lists the maximum clear spans and the framing lumber quantity required per S.F. of floor.

Design Assumptions: Dead load = joist weight plus floor weight of 5 psf plus ceiling or partition load of 10 psf.

Maximum L.L. deflection is 1/360 of the clear span.

Modulus of elasticity is 1,100,000 psi.

Fiber strength (f_b) is 1,000 psi.

10% allowance has been added to framing quantities for overlaps, waste, double joists at openings, etc. 5% added to subfloor for waste.

Maximum span is in feet and is the unsupported clear span.

| Floor Joist | | Framing Lumber | Live Load in Pounds per Square Foot | | | | | | | |
Size in Inches	Spacing C/C	B.F./S.F.	30 Span	40 Span	50 Span	60 Span	70 Span	80 Span	90 Span	100 Span
2" x 6"	12"	1.10	9'-0"	8'-6"	8'-0"	7'-8"	7'-4"	7'-0"	6'-10"	6'-7"
	16"	.83	8'-4"	7'-9"	7'-4"	7'-0"	6'-8"	6'-4"	6'-0"	—
	24"	.55	7'-2"	6'-9"	6'-2"	5'-9"	—	—	—	—
2" x 8"	12"	1.46	12'-0"	11'-2"	10'-7"	10'-0"	9'-8"	9'-4"	9'-0"	8'-8"
	16"	1.10	10'-10"	10'-2"	9'-7"	9'-2"	8'-9"	8'-4"	7'-10"	7'-6"
	24"	.74	9'-6"	8'-10"	8'-2"	7'-6"	7'-0"	6'-9"	6'-5"	6'-0"
2" x 10"	12"	1.84	15'-4"	14'-4"	13'-6"	12'-10"	12'-4"	11'-10"	11'-6"	11'-2"
	16"	1.38	13'-10"	13'-0"	12'-4"	11'-8"	11'-2"	10'-7"	10'-0"	9'-7"
	24"	.91	12'-0"	11'-4"	10'-5"	9'-9"	9'-0"	8'-7"	8'-3"	7'-10"
2" x 12"	12"	2.20	18'-7"	17'-4"	16'-5"	15'-8"	15'-0"	14'-6"	14'-0"	13'-6"
	16"	1.65	16'-10"	15'-9"	14'-10"	14'-3"	13'-7"	12'-10"	12'-3"	11'-8"
	24"	1.10	14'-9"	13'-9"	12'-8"	11'-9"	11'-2"	10'-6"	10'-0"	9'-6"
2" x 14"	12"	2.56	21'-10"	20'-6"	19'-4"	18'-5"	17'-9"	17'-0"	16'-6"	15'-10"
	16"	1.93	19'-10"	18'-7"	17'-7"	16'-9"	16'-0"	15'-2"	14'-5"	13'-9"
	24"	1.29	17'-4"	16'-3"	15'-0"	13'-10"	13'-2"	12'-4"	11'-9"	11'-3"
3" x 6"	12"	1.65	10'-9"	10'-0"	9'-6"	9'-0"	8'-8"	8'-4"	8'-0"	7'-10"
	16"	1.24	9'-9"	9'-0"	8'-8"	8'-3"	7'-10"	7'-7"	7'-4"	7'-0"
	24"	.83	8'-6"	8'-0"	7'-6"	7'-2"	6'-10"	6'-8"	6'-4"	6'-0"
3" x 8"	12"	2.20	14'-2"	13'-3"	12'-6"	11'-10"	11'-6"	11'-0"	10'-8"	10'-4"
	16"	1.65	12'-10"	12'-0"	11'-4"	10'-10"	10'-5"	10'-0"	9'-9"	9'-5"
	24"	1.10	11'-3"	10'-6"	9'-10"	9'-6"	9'-0"	8'-9"	8'-3"	7'-10"
3" x 10"	12"	2.75	18'-0"	16'-10"	16'-0"	15'-4"	14'-8"	14'-0"	13'-8"	13'-3"
	16"	2.07	16'-5"	15'-4"	14'-6"	13'-10"	13'-3"	12'-10"	12'-4"	12'-0"
	24"	1.38	14'-4"	13'-5"	12'-8"	12'-0"	11'-7"	11'-0"	10'-7"	10'-0"
3" x 12"	12"	3.30	22'-0"	20'-7"	19'-6"	18'-7"	17'-10"	17'-0"	16'-7"	16'-0"
	16"	2.48	20'-0"	18'-9"	17'-8"	16'-10"	16'-2"	15'-7"	15'-0"	14'-7"
	24"	1.65	17'-6"	16'-4"	15'-6"	14'-9"	14'-0"	13'-7"	12'-10"	12'-4"
4" x 6"	12"	2.20	—	—	—	10'-0"	9'-9"	9'-4"	9'-0"	8'-9"
	16"	1.65	—	—	—	9'-2"	8'-10"	8'-6"	8'-3"	8'-0"
	24"	1.10	—	—	—	8'-0"	7'-9"	7'-4"	7'-2"	7'-0"
4" x 8"	12"	2.93	—	—	—	13'-4"	12'-10"	12'-4"	12'-0"	11'-7"
	16"	2.20	—	—	—	12'-0"	11'-8"	11'-3"	10'-10"	10'-6"
	24"	1.47	—	—	—	10'-7"	9'-2"	9'-9"	9'-6"	9'-2"
4" x 10"	12"	3.67	—	—	—	17'-0"	16'-4"	15'-9"	15'-3"	14'-9"
	16"	2.75	—	—	—	15'-6"	14'-7"	14'-4"	13'-10"	13'-5"
	24"	1.83	—	—	—	13'-6"	13'-0"	12'-6"	12'-0"	11'-9"

Table B1010-781 Decking Material Characteristics

Description: The table below lists the maximum spans for commonly used wood decking materials for various loading conditions.

Design Assumptions: Applied load is the total load, live plus dead.

Maximum deflection is 1/180 of the clear span which is not suitable if plaster ceilings will be supported by the roof or floor.

Modulus of elasticity (E) and fiber strength (F) are as shown in table to the right.

No allowance for waste has been included.

Deck Material	Modulus of Elasticity	Fiber Strength
Cedar	1,100,000 psi	1,000 psi
Douglas Fir	1,760,000	1,200
Hemlock	1,600,000	1,200
White Spruce	1,320,000	1,200

Assume Dead Load of wood plank is 33.3 lbs. per C.F.

Table B1010-782 Maximum Spans for Wood Decking

Nominal Thickness	Type Wood	B.F. per S.F.	Total Uniform Load per S.F.							
			40	50	70	80	100	150	200	250
2"	Cedar	2.40	6.5'	6'	5.7'	5.3'	5'	4.5'	4'	3.5'
	Douglas fir	2.40	8'	7'	6.7'	6.3'	6'	5'	4.5'	4'
	Hemlock	2.40	7.5'	7'	6.5'	6'	5.5'	5'	4.5'	4'
	White spruce	2.40	7'	6.5'	6'	5.5'	5'	4.5'	4'	3.5'
3"	Cedar	3.65	11'	10'	9'	8.5'	8'	7'	6.5'	6'
	Douglas fir	3.65	13'	12'	11'	10.5'	10'	8'	7.5'	7'
	Hemlock	3.65	12'	11'	10.5'	10'	9'	8'	7.5'	7'
	White spruce	3.65	11'	10'	9.5'	9'	8'	7.5'	7'	6'
4"	Cedar	4.65	15'	14'	13'	12'	11'	10'	9'	8'
	Douglas fir	4.65	18'	17'	15'	14'	13'	11'	10.5'	10'
	Hemlock	4.65	17'	16'	14'	13.5'	13'	11'	10'	9'
	White spruce	4.65	16'	15'	13'	12.5'	12'	10'	9'	8'
6"	Cedar	6.65	23'	21'	19'	18'	17'	15'	14'	13'
	Douglas fir	6.65	24'+	24'	23'	22'	21'	18'	16'	15'
	Hemlock	6.65	24'+	24'	22'	21'	20'	17'	16'	15'
	White spruce	6.65	24'	23'	21'	20'	18'	16'	14'	13'

Table B1020-511 Maximum Roof Joist or Rafter Spans

Description: Table below lists the maximum clear spans and the framing lumber quantity required per S.F. of roof. Spans and loads are based on flat roofs.

Design Assumptions: Dead load = Joist weight plus weight of roof sheathing at 2.5 psf plus either of two alternate roofing weights.

Maximum deflection is 1/360 of the clear span.

Modulus of elasticity is 1,100,000 psi.

Fiber strength (f) is 1,000 psi.

Allowance: 10% has been added to framing quantities for overlaps, waste, double joists at openings, etc.

Maximum span is measured horizontally in feet and is the clear span.

Note: To convert to inclined measurements or to add for overhangs, multiply quantities in Table B1020-511 by factors in Table B1020-512.

Size in Inches	Center to Center Spacing	Framing B.F. per S.F.	Live Load in Pounds per Square Foot							
			Group I Covering (2.5 psf)				Group II Covering (8.0 psf)			
			15 Span	20 Span	30 Span	40 Span	20 Span	30 Span	40 Span	50 Span
2" x 4"	12"	0.73	7'-11"	7'-3"	6'-5"	5'-10"	6'-9"	6'-1"	5'-7"	5'-3"
	16"	0.55	7'-2"	6'-7"	5'-10"	5'-4"	6'-1"	5'-6"	5'-1"	4'-10"
	24"	0.37	6'-3"	5'-9"	5'-0"	4'-8"	5'-4"	4'-10"	4'-6"	4'-2"
2" x 6"	12"	1.10	12'-5"	11'-5"	10'-0"	9'-3"	10'-7"	9'-7"	8'-10"	8'-4"
	16"	0.83	11'-3"	10'-4"	9'-2"	8'-5"	9'-8"	8'-9"	8'-0"	7'-6"
	24"	0.55	9'-10"	9'-0"	8'-0"	7'-4"	8'-5"	7'-7"	7'-0"	6'-6"
2" x 8"	12"	1.46	16'-5"	15'-0"	13'-4"	12'-2"	14'-0"	12'-8"	11'-8"	11'-0"
	16"	1.10	14'-11"	13'-8"	12'-0"	11'-0"	12'-9"	11'-6"	10'-7"	11'-0"
	24"	0.74	13'-0"	12'-0"	10'-7"	9'-8"	11'-1"	10'-0"	9'-3"	10'-0"
2" x 10"	12"	1.84	20'-11"	19'-3"	17'-0"	19'-7"	17'-11"	16'-2"	14'-11"	14'-0"
	16"	1.38	19'-0"	17'-6"	15'-6"	14'-2"	16'-3"	14'-8"	13'-7"	12'-9"
	24"	0.91	16'-7"	15'-3"	13'-6"	12'-4"	14'-2"	12'-10"	11'-10"	11'-0"
2" x 12"	12"	2.20	25'-6"	23'-5"	20'-9"	18'-11"	21'-9"	19'-8"	18'-2"	17'-1"
	16"	1.65	23'-1"	21'-3"	18'-10"	17'-2"	19'-9"	17'-10"	16'-6"	15'-6"
	24"	1.10	20'-2"	18'-7"	16'-5"	15'-0"	17'-3"	15'-7"	14'-5"	13'-5"
2" x 14"	12"	2.56	30'-0"	27'-7"	24'-5"	22'-4"	25'-8"	23'-2"	21'-5"	20'-1"
	16"	1.93	27'-3"	25'-0"	22'-2"	20'-3"	23'-3"	21'-0"	19'-6"	18'-3"
	24"	1.29	23'-10"	21'-11"	19'-4"	17'-8"	20'-4"	18'-4"	17'-0"	15'-10"
3" x 6"	12"	1.65	14'-9"	13'-7"	12'-0"	11'-0"	12'-7"	11'-4"	10'-6"	9'-11"
	16"	1.24	13'-4"	12'-4"	10'-11"	9'-11"	11'-5"	10'-4"	9'-7"	9'-0"
	24"	.83	11'-9"	10'-9"	9'-6"	8'-8"	10'-0"	9'-0"	8'-4"	7'-10"
3" x 8"	12"	2.20	19'-5"	17'-11"	15'-10"	14'-6"	16'-7"	15'-0"	13'-11"	13'-0"
	16"	1.65	17'-8"	16'-9"	14'-4"	13'-2"	15'-1"	13'-7"	12'-7"	11'-10"
	24"	1.10	15'-5"	14'-2"	12'-6"	11'-6"	13'-2"	11'-11"	11'-0"	10'-4"
3" x 10"	12"	2.75	24'-10"	22'-10"	22'-2"	18'-6"	21'-9"	19'-2"	17'-9"	16'-8"
	16"	2.07	22'-6"	20'-9"	18'-4"	16'-9"	19'-9"	17'-4"	16'-1"	15'-1"
	24"	1.38	19'-8"	18'-1"	16'-0"	14'-8"	16'-10"	15'-2"	14'-0"	13'-2"
3" x 12"	12"	3.30	30'-2"	27'-9"	24'-6"	22'-5"	25'-10"	23'-4"	21'-6"	20'-3"
	16"	2.48	27'-5"	25'-3"	22'-3"	20'-4"	23'-5"	21'-2"	19'-7"	18'-4"
	24"	1.65	24'-0"	22'-0"	19'-6"	17'-10"	20'-6"	18'-6"	17'-1"	16'-0"

Table B1020-512   Factors for Converting Inclined to Horizontal

Rafters: The quantities shown in Table B1020-511 are for flat roofs. For inclined roofs using rafters, the quantities should be multiplied by the factors in Table B1020-512 to allow for the increased area of the inclined roofs.

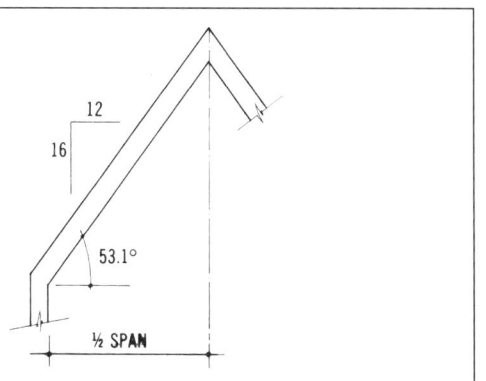

INCLINED TO HORIZONTAL SECTIONS

Roof Slope	Approx. Angle	Factor	Roof Slope	Approx. Angle	Factor
Flat	0°	1.000	12 in 12	45.0	1.414
1 in 12	4.8°	1.003	13 in 12	47.3	1.474
2 in 12	9.5°	1.014	14 in 12	49.4	1.537
3 in 12	14.0°	1.031	15 in 12	51.3	1.601
4 in 12	18.4°	1.054	16 in 12	53.1	1.667
5 in 12	22.6°	1.083	17 in 12	54.8	1.734
6 in 12	26.6°	1.118	18 in 12	56.3	1.803
7 in 12	30.3°	1.158	19 in 12	57.7	1.873
8 in 12	33.7°	1.202	20 in 12	59.0	1.943
9 in 12	36.9°	1.250	21 in 12	60.3	2.015
10 in 12	39.8°	1.302	22 in 12	61.4	2.088
11 in 12	42.5°	1.357	23 in 12	62.4	2.162

Table B1020-513 below can be used two ways:
1. Use 1/2 span with overhang on one side only.
2. Use whole span with total overhang from both sides.

Roof or Deck Overhangs: The quantities shown in table do not include cantilever overhangs. The S.F. quantities should be multiplied by the factors in Table B1020-513 to allow for the increased area of the overhang. All dimensions are horizontal. For inclined overhangs also multiply by factors in Table B1020-512 above.

Table B1020-513   Allowance Factors for Including Roof or Deck Overhangs

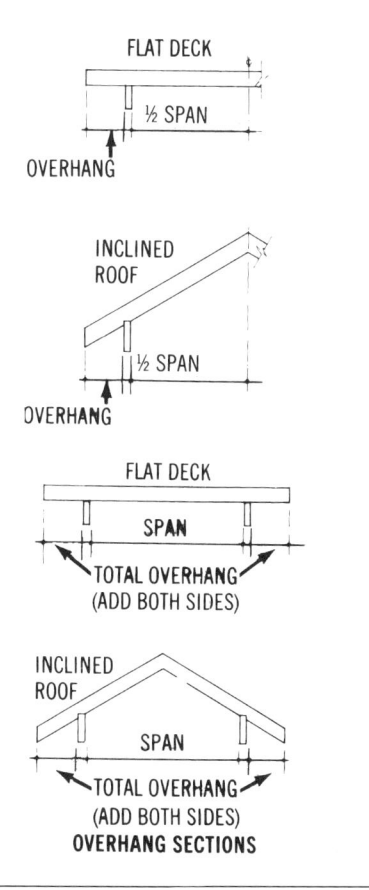

FLAT DECK
½ SPAN
OVERHANG

INCLINED ROOF
½ SPAN
OVERHANG

FLAT DECK
SPAN
TOTAL OVERHANG (ADD BOTH SIDES)

INCLINED ROOF
SPAN
TOTAL OVERHANG (ADD BOTH SIDES)
OVERHANG SECTIONS

Horizontal Span	Roof overhang measured horizontally							
	0'-6"	1'-0"	1'-6"	2'-0"	2'-6"	3'-0"	3'-6"	4'-0"
6'	1.083	1.167	1.250	1.333	1.417	1.500	1.583	1.667
7'	1.071	1.143	1.214	1.286	1.357	1.429	1.500	1.571
8'	1.063	1.125	1.188	1.250	1.313	1.375	1.438	1.500
9'	1.056	1.111	1.167	1.222	1.278	1.333	1.389	1.444
10'	1.050	1.100	1.150	1.200	1.250	1.300	1.350	1.400
11'	1.045	1.091	1.136	1.182	1.227	1.273	1.318	1.364
12'	1.042	1.083	1.125	1.167	1.208	1.250	1.292	1.333
13'	1.038	1.077	1.115	1.154	1.192	1.231	1.269	1.308
14'	1.036	1.071	1.107	1.143	1.179	1.214	1.250	1.286
15'	1.033	1.067	1.100	1.133	1.167	1.200	1.233	1.267
16'	1.031	1.063	1.094	1.125	1.156	1.188	1.219	1.250
17'	1.029	1.059	1.088	1.118	1.147	1.176	1.206	1.235
18'	1.028	1.056	1.083	1.111	1.139	1.167	1.194	1.222
19'	1.026	1.053	1.079	1.105	1.132	1.158	1.184	1.211
20'	1.025	1.050	1.075	1.100	1.125	1.150	1.175	1.200
21'	1.024	1.048	1.071	1.095	1.119	1.143	1.167	1.190
22'	1.023	1.045	1.068	1.091	1.114	1.136	1.159	1.182
23'	1.022	1.043	1.065	1.087	1.109	1.130	1.152	1.174
24'	1.021	1.042	1.063	1.083	1.104	1.125	1.146	1.167
25'	1.020	1.040	1.060	1.080	1.100	1.120	1.140	1.160
26'	1.019	1.038	1.058	1.077	1.096	1.115	1.135	1.154
27'	1.019	1.037	1.056	1.074	1.093	1.111	1.130	1.148
28'	1.018	1.036	1.054	1.071	1.089	1.107	1.125	1.143
29'	1.017	1.034	1.052	1.069	1.086	1.103	1.121	1.138
30'	1.017	1.033	1.050	1.067	1.083	1.100	1.117	1.133
32'	1.016	1.031	1.047	1.063	1.078	1.094	1.109	1.125

The "Exterior Closure" Division has many types of wall systems which are most commonly used in current United States and Canadian building construction.

Systems referenced have an illustration as well as description and cost breakdown. Total costs per square foot of each system are provided. Each system is itemized with quantities, extended and summed for material, labor, or total.

Exterior walls usually consist of several component systems. The weighted square foot cost of the enclosure is made up of individual component systems costs, prorated for their percent of wall area. For instance, a given wall system may take up 70% of the enclosure wall's composite area. The fenestration may make up the balance of 30%. The components should be factored by the proportion of area each contributes to the total.

Example: An enclosure wall is 100' long and 15' high.
   The total wall area is therefore 1,500 S.F.
   70% of the wall area is a brick & block system.
      @ $25.50/S.F. (B2010-130-1240)

   30% is a window system:

Framing	$23.00	(B2020-118-1700)
Glazing	25.85	(B2020-120-1400)
	$48.85	

Solution: Brick & block = 1,500 S.F. x .70 x $25.50 = $26,775
      Windows = 1,500 S.F. x .30 x $48.85 = $21,983
                     Total enclosure wall cost = $48,758
                           or $32.51 per S.F. of wall

The square foot unit costs include the installing contractors markup for overhead and profit. However, fees for professional services must be added along with other items. See H1010, General Conditions, for these costs. Note that interior finishes (drywall, etc.) are included elsewhere.

There may be times when the wall systems costs will need to be expressed in units of square feet of floor area.

Example: A building 25' x 25' requires a wall 100' long and 15' high.

      Cost of the wall is $32.51 per S.F. of wall.
      The floor area is 625 S.F.

Solution: Wall cost per S.F of floor area

$$\frac{100 \text{ L.F. x } 15' \text{ x } \$32.51}{625 \text{ S.F.}} = \$78.01 \text{ per S.F. of floor}$$

Concrete Block Lintels

Design Assumptions:
Bearing length each end = 8 inches
f'c = 2,000 p.s.i.
Wall load = 300 lbs. per linear foot (p.l.f.)
Floor & roof load (includes wall load) = 100 p.l.f.
Weight of 8" deep lintel (included in wall load) = 50 p.l.f.
Weight of 16" deep lintel (included in wall load) = 100 p.l.f.

Load to be supported by the lintel is shown below. The weight of the masonry above the opening can be safely assumed as the weight of a triangular section whose height is one-half the clear span of the opening. Corbelling action of the masonry above the opening may be counted upon to support the weight of the masonry outside the triangle. To the dead load of a wall must be added the uniform dead & live loads of floors & roof that frame into the wall above the opening and below the apex of the 45° triangle. Any load above the apex may be neglected.

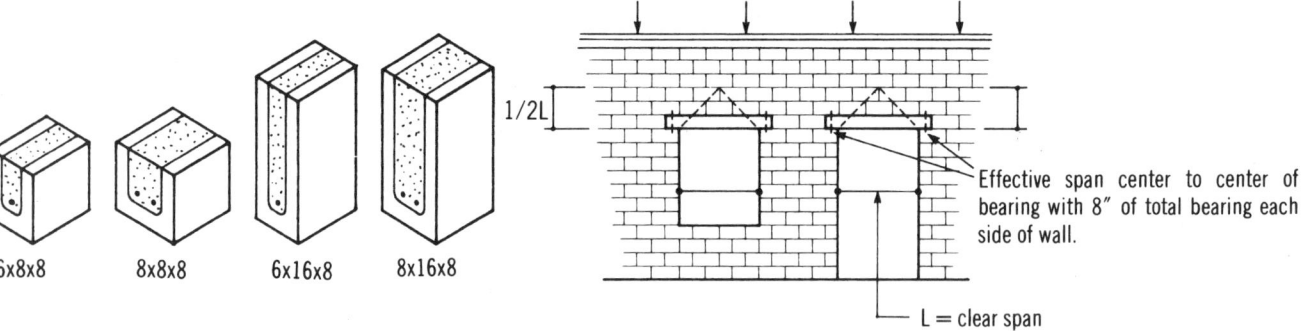

6x8x8   8x8x8   6x16x8   8x16x8

Uniform Floor & Roof Load

1/2L

Effective span center to center of bearing with 8" of total bearing each side of wall.

L = clear span

Table B2010-031   Example Using 14″ Masonry Cavity Wall with 1″ Plaster for the Exterior Closure

Total Heat Transfer is found using the equation
$Q = AU (T_2 - T_1)$ where:

Q = Heat flow, BTU per hour
A = Area, square feet
U = Overall heat transfer coefficient
$(T_2 - T_1)$ = Difference in temperature of the air on each side of the construction component in degrees Fahrenheit

Coefficients of Transmission ("U") are expressed in BTU per (hour) (square foot) (Fahrenheit degree difference in temperature between the air on two sides) and are based on 15 mph outside wind velocity.

The lower the U-value the higher the insulating value.
U = 1/R where "R" is the summation of the resistances of air films, materials and air spaces that make up the assembly.

Figure B2010-031   Example Using 14″ Masonry Cavity Wall with 1″ Plaster for the Exterior Closure

Construction	Resistance R.
1. Outside surface (15 mph wind)	0.17
2. Face Brick (4″)	0.44
3. Air space (2″, 50° mean temp, 10° diff)	1.02
4. Concrete block (8″, lightweight)	2.12
5. Plaster (1″, sand aggregate)	0.18
6. Inside Surface (still air)	0.68
Total resistance	4.61

U = 1/R = 1/4.61 = 0.22
Weight of system = 77 psf
I.S. = Initial System

Replace item 3 with 2″ smooth rigid polystyrene insulation and item 5 with 3/4″ furring and 1/2″ drywall.

Total resistance		4.61
Deduct 3. Airspace	1.02	
5. Plaster	0.18	
	1.20	
Difference 4.61 — 1.20 =		3.41
Add rigid polystyrene insulation		10.00
3/4″ air space		1.01
1/2″ gypsum board		0.45
Total resistance		14.87

U = 1/R = 1/14.87 = 0.067
Weight of system = 70 psf
M.S. = Modified System

Assume 20,000 S.F. of wall in Boston, MA (5600 degree days) Table L1030-501
Initial system effective U = Uw x M = 0.22 x 0.905 = 0.20
(M = weight correction factor from Table L1030-203)
Modified system effective U = Uw x M = 0.067 x 0.915 = 0.06
Initial systems: Q = 20,000 x 0.20 x (72° — 10°) = 248,000 BTU/hr.
Modified system: Q = 20,000 x 0.06 x (72° — 10°) = 74,400 BTU/hr.

Reduction in BTU/hr. heat loss; $\dfrac{I.S. - M.S.}{I.S.}$ x 100 = $\dfrac{248,000 - 74,400}{248,000}$ x 100 = 70%

Example of Calculating Envelope Requirements per ASHRAE 90-75

Example: For Boston, Massachusetts, based on the envelope energy considerations of ASHRAE 90-75, determine the amount of glass that could be used in a five story 50′ x 80′ office building with a 12′ floor to floor height using both systems above.

Step #1. Look up Degree Days for Boston from Table L1030-501=5600

Step #2. Look up overall "U" value for walls (buildings over three story) from Table L1030-102 U_o = 0.34

Step #3. Look up weight modification factor for each system from Table L1030-203 Mi.s. = 0.905, Mm.s. = 0.915

Step #4. Calculate effective "U" values for each system
Ui.s = 0.22 x 0.905 = 0.20,
Um.s = 0.067 x 0.915 = 0.06

Step #5. Look up R values for glass from Table L1030-401
Single glass R = 0.91, insulated glass with 1/2″ air space R = 2.04

Step #6. Calculate "U" values for glass
Single glass U = 1/R = 1/.91 = 1.10,
Insulated glass U = 1/R = 1/2.04 = 0.49

Step #7. From Table L1030-201 for single glass, enter table from left at U_o = 0.34 and proceed to the right to the intersection of the Uw = 0.20 and Uw = 0.06 lines. Project to the bottom and read 15% and 26% respectively.

Step #8. Similarly, from Table L1030-202 for insulated glass, read 28% and 47% respectively.

Step #9. Calculate area of exterior closure:
[50 + 50 + 80 + 80] x 5 x 12 = 15,600 SF

Step #10. Calculate maximum glass area for each situation:

Initial System
Single glass: A = 15,600 x .15 = 2,340 SF
Double glass: A = 15,600 x .28 = 4,368 SF

Modified System
A = 15,600 x .26 = 4,056 SF
A = 15,600 x .47 = 7,332 SF

Conclusion: In order to meet the envelope requirements of ASHRAE 90-75, the glass area is limited to the square footage shown in Step #10 for the two systems considered. If the maximum glass were used in any of the situations, then the heat loss through the walls would be based on U_o = 0.34 (Step 2).

For any system:

$$U = \frac{U \text{ Wall x Area Wall} + U \text{ Glass x Area Glass}}{\text{Total Area}}$$

SHELL

Table B2010-032   Thermal Coefficients of Exterior Closures

EXAMPLE:

Construction	Resistance R
1. Outside surface (15 MPH wind)	0.17
2. Common brick, 4"	0.80
3. Nonreflective air space, 0.75"	1.01
4. Concrete block, S&G aggregate, 8"	1.46
5. Nonreflective air space, 0.75"	1.01
6. Gypsum board, 0.5"	0.45
7. Inside surface (still air)	0.68
Total resistance	5.58

$U = 1/R = 1/5.58 = 0.18$

System weight = 97 psf

Substitution
Replace item 3 with perlite loose fill insulation and fill block cavities with the same.

Total resistance (Example)		R =	5.58
Deduct 3. Air space	1.01		
4. Concrete block	1.46		
	2.47		– 2.47
Difference			3.11
Add 0.75" perlite cavity fill			2.03
Perlite filled 8" block			2.94
Net total resistance			8.08

$U = 1/R = 1/8.08 = 0.12$

System weight = 97 psf

EXAMPLE:

Construction	Resistance R
1. Outside surface (15 MPH wind)	0.17
2. Common brick, 8"	1.60
3. Nonreflective air space, 0.75"	1.01
4. Gypsum board, 0.625"	0.56
5. Inside surface (still air)	0.68
Total resistance	4.02

$U = 1/R = 1/4.02 = 0.25$

System weight = 82 psf

Substitution
Replace item 3 with 4" blanket insulation and item 4 with 0.75" Gypsum plaster (sand agg.).

Total resistance (Example)		R =	4.02
Deduct 3. Air space	1.01		
4. Gypsum board	0.56		
	1.57		– 1.57
Difference			2.45
ADD Blanket insulation			13.00
Gypsum plaster			0.14
Net total resistance			15.59

$U = 1/R = 1/15.59 = 0.06$

System weight = 87 psf

REFERENCE TABLES

EXAMPLE:

Construction	Resistance R
1. Outside surface (15 MPH wind)	0.17
2. Cement stucco, 0.75"	0.15
3. Concrete block, 8" light weight	2.12
4. Reflective air space, 0.75"	2.77
5. Gypsum board, 0.5"	0.45
6. Inside surface (still air)	0.68
Total resistance	6.34

$U = 1/R = 1/6.34 = 0.16$

System weight = 47 psf

Substitution
Replace item 4 with 2" insulation board and item 5 with 0.75" Perlite plaster.

Total resistance (Example)		R =	6.34
Deduct 4. Air space	2.77		
5. Gypsum board	0.45		
	3.22		– 3.22
Difference			3.12
Add Insulation board 2" (polyurethane)			12.50
Perlite plaster			1.34
Net total resistance			16.96

$U = 1/R = 1/16.96 = 0.06$

System weight = 48 psf

Table B2010-112 Partially and Fully Grouted Reinforced Concrete Masonry Wall Capacities Per L.F. (Kips & In-Kips)

Thk. T (Nom.) (in)	h' (Ft.)	h'/T (in/in)	(spacing) (in O.C.)	Rebar Size (@ dL)	7.0 in-K/Ft. No (K/Ft.)	7.0 in-K/Ft. Yes (K/Ft.)	3.5 in-K/Ft. No (K/Ft.)	3.5 in-K/Ft. Yes (K/Ft.)	Without Wind No (K/Ft.)	Without Wind Yes (K/Ft.)	With Wind No 15 psf (K/Ft.)	With Wind No 30 psf (K/Ft.)	With Wind Yes 15 & 30 (K/Ft.)	Moments No (in.-K/Ft.)	Moments Yes (in.-K/Ft.)
8" Conc. Block	8'	12	48"	#8	5.10	12.55	6.25	13.70	7.75	14.90	7.45	7.45	14.90	7.55	12.20
			32"	#5	5.45	13.35	6.65	14.60	7.90	15.80	7.90	7.90	15.80	6.45	9.60
			16"	↓	6.50	15.80	7.90	17.15	9.25	18.50	9.25	9.25	18.50	7.95	12.85
			8"	↓	10.10	23.40	11.70	25.00	13.30	26.55	13.30	13.30	26.55	10.20	17.15
	12'	18	48"	#8	4.70	11.60	5.80	12.65	6.85	13.75	6.85	6.85	13.75	7.55	12.20
			32"	#5	5.05	12.35	6.15	13.45	7.30	14.60	7.30	7.30	14.60	6.45	9.60
			16"	↓	6.00	14.55	7.30	15.85	8.55	17.10	8.55	8.55	17.10	7.95	12.85
			8"	↓	9.30	21.60	10.80	23.05	12.25	24.55	12.25	12.25	24.55	10.20	17.15
	16'	24	48"	#8	3.95	9.70	4.85	10.60	5.75	11.55	5.75	—	11.55	7.55	12.20
			32"	#5	4.20	12.25	5.15	11.30	6.10	12.25	6.10	—	12.25	6.45	9.60
			16"	↓	5.05	12.20	6.10	13.30	7.15	14.35	7.15	—	14.35	7.95	12.85
			8"	↓	7.80	18.10	9.05	19.35	10.30	20.60	10.30	9.65	20.60	10.20	17.15
10" Conc. Block	8'	9.6	48"	#8	7.25	16.45	8.20	17.40	9.15	—	9.15	9.15	18.35	13.10	21.05
			32"	↓	7.80	17.60	8.80	18.60	9.80	19.55	9.80	9.80	19.55	14.55	24.15
			16"	#5	9.45	21.15	10.55	22.25	11.65	23.35	11.65	11.65	23.35	13.90	22.35
			8"	↓	14.50	31.65	15.80	32.95	17.10	34.25	17.10	17.10	34.25	18.45	30.15
	12	14.4	48"	#8	7.05	16.00	8.00	16.90	8.90	17.85	8.90	8.90	17.85	13.10	21.05
			32"	↓	7.60	17.10	8.55	18.05	9.50	19.05	9.50	9.50	19.05	14.55	24.15
			16"	#5	9.20	20.55	10.25	21.60	11.35	22.70	11.35	11.35	22.70	13.90	22.35
			8"	↓	14.10	30.75	15.35	32.05	16.65	33.00	16.65	16.65	33.30	18.35	30.15
	16'	19.2	48"	#8	6.70	15.10	7.55	16.00	8.40	16.85	8.40	8.40	16.85	13.10	21.05
			32"	↓	7.15	16.15	8.05	17.05	9.00	17.95	9.00	9.00	17.95	14.55	24.15
			16"	#5	8.70	19.40	9.70	20.40	10.70	21.45	10.70	10.70	21.45	13.90	22.35
			8"	↓	13.35	29.05	14.50	30.25	15.70	31.45	15.70	15.70	31.45	18.35	30.15
	20'	24	48"	#8	6.05	13.65	6.80	14.45	7.60	15.20	7.60	—	15.20	13.10	21.05
			32"	↓	6.45	14.60	7.30	15.40	8.10	16.25	8.10	0.70	16.25	14.55	24.15
			16"	#5	7.85	17.50	8.75	18.45	9.65	19.35	9.65	0.25	19.35	13.90	22.35
			8"	↓	12.05	26.25	13.10	27.30	14.20	28.40	14.20	13.40	28.40	18.35	30.15
12" Conc. Block	8'	8	48"	#8	9.20	20.00	10.00	20.75	10.75	21.55	10.75	10.75	21.55	15.30	24.40
			32"	↓	9.90	21.50	10.75	22.30	11.55	23.15	11.55	11.55	23.15	17.10	28.10
			16"	#5	12.10	26.05	13.00	26.95	13.90	27.85	13.90	13.90	27.85	16.20	25.90
			8"	↓	18.60	39.35	19.65	40.40	20.75	41.50	20.75	20.75	41.50	21.45	35.10
	12'	12	48"	#8	9.00	19.55	9.75	20.35	10.55	21.10	10.55	10.55	21.10	15.30	24.40
			32"	↓	9.70	21.05	10.50	21.85	11.30	22.65	11.30	11.30	22.65	17.10	28.10
			16"	#5	11.85	25.50	12.75	26.40	13.65	27.30	13.65	13.65	27.30	16.20	25.90
			8"	↓	18.20	38.50	19.25	39.55	20.30	40.60	20.30	20.30	40.60	21.45	35.10
	16'	16	48"	#8	8.65	18.75	9.35	19.50	10.10	20.20	10.10	10.10	20.20	15.30	24.40
			32"	↓	9.30	20.15	10.05	20.90	10.85	21.70	10.85	10.85	21.70	17.10	28.10
			16"	#5	11.35	24.40	12.20	25.25	13.05	26.15	13.05	13.05	26.15	16.20	25.90
			8"	↓	17.45	36.90	18.45	37.90	19.45	38.90	19.45	19.45	38.90	21.45	35.10
	24'	24	48"	#8	7.05	15.35	7.65	15.95	8.30	16.55	8.30	—	16.55	15.30	24.40
			32"	↓	7.60	16.55	8.25	17.15	8.90	17.80	8.90	—	17.80	17.10	28.10
			16"	#5	9.30	20.00	10.00	20.70	10.70	21.45	10.40*	—	21.45	16.20	25.90
			8"	↓	14.30	30.25	15.10	31.10	15.95	31.90	15.55*	15.15	31.90	21.45	35.10

*Zone 3 only

SHELL

REFERENCE TABLES

Table B2010-114 Fully Grouted Reinforced Masonry Wall Capacities Per L.F. (Kips & In-Kips)

Thk. T (Nom.) (in)	h' (Ft.)	h'/T (in/in)	Type Wall Brick Grout Conc. M.U.	Rebar Size and Spacing (in. O.C.)	Eccentric 7.0 in-K/Ft. No (K/Ft.)	Eccentric 7.0 in-K/Ft. Yes (K/Ft.)	Eccentric 3.5 in-K/Ft. No (K/Ft.)	Eccentric 3.5 in-K/Ft. Yes (K/Ft.)	Without Wind or Ecc. No (K/Ft.)	Without Wind or Ecc. Yes (K/Ft.)	With Wind No 15 psf (K/Ft.)	With Wind No 30 psf (K/Ft.)	With Wind Yes 15 & 30 (K/Ft.)	Moment No (in.-K/Ft.)	Moment Yes (in.-K/Ft.)
8"	8'	12	4"0"4"	#5@32	10.10	23.35	11.70	24.50	13.30	26.55	13.30	13.30	26.55	6.90	9.65
			Solid CMU		12.75	28.70	14.35	30.30	15.95	31.90	15.95	15.95	31.90	7.80	9.75
	12'	18	4"0"4"		9.30	21.60	10.80	23.05	12.25	24.55	12.25	12.25	24.55	6.90	9.65
			Solid CMU		16.65	26.50	18.15	28.00	14.70	29.45	14.70	14.70	29.45	7.80	9.75
	16'	24	4"0"4"		7.80	18.10	9.05	19.35	10.30	20.60	10.30	9.65	20.60	6.90	9.65
			Solid CMU	↓	9.85	22.20	11.10	23.45	12.35	24.70	12.35	12.35	12.70	7.80	9.75
10"	8'	9.6	4"0"6"	#5@24	14.45	31.50	15.75	32.80	17.05	34.10	17.05	17.05	34.10	11.20	16.20
			Solid CMU		17.85	38.35	19.15	39.65	20.45	40.90	20.45	20.45	40.90	12.65	16.35
			4"2"4"		14.45	31.50	15.75	32.80	17.05	34.10	17.05	17.05	34.10	11.20	16.20
	12'	14.4	4"0"6"		13.90	30.35	15.15	31.60	16.40	32.80	16.40	16.40	32.80	11.20	16.20
			Solid CMU		17.20	36.90	18.45	38.15	19.70	39.40	19.70	19.70	39.40	12.65	16.35
			4"2"4"		13.90	30.35	15.15	31.60	16.40	32.80	16.40	16.40	32.80	11.20	16.20
	16'	19.2	4"0"6"		12.85	28.05	14.00	29.20	15.15	30.35	15.15	15.15	30.35	11.20	16.20
			Solid CMU		15.90	34.10	17.05	35.25	18.20	36.40	13.20	13.20	36.40	12.65	16.35
			4"2"4"		12.85	23.05	14.00	29.20	15.15	30.35	15.15	15.15	30.35	11.20	16.20
	20'	24	4"0"6"		11.15	24.25	12.10	25.25	13.10	26.20	13.10	12.40	26.20	11.20	16.20
			Solid CMU		13.75	29.50	14.75	30.50	15.75	31.50	15.75	15.75	31.50	12.65	16.35
			4"2"4"	↓	11.15	24.25	12.10	25.25	13.10	26.20	13.10	12.40	26.20	11.20	16.20
12"	8'	8	4"0"8"	#8@48	18.60	39.35	19.65	40.40	20.75	41.50	20.75	20.75	41.50	16.65	24.90
			Solid CMU		22.75	47.65	23.80	48.70	24.90	49.80	24.90	24.90	49.80	18.80	25.10
			4"2"6"	#5@20	18.60	39.35	19.65	40.40	20.75	41.50	20.75	20.75	41.50	11.90	18.75
	12'	12	4"0"8"	#8@48	18.20	38.50	19.25	39.55	20.30	40.60	20.30	20.30	40.60	16.65	24.90
			Solid CMU	↓	22.25	46.65	23.30	47.70	24.35	48.75	24.35	24.35	48.75	18.80	25.10
			4"2"6"	#5@20	18.20	38.50	19.25	39.55	20.30	40.60	20.30	20.30	40.60	11.90	18.75
	16'	16	4"0"8"	#8@48	17.45	36.90	18.45	37.90	19.45	38.90	19.45	19.45	38.90	16.65	24.90
			Solid CMU	↓	21.30	44.70	22.35	45.70	23.35	46.70	23.25	23.35	46.70	18.80	25.10
			4"2"6"	#5@20	17.45	36.90	18.45	37.90	19.45	38.90	19.45	19.45	38.90	11.90	18.75
	24'	24	4"0"8"	#8@48	14.30	30.25	15.10	31.10	15.95	31.90	15.95	15.15	31.90	16.65	24.90
			Solid CMU	↓	17.15	36.65	18.30	37.45	19.15	38.30	19.15	19.15	38.30	18.80	25.10
			4"2"6"	#5@20	14.30	30.25	15.10	31.10	15.95	31.90	15.95	15.15	31.90	11.90	18.75
16"	8'	6	4"0"12"	#8@32	26.40	54.40	27.20	55.20	28.00	56.05	28.00	28.00	56.05	31.20	49.10
			Solid CMU	↓	32.00	65.65	32.80	66.45	33.60	67.25	33.60	33.60	67.25	35.25	50.40
			4"2"10"	#5@15	26.40	54.40	27.20	55.20	28.00	56.05	28.00	28.00	56.05	14.75	24.10
	12'	9	4"0"12"	#8@32	26.15	53.95	26.95	54.75	27.75	55.55	27.75	27.75	55.55	31.20	49.10
			Solid CMU	↓	31.70	66.05	32.50	65.85	33.30	66.65	33.00	33.30	66.65	35.25	50.40
			4"2"10"	#5@15	26.15	53.95	26.95	54.75	27.75	55.55	27.75	27.75	55.55	14.75	24.10
	16'	12	4"0"12"	#8@32	25.70	53.05	26.50	53.80	28.30	54.60	27.30	27.30	54.60	31.20	49.10
			Solid CMU		31.20	63.95	31.95	64.75	32.75	65.50	32.75	32.75	65.50	35.25	50.40
			4"2"10"	↓	25.70	53.05	26.50	53.80	27.30	54.60	27.30	27.30	59.60	14.75	24.10
	32'	24	4"0"12"	#8@32	20.35	41.95	20.95	42.55	21.60	43.20	18.75*	18.75*	43.20	31.20	49.10
			Solid CMU	↓	24.65	50.60	25.30	51.20	25.90	51.85	24.55*	24.55*	51.85	35.25	50.40
			4"2"10"	#5@15	20.35	41.95	20.95	42.55	21.60	43.20	18.75*	18.75*	43.20	14.75	24.10

*Zone 3 Only

Table B2010-116   Unreinforced Masonry Wall Capacities Per L.F. (Kips & In-Kips)

Thk. T (Nom.) (in)	Earthquake Zones 0 & 1 Only Length Or Height h' (Ft.)	h'/t (in/in)	Type of Wall	Eccentric Loads 7.0 (K/Ft.)	3.5 (K/Ft.)	Without Wind or Eccentric Loads (K/Ft.)	With Wind 15 psf (K/Ft.)	30 psf (K/Ft.)	Not Wind or Earthquake Inspection No (in-K/Ft.)	Yes (in-K/Ft.)	Wind or Earthquake Inspection No (in-K/Ft.)	Yes (in-K/Ft.)
8"	8'	12	Solid Brick	4.85	7.50	10.15	10.15	10.15	1.15	2.30	1.55	3.10
			Solid CM Units	8.85	11.50	14.15	14.15	14.15	.70	1.40	.95	1.85
			Hollow CM Units	7.95	10.60	13.30	13.30	13.30	.70	1.40	.60	1.25
	10'	15	Solid Brick	4.70	7.30	9.85	9.85	9.85	1.15	2.30	1.55	3.10
			Solid CM Units	8.55	11.15	13.75	13.75	13.75	.70	1.40	.90	1.85
			Hollow CM Units	—	4.45	6.30	6.30	6.05	.45	.95	.60	1.25
	12'	18	Solid Brick	—	6.95	9.40	9.40	7.95	1.15	2.30	1.55	3.10
			Solid CM Units	8.15	10.60	13.10	13.10	12.90	.70	1.40	.90	1.85
			Hollow CM Units	—	4.25	6.00	6.00	4.75	.45	.95	.60	1.25
12"	8'	8	Solid Brick	12.30	14.10	15.90	15.90	15.90	2.70	5.40	3.60	7.20
			Solid CM Units	18.50	20.30	22.10	22.10	22.10	1.60	3.20	2.15	4.30
			Hollow CM Units	7.85	9.10	10.40	10.40	10.40	1.10	2.25	1.50	3.00
			Brick & Hollow CMU	5.50	7.80	10.05	10.05	10.05	1.35	2.70	1.80	3.60
	12'	12	Solid Brick	12.05	13.80	15.55	15.55	15.55	2.70	5.40	3.60	7.20
			Solid CM Units	13.45	14.75	16.05	16.05	16.05	1.60	3.20	2.15	4.30
			Hollow CM Units	7.65	8.90	10.15	10.15	10.15	1.10	2.25	1.50	3.00
			Brick & Hollow CMU	5.40	7.65	9.85	9.85	9.05	1.35	2.70	1.80	3.60
	16'	16	Solid Brick	11.55	13.20	14.90	14.90	14.35	2.70	5.40	3.60	7.20
			Solid CM Units	17.35	19.05	20.70	20.70	20.70	1.60	3.20	2.15	4.30
			Hollow CM Units	7.35	8.55	9.75	9.75	9.05	1.10	2.25	1.50	3.00
			Brick & Hollow CMU	5.20	7.30	9.45	9.10	—	1.35	2.70	1.80	3.60
16"	12'	9	Solid Brick	18.60	19.95	21.30	21.30	21.30	4.85	9.75	6.50	13.00
			Solid CM Units	26.95	28.30	29.60	29.60	29.60	2.90	5.85	3.90	7.80
			Hollow CM Units	10.85	12.10	13.30	13.30	13.30	1.50	3.05	2.00	4.05
			Brick & Hollow CMU	9.30	11.10	12.85	12.85	12.85	2.20	4.40	2.90	5.85
	16'	12	Solid Brick	18.30	19.60	20.90	20.90	20.90	4.85	9.75	6.50	13.00
			Solid CM Units	26.50	27.80	29.10	29.10	29.10	2.90	5.85	3.90	7.80
			Hollow CM Units	10.70	11.90	13.10	13.10	13.10	1.50	3.05	2.00	4.05
			Brick & Hollow CMU	9.15	10.90	12.65	12.65	11.10	2.20	4.40	2.90	5.85
	20'	15	Solid Brick	17.80	19.05	20.30	20.30	20.30	4.85	9.75	6.50	13.00
			Solid CM Units	25.75	27.00	28.25	28.25	28.25	2.90	5.85	3.90	7.80
			Hollow CM Units	10.40	11.55	12.70	12.70	10.95	1.50	3.05	2.00	4.05
			Brick & Hollow CMU	8.90	10.60	12.30	12.00	—	2.20	4.40	2.90	5.85
20"	12'	7.2	Solid Brick	24.80	25.85	26.90	26.90	26.90	7.70	15.40	10.25	20.50
			Solid CM Units	35.30	36.40	37.45	37.45	37.45	4.60	9.20	6.15	12.30
			Hollow CM Units	16.15	17.15	18.15	18.15	18.15	2.50	5.05	3.35	6.75
			Brick & Hollow CMU	16.10	17.25	18.45	18.45	18.45	4.05	8.10	5.40	10.80
	16'	9.6	Solid Brick	24.55	25.60	26.70	26.70	26.70	7.70	15.40	10.25	20.50
			Solid CM Units	35.00	36.10	37.15	37.15	37.15	4.60	9.20	6.15	12.30
			Hollow CM Units	16.00	17.00	18.00	18.00	18.00	2.50	5.05	3.35	6.75
			Brick & Hollow CMU	15.95	17.10	18.30	18.30	18.30	4.05	8.10	5.40	10.80
	24'	14.4	Solid Brick	23.70	24.70	25.75	25.75	25.75	7.70	15.40	10.25	20.50
			Solid CM Units	33.80	34.80	35.80	35.80	35.80	4.60	9.20	6.15	12.30
			Hollow CM Units	15.45	16.40	17.40	17.40	16.05	2.50	5.05	3.35	6.75
			Brick & Hollow CMU	15.40	16.50	17.65	17.65	15.25	4.05	8.10	5.40	10.80

Table B2010-202 Mix Proportions by Volume, Compressive Strength of Mortar

Where Used	Mortar Type	Allowable Proportions by Volume				Compressive Strength @ 28 days
		Portland Cement	Masonry Cement	Hydrated Lime	Masonry Sand	
Plain Masonry	M	1	1	—	6	
		1	—	1/4	3	2500 psi
	S	1/2	1	—	4	
		1	—	1/4 to 1/2	4	1800 psi
	N	—	1	—	3	
		1	—	1/2 to 1-1/4	6	750 psi
	O	—	1	—	3	
		1	—	1-1/4 to 2-1/2	9	350 psi
	K	1	—	2-1/2 to 4	12	75 psi
Reinforced Masonry	PM	1	1	—	6	2500 psi
	PL	1	—	1/4 to 1/2	4	2500 psi

Note: The total aggregate should be between 2.25 to 3 times the sum of the cement and lime used.

Table B2010-203 Volume and Cost Per S.F. of Grout Fill for Concrete Block Walls

Center to Center Spacing Grouted Cores	6" C.M.U. Per S.F.				8" C.M.U. Per S.F.				12" C.M.U. Per S.F.			
	Volume in C.F.		Cost		Volume in C.F.		Cost		Volume in C.F.		Cost	
	40% Solid	75% Solid	40% Solid	75% Solid	40% Solid	75% Solid	40% Solid	75% Solid	40% Solid	75% Solid	40% Solid	75% Solid
All cores grouted solid	.27	.11	$2.40	$.98	.36	.15	$3.20	$1.34	.55	.23	$4.90	$2.05
cores grouted 16" O.C.	.14	.06	1.25	.53	.18	.08	1.60	.71	.28	.12	2.49	1.07
cores grouted 24" O.C.	.09	.04	.80	.36	.12	.05	1.07	.45	.18	.08	1.60	.71
cores grouted 32" O.C.	.07	.03	.62	.27	.09	.04	.80	.36	.14	.06	1.25	.53
cores grouted 40" O.C.	.05	.02	.45	.18	.07	.03	.62	.27	.11	.05	.98	.45
cores grouted 48" O.C.	.04	.02	.36	.18	.06	.03	.53	.27	.09	.04	.80	.36

Installed costs, including O&P, based on High-Lift Grouting method

Low-Lift Grouting is used when the wall is built to a maximum height of 5'. The grout is pumped or poured into the cores of the concrete block. The operation is repeated after each five additional feet of wall height has been completed.

High-Lift Grouting is used when the wall has been built to the full-story height. Two advantages are: the vertical reinforcing steel can be placed after the wall is completed; and the grout can be supplied by a ready-mix concrete supplier so that it may be pumped in a continuous operation.

B20 Exterior Enclosure | RB2010-430 | Pre-Engin. Steel Building

SHELL

Pre-engineered Steel Buildings

These buildings are manufactured by many companies and normally erected by franchised dealers throughout the U.S. The four basic types are: Rigid Frames, Truss type, Post and Beam and the Sloped Beam type. Most popular roof slope is low pitch of 1″ in 12″. The minimum economical area of these buildings is about 3000 S.F. of floor area. Bay sizes are usually 20′ to 24′ but can go as high as 30′ with heavier girts and purlins. Eave heights are usually 12′ to 24′ with 18′ to 20′ most typical.

Material prices generally do not include floors, foundations, interior finishes or utilities. Costs assume at least three bays of 24′ each, a 1″ in 12″ roof slope, and they are based on 30 psf roof load and 20 psf wind load (wind load is a function of wind speed, building height, and terrain characteristics; this should be determined by a Registered Structural Engineer) and no unusual requirements. Costs include the structural frame, 26 ga. non-insulated colored corrugated or ribbed roofing or siding panels, fasteners, closures, trim and flashing but no allowance for insulation, doors, windows, skylights, gutters or downspouts. Very large projects would generally cost less for materials than the prices shown.

Conditions at the site, weather, shape and size of the building, and labor availability will affect the erection cost of the building.

Table B3010-011   Thermal Coefficients for Roof Systems

EXAMPLE:

Construction	Resistance R
1. Outside surface (15 MPH wind)	0.17
2. Built up roofing .375" thick	0.33
3. Rigid roof insulation, 2"	5.26
4. Metal decking	0.00
5. Air space (non reflective)	1.14
6. Structural members	0.00
7. Gypsum board ceiling 5/8" thick	0.56
8. Inside surface (still air)	0.61
Total resistance	8.07

$U = 1/R = 1/8.07 =$ 0.12

Substitution

Replace item 3 with Perlite lightweight concrete and item 7 with acoustical tile.

Total resistance (Example)		**R =**	8.07
Deduct 3. Rigid insulation	5.26		
7. Gypsum board ceiling	0.05		
	5.31		-5.31
Difference			2.76
Add Perlite concrete, 4"			4.32
Acoustical tile, 3/4"			1.56
		Total resistance	8.64

$U = 1/R = 1/8.64 =$ 0.12

EXAMPLE

Construction	Resistance R
1. Outside surface (15 MPH wind)	0.17
2. Built up roofing .375" thick	0.33
3. Rigid roof insulation, 2" thick	1.20
4. Form bd., 1/2" thick	0.45
5. Air space (non reflective)	1.14
6. Structural members	0.00
7. Gypsum board ceiling 3/4" thick	0.14
8. Inside surface (still air)	0.61
Total resistance	4.04

$U = 1/R = 1/4.04 =$ 0.25

(for modified system see next page)

Table B3010-011 Thermal Coefficients for Roof Systems

Substitution
Replace item 3 with Tectum board and item 7 with acoustical tile.

Total resistance (Example)		R = 4.04
Deduct 3. Gypsum concrete	1.20	
4. Form bd.	0.45	
7. Plaster	0.14	
	1.79	<-1.79>
Difference		2.25
Add Tectum bd. 3" thick		5.25
Acoustical tile, 3/4"		1.56
Total resistance		9.06

U = 1/R = 1/9.06 = 0.11

(for limited system see previous page)

EXAMPLE:

Construction	Resistance R
1. Outside surface (15 MPH wind)	0.17
2. Asphalt shingle roof	0.44
3. Felt building paper	0.06
4. Plywood sheathing, 5/8" thick	0.78
5. Reflective air space, 3-1/2"	2.17
6. Gypsum board, 1/2" foil backed	0.45
7. Inside surface (still air)	0.61
Total resistance	4.68

U = 1/R = 1/4.68 = 0.21

Substitution
Replace item 2 with slate, item 5 with insulation and item 6 with gypsum plaster

Total resistance (Example)		R = 4.68
Deduct 2. Asphalt shingles	0.44	
5. Air space	2.17	
6. Gypsum board	0.45	
	3.06	<-3.06>
Difference		1.62
Add Slate, 1/4" thick		0.03
Insulation, 6" blanket		19.00
Gypsum sand plaster 3/4"		0.14
Total resistance		20.79

U = 1/R = 1/20.79 = 0.05

REFERENCE TABLES

Table C2010-101 Typical Range of Risers for Various Story Heights

General Design: See Table L1010-101 for code requirements.
Maximum height between landings is 12′; usual stair angle is 20°
to 50° with 30° to 35° best. Usual relation of riser to treads is:

    Riser + tread = 17.5
    2x (Riser) + tread = 25
    Riser x tread = 70 or 75

Maximum riser height is 7″ for commercial, 8-1/4″ for residential.
Usual riser height is 6-1/2″ to 7-1/4″.
Minimum tread width is 11″ for commercial, 9″ for residential.

Story Height	Minimum Risers	Maximum Riser Height	Tread Width	Maximum Risers	Minimum Riser Height	Tread Width	Average Risers	Average Riser Height	Tread Width
7′-6″	12	7.50″	10.00″	14	6.43″	11.07″	13	6.92″	10.58″
8′-0″	13	7.38	10.12	15	6.40	11.10	14	6.86	10.64
8′-6″	14	7.29	10.21	16	6.38	11.12	15	6.80	10.70
9′-0″	15	7.20	10.30	17	6.35	11.15	16	6.75	10.75
9′-6″	16	7.13	10.37	18	6.33	11.17	17	6.71	10.79
10′-0″	16	7.50	10.00	19	6.32	11.18	18	6.67	10.83
10′-6″	17	7.41	10.09	20	6.30	11.20	18	7.00	10.50
11′-0″	18	7.33	10.17	21	6.29	11.21	19	6.95	10.55
11′-6″	19	7.26	10.24	22	6.27	11.23	20	6.90	10.60
12′-0″	20	7.20	10.30	23	6.26	11.24	21	6.86	10.64
12′-6″	20	7.50	10.00	24	6.25	11.25	22	6.82	10.68
13′-0″	21	7.43	10.07	25	6.24	11.26	22	7.09	10.41
13′-6″	22	7.36	10.14	25	6.48	11.02	23	7.04	10.46
14′-0″	23	7.30	10.20	26	6.46	11.04	24	7.00	10.50

For information about Means Estimating Seminars, see yellow pages 12 and 13 in back of book

Table D1010-011   Elevator Hoistway Sizes

Elevator Type	Floors	Building Type	Capacity Lbs.	Capacity Passengers	Entry	Hoistway Width	Hoistway Depth	S.F. Area per Floor
Hydraulic	5	Apt./Small office	1500	10	S	6'-7"	4'-6"	29.6
			2000	13	S	7'-8"	4'-10"	37.4
	7	Average office/Hotel	2500	16	S	8'-4"	5'-5"	45.1
			3000	20	S	8'-4"	5'-11"	49.3
		Large office/Store	3500	23	S	8'-4"	6'-11"	57.6
		Freight light duty	2500		D	7'-2"	7'-10"	56.1
		Heavy duty	5000		D	10'-2"	10'-10"	110.1
			7500		D	10'-2"	12'-10"	131
			5000		D	10'-2"	10'-10"	110.1
			7500		D	10'-2"	12'-10"	131
			10000		D	10'-4"	14'-10"	153.3
		Hospital	3500		D	6'-10"	9'-2"	62.7
			4000		D	7'-4"	9'-6"	69.6
Electric Traction, high speed	High Rise	Apt./Small office	2000	13	S	7'-8"	5'-10"	44.8
			2500	16	S	8'-4"	6'-5"	54.5
			3000	20	S	8'-4"	6'-11"	57.6
			3500	23	S	8'-4"	7'-7"	63.1
		Store	3500	23	S	9'-5"	6'-10"	64.4
		Large office	4000	26	S	9'-4"	7'-6"	70
		Hospital	3500		D	7'-6"	9'-2"	69.4
			4000		D	7'-10"	9'-6"	58.8
Geared, low speed		Apartment	1200	8	S	6'-4"	5'-3"	33.2
			2000	13	S	7'-8"	5'-8"	43.5
			2500	16	S	8'-4"	6'-3"	52
		Office	3000	20	S	8'-4"	6'-9"	56
		Store	3500	23	S	9'-5"	6'-10"	64.4
						Add 4" width for multiple units		

Figure D1010-012 Elevator Passenger Capacity During Peak Periods

Figure D1010-013 Speed & Travel Chart

All speeds are available with any capacity car

Table D1010-014 Elevator Speed vs. Height Requirements

Building Type and Elevator Capacities	Travel Speeds in Feet per Minute								
	100 fpm	200 fpm	250 fpm	350 fpm	400 fpm	500 fpm	700 fpm	800 fpm	1000 fpm
Apartments 1200 lb. to 2500 lb.	to 70	to 100	to 125	to 150	to 175	to 250	to 350		
Department Stores 1200 lb. to 2500 lb.		100		125	175	250	350		
Hospitals 3500 lb. to 4000 lb.	70	100	125	150	175	250	350		
Office Buildings 2000 lb. to 4000 lb.		100	125	150		175	250	to 350	over 350

Note: Vertical transportation capacity may be determined by code occupancy requirements of the number of square feet per person divided into the total square feet of building type. If we are contemplating an office building, we find that the Occupancy Code Requirement is 100 S.F. per person. For a 20,000 S.F. building, we would have a legal capacity of two hundred people. Elevator handling capacity is subject to the five minute evacuation recommendation, but it may vary from 11 to 18%. Speed required is a function of the travel height, number of stops and capacity of the elevator.

Table D2010-031 Plumbing Approximations for Quick Estimating

Water Control

Water Meter; Backflow Preventer,

Shock Absorbers; Vacuum

Breakers; Mixer .... 10 to 15% of Fixtures

Pipe And Fittings .... 30 to 60% of Fixtures

> **Note:** Lower percentage for compact buildings or larger buildings with plumbing in one area.
> Larger percentage for large buildings with plumbing spread out.
> In extreme cases pipe may be more than 100% of fixtures.
> Percentages **do not** include special purpose or process piping.

Plumbing Labor

1 & 2 Story Residential ....Rough-in Labor = 80% of Materials

Apartment Buildings .... Rough-in Labor = 90 to 100% of Materials

Labor for handling and placing fixtures is approximately 25 to 30% of fixtures

Quality/Complexity Multiplier (for all installations)

Economy installation, add ....0 to 5%

Good quality, medium complexity, add ....5 to 15%

Above average quality and complexity, add ....15 to 25%

Table D2010-032 Pipe Material Considerations

1. Malleable fittings should be used for gas service.
2. Malleable fittings are used where there are stresses/strains due to expansion and vibration.
3. Cast fittings may be broken as an aid to disassembling of heating lines frozen by long use, temperature and minerals.
4. Cast iron pipe is extensively used for underground and submerged service.
5. Type M (light wall) copper tubing is available in hard temper only and is used for nonpressure and less severe applications than K and L.
6. Type L (medium wall) copper tubing, available hard or soft for interior service.
7. Type K (heavy wall) copper tubing, available in hard or soft temper for use where conditions are severe. For underground and interior service.
8. Hard drawn tubing requires fewer hangers or supports but should not be bent. Silver brazed fittings are recommended, however soft solder is normally used.
9. Type DMV (very light wall) copper tubing designed for drainage, waste and vent plus other non-critical pressure services.

Table D2010-033 Domestic/Imported Pipe and Fittings Cost

The prices shown in this publication for steel/cast iron pipe and steel, cast iron, malleable iron fittings are based on domestic production sold at the normal trade discounts. The above listed items of foreign manufacture may be available at prices of 1/3 to 1/2 those shown. Some imported items after minor machining or finishing operations are being sold as domestic to further complicate the system.

Caution: Most pipe prices in this book also include a coupling and pipe hangers which for the larger sizes can add significantly to the per foot cost and should be taken into account when comparing "book cost" with quoted supplier's cost.

Table D2010-401 Minimum Plumbing Fixture Requirements

Minimum Plumbing Fixture Requirements

Type of Building or Occupancy (2)	Water Closets (14) (Fixtures per Person)		Urinals (5,10) (Fixtures per Person)		Lavatories (Fixtures per Person)		Bathtubs or Showers (Fixtures per Person)	Drinking Fountains (Fixtures per Person) (3, 13)
	Male	Female	Male	Female	Male	Female		
Assembly Places-Theatres, Auditoriums, Convention Halls, etc.-for permanent employee use	1: 1 - 15 2: 16 - 35 3: 36 - 55 Over 55, add 1 fixture for each additional 40 persons	1: 1 - 15 2: 16 - 35 3: 36 - 55	0: 1 - 9 1: 10 - 50 Add one fixture for each additional 50 males		1 per 40	1 per 40		
Assembly Places-Theatres, Auditoriums, Convention Halls, etc. - for public use	1: 1 - 100 2: 101 - 200 3: 201 - 400 Over 400, add 1 fixture for each additional 500 males and 1 for each additional 125 females	3: 1 - 50 4: 51 - 100 8: 101 - 200 11: 201 - 400	1: 1 - 100 2: 101 - 200 3: 201 - 400 4: 401 - 600 Over 600, add 1 fixture for each additional 300 males		1: 1 - 200 2: 201 - 400 3: 401 - 750 Over 750, add 1 fixture for each additional 500 persons	1: 1 - 200 2: 201 - 400 3: 401 - 750		1: 1 - 150 2: 151 - 400 3: 401 - 750 Over 750, add one fixture for each additional 500 persons
Dormitories (9) School or Labor	1 per 10 Add 1 fixture for each additional 25 males (over 10) and 1 for each additional 20 females (over 8)	1 per 8	1 per 25 Over 150, add 1 fixture for each additional 50 males		1 per 12 Over 12 add 1 fixture for each additional 20 males and 1 for each 15 additional females	1 per 12	1 per 8 For females add 1 bathtub per 30. Over 150, add 1 per 20	1 per 150 (12)
Dormitories- for Staff Use	1: 1 - 15 2: 16 - 35 3: 36 - 55 Over 55, add 1 fixture for each additional 40 persons	1: 1 - 15 3: 16 - 35 4: 36 - 55	1 per 50		1 per 40	1 per 40	1 per 8	
Dwellings: Single Dwelling / Multiple Dwelling or Apartment House	1 per dwelling 1 per dwelling or apartment unit				1 per dwelling 1 per dwelling or apartment unit		1 per dwelling 1 per dwelling or apartment unit	
Hospital Waiting rooms	1 per room				1 per room			1 per 150 (12)
Hospitals- for employee use	1: 1 - 15 2: 16 - 35 3: 36 - 55 Over 55, add 1 fixture for each additional 40 persons	1: 1 - 15 3: 16 - 35 4: 36 - 55	0: 1 - 9 1: 10 - 50 Add 1 fixture for each additional 50 males		1 per 40	1 per 40		
Hospitals: Individual Room / Ward Room	1 per room 1 per 8 patients				1 per room 1 per 10 patients		1 per room 1 per 20 patients	1 per 150 (12)
Industrial (6) Warehouses Workshops, Foundries and similar establishments- for employee use	1: 1 -10 2: 11 - 25 3: 26 - 50 4: 51 - 75 5: 76 - 100 Over 100, add 1 fixture for each additional 30 persons	1: 1 -10 2: 11 - 25 3: 26 - 50 4: 51 - 75 5: 76 - 100			Up to 100, per 10 persons Over 100, 1 per 15 persons (7, 8)		1 shower for each 15 persons exposed to excessive heat or to skin contamination with poisonous, infectious or irritating material	1 per 150 (12)
Institutional - Other than Hospitals or Penal Institutions (on each occupied floor)	1 per 25	1 per 20	0: 1 - 9 1: 10 - 50 Add 1 fixture for each additional 50 males		1 per 10	1 per 10	1 per 8	1 per 150 (12)
Institutional - Other than Hospitals or Penal Institutions (on each occupied floor)- for employee use	1: 1 - 15 2: 16 - 35 3: 36 - 55 Over 55, add 1 fixture for each additional 40 persons	1: 1 - 15 3: 16 - 35 4: 36 - 55	0: 1 - 9 1: 10 - 50 Add 1 fixture for each additional 50 males		1 per 40	1 per 40	1 per 8	1 per 150 (12)
Office or Public Buildings	1: 1 - 100 2: 101 - 200 3: 201 - 400 Over 400, add 1 fixture for each additional 500 males and 1 for each additional 150 females	3: 1 - 50 4: 51 - 100 8: 101 - 200 11: 201 - 400	1: 1 - 100 2: 101 - 200 3: 201 - 400 4: 401 - 600 Over 600, add 1 fixture for each additional 300 males		1: 1 - 200 2: 201 - 400 3: 401 - 750 Over 750, add 1 fixture for each additional 500 persons	1: 1 - 200 2: 201 - 400 3: 401 - 750		1 per 150 (12)
Office or Public Buildings - for employee use	1: 1 - 15 2: 16 - 35 3: 36 - 55 Over 55, add 1 fixture for each additional 40 persons	1: 1 - 15 3: 16 - 35 4: 36 - 55	0: 1 - 9 1: 10 - 50 Add 1 fixture for each additional 50 males		1 per 40	1 per 40		

Table D2010-401 Minimum Plumbing Fixture Requirements (cont.)

Minimum Plumbing Fixture Requirements

Type of Building or Occupancy	Water Closets (14) (Fixtures per Person)		Urinals (5, 10) (Fixtures per Person)		Lavatories (Fixtures per Person)		Bathtubs or Showers (Fixtures per Person)	Drinking Fountains (Fixtures per Person) (3, 13)
	Male	Female	Male	Female	Male	Female		
Penal Institutions - for employee use	1: 1 - 15 2: 16 - 35 3: 36 - 55	1: 1 - 15 3: 16 - 35 4: 36 - 55	0: 1 - 9 1: 10 - 50		1 per 40	1 per 40		1 per 150 (12)
	Over 55, add 1 fixture for each additional 40 persons		Add 1 fixture for each additional 50 males					
Penal Institutions - for prison use Cell	1 per cell				1 per cell			1 per cellblock floor
Exercise room	1 per exercise room		1 per exercise room		1 per exercise room			1 per exercise room
Restaurants, Pubs and Lounges (11)	1: 1 - 50 2: 51 - 150 3: 151 - 300	1: 1 - 50 2: 51 - 150 4: 151 - 300	1: 1 - 150		1: 1 - 150 2: 151 - 200 3: 201 - 400	1: 1 - 150 2: 151 - 200 3: 201 - 400		
	Over 300, add 1 fixture for each additional 200 persons		Over 150, add 1 fixture for each additional 150 males		Over 400, add 1 fixture for each additional 400 persons			
Schools - for staff use All Schools	1: 1 - 15 2: 16 - 35 3: 36 - 55	1: 1 - 15 3: 16 - 35 4: 36 - 55	1 per 50		1 per 40	1 per 40		
	Over 55, add 1 fixture for each additional 40 persons							
Schools - for student use: Nursery	1: 1 - 20 2: 21 - 50	1: 1 - 20 2: 21 - 50			1: 1 - 25 2: 26 - 50	1: 1 - 25 2: 26 - 50		1 per 150 (12)
	Over 50, add 1 fixture for each additional 50 persons				Over 50, add 1 fixture for each additional 50 persons			
Elementary	1 per 30	1 per 25	1 per 75		1 per 35	1 per 35		1 per 150 (12)
Secondary	1 per 40	1 per 30	1 per 35		1 per 40	1 per 40		1 per 150 (12)
Others (Colleges, Universities, Adult Centers, etc.	1 per 40	1 per 30	1 per 35		1 per 40	1 per 40		1 per 150 (12)
Worship Places Educational and Activities Unit	1 per 150	1 per 75	1 per 150		1 per 2 water closets			1 per 150 (12)
Worship Places Principal Assembly Place	1 per 150	1 per 75	1 per 150		1 per 2 water closets			1 per 150 (12)

Notes:
1. The figures shown are based upon one (1) fixture being the minimum required for the number of persons indicated or any fraction thereof.
2. Building categories not shown on this table shall be considered separately by the Administrative Authority.
3. Drinking fountains shall not be installed in toilet rooms.
4. Laundry trays. One (1) laundry tray or one (1) automatic washer standpipe for each dwelling unit or one (1) laundry trays or one (1) automatic washer standpipes, or combination thereof, for each twelve (12) apartments. Kitchen sinks, one (1) for each dwelling or apartment unit.
5. For each urinal added in excess of the minimum required, one water closet may be deducted. The number of water closets shall not be reduced to less than two-thirds (2/3) of the minimum requirement.
6. As required by ANSI Z4.1-1968, Sanitation in Places of Employment.
7. Where there is exposure to skin contamination with poisonous, infectious, or irritating materials, provide one (1) lavatory for each five (5) persons.
8. Twenty-four (24) lineal inches of wash sink or eighteen (18) inches of a circular basin, when provided with water outlets for such space shall be considered equivalent to one (1) lavatory.
9. Laundry trays, one (1) for each fifty (50) persons. Service sinks, one (1) for each hundred (100) persons.
10. General. In applying this schedule of facilities, consideration shall be given to the accessibility of the fixtures. Conformity purely on a numerical basis may not result in an installation suited to the need of the individual establishment. For example, schools should be provided with toilet facilities on each floor having classrooms.
    a. Surrounding materials, wall and floor space to a point two (2) feet in front of urinal lip and four (4) feet above the floor, and at least two (2) feet to each side of the urinal shall be lined with non-absorbent materials.
    b. Trough urinals shall be prohibited.
11. A restaurant is defined as a business which sells food to be consumed on the premises.
    a. The number of occupants for a drive-in restaurant shall be considered as equal to the number of parking stalls.
    b. Employee toilet facilities shall not to be included in the above restaurant requirements. Hand washing facilities shall be available in the kitchen for employees.
12. Where food is consumed indoors, water stations may be substituted for drinking fountains. Offices, or public buildings for use by more than six (6) persons shall have one (1) drinking fountain for the first one hundred fifty (150) persons and one additional fountain for each three hundred (300) persons thereafter.
13. There shall be a minimum of one (1) drinking fountain per occupied floor in schools, theaters, auditoriums, dormitories, offices of public building.
14. The total number of water closets for females shall be at least equal to the total number of water closets and urinals required for males.

Table D2010-403   Water Cooler Application

Type of Service	Requirement
Office, School or Hospital	12 persons per gallon per hour
Office, Lobby or Department Store	4 or 5 gallons per hour per fountain
Light manufacturing	7 persons per gallon per hour
Heavy manufacturing	5 persons per gallon per hour
Hot heavy manufacturing	4 persons per gallon per hour
Hotel	.08 gallons per hour per room
Theater	1 gallon per hour per 100 seats

Table D2020-101   Hot Water Consumption Rates

Type of Building	Size Factor	Maximum Hourly Demand	Average Day Demand
Apartment Dwellings	No. of Apartments: Up to 20 21 to 50 51 to 75 76 to 100 101 to 200 201 up	 12.0 Gal. per apt. 10.0 Gal. per apt. 8.5 Gal. per apt. 7.0 Gal. per apt. 6.0 Gal. per apt. 5.0 Gal. per apt.	 42.0 Gal. per apt. 40.0 Gal. per apt. 38.0 Gal. per apt. 37.0 Gal. per apt. 36.0 Gal. per apt. 35.0 Gal. per apt.
Dormitories	Men Women	3.8 Gal. per man 5.0 Gal. per woman	13.1 Gal. per man 12.3 Gal. per woman
Hospitals	Per bed	23.0 Gal. per patient	90.0 Gal. per patient
Hotels	Single room with bath Double room with bath	17.0 Gal. per unit 27.0 Gal. per unit	50.0 Gal. per unit 80.0 Gal. per unit
Motels	No. of units: Up to 20 21 to 100 101 Up	 6.0 Gal. per unit 5.0 Gal. per unit 4.0 Gal. per unit	 20.0 Gal. per unit 14.0 Gal. per unit 10.0 Gal. per unit
Nursing Homes		4.5 Gal. per bed	18.4 Gal. per bed
Office buildings		0.4 Gal. per person	1.0 Gal. per person
Restaurants	Full meal type Drive-in snack type	1.5 Gal./max. meals/hr. 0.7 Gal./max. meals/hr.	2.4 Gal. per meal 0.7 Gal. per meal
Schools	Elementary Secondary & High	0.6 Gal. per student 1.0 Gal. per student	0.6 Gal. per student 1.8 Gal. per student

For evaluation purposes, recovery rate and storage capacity are inversely proportional. Water heaters should be sized so that the maximum hourly demand anticipated can be met in addition to allowance for the heat loss from the pipes and storage tank.

Table D2020-102   Fixture Demands in Gallons Per Fixture Per Hour

Table below is based on 140°F final temperature except for dishwashers in public places (*) where 180°F water is mandatory.

Fixture	Apartment House	Club	Gym	Hospital	Hotel	Indust. Plant	Office	Private Home	School
Bathtubs	20	20	30	20	20			20	
Dishwashers, automatic	15	50-150*		50-150*	50-200*	20-100*		15	20-100*
Kitchen sink	10	20		20	30	20	20	10	20
Laundry, stationary tubs	20	28		28	28			20	
Laundry, automatic wash	75	75		100	150			75	
Private lavatory	2	2	2	2	2	2	2	2	2
Public lavatory	4	6	8	6	8	12	6		15
Showers	30	150	225	75	75	225	30	30	225
Service sink	20	20		20	30	20	20	15	20
Demand factor	0.30	0.30	0.40	0.25	0.25	0.40	0.30	0.30	0.40
Storage capacity factor	1.25	0.90	1.00	0.60	0.80	1.00	2.00	0.70	1.00

To obtain the probable maximum demand multiply the total demands for the fixtures (gal./fixture/hour) by the demand factor. The heater should have a heating capacity in gallons per hour equal to this maximum. The storage tank should have a capacity in gallons equal to the probable maximum demand multiplied by the storage capacity factor.

REFERENCE TABLES

Table D3010-601   Collector Tilt for Domestic Hot Water

Optimum collector tilt is usually equal to the site latitude. Variations of plus or minus 10 degrees are acceptable and orientation of 20 degrees on either side of true south is acceptable; however, local climate and collector type may influence the choice between east or west deviations.

Flat plate collectors consist of a number of components as follows: Insulation to reduce heat loss through the bottom and sides of the collector. The enclosure which contains all the components in this assembly is usually weatherproof and prevents dust, wind and water from coming in contact with the absorber plate. The cover plate usually consists of one or more layers of a variety of glass or plastic and reduces the reradiation. It creates an air space which traps the heat by reducing radiation losses between the cover plate and the absorber plate.

The absorber plate must have a good thermal bond with the fluid passages.

The absorber plate is usually metallic and treated with a surface coating which improves absorptivity. Black or dark paints or selective coatings are used for this purpose, and the design of this passage and plate combination helps determine a solar system's effectiveness.

Heat transfer fluid passage tubes are attached above and below or integral with an absorber plate for the purpose of transferring thermal energy from the absorber plate to a heat transfer medium. The heat exchanger is a device for transferring thermal energy from one fluid to another. The rule of thumb of space heating sizing is one S.F. of collector per 2.5 S.F. of floor space.

For domestic hot water the rule of thumb is 3/4 S.F. of collector for one gallon of water used per day, on an average use of twenty-five gallons per day per person, plus ten gallons per dishwasher or washing machine.

Table D3020-011
Heating Systems

The basic function of a heating system is to bring an enclosed volume up to a desired temperature and then maintain that temperature within a reasonable range. To accomplish this, the selected system must have sufficient capacity to offset transmission losses resulting from the temperature difference on the interior and exterior of the enclosing walls in addition to losses due to cold air infiltration through cracks, crevices and around doors and windows. The amount of heat to be furnished is dependent upon the building size, construction, temperature difference, air leakage, use, shape, orientation and exposure. Air circulation is also an important consideration. Circulation will prevent stratification which could result in heat losses through uneven temperatures at various levels. For example, the most

Heat Transmission

Heat transfer is an important parameter to be considered during selection of the exterior wall style, material and window area. A high rate of transfer will permit greater heat loss during the wintertime with the resultant increase in heating energy costs and a greater rate of heat gain in the summer with proportionally greater cooling cost. Several terms are used to describe various aspects of heat transfer. However, for general estimating purposes this book lists U valves for systems of construction materials. U is the "overall heat transfer coefficient." It is defined as the heat flow per hour through one square foot when the temperature difference in the air on either side of the structure wall, roof, ceiling or floor is one degree Fahrenheit. The structural segment may be a single homogeneous material or a composite.

efficient use of unit heaters can usually be achieved by circulating the space volume through the total number of units once every 20 minutes or 3 times an hour. This general rule must, of course, be adapted for special cases such as large buildings with low ratios of heat transmitting surface to cubical volume. The type of occupancy of a building will have considerable bearing on the number of heat transmitting units and the location selected. It is axiomatic, however, that the basis of any successful heating system is to provide the maximum amount of heat at the points of maximum heat loss such as exposed walls, windows, and doors. Large roof areas, wind direction, and wide doorways create problems of excessive heat loss and require special consideration and treatment.

Total heat transfer is found using the following equation:

$Q = AU(T_2 - T_1)$ where
- Q = Heat flow, BTU per hour
- A = Area, square feet
- U = Overall heat transfer coefficient
- $(T_2 - T_1)$ = Difference in temperature of air on each side of the construction component. (Also abbreviated TD)

Note that heat can flow through all surfaces of any building and this flow is in addition to heat gain or loss due to ventilation, infiltration and generation (appliances, machinery, people).

Table D3020-012 Heating Approximations for Quick Estimating

Oil Piping & Boiler Room Piping:

Small System . . . . 20 to 30% of Boiler

Complex System
with Pumps, Headers, Etc. . . . . 80 to 110% of Boiler

Breeching With Insulation:

Small . . . . 10 to 15% of Boiler

Large . . . . 15 to 25% of Boiler

Coils: . . . . 15 to 30% of Containing Unit

Balancing (Independent) . . . . 1/2% of H.V.A.C. Estimating

Quality/Complexity Adjustment: For all heating installations add these adjustments to the estimate to more closely allow for the equipment and conditions of the particular job under consideration.

Economy installation, add . . . . 0 to 5% of System

Good quality, medium complexity, add . . . . 5 to 15% of System

Above average quality and complexity, add . . . . 15 to 25% of System

Table D3020-021 Factor for Determining Heat Loss for Various Types of Buildings

General: While the most accurate estimates of heating requirements would naturally be based on detailed information about the building being considered, it is possible to arrive at a reasonable approximation using the following procedure:

1. Calculate the cubic volume of the room or building.
2. Select the appropriate factor from Table D3020-021 below. Note that the factors apply only to inside temperatures listed in the first column and to 0°F outside temperature.

3. If the building has bad north and west exposures, multiply the heat loss factor by 1.1.
4. If the outside design temperature is other than 0°F, multiply the factor from Table D3020-021 by the factor from Table D3020-022.
5. Multiply the cubic volume by the factor selected from Table D3020-021. This will give the estimated BTUH heat loss which must be made up to maintain inside temperature.

Building Type	Conditions	Qualifications	Loss Factor*
Factories & Industrial Plants General Office Areas at 70°F	One Story	Skylight in Roof	6.2
		No Skylight in Roof	5.7
	Multiple Story	Two Story	4.6
		Three Story	4.3
		Four Story	4.1
		Five Story	3.9
		Six Story	3.6
	All Walls Exposed	Flat Roof	6.9
		Heated Space Above	5.2
	One Long Warm Common Wall	Flat Roof	6.3
		Heated Space Above	4.7
	Warm Common Walls on Both Long Sides	Flat Roof	5.8
		Heated Space Above	4.1
Warehouses at 60°F	All Walls Exposed	Skylights in Roof	5.5
		No Skylight in Roof	5.1
		Heated Space Above	4.0
	One Long Warm Common Wall	Skylight in Roof	5.0
		No Skylight in Roof	4.9
		Heated Space Above	3.4
	Warm Common Walls on Both Long Sides	Skylight in Roof	4.7
		No Skylight in Roof	4.4
		Heated Space Above	3.0

*Note: This table tends to be conservative particularly for new buildings designed for minimum energy consumption.

Table D3020-022 Outside Design Temperature Correction Factor (for Degrees Fahrenheit)

Outside Design Temperature	50	40	30	20	10	0	−10	−20	−30
Correction Factor	0.29	0.43	0.57	0.72	0.86	1.00	1.14	1.28	1.43

Figure D3020-023 and D3020-024 provide a way to calculate heat transmission of various construction materials from their U values and the TD (Temperature Difference).

1. From the exterior Enclosure Division or elsewhere, determine U values for the construction desired.
2. Determine the coldest design temperature. The difference between this temperature and the desired interior temperature is the TD (temperature difference).

3. Enter Figure D3020-023 or D3020-024 at correct U value. Cross horizontally to the intersection with appropriate TD. Read transmission per square foot from bottom of figure.
4. Multiply this value of BTU per hour transmission per square foot of area by the total surface area of that type of construction.

Table D3020-023 Transmission of Heat

Table D3020-024 Transmission of Heat (Low Rate)

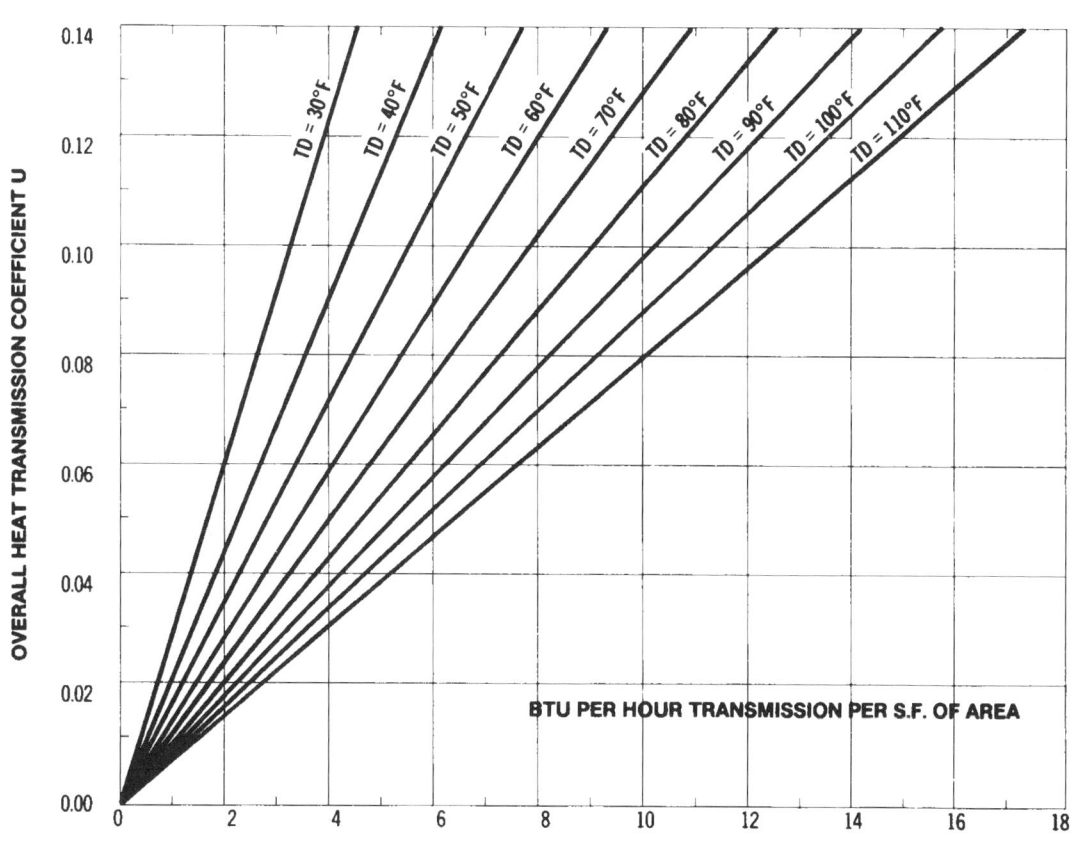

Table D3030-012 Air Conditioning Requirements

BTU's per hour per S.F. of floor area and S.F. per ton of air conditioning.

Type of Building	BTU per S.F.	S.F. per Ton	Type of Building	BTU per S.F.	S.F. per Ton	Type of Building	BTU per S.F.	S.F. per Ton
Apartments, Individual	26	450	Dormitory, Rooms	40	300	Libraries	50	240
Corridors	22	550	Corridors	30	400	Low Rise Office, Exterior	38	320
Auditoriums & Theaters	40	300/18*	Dress Shops	43	280	Interior	33	360
Banks	50	240	Drug Stores	80	150	Medical Centers	28	425
Barber Shops	48	250	Factories	40	300	Motels	28	425
Bars & Taverns	133	90	High Rise Office—Ext. Rms.	46	263	Office (small suite)	43	280
Beauty Parlors	66	180	Interior Rooms	37	325	Post Office, Individual Office	42	285
Bowling Alleys	68	175	Hospitals, Core	43	280	Central Area	46	260
Churches	36	330/20*	Perimeter	46	260	Residences	20	600
Cocktail Lounges	68	175	Hotel, Guest Rooms	44	275	Restaurants	60	200
Computer Rooms	141	85	Corridors	30	400	Schools & Colleges	46	260
Dental Offices	52	230	Public Spaces	55	220	Shoe Stores	55	220
Dept. Stores, Basement	34	350	Industrial Plants, Offices	38	320	Shop'g. Ctrs., Supermarkets	34	350
Main Floor	40	300	General Offices	34	350	Retail Stores	48	250
Upper Floor	30	400	Plant Areas	40	300	Specialty	60	200

*Persons per ton
12,000 BTU = 1 ton of air conditioning

Table D3030-013 Psychrometric Table

Dewpoint or Saturation Temperature (F)

100	32	35	40	45	50	55	60	65	70	75	80	85	90	95	100
90	30	33	37	42	47	52	57	62	67	72	77	82	87	92	97
80	27	30	34	39	44	49	54	58	64	68	73	78	83	88	93
70	24	27	31	36	40	45	50	55	60	64	69	74	79	84	88
60	20	24	28	32	36	41	46	51	55	60	65	69	74	79	83
50	16	20	24	28	33	36	41	46	50	55	60	64	69	73	78
40	12	15	18	23	27	31	35	40	45	49	53	58	62	67	71
30	8	10	14	18	21	25	29	33	37	42	46	50	54	59	62
20	6	7	8	9	13	16	20	24	28	31	35	40	43	48	52
10	4	4	5	5	6	8	9	10	13	17	20	24	27	30	34
	32	35	40	45	50	55	60	65	70	75	80	85	90	95	100

Dry bulb temperature (F)

This table shows the relationship between RELATIVE HUMIDITY, DRY BULB TEMPERATURE AND DEWPOINT. As an example, assume that the thermometer in a room reads 75°F, and we know that the relative humidity is 50%. The chart shows the dewpoint temperature to be 55°. That is, any surface colder than 55°F will "sweat" or collect condensing moisture. This surface could be the outside of an uninsulated chilled water pipe in the summertime, or the inside surface of a wall or deck in the wintertime. After determining the extreme ambient parameters, the table at the left is useful in determining which surfaces need insulation or vapor barrier protection.

Table D3030-014   Recommended Ventilation Air Changes

Table below lists range of time in minutes per change for various types of facilities.

Assembly Halls	2-10	Dance Halls	2-10	Laundries	1-3
Auditoriums	2-10	Dining Rooms	3-10	Markets	2-10
Bakeries	2-3	Dry Cleaners	1-5	Offices	2-10
Banks	3-10	Factories	2-5	Pool Rooms	2-5
Bars	2-5	Garages	2-10	Recreation Rooms	2-10
Beauty Parlors	2-5	Generator Rooms	2-5	Sales Rooms	2-10
Boiler Rooms	1-5	Gymnasiums	2-10	Theaters	2-8
Bowling Alleys	2-10	Kitchens-Hospitals	2-5	Toilets	2-5
Churches	5-10	Kitchens-Restaurant	1-3	Transformer Rooms	1-5

CFM air required for changes = Volume of room in cubic feet ÷ Minutes per change.

Table D3030-015   Ductwork

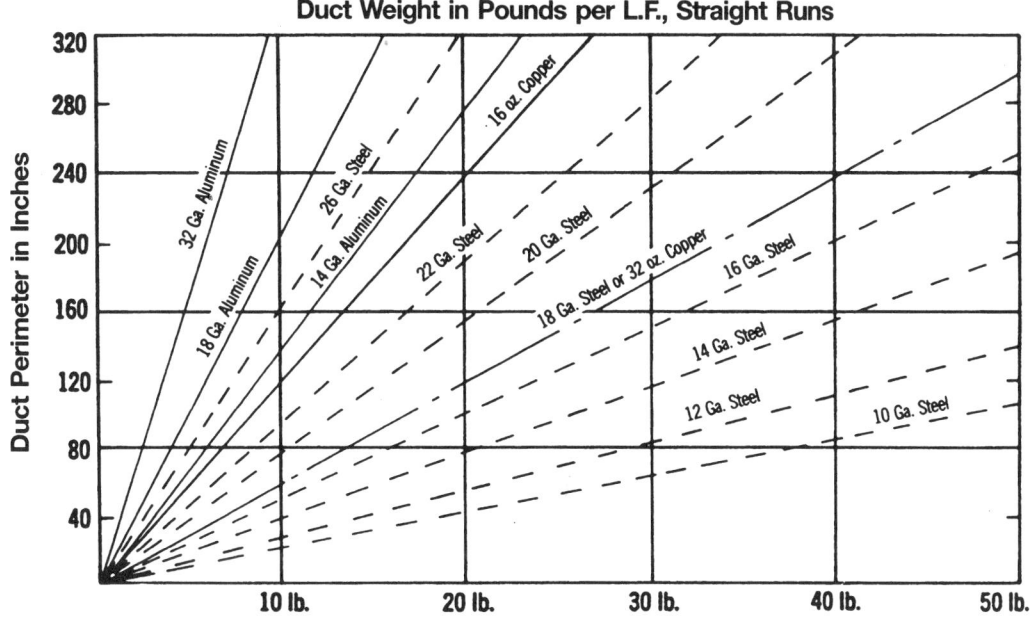

Duct Weight in Pounds per L.F., Straight Runs

Add to the above for fittings; 90° elbow is 3 L.F.; 45° elbow is 2.5 L.F.; offset is 4 L.F.; transition offset is 6 L.F.; square-to-round transition is 4 L.F.; 90° reducing elbow is 5 L.F. For bracing and waste, add 20% to aluminum and copper, 15% to steel.

Table D3030-017   Ductwork Packages (per Ton of Cooling)

System	Sheet Metal	Insulation	Diffusers	Return Register
Roof Top Unit Single Zone	120 Lbs.	52 S.F.	1	1
Roof Top Unit Multizone	240 Lbs.	104 S.F.	2	1
Self-contained Air or Water Cooled	108 Lbs.	—	2	—
Split System Air Cooled	102 Lbs.	—	2	—

Systems reflect most common usage. Refer to system graphics for duct layout.

Table D3030-018   Quality/Complexity Adjustment for Air Conditioning Systems

Economy installation, add	0 to 5%
Good quality, medium complexity, add	5 to 15%
Above average quality and complexity, add	15 to 25%

Add the above adjustments to the estimate to more closely allow for the equipment and conditions of the particular job under consideration.

Table D3030-019 Sheet Metal Calculator (Weight in Lb./Ft. of Length)

Gauge	26	24	22	20	18	16	Gauge	26	24	22	20	18	16
Wt.-Lb./S.F.	.906	1.156	1.406	1.656	2.156	2.656	Wt.-Lb./S.F.	.906	1.156	1.406	1.656	2.156	2.656
SMACNA Max. Dimension – Long Side		30″	54″	84″	85″ Up		SMACNA Max. Dimension – Long Side		30″	54″	84″	85″ Up	
Sum-2 sides							Sum-2 Sides						
2	.3	.40	.50	.60	.80	.90	56	9.3	12.0	14.0	16.2	21.3	25.2
3	.5	.65	.80	.90	1.1	1.4	57	9.5	12.3	14.3	16.5	21.7	25.7
4	.7	.85	1.0	1.2	1.5	1.8	58	9.7	12.5	14.5	16.8	22.0	26.1
5	.8	1.1	1.3	1.5	1.9	2.3	59	9.8	12.7	14.8	17.1	22.4	26.6
6	1.0	1.3	1.5	1.7	2.3	2.7	60	10.0	12.9	15.0	17.4	22.8	27.0
7	1.2	1.5	1.8	2.0	2.7	3.2	61	10.2	13.1	15.3	17.7	23.2	27.5
8	1.3	1.7	2.0	2.3	3.0	3.6	62	10.3	13.3	15.5	18.0	23.6	27.9
9	1.5	1.9	2.3	2.6	3.4	4.1	63	10.5	13.5	15.8	18.3	24.0	28.4
10	1.7	2.2	2.5	2.9	3.8	4.5	64	10.7	13.7	16.0	18.6	24.3	28.8
11	1.8	2.4	2.8	3.2	4.2	5.0	65	10.8	13.9	16.3	18.9	24.7	29.3
12	2.0	2.6	3.0	3.5	4.6	5.4	66	11.0	14.1	16.5	19.1	25.1	29.7
13	2.2	2.8	3.3	3.8	4.9	5.9	67	11.2	14.3	16.8	19.4	25.5	30.2
14	2.3	3.0	3.5	4.1	5.3	6.3	68	11.3	14.6	17.0	19.7	25.8	30.6
15	2.5	3.2	3.8	4.4	5.7	6.8	69	11.5	14.8	17.3	20.0	26.2	31.1
16	2.7	3.4	4.0	4.6	6.1	7.2	70	11.7	15.0	17.5	20.3	26.6	31.5
17	2.8	3.7	4.3	4.9	6.5	7.7	71	11.8	15.2	17.8	20.6	27.0	32.0
18	3.0	3.9	4.5	5.2	6.8	8.1	72	12.0	15.4	18.0	20.9	27.4	32.4
19	3.2	4.1	4.8	5.5	7.2	8.6	73	12.2	15.6	18.3	21.2	27.7	32.9
20	3.3	4.3	5.0	5.8	7.6	9.0	74	12.3	15.8	18.5	21.5	28.1	33.3
21	3.5	4.5	5.3	6.1	8.0	9.5	75	12.5	16.1	18.8	21.8	28.5	33.8
22	3.7	4.7	5.5	6.4	8.4	9.9	76	12.7	16.3	19.0	22.0	28.9	34.2
23	3.8	5.0	5.8	6.7	8.7	10.4	77	12.8	16.5	19.3	22.3	29.3	34.7
24	4.0	5.2	6.0	7.0	9.1	10.8	78	13.0	16.7	19.5	22.6	29.6	35.1
25	4.2	5.4	6.3	7.3	9.5	11.3	79	13.2	16.9	19.8	22.9	30.0	35.6
26	4.3	5.6	6.5	7.5	9.9	11.7	80	13.3	17.1	20.0	23.2	30.4	36.0
27	4.5	5.8	6.8	7.8	10.3	12.2	81	13.5	17.3	20.3	23.5	30.8	36.5
28	4.7	6.0	7.0	8.1	10.6	12.6	82	13.7	17.5	20.5	23.8	31.2	36.9
29	4.8	6.2	7.3	8.4	11.0	13.1	83	13.8	17.8	20.8	24.1	31.5	37.4
30	5.0	6.5	7.5	8.7	11.4	13.5	84	14.0	18.0	21.0	24.4	31.9	37.8
31	5.2	6.7	7.8	9.0	11.8	14.0	85	14.2	18.2	21.3	24.7	32.3	38.3
32	5.3	6.9	8.0	9.3	12.2	14.4	86	14.3	18.4	21.5	24.9	32.7	38.7
33	5.5	7.1	8.3	9.6	12.5	14.9	87	14.5	18.6	21.8	25.2	33.1	39.2
34	5.7	7.3	8.5	9.9	12.9	15.3	88	14.7	18.8	22.0	25.5	33.4	39.6
35	5.8	7.5	8.8	10.2	13.3	15.8	89	14.8	19.0	22.3	25.8	33.8	40.1
36	6.0	7.8	9.0	10.4	13.7	16.2	90	15.0	19.3	22.5	26.1	34.2	40.5
37	6.2	8.0	9.3	10.7	14.1	16.7	91	15.2	19.5	22.8	26.4	34.6	41.0
38	6.3	8.2	9.5	11.0	14.4	17.1	92	15.3	19.7	23.0	26.7	35.0	41.4
39	6.5	8.4	9.8	11.3	14.8	17.6	93	15.5	19.9	23.3	27.0	35.3	41.9
40	6.7	8.6	10.0	11.6	15.2	18.0	94	15.7	20.1	23.5	27.3	35.7	42.3
41	6.8	8.8	10.3	11.9	15.6	18.5	95	15.8	20.3	23.8	27.6	36.1	42.8
42	7.0	9.0	10.5	12.2	16.0	18.9	96	16.0	20.5	24.0	27.8	36.5	43.2
43	7.2	9.2	10.8	12.5	16.3	19.4	97	16.2	20.8	24.3	28.1	36.9	43.7
44	7.3	9.5	11.0	12.8	16.7	19.8	98	16.3	21.0	24.5	28.4	37.2	44.1
45	7.5	9.7	11.3	13.1	17.1	20.3	99	16.5	21.2	24.8	28.7	37.6	44.6
46	7.7	9.9	11.5	13.3	17.5	20.7	100	16.7	21.4	25.0	29.0	38.0	45.0
47	7.8	10.1	11.8	13.6	17.9	21.2	101	16.8	21.6	25.3	29.3	38.4	45.5
48	8.0	10.3	12.0	13.9	18.2	21.6	102	17.0	21.8	25.5	29.6	38.8	45.9
49	8.2	10.5	12.3	14.2	18.6	22.1	103	17.2	22.0	25.8	29.9	39.1	46.4
50	8.3	10.7	12.5	14.5	19.0	22.5	104	17.3	22.3	26.0	30.2	39.5	46.8
51	8.5	11.0	12.8	14.8	19.4	23.0	105	17.5	22.5	26.3	30.5	39.9	47.3
52	8.7	11.2	13.0	15.1	19.8	23.4	106	17.7	22.7	26.5	30.7	40.3	47.7
53	8.8	11.4	13.3	15.4	20.1	23.9	107	17.8	22.9	26.8	31.0	40.7	48.2
54	9.0	11.6	13.5	15.7	20.5	24.3	108	18.0	23.1	27.0	31.3	41.0	48.6
55	9.2	11.8	13.8	16.0	20.9	24.8	109	18.2	23.3	27.3	31.6	41.4	49.1
							110	18.3	23.5	27.5	31.9	41.8	49.5

Example: If duct is 34″ x 20″ x 15′ long, 34″ is greater than 30″ maximum, for 24 ga. so must be 22 ga. 34″ + 20″ = 54″ going across from 54″ find 13.5 lb. per foot. 13.5 x 15′ = 202.5 lbs.

For S.F. of surface area 202.5 ÷ 1.406 = 144 S.F.
Note: Figures include an allowance for scrap.

Table D4010-102 System Classification

System Classification
Rules for installation of sprinkler systems vary depending on the classification of occupancy falling into one of three categories as follows:

Light Hazard Occupancy
The protection area allotted per sprinkler should not exceed 225 S.F., with the maximum distance between lines and sprinklers on lines being 15'. The sprinklers do not need to be staggered. Branch lines should not exceed eight sprinklers on either side of a cross main. Each large area requiring more than 100 sprinklers and without a sub-dividing partition should be supplied by feed mains or risers sized for ordinary hazard occupancy.
Maximum system area = 52,000 S.F.

Included in this group are:

Churches	Nursing Homes
Clubs	Offices
Educational	Residential
Hospitals	Restaurants
Institutional	Theaters and Auditoriums
Libraries	(except stages and prosceniums)
(except large stack rooms)	Unused Attics
Museums	

Ordinary Hazard Occupancy
The protection area allotted per sprinkler shall not exceed 130 S.F. of noncombustible ceiling and 130 S.F. of combustible ceiling. The maximum allowable distance between sprinkler lines and sprinklers on line is 15'. Sprinklers shall be staggered if the distance between heads exceeds 12'. Branch lines should not exceed eight sprinklers on either side of a cross main.
Maximum system area = 52,000 S.F.

Included in this group are:

Group 1
Automotive Parking and Showrooms
Bakeries
Beverage manufacturing
Canneries
Dairy Products Manufacturing/Processing
Electronic Plans
Glass and Glass Products Manufacturing
Laundries
Restaurant Service Areas

Group 2
Cereal Mills
Chemical Plants—Ordinary
Confectionery Products
Distilleries
Dry Cleaners
Feed Mills
Horse Stables
Leather Goods Manufacturing
Libraries—Large Stack Room Areas
Machine Shops
Metal Working
Mercantile
Paper and Pulp Mills
Paper Process Plants
Piers and Wharves
Post Offices
Printing and Publishing
Repair Garages
Stages
Textile Manufacturing
Tire Manufacturing
Tobacco Products Manufacturing
Wood Machining
Wood Product Assembly

Extra Hazard Occupancy
The protection area allotted per sprinkler shall not exceed 100 S.F. of noncombustible ceiling and 100 S.F. of combustible ceiling. The maximum allowable distance between lines and between sprinklers on lines is 12'. Sprinklers on alternate lines shall be staggered if the distance between sprinklers on lines exceeds 8'. Branch lines should not exceed six sprinklers on either side of a cross main.
Maximum system area:
   Design by pipe schedule = 25,000 S.F.
   Design by hydraulic calculation = 40,000 S.F.

Included in this group are:

Group 1
Aircraft hangars
Combustible Hydraulic Fluid Use Area
Die Casting
Metal Extruding
Plywood/Particle Board Manufacturing
Printing (inkls with flash points < 100 degrees F
Rubber Reclaiming, Compounding, Drying, Milling, Vulcanizing
Saw Mills
Textile Picking, Opening, Blending, Garnetting, Carding, Combing of Cotton, Synthetics, Wood Shoddy, or Burlap
Upholstering with Plastic Foams

Group 2
Asphalt Saturating
Flammable Liquids Spraying
Flow Coating
Manufactured/Modular Home Building Assemblies (where finished enclosure is present and has combustible interiors)
Open Oil Quenching
Plastics Processing
Solvent Cleaning
Varnish and Paint Dipping

REFERENCE TABLES

Table D4020-301 Standpipe Systems

The basis for standpipe system design is National Fire Protection Association NFPA 14, however, the authority having jurisdiction should be consulted for special conditions, local requirements and approval.

Standpipe systems, properly designed and maintained, are an effective and valuable time saving aid for extinguishing fires, especially in the upper stories of tall buildings, the interior of large commercial or industrial malls, or other areas where construction features or access make the laying of temporary hose lines time consuming and/or hazardous. Standpipes are frequently installed with automatic sprinkler systems for maximum protection.

There are three general classes of service for standpipe systems:
Class I for use by fire departments and personnel with special training for heavy streams (2-1/2″ hose connections).
Class II for use by building occupants until the arrival of the fire department (1-1/2″ hose connector with hose).

Class III for use by either fire departments and trained personnel or by the building occupants (both 2-1/2″ and 1-1/2″ hose connections or one 2-1/2″ hose valve with an easily removable 2-1/2″ by 1-1/2″ adapter).

Standpipe systems are also classified by the way water is supplied to the system. The four basic types are:
Type 1: Wet standpipe system having supply valve open and water pressure maintained at all times.
Type 2: Standpipe system so arranged through the use of approved devices as to admit water to the system automatically by opening a hose valve.
Type 3: Standpipe system arranged to admit water to the system through manual operation of approved remote control devices located at each hose station.
Type 4: Dry standpipe having no permanent water supply.

Table D4020-302 NFPA 14 Basic Standpipe Design

Class	Design-Use	Pipe Size Minimums	Water Supply Minimums
Class I	2 1/2″ hose connection on each floor All areas within 150′ of an exit in every exit stairway Fire Department Trained Personnel	Height to 100′, 4″ dia. Heights above 100′, 6″ dia. (275′ max. except with pressure regulators 400′ max.)	For each standpipe riser 500 GPM flow For common supply pipe allow 500 GPM for first standpipe plus 250 GPM for each additional standpipe (2500 GPM max. total) 30 min. duration 65 PSI at 500 GPM
Class II	1 1/2″ hose connection with hose on each floor All areas within 130′ of hose connection measured along path of hose travel Occupant personnel	Height to 50′, 2″ dia. Height above 50′, 2 1/2″ dia.	For each standpipe riser 100 GPM flow For multiple riser common supply pipe 100 GPM 300 min. duration, 65 PSI at 100 GPM
Class III	Both of above. Class I valved connections will meet Class III with additional 2 1/2″ by 1 1/2″ adapter and 1 1/2″ hose.	Same as Class I	Same as Class I

*Note: Where 2 or more standpipes are installed in the same building or section of building they shall be interconnected at the bottom.

Combined Systems

Combined systems are systems where the risers supply both automatic sprinklers and 2-1/2″ hose connection outlets for fire department use. In such a system the sprinkler spacing pattern shall be in accordance with NFPA 13 while the risers and supply piping will be sized in accordance with NFPA 14. When the building is completely sprinklered the risers may be sized by hydraulic calculation. The minimum size riser for buildings not completely sprinklered is 6″.
The minimum water supply of a completely sprinklered, light hazard, high-rise occupancy building will be 500 GPM while the supply required for other types of completely sprinklered high-rise buildings is 1000 GPM.

General System Requirements

1. Approved valves will be provided at the riser for controlling branch lines to hose outlets.
2. A hose valve will be provided at each outlet for attachment of hose.
3. Where pressure at any standpipe outlet exceeds 100 PSI a pressure reducer must be installed to limit the pressure to 100 PSI. Note that

the pressure head due to gravity in 100′ of riser is 43.4 PSI. This must be overcome by city pressure, fire pumps, or gravity tanks to provide adequate pressure at the top of the riser.

4. Each hose valve on a wet system having linen hose shall have an automatic drip connection to prevent valve leakage from entering the hose.
5. Each riser will have a valve to isolate it from the rest of the system.
6. One or more fire department connections as an auxiliary supply shall be provided for each Class I or Class III standpipe system. In buildings having two or more zones, a connection will be provided for each zone.
7. There will be no shutoff valve in the fire department connection, but a check valve will be located in the line before it joins the system.
8. All hose connections street side will be identified on a cast plate or fitting as to purpose.

Table D4020-303 Quality/Complexity Adjustment for Sprinkler/Standpipe Systems

Economy installation, add	0 to 5%
Good quality, medium complexity, add	5 to 15%
Above average quality and complexity, add	15 to 25%

Figure D5010-111 Typical Overhead Service Entrance

Figure D5010-112 Typical Commercial Electric System

A = 1" Conduit w/1-#1/0 Wire XHHW
B = 2-3" Conduits w/4-350kcmil XHHW in Each
C = 1¼" Conduit w/4 #3 THHN
D = 1¼" Conduit w/3 #1 XHHW
E = 1" Conduit w/4 #6 THHN
F = ½" Conduit w/2 #12 THHN

Figure D5010-113 Preliminary Procedure

1. Determine building size and use.
2. Develop total load in watts
   a. Lighting
   b. Receptacle
   c. Air Conditioning
   d. Elevator
   e. Other power requirements
3. Determine best voltage available from utility company.
4. Determine cost from tables for loads (a) thru (e) above.
5. Determine size of service from formulas (D5010-116).
6. Determine costs for service, panels, and feeders from tables.

Figure D5010-114 Office Building 90′ x 210′, 3 story, w/garage

Garage Area = 18,900 S.F.
Office Area = 56,700 S.F.
Elevator = 2 @ 125 FPM

Tables	Power Required	Watts
D5010-1151	Garage Lighting .5 Watt/S.F.	9,450
	Office Lighting 3 Watts/S.F.	170,100
D5010-1151	Office Receptacles 2 Watts/S.F.	113,400
D5010-1151, R16050-060	Low Rise Office A.C. 4.3 Watts/S.F.	243,810
D5010-1152, 1153	Elevators - 2 @ 20 HP = 2 @ 17,404 Watts/Ea.	34,808
D5010-1151	Misc. Motors + Power 1.2 Watts/S.F.	68,040
	Total	639,608 Watts

Voltage Available
277/480V, 3 Phase, 4 Wire

Formula

D5010-116

$$\text{Amperes} = \frac{\text{Watts}}{\text{Volts x Power Factor x 1.73}} = \frac{639,608}{480V \text{ x } .8 \text{ x } 1.73} = 963 \text{ Amps}$$

Use 1200 Amp Service

System	Description	Unit Cost	Unit	Total
D5020 210 0200	Garage Lighting (Interpolated)	$1.10	S.F.	$ 20,790
D5020 210 0280	Office Lighting	6.55	S.F.	371,385
D5020 115 0880	Receptacle - Undercarpet	3.16	S.F.	179,172
D5020 140 0280	Air Conditioning	.51	S.F.	28,917
D5020 135 0320	Misc. Pwr.	.28	S.F.	15,876
D5020 145 2120	Elevators - 2 @ 20HP	2,350.00	Ea.	4,700
D5010 120 0480	Service - 1200 Amp (add 25% for 277/480V)	17,025.00	Ea.	21,281
D5010 230 0480	Feeder - Assume 200 Ft.	233.00	Ft.	46,600
D5010 240 0320	Panels - 1200 Amp (add 20% for 277/480V)	25,525.00	Ea.	30,630
D5030 910 0400	Fire Detection	26,950.00	Ea.	26,950
	Total			$746,301
	or			$746,300

Table D5010-1151 Nominal Watts Per S.F. for Electric Systems for Various Building Types

Type Construction	1. Lighting	2. Devices	3. HVAC	4. Misc.	5. Elevator	Total Watts
Apartment, luxury high rise	2	2.2	3	1		
Apartment, low rise	2	2	3	1		
Auditorium	2.5	1	3.3	.8		
Bank, branch office	3	2.1	5.7	1.4		
Bank, main office	2.5	1.5	5.7	1.4		
Church	1.8	.8	3.3	.8		
College, science building	3	3	5.3	1.3		
College, library	2.5	.8	5.7	1.4		
College, physical education center	2	1	4.5	1.1		
Department store	2.5	.9	4	1		
Dormitory, college	1.5	1.2	4	1		
Drive-in donut shop	3	4	6.8	1.7		
Garage, commercial	.5	.5	0	.5		
Hospital, general	2	4.5	5	1.3		
Hospital, pediatric	3	3.8	5	1.3		
Hotel, airport	2	1	5	1.3		
Housing for the elderly	2	1.2	4	1		
Manufacturing, food processing	3	1	4.5	1.1		
Manufacturing, apparel	2	1	4.5	1.1		
Manufacturing, tools	4	1	4.5	1.1		
Medical clinic	2.5	1.5	3.2	1		
Nursing home	2	1.6	4	1		
Office building, hi rise	3	2	4.7	1.2		
Office building, low rise	3	2	4.3	1.2		
Radio-TV studio	3.8	2.2	7.6	1.9		
Restaurant	2.5	2	6.8	1.7		
Retail store	2.5	.9	5.5	1.4		
School, elementary	3	1.9	5.3	1.3		
School, junior high	3	1.5	5.3	1.3		
School, senior high	2.3	1.7	5.3	1.3		
Supermarket	3	1	4	1		
Telephone exchange	1	.6	4.5	1.1		
Theater	2.5	1	3.3	.8		
Town Hall	2	1.9	5.3	1.3		
U.S. Post Office	3	2	5	1.3		
Warehouse, grocery	1	.6	0	.5		

Rule of Thumb: 1 KVA = 1 HP (Single Phase)

Three Phase:

$$\text{Watts} = 1.73 \times \text{Volts} \times \text{Current} \times \text{Power Factor} \times \text{Efficiency}$$

$$\text{Horsepower} = \frac{\text{Volts} \times \text{Current} \times 1.73 \times \text{Power Factor}}{746 \text{ Watts}}$$

Table D5010-1152 Horsepower Requirements for Elevators with 3 Phase Motors

Type	Maximum Travel Height in Ft.	Travel Speeds in FPM	Capacity of Cars in Lbs.		
			1200	1500	1800
Hydraulic	70	70	10	15	15
		85	15	15	15
		100	15	15	20
		110	20	20	20
		125	20	20	20
		150	25	25	25
		175	25	30	30
		200	30	30	40
Geared Traction	300	200			
		350			
			2000	2500	3000
Hydraulic	70	70	15	20	20
		85	20	20	25
		100	20	25	30
		110	20	25	30
		125	25	30	40
		150	30	40	50
		175	40	50	50
		200	40	50	60
Geared Traction	300	200	10	10	15
		350	15	15	23
			3500	4000	4500
Hydraulic	70	70	20	25	30
		85	25	30	30
		100	30	40	40
		110	40	40	50
		125	40	50	50
		150	50	50	60
		175	60		
		200	60		
Geared Traction	300	200	15		23
		350	23		35

The power factor of electric motors varies from 80% to 90% in larger size motors. The efficiency likewise varies from 80% on a small motor to 90% on a large motor.

Table D5010-1153 Watts per Motor

90% Power Factor & Efficiency @ 200 or 460V			
HP	Watts	HP	Watts
10	9024	30	25784
15	13537	40	33519
20	17404	50	41899
25	21916	60	49634

Table D5010-116 Electrical Formulas

Ohm's Law

Ohm's Law is a method of explaining the relation existing between voltage, current, and resistance in an electrical circuit. It is practically the basis of all electrical calculations. The term "electromotive force" is often used to designate pressure in volts. This formula can be expressed in various forms.

To find the current in amperes:

$$\text{Current} = \frac{\text{Voltage}}{\text{Resistance}} \quad \text{or} \quad \text{Amperes} = \frac{\text{Volts}}{\text{Ohms}} \quad \text{or} \quad I = \frac{E}{R}$$

The flow of current in amperes through any circuit is equal to the voltage or electromotive force divided by the resistance of that circuit.

To find the pressure or voltage:

$$\text{Voltage} = \text{Current} \times \text{Resistance} \quad \text{or} \quad \text{Volts} = \text{Amperes} \times \text{Ohms}$$
$$\text{or} \quad E = I \times R$$

The voltage required to force a current through a circuit is equal to the resistance of the circuit multiplied by the current.

To find the resistance:

$$\text{Resistance} = \frac{\text{Voltage}}{\text{Current}} \quad \text{or} \quad \text{Ohms} = \frac{\text{Volts}}{\text{Amperes}} \quad \text{or} \quad R = \frac{E}{I}$$

The resistance of a circuit is equal to the voltage divided by the current flowing through that circuit.

Power Formulas

One horsepower = 746 watts   One kilowatt = 1000 watts

The power factor of electric motors varies from 80% to 90% in the larger size motors.

Single-Phase Alternating Current Circuits

Power in Watts = Volts x Amperes x Power Factor

To find current in amperes:

$$\text{Current} = \frac{\text{Watts}}{\text{Volts} \times \text{Power Factor}} \quad \text{or}$$

$$\text{Amperes} = \frac{\text{Watts}}{\text{Volts} \times \text{Power Factor}} \quad \text{or} \quad I = \frac{W}{E \times PF}$$

To find current of a motor, single phase:

$$\text{Current} = \frac{\text{Horsepower} \times 746}{\text{Volts} \times \text{Power Factor} \times \text{Efficiency}} \quad \text{or}$$

$$I = \frac{HP \times 746}{E \times PF \times \text{Eff.}}$$

To find horsepower of a motor, single phase:

$$\text{Horsepower} = \frac{\text{Volts} \times \text{Current} \times \text{Power Factor} \times \text{Efficiency}}{746 \text{ Watts}}$$

$$HP = \frac{E \times I \times PF \times \text{Eff.}}{746}$$

To find power in watts of a motor, single phase:

Watts = Volts x Current x Power Factor x Efficiency   or
Watts = E x I x PF x Eff.

To find single phase kVA:

$$1 \text{ Phase kVA} = \frac{\text{Volts} \times \text{Amps}}{1000}$$

Three-Phase Alternating Current Circuits

Power in Watts = Volts x Amperes x Power Factor x 1.73

To find current in amperes in each wire:

$$\text{Current} = \frac{\text{Watts}}{\text{Voltage} \times \text{Power Factor} \times 1.73} \quad \text{or}$$

$$\text{Amperes} = \frac{\text{Watts}}{\text{Volts} \times \text{Power Factor} \times 1.73} \quad \text{or} \quad I = \frac{W}{E \times PF \times 1.73}$$

To find current of a motor, 3 phase:

$$\text{Current} = \frac{\text{Horsepower} \times 746}{\text{Volts} \times \text{Power Factor} \times \text{Efficiency} \times 1.73} \quad \text{or}$$

$$I = \frac{HP \times 746}{E \times PF \times \text{Eff.} \times 1.73}$$

To find horsepower of a motor, 3 phase:

$$\text{Horsepower} = \frac{\text{Volts} \times \text{Current} \times 1.73 \times \text{Power Factor}}{746 \text{ Watts}}$$

$$HP = \frac{E \times I \times 1.73 \times PF}{746}$$

To find power in watts of a motor, 3 phase:

Watts = Volts x Current x 1.73 x Power Factor x Efficiency   or
Watts = E x I x 1.73 x PF x Eff.

To find 3 phase kVA:

$$3 \text{ phase kVA} = \frac{\text{Volts} \times \text{Amps} \times 1.73}{1000} \quad \text{or}$$

$$\text{kVA} = \frac{V \times A \times 1.73}{1000}$$

Power Factor (PF) is the percentage ratio of the measured watts (effective power) to the volt-amperes (apparent watts)

$$\text{Power Factor} = \frac{\text{Watts}}{\text{Volts} \times \text{Amperes}} \times 100\%$$

A **Conceptual Estimate** of the costs for a building when final drawings are not available can be quickly figured by using **Table D5010-117 Cost Per S.F. for Electrical Systems for Various Building Types.** The following definitions apply to this table.

1. **Service and Distribution:** This system includes the incoming primary feeder from the power company, main building transformer, metering arrangement, switchboards, distribution panel boards, stepdown transformers, and power and lighting panels. Items marked (*) include the cost of the primary feeder and transformer. In all other projects the cost of the primary feeder and transformer is paid for by the local power company.

2. **Lighting:** Includes all interior fixtures for decor, illumination, exit and emergency lighting. Fixtures for exterior building lighting are included, but parking area lighting is not included unless mentioned. See also Section D5020 for detailed analysis of lighting requirements and costs.

3. **Devices:** Includes all outlet boxes, receptacles, switches for lighting control, dimmers and cover plates.

4. **Equipment Connections:** Includes all materials and equipment for making connections for heating, ventilating and air conditioning, food service and other motorized items requiring connections.

5. **Basic Materials:** This category includes all disconnect power switches not part of service equipment, raceways for wires, pull boxes, junction boxes, supports, fittings, grounding materials, wireways, busways, and wire and cable systems.

6. **Special Systems:** Includes installed equipment only for the particular system such as fire detection and alarm, sound, emergency generator and others as listed in the table.

Figure 9.0-D5010-117 Cost per S.F. for Electric Systems for Various Building Types

Type Construction	1. Service & Distrib.	2. Lighting	3. Devices	4. Equipment Connections	5. Basic Materials	6. Special Systems Fire Alarm & Detection	Lightning Protection	Master TV Antenna
Apartment, luxury high rise	$1.71	$1.09	$ .81	$1.05	$ 2.83	$ .52		$.36
Apartment, low rise	.99	.91	.74	.89	1.65	.44		
Auditorium	2.17	5.50	.67	1.50	3.39	.67		
Bank, branch office	2.61	6.11	1.09	1.54	3.21	1.92		
Bank, main office	1.96	3.32	.35	.65	3.48	.99		
Church	1.36	3.37	.42	.39	1.67	.99		
* College, science building	2.81	4.47	1.60	1.36	4.38	.87		
* College, library	1.84	2.48	.32	.71	2.08	.99		
* College, physical education center	2.89	3.46	.44	.58	1.62	.56		
Department store	.96	2.39	.32	1.03	2.79	.44		
* Dormitory, college	1.30	3.07	.33	.65	2.74	.70		.44
Drive-in donut shop	3.63	9.18	1.57	1.55	4.48	—		
Garage, commercial	.49	1.24	.24	.48	.96	—		
* Hospital, general	7.02	4.73	1.79	1.21	5.54	.61	$.20	
* Hospital, pediatric	6.15	7.04	1.55	4.46	10.29	.69		.53
* Hotel, airport	2.76	3.87	.33	.60	4.07	.57	.32	.49
Housing for the elderly	.78	.92	.43	1.17	3.47	.69		.43
Manufacturing, food processing	1.75	4.83	.30	2.22	3.79	.43		
Manufacturing, apparel	1.16	2.53	.36	.88	2.03	.38		
Manufacturing, tools	2.67	5.90	.34	1.04	3.39	.45		
Medical clinic	1.38	2.24	.54	1.74	2.94	.68		
Nursing home	1.82	3.82	.54	.47	3.39	.94		.36
Office building	2.44	5.17	.30	.88	3.55	.50	.29	
Radio-TV studio	1.75	5.33	.81	1.62	4.19	.64		
Restaurant	6.53	5.06	.98	2.46	5.04	.38		
Retail store	1.38	2.69	.33	.60	1.58	—		
School, elementary	2.35	4.77	.66	.60	4.23	.60		.28
School, junior high	1.43	3.99	.33	1.09	3.36	.70		
* School, senior high	1.54	3.13	.57	1.42	3.74	.62		
Supermarket	1.57	2.70	.39	2.37	3.24	.31		
* Telephone exchange	4.08	1.49	.37	1.32	2.52	1.15		
Theater	3.01	3.69	.68	2.06	3.29	.85		
Town Hall	1.85	2.92	.68	.74	4.35	.56		
* U.S. Post Office	5.40	3.75	.69	1.13	3.11	.56		
Warehouse, grocery	1.01	1.63	.25	.60	2.31	.36		

*Includes cost of primary feeder and transformer. Cont'd on next page.

COST ASSUMPTIONS:

Each of the projects analyzed in Table D5010-117 was bid within the last 10 years in the northeastern part of the United States. Bid prices have been adjusted to Jan. 1 levels. The list of projects is by no means all-inclusive, yet by carefully examining the various systems for a particular building type, certain cost relationships will emerge. The use of Square Foot/Cubic Costs, in the Reference section, along with the Project Size Modifier, should produce a budget S.F. cost for the electrical portion of a job that is consistent with the amount of design information normally available at the conceptual estimate stage.

Figure D5010-117 Cost per S.F. for Electric Systems for Various Building Types (cont.)

Type Construction	6. Special Systems (cont.)						
	Intercom Systems	Sound Systems	Closed Circuit TV	Snow Melting	Emergency Generator	Security	Master Clock Sys.
Apartment, luxury high rise	$ .68						
Apartment, low rise	.49						
Auditorium		$1.78	$ .82		$1.26		
Bank, branch office	.94		1.88			$1.59	
Bank, main office	.53		.41		1.02	.87	$.38
Church	.66	.41				.26	
* College, science building	.69	.88			1.42	.45	.43
* College, library		.82			.66	.44	
* College, physical education center		.91				.46	
Department store					.31		
* Dormitory, college	.92						
Drive-in donut shop							.20
Garage, commercial							.17
Hospital, general	.70		.30		1.74		
* Hospital, pediatric	4.75	.48	.52		1.09		
* Hotel, airport	.69				.65		
Housing for the elderly	.83						
Manufacturing, food processing		.32			2.26		
Manufacturing apparel		.42					
Manufacturing, tools		.53		$.35			
Medical clinic							
Nursing home	1.58				.58		
Office Building		.28			.58	.31	.17
Radio-TV studio	.93				1.43		.66
Restaurant		.42					
Retail Store							
School, elementary		.30					.30
School, junior high		.76			.48		.52
* School, senior high	.63		.42		.66	.38	.39
Supermarket		.33			.61	.42	
* Telephone exchange					5.75	.24	
Theater		.61					
Town Hall							.30
* U.S. Post Office	.61			.16	.64		
Warehouse, grocery	.40						

*Includes cost of primary feeder and transformer. Cont'd on next page.

General: Variations in the following square foot costs are due to the type of structural systems of the buildings, geographical location, local electrical codes, designer's preference for specific materials and equipment, and the owner's particular requirements.

Figure D5010-117 Cost per S.F. for Total Electric Systems for Various Building Types (cont.)

Type Construction	Basic Description	Total Floor Area in Square Feet	Total Cost per Square Foot for Total Electric Systems
Apartment building, luxury high rise	All electric, 18 floors, 86 1 B.R., 34 2 B.R.	115,000	$ 9.05
Apartment building, low rise	All electric, 2 floors, 44 units, 1 & 2 B.R.	40,200	6.11
Auditorium	All electric, 1,200 person capacity	28,000	17.76
Bank, branch office	All electric, 1 floor	2,700	20.89
Bank, main office	All electric, 8 floors	54,900	13.96
Church	All electric, incl. Sunday school	17,700	9.53
*College, science building	All electric, 3-1/2 floors, 47 rooms	27,500	19.36
*College, library	All electric	33,500	10.34
*College, physical education center	All electric	22,000	10.92
Department store	Gas heat, 1 floor	85,800	8.24
*Dormitory, college	All electric, 125 rooms	63,000	10.15
Drive-in donut shop	Gas heat, incl. parking area lighting	1,500	20.61
Garage, commercial	All electric	52,300	3.58
*Hospital, general	Steam heat, 4-story garage, 300 beds	540,000	23.84
*Hospital, pediatric	Steam heat, 6 stories	278,000	37.55
Hotel, airport	All electric, 625 guest rooms	536,000	14.35
Housing for the elderly	All electric, 7 floors, 100 1 B.R. units	67,000	8.72
Manufacturing, food processing	Electric heat, 1 floor	9,600	15.90
Manufacturing, apparel	Electric heat, 1 floor	28,000	7.76
Manufacturing, tools	Electric heat, 2 floors	42,000	14.67
Medical clinic	Electric heat, 2 floors	22,700	9.52
Nursing home	Gas heat, 3 floors, 60 beds	21,000	13.50
Office building	All electric, 15 floors	311,200	14.47
Radio-TV studio	Electric heat, 3 floors	54,000	17.36
Restaurant	All electric	2,900	20.87
Retail store	All electric	3,000	6.58
School, elementary	All electric, 1 floor	39,500	14.09
School, junior high	All electric, 1 floor	49,500	12.66
*School, senior high	All electric, 1 floor	158,300	13.50
Supermarket	Gas heat	30,600	11.94
*Telephone exchange	Gas heat, 300 kW emergency generator	24,800	16.92
Theater	Electric heat, twin cinema	14,000	14.19
Town Hall	All electric	20,000	11.40
*U.S. Post Office	All electric	495,000	16.05
Warehouse, grocery	All electric	96,400	6.56

*Includes cost of primary feeder and transformer.

D5010-120 Electric Circuit Voltages

General: The following method provides the user with a simple non-technical means of obtaining comparative costs of wiring circuits. The circuits considered serve the electrical loads of motors, electric heating, lighting and transformers, for example, that require low voltage 60 Hertz alternating current.

The method used here is suitable only for obtaining estimated costs. It is **not** intended to be used as a substitute for electrical engineering design applications.

Conduit and wire circuits can represent from twenty to thirty percent of the total building electrical cost. By following the described steps and using the tables the user can translate the various types of electric circuits into estimated costs.

Wire Size: Wire size is a function of the electric load which is usually listed in one of the following units:

1. Amperes (A)
2. Watts (W)
3. Kilowatts (kW)
4. Volt amperes (VA)
5. Kilovolt amperes (kVA)
6. Horsepower (HP)

These units of electric load must be converted to amperes in order to obtain the size of wire necessary to carry the load. To convert electric load units to amperes one must have an understanding of the voltage classification of the power source and the voltage characteristics of the electrical equipment or load to be energized. The seven A.C. circuits commonly used are illustrated in Figures D5010-121 thru D5010-127 showing the tranformer load voltage and the point of use voltage at the point on the circuit where the load is connected. The difference between the source and point of use

voltages is attributed to the circuit voltage drop and is considered to be approximately 4%.

Motor Voltages: Motor voltages are listed by their point of use voltage and not the power source voltage.

For example: 460 volts instead of 480 volts
200 instead of 208 volts
115 volts instead of 120 volts

Lighting and Heating Voltages: Lighting and heating equipment voltages are listed by the power source voltage and not the point of wire voltage.

For example: 480, 277, 120 volt lighting
480 volt heating or air conditioning unit
208 volt heating unit

Transformer Voltages: Transformer primary (input) and secondary (output) voltages are listed by the power source voltage.

For example: Single phase 10 kVA
Primary 240/480 volts
Secondary 120/240 volts

In this case, the primary voltage may be 240 volts with a 120 volts secondary or may be 480 volts with either a 120V or a 240V secondary.

For example: Three phase 10 kVA
Primary 480 volts
Secondary 208Y/120 volts

In this case the transformer is suitable for connection to a circuit with a 3 phase 3 wire or 3 phase 4 wire circuit with a 480 voltage. This application will provide a secondary circuit of 3 phase 4 wire with 208 volts between phase wires and 120 volts between any phase wire and the neutral (white) wire.

Figure D5010-121

3 Wire, 1 Phase, 120/240 Volt System

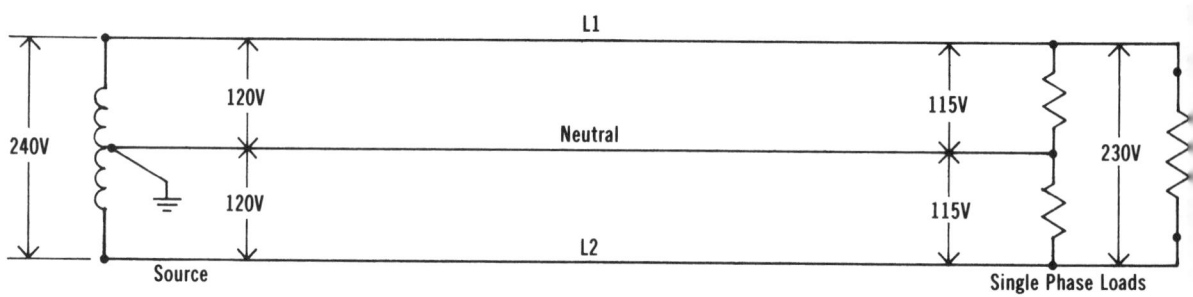

Figure D5010-122

4 wire, 3 Phase, 208Y/120 Volt System

Figure D5010-123 Electric Circuit Voltages (cont.)

3 Wire, 3 Phase 240 Volt System

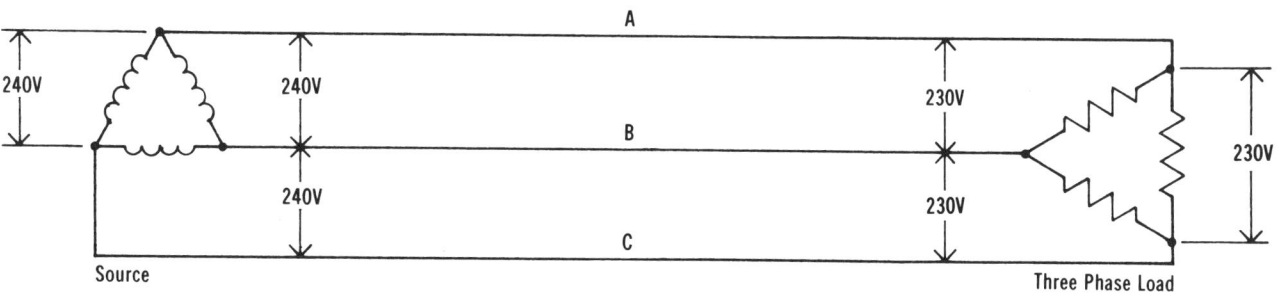

Figure D5010-124

4 Wire, 3 Phase, 240/120 Volt System

Figure D5010-125

3 Wire, 3 Phase 480 Volt System

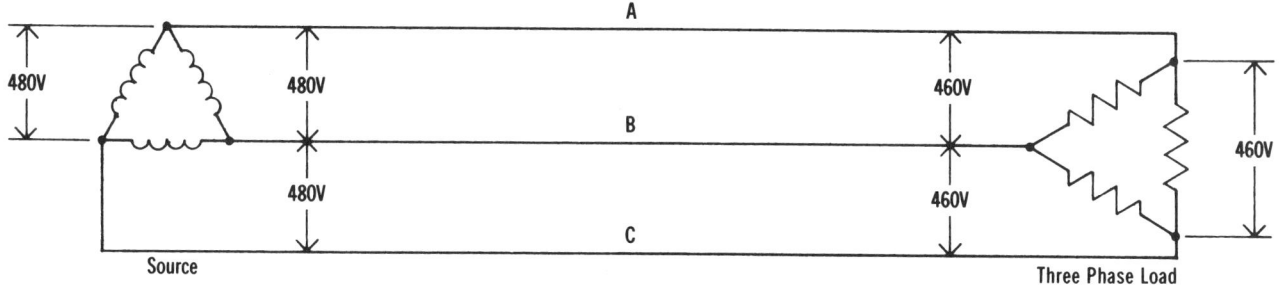

Figure D5010-126

4 Wire, 3 Phase, 480Y/277 Volt System

Figure D5010-127 Electric Circuit Voltages (cont.)

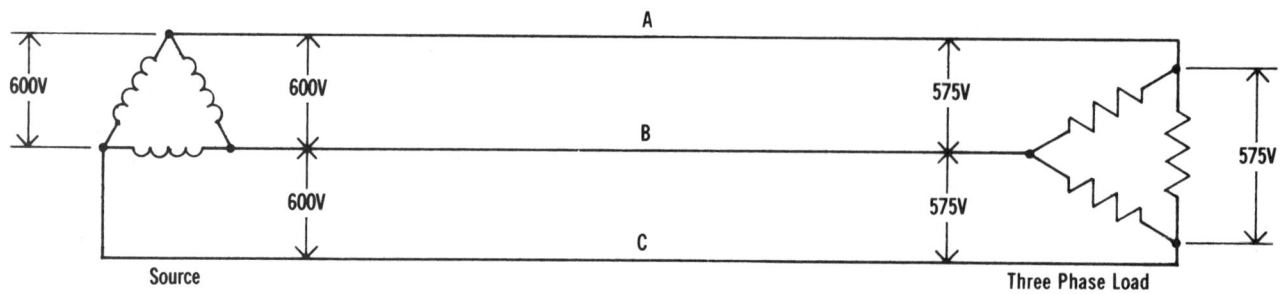

3 Wire, 3 Phase, 600 Volt System

Maximum Circuit Length: Table D5010-130 indicates typical maximum installed length a circuit can have and still maintain an adequate voltage level at the point of use. The circuit length is similar to the conduit length.

If the circuit length for an ampere load and a copper wire size exceeds the length obtained from Table D5010-130, use the next largest wire size to compensate voltage drop.

Example: A 130 ampere load at 480 volts, 3 phase, 3 wire with No. 1 wire can be run a maximum of 555 L.F. and provide satisfactory operation. If the same load is to be wired at the end of a 625 L.F. circuit, then a larger wire must be used.

Table D5010-130   Maximum Circuit Length (approximate) for Various Power Requirements Assuming THW, Copper Wire @ 75° C, Based Upon a 4% Voltage Drop

| Amperes | Wire Size | Maximum Circuit Length in Feet | | | | |
| | | 2 Wire, 1 Phase | | 3 Wire, 3 Phase | | |
		120V	240V	240V	480V	600V
15	14*	50	105	120	240	300
	14	50	100	120	235	295
20	12*	60	125	145	290	360
	12	60	120	140	280	350
30	10*	65	130	155	305	380
	10	65	130	150	300	375
50	8	60	125	145	285	355
65	6	75	150	175	345	435
85	4	90	185	210	425	530
115	2	110	215	250	500	620
130	1	120	240	275	555	690
150	1/0	130	260	305	605	760
175	2/0	140	285	330	655	820
200	3/0	155	315	360	725	904
230	4/0	170	345	395	795	990
255	250	185	365	420	845	1055
285	300	195	395	455	910	1140
310	350	210	420	485	975	1220
380	500	245	490	565	1130	1415

*Solid Conductor

Note: The circuit length is the one-way distance between the origin and the load.

RD5010-141 Minimum Copper and Aluminum Wire Size Allowed for Various Types of Insulation

Minimum Wire Sizes

Amperes	Copper THW THWN or XHHW	Copper THHN XHHW *	Aluminum THW XHHW	Aluminum THHN XHHW *	Amperes	Copper THW THWN or XHHW	Copper THHN XHHW *	Aluminum THW XHHW	Aluminum THHN XHHW *
15A	#14	#14	#12	#12	195	3/0	2/0	250kcmil	4/0
20	#12	#12	#10	#10	200	3/0	3/0	250kcmil	4/0
25	#10	#10	#10	#10	205	4/0	3/0	250kcmil	4/0
30	#10	#10	#8	#8	225	4/0	3/0	300kcmil	250kcmil
40	#8	#8	#8	#8	230	4/0	4/0	300kcmil	250kcmil
45	#8	#8	#6	#8	250	250kcmil	4/0	350kcmil	300kcmil
50	#8	#8	#6	#6	255	250kcmil	4/0	400kcmil	300kcmil
55	#6	#8	#4	#6	260	300kcmil	4/0	400kcmil	350kcmil
60	#6	#6	#4	#6	270	300kcmil	250kcmil	400kcmil	350kcmil
65	#6	#6	#4	#4	280	300kcmil	250kcmil	500kcmil	350kcmil
75	#4	#6	#3	#4	285	300kcmil	250kcmil	500kcmil	400kcmil
85	#4	#4	#2	#3	290	350kcmil	250kcmil	500kcmil	400kcmil
90	#3	#4	#2	#2	305	350kcmil	300kcmil	500kcmil	400kcmil
95	#3	#4	#1	#2	310	350kcmil	300kcmil	500kcmil	500kcmil
100	#3	#3	#1	#2	320	400kcmil	300kcmil	600kcmil	500kcmil
110	#2	#3	1/0	#1	335	400kcmil	350kcmil	600kcmil	500kcmil
115	#2	#2	1/0	#1	340	500kcmil	350kcmil	600kcmil	500kcmil
120	#1	#2	1/0	1/0	350	500kcmil	350kcmil	700kcmil	500kcmil
130	#1	#2	2/0	1/0	375	500kcmil	400kcmil	700kcmil	600kcmil
135	1/0	#1	2/0	1/0	380	500kcmil	400kcmil	750kcmil	600kcmil
150	1/0	#1	3/0	2/0	385	600kcmil	500kcmil	750kcmil	600kcmil
155	2/0	1/0	3/0	3/0	420	600kcmil	500kcmil		700kcmil
170	2/0	1/0	4/0	3/0	430		500kcmil		750kcmil
175	2/0	2/0	4/0	3/0	435		600kcmil		750kcmil
180	3/0	2/0	4/0	4/0	475		600kcmil		

Dry Locations Only

Notes:
1. Size #14 to 4/0 is in AWG units (American Wire Gauge).
2. Size 250 to 750 is in kcmil units (Thousand Circular Mils).
3. Use next higher ampere value if exact value is not listed in table.
4. For loads that operate continuously increase ampere value by 25% to obtain proper wire size.
5. Refer to Table D5010-130 for the maximum circuit length for the various size wires.
6. Table D5010-141 has been written for estimating purpose only, based on ambient temperature of 30 °C (86°F); for ambient temperature other than 30 °C (86°F), ampacity correction factors will be applied.

Table D5010-142 Conductors in Conduit

Table below lists maximum number of conductors for various sized conduit using THW, TW or THWN insulations.

Copper Wire Size	1/2" TW	1/2" THW	1/2" THWN	3/4" TW	3/4" THW	3/4" THWN	1" TW	1" THW	1" THWN	1-1/4" TW	1-1/4" THW	1-1/4" THWN	1-1/2" TW	1-1/2" THW	1-1/2" THWN	2" TW	2" THW	2" THWN	2-1/2" TW	2-1/2" THW	2-1/2" THWN	3" THW	3" THWN	3-1/2" THW	3-1/2" THWN	4" THW	4" THWN
#14	9	6	13	15	10	24	25	16	39	44	29	69	60	40	94	99	65	154	142	93		143		192			
#12	7	4	10	12	8	18	19	13	29	35	24	51	47	32	70	78	53	114	111	76	164	117		157			
#10	5	4	6	9	6	11	15	11	18	26	19	32	36	26	44	60	43	73	85	61	104	95	160	127		163	
#8	2	1	3	4	3	5	7	5	9	12	10	16	17	13	22	28	22	36	40	32	51	49	79	66	106	85	136
#6		1	1		2	4		4	6		7	11		10	15		16	26		23	37	36	57	48	76	62	98
#4		1	1		1	2		3	4		5	7		7	9		12	16		17	22	27	35	36	47	47	60
#3		1	1		1	1		2	3		4	6		6	8		10	13		15	19	23	29	31	39	40	51
#2		1	1		1	1		2	3		4	5		5	7		9	11		13	16	20	25	27	33	34	43
#1					1	1		1	1		3	3		4	5		6	8		9	12	14	18	19	25	25	32
1/0					1	1		1	1		2	3		3	4		5	7		8	10	12	15	16	21	21	27
2/0					1	1		1	1		1	2		3	3		5	6		7	8	10	13	14	17	18	22
3/0					1	1		1	1		1	1		2	3		4	5		6	7	9	11	12	14	15	18
4/0						1		1	1		1	1		1	2		3	4		5	6	7	9	10	12	13	15
250 MCM								1	1		1	1		1	1		2	3		4	4	6	7	8	10	10	12
300								1	1		1	1		1	1		2	3		3	4	5	6	7	8	9	11
350									1		1	1		1	1		1	2		3	3	4	5	6	7	8	9
400											1	1		1	1		1	1		2	3	4	5	5	6	7	8
500											1	1		1	1		1	1		1	2	3	4	4	5	6	7
600												1		1	1		1	1		1	1	3	3	4	4	5	5
700														1	1		1	1		1	1	2	3	3	4	4	5
750														1	1		1	1		1	1	2	2	3	3	4	4

Table D5010-146A Metric Equivalent, Conduit

U.S. vs. European Conduit – Approximate Equivalents

United States Trade Size	United States Inside Diameter Inch/MM	European Trade Size	European Inside Diameter MM
1/2	.622/15.8	11	16.4
3/4	.824/20.9	16	19.9
1	1.049/26.6	21	25.5
1¼	1.380/35.0	29	34.2
1½	1.610/40.9	36	44.0
2	2.067/52.5	42	51.0
2½	2.469/62.7		
3	3.068/77.9		
3½	3.548/90.12		
4	4.026/102.3		
5	5.047/128.2		
6	6.065/154.1		

Table D5010-146B Metric Equivalent, Wire

U.S. vs. European Wire – Approximate Equivalents

United States Size AWG or MCM	United States Area Cir. Mils.(CM) MM²	European Size MM²	European Area Cir. Mils.
18	1620/.82	.75	1480
16	2580/1.30	1.0	1974
14	4110/2.08	1.5	2961
12	6530/3.30	2.5	4935
10	10,380/5.25	4	7896
8	16,510/8.36	6	11,844
6	26,240/13.29	10	19,740
4	41,740/21.14	16	31,584
3	52,620/26.65	25	49,350
2	66,360/33.61	–	–
1	83,690/42.39	35	69,090
1/0	105,600/53.49	50	98,700
2/0	133,100/67.42	–	–
3/0	167,800/85.00	70	138,180
4/0	211,600/107.19	95	187,530
250	250,000/126.64	120	236,880
300	300,000/151.97	150	296,100
350	350,000/177.30	–	–
400	400,000/202.63	185	365,190
500	500,000/253.29	240	473,760
600	600,000/303.95	300	592,200
700	700,000/354.60	–	–
750	750,000/379.93	–	–

SERVICES

REFERENCE TABLES

D5010-147 Concrete for Conduit Encasement

Table below lists C.Y. of concrete for 100 L.F. of trench. Conduits separation center to center should meet 7.5" (N.E.C.).

Number of Conduits	1	2	3	4	6	8	9	Number of Conduits
Trench Dimension	11.5" x 11.5"	11.5" x 19"	11.5" x 27"	19" x 19"	19" x 27"	19" x 38"	27" x 27"	Trench Dimension
Conduit Diameter 2.0"	3.29	5.39	7.64	8.83	12.51	17.66	17.72	Conduit Diameter 2.0"
2.5"	3.23	5.29	7.49	8.62	12.19	17.23	17.25	2.5"
3.0"	3.15	5.13	7.24	8.29	11.71	16.59	16.52	3.0"
3.5"	3.08	4.97	7.02	7.99	11.26	15.98	15.84	3.5"
4.0"	2.99	4.80	6.76	7.65	10.74	15.30	15.07	4.0"
5.0"	2.78	4.37	6.11	6.78	9.44	13.57	13.12	5.0"
6.0"	2.52	3.84	5.33	5.74	7.87	11.48	10.77	6.0"

Table D5010-148 Size Required and Weight (Lbs./1000 L.F.) of Aluminum and Copper THW Wire by Ampere Load

Amperes	Copper Size	Aluminum Size	Copper Weight	Aluminum Weight
15	14	12	24	11
20	12	10	33	17
30	10	8	48	39
45	8	6	77	52
65	6	4	112	72
85	4	2	167	101
100	3	1	205	136
115	2	1/0	252	162
130	1	2/0	324	194
150	1/0	3/0	397	233
175	2/0	4/0	491	282
200	3/0	250	608	347
230	4/0	300	753	403
255	250	400	899	512
285	300	500	1068	620
310	350	500	1233	620
335	400	600	1396	772
380	500	750	1732	951

Table D5010-150 Transformer Weight (Lbs.) by kVA

Oil Filled 3 Phase 5/15 kV To 480/277			
kVA	Lbs.	kVA	Lbs.
150	1800	1000	6200
300	2900	1500	8400
500	4700	2000	9700
750	5300	3000	15000

Dry 240/480 To 120/240 Volt			
1 Phase		3 Phase	
kVA	Lbs.	kVA	Lbs.
1	23	3	90
2	36	6	135
3	59	9	170
5	73	15	220
7.5	131	30	310
10	149	45	400
15	205	75	600
25	255	112.5	950
37.5	295	150	1140
50	340	225	1575
75	550	300	1870
100	670	500	2850
167	900	750	4300

Table D5010-149 Weight (Lbs./L.F.) of 4 Pole Aluminum and Copper Bus Duct by Ampere Load

Amperes	Aluminum Feeder	Copper Feeder	Aluminum Plug-In	Copper Plug-In
225			7	7
400			8	13
600	10	10	11	14
800	10	19	13	18
1000	11	19	16	22
1350	14	24	20	30
1600	17	26	25	39
2000	19	30	29	46
2500	27	43	36	56
3000	30	48	42	73
4000	39	67		
5000		78		

Table D5010-151 Generator Weight (Lbs.) by kW

3 Phase 4 Wire 277/480 Volt			
Gas		Diesel	
kW	Lbs.	kW	Lbs.
7.5	600	30	1800
10	630	50	2230
15	960	75	2250
30	1500	100	3840
65	2350	125	4030
85	2570	150	5500
115	4310	175	5650
170	6530	200	5930
		250	6320
		300	7840
		350	8220
		400	10750
		500	11900

Table D5010-152A   Conduit Weight Comparisons (Lbs. per 100 ft.) Empty

Type	1/2"	3/4"	1"	1-1/4"	1-1/2"	2"	2-1/2"	3"	3-1/2"	4"	5"	6"
Rigid Aluminum	28	37	55	72	89	119	188	246	296	350	479	630
Rigid Steel	79	105	153	201	249	332	527	683	831	972	1314	1745
Intermediate Steel (IMC)	60	82	116	150	182	242	401	493	573	638		
Electrical Metallic Tubing (EMT)	29	45	65	96	111	141	215	260	365	390		
Polyvinyl Chloride, Schedule 40	16	22	32	43	52	69	109	142	170	202	271	350
Polyvinyl Chloride Encased Burial						38		67	88	105	149	202
Fibre Duct Encased Burial						127		164	180	206	400	511
Fibre Duct Direct Burial						150		251	300	354		
Transite Encased Burial						160		240	290	330	450	550
Transite Direct Burial						220		310		400	540	640

Table D5010-152B   Conduit Weight Comparisons (Lbs. per 100 ft.) with Maximum Cable Fill*

Type	1/2"	3/4"	1"	1-1/4"	1-1/2"	2"	2-1/2"	3"	3-1/2"	4"	5"	6"
Rigid Galvanized Steel (RGS)	104	140	235	358	455	721	1022	1451	1749	2148	3083	4343
Intermediate Steel (IMC)	84	113	186	293	379	611	883	1263	1501	1830		
Electrical Metallic Tubing (EMT)	54	116	183	296	368	445	641	930	1215	1540		

*Conduit & Heaviest Conductor Combination

D5010-153   Weight Comparisons of Common Size Cast Boxes in Lbs.

Size NEMA 4 or 9	Cast Iron	Cast Aluminum	Size NEMA 7	Cast Iron	Cast Aluminum
6" x 6" x 6"	17	7	6" x 6" x 6"	40	15
8" x 6" x 6"	21	8	8" x 6" x 6"	50	19
10" x 6" x 6"	23	9	10" x 6" x 6"	55	21
12" x 12" x 6"	52	20	12" x 6" x 6"	100	37
16" x 16" x 6"	97	36	16" x 16" x 6"	140	52
20" x 20" x 6"	133	50	20" x 20" x 6"	180	67
24" x 18" x 8"	149	56	24" x 18" x 8"	250	93
24" x 24" x 10"	238	88	24" x 24" x 10"	358	133
30" x 24" x 12"	324	120	30" x 24" x 10"	475	176
36" x 36" x 12"	500	185	30" x 24" x 12"	510	189

General: Lighting and electric heating loads are expressed in watts and kilowatts.

Cost Determination:

The proper ampere values can be obtained as follows:

1. Convert watts to kilowatts
   (watts 1000 ÷ kilowatts)
2. Determine voltage rating of equipment.
3. Determine whether equipment is single phase or three phase.
4. Refer to Table D5010-158 to find ampere value from kW, Ton and Btu/hr. values.
5. Determine type of wire insulation – TW, THW, THWN.
6. Determine if wire is copper or aluminum.
7. Refer to Table D5010-141 to obtain copper or aluminum wire size from ampere values.

Notes:

1. Phase refers to single phase, 2 wire circuits.
2. Phase refers to three phase, 3 wire circuits.
3. For circuits which operate continuously for 3 hours or more, multiply the ampere values by 1.25 for a given kW requirement.
4. For kW ratings not listed, add ampere values.

8. Next refer to Table D5010-142 for the proper conduit size to accommodate the number and size of wires in each particular case.
9. Next refer to unit price data for the per linear foot cost of the conduit.
10. Next refer to unit price data for the per linear foot cost of the wire. Multiply cost of wire per L.F. x number of wires in the circuits to obtain total wire cost per L.F.
11. Add values obtained in Step 9 and 10 for total cost per linear foot for conduit and wire x length of circuit = Total Cost.

For example: Find the ampere value of
9 kW at 208 volt, single phase.

$$\begin{array}{l} 4\ kW = 19.2A \\ 5\ kW = 24.0A \\ \hline 9\ kW = 43.2A \end{array}$$

5. "Length of Circuit" refers to the one way distance of the run, not to the total sum of wire lengths.

Table D5010-158 Ampere Values as Determined by kW Requirements, BTU/HR or Ton, Voltage and Phase Values

			120V	208V		240V		277V	480V
			1 Phase	1 Phase	3 Phase	1 Phase	3 Phase	1 Phase	3 Phase
kW	Ton	BTU/HR							
0.5	.1422	1,707	4.2A	2.4A	1.4A	2.1A	1.2A	1.8A	0.6A
0.75	.2133	2,560	6.2	3.6	2.1	3.1	1.9	2.7	.9
1.0	.2844	3,413	8.3	4.9	2.8	4.2	2.4	3.6	1.2
1.25	.3555	4,266	10.4	6.0	3.5	5.2	3.0	4.5	1.5
1.5	.4266	5,120	12.5	7.2	4.2	6.3	3.1	5.4	1.8
2.0	.5688	6,826	16.6	9.7	5.6	8.3	4.8	7.2	2.4
2.5	.7110	8,533	20.8	12.0	7.0	10.4	6.1	9.1	3.1
3.0	.8532	10,239	25.0	14.4	8.4	12.5	7.2	10.8	3.6
4.0	1.1376	13,652	33.4	19.2	11.1	16.7	9.6	14.4	4.8
5.0	1.4220	17,065	41.6	24.0	13.9	20.8	12.1	18.1	6.1
7.5	2.1331	25,598	62.4	36.0	20.8	31.2	18.8	27.0	9.0
10.0	2.8441	34,130	83.2	48.0	27.7	41.6	24.0	36.5	12.0
12.5	3.5552	42,663	104.2	60.1	35.0	52.1	30.0	45.1	15.0
15.0	4.2662	51,195	124.8	72.0	41.6	62.4	37.6	54.0	18.0
20.0	5.6883	68,260	166.4	96.0	55.4	83.2	48.0	73.0	24.0
25.0	7.1104	85,325	208.4	120.2	70.0	104.2	60.0	90.2	30.0
30.0	8.5325	102,390		144.0	83.2	124.8	75.2	108.0	36.0
35.0	9.9545	119,455		168.0	97.1	145.6	87.3	126.0	42.1
40.0	11.3766	136,520		192.0	110.8	166.4	96.0	146.0	48.0
45.0	12.7987	153,585			124.8	187.5	112.8	162.0	54.0
50.0	14.2208	170,650			140.0	208.4	120.0	180.4	60.0
60.0	17.0650	204,780			166.4		150.4	216.0	72.0
70.0	19.9091	238,910			194.2		174.6		84.2
80.0	22.7533	273,040			221.6		192.0		96.0
90.0	25.5975	307,170					225.6		108.0
100.0	28.4416	341,300							120.0

REFERENCE TABLES

495

General: Control transformers are listed in VA. Step-down and power transformers are listed in kVA.

Cost Determination:

1. Convert VA to kVA. Volt amperes (VA) ÷ 1000 = Kilovolt amperes (kVA).
2. Determine voltage rating of equipment.
3. Determine whether equipment is single phase or three phase.
4. Refer to Table D5010-161 to find ampere value from kVA value.
5. Determine type of wire insulation – TW, THW, THWN.
6. Determine if wire is copper or aluminum.
7. Refer to Table D5010-161 to obtain copper or aluminum wire size from ampere values.
8. Next refer to Table D5010-142 for the proper conduit size to accommodate the number and size of wires in each particular case.
9. Next refer to unit price data for the per linear foot cost of the conduit.
10. Next refer to unit price data for the per linear foot cost of the wire. Multiply cost of wire per L.F. x number of wires in the circuits to obtain total wire cost.
11. Add values obtained in Step 9 and 10 for total cost per linear foot for conduit and wire x length of circuit = Total Cost.

Example: A transformer rated 10 kVA 480 volts primary, 240 volts secondary, 3 phase has the capacity to furnish the following:

1. Primary amperes = 10 kVA x 1.20 = 12 amperes (from Table D5010-161)
2. Secondary amperes = 10 kVA x 2.40 = 24 amperes (from Table D5010-161)

Note: Transformers can deliver generally 125% of their rated kVA. For instance, a 10 kVA rated transformer can safely deliver 12.5 kVA.

Table D5010-161 Multiplier Values for kVA to Amperes Determined by Voltage and Phase Values

Volts	Multiplier for Circuits	
	2 Wire, 1 Phase	3 Wire, 3 Phase
115	8.70	
120	8.30	
230	4.30	2.51
240	4.16	2.40
200	5.00	2.89
208	4.80	2.77
265	3.77	2.18
277	3.60	2.08
460	2.17	1.26
480	2.08	1.20
575	1.74	1.00
600	1.66	0.96

General: Motors can be powered by any of the seven systems shown in Figure D5010-121 thru Figure D5010-127 provided the motor voltage characteristics are compatible with the power system characteristics.

Cost Determination:

Motor Amperes for the various size H.P. and voltage are listed in Table D5010-171. To find the amperes, locate the required H.P. rating and locate the amperes under the appropriate circuit characteristics.

For example:

A. 100 H.P., 3 phase, 460 volt motor = 124 amperes (Table D5010-171)

B. 10 H.P., 3 phase, 200 volt motor = 32.2 amperes (Table D5010-171)

Motor Wire Size: After the amperes are found in Table D5010-171 the amperes must be increased 25% to compensate for power losses. Next refer to Table D5010-141. Find the appropriate insulation column for copper or aluminum wire to determine the proper wire size.

For example:

A. 100 H.P., 3 phase, 460 volt motor has an ampere value of 124 amperes from Table D5010-171

B. 124A x 1.25 = 155 amperes

C. Refer to Table D5010-141 for THW or THWN wire insulations to find the proper wire size. For a 155 ampere load using copper wire a size 2/0 wire is needed.

D. For the 3 phase motor three wires of 2/0 size are required.

Conduit Size: To obtain the proper conduit size for the wires and type of insulation used, refer to Table D5010-142.

For example: For the 100 H.P., 460V, 3 phase motor, it was determined that three 2/0 wires are required. Assuming THWN insulated copper wire, use Table D5010-142 to determine that three 2/0 wires require 1-1/2" conduit.

Material Cost of the conduit and wire system depends on:

1. Wire size required
2. Copper or aluminum wire
3. Wire insulation type selected
4. Steel or plastic conduit
5. Type of conduit raceway selected.

Labor Cost of the conduit and wire system depends on:

1. Type and size of conduit
2. Type and size of wires installed
3. Location and height of installation in building or depth of trench
4. Support system for conduit.

Table D5010-171 Ampere Values Determined by Horsepower, Voltage and Phase Values

H.P.	Single Phase		Three Phase			
	115V	230V	200 V	230V	460V	575V
1/6	4.4A	2.2A				
1/4	5.8	2.9				
1/3	7.2	3.6				
1/2	9.8	4.9	2.3A	2.0A	1.0A	0.8A
3/4	13.8	6.9	3.2	2.8	1.4	1.1
1	16	8	4.1	3.6	1.8	1.4
1-1/2	20	10	6.0	5.2	2.6	2.1
2	24	12	7.8	6.8	3.4	2.7
3	34	17	11.0	9.6	4.8	3.9
5			17.5	15.2	7.6	6.1
7-1/2			25.3	22	11	9
10			32.2	28	14	11
15			48.3	42	21	17
20			62.1	54	27	22
25			78.2	68	34	27
30			92.0	80	40	32
40			119.6	104	52	41
50			149.5	130	65	52
60			177	154	77	62
75			221	192	96	77
100			285	248	124	99
125			359	312	156	125
150			414	360	180	144
200			552	480	240	192

Cost Determination (cont.)

Magnetic starters, switches, and motor connection:

To complete the cost picture from H.P. to Costs additional items must be added to the cost of the conduit and wire system to arrive at a total cost.

1. Assembly D5020 160 Magnetic Starters Installed Cost lists the various size starters for single phase and three phase motors.
2. Assembly D5020 165 Heavy Duty Safety Switches Installed Cost lists safety switches required at the beginning of a motor circuit and also one required in the vicinity of the motor location.
3. Assembly D5020 170 Motor Connection lists the various costs for single and three phase motors.

Worksheet to obtain total motor wiring costs:

It is assumed that the motors or motor driven equipment are furnished an installed under other sections for this estimate and the following work is done under this section:

1. Conduit
2. Wire (add 10% for additional wire beyond conduit ends for connection to switches, boxes, starters, etc.)
3. Starters
4. Safety switches
5. Motor connections

Table D5010-172

Worksheet for Motor Circuits				Cost	
Item	Type	Size	Quantity	Unit	Total
Wire					
Conduit					
Switch					
Starter					
Switch					
Motor Connection					
Other					
Total Cost					

Table D5010-173 Maximum Horsepower for Starter Size by Voltage

Starter	Maximum HP (3φ)			
Size	208V	240V	480V	600V
00	1½	1½	2	2
0	3	3	5	5
1	7½	7½	10	10
2	10	15	25	25
3	25	30	50	50
4	40	50	100	100
5		100	200	200
6		200	300	300
7		300	600	600
8		450	900	900
8L		700	1500	1500

General: The cost of the lighting portion of the electrical costs is dependent upon:
1. The footcandle requirement of the proposed building.
2. The type of fixtures required.
3. The ceiling heights of the building.
4. Reflectance value of ceilings, walls and floors.
5. Fixture efficiencies and spacing vs. mounting height ratios.

Footcandle Requirements: See Table D5020-204 for Footcandle and Watts per S.F. determination.

Table D5020-201 IESNA* Recommended Illumination Levels in Footcandles

Commercial Buildings			Industrial Buildings		
Type	Description	Footcandles	Type	Description	Footcandles
Bank	Lobby	50	Assembly Areas	Rough bench & machine work	50
	Customer Areas	70		Medium bench & machine work	100
	Teller Stations	150		Fine bench & machine work	500
	Accounting Areas	150	Inspection Areas	Ordinary	50
Offices	Routine Work	100		Difficult	100
	Accounting	150		Highly Difficult	200
	Drafting	200	Material Handling	Loading	20
	Corridors, Halls, Washrooms	30		Stock Picking	30
Schools	Reading or Writing	70		Packing, Wrapping	50
	Drafting, Labs, Shops	100	Stairways	Service Areas	20
	Libraries	70	Washrooms	Service Areas	20
	Auditoriums, Assembly	15	Storage Areas	Inactive	5
	Auditoriums, Exhibition	30		Active, Rough, Bulky	10
Stores	Circulation Areas	30		Active, Medium	20
	Stock Rooms	30		Active, Fine	50
	Merchandise Areas, Service	100	Garages	Active Traffic Areas	20
	Self-Service Areas	200		Service & Repair	100

*IESNA - Illuminating Engineering Society of North America

Table D5020-202 General Lighting Loads by Occupancies

Type of Occupancy	Unit Load per S.F. (Watts)
Armories and Auditoriums	1
Banks	5
Barber Shops and Beauty Parlors	3
Churches	1
Clubs	2
Court Rooms	2
*Dwelling Units	3
Garages — Commercial (storage)	½
Hospitals	2
*Hotels and Motels, including apartment houses without provisions for cooking by tenants	2
Industrial Commercial (Loft) Buildings	2
Lodge Rooms	1½
Office Buildings	5
Restaurants	2
Schools	3
Stores	3
Warehouses (storage)	¼
*In any of the above occupancies except one-family dwellings and individual dwelling units of multi-family dwellings:	
Assembly Halls and Auditoriums	1
Halls, Corridors, Closets	½
Storage Spaces	¼

Table D5020-203 Lighting Limit (Connected Load) for Listed Occupancies: New Building Proposed Energy Conservation Guideline

Type of Use	Maximum Watts per S.F.
Interior	
Category A: Classrooms, office areas, automotive mechanical areas, museums, conference rooms, drafting rooms, clerical areas, laboratories, merchandising areas, kitchens, examining rooms, book stacks, athletic facilities.	3.00
Category B: Auditoriums, waiting areas, spectator areas, restrooms, dining areas, transportation terminals, working corridors in prisons and hospitals, book storage areas, active inventory storage, hospital bedrooms, hotel and motel bedrooms, enclosed shopping mall concourse areas, stairways.	1.00
Category C: Corridors, lobbies, elevators, inactive storage areas.	0.50
Category D: Indoor parking.	0.25
Exterior	
Category E: Building perimeter: wall-wash, facade, canopy.	5.00 (per linear foot)
Category F: Outdoor parking.	0.10

Table D5020-204   Procedure for Calculating Footcandles and Watts Per Square Foot

1. Initial footcandles = No. of fixtures × lamps per fixture × lumens per lamp × coefficient of utilization ÷ square feet
2. Maintained footcandles = initial footcandles × maintenance factor
3. Watts per square foot = No. of fixtures × lamps × (lamp watts + ballast watts) ÷ square feet

Example: To find footcandles and watts per S.F. for an office 20' x 20' with 11 fluorescent fixtures each having 4–40 watt C.W. lamps.

Based on good reflectance and clean conditions:

Lumens per lamp = 40 watt cool white at 3150 lumens per lamp RD5020-250, Table RD5020-251

Coefficient of utilization = .42 (varies from .62 for light colored areas to .27 for dark)

Maintenance factor = .75 (varies from .80 for clean areas with good maintenance to .50 for poor)

Ballast loss = 8 watts per lamp. (Varies with manufacturer. See manufacturers' catalog.)

1. Initial footcandles:

$$\frac{11 \times 4 \times 3150 \times .42}{400} = \frac{58,212}{400} = 145 \text{ footcandles}$$

2. Maintained footcandles:

$$145 \times .75 = 109 \text{ footcandles}$$

3. Watts per S.F.

$$\frac{11 \times 4 (40 + 8)}{400} = \frac{2,112}{400} = 5.3 \text{ watts per S.F.}$$

Table D5020-205   Approximate Watts Per Square Foot for Popular Fixture Types

Due to the many variables involved, use for preliminary estimating only:

  a. Fluorescent – industrial System D5020 208
  b. Fluorescent – lens unit System D5020 208 Fixture types B & C
  c. Fluorescent – louvered unit
  d. Incandescent – open reflector System D5020 214, Type D
  e. Incandescent – lens unit System D5020 214, Type A
  f. Incandescent – down light System D5020 214, Type B

REFERENCE TABLES

Table D5020-241   Comparison - Cost of Operation of High Intensity Discharge Lamps

Lamp Type	Wattage	Life (Hours)	1 Circuit Wattage	Average Initial Lumens	2 L.L.D.	3 Mean Lumens
M.V.	100 DX	24,000	125	4,000	61%	2,440
L.P.S.	SOX-35	18,000	65	4,800	100%	4,800
H.P.S.	LU-70	12,000	84	5,800	90%	5,220
M.H.	No Equivalent					
M.V.	175 DX	24,000	210	8,500	66%	5,676
M.V.	250 DX	24,000	295	13,000	66%	7,986
L.P.S.	SOX-55	18,000	82	8,000	100%	8,000
H.P.S.	LU-100	12,000	120	9,500	90%	8,550
M.H.	No Equivalent					
M.V.	400 DX	24,000	465	24,000	64%	14,400
L.P.S.	SOX-90	18,000	141	13,500	100%	13,500
H.P.S.	LU-150	16,000	188	16,000	90%	14,400
M.H.	MH-175	7,500	210	14,000	73%	10,200
M.V.	No Equivalent					
L.P.S.	SOX-135	18,000	147	22,500	100%	22,500
H.P.S.	LU-250	20,000	310	25,500	92%	23,205
M.H.	MH-250	7,500	295	20,500	78%	16,000
M.V.	1000 DX	24,000	1,085	63,000	61%	37,820
L.P.S.	SOX-180	18,000	248	33,000	100%	33,000
H.P.S.	LU-400	20,000	480	50,000	90%	45,000
M.H.	MH-400	15,000	465	34,000	72%	24,600
H.P.S.	LU-1000	15,000	1,100	140,000	91%	127,400

1. Includes ballast losses and average lamp watts
2. Lamp lumen depreciation (% of initial light output at 70% rated life)
3. Lamp lumen output at 70% rated life (L.L.D. x initial)

M.V.   = Mercury Vapor

L.P.S. = Low pressure sodium

H.P.S. = High pressure sodium

L.H.   = Metal halide

Table D5020-242   For Other than Regular Cool White (CW) Lamps

Multiply Material Costs as Follows:					
Regular Lamps	Cool white deluxe (CWX)	x 1.35	Energy Saving Lamps	Cool white (CW/ES)	x 1.35
	Warm white deluxe (WWX)	x 1.35		Cool white deluxe (CWX/ES)	x 1.65
	Warm white (WW)	x 1.30		Warm white (WW/ES)	x 1.55
	Natural (N)	x 2.05		Warm white deluxe (WWX/ES)	x 1.65

D50 Electrical   RD5020-250   Lighting

Table D5020-251   Lamp Comparison Chart with Enclosed Floodlight, Ballast, & Lamp for Pole

Type	Watts	Initial Lumens	Lumens per Watt	Lumens @ 40% Life	Life (Hours)
Incandescent	150	2,880	19	85	750
	300	6,360	21	84	750
	500	10,850	22	80	1,000
	1,000	23,740	24	80	1,000
	1,500	34,400	23	80	1,000
Tungsten	500	10,950	22	97	2,000
Halogen	1,500	35,800	24	97	2,000
Fluorescent	40	3,150	79	88	20,000
Cool	110	9,200	84	87	12,000
White	215	16,000	74	81	12,000
Deluxe	250	12,100	48	86	24,000
Mercury	400	22,500	56	85	24,000
	1,000	63,000	63	75	24,000
Metal	175	14,000	80	77	7,500
Halide	400	34,000	85	75	15,000
	1,000	100,000	100	83	10,000
	1,500	155,000	103	92	1,500
High	70	5,800	83	90	20,000
Pressure	100	9,500	95	90	20,000
Sodium	150	16,000	107	90	24,000
	400	50,000	125	90	24,000
	1,000	140,000	140	90	24,000
Low	55	4,600	131	98	18,000
Pressure	90	12,750	142	98	18,000
Sodium	180	33,000	183	98	18,000

Color: High Pressure Sodium — Slightly Yellow
   Low Pressure Sodium — Yellow
   Mercury Vapor — Green-Blue
   Metal Halide — Blue White

Note: Pole not included.

D50 Electrical   RD5020-600   Air Conditioning

Table D5020-602   Central Air Conditioning Watts per S.F., BTU's per Hour per S.F. of Floor Area and S.F. per Ton of Air Conditioning

Type Building	Watts per S.F.	BTUH per S.F.	S.F. per Ton	Type Building	Watts per S.F.	BTUH per S.F.	S.F. per Ton	Type Building	Watts per S.F.	BTUH per S.F.	S.F. per Ton
Apartments, Individual	3	26	450	Dormitory, Rooms	4.5	40	300	Libraries	5.7	50	240
Corridors	2.5	22	550	Corridors	3.4	30	400	Low Rise Office, Ext.	4.3	38	320
Auditoriums & Theaters	3.3	40	300/18*	Dress Shops	4.9	43	280	Interior	3.8	33	360
Banks	5.7	50	240	Drug Stores	9	80	150	Medical Centers	3.2	28	425
Barber Shops	5.5	48	250	Factories	4.5	40	300	Motels	3.2	28	425
Bars & Taverns	15	133	90	High Rise Off.-Ext. Rms.	5.2	46	263	Office (small suite)	4.9	43	280
Beauty Parlors	7.6	66	180	Interior Rooms	4.2	37	325	Post Office, Int. Office	4.9	42	285
Bowling Alleys	7.8	68	175	Hospitals, Core	4.9	43	280	Central Area	5.3	46	260
Churches	3.3	36	330/20*	Perimeter	5.3	46	260	Residences	2.3	20	600
Cocktail Lounges	7.8	68	175	Hotels, Guest Rooms	5	44	275	Restaurants	6.8	60	200
Computer Rooms	16	141	85	Public Spaces	6.2	55	220	Schools & Colleges	5.3	46	260
Dental Offices	6	52	230	Corridors	3.4	30	400	Shoe Stores	6.2	55	220
Dept. Stores, Basement	4	34	350	Industrial Plants, Offices	4.3	38	320	Shop'g. Ctrs., Sup. Mkts.	4	34	350
Main Floor	4.5	40	300	General Offices	4	34	350	Retail Stores	5.5	48	250
Upper Floor	3.4	30	400	Plant Areas	4.5	40	300	Specialty Shops	6.8	60	200

* Persons per ton   12,000 BTUH = 1 ton of air conditioning

For information about Means Estimating Seminars, see yellow pages 12 and 13 in back of book   503

The Building Site Work Division is divided into five sections:

G10 Site Preparation
G20 Site Improvements
G30 Site Mechanical Utilities
G40 Site Electrical Utilities
G90 Other Site Construction

Clear and Grub, Demolition and Landscaping must be added to these costs as they apply to individual sites.

The Site Preparation section includes typical bulk excavation, trench excavation, and backfill. To this must be added sheeting, dewatering, drilling and blasting, traffic control, and any unusual items relating to a particular site.

The Site Improvements section, including Parking Lot and Roadway construction, has quantities and prices for typical designs with gravel base and asphaltic concrete pavement. Different climate, soil conditions, material availability, and owner's requirements can easily be adapted to these parameters.

The Site Electrical Utilities sections tabulate the installed price for pipe. Excavation and backfill costs for the pipe are derived from the Site Preparation section.

By following the examples and illustrations of typical site work conditions in this division and adapting them to your particular site, an accurate price, including O&P, can be determined.

Table G1010-011 Site Preparation

Description	Unit	Total Costs
Clear and Grub Average	Acre	$7,700
Bulk Excavation with Front End Loader	C.Y.	1.72
Spread and Compact Dumped Material	C.Y.	2.19
Fine Grade and Seed	S.Y.	2.59

Bulk excavation on building jobs today is done entirely by machine. Hand work is used only for squaring out corners, fine grading, cutting out trenches and footings below the general excavating grade, and making minor miscellaneous cuts. Hand trimming of material other than sand and gravel is generally done with picks and shovels.

The following figures on machine excavating apply to basement pits, large footings and other excavation for buildings. On grading excavation this performance can be bettered by 33% to 50%. The figures presume loading directly into trucks. Allowance has been made for idle time due to truck delay and moving about on the job. Figures include no water problems or slow down due to sheeting and bracing operations. They also do not include truck spotters or labor for cleanup.

Table G1030-401 Bulk Excavation Cost, 15' Deep

1½ C.Y. Hydraulic Excavator Backhoe, Four Tandem Trucks with a Two Mile round Trip Haul Distance

Item			Daily Cost	Labor & Equipment	Unit Price (C.Y.)
Equipment operator @	$ 57.50	per hr. x 8 hrs.	$ 460.00		
Laborer @	$ 42.65	per hr. x 8 hrs.	341.20	$2,186.80 Labor	$3.04 Labor
Four truck drivers @	$ 43.30	per hr. x 8 hrs.	1,385.60		
1½ C.Y. backhoe @	$722.80	per day	$ 722.80	$2,629.20 Equipment	$3.65 Equipment
Four 16 ton trucks @	$476.60	per day	1,906.40		
Production = 720 C.Y. per day			$4,816.00	$4,816.00 Labor & Equip.	$6.69 Labor & Equip.

Table G1030-402 Total Cost Per C.Y. Excavation for Various Sized Jobs

No. of C.Y. to be Excavated	Cost Per Cubic Yard	Mobilizing & Demobilizing Cost*	Total Cost Per Cubic Yard
500 C.Y.	$6.69	$1.22	$7.91
1,000		.61	7.30
1,500		.41	7.10
2,000		.31	7.00
3,000		.20	6.89
5,000		.12	6.81
10,000		.06	6.75
15,000		.04	6.73

*$305.00 for Mobilizing or Demobilizing 1½ C.Y. Hydraulic Excavator

Table G1030-403 Trench Bottom Widths for Various Outside Diameters of Buried Pipe

Outside Diameter	Trench Bottom Width
24 inches	4.1 feet
30	4.9
36	5.6
42	6.3
48	6.3
60	8.5
72	10.0
84	11.4

SINGLE UNIT

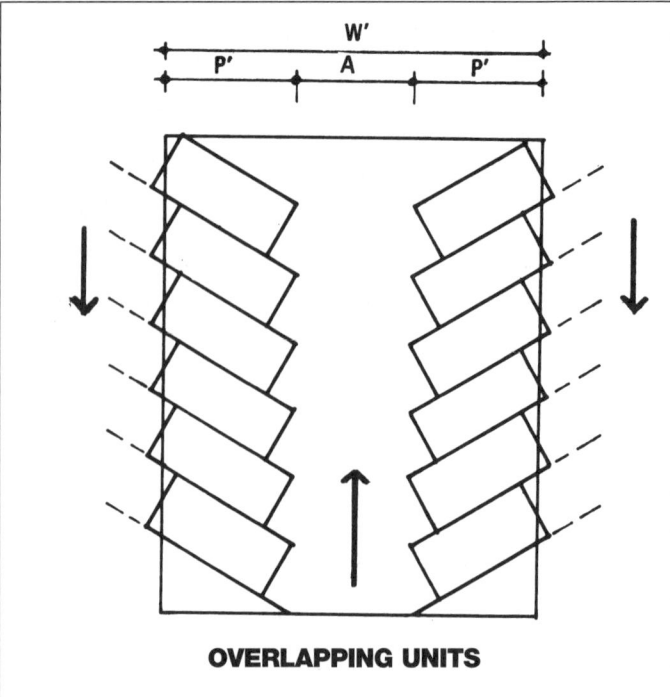

OVERLAPPING UNITS

Table G2020-502 Layout Data Based on 9' x 19' Parking Stall Size

φ	P	A	C	W	N	G	D	L	P'	W'
Angle of Stall	Parking Depth	Aisle Width	Curb Length	Width Overall	Net Car Area	Gross Car Area	Distance Last Car	Lost Area	Parking Depth	Width Overall
90°	19'.	24'	9'.	62'.	171 S.F.	171 S.F.	9'.	0	19'.	62'.
60°	21'.8	18'	10.4'	60'.	171 S.F.	217 S.F.	7.8'	205 S.F.	18.8'	55.5'
45°	19.8'	13'	12.8'	52.7'	171 S.F.	252 S.F.	6.4'	286 S.F.	16.6'	46.2'

Note: *Square foot per car areas do not include the area of the travel lane.*

90° Stall Angle: The main reason for use of this stall angle is to achieve the highest car capacity. This may be sound reasoning for employee lots with all day parking, but in most (in & out) lots there is difficulty in entering the stalls and no traffic lane direction. This may outweigh the advantage of high capacity.

60° Stall Angle: This layout is used most often due to the ease of entering and backing out, also the traffic aisle may be smaller.

45° Stall Angle: Requires a small change of direction from the traffic aisle to the stall, so the aisle may be reduced in width.

162'-6"

13 SPACES

CATCH BASIN

9 SPACES

5 SPACES

120'-0'

9 SPACES

MAN HOLE

CATCH BASIN

14 SPACES

PARKING LOT PLAN (50 CAR)

Preliminary Design Data: The space required for parking and maneuvering is between 300 and 400 S.F. per car, depending upon engineering layout and design.

Ninety degree (90°) parking, with a central driveway and two rows of parked cars, will provide the best economy.

Diagonal parking is easier than 90° for the driver and reduces the necessary driveway width, but requires more total space.

Table H1010-101 General Contractor's Overhead

There are two distinct types of overhead on a construction project: Project Overhead and Main Office Overhead. Project Overhead includes those costs at a construction site not directly associated with the installation of construction materials. Examples of Project Overhead costs include the following:

1. Superintendent
2. Construction office and storage trailers
3. Temporary sanitary facilities
4. Temporary utilities
5. Security fencing
6. Photographs
7. Clean up
8. Performance and payment bonds

The above Project Overhead items are also referred to as General Requirements and therefore are estimated in Division 1. Division 1 is the first division listed in the CSI MasterFormat but it is usually the last division estimated. The sum of the costs in Divisions 1 through 16 is referred to as the sum of the direct costs.

All construction projects also include indirect costs. The primary components of indirect costs are the contractor's Main Office Overhead and profit. The amount of the Main Office Overhead expense varies depending on the the following:

1. Owner's compensation
2. Project managers and estimator's wages
3. Clerical support wages
4. Office rent and utilities
5. Corporate legal and accounting costs
6. Advertising
7. Automobile expenses
8. Association dues
9. Travel and entertainment expenses

These costs are usually calculated as a percentage of annual sales volume. This percentage can range from 35% for a small contractor doing less than $500,000 to 5% for a large contractor with sales in excess of $100 million.

Table H1010-102 Main Office Expense

A General Contractor's main office expense consists of many items not detailed in the front portion of the book. The percentage of main office expense declines with increased annual volume of the contractor. Typical main office expense ranges from 2% to 20% with the median about 7.2% of total volume. This equals about 7.7% of direct costs. The following are approximate percentages of total overhead for different items usually included in a General Contractor's main office overhead. With different accounting procedures, these percentages may vary.

Item	Typical Range			Average
Managers', clerical and estimators' salaries	40 %	to	55 %	48%
Profit sharing, pension and bonus plans	2	to	20	12
Insurance	5	to	8	6
Estimating and project management (not including salaries)	5	to	9	7
Legal, accounting and data processing	0.5	to	5	3
Automobile and light truck expense	2	to	8	5
Depreciation of overhead capital expenditures	2	to	6	4
Maintenance of office equipment	0.1	to	1.5	1
Office rental	3	to	5	4
Utilities including phone and light	1	to	3	2
Miscellaneous	5	to	15	8
Total				100%

Table H1010-201   Architectural Fees

Tabulated below are typical percentage fees by project size, for good professional architectural service. Fees may vary from those listed depending upon degree of design difficulty and economic conditions in any particular area.

Rates can be interpolated horizontally and vertically. Various portions of the same project requiring different rates should be adjusted

proportionately. For alterations, add 50% to the fee for the first $500,000 of project cost and add 25% to the fee for project cost over $500,000.

Architectural fees tabulated below include Structural, Mechanical and Electrical Engineering Fees. They do not include the fees for special consultants such as kitchen planning, security, acoustical, interior design, etc.

Building Types	Total Project Size in Thousands of Dollars						
	100	250	500	1,000	5,000	10,000	50,000
Factories, garages, warehouses, repetitive housing	9.0%	8.0%	7.0%	6.2%	5.3%	4.9%	4.5%
Apartments, banks, schools, libraries, offices, municipal buildings	12.2	12.3	9.2	8.0	7.0	6.6	6.2
Churches, hospitals, homes, laboratories, museums, research	15.0	13.6	12.7	11.9	9.5	8.8	8.0
Memorials, monumental work, decorative furnishings	—	16.0	14.5	13.1	10.0	9.0	8.3

Table H1010-202   Engineering Fees

Typical **Structural Engineering Fees** based on type of construction and total project size. These fees are included in Architectural Fees.

Type of Construction	Total Project Size (in thousands of dollars)			
	$500	$500-$1,000	$1,000-$5,000	Over $5000
Industrial buildings, factories & warehouses	Technical payroll times 2.0 to 2.5	1.60%	1.25%	1.00%
Hotels, apartments, offices, dormitories, hospitals, public buildings, food stores		2.00%	1.70%	1.20%
Museums, banks, churches and cathedrals		2.00%	1.75%	1.25%
Thin shells, prestressed concrete, earthquake resistive		2.00%	1.75%	1.50%
Parking ramps, auditoriums, stadiums, convention halls, hangars & boiler houses		2.50%	2.00%	1.75%
Special buildings, major alterations, underpinning & future expansion	↓	Add to above 0.5%	Add to above 0.5%	Add to above 0.5%

For complex reinforced concrete or unusually complicated structures, add 20% to 50%.

Table H1010-203   Mechanical and Electrical Fees

Typical **Mechanical and Electrical Engineering Fees** based on the size of the subcontract. The fee structure for both are shown below. These fees are included in Architectural Fees.

Type of Construction	Subcontract Size							
	$25,000	$50,000	$100,000	$225,000	$350,000	$500,000	$750,000	$1,000,000
Simple structures	6.4%	5.7%	4.8%	4.5%	4.4%	4.3%	4.2%	4.1%
Intermediate structures	8.0	7.3	6.5	5.6	5.1	5.0	4.9	4.8
Complex structures	10.1	9.0	9.0	8.0	7.5	7.5	7.0	7.0

For renovations, add 15% to 25% to applicable fee.

Table H1010-204   Construction Time Requirements

Table at left is average construction time in months for different types of building projects. Table at right is the construction time in months for different size projects. Design time runs 25% to 40% of construction time.

Type Building	Construction Time	Project Value	Construction Time
Industrial Buildings	12 Months	Under $1,400,000	10 Months
Commercial Buildings	15 Months	Up to $3,800,000	15 Months
Research & Development	18 Months	Up to $19,000,000	21 Months
Institutional Buildings	20 Months	over $19,000,000	28 Months

Table H1010-301 Builder's Risk Insurance

Builder's Risk Insurance is insurance on a building during construction. Premiums are paid by the owner or the contractor. Blasting, collapse and underground insurance would raise total insurance costs above those listed. Floater policy for materials delivered to the job runs $.75 to $1.25 per $100 value. Contractor equipment insurance runs $.50 to $1.50 per $100 value. Insurance for miscellaneous tools to $1,500 value runs from $3.00 to $7.50 per $100 value.

Tabulated below are New England Builder's Risk insurance rates in dollars per $100 value for $1,000 deductible. For $25,000 deductible, rates can be reduced 13% to 34%. On contracts over $1,000,000, rates may be lower than those tabulated. Policies are written annually for the total completed value in place. For "all risk" insurance (excluding flood, earthquake and certain other perils) add $.025 to total rates below.

Coverage	Frame Construction (Class 1)			Brick Construction (Class 4)			Fire Resistive (Class 6)		
	Range		Average	Range		Average	Range		Average
Fire Insurance	$.350 to $.850		$.600	$.158 to $.189		$.174	$.052 to $.080		$.070
Extended Coverage	.115 to .200		.158	.080 to .105		.101	.081 to .105		.100
Vandalism	.012 to .016		.014	.008 to .011		.011	.008 to .011		.010
Total Annual Rate	$.477 to $1.066		$.772	$.246 to $.305		$.286	$.141 to $.196		$.180

Table H1010-302 Performance Bond

This table shows the cost of a Performance Bond for a construction job scheduled to be completed in 12 months. Add 1% of the premium cost per month for jobs requiring more than 12 months to complete. The rates are "standard" rates offered to contractors that the bonding company considers financially sound and capable of doing the work. Preferred rates

are offered by some bonding companies based upon financial strength of the contractor. Actual rates vary from contractor to contractor and from bonding company to bonding company. Contractors should prequalify through a bonding agency before submitting a bid on a contract that requires a bond.

Contract Amount	Building Construction Class B Projects	Highways & Bridges	
		Class A New Construction	Class A-1 Highway Resurfacing
First $ 100,000 bid	$25.00 per M	$15.00 per M	$9.40 per M
Next 400,000 bid	$ 2,500 plus $15.00 per M	$ 1,500 plus $10.00 per M	$ 940 plus $7.20 per M
Next 2,000,000 bid	8,500 plus 10.00 per M	5,500 plus 7.00 per M	3,820 plus 5.00 per M
Next 2,500,000 bid	28,500 plus 7.50 per M	19,500 plus 5.50 per M	15,820 plus 4.50 per M
Next 2,500,000 bid	47,250 plus 7.00 per M	33,250 plus 5.00 per M	28,320 plus 4.50 per M
Over 7,500,000 bid	64,750 plus 6.00 per M	45,750 plus 4.50 per M	39,570 plus 4.00 per M

Table H1010-401 Workers' Compensation Insurance Rates by Trade

The table below tabulates the national averages for Workers' Compensation insurance rates by trade and type of building. The average "Insurance Rate" is multiplied by the "% of Building Cost" for each trade. This produces the "Workers' Compensation Cost" by % of total labor cost, to be added for each trade by building type to determine the weighted average Workers' Compensation rate for the building types analyzed.

Trade	Insurance Rate (% Labor Cost)		% of Building Cost			Workers' Compensation		
	Range	Average	Office Bldgs.	Schools & Apts.	Mfg.	Office Bldgs.	Schools & Apts.	Mfg.
Excavation, Grading, etc.	4.2 % to 19.5%	10.6%	4.8%	4.9%	4.5%	.51%	.52%	.48%
Piles & Foundations	7.7 to 76.7	22.1	7.1	5.2	8.7	1.57	1.15	1.92
Concrete	5.2 to 35.8	15.8	5.0	14.8	3.7	.79	2.34	.58
Masonry	5.1 to 28.3	14.9	6.9	7.5	1.9	1.03	1.12	.28
Structural Steel	7.8 to 103.8	39.7	10.7	3.9	17.6	4.25	1.55	6.99
Miscellaneous & Ornamental Metals	4.9 to 24.8	12.2	2.8	4.0	3.6	.34	.49	.44
Carpentry & Millwork	7.0 to 53.2	18.4	3.7	4.0	0.5	.68	.74	.09
Metal or Composition Siding	5.3 to 32.1	17.3	2.3	0.3	4.3	.40	.05	.74
Roofing	7.8 to 78.9	32.4	2.3	2.6	3.1	.75	.84	1.00
Doors & Hardware	4.5 to 24.9	11.8	0.9	1.4	0.4	.11	.17	.05
Sash & Glazing	4.6 to 33.7	13.9	3.5	4.0	1.0	.49	.56	.14
Lath & Plaster	4.3 to 40.2	14.0	3.3	6.9	0.8	.46	.97	.11
Tile, Marble & Floors	2.4 to 31.8	9.5	2.6	3.0	0.5	.25	.29	.05
Acoustical Ceilings	2.4 to 29.7	11.4	2.4	0.2	0.3	.27	.02	.03
Painting	4.5 to 33.7	13.3	1.5	1.6	1.6	.20	.21	.21
Interior Partitions	7.0 to 53.2	18.4	3.9	4.3	4.4	.72	.79	.81
Miscellaneous Items	2.6 to 137.6	17.3	5.2	3.7	9.7	.90	.64	1.68
Elevators	2.8 to 42.8	6.9	2.1	1.1	2.2	.14	.08	.15
Sprinklers	3.2 to 22.1	8.6	0.5	—	2.0	.04	—	.17
Plumbing	2.7 to 12.4	8.2	4.9	7.2	5.2	.40	.59	.43
Heat., Vent., Air Conditioning	4.0 to 27.2	11.8	13.5	11.0	12.9	1.59	1.30	1.52
Electrical	2.6 to 12.7	6.8	10.1	8.4	11.1	.69	.57	.75
Total	2.4 % to 137.6%	—	100.0%	100.0%	100.0%	16.58%	14.99%	18.62%
		Overall Weighted Average 16.73%						

Table H1010-402 Workers' Compensation Insurance Rates by States

The table below lists the weighted average Workers' Compensation base rate for each state with a factor comparing this with the national average of 16.3%.

State	Weighted Average	Factor	State	Weighted Average	Factor	State	Weighted Average	Factor
Alabama	25.7%	158	Kentucky	20.5%	126	North Dakota	13.1%	80
Alaska	24.0	147	Louisiana	28.2	173	Ohio	14.5	89
Arizona	7.5	46	Maine	20.9	128	Oklahoma	14.7	90
Arkansas	13.7	84	Maryland	19.1	117	Oregon	13.0	80
California	18.8	115	Massachusetts	13.4	82	Pennsylvania	14.3	88
Colorado	10.9	67	Michigan	17.4	107	Rhode Island	21.1	129
Connecticut	22.4	137	Minnesota	25.5	156	South Carolina	16.8	103
Delaware	17.3	106	Mississippi	17.9	110	South Dakota	17.1	105
District of Columbia	17.1	105	Missouri	17.7	109	Tennessee	18.1	111
Florida	23.3	143	Montana	20.4	125	Texas	12.7	78
Georgia	22.9	140	Nebraska	22.9	140	Utah	12.3	75
Hawaii	16.9	104	Nevada	13.6	83	Vermont	25.0	153
Idaho	11.7	72	New Hampshire	22.6	139	Virginia	13.5	83
Illinois	18.8	115	New Jersey	12.2	75	Washington	10.9	67
Indiana	7.1	44	New Mexico	17.7	109	West Virginia	13.2	81
Iowa	12.3	75	New York	14.4	88	Wisconsin	14.7	90
Kansas	8.9	55	North Carolina	15.4	94	Wyoming	9.1	56
			Weighted Average for U.S. is 16.7% of payroll = 100%					

Rates in the following table are the base or manual costs per $100 of payroll for Workers' Compensation in each state. Rates are usually applied to straight time wages only and not to premium time wages and bonuses.

The weighted average skilled worker rate for 35 trades is 16.3%. For bidding purposes, apply the full value of Workers' Compensation directly to total labor costs, or if labor is 38%, materials 42% and overhead and profit 20% of total cost, carry 38/80 x 16.3% =7.7% of cost (before overhead and profit) into overhead. Rates vary not only from state to state but also with the experience rating of the contractor.

Rates are the most current available at the time of publication.

Table H1010-402 Workers' Compensation Insurance Rates by Trade and State (cont.)

State	Carpentry — 3 stories or less	Carpentry — interior cab. work	Carpentry — general	Concrete Work — NOC	Concrete Work — flat (flr., sdwk.)	Electrical Wiring — inside	Excavation — earth NOC	Excavation — rock	Glaziers	Insulation Work	Lathing	Masonry	Painting & Decorating	Pile Driving	Plastering	Plumbing	Roofing	Sheet Metal Work (HVAC)	Steel Erection — door & sash	Steel Erection — inter., ornam.	Steel Erection — structure	Steel Erection — NOC	Tile Work — (interior ceramic)	Waterproofing	Wrecking
	5651	5437	5403	5213	5221	5190	6217	6217	5462	5479	5443	5022	5474	6003	5480	5183	5551	5538	5102	5102	5040	5057	5348	9014	5701
AL	29.57	16.77	31.62	11.24	10.07	8.55	12.90	12.90	26.88	17.64	12.34	23.46	32.15	38.02	32.13	11.79	65.04	21.69	19.37	19.37	56.15	31.21	12.21	6.54	56.15
AK	21.71	15.70	15.84	13.63	12.88	13.24	20.07	20.07	26.84	31.05	11.29	31.31	21.16	51.97	26.78	12.29	41.90	14.82	12.85	12.85	50.64	25.20	9.35	9.63	50.64
AZ	6.88	5.05	11.62	5.56	3.49	3.85	4.74	4.74	7.40	7.69	4.27	5.37	4.87	9.83	5.20	3.90	12.19	5.65	9.50	9.50	18.98	8.11	2.04	2.34	50.31
AR	11.87	9.14	15.06	11.65	6.81	5.31	9.16	9.16	11.83	30.45	7.61	10.44	10.67	16.93	12.24	5.61	22.08	12.99	7.43	7.43	36.81	16.98	6.36	4.67	38.24
CA	28.28	9.07	28.28	13.18	13.18	9.93	7.88	7.88	16.37	23.34	10.90	14.55	19.45	21.18	18.12	11.59	39.79	15.33	13.55	13.55	23.93	21.91	7.74	19.45	21.91
CO	14.58	6.77	10.47	11.38	6.17	4.52	8.41	8.41	7.33	10.76	4.73	10.57	8.54	13.61	8.00	7.15	19.17	10.23	6.98	6.98	27.39	12.94	7.07	4.70	27.39
CT	18.69	17.64	27.54	26.02	12.53	8.70	10.59	10.59	16.86	19.70	19.27	24.01	18.88	25.92	24.21	10.86	43.51	13.48	16.02	16.02	61.01	29.58	9.39	6.44	50.86
DE	17.74	17.74	14.29	14.46	11.19	7.77	11.13	11.13	13.41	14.29	15.42	14.16	18.57	23.83	15.42	9.28	31.57	11.76	14.60	14.60	33.94	14.60	11.81	14.16	33.94
DC	12.26	11.99	12.28	14.08	16.31	7.27	9.99	9.99	23.64	10.47	9.64	19.37	10.29	17.95	12.01	14.00	18.86	8.55	17.12	17.12	46.07	20.00	28.91	4.70	46.07
FL	24.98	19.24	28.32	24.68	12.25	10.16	12.77	12.77	18.92	17.94	13.27	20.39	19.51	47.98	24.53	10.82	37.58	16.40	13.31	13.31	56.10	32.45	10.28	8.90	56.10
GA	32.05	17.69	24.16	15.16	10.44	8.97	16.86	16.86	15.69	22.02	17.16	20.30	17.13	31.16	19.94	10.77	46.74	17.52	13.97	13.97	44.46	48.01	9.64	8.55	44.46
HI	17.38	11.62	28.20	13.74	12.27	7.10	7.73	7.73	20.55	21.19	10.94	17.87	11.33	20.15	15.61	6.06	33.24	8.02	11.11	11.11	31.71	21.70	9.78	11.54	31.71
ID	11.92	6.31	13.47	10.64	5.99	5.25	6.90	6.90	9.58	8.14	6.31	9.91	8.88	15.41	9.45	5.99	29.54	9.11	7.94	7.94	33.81	14.13	7.27	4.09	33.81
IL	17.63	14.46	17.30	27.55	11.47	19.13	9.77	9.77	17.62	12.10	9.88	16.21	10.52	28.78	11.82	9.62	27.47	13.74	15.18	15.18	52.69	23.84	13.78	4.43	52.69
IN	5.29	14.84	6.93	5.29	13.20	2.58	4.34	4.34	4.67	4.81	2.49	5.20	5.25	8.82	4.02	2.73	11.74	4.35	5.20	5.20	18.17	8.49	3.31	2.82	18.17
IA	11.94	5.77	9.18	13.10	7.10	4.40	5.19	5.19	12.22	8.84	5.48	9.22	8.30	10.69	8.48	5.61	17.71	6.88	8.97	8.97	42.71	42.83	4.81	3.70	31.77
KS	9.72	8.65	10.24	7.86	5.92	3.32	4.15	4.15	7.75	7.72	4.21	8.13	6.79	10.74	7.35	5.65	18.03	6.03	7.26	7.26	22.40	11.92	5.40	3.31	23.40
KY	19.74	12.89	18.65	24.60	7.70	8.10	15.00	15.00	21.87	24.34	11.81	10.91	14.25	22.06	15.53	8.42	54.39	17.05	14.51	14.51	50.29	28.09	12.52	5.04	50.26
LA	23.14	24.93	53.17	26.43	15.61	9.88	17.49	17.49	20.14	21.43	24.51	28.33	29.58	31.25	22.37	8.64	77.12	21.20	18.98	18.98	51.76	24.21	13.77	13.78	66.41
ME	13.43	10.42	43.59	24.29	11.12	3.72	12.56	12.56	13.35	15.39	14.03	17.26	16.07	31.44	17.95	9.07	32.07	10.07	15.08	15.08	37.84	61.81	11.66	5.92	37.84
MD	13.71	5.95	10.55	18.82	8.70	5.15	12.91	12.91	25.52	20.07	11.37	13.31	9.37	26.22	14.21	8.58	48.72	12.74	13.66	13.66	56.25	37.81	8.03	7.14	26.80
MA	9.93	6.07	16.09	17.95	8.10	3.69	6.49	6.49	7.55	13.48	5.76	13.04	7.54	14.31	5.34	4.55	38.25	7.47	12.35	12.35	35.13	27.39	9.24	3.32	28.82
MI	19.68	11.72	21.53	19.00	8.83	5.94	9.96	9.96	11.76	13.31	13.59	17.06	13.34	36.96	15.69	7.35	33.24	9.51	9.92	9.92	36.96	27.43	10.97	5.38	39.06
MN	19.53	19.80	41.48	14.65	15.73	7.53	14.95	14.95	15.68	11.90	22.05	19.18	16.95	24.63	22.05	10.88	70.18	14.23	10.58	10.58	104.13	33.30	14.55	6.68	132.92
MS	16.68	14.87	24.71	12.67	9.14	5.85	12.50	12.50	13.10	15.77	8.50	11.88	13.12	19.91	28.30	9.82	38.94	20.61	10.87	10.87	32.50	30.23	10.01	6.49	32.50
MO	24.01	11.40	15.92	15.49	10.23	6.90	9.30	9.30	11.03	17.88	11.24	14.55	13.46	16.69	15.51	9.66	31.18	13.82	11.26	11.26	53.28	36.77	8.63	6.32	53.28
MT	21.18	10.79	19.52	12.88	10.39	6.82	19.69	19.69	13.67	20.72	22.29	15.50	10.31	60.73	12.40	10.22	45.28	12.09	12.45	12.45	40.26	22.93	6.26	5.14	40.26
NE	23.85	14.32	22.60	27.52	12.77	9.52	18.07	18.07	21.82	33.02	12.77	23.10	17.70	23.25	17.17	11.13	38.42	17.27	13.90	13.90	60.90	30.50	10.52	7.63	50.40
NV	20.81	8.10	10.79	9.56	7.96	6.16	10.76	10.76	12.49	10.61	9.45	9.87	9.00	16.26	9.36	7.27	18.09	21.51	10.59	10.59	24.87	29.06	6.55	5.43	4.09
NH	27.93	14.04	25.82	31.58	12.51	7.80	18.33	18.33	13.55	38.86	9.24	22.82	12.51	20.88	15.76	11.83	49.65	16.44	13.62	13.62	52.78	22.94	14.95	6.55	52.78
NJ	12.63	9.48	12.63	10.77	8.25	4.50	8.33	8.33	7.06	12.93	11.17	12.87	10.95	13.78	11.17	6.14	32.47	6.17	10.72	10.72	22.56	12.00	4.75	4.81	26.14
NM	29.00	8.32	20.65	18.60	11.34	7.16	10.46	10.46	14.57	12.61	8.59	12.43	13.86	21.32	11.32	8.92	30.32	10.81	19.90	19.90	40.88	27.34	8.55	7.65	40.88
NY	17.66	6.98	15.91	17.16	13.57	6.32	10.02	10.02	12.02	9.09	5.11	19.20	13.75	16.81	8.43	8.53	33.44	17.47	9.63	9.63	18.80	19.14	9.43	6.09	11.29
NC	16.52	12.25	14.99	13.22	7.32	8.85	9.45	9.45	11.16	11.72	10.68	9.98	10.17	15.13	14.39	8.39	24.40	12.94	7.40	7.40	68.39	20.27	6.05	4.83	68.39
ND	10.49	10.49	10.49	5.94	5.94	4.15	5.93	5.93	10.49	10.49	9.25	8.19	6.57	21.54	9.25	5.46	20.98	5.46	21.45	21.45	21.54	21.54	10.49	20.98	14.48
OH	7.48	9.85	9.98	11.81	11.78	6.16	9.02	9.02	8.65	14.76	38.58	12.85	13.30	23.94	3.76	7.06	24.80	9.48	10.09	10.09	31.53	15.16	9.91	5.99	31.53
OK	13.44	9.11	12.70	12.65	6.63	5.82	10.25	10.25	16.10	19.50	8.46	10.98	9.09	20.29	12.21	7.24	22.32	9.49	13.78	13.78	39.78	24.01	6.95	5.88	39.78
OR	16.72	9.16	15.83	14.23	8.06	4.50	9.48	9.48	14.13	8.79	7.41	14.91	11.59	15.81	11.05	5.30	22.92	9.68	9.53	9.53	29.02	11.69	10.63	4.52	29.02
PA	13.48	13.48	12.50	14.67	10.14	6.58	8.34	8.34	9.68	12.50	11.87	12.38	14.03	17.07	11.87	7.29	27.70	8.07	15.29	15.29	24.67	15.29	8.67	12.38	24.67
RI	19.53	11.65	18.07	18.23	16.24	4.43	10.38	10.38	12.85	22.78	11.97	25.11	24.13	37.66	17.25	8.52	33.92	10.29	14.07	14.07	59.49	37.50	14.36	7.70	78.79
SC	22.26	17.21	19.93	13.51	7.81	9.42	11.95	11.95	14.32	11.08	8.53	13.27	16.47	18.32	18.15	8.83	43.19	10.97	11.47	11.47	24.55	27.29	8.90	5.98	24.55
SD	21.00	17.87	17.14	19.81	7.46	5.25	15.58	15.58	11.54	12.77	8.94	7.73	12.16	23.97	11.77	10.03	22.29	11.34	11.11	11.11	62.70	23.72	7.24	4.06	62.70
TN	30.13	13.71	22.70	18.06	9.32	8.15	13.14	13.14	11.29	13.81	12.27	15.91	13.61	16.19	16.22	10.57	36.23	13.55	12.52	12.52	42.02	23.33	9.07	6.20	42.02
TX	13.47	9.95	13.47	11.11	8.00	7.76	8.79	8.79	9.83	12.47	6.78	12.04	8.63	16.18	8.63	7.35	19.58	16.38	9.69	9.69	31.53	13.20	6.81	7.37	14.13
UT	9.45	6.23	10.52	7.68	7.99	7.05	6.38	6.38	9.55	10.95	12.32	13.75	15.29	17.24	10.60	6.41	26.27	6.30	8.99	8.99	23.51	23.51	6.06	5.83	26.53
VT	24.65	11.60	22.01	36.23	15.66	8.18	13.75	13.75	25.11	29.18	12.53	23.26	13.86	23.35	18.10	11.77	42.54	15.20	15.81	15.81	72.09	57.39	9.47	12.14	72.09
VA	13.36	8.95	11.28	12.51	5.79	5.62	7.58	7.58	9.32	10.10	17.52	9.64	10.83	15.93	9.09	6.69	21.06	8.52	11.01	11.01	49.54	21.40	6.14	3.14	49.54
WA	10.29	10.29	10.29	7.89	7.89	3.28	8.11	8.11	12.02	10.01	10.29	9.82	9.08	18.17	11.05	4.63	18.54	4.64	12.96	12.96	8.65	8.65	10.03	10.23	8.65
WV	12.30	12.30	12.30	27.37	27.37	5.86	8.90	8.90	6.99	13.15	13.15	11.94	12.35	13.34	12.35	5.48	14.63	6.99	10.89	10.89	10.89	10.89	13.15	4.46	48.62
WI	12.40	11.36	16.86	12.29	9.90	4.24	7.44	7.44	13.31	11.62	8.09	18.99	14.32	13.64	12.11	5.42	34.72	7.16	9.18	9.18	37.88	24.06	13.99	5.52	37.88
WY	8.27	8.27	8.27	8.27	8.27	8.27	8.27	8.27	8.27	8.27	8.27	8.27	8.27	8.27	8.27	8.27	8.27	8.27	8.27	8.27	8.27	8.27	8.27	8.27	8.27
AVG.	17.27	11.81	18.39	15.82	10.29	6.84	10.63	10.63	13.87	15.76	11.44	14.92	13.29	22.07	14.00	8.22	32.40	11.76	12.19	12.19	39.65	24.35	9.52	7.04	40.06

Table H1010-404 Workers' Compensation (cont.) (Canada in Canadian dollars)

Province		Alberta	British Columbia	Manitoba	Ontario	New Brunswick	Newfndld. & Labrador	Northwest Territories	Nova Scotia	Prince Edward Island	Quebec	Saskat-chewan	Yukon
Carpentry—3 stories or less	Rate	9.71	7.19	4.57	4.62	4.83	9.78	4.61	7.67	6.04	15.41	7.33	5.58
	Code	42143	721028	40102	723	422	4226	4-41	4226	401	80110	B1317	202
Carpentry—interior cab. work	Rate	1.60	6.16	4.57	4.62	4.31	6.37	4.61	5.74	4.05	15.41	3.68	5.58
	Code	42133	721021	40102	723	427	4270	4-41	4274	402	80110	B11-27	202
CARPENTRY—general	Rate	9.71	7.19	4.57	4.62	4.83	6.37	4.61	7.67	6.04	15.41	7.33	5.58
	Code	42143	721028	40102	723	422	4299	4-41	4226	401	80110	B1317	202
CONCRETE WORK—NOC	Rate	5.96	7.53	7.60	15.25	4.83	9.78	4.61	4.83	6.04	16.51	7.33	5.58
	Code	42104	721010	40110	748	422	4224	4-41	4224	401	80100	B13-14	203
CONCRETE WORK—flat (flr. sidewalk)	Rate	5.96	7.53	7.60	15.25	4.83	9.78	4.61	4.83	6.04	16.51	7.33	5.58
	Code	42104	721010	40110	748	422	4224	4-41	4224	401	80100	B13-14	203
ELECTRICAL Wiring—inside	Rate	2.50	2.17	2.52	2.94	2.35	3.25	4.13	2.23	4.05	7.64	3.68	5.58
	Code	42124	721019	40203	704	426	4261	4-46	4261	402	80170	B11-05	206
EXCAVATION—earth NOC	Rate	3.31	3.85	3.84	4.20	3.16	5.53	3.71	4.11	4.25	8.36	4.37	5.58
	Code	40604	721031	40706	711	421	4214	4-43	4214	404	80030	R11-06	207
EXCAVATION—rock	Rate	3.31	3.85	3.84	4.20	3.16	5.53	3.71	4.11	4.25	8.36	4.37	5.58
	Code	40604	721031	40706	711	421	4214	4-43	4214	404	80030	R11-06	207
GLAZIERS	Rate	4.17	4.07	4.57	8.12	5.58	7.07	4.61	7.67	4.05	14.36	7.33	5.58
	Code	42121	715020	40109	751	423	4233	4-41	4233	402	80150	B13-04	212
INSULATION WORK	Rate	3.33	5.88	4.57	8.12	5.58	7.07	4.61	7.67	6.04	15.41	6.15	5.58
	Code	42184	721029	40102	751	423	4234	4-41	4234	401	80110	B12-07	202
LATHING	Rate	6.71	8.99	4.57	4.62	4.31	6.37	4.61	5.74	4.05	15.41	7.33	5.58
	Code	42135	721033	40102	723	427	4279	4-41	4271	402	80110	B13-16	202
MASONRY	Rate	5.96	8.99	4.57	11.44	5.58	7.07	4.61	7.67	6.04	16.51	7.33	5.58
	Code	42102	721037	40102	741	423	4231	4-41	4231	401	80100	B13-18	202
PAINTING & DECORATING	Rate	5.20	5.88	3.45	6.41	4.31	6.37	4.61	5.74	4.05	15.41	6.15	5.58
	Code	42111	721041	40105	719	427	4275	4-41	4275	402	80110	B12-01	202
PILE DRIVING	Rate	5.96	5.63	3.84	5.84	4.83	9.78	3.71	4.83	6.04	8.36	7.33	5.58
	Code	42159	722004	40706	732	422	4221	4-43	4221	401	80030	B13-10	202
PLASTERING	Rate	6.71	8.99	5.13	6.41	4.31	6.37	4.61	5.74	4.05	15.41	6.15	5.58
	Code	42135	721042	40108	719	427	4271	4-41	4271	402	80110	B12-21	202
PLUMBING	Rate	2.50	4.67	2.88	3.67	3.09	3.87	4.13	2.23	4.05	7.61	3.68	5.58
	Code	42122	721043	40204	707	424	4241	4-46	4241	402	80160	B11-01	214
ROOFING	Rate	9.88	10.10	6.52	11.60	10.25	9.78	4.61	9.49	6.04	22.49	7.33	5.58
	Code	42118	721036	40403	728	430	4236	4-41	4236	401	80130	B13-20	202
SHEET METAL WORK (HVAC)	Rate	2.50	4.67	6.52	3.67	3.09	3.87	4.13	3.23	4.05	7.61	3.68	7.39
	Code	42117	721043	40402	707	424	4244	4-46	4244	402	80160	B11-07	208
STEEL ERECTION—door & sash	Rate	3.33	13.44	9.89	15.25	4.83	9.78	4.61	7.67	6.04	29.92	7.33	5.58
	Code	42106	722005	40502	748	422	4227	4-41	4227	401	80080	B13-22	202
STEEL ERECTION—inter., ornam.	Rate	3.33	13.44	9.89	15.25	4.83	9.78	4.61	7.67	6.04	29.92	7.33	5.58
	Code	42106	722005	40502	748	422	4227	4-41	4227	401	80080	B13-22	202
STEEL ERECTION—structure	Rate	3.33	13.44	9.89	15.25	4.83	9.78	4.61	7.67	6.04	29.92	7.33	5.58
	Code	42106	722005	40502	748	422	4227	4-41	4227	401	80080	B13-22	202
STEEL ERECTION—NOC	Rate	3.33	13.44	9.89	15.25	4.83	9.78	4.61	7.67	6.04	29.92	7.33	5.58
	Code	42106	722005	40502	748	422	4227	4-41	4227	401	80080	B13-22	202
TILE WORK—inter. (ceramic)	Rate	4.62	5.67	2.17	6.41	4.31	6.37	4.61	5.74	4.05	15.41	7.33	5.58
	Code	42113	721054	40103	719	427	4276	4-41	4276	402	80110	B13-01	202
WATERPROOFING	Rate	5.20	5.88	4.57	4.62	5.58	6.37	4.61	7.67	4.05	22.49	6.15	5.58
	Code	42139	721016	40102	723	423	4299	4-41	4239	402	80130	B12-17	202
WRECKING	Rate	3.31	5.91	6.46	15.25	3.16	5.53	3.71	4.11	6.04	15.41	7.33	5.58
	Code	40604	721005	40106	748	421	4211	4-43	4211	401	80110	B13-09	202

GENERAL CONDITIONS & PROFIT

REFERENCE TABLES

| General Conditions | RH1020-300 | Overhead & Miscellaneous Data |

Unemployment Taxes and Social Security Taxes

State Unemployment Tax rates vary not only from state to state, but also with the experience rating of the contractor. The Federal Unemployment Tax rate is 6.2% of the first $7,000 of wages. This is reduced by a credit of up to 5.4% for timely payment to the state. The minimum Federal Unemployment Tax is 0.8% after all credits.

Social Security (FICA) for 2006 is estimated at time of publication to be 7.65% of wages up to $90,000.

| General Conditions | RH1020-400 | Overtime |

Overtime

One way to improve the completion date of a project or eliminate negative float from a schedule is to compress activity duration times. This can be achieved by increasing the crew size or working overtime with the proposed crew.

To determine the costs of working overtime to compress activity duration

times, consider the following examples. Below is an overtime efficiency and cost chart based on a five, six, or seven day week with an eight through twelve hour day. Payroll percentage increases for time and one half and double time are shown for the various working days.

| Days per Week | Hours per Day | Production Efficiency | | | | | Payroll Cost Factors | |
		1 Week	2 Weeks	3 Weeks	4 Weeks	Average 4 Weeks	@ 1-1/2 Times	@ 2 Times
	8	100%	100%	100%	100%	100 %	100 %	100 %
	9	100	100	95	90	96.25	105.6	111.1
5	10	100	95	90	85	91.25	110.0	120.0
	11	95	90	75	65	81.25	113.6	127.3
	12	90	85	70	60	76.25	116.7	133.3
	8	100	100	95	90	96.25	108.3	116.7
	9	100	95	90	85	92.50	113.0	125.9
6	10	95	90	85	80	87.50	116.7	133.3
	11	95	85	70	65	78.75	119.7	139.4
	12	90	80	65	60	73.75	122.2	144.4
	8	100	95	85	75	88.75	114.3	128.6
	9	95	90	80	70	83.75	118.3	136.5
7	10	90	85	75	65	78.75	121.4	142.9
	11	85	80	65	60	72.50	124.0	148.1
	12	85	75	60	55	68.75	126.2	152.4

| General Conditions | RH1030-100 | General |

Sales Tax by State

State sales tax on materials is tabulated below (5 states have no sales tax). Many states allow local jurisdictions, such as a county or city, to levy additional sales tax.

Some projects may be sales tax exempt, particularly those constructed with public funds.

State	Tax (%)	State	Tax (%)	State	Tax (%)	State	Tax (%)
Alabama	4	Illinois	6.25	Montana	0	Rhode Island	7
Alaska	0	Indiana	6	Nebraska	5.5	South Carolina	5
Arizona	5.6	Iowa	5	Nevada	6.5	South Dakota	4
Arkansas	6	Kansas	5.3	New Hampshire	0	Tennessee	7
California	6.25	Kentucky	6	New Jersey	6	Texas	6.25
Colorado	2.9	Louisiana	4	New Mexico	5	Utah	4.75
Connecticut	6	Maine	5	New York	4.25	Vermont	6
Delaware	0	Maryland	5	North Carolina	4.5	Virginia	4
District of Columbia	5.75	Massachusetts	5	North Dakota	5	Washington	6.5
Florida	6	Michigan	6	Ohio	6	West Virginia	6
Georgia	4	Minnesota	6.5	Oklahoma	4.5	Wisconsin	5
Hawaii	4	Mississippi	7	Oregon	0	Wyoming	4
Idaho	6	Missouri	4.225	Pennsylvania	6	Average	4.83 %

Table H1040-101 Steel Tubular Scaffolding

On new construction, tubular scaffolding is efficient up to 60' high or five stories. Above this it is usually better to use a hung scaffolding if construction permits. Swing scaffolding operations may interfere with tenants. In this case, the tubular is more practical at all heights.

In repairing or cleaning the front of an existing building the cost of tubular scaffolding per S.F. of building front increases as the height increases above the first tier. The first tier cost is relatively high due to leveling and alignment.

The minimum efficient crew for erection is three workers. For heights over 50', a crew of four is more efficient. Use two or more on top and two at the bottom for handing up or hoisting. Four workers can erect and dismantle about nine frames per hour up to five stories. From five to eight stories they will average six frames per hour. With 7' horizontal spacing this will run about 400 S.F. and 265 S.F. of wall surface, respectively. Time

for placing planks must be added to the above. On heights above 50', five planks can be placed per labor-hour.

The cost per 1,000 S.F. of building front in the table below was developed by pricing the materials required for a typical tubular scaffolding system eleven frames long and two frames high. Planks were figured five wide for standing plus two wide for materials.

Frames are 5' wide and usually spaced 7' O.C. horizontally. Sidewalk frames are 6' wide. Rental rates will be lower for jobs over three months' duration.

For jobs under twenty-five frames, add 50% to rental cost. These figures do not include accessories which are listed separately below. Large quantities for long periods can reduce rental rates by 20%.

Item	Unit	Monthly Rent	Per 1,000 S. F. of Building Front	
			No. of Pieces	Rental per Month
5' Wide Standard Frame, 6'-4" High	Ea.	$ 3.94	24	$ 94.56
Leveling Jack & Plate		1.58	24	37.92
Cross Brace		.63	44	27.72
Side Arm Bracket, 21"		1.58	12	18.96
Guardrail Post		1.05	12	12.60
Guardrail, 7' section		.79	22	17.38
Stairway Section		10.50	2	21.00
Stairway Starter Bar		.11	1	.11
Stairway Inside Handrail		5.25	2	10.50
Stairway Outside Handrail		5.25	2	10.50
Walk-Thru Frame Guardrail		2.10	2	4.20
			Total	$255.45
			Per C.S.F., 1 Use/Mo.	$ 25.55

Scaffolding is often used as falsework over 15' high during construction of cast-in-place concrete beams and slabs. Two-feet wide scaffolding is generally used for heavy beam construction. The span between frames depends upon the load to be carried with a maximum span of 5'.

Heavy duty shoring frames with a capacity of 10,000#/leg can be spaced up to 10' O. C. depending upon form support design and loading.

Scaffolding used as horizontal shoring requires less than half the material required with conventional shoring.

On new construction, erection is done by carpenters.

Rolling towers supporting horizontal shores can reduce labor and speed the job. For maintenance work, catwalks with spans up to 70' can be supported by the rolling towers.

General: The following information on current city cost indexes is calculated for over 930 zip code locations in the United States and Canada. Index figures for both material and installation are based upon the 30 major city average of 100 and represent the cost relationship on July 1, 2005.

In addition to index adjustment, the user should consider:
1. productivity
2. management efficiency
3. competitive conditions
4. automation
5. restrictive union practices
6. unique local requirements
7. regional variations due to specific building codes

The weighted-average index is calculated from about 72 materials and equipment types and 21 building trades. The component contribution of these in a model building is tabulated in Table J1010-011 below.

If the systems component distribution of a building is unknown, the weighted-average index can be used to adjust the cost for any city.

Table J1010-011 Labor, Material, and Equipment Cost Distribution for Weighted Average Listed by System Division

Division No.	Building System	Percentage	Division No.	Building System	Percentage
A	Substructure	6.3%	D10	Services: Conveying	3.7%
B10	Shell: Superstructure	18.5	D20-40	Mechanical	21.2
B20	Exterior Closure	11.7	D50	Electrical	13.1
B30	Roofing	2.5	E	Equipment & Furnishings	2.0
C	Interior Construction	16.6	G	Site Work	4.4
				Total weighted average (Div. A-G)	100.0%

How to Use the Component Indexes: Table J1010-012 below shows how the average costs obtained from this book for each division should be adjusted for your particular building in your city. The example is a building adjusted for New York, NY. Indexes for other cities are tabulated in Division RJ1030. These indexes should also be used when compiling data on the form in Division K1010.

Table J1010-012 Adjustment of "Book" Costs to a Particular City

Systems Division Number	System Description	New York, NY City Cost Index	Cost Factor	Book Cost	Adjusted Cost	City Cost Index	Cost Factor	Book Cost	Adjusted Cost
A	Substructure	145.5	1.455	$ 88,300	$ 128,500				
B10	Shell: Superstructure	128.2	1.282	244,000	312,800				
B20	Exterior Closure	133.2	1.332	165,300	220,200				
B30	Roofing	132.6	1.326	36,500	48,400				
C	Interior Construction	135.3	1.353	235,400	318,500				
D10	Services: Conveying	110.1	1.101	47,100	51,900				
D20-40	Mechanical	126.9	1.269	292,200	370,800				
D50	Electrical	144.6	1.446	163,500	236,400				
E	Equipment	105.4	1.054	24,700	26,000				
G	Site Work	130.9	1.309	69,500	91,000				
Total Cost, (Div. A-G)				$1,366,500	$1,804,500				

Alternate Method: Use the weighted-average index for your city rather than the component indexes. **For example:**

(Total weighted-average index A-G)/100 x (total cost) = Adjusted city cost
New York, NY 1.319 x $1,366,500 = $1,802,400

COST INDEXES

City to City: To convert known or estimated costs from one city to another, cost indexes can be used as follows:

$$\text{Unknown City Cost} = \text{Known Cost} \times \frac{\text{Unknown City Index}}{\text{Known City Index}}$$

For example: If the building cost in Boston, MA, is $2,000,000, how much would a duplicated building cost in Los Angeles, CA?

$$\text{L.A. Cost} = \$2,000,000 \times \frac{\text{(Los Angeles) } 106.8}{\text{(Boston) } 115.6} = \$1,848,000$$

The table below lists both the Means City Cost Index based on January 1, 1993 = 100 as well as the computed value of an index based on January 1, 2006 costs. Since the January 1, 2006 figure is estimated, space is left to write in the actual index figures as they become available through either the quarterly *Means Construction Cost Indexes* or as printed in the *Engineering News-Record*. To compute the actual index based on January 1, 2006 = 100, divide the Quarterly City Cost Index for a particular year by the actual January 1, 2006 Quarterly City Cost Index. Space has been left to advance the index figures as the year progresses.

Table J1020-011   Historical Cost Indexes

Year	Historical Cost Index Jan. 1, 1993 = 100 Est.	Actual	Current Index Based on Jan. 1, 2006 = 100 Est.	Actual	Year	Historical Cost Index Jan. 1, 1993 = 100 Actual	Current Index Based on Jan. 1, 2006 = 100 Est.	Actual	Year	Historical Cost Index Jan. 1, 1993 = 100 Actual	Current Index Based on Jan. 1, 2006 = 100 Est.	Actual
Oct. 2006					July 1991	96.8	62.1		July 1973	37.7	24.2	
July 2006					1990	94.3	60.5		1972	34.8	22.3	
April 2006					1989	92.1	59.1		1971	32.1	20.6	
Jan. 2006	155.9		100.0	100.0	1988	89.9	57.6		1970	28.7	18.4	
July 2005		151.6	97.2		1987	87.7	56.2		1969	26.9	17.3	
2004		143.7	92.2		1986	84.2	54.0		1968	24.9	16.0	
2003		132.0	84.7		1985	82.6	53.0		1967	23.5	15.1	
2002		128.7	82.6		1984	82.0	52.6		1966	22.7	14.6	
2001		125.1	80.2		1983	80.2	51.4		1965	21.7	13.9	
2000		120.9	77.5		1982	76.1	48.8		1964	21.2	13.6	
1999		117.6	75.4		1981	70.0	44.9		1963	20.7	13.3	
1998		115.1	73.8		1980	62.9	40.3		1962	20.2	13.0	
1997		112.8	72.4		1979	57.8	37.1		1961	19.8	12.7	
1996		110.2	70.7		1978	53.5	34.3		1960	19.7	12.6	
1995		107.6	69.0		1977	49.5	31.8		1959	19.3	12.4	
1994		104.4	67.0		1976	46.9	30.1		1958	18.8	12.1	
1993		101.7	65.2		1975	44.8	28.7		1957	18.4	11.8	
1992		99.4	63.8		1974	41.4	26.6		1956	17.6	11.3	

To find the **current cost** from a project built previously in either the same city or a different city, the following formula is used:

$$\text{Present Cost (City X)} = \frac{\text{Current HCI} \times \text{CCI (City X)}}{\text{Previous HCI} \times \text{CCI (City Y)}} \times \text{Former Cost (City Y)}$$

For example: Find the construction cost of a building to be built in San Francisco, CA, as of January 1, 2006 when the identical building cost $500,000 in Boston, MA, on July 1, 1968.

$$\text{Jan. 1, 2006 (San Francisco)} = \frac{\text{(San Francisco) } 155.9 \times 121.7}{\text{(Boston) } 24.9 \times 115.6} \times \$500,000 = \$3,295,500$$

Note: The City Cost Indexes for Canada can be used to convert U.S. national averages to local costs in Canadian dollars.

To Project Future Construction Costs: Using the results of the last five years average percentage increase as a basis, an average increase of 2.7% could be used.

The historical index figures above are compiled from the Means Construction Index Service.

ALABAMA

DIV. NO.	BUILDING SYSTEMS	BIRMINGHAM			HUNTSVILLE			MOBILE			MONTGOMERY			TUSCALOOSA		
		MAT.	INST.	TOTAL	MAT.	INST.	TOTAL	MAT.	INST.	TOTAL	MAT.	INST.	TOTAL	MAT.	INST.	TOTAL
A	Substructure	91.7	78.3	83.4	88.5	73.8	79.5	94.0	60.7	73.5	94.6	62.4	74.8	90.8	60.2	71.9
B10	Shell: Superstructure	93.0	84.6	89.7	93.4	79.4	87.8	93.3	65.7	82.2	93.3	73.4	85.3	93.3	71.2	84.4
B20	Exterior Closure	91.4	77.8	84.7	91.3	68.5	80.1	91.6	53.1	72.8	91.2	41.1	66.7	90.8	44.6	68.2
B30	Roofing	96.7	79.4	90.0	96.5	75.0	88.2	96.5	69.7	86.1	96.2	65.2	84.2	96.5	62.5	83.3
C	Interior Construction	99.2	69.4	86.6	98.9	63.5	83.9	101.5	51.0	80.1	101.0	44.9	77.2	98.9	40.0	73.9
D10	Services: Conveying	100.0	87.1	96.3	100.0	86.0	96.0	100.0	66.3	90.4	100.0	80.9	94.5	100.0	79.6	94.2
D20 - 40	Mechanical	99.9	67.0	86.3	99.9	67.2	86.3	99.9	68.2	86.7	99.9	40.2	75.1	99.9	35.4	73.1
D50	Electrical	97.5	67.0	82.7	97.5	68.8	83.6	97.5	50.1	74.4	96.7	67.4	82.5	97.0	67.0	82.4
E	Equipment & Furnishings	100.0	73.3	98.4	100.0	66.3	98.0	100.0	50.7	97.0	100.0	47.5	96.8	100.0	38.3	96.2
G	Site Work	84.6	92.7	90.4	82.8	92.7	89.8	94.3	85.4	88.0	92.9	86.0	88.0	83.3	91.4	89.0
A-G	WEIGHTED AVERAGE	96.5	75.2	87.0	96.3	72.0	85.5	97.2	61.1	81.2	97.0	57.1	79.3	96.3	55.4	78.1

ALASKA / ARIZONA

DIV. NO.	BUILDING SYSTEMS	ANCHORAGE			FAIRBANKS			JUNEAU			FLAGSTAFF			MESA/TEMPE		
		MAT.	INST.	TOTAL	MAT.	INST.	TOTAL	MAT.	INST.	TOTAL	MAT.	INST.	TOTAL	MAT.	INST.	TOTAL
A	Substructure	174.7	118.5	140.1	153.7	120.2	133.0	167.9	118.5	137.5	94.0	79.1	84.8	97.8	76.4	84.6
B10	Shell: Superstructure	139.1	107.8	126.5	134.9	109.2	124.6	137.8	107.7	125.7	94.5	70.3	84.8	95.3	69.0	84.7
B20	Exterior Closure	166.7	119.4	143.5	155.5	119.3	137.8	163.6	119.4	142.0	121.9	65.5	94.3	103.7	55.3	80.1
B30	Roofing	173.1	115.3	150.7	170.5	118.9	150.5	171.0	115.3	149.4	99.7	69.5	88.0	99.1	63.0	85.1
C	Interior Construction	138.9	116.0	129.2	137.2	120.6	130.2	136.9	116.0	128.0	99.6	63.8	84.4	98.5	62.9	83.4
D10	Services: Conveying	100.0	115.4	104.4	100.0	115.4	104.4	100.0	115.4	104.4	100.0	83.1	95.2	100.0	78.2	93.8
D20 - 40	Mechanical	100.5	108.7	103.9	100.5	116.9	107.3	100.5	105.5	102.6	100.2	76.8	90.5	100.2	70.0	87.7
D50	Electrical	148.4	114.2	131.8	150.2	114.2	132.7	150.2	114.2	132.7	99.5	64.1	82.3	91.4	64.1	78.2
E	Equipment & Furnishings	100.0	113.7	100.8	100.0	118.7	101.1	100.0	113.7	100.8	100.0	67.9	98.0	100.0	67.2	98.0
G	Site Work	145.3	134.3	137.5	129.5	134.3	132.9	141.4	134.3	136.3	80.7	100.8	95.0	86.7	103.9	98.9
A-G	WEIGHTED AVERAGE	133.7	114.7	125.3	130.3	117.6	124.6	132.5	114.0	124.3	100.4	72.3	87.9	97.8	68.9	85.0

ARIZONA / ARKANSAS

DIV. NO.	BUILDING SYSTEMS	PHOENIX			PRESCOTT			TUCSON			FORT SMITH			JONESBORO		
		MAT.	INST.	TOTAL	MAT.	INST.	TOTAL	MAT.	INST.	TOTAL	MAT.	INST.	TOTAL	MAT.	INST.	TOTAL
A	Substructure	97.7	80.6	87.2	92.2	70.1	78.6	95.4	80.3	86.1	89.2	62.9	73.0	87.2	62.1	71.8
B10	Shell: Superstructure	96.5	71.7	86.5	94.2	62.3	81.4	95.5	71.0	85.6	95.5	62.6	82.3	89.3	62.6	78.6
B20	Exterior Closure	97.7	66.5	82.5	107.6	57.0	82.9	98.4	65.6	82.3	91.1	54.6	73.3	89.2	47.6	68.9
B30	Roofing	99.1	69.4	87.5	98.3	60.6	83.7	100.7	65.5	87.1	99.6	44.9	78.4	101.6	50.3	81.7
C	Interior Construction	100.4	67.2	86.3	98.2	53.0	79.1	96.8	64.7	83.2	95.6	47.8	75.3	92.0	46.3	72.6
D10	Services: Conveying	100.0	83.6	95.3	100.0	82.4	95.0	100.0	83.6	95.3	100.0	76.7	93.3	100.0	51.9	86.3
D20 - 40	Mechanical	100.2	76.9	90.6	100.2	75.4	89.9	100.2	70.5	87.9	100.1	41.6	75.8	100.2	41.5	75.8
D50	Electrical	100.5	64.1	82.8	99.2	64.1	82.1	93.4	62.0	78.1	95.2	66.6	81.3	102.1	46.6	75.1
E	Equipment & Furnishings	100.0	69.1	98.1	100.0	53.5	97.2	100.0	69.0	98.1	100.0	40.4	96.4	100.0	46.5	96.7
G	Site Work	87.1	104.7	99.6	69.7	100.0	91.2	83.9	104.4	98.5	79.4	83.5	82.3	100.7	98.7	99.3
A-G	WEIGHTED AVERAGE	98.8	73.6	87.6	98.3	66.8	84.3	97.1	71.2	85.6	95.9	56.9	78.6	95.2	53.4	76.6

ARKANSAS / CALIFORNIA

DIV. NO.	BUILDING SYSTEMS	LITTLE ROCK			PINE BLUFF			TEXARKANA			ANAHEIM			BAKERSFIELD		
		MAT.	INST.	TOTAL	MAT.	INST.	TOTAL	MAT.	INST.	TOTAL	MAT.	INST.	TOTAL	MAT.	INST.	TOTAL
A	Substructure	88.7	68.1	76.1	83.1	68.0	73.8	90.5	52.2	66.9	104.3	116.4	111.7	104.3	115.1	111.0
B10	Shell: Superstructure	92.1	66.3	81.7	92.1	66.1	81.6	87.8	50.2	72.7	109.4	108.8	109.2	104.5	108.1	105.9
B20	Exterior Closure	90.2	55.4	73.2	101.9	55.3	79.2	89.0	34.2	62.2	101.6	110.8	106.1	108.6	109.4	109.0
B30	Roofing	98.0	47.6	78.4	98.0	47.6	78.4	99.2	41.5	76.8	102.7	117.6	108.5	99.2	110.6	103.6
C	Interior Construction	95.7	61.2	81.1	90.7	61.2	78.2	94.9	39.2	71.3	108.3	117.3	112.1	108.2	114.6	110.9
D10	Services: Conveying	100.0	76.7	93.3	100.0	76.7	93.3	100.0	36.1	81.7	100.0	111.7	103.3	100.0	110.4	103.0
D20 - 40	Mechanical	100.1	57.9	82.6	100.1	43.9	76.8	100.1	36.7	73.8	100.2	107.3	103.1	100.2	96.2	98.6
D50	Electrical	94.9	68.4	82.0	94.6	68.4	81.8	96.3	38.8	68.4	89.5	105.3	97.2	89.6	97.2	93.3
E	Equipment & Furnishings	100.0	60.9	97.6	100.0	60.9	97.6	100.0	40.2	96.4	100.0	121.8	101.3	100.0	121.1	101.3
G	Site Work	79.3	83.5	82.3	81.3	83.5	82.9	95.0	84.3	87.4	104.2	109.2	107.7	107.8	106.3	106.7
A-G	WEIGHTED AVERAGE	95.1	63.8	81.2	95.2	60.9	80.0	94.6	44.0	72.1	102.6	110.6	106.1	102.3	105.9	103.9

CALIFORNIA

DIV. NO.	BUILDING SYSTEMS	FRESNO			LOS ANGELES			OAKLAND			OXNARD			REDDING		
		MAT.	INST.	TOTAL	MAT.	INST.	TOTAL	MAT.	INST.	TOTAL	MAT.	INST.	TOTAL	MAT.	INST.	TOTAL
A	Substructure	107.3	115.3	112.2	97.6	115.6	108.7	131.8	120.0	124.5	108.9	114.9	112.6	117.8	114.2	115.6
B10	Shell: Superstructure	110.0	109.2	109.7	108.8	107.9	108.5	109.6	115.1	111.8	105.6	108.6	106.8	111.4	108.4	110.2
B20	Exterior Closure	110.4	112.3	111.3	103.2	116.0	109.5	132.8	124.5	128.7	110.5	109.5	110.0	115.7	110.7	113.3
B30	Roofing	95.9	112.3	102.2	100.8	118.5	107.7	110.2	130.0	117.9	105.4	115.9	109.5	106.0	110.1	107.6
C	Interior Construction	111.9	123.5	116.8	104.6	117.1	109.9	111.8	132.9	120.8	107.6	116.8	111.5	109.7	121.6	114.8
D10	Services: Conveying	100.0	125.3	107.2	100.0	110.8	103.1	100.0	127.7	107.9	100.0	112.0	103.4	100.0	124.2	106.9
D20 - 40	Mechanical	100.2	109.1	103.9	100.1	107.2	103.1	100.3	136.0	115.1	100.2	107.3	103.2	100.2	105.6	102.5
D50	Electrical	88.4	96.6	92.4	97.0	114.2	105.3	103.4	138.9	120.6	95.4	106.8	100.9	98.1	104.2	101.1
E	Equipment & Furnishings	100.0	124.6	101.5	100.0	122.1	101.3	100.0	136.3	102.2	100.0	121.6	101.3	100.0	124.7	101.5
G	Site Work	109.5	106.4	107.3	100.9	107.9	105.9	154.0	104.6	119.0	109.4	104.0	105.6	113.8	105.2	107.7
A-G	WEIGHTED AVERAGE	104.1	110.9	107.1	102.5	112.1	106.8	110.7	127.1	118.0	103.8	110.0	106.5	106.6	110.3	108.3

CALIFORNIA

DIV. NO.	BUILDING SYSTEMS	RIVERSIDE			SACRAMENTO			SAN DIEGO			SAN FRANCISCO			SAN JOSE		
		MAT.	INST.	TOTAL	MAT.	INST.	TOTAL	MAT.	INST.	TOTAL	MAT.	INST.	TOTAL	MAT.	INST.	TOTAL
A	Substructure	105.9	115.8	112.0	113.5	116.9	115.6	109.4	106.5	107.6	135.4	122.3	127.4	123.9	118.7	120.7
B10	Shell: Superstructure	110.1	108.8	109.5	103.1	109.2	105.6	108.7	103.0	106.4	114.9	116.6	115.6	107.3	115.9	110.8
B20	Exterior Closure	98.6	110.4	104.4	119.1	112.9	116.0	105.9	107.8	106.8	133.6	127.5	130.6	126.9	124.7	125.8
B30	Roofing	102.9	117.2	108.4	110.6	114.4	112.1	107.8	104.2	106.4	113.4	131.4	120.4	101.9	131.0	113.2
C	Interior Construction	107.1	117.3	111.5	116.9	122.4	119.2	106.2	107.2	106.6	115.5	134.4	123.5	102.7	132.7	115.5
D10	Services: Conveying	100.0	111.7	103.3	100.0	126.2	107.5	100.0	111.6	103.3	100.0	128.3	108.1	100.0	126.9	107.7
D20 - 40	Mechanical	100.1	107.3	103.1	100.2	106.1	101.8	100.3	104.6	101.8	100.3	148.6	120.3	100.2	138.0	119.3
D50	Electrical	89.6	99.7	94.5	97.7	106.1	101.8	98.9	96.4	97.7	103.4	150.5	126.3	103.3	140.8	121.5
E	Equipment & Furnishings	100.0	121.5	101.3	100.0	125.0	101.5	100.0	105.9	100.4	100.0	137.0	102.3	100.0	136.5	102.2
G	Site Work	102.1	106.8	105.5	121.1	111.6	114.3	100.9	100.7	100.7	155.4	111.5	124.3	144.7	99.7	112.8
A-G	WEIGHTED AVERAGE	102.2	109.5	105.5	106.6	113.0	109.4	104.0	103.9	104.0	112.8	132.9	121.7	107.3	127.5	116.3

COST INDEXES

DIV. NO.	BUILDING SYSTEMS	CALIFORNIA									COLORADO					
		SANTA BARBARA			STOCKTON			VALLEJO			COLORADO SPRINGS			DENVER		
		MAT.	INST.	TOTAL	MAT.	INST.	TOTAL	MAT.	INST.	TOTAL	MAT.	INST.	TOTAL	MAT.	INST.	TOTAL
A	Substructure	108.7	115.4	112.8	108.7	115.2	112.7	117.3	120.3	119.1	98.0	88.7	92.3	95.0	91.4	92.7
B10	Shell: Superstructure	105.9	108.5	107.0	107.2	109.4	108.1	106.7	112.8	109.2	97.7	88.0	93.8	99.4	88.3	94.9
B20	Exterior Closure	108.8	110.5	109.6	110.4	112.8	111.6	102.7	125.2	113.7	111.0	83.2	97.4	107.4	85.0	96.4
B30	Roofing	100.3	114.5	105.8	105.6	111.6	107.9	113.2	126.1	118.2	105.9	85.1	97.8	105.4	80.9	95.9
C	Interior Construction	108.2	115.6	111.3	107.9	122.3	114.0	118.1	132.4	124.2	98.2	83.2	91.8	99.7	87.6	94.6
D10	Services: Conveying	100.0	112.0	103.4	100.0	125.3	107.2	100.0	125.7	107.4	100.0	90.1	97.2	100.0	90.2	97.2
D20 - 40	Mechanical	100.2	107.3	103.2	100.2	104.4	101.9	100.2	122.9	109.6	100.1	79.9	91.7	100.1	87.4	94.8
D50	Electrical	86.7	110.0	98.0	97.9	123.1	110.1	93.0	123.5	107.9	99.6	88.6	94.3	101.5	94.9	98.3
E	Equipment & Furnishings	100.0	121.7	101.3	100.0	125.8	101.6	100.0	135.4	102.2	100.0	87.4	99.2	100.0	87.8	99.3
G	Site Work	109.4	106.3	107.2	107.3	105.7	106.2	119.2	111.6	113.8	93.9	96.9	96.0	92.4	106.2	102.2
A-G	**WEIGHTED AVERAGE**	102.6	110.5	106.1	104.4	113.5	108.4	105.5	122.2	112.9	100.4	85.7	93.9	100.6	89.9	95.8

DIV. NO.	BUILDING SYSTEMS	COLORADO												CONNECTICUT		
		FORT COLLINS			GRAND JUNCTION			GREELEY			PUEBLO			BRIDGEPORT		
		MAT.	INST.	TOTAL	MAT.	INST.	TOTAL	MAT.	INST.	TOTAL	MAT.	INST.	TOTAL	MAT.	INST.	TOTAL
A	Substructure	104.5	85.7	93.0	111.5	86.8	96.3	92.8	67.5	77.2	102.8	88.9	94.2	107.7	117.5	113.7
B10	Shell: Superstructure	98.2	81.6	91.5	100.0	81.4	92.5	95.7	68.6	84.8	99.1	88.8	94.9	102.1	123.4	110.7
B20	Exterior Closure	118.0	64.1	91.6	129.2	65.9	98.3	108.6	46.5	78.3	105.9	82.5	94.4	109.5	127.6	118.3
B30	Roofing	105.9	73.1	93.1	99.7	68.7	87.6	105.2	61.7	88.3	99.2	84.7	93.6	101.9	125.8	111.1
C	Interior Construction	97.1	77.8	88.9	103.6	80.6	93.9	96.5	50.3	76.9	100.7	84.9	94.0	104.5	124.0	112.8
D10	Services: Conveying	100.0	88.4	96.7	100.0	90.1	97.2	100.0	83.9	95.4	100.0	91.2	97.5	100.0	114.5	104.1
D20 - 40	Mechanical	100.1	83.1	93.0	100.1	72.7	88.7	100.1	77.5	90.7	100.1	71.2	88.1	100.1	111.8	104.9
D50	Electrical	96.4	94.2	95.3	90.9	63.9	77.8	96.4	94.1	95.3	91.8	80.6	86.3	97.7	109.6	103.5
E	Equipment & Furnishings	100.0	85.3	99.1	100.0	83.8	99.0	100.0	48.6	96.9	100.0	86.9	99.2	100.0	121.2	101.3
G	Site Work	103.3	99.3	100.5	123.2	102.0	108.1	91.1	98.0	96.0	115.1	96.5	101.9	102.0	105.1	104.2
A-G	**WEIGHTED AVERAGE**	101.1	82.4	92.8	103.6	76.8	91.7	98.7	70.4	86.1	100.0	83.1	92.5	102.4	117.8	109.2

DIV. NO.	BUILDING SYSTEMS	CONNECTICUT														
		BRISTOL			HARTFORD			NEW BRITAIN			NEW HAVEN			NORWALK		
		MAT.	INST.	TOTAL	MAT.	INST.	TOTAL	MAT.	INST.	TOTAL	MAT.	INST.	TOTAL	MAT.	INST.	TOTAL
A	Substructure	103.4	117.3	112.0	104.1	117.3	112.2	104.5	117.3	112.4	105.5	117.6	113.0	106.6	119.9	114.8
B10	Shell: Superstructure	101.2	123.0	110.0	105.3	123.0	112.4	98.7	123.0	108.5	99.1	123.3	108.8	101.9	124.6	111.1
B20	Exterior Closure	105.7	124.2	114.7	105.7	124.2	114.7	105.7	124.2	114.7	120.1	127.5	123.8	105.8	127.4	116.4
B30	Roofing	102.0	121.4	109.6	100.8	121.4	108.8	102.0	122.6	110.0	102.1	122.6	110.0	102.0	127.2	111.8
C	Interior Construction	104.5	123.6	112.6	104.5	123.6	112.6	104.5	123.6	112.6	104.5	124.0	112.8	104.5	124.0	112.8
D10	Services: Conveying	100.0	114.5	104.1	100.0	114.5	104.1	100.0	114.5	104.1	100.0	114.5	104.1	100.0	114.5	104.1
D20 - 40	Mechanical	100.1	111.8	104.9	100.1	111.8	104.9	100.1	111.8	104.9	100.1	111.8	104.9	100.1	111.9	105.0
D50	Electrical	97.7	105.8	101.6	97.3	110.7	103.8	97.7	105.8	101.7	97.6	109.7	103.5	97.7	109.6	103.5
E	Equipment & Furnishings	100.0	121.0	101.3	100.0	121.2	101.3	100.0	121.0	101.3	100.0	121.2	101.3	100.0	121.2	101.3
G	Site Work	101.0	105.1	103.9	101.4	105.1	104.0	101.2	105.1	104.0	101.0	106.0	104.5	101.7	105.1	104.1
A-G	**WEIGHTED AVERAGE**	101.6	116.6	108.2	102.3	117.3	109.0	101.1	116.6	108.0	102.8	117.8	109.4	101.9	118.2	109.1

DIV. NO.	BUILDING SYSTEMS	CONNECTICUT						D.C.			DELAWARE			FLORIDA		
		STAMFORD			WATERBURY			WASHINGTON			WILMINGTON			DAYTONA BEACH		
		MAT.	INST.	TOTAL	MAT.	INST.	TOTAL	MAT.	INST.	TOTAL	MAT.	INST.	TOTAL	MAT.	INST.	TOTAL
A	Substructure	107.7	120.0	115.3	107.6	117.5	113.7	109.6	86.4	95.3	87.0	99.4	94.6	95.3	75.3	83.0
B10	Shell: Superstructure	102.1	124.9	111.3	102.2	123.3	110.7	105.8	98.2	102.8	99.1	105.1	101.5	94.2	83.7	90.0
B20	Exterior Closure	105.9	127.5	116.4	105.9	127.5	116.5	102.5	85.1	94.0	113.2	95.2	104.4	92.7	65.2	79.2
B30	Roofing	102.0	127.3	111.8	102.0	122.6	110.0	103.6	84.2	96.1	101.8	106.6	103.7	99.4	67.5	87.0
C	Interior Construction	104.6	124.0	112.8	104.5	124.3	112.9	103.6	85.4	95.9	97.7	99.2	98.3	103.8	69.5	89.3
D10	Services: Conveying	100.0	114.5	104.1	100.0	114.5	104.1	100.0	98.3	99.5	100.0	92.2	97.8	100.0	83.3	95.2
D20 - 40	Mechanical	100.1	111.9	105.0	100.1	111.8	104.9	99.9	91.4	96.4	100.3	113.6	105.8	99.9	70.3	87.6
D50	Electrical	97.7	150.4	123.3	97.2	109.6	103.2	99.2	98.3	98.7	96.2	102.2	99.1	96.6	65.4	81.4
E	Equipment & Furnishings	100.0	123.1	101.4	100.0	121.1	101.3	100.0	81.4	98.9	100.0	99.6	100.0	100.0	69.5	98.1
G	Site Work	102.4	105.1	104.3	101.6	105.1	104.1	103.9	89.2	93.5	87.5	111.0	104.2	111.0	86.1	93.4
A-G	**WEIGHTED AVERAGE**	102.0	124.0	111.8	101.9	117.7	108.9	102.5	92.2	97.5	99.7	103.9	101.5	98.3	72.8	87.0

DIV. NO.	BUILDING SYSTEMS	FLORIDA														
		FORT LAUDERDALE			JACKSONVILLE			MELBOURNE			MIAMI			ORLANDO		
		MAT.	INST.	TOTAL	MAT.	INST.	TOTAL	MAT.	INST.	TOTAL	MAT.	INST.	TOTAL	MAT.	INST.	TOTAL
A	Substructure	95.7	69.2	79.4	95.8	61.9	74.9	107.3	78.4	89.5	94.0	69.3	78.8	100.1	75.3	84.8
B10	Shell: Superstructure	94.5	77.9	87.8	94.0	66.5	83.0	103.2	85.6	96.1	98.2	78.1	90.1	99.9	83.0	93.1
B20	Exterior Closure	91.2	65.6	78.7	92.5	51.2	72.3	94.6	75.0	85.0	90.5	66.1	78.6	92.2	64.9	78.9
B30	Roofing	99.5	74.9	89.9	99.7	58.6	83.7	99.8	76.3	90.7	103.3	71.1	90.8	99.6	67.8	87.3
C	Interior Construction	102.3	62.7	85.5	103.8	49.7	80.9	102.2	73.9	90.2	103.5	62.8	86.2	104.0	69.4	89.3
D10	Services: Conveying	100.0	91.4	97.6	100.0	80.6	94.5	100.0	87.6	96.4	100.0	91.4	97.6	100.0	83.3	95.2
D20 - 40	Mechanical	99.9	67.2	86.3	99.9	49.5	79.0	99.9	75.9	89.9	99.9	73.8	89.1	99.9	65.8	85.8
D50	Electrical	96.6	73.4	85.3	96.4	66.9	82.0	97.5	72.1	85.2	97.5	71.6	84.9	97.3	43.5	71.2
E	Equipment & Furnishings	100.0	64.4	97.8	100.0	51.3	97.0	100.0	70.4	98.2	100.0	64.3	97.8	100.0	70.7	98.2
G	Site Work	99.0	73.2	80.7	111.2	86.7	93.8	119.3	86.8	96.2	98.5	73.0	80.4	111.9	85.8	93.4
A-G	**WEIGHTED AVERAGE**	97.7	70.3	85.5	98.3	59.6	81.1	100.8	77.8	90.6	98.7	71.4	86.6	99.7	68.6	85.9

DIV. NO.	BUILDING SYSTEMS	FLORIDA														
		PANAMA CITY			PENSACOLA			ST. PETERSBURG			TALLAHASSEE			TAMPA		
		MAT.	INST.	TOTAL	MAT.	INST.	TOTAL	MAT.	INST.	TOTAL	MAT.	INST.	TOTAL	MAT.	INST.	TOTAL
A	Substructure	101.1	45.9	67.1	100.5	59.7	75.4	102.9	59.3	76.1	98.0	53.9	70.9	101.3	74.7	84.9
B10	Shell: Superstructure	94.9	49.2	76.5	94.4	65.1	82.6	98.2	65.1	84.9	89.5	60.7	77.9	98.1	83.0	92.0
B20	Exterior Closure	99.8	28.9	65.1	97.6	50.4	74.5	115.3	48.9	82.8	93.6	40.5	67.6	93.2	72.5	83.1
B30	Roofing	99.9	30.7	73.0	99.7	51.4	80.9	99.3	47.8	79.3	99.7	46.5	79.0	99.6	58.1	83.5
C	Interior Construction	102.3	24.9	69.5	100.9	49.8	79.2	101.9	46.6	78.5	103.2	37.3	75.2	103.9	68.1	88.7
D10	Services: Conveying	100.0	45.3	84.4	100.0	47.8	85.1	100.0	51.0	86.0	100.0	78.4	93.8	100.0	85.5	95.9
D20 - 40	Mechanical	99.9	24.3	68.5	99.9	48.2	78.4	99.9	48.0	78.4	99.9	38.5	74.4	99.9	71.3	88.0
D50	Electrical	95.6	32.6	64.9	98.7	62.5	81.1	96.9	47.0	72.6	97.0	40.3	69.4	96.5	47.0	72.4
E	Equipment & Furnishings	100.0	26.6	95.5	100.0	50.1	97.0	100.0	45.4	96.7	100.0	36.1	96.1	100.0	73.3	98.4
G	Site Work	124.6	83.2	95.2	122.5	85.8	96.4	112.5	85.1	93.1	112.4	85.2	93.1	112.5	85.8	93.5
A-G	**WEIGHTED AVERAGE**	99.4	36.8	71.6	99.1	57.2	80.5	101.6	54.2	80.5	97.6	48.2	75.6	99.4	70.8	86.7

COST INDEXES

GEORGIA

DIV. NO.	BUILDING SYSTEMS	ALBANY MAT.	INST.	TOTAL	ATLANTA MAT.	INST.	TOTAL	AUGUSTA MAT.	INST.	TOTAL	COLUMBUS MAT.	INST.	TOTAL	MACON MAT.	INST.	TOTAL
A	Substructure	96.8	59.0	73.5	99.3	82.3	88.8	95.5	63.1	75.5	96.6	66.0	77.8	96.3	66.6	78.0
B10	Shell: Superstructure	94.4	70.6	84.8	93.4	79.4	87.7	91.6	61.1	79.3	94.0	77.8	87.5	90.6	75.2	84.4
B20	Exterior Closure	92.4	42.2	67.9	96.5	69.9	83.5	93.9	42.0	68.5	92.1	57.8	75.3	97.1	42.2	70.3
B30	Roofing	96.5	57.1	81.2	101.9	75.5	91.6	101.4	50.0	81.5	96.3	61.7	82.9	94.9	59.4	81.1
C	Interior Construction	102.2	42.0	76.7	99.1	76.8	89.6	96.1	48.0	75.7	102.1	55.7	82.4	96.2	51.5	77.2
D10	Services: Conveying	100.0	82.0	94.8	100.0	84.1	95.5	100.0	64.8	89.9	100.0	82.0	94.8	100.0	80.0	94.3
D20 - 40	Mechanical	99.9	47.9	78.3	100.1	78.1	91.0	100.1	63.8	85.0	99.9	48.3	78.5	99.9	40.4	75.2
D50	Electrical	92.2	65.0	78.9	95.5	80.1	88.0	95.7	48.1	72.6	92.6	49.7	71.7	90.3	66.4	78.7
E	Equipment & Furnishings	100.0	42.4	96.5	100.0	78.2	98.7	100.0	50.9	97.0	100.0	56.2	97.3	100.0	56.0	97.3
G	Site Work	99.3	75.5	82.4	98.8	95.9	96.7	95.5	92.5	93.4	99.2	76.3	83.0	99.5	93.6	95.3
A-G	WEIGHTED AVERAGE	97.2	56.3	79.1	97.7	79.0	89.4	96.3	57.5	79.0	97.2	60.4	80.8	95.7	59.1	79.4

GEORGIA / HAWAII / IDAHO

DIV. NO.	BUILDING SYSTEMS	SAVANNAH MAT.	INST.	TOTAL	VALDOSTA MAT.	INST.	TOTAL	HONOLULU MAT.	INST.	TOTAL	BOISE MAT.	INST.	TOTAL	LEWISTON MAT.	INST.	TOTAL
A	Substructure	95.1	60.7	73.9	96.0	54.0	70.1	168.6	127.4	143.3	95.2	87.7	90.6	102.7	85.7	92.3
B10	Shell: Superstructure	90.7	69.1	82.0	92.7	58.9	79.1	148.8	122.7	138.3	99.0	79.1	91.0	95.1	83.4	90.4
B20	Exterior Closure	93.6	53.0	73.8	99.5	47.5	74.1	120.8	128.4	124.5	121.4	72.3	97.4	129.4	82.3	106.4
B30	Roofing	96.6	52.6	79.5	96.4	59.7	82.2	107.6	129.4	116.1	92.6	80.6	88.0	136.2	79.2	114.1
C	Interior Construction	103.6	50.0	80.9	97.8	40.7	73.6	121.3	142.4	130.2	96.8	74.2	87.2	137.3	71.6	109.4
D10	Services: Conveying	100.0	68.4	91.0	100.0	68.4	91.0	100.0	120.1	105.8	100.0	80.8	94.5	100.0	85.8	95.9
D20 - 40	Mechanical	99.9	46.1	75.3	99.9	40.6	75.3	100.2	116.5	107.0	100.1	75.5	89.9	101.2	89.1	96.2
D50	Electrical	94.1	61.9	78.4	90.8	29.5	61.0	108.9	123.3	115.9	96.0	77.0	86.8	81.3	83.6	82.4
E	Equipment & Furnishings	100.0	49.2	96.9	100.0	39.9	96.3	100.0	145.6	102.8	100.0	81.2	98.9	100.0	63.2	97.8
G	Site Work	99.6	76.4	83.2	108.7	75.7	85.3	139.4	106.6	116.1	79.8	102.4	95.9	84.8	96.7	93.2
A-G	WEIGHTED AVERAGE	97.0	57.7	79.6	96.9	47.6	75.0	120.6	124.8	122.5	100.2	78.9	90.7	107.4	83.6	96.8

IDAHO / ILLINOIS

DIV. NO.	BUILDING SYSTEMS	POCATELLO MAT.	INST.	TOTAL	TWIN FALLS MAT.	INST.	TOTAL	CHICAGO MAT.	INST.	TOTAL	DECATUR MAT.	INST.	TOTAL	EAST ST. LOUIS MAT.	INST.	TOTAL
A	Substructure	95.3	87.5	90.5	97.9	59.2	74.1	108.2	127.3	120.0	98.2	100.7	99.8	96.2	104.3	101.2
B10	Shell: Superstructure	105.2	78.4	94.4	105.7	59.4	87.1	100.5	131.2	112.9	98.9	104.4	101.1	95.8	112.5	102.5
B20	Exterior Closure	116.3	63.4	90.5	124.8	39.7	83.2	100.9	135.9	118.0	85.6	105.9	95.5	78.6	112.9	95.4
B30	Roofing	92.8	72.0	84.7	93.6	45.4	74.9	101.4	130.6	112.8	94.8	100.4	97.0	89.5	92.0	90.9
C	Interior Construction	97.0	74.4	87.4	97.7	38.5	72.6	98.1	136.4	114.3	96.2	102.4	98.8	90.7	109.7	98.8
D10	Services: Conveying	100.0	80.8	94.5	100.0	47.1	84.9	100.0	122.1	106.3	100.0	87.1	96.3	100.0	87.2	96.3
D20 - 40	Mechanical	100.1	75.4	89.8	100.1	62.1	84.3	100.0	126.1	110.8	100.0	102.6	101.1	99.9	92.3	96.8
D50	Electrical	91.1	76.1	83.8	83.3	69.9	76.8	98.8	117.1	107.7	98.7	89.1	94.0	95.9	89.5	92.8
E	Equipment & Furnishings	100.0	81.1	98.9	100.0	35.4	96.1	100.0	135.1	102.1	100.0	98.9	99.9	100.0	110.1	100.6
G	Site Work	80.1	102.4	95.9	86.5	100.3	96.3	98.7	90.8	93.1	90.6	95.8	94.3	105.2	96.0	98.7
A-G	WEIGHTED AVERAGE	100.3	77.3	90.1	100.8	57.7	81.7	100.1	126.3	111.7	97.0	101.1	98.8	94.5	102.0	97.8

ILLINOIS / INDIANA

DIV. NO.	BUILDING SYSTEMS	JOLIET MAT.	INST.	TOTAL	PEORIA MAT.	INST.	TOTAL	ROCKFORD MAT.	INST.	TOTAL	SPRINGFIELD MAT.	INST.	TOTAL	ANDERSON MAT.	INST.	TOTAL
A	Substructure	108.2	121.6	116.4	98.0	101.2	100.0	99.6	107.3	104.4	95.1	100.4	98.4	97.6	84.8	89.7
B10	Shell: Superstructure	98.8	122.2	108.2	98.5	104.5	100.9	98.9	116.9	106.2	100.0	103.4	101.3	93.4	85.0	90.0
B20	Exterior Closure	101.3	127.6	114.1	105.0	109.6	107.3	92.8	108.0	100.3	86.4	106.4	96.2	94.0	85.1	89.7
B30	Roofing	100.7	123.4	109.5	94.9	104.5	98.6	98.0	115.0	104.6	94.4	101.5	97.2	100.3	81.6	93.0
C	Interior Construction	96.6	136.9	113.7	96.2	103.6	99.4	96.3	112.7	103.2	96.4	102.3	98.9	93.6	81.5	88.5
D10	Services: Conveying	100.0	118.0	105.1	100.0	87.0	96.3	100.0	95.1	98.6	100.0	87.5	96.4	100.0	92.3	97.8
D20 - 40	Mechanical	100.1	113.4	105.6	100.0	98.4	99.3	100.1	101.9	100.8	100.0	101.9	100.8	99.9	78.8	91.2
D50	Electrical	97.1	103.8	100.4	98.0	86.6	92.4	98.1	107.4	102.6	98.6	86.9	92.9	85.1	90.0	87.8
E	Equipment & Furnishings	100.0	139.8	102.4	100.0	101.0	100.1	100.0	117.0	101.0	100.0	98.9	99.9	100.0	80.3	98.8
G	Site Work	98.6	89.6	92.2	98.1	94.9	95.8	97.4	95.5	96.1	95.4	95.8	95.7	90.3	96.3	94.5
A-G	WEIGHTED AVERAGE	99.4	118.5	107.9	99.1	99.9	99.4	98.1	107.9	102.4	97.3	99.8	98.4	94.9	84.8	90.4

INDIANA

DIV. NO.	BUILDING SYSTEMS	BLOOMINGTON MAT.	INST.	TOTAL	EVANSVILLE MAT.	INST.	TOTAL	FORT WAYNE MAT.	INST.	TOTAL	GARY MAT.	INST.	TOTAL	INDIANAPOLIS MAT.	INST.	TOTAL
A	Substructure	94.6	83.9	88.0	93.2	96.2	95.0	101.2	83.2	90.1	100.4	105.1	103.3	95.5	88.0	90.9
B10	Shell: Superstructure	95.1	77.8	88.1	88.7	85.5	87.4	94.2	83.1	89.7	93.9	105.2	98.4	94.3	80.9	88.9
B20	Exterior Closure	102.0	80.3	91.4	99.9	85.7	93.0	94.1	83.8	89.1	94.7	107.2	100.8	101.6	83.9	92.9
B30	Roofing	91.3	85.2	88.9	99.3	88.8	92.8	100.1	86.3	94.7	99.5	105.5	101.9	98.5	86.9	94.0
C	Interior Construction	98.8	83.1	92.1	94.1	83.0	89.4	93.5	77.9	86.9	93.2	108.4	99.6	98.4	85.9	93.1
D10	Services: Conveying	100.0	91.7	97.6	100.0	97.7	99.3	100.0	91.2	97.5	100.0	92.4	97.8	100.0	93.6	98.2
D20 - 40	Mechanical	99.7	84.9	93.6	99.9	86.8	94.5	99.9	81.2	92.2	99.9	101.3	100.5	99.9	87.7	94.9
D50	Electrical	101.6	88.0	95.0	95.0	87.6	91.4	86.2	82.9	84.6	94.4	104.1	99.1	104.4	90.6	97.7
E	Equipment & Furnishings	100.0	80.9	98.8	100.0	79.7	98.8	100.0	77.1	98.6	100.0	106.9	100.4	100.0	84.6	99.1
G	Site Work	78.3	95.9	90.8	83.4	130.4	116.8	92.0	96.4	95.1	90.9	99.1	96.7	89.3	100.3	97.1
A-G	WEIGHTED AVERAGE	98.2	84.1	91.9	95.3	90.0	93.0	95.4	83.1	89.9	96.2	104.3	99.8	98.8	87.2	93.7

INDIANA / IOWA

DIV. NO.	BUILDING SYSTEMS	MUNCIE MAT.	INST.	TOTAL	SOUTH BEND MAT.	INST.	TOTAL	TERRE HAUTE MAT.	INST.	TOTAL	CEDAR RAPIDS MAT.	INST.	TOTAL	COUNCIL BLUFFS MAT.	INST.	TOTAL
A	Substructure	98.6	84.4	89.9	96.3	85.9	89.9	91.5	96.0	94.3	103.1	84.0	91.3	105.5	72.6	85.2
B10	Shell: Superstructure	96.8	85.0	92.1	93.1	92.3	92.8	89.0	85.9	87.7	91.6	87.1	89.8	94.8	77.7	87.9
B20	Exterior Closure	98.4	85.1	91.9	88.9	85.3	87.1	107.5	85.5	96.7	100.3	81.2	91.0	100.9	74.7	88.1
B30	Roofing	93.6	81.6	89.0	99.3	88.0	94.9	95.5	84.7	91.3	99.0	82.5	92.6	98.3	64.5	85.2
C	Interior Construction	94.3	81.4	88.8	89.9	81.8	86.5	94.4	85.2	90.5	106.4	76.2	93.6	100.0	55.9	81.3
D10	Services: Conveying	100.0	91.9	97.7	100.0	91.6	97.6	100.0	96.4	99.0	100.0	92.0	97.7	100.0	71.6	91.9
D20 - 40	Mechanical	99.7	78.7	91.0	99.9	82.8	92.8	99.9	76.9	90.4	100.3	85.9	94.3	100.3	79.0	91.5
D50	Electrical	89.7	74.6	82.4	99.2	86.5	92.9	93.5	88.0	90.9	91.1	85.5	90.0	101.0	85.5	93.4
E	Equipment & Furnishings	100.0	79.5	98.8	100.0	81.2	98.9	100.0	80.9	98.8	100.0	80.1	98.8	100.0	54.5	97.2
G	Site Work	78.9	95.6	90.8	90.2	95.8	94.2	84.8	130.4	117.2	89.3	94.6	93.0	93.6	91.3	91.9
A-G	WEIGHTED AVERAGE	96.2	82.4	90.1	95.3	86.8	91.5	96.0	88.3	92.6	98.9	84.4	92.5	99.3	75.2	88.6

IOWA

DIV. NO.	BUILDING SYSTEMS	DAVENPORT MAT.	INST.	TOTAL	DES MOINES MAT.	INST.	TOTAL	DUBUQUE MAT.	INST.	TOTAL	SIOUX CITY MAT.	INST.	TOTAL	WATERLOO MAT.	INST.	TOTAL
A	Substructure	100.3	96.6	98.0	101.9	84.2	91.0	99.3	82.7	89.0	103.6	72.0	84.2	102.9	64.2	79.1
B10	Shell: Superstructure	91.1	98.7	94.1	91.9	86.0	89.5	88.8	85.4	87.4	91.5	74.6	84.7	91.7	73.3	84.3
B20	Exterior Closure	98.4	91.7	95.1	96.2	79.6	88.1	99.5	76.1	88.1	97.0	60.6	79.2	96.8	62.9	80.2
B30	Roofing	98.5	93.0	96.3	99.2	78.8	91.3	98.6	69.9	87.4	98.5	58.2	82.8	98.2	49.5	79.3
C	Interior Construction	104.0	91.8	98.8	103.0	74.1	90.8	102.5	65.0	86.6	104.7	67.4	88.9	102.5	50.1	80.3
D10	Services: Conveying	100.0	95.6	98.7	100.0	92.5	97.9	100.0	90.7	97.3	100.0	90.4	97.3	100.0	85.5	95.8
D20 - 40	Mechanical	100.3	95.3	98.2	100.3	80.2	92.0	100.3	78.9	91.4	100.3	82.4	92.9	100.3	45.2	77.4
D50	Electrical	94.7	93.0	93.9	96.6	81.2	89.1	99.6	68.6	84.6	96.1	78.0	87.3	96.1	67.0	82.0
E	Equipment & Furnishings	100.0	91.8	99.5	100.0	80.4	98.8	100.0	66.8	98.0	100.0	69.1	98.1	100.0	47.9	96.8
G	Site Work	87.5	99.0	95.7	81.9	99.8	94.6	87.7	91.5	90.4	95.9	95.3	95.5	88.1	94.4	92.5
A-G	WEIGHTED AVERAGE	97.9	94.9	96.5	97.8	82.2	90.9	97.8	77.2	88.7	98.4	74.9	88.0	97.8	62.0	81.9

KANSAS

DIV. NO.	BUILDING SYSTEMS	DODGE CITY MAT.	INST.	TOTAL	KANSAS CITY MAT.	INST.	TOTAL	SALINA MAT.	INST.	TOTAL	TOPEKA MAT.	INST.	TOTAL	WICHITA MAT.	INST.	TOTAL
A	Substructure	112.2	69.8	86.1	93.6	93.3	93.4	101.0	63.6	78.0	94.5	64.9	76.3	91.7	69.3	77.9
B10	Shell: Superstructure	98.9	72.6	88.3	99.6	95.2	97.8	96.4	72.6	86.8	99.7	77.3	90.7	98.9	75.6	89.6
B20	Exterior Closure	108.3	58.9	84.2	101.9	96.3	99.2	109.8	52.6	81.9	97.3	65.2	81.6	94.4	66.8	80.9
B30	Roofing	95.5	60.8	82.0	93.3	99.3	95.6	95.1	56.7	80.2	95.0	68.5	84.7	94.7	64.1	82.8
C	Interior Construction	95.1	64.2	82.0	93.6	93.3	93.5	94.4	45.7	73.7	95.1	46.2	74.3	95.0	60.7	80.5
D10	Services: Conveying	100.0	47.7	85.0	100.0	78.7	93.9	100.0	64.1	89.7	100.0	69.0	91.1	100.0	67.5	90.7
D20 - 40	Mechanical	100.0	64.3	85.2	99.9	95.9	98.2	100.0	64.2	85.2	100.0	71.6	88.2	100.0	65.4	85.7
D50	Electrical	99.4	74.8	87.4	104.3	106.8	105.5	99.1	70.8	85.4	103.0	78.7	91.2	101.3	70.8	86.5
E	Equipment & Furnishings	100.0	73.9	98.4	100.0	94.2	99.6	100.0	48.7	96.9	100.0	41.4	96.4	100.0	55.2	97.3
G	Site Work	112.1	92.6	98.2	94.6	91.2	92.2	102.3	93.0	95.7	95.0	90.3	91.7	95.9	93.0	93.8
A-G	WEIGHTED AVERAGE	100.4	68.4	86.2	98.9	96.1	97.7	99.2	63.9	83.5	98.7	69.3	85.6	97.9	69.5	85.3

KENTUCKY / LOUISIANA

DIV. NO.	BUILDING SYSTEMS	BOWLING GREEN MAT.	INST.	TOTAL	LEXINGTON MAT.	INST.	TOTAL	LOUISVILLE MAT.	INST.	TOTAL	OWENSBORO MAT.	INST.	TOTAL	ALEXANDRIA MAT.	INST.	TOTAL
A	Substructure	84.6	85.2	85.0	92.2	83.8	87.0	89.5	87.6	88.4	87.5	86.2	86.7	93.9	56.8	71.1
B10	Shell: Superstructure	90.7	79.4	86.1	93.9	83.1	89.6	99.1	86.7	94.1	84.9	79.1	82.5	87.2	62.0	77.0
B20	Exterior Closure	100.8	77.4	89.3	96.0	52.6	74.8	96.7	82.5	89.7	106.7	56.0	81.9	100.4	53.4	77.4
B30	Roofing	85.6	80.5	83.6	95.8	82.0	90.4	85.8	81.3	84.1	95.4	69.9	85.5	99.6	52.2	81.2
C	Interior Construction	93.0	78.3	86.8	95.5	66.3	83.1	95.2	81.5	89.4	91.0	66.4	80.6	94.7	47.2	74.6
D10	Services: Conveying	100.0	57.5	87.8	100.0	77.5	93.6	100.0	79.5	94.1	100.0	99.5	99.9	100.0	70.3	91.5
D20 - 40	Mechanical	99.9	86.6	94.4	99.9	47.0	78.0	99.9	84.6	93.6	99.9	51.7	79.9	100.1	62.5	84.5
D50	Electrical	93.9	85.5	89.8	94.4	49.5	72.6	94.4	85.5	90.1	93.5	84.1	88.9	92.0	60.4	76.6
E	Equipment & Furnishings	100.0	83.2	99.0	100.0	71.4	98.3	100.0	82.9	99.0	100.0	62.1	97.7	100.0	43.0	96.5
G	Site Work	70.4	100.4	91.7	79.2	102.2	95.6	67.8	100.2	90.8	83.5	129.6	116.2	101.6	85.5	90.2
A-G	WEIGHTED AVERAGE	94.6	82.7	89.3	96.0	65.7	82.5	96.4	85.4	91.5	94.4	73.5	85.1	95.4	59.5	79.5

LOUISIANA

DIV. NO.	BUILDING SYSTEMS	BATON ROUGE MAT.	INST.	TOTAL	LAKE CHARLES MAT.	INST.	TOTAL	MONROE MAT.	INST.	TOTAL	NEW ORLEANS MAT.	INST.	TOTAL	SHREVEPORT MAT.	INST.	TOTAL
A	Substructure	89.6	64.2	74.0	92.8	64.4	75.3	93.6	57.7	71.5	92.7	74.6	81.6	94.3	59.6	73.0
B10	Shell: Superstructure	93.0	65.2	81.8	89.4	65.3	79.7	87.1	61.6	76.8	98.9	73.2	88.5	85.4	62.5	76.2
B20	Exterior Closure	101.2	54.8	78.5	100.1	56.1	78.6	98.0	58.1	78.5	100.8	64.0	82.8	93.5	51.0	72.7
B30	Roofing	99.0	58.5	83.2	101.2	58.0	84.4	99.6	54.4	82.1	102.3	62.9	87.0	98.7	54.4	81.5
C	Interior Construction	102.5	60.5	84.7	102.6	57.6	83.5	94.6	44.3	73.2	103.2	65.5	87.2	96.1	49.1	76.1
D10	Services: Conveying	100.0	78.8	93.9	100.0	78.9	94.0	100.0	63.6	89.6	100.0	80.7	94.5	100.0	73.4	92.4
D20 - 40	Mechanical	100.0	58.0	82.6	100.0	63.5	84.8	100.1	53.8	80.9	100.0	65.4	85.6	100.1	62.4	84.4
D50	Electrical	94.0	63.7	79.3	94.0	66.2	80.4	93.7	61.3	78.0	95.1	69.0	82.4	92.6	70.2	81.7
E	Equipment & Furnishings	100.0	62.4	97.7	100.0	61.8	97.7	100.0	43.5	96.6	100.0	69.9	98.2	100.0	47.4	96.8
G	Site Work	106.4	85.0	91.2	108.0	84.8	91.5	101.6	85.0	89.8	110.0	87.6	94.1	100.4	84.8	89.3
A-G	WEIGHTED AVERAGE	98.1	62.9	82.5	97.5	64.1	82.7	95.3	57.9	78.7	99.8	69.7	86.4	94.6	61.3	79.8

MAINE / MARYLAND

DIV. NO.	BUILDING SYSTEMS	AUGUSTA MAT.	INST.	TOTAL	BANGOR MAT.	INST.	TOTAL	LEWISTON MAT.	INST.	TOTAL	PORTLAND MAT.	INST.	TOTAL	BALTIMORE MAT.	INST.	TOTAL
A	Substructure	86.1	79.0	81.7	85.6	88.1	87.1	97.4	88.1	91.7	92.0	88.1	89.6	97.5	82.6	88.3
B10	Shell: Superstructure	87.7	78.2	83.9	87.0	83.9	85.8	92.3	83.9	88.9	92.2	83.9	88.9	100.7	89.3	96.1
B20	Exterior Closure	107.6	51.8	80.3	115.5	59.3	88.0	104.7	59.3	82.5	103.6	59.3	81.9	97.3	76.7	87.2
B30	Roofing	102.3	54.9	83.9	102.1	61.1	86.2	102.0	61.1	86.1	101.8	61.1	86.0	97.8	80.1	90.9
C	Interior Construction	101.3	57.6	82.7	100.2	80.7	91.9	103.8	80.7	94.0	103.8	80./	94.0	95.4	76.6	87.4
D10	Services: Conveying	100.0	59.4	88.4	100.0	77.2	93.5	100.0	77.2	93.5	100.0	77.2	93.5	100.0	87.2	96.3
D20 - 40	Mechanical	100.1	76.0	90.1	100.1	77.1	90.5	100.1	77.1	90.5	100.1	77.1	90.5	100.0	86.1	94.2
D50	Electrical	97.6	87.8	92.9	95.9	85.0	90.6	97.7	85.0	91.5	97.8	85.0	91.6	105.1	91.2	98.4
E	Equipment & Furnishings	100.0	61.8	97.7	100.0	93.7	99.6	100.0	93.7	99.6	100.0	93.7	99.6	100.0	76.8	98.6
G	Site Work	84.2	102.6	97.2	84.1	100.5	95.7	82.6	100.5	95.3	81.2	100.5	94.9	98.2	92.6	94.2
A-G	WEIGHTED AVERAGE	97.4	73.2	86.7	97.7	79.8	89.8	98.9	79.8	90.4	98.5	79.8	90.2	99.5	84.7	92.9

MARYLAND / MASSACHUSETTS

DIV. NO.	BUILDING SYSTEMS	HAGERSTOWN MAT.	INST.	TOTAL	BOSTON MAT.	INST.	TOTAL	BROCKTON MAT.	INST.	TOTAL	FALL RIVER MAT.	INST.	TOTAL	HYANNIS MAT.	INST.	TOTAL
A	Substructure	84.1	75.5	78.8	105.6	133.7	122.9	102.5	126.8	117.5	100.4	123.8	114.8	90.9	123.5	111.0
B10	Shell: Superstructure	94.5	81.6	89.3	104.0	133.8	116.0	100.8	128.1	111.8	100.3	122.0	109.0	95.2	121.5	105.7
B20	Exterior Closure	94.0	78.3	86.3	116.6	144.9	130.4	114.3	138.9	126.3	114.2	137.4	125.5	110.9	137.5	123.9
B30	Roofing	97.2	70.9	87.0	102.4	140.1	117.0	102.3	132.0	113.8	102.2	126.9	111.8	101.8	128.0	112.0
C	Interior Construction	92.5	73.3	84.4	101.6	144.2	119.6	100.3	131.7	113.6	100.2	130.0	112.9	95.7	129.8	110.2
D10	Services: Conveying	100.0	90.4	97.3	100.0	117.8	105.1	100.0	116.1	104.6	100.0	117.1	104.9	100.0	116.1	104.6
D20 - 40	Mechanical	100.0	90.0	95.9	100.1	128.8	112.0	100.1	110.1	104.2	100.1	109.9	104.2	100.1	109.8	104.1
D50	Electrical	101.7	84.3	93.2	96.8	120.4	108.3	95.8	103.4	99.5	95.4	103.4	99.3	94.0	103.4	98.6
E	Equipment & Furnishings	100.0	76.9	98.6	100.0	137.5	102.3	100.0	121.8	101.3	100.0	120.4	101.2	100.0	120.5	101.2
G	Site Work	88.2	88.1	88.1	90.1	109.1	103.6	87.7	105.1	100.0	86.6	105.3	99.8	85.2	105.1	99.3
A-G	WEIGHTED AVERAGE	96.2	81.9	89.8	102.5	132.0	115.6	101.1	121.0	110.0	100.9	119.1	109.0	98.1	119.0	107.4

MASSACHUSETTS

DIV. NO.	BUILDING SYSTEMS	LAWRENCE MAT.	INST.	TOTAL	LOWELL MAT.	INST.	TOTAL	NEW BEDFORD MAT.	INST.	TOTAL	PITTSFIELD MAT.	INST.	TOTAL	SPRINGFIELD MAT.	INST.	TOTAL
A	Substructure	102.8	124.8	116.4	97.0	124.9	114.1	98.5	123.8	114.1	98.6	107.6	104.1	99.4	111.7	107.0
B10	Shell: Superstructure	98.8	124.3	109.1	97.4	122.8	107.6	99.9	122.0	108.8	97.6	105.7	100.9	99.9	109.7	103.9
B20	Exterior Closure	114.0	138.6	126.0	102.3	138.2	119.8	114.1	137.4	125.5	102.6	113.1	107.7	102.4	115.7	108.9
B30	Roofing	102.2	131.9	113.7	102.0	131.0	113.3	102.2	126.9	111.8	102.1	108.7	104.6	102.0	111.2	105.6
C	Interior Construction	100.1	130.3	112.9	103.7	130.5	115.1	100.2	130.0	112.9	103.7	104.4	104.0	103.7	111.3	106.9
D10	Services: Conveying	100.0	116.1	104.6	100.0	116.1	104.6	100.0	117.1	104.9	100.0	105.9	101.7	100.0	107.1	102.0
D20 - 40	Mechanical	100.1	111.4	104.8	100.1	124.6	110.2	100.1	109.9	104.2	100.1	97.0	98.8	100.1	98.4	99.4
D50	Electrical	97.3	120.4	108.5	97.7	120.4	108.7	96.7	103.4	100.0	97.7	93.7	95.8	97.7	93.7	95.8
E	Equipment & Furnishings	100.0	121.3	101.3	100.0	121.4	101.3	100.0	120.4	101.2	100.0	96.4	99.8	100.0	105.8	100.4
G	Site Work	88.4	105.1	100.2	87.0	105.1	99.9	86.3	105.3	99.7	88.3	103.8	99.3	87.6	103.9	99.2
A-G	WEIGHTED AVERAGE	100.9	120.6	110.5	99.7	125.0	111.0	100.8	119.1	109.0	99.9	103.0	101.3	100.4	105.9	102.8

MASSACHUSETTS / MICHIGAN

DIV. NO.	BUILDING SYSTEMS	WORCESTER MAT.	INST.	TOTAL	ANN ARBOR MAT.	INST.	TOTAL	DEARBORN MAT.	INST.	TOTAL	DETROIT MAT.	INST.	TOTAL	FLINT MAT.	INST.	TOTAL
A	Substructure	97.0	126.1	114.9	90.9	111.0	103.3	89.6	116.5	106.2	96.9	116.6	109.0	90.2	99.1	95.7
B10	Shell: Superstructure	99.5	125.7	110.0	100.2	121.5	108.8	100.0	125.6	110.3	104.9	114.0	108.6	100.6	112.2	105.3
B20	Exterior Closure	102.3	138.8	120.1	97.7	105.0	101.3	97.6	121.4	109.3	97.3	120.8	108.8	99.7	92.5	95.2
B30	Roofing	102.0	125.3	111.1	102.8	110.3	105.7	101.9	125.6	111.1	99.5	125.6	109.6	100.8	90.5	96.8
C	Interior Construction	103.7	130.3	115.0	94.5	116.7	103.9	94.4	123.1	106.6	95.7	123.1	107.3	94.3	91.7	93.2
D10	Services: Conveying	100.0	109.3	102.7	100.0	108.6	102.4	100.0	111.2	103.2	100.0	111.2	103.2	100.0	89.5	97.0
D20 - 40	Mechanical	100.1	108.1	103.4	100.0	99.6	99.8	100.0	117.5	107.3	100.1	120.0	108.3	100.0	99.9	99.9
D50	Electrical	97.7	106.8	102.1	94.1	85.8	90.0	94.1	119.2	106.3	95.8	119.1	107.1	94.1	108.0	100.9
E	Equipment & Furnishings	100.0	120.7	101.3	100.0	119.8	101.2	100.0	123.1	101.4	100.0	123.1	101.4	100.0	95.7	99.7
G	Site Work	87.0	105.1	99.8	79.5	94.8	90.4	79.2	95.2	90.6	91.4	97.6	95.8	71.6	93.7	87.3
A-G	WEIGHTED AVERAGE	100.1	120.0	109.0	97.4	105.9	101.2	97.3	118.9	106.9	99.1	117.6	107.3	97.2	99.9	98.4

MICHIGAN

DIV. NO.	BUILDING SYSTEMS	GRAND RAPIDS MAT.	INST.	TOTAL	KALAMAZOO MAT.	INST.	TOTAL	LANSING MAT.	INST.	TOTAL	MUSKEGON MAT.	INST.	TOTAL	SAGINAW MAT.	INST.	TOTAL
A	Substructure	92.9	85.5	88.4	94.0	89.9	91.5	92.1	98.1	95.8	92.7	87.7	89.6	89.2	97.9	94.6
B10	Shell: Superstructure	94.0	80.9	88.8	93.0	85.9	90.2	99.2	111.3	104.1	91.0	84.3	88.3	100.2	110.8	104.4
B20	Exterior Closure	97.0	54.8	76.4	100.0	82.8	91.6	94.8	94.0	94.4	94.3	77.4	86.0	98.3	89.1	93.8
B30	Roofing	91.3	63.4	80.4	90.9	83.3	87.9	100.8	91.0	97.0	90.0	76.3	84.7	101.2	89.4	96.6
C	Interior Construction	94.5	68.1	83.3	93.3	82.5	88.7	96.8	94.4	95.8	91.7	79.5	86.5	94.0	89.3	92.0
D10	Services: Conveying	100.0	97.9	99.4	100.0	100.3	100.1	100.0	91.3	97.5	100.0	98.8	99.7	100.0	91.3	97.5
D20 - 40	Mechanical	99.9	55.0	81.3	99.9	84.6	93.6	100.0	90.0	95.8	99.8	86.5	94.3	100.0	92.1	96.7
D50	Electrical	94.8	59.8	77.8	93.7	88.6	91.2	92.3	86.1	89.3	94.2	82.5	88.5	93.4	81.2	87.5
E	Equipment & Furnishings	100.0	71.8	98.3	100.0	81.3	98.9	100.0	94.9	99.7	100.0	78.0	98.7	100.0	94.5	99.7
G	Site Work	83.4	87.7	86.5	83.9	88.1	86.9	84.7	93.3	90.8	81.4	87.9	86.0	73.5	93.3	87.5
A-G	WEIGHTED AVERAGE	96.0	68.1	83.6	95.8	85.8	91.4	97.2	92.6	95.3	94.4	83.6	89.6	97.0	93.4	95.4

MINNESOTA

DIV. NO.	BUILDING SYSTEMS	DULUTH MAT.	INST.	TOTAL	MINNEAPOLIS MAT.	INST.	TOTAL	ROCHESTER MAT.	INST.	TOTAL	SAINT PAUL MAT.	INST.	TOTAL	ST. CLOUD MAT.	INST.	TOTAL
A	Substructure	103.6	112.3	109.0	102.6	126.6	117.4	101.2	109.2	106.1	101.8	123.7	115.3	93.7	123.1	111.8
B10	Shell: Superstructure	95.8	122.0	106.4	97.8	136.3	113.3	95.2	124.9	107.2	94.4	134.9	110.7	91.0	133.0	107.9
B20	Exterior Closure	101.1	117.5	109.1	102.9	136.4	119.2	100.7	118.4	109.4	108.8	136.2	122.2	97.0	128.7	112.5
B30	Roofing	100.6	122.4	109.0	100.3	138.3	115.1	100.4	106.5	102.8	100.2	137.4	114.7	98.2	122.3	107.6
C	Interior Construction	100.1	123.6	110.1	101.7	138.6	117.4	100.0	109.3	103.9	97.5	135.4	113.6	91.0	134.8	109.6
D10	Services: Conveying	100.0	99.8	99.9	100.0	105.1	101.4	100.0	98.4	99.5	100.0	104.6	101.3	100.0	103.2	100.9
D20 - 40	Mechanical	100.1	114.6	106.1	100.1	119.3	108.1	100.1	106.4	102.7	100.1	115.7	106.6	99.7	117.9	107.2
D50	Electrical	103.5	100.6	102.1	105.3	114.3	109.7	103.5	87.2	95.6	104.1	114.2	109.0	101.4	114.2	107.7
E	Equipment & Furnishings	100.0	122.9	101.4	100.0	137.3	102.3	100.0	109.9	100.6	100.0	132.6	102.0	100.0	132.4	102.0
G	Site Work	83.1	102.5	96.9	83.0	108.8	101.3	82.2	101.8	96.1	85.3	103.2	98.0	81.2	105.3	98.3
A-G	WEIGHTED AVERAGE	99.5	114.5	106.2	100.5	126.7	112.2	99.2	108.5	103.4	99.7	124.5	110.7	95.7	123.3	108.0

MISSISSIPPI / MISSOURI

DIV. NO.	BUILDING SYSTEMS	BILOXI MAT.	INST.	TOTAL	GREENVILLE MAT.	INST.	TOTAL	JACKSON MAT.	INST.	TOTAL	MERIDIAN MAT.	INST.	TOTAL	CAPE GIRARDEAU MAT.	INST.	TOTAL
A	Substructure	102.0	56.7	74.1	102.8	49.8	70.2	100.3	52.6	70.9	97.7	50.5	68.6	89.3	84.2	86.1
B10	Shell: Superstructure	92.9	65.9	82.0	90.9	56.3	76.9	92.6	59.2	79.1	89.9	57.0	76.7	91.3	95.6	93.0
B20	Exterior Closure	92.2	40.8	67.1	115.1	40.9	78.8	93.2	41.2	67.8	90.5	32.1	62.0	101.3	82.3	92.0
B30	Roofing	96.4	47.3	77.3	96.5	41.1	74.9	96.2	42.5	75.3	96.0	41.1	74.7	92.6	80.7	88.0
C	Interior Construction	102.5	41.4	76.6	99.1	34.3	71.6	102.7	39.2	75.8	98.6	32.4	70.5	92.0	75.8	85.1
D10	Services: Conveying	100.0	53.0	86.5	100.0	52.3	86.4	100.0	52.3	86.4	100.0	52.4	86.4	100.0	65.5	90.1
D20 - 40	Mechanical	99.9	53.9	80.8	99.9	34.3	72.7	99.8	34.0	72.5	99.9	34.0	72.6	100.1	100.9	100.5
D50	Electrical	97.0	61.0	79.5	96.3	37.1	67.6	97.8	37.1	68.3	96.0	61.3	79.1	101.4	114.6	107.8
E	Equipment & Furnishings	100.0	41.6	96.4	100.0	33.2	95.9	100.0	39.0	96.3	100.0	33.1	95.9	100.0	76.3	98.6
G	Site Work	101.4	85.6	90.1	105.4	85.3	91.2	98.1	85.3	89.0	97.5	85.3	88.8	94.0	92.8	93.2
A-G	WEIGHTED AVERAGE	97.8	55.5	79.0	99.3	44.6	75.0	97.8	46.1	74.8	96.0	46.7	74.1	96.4	92.3	94.6

MISSOURI

DIV. NO.	BUILDING SYSTEMS	COLUMBIA MAT.	INST.	TOTAL	JOPLIN MAT.	INST.	TOTAL	KANSAS CITY MAT.	INST.	TOTAL	SPRINGFIELD MAT.	INST.	TOTAL	ST. JOSEPH MAT.	INST.	TOTAL
A	Substructure	93.2	86.8	89.2	105.8	80.0	89.9	98.9	105.9	103.2	101.2	82.2	89.5	98.7	94.8	96.3
B10	Shell: Superstructure	92.7	102.7	96.7	101.0	82.3	93.5	105.6	111.7	108.1	98.3	93.5	96.4	102.4	100.3	101.5
B20	Exterior Closure	105.3	91.3	98.5	93.4	65.1	79.6	97.1	107.6	102.2	92.2	83.7	88.1	96.9	95.1	96.0
B30	Roofing	89.5	84.5	87.6	93.6	72.8	85.5	92.3	108.3	98.5	94.2	77.3	87.6	92.8	94.5	93.5
C	Interior Construction	91.9	73.7	84.2	95.8	67.7	83.9	96.6	106.2	100.7	96.6	70.9	85.7	96.4	84.4	91.3
D10	Services: Conveying	100.0	94.1	98.3	100.0	77.7	93.6	100.0	96.1	98.9	100.0	90.5	97.3	100.0	93.3	98.1
D20 - 40	Mechanical	99.9	101.8	100.7	100.1	57.5	82.4	99.9	105.4	102.2	100.0	69.7	87.4	100.1	92.4	96.9
D50	Electrical	99.0	92.5	95.8	96.8	62.1	80.0	108.3	109.4	108.8	102.4	67.1	85.2	105.3	93.2	99.4
E	Equipment & Furnishings	100.0	67.5	98.0	100.0	72.5	98.3	100.0	106.3	100.4	100.0	72.6	98.3	100.0	83.6	99.0
G	Site Work	103.5	94.1	96.8	104.6	98.6	100.4	99.4	98.0	98.4	97.8	93.6	94.8	101.0	93.5	95.7
A-G	WEIGHTED AVERAGE	97.1	92.4	95.0	98.6	70.5	86.1	100.9	106.8	103.5	98.4	78.7	89.6	99.9	93.3	97.0

DIV. NO.	BUILDING SYSTEMS	MISSOURI			MONTANA										
		ST. LOUIS			BILLINGS			BUTTE			GREAT FALLS			HELENA		
		MAT.	INST.	TOTAL	MAT.	INST.	TOTAL	MAT.	INST.	TOTAL	MAT.	INST.	TOTAL	MAT.	INST.	TOTAL
A	Substructure	89.1	104.8	98.8	108.8	75.0	88.0	112.6	74.2	88.9	116.9	71.7	89.1	119.3	72.8	90.7
B10	Shell: Superstructure	95.5	115.6	103.6	105.1	77.4	95.1	105.1	76.7	91.5	105.8	75.6	93.7	105.7	72.5	92.3
B20	Exterior Closure	91.7	113.4	102.3	107.2	69.8	88.9	104.2	65.1	85.1	108.9	72.4	91.0	106.6	65.4	86.5
B30	Roofing	92.6	106.3	97.9	98.9	66.9	86.5	98.4	66.7	86.1	99.0	66.4	86.3	99.0	68.6	87.2
C	Interior Construction	94.6	103.9	98.5	103.2	61.5	85.5	99.5	56.5	81.3	103.6	61.4	85.7	103.0	61.4	85.4
D10	Services: Conveying	100.0	99.2	99.8	100.0	67.7	90.8	100.0	66.7	90.5	100.0	69.1	91.2	100.0	55.3	87.2
D20 - 40	Mechanical	100.1	108.3	103.5	100.3	76.5	90.4	100.3	69.5	87.5	100.3	78.2	91.1	100.3	73.9	89.4
D50	Electrical	104.3	114.6	109.3	94.2	78.7	86.6	100.9	75.6	88.6	94.2	73.3	84.0	94.2	74.6	84.7
E	Equipment & Furnishings	100.0	103.7	100.2	100.0	63.9	97.8	100.0	61.9	97.7	100.0	63.7	97.8	100.0	63.9	97.8
G	Site Work	93.3	99.1	95.2	87.8	95.8	93.5	94.2	94.5	94.4	97.7	95.3	96.0	99.6	95.2	96.5
A-G	WEIGHTED AVERAGE	97.0	109.0	102.3	102.1	74.5	89.8	101.2	70.9	87.7	102.7	73.8	89.9	102.5	71.5	88.7

DIV. NO.	BUILDING SYSTEMS	MONTANA			NEBRASKA										
		MISSOULA			GRAND ISLAND			LINCOLN			NORTH PLATTE			OMAHA		
		MAT.	INST.	TOTAL	MAT.	INST.	TOTAL	MAT.	INST.	TOTAL	MAT.	INST.	TOTAL	MAT.	INST.	TOTAL
A	Substructure	91.6	71.4	79.2	111.4	65.6	83.2	102.3	65.1	79.4	111.7	62.1	81.2	104.9	78.3	88.5
B10	Shell: Superstructure	95.0	75.8	87.3	97.4	70.5	86.6	99.8	70.7	88.1	97.6	67.3	85.4	100.1	76.0	90.4
B20	Exterior Closure	109.4	59.9	85.2	101.0	50.0	76.1	96.0	64.6	80.6	93.9	43.0	69.0	98.2	75.9	87.3
B30	Roofing	97.7	69.3	86.7	94.6	60.6	81.4	94.9	60.8	81.7	94.5	48.1	76.5	89.3	69.6	81.7
C	Interior Construction	99.6	58.1	82.0	92.0	45.3	72.2	96.0	43.2	73.6	91.9	45.1	72.0	107.0	68.8	90.8
D10	Services: Conveying	100.0	50.5	85.9	100.0	72.5	92.1	100.0	72.5	92.1	100.0	52.7	86.5	100.0	70.6	91.6
D20 - 40	Mechanical	100.3	69.3	87.4	100.0	74.8	89.6	100.0	74.9	89.6	100.0	73.8	89.1	99.8	81.1	92.0
D50	Electrical	99.0	78.4	89.0	91.0	70.3	80.9	96.6	70.3	83.8	92.9	43.0	68.7	90.6	86.3	88.5
E	Equipment & Furnishings	100.0	58.5	97.5	100.0	45.4	96.7	100.0	40.2	96.4	100.0	43.5	96.6	100.0	75.4	98.5
G	Site Work	77.3	94.3	89.4	101.5	91.3	94.3	92.8	91.3	91.7	102.4	89.4	93.1	80.4	90.2	87.4
A-G	WEIGHTED AVERAGE	98.9	70.1	86.1	97.5	65.6	83.3	98.2	67.1	84.4	97.1	58.8	80.1	99.3	78.2	89.9

DIV. NO.	BUILDING SYSTEMS	NEVADA									NEW HAMPSHIRE					
		CARSON CITY			LAS VEGAS			RENO			MANCHESTER			NASHUA		
		MAT.	INST.	TOTAL	MAT.	INST.	TOTAL	MAT.	INST.	TOTAL	MAT.	INST.	TOTAL	MAT.	INST.	TOTAL
A	Substructure	104.0	100.7	102.0	99.1	109.4	105.5	104.3	101.3	102.5	102.8	84.2	91.3	97.9	84.2	89.5
B10	Shell: Superstructure	96.2	101.4	98.3	95.5	108.7	100.8	96.5	102.7	99.0	100.7	80.6	92.6	99.6	80.6	91.9
B20	Exterior Closure	117.0	85.2	101.4	112.9	101.9	107.5	116.8	85.3	101.4	104.8	88.4	96.8	104.9	88.4	96.9
B30	Roofing	99.7	91.4	96.5	103.2	103.6	103.4	99.7	91.4	96.5	101.9	85.5	95.5	102.2	85.5	95.7
C	Interior Construction	95.4	94.2	94.9	96.4	108.4	101.5	96.7	94.4	95.8	103.8	70.0	89.5	103.9	70.0	89.5
D10	Services: Conveying	100.0	108.5	102.4	100.0	102.6	100.7	100.0	108.5	102.4	100.0	77.7	93.6	100.0	77.7	93.6
D20 - 40	Mechanical	100.1	93.0	97.2	100.1	107.1	103.0	100.1	93.1	97.2	100.1	90.3	96.0	100.1	90.3	96.0
D50	Electrical	91.0	110.9	100.7	93.0	108.8	100.6	91.0	110.9	100.7	97.8	72.7	85.6	97.6	72.7	85.5
E	Equipment & Furnishings	100.0	103.2	100.2	100.0	108.2	100.5	100.0	103.2	100.2	100.0	54.3	97.2	100.0	54.3	97.2
G	Site Work	63.7	104.1	92.3	63.0	106.3	93.7	63.2	104.1	92.2	88.1	100.0	96.5	89.0	100.0	96.8
A-G	WEIGHTED AVERAGE	98.6	97.9	98.3	98.3	107.1	102.2	98.9	98.2	98.6	100.9	82.4	92.7	100.6	82.4	92.5

DIV. NO.	BUILDING SYSTEMS	NEW HAMPSHIRE			NEW JERSEY										
		PORTSMOUTH			CAMDEN			ELIZABETH			JERSEY CITY			NEWARK		
		MAT.	INST.	TOTAL	MAT.	INST.	TOTAL	MAT.	INST.	TOTAL	MAT.	INST.	TOTAL	MAT.	INST.	TOTAL
A	Substructure	88.4	80.7	83.6	90.9	118.7	108.1	93.0	122.0	110.9	90.9	121.8	109.9	101.3	122.1	114.1
B10	Shell: Superstructure	94.2	77.8	87.6	97.8	111.9	103.5	95.8	116.9	104.3	97.8	115.3	104.9	99.7	116.9	106.6
B20	Exterior Closure	100.5	78.2	89.6	101.1	120.2	110.5	112.0	125.8	118.7	101.1	125.5	113.1	105.7	125.8	115.5
B30	Roofing	101.8	89.1	96.9	101.7	123.0	110.0	102.5	134.3	114.9	101.7	134.3	114.4	101.9	134.3	114.5
C	Interior Construction	100.8	62.8	84.6	104.0	124.2	112.6	105.4	129.2	115.5	104.0	129.2	114.7	107.9	129.2	116.9
D10	Services: Conveying	100.0	73.3	92.4	100.0	115.3	104.4	100.0	116.9	104.8	100.0	116.9	104.8	100.0	116.9	104.8
D20 - 40	Mechanical	100.1	84.8	93.7	100.1	116.6	106.9	100.1	121.9	109.1	100.1	125.9	110.8	100.1	124.7	110.3
D50	Electrical	95.9	72.7	84.6	98.0	114.8	106.2	95.9	135.0	114.9	99.5	137.6	118.0	99.4	137.6	118.0
E	Equipment & Furnishings	100.0	53.8	97.2	100.0	124.7	101.5	100.0	130.5	101.9	100.0	130.5	101.9	100.0	130.6	101.9
G	Site Work	84.0	99.0	94.6	91.3	105.2	101.2	105.7	105.6	105.6	91.3	105.0	101.0	109.0	105.6	106.6
A-G	WEIGHTED AVERAGE	97.8	78.0	89.0	99.6	116.8	107.2	100.8	123.6	110.9	99.8	124.5	110.8	102.1	124.6	112.1

DIV. NO.	BUILDING SYSTEMS	NEW JERSEY						NEW MEXICO								
		PATERSON			TRENTON			ALBUQUERQUE			FARMINGTON			LAS CRUCES		
		MAT.	INST.	TOTAL	MAT.	INST.	TOTAL	MAT.	INST.	TOTAL	MAT.	INST.	TOTAL	MAT.	INST.	TOTAL
A	Substructure	105.8	121.8	115.7	100.3	120.2	112.6	100.7	80.7	88.4	103.2	80.7	89.4	94.5	69.8	79.3
B10	Shell: Superstructure	96.9	116.7	104.9	95.8	113.3	102.8	100.6	79.8	92.2	99.3	79.8	91.5	97.1	69.6	86.0
B20	Exterior Closure	103.7	125.7	114.5	100.9	120.5	110.5	111.5	66.6	89.6	116.1	66.6	91.9	92.8	60.9	77.2
B30	Roofing	102.3	125.5	111.3	100.6	122.5	109.1	98.9	74.4	89.4	99.6	74.4	89.8	84.3	67.7	77.8
C	Interior Construction	107.7	129.2	116.8	104.1	127.3	113.9	96.8	67.4	84.3	96.6	67.4	84.2	99.1	65.7	85.0
D10	Services: Conveying	100.0	116.9	104.8	100.0	115.1	104.3	100.0	71.8	91.9	100.0	71.8	91.9	100.0	67.6	90.7
D20 - 40	Mechanical	100.1	125.9	110.8	100.1	127.2	111.3	100.1	72.2	88.5	100.1	72.2	88.5	100.3	71.6	88.4
D50	Electrical	99.5	135.0	116.8	98.1	134.7	115.9	84.0	78.1	81.1	82.4	78.1	80.3	84.7	58.9	72.2
E	Equipment & Furnishings	100.0	130.4	101.8	100.0	130.5	101.9	100.0	70.0	98.2	100.0	70.0	98.2	100.0	68.3	98.1
G	Site Work	104.3	105.0	104.8	93.3	105.5	102.0	79.4	111.1	101.9	85.5	111.1	103.7	89.7	86.5	87.4
A-G	WEIGHTED AVERAGE	101.4	124.2	111.5	99.6	122.6	109.8	98.5	76.2	88.6	98.7	76.2	88.7	95.8	67.9	83.4

DIV. NO.	BUILDING SYSTEMS	NEW MEXICO						NEW YORK								
		ROSWELL			SANTA FE			ALBANY			BINGHAMTON			BUFFALO		
		MAT.	INST.	TOTAL	MAT.	INST.	TOTAL	MAT.	INST.	TOTAL	MAT.	INST.	TOTAL	MAT.	INST.	TOTAL
A	Substructure	98.6	79.2	86.6	98.7	80.7	87.6	90.8	99.4	96.1	103.2	88.5	94.2	109.8	109.5	109.6
B10	Shell: Superstructure	98.2	76.1	89.3	98.4	79.8	90.9	95.5	102.9	98.5	95.5	101.7	98.0	102.4	104.0	103.1
B20	Exterior Closure	120.2	66.1	93.7	110.1	66.6	88.9	99.4	96.9	98.2	98.7	89.1	94.0	99.7	116.8	108.1
B30	Roofing	99.9	74.4	90.0	99.1	74.4	89.5	93.8	91.9	93.1	102.7	86.1	96.3	97.7	109.4	102.3
C	Interior Construction	95.1	66.7	83.1	94.8	67.4	83.2	97.1	91.6	94.8	95.0	81.4	89.2	94.4	115.0	103.1
D10	Services: Conveying	100.0	71.8	91.9	100.0	71.8	91.9	100.0	96.3	98.9	100.0	94.9	98.5	100.0	106.3	101.8
D20 - 40	Mechanical	100.1	72.0	88.4	100.1	72.2	88.5	100.3	91.8	96.8	100.5	84.5	93.9	99.8	99.6	99.7
D50	Electrical	83.1	78.1	80.7	84.0	78.1	81.1	103.6	94.1	99.0	101.6	87.4	94.7	100.2	99.2	99.7
E	Equipment & Furnishings	100.0	70.0	98.2	100.0	70.0	98.2	100.0	92.0	99.5	100.0	79.1	98.7	100.0	113.8	100.8
G	Site Work	89.5	111.1	104.8	79.5	111.1	101.9	71.4	106.5	96.3	94.2	89.2	90.6	96.9	93.7	94.6
A-G	WEIGHTED AVERAGE	98.7	75.3	88.3	97.5	76.2	88.0	97.8	96.4	97.2	98.5	88.8	94.2	99.8	105.7	102.4

COST INDEXES

523

NEW YORK

DIV. NO.	BUILDING SYSTEMS	HICKSVILLE			NEW YORK			RIVERHEAD			ROCHESTER			SCHENECTADY		
		MAT.	INST.	TOTAL	MAT.	INST.	TOTAL	MAT.	INST.	TOTAL	MAT.	INST.	TOTAL	MAT.	INST.	TOTAL
A	Substructure	99.1	153.0	132.3	118.4	162.4	145.5	100.9	153.0	133.0	104.8	101.5	102.7	95.9	99.0	97.8
B10	Shell: Superstructure	101.5	148.5	120.4	107.0	159.7	128.2	102.5	148.5	121.1	98.8	103.8	100.9	99.0	102.7	99.3
B20	Exterior Closure	109.4	158.9	133.6	108.6	158.8	133.2	111.6	158.9	134.8	106.7	99.6	103.2	99.9	95.8	97.9
B30	Roofing	109.5	147.2	124.2	114.4	161.3	132.6	110.1	147.2	124.5	94.6	98.4	96.1	93.9	91.3	92.9
C	Interior Construction	95.2	142.8	115.4	102.5	179.7	135.3	96.3	142.8	116.0	96.8	102.2	99.1	96.9	91.3	94.6
D10	Services: Conveying	100.0	135.1	110.0	100.0	135.4	110.1	100.0	135.1	110.0	100.0	100.2	100.0	100.0	95.7	98.8
D20 - 40	Mechanical	99.8	152.0	121.5	100.2	164.6	126.9	99.9	152.0	121.5	99.8	94.7	97.7	100.3	91.1	96.5
D50	Electrical	102.0	155.1	127.8	109.9	181.3	144.6	103.5	155.1	128.6	103.9	94.9	99.5	101.1	94.1	97.7
E	Equipment & Furnishings	100.0	155.3	103.4	100.0	189.1	105.4	100.0	155.3	103.4	100.0	103.0	100.2	100.0	92.0	99.5
G	Site Work	113.8	131.8	126.6	135.0	129.2	130.9	114.9	131.8	126.9	74.6	107.3	97.8	71.6	106.4	96.3
A-G	WEIGHTED AVERAGE	101.2	149.5	122.7	106.0	164.4	131.9	102.2	149.5	123.2	99.8	99.7	99.8	98.1	95.9	97.1

NEW YORK

DIV. NO.	BUILDING SYSTEMS	SYRACUSE			UTICA			WATERTOWN			WHITE PLAINS			YONKERS		
		MAT.	INST.	TOTAL	MAT.	INST.	TOTAL	MAT.	INST.	TOTAL	MAT.	INST.	TOTAL	MAT.	INST.	TOTAL
A	Substructure	98.7	97.0	97.7	91.7	92.0	91.9	100.1	92.1	95.2	106.1	139.2	126.5	116.6	139.1	130.5
B10	Shell: Superstructure	97.1	100.2	98.3	94.5	95.1	94.7	95.3	95.2	95.3	94.9	138.8	112.6	103.7	138.9	117.8
B20	Exterior Closure	102.2	97.8	100.1	99.2	90.5	95.0	106.9	99.8	98.5	101.3	134.6	117.6	108.6	134.6	121.3
B30	Roofing	102.2	95.6	99.7	93.9	91.7	93.0	94.1	91.7	93.2	114.9	141.2	125.1	115.3	141.2	125.3
C	Interior Construction	95.7	88.9	92.8	96.9	83.5	91.2	95.3	87.4	92.0	94.2	146.1	116.2	100.5	148.3	120.7
D10	Services: Conveying	100.0	97.6	99.3	100.0	94.1	98.3	100.0	76.5	93.3	100.0	132.1	109.2	100.0	132.1	109.2
D20 - 40	Mechanical	100.3	89.0	95.6	100.3	87.2	94.9	100.3	74.1	89.5	100.5	131.7	113.4	100.5	131.7	113.4
D50	Electrical	101.1	92.2	96.8	97.7	91.8	94.8	101.1	91.4	96.4	99.0	147.1	122.4	106.4	147.1	126.2
E	Equipment & Furnishings	100.0	87.6	99.2	100.0	83.3	99.0	100.0	92.0	99.5	100.0	140.3	102.5	100.0	140.2	102.4
G	Site Work	92.8	106.9	102.8	69.5	105.3	94.9	77.3	107.2	98.5	126.7	125.8	126.0	134.4	125.6	128.1
A-G	WEIGHTED AVERAGE	99.0	94.7	97.1	96.9	90.9	94.2	98.5	88.5	94.1	99.5	138.2	116.7	104.5	138.5	119.6

NORTH CAROLINA

DIV. NO.	BUILDING SYSTEMS	ASHEVILLE			CHARLOTTE			DURHAM			FAYETTEVILLE			GREENSBORO		
		MAT.	INST.	TOTAL	MAT.	INST.	TOTAL	MAT.	INST.	TOTAL	MAT.	INST.	TOTAL	MAT.	INST.	TOTAL
A	Substructure	96.1	52.9	69.5	98.4	53.0	70.5	96.6	56.5	71.9	93.7	56.5	70.8	96.1	56.4	71.6
B10	Shell: Superstructure	93.1	62.3	80.7	95.9	62.3	82.4	97.2	64.9	84.2	96.3	64.9	83.7	97.5	64.8	84.3
B20	Exterior Closure	93.1	42.6	68.4	95.7	42.6	69.8	94.3	43.2	69.3	94.8	43.2	69.6	94.2	43.2	69.2
B30	Roofing	101.2	46.2	79.8	101.2	48.1	80.6	101.7	46.9	80.4	101.4	47.7	80.5	101.7	46.9	80.4
C	Interior Construction	95.3	43.3	73.2	99.1	43.3	75.4	99.1	43.8	75.6	95.5	43.8	73.5	99.1	43.8	75.6
D10	Services: Conveying	100.0	61.5	89.0	100.0	61.5	89.0	100.0	79.3	94.1	100.0	79.3	94.1	100.0	61.5	89.0
D20 - 40	Mechanical	100.1	42.6	76.2	100.1	42.6	76.2	100.1	42.8	76.3	100.1	42.8	76.3	100.1	42.7	76.3
D50	Electrical	97.8	39.4	69.4	99.6	38.5	69.9	97.3	37.7	68.3	94.7	37.7	67.0	97.9	37.7	68.6
E	Equipment & Furnishings	100.0	42.3	96.5	100.0	42.3	96.5	100.0	42.2	96.5	100.0	42.2	96.5	100.0	42.2	96.5
G	Site Work	103.2	71.4	80.6	103.3	71.4	80.7	103.0	81.5	87.7	101.2	81.5	87.2	102.9	81.5	87.7
A-G	WEIGHTED AVERAGE	96.8	48.9	75.5	98.6	48.9	76.5	98.4	50.7	77.2	97.1	50.8	76.5	98.5	50.3	77.1

NORTH CAROLINA / NORTH DAKOTA

DIV. NO.	BUILDING SYSTEMS	RALEIGH			WILMINGTON			WINSTON-SALEM			BISMARCK			FARGO		
		MAT.	INST.	TOTAL	MAT.	INST.	TOTAL	MAT.	INST.	TOTAL	MAT.	INST.	TOTAL	MAT.	INST.	TOTAL
A	Substructure	100.4	56.5	73.4	96.2	54.2	70.3	98.1	55.3	71.7	102.5	66.7	80.5	103.8	67.3	81.3
B10	Shell: Superstructure	96.5	64.9	83.8	92.8	64.9	81.6	95.8	62.3	82.3	94.8	70.1	84.8	97.0	69.7	86.0
B20	Exterior Closure	95.7	43.2	70.0	88.2	43.2	66.2	94.3	42.8	69.1	99.3	64.1	82.1	109.9	50.1	80.7
B30	Roofing	101.5	46.9	80.3	101.2	47.7	80.4	101.7	46.9	80.4	99.3	54.0	81.7	99.9	54.7	82.3
C	Interior Construction	97.8	43.8	74.9	95.7	43.8	73.6	99.1	43.3	75.5	107.3	51.9	83.8	107.2	50.4	83.1
D10	Services: Conveying	100.0	79.3	94.1	100.0	79.3	94.1	100.0	61.5	89.0	100.0	72.6	92.2	100.0	72.6	92.2
D20 - 40	Mechanical	100.1	42.8	76.3	100.1	42.8	76.2	100.1	42.7	76.2	100.5	65.3	85.9	100.5	69.3	87.6
D50	Electrical	97.9	37.7	68.6	98.3	37.7	68.8	97.9	37.7	68.6	93.6	69.7	82.0	93.5	68.7	81.4
E	Equipment & Furnishings	100.0	42.2	96.5	100.0	42.2	96.5	100.0	42.2	96.5	100.0	46.3	96.7	100.0	46.8	96.8
G	Site Work	104.2	81.5	88.1	104.2	71.4	80.9	103.3	81.4	87.8	89.9	96.4	94.5	89.6	96.4	94.4
A-G	WEIGHTED AVERAGE	98.4	50.7	77.2	96.4	49.9	75.7	98.2	49.6	76.6	99.3	66.6	84.8	100.9	65.2	85.1

NORTH DAKOTA / OHIO

DIV. NO.	BUILDING SYSTEMS	GRAND FORKS			MINOT			AKRON			CANTON			CINCINNATI		
		MAT.	INST.	TOTAL	MAT.	INST.	TOTAL	MAT.	INST.	TOTAL	MAT.	INST.	TOTAL	MAT.	INST.	TOTAL
A	Substructure	104.3	63.3	79.0	104.3	68.7	82.4	97.9	101.2	99.9	98.5	94.7	96.2	87.8	92.9	91.0
B10	Shell: Superstructure	94.8	66.1	83.3	94.8	71.8	85.6	91.5	89.7	90.8	91.6	82.1	87.8	88.5	88.0	88.3
B20	Exterior Closure	105.4	64.3	85.3	104.3	65.5	85.3	95.7	101.4	98.5	95.4	89.0	92.3	90.7	93.2	92.0
B30	Roofing	99.9	52.6	81.5	99.6	54.9	82.3	99.5	100.6	99.9	100.0	96.6	98.7	94.1	96.5	95.0
C	Interior Construction	107.1	42.7	79.8	106.5	57.3	85.6	101.7	97.3	99.8	99.0	84.4	92.8	98.0	89.4	94.3
D10	Services: Conveying	100.0	45.5	84.4	100.0	72.6	92.2	100.0	85.6	95.9	100.0	73.6	92.5	100.0	93.6	98.2
D20 - 40	Mechanical	100.5	45.2	77.5	100.5	63.4	85.1	99.9	102.0	100.8	99.9	86.3	94.2	99.8	88.5	95.1
D50	Electrical	96.2	69.7	83.3	99.2	70.9	85.4	98.2	94.9	96.6	97.4	91.0	94.3	98.1	81.5	90.1
E	Equipment & Furnishings	100.0	42.3	96.5	100.0	54.7	97.2	100.0	94.5	99.7	100.0	84.1	99.0	100.0	86.7	99.2
G	Site Work	97.7	93.9	95.0	95.3	96.4	96.1	101.9	107.1	105.6	102.1	107.1	105.7	75.0	108.8	99.0
A-G	WEIGHTED AVERAGE	100.5	59.2	82.2	100.6	67.8	86.1	97.8	98.0	97.9	97.3	88.4	93.4	94.9	90.2	92.8

OHIO

DIV. NO.	BUILDING SYSTEMS	CLEVELAND			COLUMBUS			DAYTON			LORAIN			SPRINGFIELD		
		MAT.	INST.	TOTAL	MAT.	INST.	TOTAL	MAT.	INST.	TOTAL	MAT.	INST.	TOTAL	MAT.	INST.	TOTAL
A	Substructure	96.9	105.4	102.1	86.7	90.8	89.3	82.9	89.1	86.7	95.0	104.1	100.6	85.3	91.0	88.8
B10	Shell: Superstructure	92.4	93.8	93.0	94.4	83.5	90.0	87.0	79.8	84.1	91.3	92.5	91.8	87.4	81.4	85.0
B20	Exterior Closure	96.5	105.6	100.9	98.4	90.0	94.3	90.8	87.7	89.3	93.6	96.6	96.6	90.5	87.0	88.8
B30	Roofing	98.2	114.1	104.4	98.1	93.2	96.2	99.3	89.8	95.6	99.9	108.3	103.2	99.2	90.8	95.9
C	Interior Construction	96.9	103.5	99.7	98.8	87.0	93.8	98.5	79.7	90.5	99.0	105.4	101.7	97.7	85.0	92.3
D10	Services: Conveying	100.0	106.0	101.7	100.0	94.0	98.3	100.0	92.7	97.9	100.0	103.1	100.9	100.0	92.7	97.9
D20 - 40	Mechanical	99.9	108.1	103.3	100.0	91.7	96.6	100.8	88.0	95.5	99.9	96.8	96.1	100.8	88.0	95.4
D50	Electrical	98.2	107.1	102.6	98.1	88.4	93.4	96.1	88.3	92.3	97.5	88.6	93.2	96.1	87.0	91.7
E	Equipment & Furnishings	100.0	99.7	100.0	100.0	85.0	99.1	100.0	76.7	98.6	100.0	103.4	100.2	100.0	84.4	99.1
G	Site Work	101.9	107.2	105.6	86.3	102.0	97.4	73.9	108.6	98.5	101.2	106.4	104.9	74.5	107.7	98.0
A-G	WEIGHTED AVERAGE	97.2	104.3	100.4	97.3	89.6	93.9	94.6	87.0	91.2	96.9	97.1	97.0	94.6	87.9	91.6

OHIO / OKLAHOMA

DIV. NO.	BUILDING SYSTEMS	OHIO TOLEDO MAT.	INST.	TOTAL	YOUNGSTOWN MAT.	INST.	TOTAL	OKLAHOMA ENID MAT.	INST.	TOTAL	LAWTON MAT.	INST.	TOTAL	MUSKOGEE MAT.	INST.	TOTAL
A	Substructure	86.6	100.3	95.0	97.4	98.4	98.0	94.6	59.9	73.2	92.8	65.4	75.9	87.7	49.1	63.9
B10	Shell: Superstructure	94.2	93.6	93.9	91.4	87.8	90.0	90.6	58.5	77.7	93.1	62.6	80.8	89.7	48.7	73.2
B20	Exterior Closure	102.9	101.4	102.2	95.1	95.6	95.3	95.6	60.7	78.5	91.1	61.2	76.5	96.7	53.2	75.4
B30	Roofing	100.2	108.7	103.5	100.1	99.6	99.9	99.9	58.9	84.0	99.6	61.3	84.8	99.5	43.7	77.8
C	Interior Construction	97.5	96.0	96.9	99.0	90.2	95.2	93.9	43.9	72.7	96.3	55.8	79.1	95.6	36.9	70.7
D10	Services: Conveying	100.0	86.5	96.1	100.0	85.5	95.8	100.0	62.9	89.4	100.0	62.9	89.4	100.0	62.6	89.3
D20 - 40	Mechanical	100.0	102.9	101.2	99.9	92.5	95.8	100.1	65.3	85.6	100.1	65.3	85.6	100.1	29.2	70.6
D50	Electrical	97.9	105.7	101.7	97.5	93.1	95.4	94.9	71.5	83.5	96.4	71.5	84.3	94.5	34.3	65.2
E	Equipment & Furnishings	100.0	94.9	99.7	100.0	87.5	99.2	100.0	37.9	96.2	100.0	56.0	97.3	100.0	35.5	96.1
G	Site Work	85.6	103.0	98.0	101.8	107.5	105.8	107.5	89.4	94.6	102.7	91.0	94.4	94.4	84.9	87.6
A-G	WEIGHTED AVERAGE	97.6	100.0	98.7	97.2	93.3	95.5	96.0	62.0	80.9	96.4	65.3	82.6	95.6	44.0	72.7

OKLAHOMA / OREGON

DIV. NO.	BUILDING SYSTEMS	OKLAHOMA CITY MAT.	INST.	TOTAL	TULSA MAT.	INST.	TOTAL	OREGON EUGENE MAT.	INST.	TOTAL	MEDFORD MAT.	INST.	TOTAL	PORTLAND MAT.	INST.	TOTAL
A	Substructure	97.5	63.8	76.7	93.1	61.7	73.8	106.6	104.4	105.3	109.0	104.3	106.1	111.8	104.6	107.4
B10	Shell: Superstructure	95.6	60.9	81.7	92.8	68.0	82.8	96.9	100.4	98.3	96.5	100.2	98.0	99.0	100.7	99.7
B20	Exterior Closure	91.9	62.6	77.6	90.8	63.5	77.5	110.0	103.4	106.8	115.2	103.4	109.4	110.2	105.6	107.9
B30	Roofing	98.9	61.0	84.2	99.5	57.4	83.1	105.2	94.8	101.2	105.8	91.7	100.3	105.1	101.3	103.6
C	Interior Construction	96.3	49.4	76.4	96.0	49.4	76.2	104.6	101.6	103.3	106.9	100.7	104.3	103.4	101.6	102.6
D10	Services: Conveying	100.0	63.9	89.7	100.0	64.7	89.9	100.0	106.5	101.9	100.0	106.5	101.9	100.0	106.5	101.9
D20 - 40	Mechanical	100.1	66.4	86.1	100.1	60.3	83.6	100.2	106.1	102.7	100.2	106.1	102.7	100.2	108.9	103.8
D50	Electrical	95.8	71.5	84.0	96.4	44.7	71.3	98.2	100.7	99.4	101.5	90.2	96.0	98.6	105.4	100.3
E	Equipment & Furnishings	100.0	45.8	96.7	100.0	45.7	96.7	100.0	105.4	100.3	100.0	104.9	100.3	100.0	105.6	100.3
G	Site Work	104.1	91.6	95.2	100.2	85.9	90.1	108.6	104.8	105.9	117.5	104.8	108.5	109.9	104.8	106.3
A-G	WEIGHTED AVERAGE	97.1	64.3	82.6	96.2	59.9	80.1	101.7	102.8	102.2	103.3	101.1	102.3	102.2	104.6	103.3

OREGON / PENNSYLVANIA

DIV. NO.	BUILDING SYSTEMS	OREGON SALEM MAT.	INST.	TOTAL	PENNSYLVANIA ALLENTOWN MAT.	INST.	TOTAL	ALTOONA MAT.	INST.	TOTAL	ERIE MAT.	INST.	TOTAL	HARRISBURG MAT.	INST.	TOTAL
A	Substructure	109.3	104.5	106.3	92.9	106.7	101.4	98.0	92.2	94.4	98.6	95.0	96.4	96.5	95.4	95.8
B10	Shell: Superstructure	97.9	100.6	99.0	96.2	115.2	103.9	91.3	102.4	95.7	92.5	103.8	97.0	98.5	106.2	101.6
B20	Exterior Closure	112.3	105.6	109.0	99.9	101.0	100.4	92.2	87.0	89.7	86.0	95.9	90.8	98.4	91.9	95.2
B30	Roofing	105.1	97.5	102.1	102.2	119.0	108.7	101.1	92.9	97.9	100.9	98.7	100.0	105.3	111.1	107.5
C	Interior Construction	104.1	101.6	103.0	94.2	108.0	100.1	90.4	83.9	87.6	92.2	93.6	92.8	93.7	88.6	91.5
D10	Services: Conveying	100.0	106.5	101.9	100.0	104.7	101.3	100.0	99.3	99.8	100.0	104.2	101.2	100.0	99.0	99.7
D20 - 40	Mechanical	100.2	106.1	102.7	100.3	109.5	104.1	99.8	90.4	95.9	99.8	94.0	97.4	100.3	94.0	97.7
D50	Electrical	98.3	100.7	99.5	100.7	99.5	100.1	89.8	105.0	97.2	91.3	90.6	91.0	99.7	86.8	93.4
E	Equipment & Furnishings	100.0	105.4	100.3	100.0	113.7	100.8	100.0	87.9	99.3	100.0	94.7	99.7	100.0	86.6	99.2
G	Site Work	108.4	104.8	105.9	91.5	106.6	102.2	94.3	106.5	102.9	91.0	107.3	102.5	83.7	105.4	99.1
A-G	WEIGHTED AVERAGE	102.2	103.2	102.6	98.0	107.4	102.2	94.4	94.5	94.5	94.4	96.7	95.4	98.2	95.3	96.9

PENNSYLVANIA

DIV. NO.	BUILDING SYSTEMS	PHILADELPHIA MAT.	INST.	TOTAL	PITTSBURGH MAT.	INST.	TOTAL	READING MAT.	INST.	TOTAL	SCRANTON MAT.	INST.	TOTAL	YORK MAT.	INST.	TOTAL
A	Substructure	100.9	124.3	115.3	99.3	102.1	101.0	84.1	98.5	92.9	95.4	97.4	96.6	88.7	95.3	92.8
B10	Shell: Superstructure	103.9	130.8	114.8	97.2	112.9	103.5	97.6	107.7	101.6	98.3	110.4	103.2	94.5	106.1	99.2
B20	Exterior Closure	107.4	135.0	120.9	90.3	103.2	96.6	108.2	94.2	101.4	100.1	96.6	98.4	101.3	91.9	96.7
B30	Roofing	102.8	137.2	116.2	101.9	99.8	101.1	102.2	113.3	106.5	102.1	104.3	103.0	100.8	111.1	104.8
C	Interior Construction	98.5	138.1	115.3	94.3	101.9	97.5	95.6	91.6	93.9	95.7	90.9	93.7	91.6	88.6	90.4
D10	Services: Conveying	100.0	124.3	107.0	100.0	103.7	101.1	100.0	102.3	100.6	100.0	104.2	101.2	100.0	99.0	99.7
D20 - 40	Mechanical	100.1	133.4	113.9	100.1	98.8	99.6	100.3	107.9	103.5	100.3	86.2	94.5	100.3	94.0	97.7
D50	Electrical	96.1	138.2	116.6	95.9	105.1	100.3	97.9	93.5	95.7	100.7	91.7	96.3	94.5	86.8	90.7
E	Equipment & Furnishings	100.0	136.6	102.2	100.0	98.5	99.9	100.0	86.3	99.2	100.0	85.8	99.1	100.0	86.6	99.2
G	Site Work	100.9	95.3	96.9	98.6	109.0	106.0	97.7	111.0	107.1	92.1	106.7	102.4	82.9	105.4	98.8
A-G	WEIGHTED AVERAGE	101.0	131.1	114.4	97.0	104.2	100.2	98.8	100.9	99.7	98.8	96.3	97.7	96.2	95.3	95.8

PUERTO RICO / RHODE ISLAND / SOUTH CAROLINA

DIV. NO.	BUILDING SYSTEMS	PUERTO RICO SAN JUAN MAT.	INST.	TOTAL	RHODE ISLAND PROVIDENCE MAT.	INST.	TOTAL	SOUTH CAROLINA CHARLESTON MAT.	INST.	TOTAL	COLUMBIA MAT.	INST.	TOTAL	FLORENCE MAT.	INST.	TOTAL
A	Substructure	124.9	39.1	72.0	93.2	112.1	104.8	90.4	55.4	68.8	87.8	56.0	68.2	82.7	56.1	66.3
B10	Shell: Superstructure	115.9	26.4	79.9	98.8	112.4	104.2	92.0	64.1	80.8	91.7	64.6	80.8	88.5	64.6	78.9
B20	Exterior Closure	164.4	17.4	92.6	111.8	125.6	118.5	97.8	38.4	68.7	97.2	39.2	68.9	97.5	38.4	68.6
B30	Roofing	147.8	24.3	99.9	101.0	114.4	106.2	101.3	43.7	78.9	101.3	43.2	78.7	101.5	44.0	79.2
C	Interior Construction	206.4	18.3	126.6	99.9	120.0	108.5	99.3	38.2	73.4	99.2	40.1	74.2	94.7	40.1	71.6
D10	Services: Conveying	100.0	19.8	77.1	100.0	94.7	98.5	100.0	68.9	91.1	100.0	68.9	91.1	100.0	68.9	91.1
D20 - 40	Mechanical	103.7	15.3	67.0	100.1	109.6	104.0	100.1	43.0	76.4	100.1	36.2	73.6	100.1	36.2	73.6
D50	Electrical	129.8	15.3	74.2	96.5	105.4	100.8	97.3	32.4	65.8	97.9	34.2	66.9	95.5	19.5	58.6
E	Equipment & Furnishings	100.0	18.3	95.0	100.0	115.1	100.9	100.0	35.6	96.1	100.0	38.6	96.3	100.0	37.9	96.2
G	Site Work	116.8	92.2	99.3	82.1	104.1	97.7	94.2	80.4	84.4	93.4	80.4	84.1	104.0	80.4	87.2
A-G	WEIGHTED AVERAGE	135.2	25.5	86.5	99.9	112.7	105.6	97.3	47.9	75.3	97.1	47.3	75.0	95.4	45.2	73.1

SOUTH CAROLINA / SOUTH DAKOTA

DIV. NO.	BUILDING SYSTEMS	GREENVILLE MAT.	INST.	TOTAL	SPARTANBURG MAT.	INST.	TOTAL	SOUTH DAKOTA ABERDEEN MAT.	INST.	TOTAL	PIERRE MAT.	INST.	TOTAL	RAPID CITY MAT.	INST.	TOTAL
A	Substructure	83.3	53.3	64.8	83.7	53.3	65.0	99.7	56.6	73.1	97.5	57.8	73.1	96.7	55.5	71.4
B10	Shell: Superstructure	89.5	59.2	77.3	89.8	59.2	77.5	99.3	57.0	82.3	98.8	61.2	83.6	100.2	59.8	84.0
B20	Exterior Closure	94.2	37.7	66.6	95.1	37.7	67.0	100.7	56.0	78.9	99.2	59.6	79.8	99.3	53.9	77.1
B30	Roofing	101.5	44.0	79.2	101.6	44.0	79.3	98.3	48.4	78.9	98.3	46.6	78.3	98.7	44.2	77.5
C	Interior Construction	96.1	39.3	72.0	96.6	39.3	72.3	101.8	41.0	76.0	102.4	40.8	76.3	104.0	43.3	78.3
D10	Services: Conveying	100.0	68.9	91.1	100.0	68.9	91.1	100.0	56.0	87.4	100.0	67.5	90.7	100.0	65.5	90.1
D20 - 40	Mechanical	100.1	36.2	73.6	100.1	36.2	73.6	100.2	38.6	74.6	100.2	37.0	73.9	100.2	34.5	72.9
D50	Electrical	97.4	31.1	65.2	97.4	31.1	65.2	96.7	59.3	78.5	93.0	54.7	74.4	93.5	54.8	74.7
E	Equipment & Furnishings	100.0	38.4	96.3	100.0	38.4	96.3	100.0	37.7	96.2	100.0	39.5	96.3	100.0	35.4	96.1
G	Site Work	99.2	80.0	85.6	99.0	80.0	85.5	86.7	93.3	91.4	85.0	93.3	90.9	84.8	93.1	90.7
A-G	WEIGHTED AVERAGE	95.7	45.4	73.4	95.9	45.4	73.5	99.5	53.1	78.9	98.8	53.5	78.7	99.4	52.2	78.4

COST INDEXES

DIV. NO.	BUILDING SYSTEMS	SOUTH DAKOTA SIOUX FALLS MAT.	INST.	TOTAL	TENNESSEE CHATTANOOGA MAT.	INST.	TOTAL	JACKSON MAT.	INST.	TOTAL	JOHNSON CITY MAT.	INST.	TOTAL	KNOXVILLE MAT.	INST.	TOTAL
A	Substructure	98.6	59.4	74.5	98.9	58.6	74.1	97.4	53.7	70.5	86.2	57.4	68.5	93.1	56.3	70.4
B10	Shell: Superstructure	100.5	62.2	85.0	94.6	62.0	81.5	92.0	57.9	78.3	88.9	62.6	78.3	94.1	62.1	81.3
B20	Exterior Closure	97.0	59.6	78.7	94.5	45.6	70.6	104.0	36.4	71.0	114.6	45.7	80.9	82.3	45.7	64.4
B30	Roofing	98.3	48.0	78.8	97.8	50.9	79.6	95.1	40.5	73.9	92.1	48.0	75.0	90.2	48.2	73.9
C	Interior Construction	104.0	47.8	80.2	99.1	46.5	76.8	97.7	35.2	71.2	97.0	47.8	76.2	93.2	47.8	73.9
D10	Services: Conveying	100.0	67.5	90.7	100.0	52.4	86.4	100.0	50.5	85.8	100.0	61.1	88.9	100.0	61.1	88.9
D20 - 40	Mechanical	100.2	36.0	73.6	99.9	44.8	77.1	100.0	55.5	81.5	99.8	59.0	82.9	99.8	59.0	82.9
D50	Electrical	92.9	71.9	82.7	104.2	52.8	79.2	100.9	51.1	76.7	92.5	51.1	72.4	97.9	59.7	79.3
E	Equipment & Furnishings	100.0	40.9	96.4	100.0	46.5	96.7	100.0	37.2	96.2	100.0	47.4	96.8	100.0	47.8	96.8
G	Site Work	85.9	95.3	92.5	101.1	95.9	97.4	99.0	95.2	96.3	110.3	64.4	92.0	89.5	84.6	86.0
A-G	WEIGHTED AVERAGE	99.2	57.3	80.6	98.6	54.1	78.8	98.3	51.7	77.6	97.3	56.4	79.1	94.7	57.5	78.2

DIV. NO.	BUILDING SYSTEMS	TENNESSEE MEMPHIS MAT.	INST.	TOTAL	NASHVILLE MAT.	INST.	TOTAL	TEXAS ABILENE MAT.	INST.	TOTAL	AMARILLO MAT.	INST.	TOTAL	AUSTIN MAT.	INST.	TOTAL
A	Substructure	93.0	74.9	81.8	94.0	75.1	82.4	94.3	54.4	69.7	97.5	60.4	74.6	90.1	61.2	72.3
B10	Shell: Superstructure	94.5	82.7	89.1	95.0	80.2	89.1	95.7	56.1	79.8	96.5	60.5	82.0	94.6	60.2	80.7
B20	Exterior Closure	89.4	76.7	83.2	87.1	67.3	77.4	91.1	54.0	73.0	92.0	51.6	72.2	81.9	56.4	69.4
B30	Roofing	91.8	72.5	84.3	95.8	65.8	84.1	99.6	47.5	79.4	101.7	47.1	80.5	88.7	52.2	74.5
C	Interior Construction	98.1	63.3	83.3	100.9	68.0	86.9	94.5	47.6	74.6	94.5	54.6	77.6	92.4	54.8	76.5
D10	Services: Conveying	100.0	82.3	94.9	100.0	81.4	94.7	100.0	78.0	93.7	100.0	62.7	89.3	100.0	62.3	89.2
D20 - 40	Mechanical	99.9	75.5	87.9	99.8	71.0	87.9	100.1	45.1	77.3	100.1	51.5	79.9	99.8	59.5	83.1
D50	Electrical	98.3	76.7	87.8	102.3	69.4	86.3	96.5	48.8	73.3	97.0	60.0	79.0	99.5	69.8	85.0
E	Equipment & Furnishings	100.0	67.9	98.0	100.0	66.1	97.9	100.0	41.9	96.5	100.0	56.7	97.4	100.0	60.9	97.6
G	Site Work	93.7	91.4	92.0	94.2	98.5	97.3	100.7	85.2	89.7	101.0	86.0	90.4	90.4	84.8	86.4
A-G	WEIGHTED AVERAGE	96.6	76.2	87.5	97.5	73.7	86.9	96.6	53.3	77.4	97.2	58.0	79.8	94.6	61.7	80.0

DIV. NO.	BUILDING SYSTEMS	TEXAS BEAUMONT MAT.	INST.	TOTAL	CORPUS CHRISTI MAT.	INST.	TOTAL	DALLAS MAT.	INST.	TOTAL	EL PASO MAT.	INST.	TOTAL	FORT WORTH MAT.	INST.	TOTAL
A	Substructure	90.7	60.7	72.2	105.5	52.8	73.1	100.1	65.3	78.7	95.7	54.0	70.0	93.5	63.2	74.9
B10	Shell: Superstructure	98.1	58.7	82.3	96.5	61.5	82.4	94.0	71.3	84.9	95.9	54.3	79.1	93.1	62.9	81.0
B20	Exterior Closure	93.7	58.2	76.3	81.8	50.7	66.6	93.0	61.4	77.6	90.5	51.1	71.2	92.1	60.4	76.6
B30	Roofing	105.2	57.0	86.5	91.1	44.7	73.1	99.5	63.4	85.5	99.2	52.0	80.9	99.9	53.2	81.8
C	Interior Construction	94.7	55.0	77.9	98.9	40.3	74.0	101.4	59.0	83.4	94.5	50.5	75.8	95.1	56.8	78.8
D10	Services: Conveying	100.0	80.9	94.5	100.0	79.4	94.1	100.0	81.5	94.7	100.0	61.4	89.0	100.0	81.0	94.6
D20 - 40	Mechanical	100.0	66.1	85.9	99.8	46.0	77.5	99.8	67.8	86.5	100.1	34.4	72.8	100.1	59.7	83.3
D50	Electrical	94.4	72.5	83.7	97.6	54.3	76.6	96.1	65.5	81.2	95.7	56.2	76.5	95.7	63.8	80.2
E	Equipment & Furnishings	100.0	54.0	97.2	100.0	38.8	96.3	100.0	60.2	97.6	100.0	50.1	97.0	100.0	60.0	97.6
G	Site Work	96.8	87.1	89.9	120.8	79.7	91.6	124.0	86.0	97.1	101.2	84.8	89.5	100.4	85.6	89.9
A-G	WEIGHTED AVERAGE	97.1	64.1	82.4	97.3	53.1	77.7	98.3	67.1	84.4	96.6	51.6	76.6	96.2	62.8	81.4

DIV. NO.	BUILDING SYSTEMS	TEXAS HOUSTON MAT.	INST.	TOTAL	LAREDO MAT.	INST.	TOTAL	LUBBOCK MAT.	INST.	TOTAL	ODESSA MAT.	INST.	TOTAL	SAN ANTONIO MAT.	INST.	TOTAL
A	Substructure	96.1	72.6	81.7	85.5	57.9	68.5	99.3	55.7	72.4	94.4	54.0	69.5	85.9	68.2	75.0
B10	Shell: Superstructure	101.4	79.6	92.6	94.0	56.2	78.8	98.8	64.6	85.0	95.3	55.2	79.2	94.6	64.6	82.5
B20	Exterior Closure	91.2	67.4	79.6	85.8	53.1	69.3	92.0	50.9	71.9	91.1	50.0	71.0	85.7	63.9	75.0
B30	Roofing	99.4	67.1	86.9	87.8	48.6	72.5	91.8	49.2	75.3	99.6	44.0	78.0	87.8	64.4	78.7
C	Interior Construction	101.9	66.9	87.1	92.1	41.4	70.6	100.4	40.8	75.1	94.5	39.4	71.1	93.2	56.6	77.7
D10	Services: Conveying	100.0	84.9	95.7	100.0	76.2	93.2	100.0	77.7	93.6	100.0	62.1	89.2	100.0	80.9	94.5
D20 - 40	Mechanical	99.9	72.4	88.5	99.8	41.5	75.6	99.7	48.3	78.3	100.1	37.7	74.2	99.8	72.4	88.4
D50	Electrical	97.0	71.0	84.3	99.3	62.8	81.5	95.1	45.9	71.2	96.5	42.7	70.4	99.5	62.8	81.7
E	Equipment & Furnishings	100.0	68.7	98.1	100.0	39.1	96.3	100.0	41.9	96.5	100.0	39.7	96.3	100.0	57.7	97.4
G	Site Work	120.2	85.2	95.3	90.4	84.6	86.3	126.0	82.9	95.4	100.7	85.5	89.9	90.0	89.5	89.7
A-G	WEIGHTED AVERAGE	99.6	73.0	87.7	94.6	53.7	76.5	98.6	53.5	78.6	96.6	48.5	75.2	95.0	67.0	82.5

DIV. NO.	BUILDING SYSTEMS	TEXAS WACO MAT.	INST.	TOTAL	WICHITA FALLS MAT.	INST.	TOTAL	UTAH LOGAN MAT.	INST.	TOTAL	OGDEN MAT.	INST.	TOTAL	PROVO MAT.	INST.	TOTAL
A	Substructure	88.7	56.9	69.1	92.3	56.1	70.0	92.8	75.2	81.9	93.2	75.2	82.1	94.2	74.7	82.2
B10	Shell: Superstructure	94.5	56.6	79.3	95.2	57.5	80.1	97.9	70.8	87.0	98.7	70.8	87.4	97.1	70.8	86.5
B20	Exterior Closure	92.8	57.3	75.4	93.0	58.9	76.3	121.4	60.8	91.8	107.1	60.8	84.5	123.4	60.8	92.8
B30	Roofing	100.1	48.7	80.1	100.1	51.2	81.1	99.4	69.0	87.6	98.2	69.0	86.9	101.2	69.0	88.7
C	Interior Construction	95.1	38.9	71.3	95.2	49.3	75.7	93.1	58.2	78.3	92.5	58.2	77.9	95.2	59.4	80.0
D10	Services: Conveying	100.0	81.0	94.6	100.0	62.7	89.3	100.0	72.5	92.1	100.0	72.5	92.1	100.0	72.5	92.1
D20 - 40	Mechanical	100.1	55.5	81.6	100.1	51.3	79.8	100.1	71.2	88.1	100.1	71.2	88.1	100.1	71.2	88.1
D50	Electrical	96.5	67.1	82.2	99.6	65.6	83.1	91.0	74.5	83.0	91.5	74.5	83.3	91.2	74.5	83.1
E	Equipment & Furnishings	100.0	38.6	96.3	100.0	42.7	96.5	100.0	58.7	97.5	100.0	58.7	97.5	100.0	58.7	97.5
G	Site Work	99.3	85.6	89.6	100.2	85.2	89.5	89.7	100.5	97.4	79.1	100.5	94.3	86.7	98.5	95.1
A-G	WEIGHTED AVERAGE	96.4	57.5	79.1	97.1	57.9	79.7	99.1	70.5	86.4	97.4	70.5	85.5	99.6	70.5	86.7

DIV. NO.	BUILDING SYSTEMS	UTAH SALT LAKE CITY MAT.	INST.	TOTAL	VERMONT BURLINGTON MAT.	INST.	TOTAL	RUTLAND MAT.	INST.	TOTAL	VIRGINIA ALEXANDRIA MAT.	INST.	TOTAL	ARLINGTON MAT.	INST.	TOTAL
A	Substructure	97.0	75.2	83.6	98.5	66.2	78.6	96.7	66.2	77.9	95.9	81.2	86.9	97.8	74.9	83.7
B10	Shell: Superstructure	102.0	70.8	89.4	100.7	60.6	84.6	99.6	60.7	83.9	100.2	89.2	95.8	99.2	79.6	91.3
B20	Exterior Closure	130.6	60.8	96.4	105.1	58.1	82.1	99.4	58.1	79.2	99.8	76.1	88.2	109.7	70.2	90.4
B30	Roofing	100.9	69.0	88.5	101.7	53.4	83.0	101.5	54.8	83.4	100.3	81.9	93.1	101.8	71.2	89.9
C	Interior Construction	92.5	59.4	78.5	102.4	51.4	80.8	102.3	51.4	80.7	99.5	78.5	90.6	97.7	72.1	86.8
D10	Services: Conveying	100.0	72.5	92.1	100.0	98.6	99.6	100.0	98.6	99.6	100.0	88.3	96.6	100.0	67.2	90.6
D20 - 40	Mechanical	100.2	71.2	88.2	100.1	66.1	86.0	100.1	66.1	86.0	99.9	86.3	94.3	100.1	82.7	92.8
D50	Electrical	95.0	74.5	85.0	98.3	68.7	83.9	97.6	68.7	83.6	98.3	95.1	96.8	95.8	95.2	95.5
E	Equipment & Furnishings	100.0	58.7	97.5	100.0	48.1	96.8	100.0	48.1	96.8	100.0	75.9	98.5	100.0	75.4	98.5
G	Site Work	79.3	100.4	94.3	81.7	98.8	93.9	81.5	98.8	93.8	115.4	85.7	94.4	125.3	81.9	94.5
A-G	WEIGHTED AVERAGE	101.3	70.7	87.7	100.5	64.9	84.7	99.5	64.9	84.1	99.9	84.9	93.3	100.5	79.3	91.1

COST INDEXES

VIRGINIA

DIV. NO.	BUILDING SYSTEMS	NEWPORT NEWS			NORFOLK			PORTSMOUTH			RICHMOND			ROANOKE		
		MAT.	INST.	TOTAL	MAT.	INST.	TOTAL	MAT.	INST.	TOTAL	MAT.	INST.	TOTAL	MAT.	INST.	TOTAL
A	Substructure	97.9	66.8	78.7	100.0	67.2	79.8	96.1	66.5	77.9	101.8	65.4	79.4	104.7	55.2	74.2
B10	Shell: Superstructure	100.6	76.9	91.0	100.6	77.0	91.1	98.8	76.0	89.6	103.0	75.3	91.9	102.0	65.9	87.5
B20	Exterior Closure	98.4	55.9	77.6	101.4	56.7	79.6	100.7	56.5	79.1	97.8	51.6	75.2	98.5	42.3	71.0
B30	Roofing	101.3	49.4	81.2	101.1	49.4	81.0	101.3	49.4	81.1	100.9	54.0	82.7	101.3	44.8	79.4
C	Interior Construction	99.2	55.9	80.8	99.1	57.9	81.6	97.1	57.9	80.4	99.1	57.5	81.5	99.1	36.6	72.6
D10	Services: Conveying	100.0	80.6	94.4	100.0	80.6	94.4	100.0	80.9	94.5	100.0	80.7	94.5	100.0	76.8	93.4
D20 - 40	Mechanical	100.1	63.4	84.8	100.1	63.6	84.9	100.1	63.6	84.9	100.1	67.4	86.5	100.1	39.3	74.9
D50	Electrical	97.7	59.2	79.0	97.8	63.1	80.9	96.2	63.1	80.1	98.4	71.5	85.3	97.7	34.2	66.8
E	Equipment & Furnishings	100.0	61.1	97.6	100.0	61.3	97.6	100.0	61.3	97.6	100.0	56.7	97.4	100.0	34.4	96.0
G	Site Work	106.6	84.0	90.6	105.8	85.2	91.2	104.9	83.4	89.6	107.3	84.4	91.1	105.0	80.2	87.4
A-G	WEIGHTED AVERAGE	99.6	64.7	84.1	100.0	65.8	84.8	98.9	65.5	84.0	100.3	66.7	85.4	100.2	48.1	77.1

WASHINGTON

DIV. NO.	BUILDING SYSTEMS	EVERETT			RICHLAND			SEATTLE			SPOKANE			TACOMA		
		MAT.	INST.	TOTAL	MAT.	INST.	TOTAL	MAT.	INST.	TOTAL	MAT.	INST.	TOTAL	MAT.	INST.	TOTAL
A	Substructure	98.4	105.0	102.5	111.9	84.5	95.0	100.7	105.4	103.6	115.3	84.5	96.4	99.3	105.7	103.3
B10	Shell: Superstructure	104.8	94.5	100.6	98.1	82.1	91.7	106.0	96.0	102.0	101.3	82.1	93.6	105.7	94.9	101.4
B20	Exterior Closure	111.6	102.2	107.0	108.1	81.7	95.2	110.2	102.8	106.6	108.8	81.7	95.6	110.0	102.7	106.5
B30	Roofing	104.6	98.0	102.0	140.1	80.8	117.1	104.6	100.5	103.0	137.9	80.7	115.7	104.2	98.9	102.2
C	Interior Construction	106.2	100.3	103.7	117.3	73.3	98.6	106.1	100.6	103.8	118.2	78.5	101.3	105.5	100.6	103.4
D10	Services: Conveying	100.0	103.0	100.9	100.0	83.6	95.3	100.0	103.5	101.0	100.0	83.6	95.3	100.0	103.5	101.0
D20 - 40	Mechanical	99.9	101.9	100.7	100.5	98.5	99.7	99.9	112.2	105.0	100.5	87.7	95.2	100.0	100.6	100.3
D50	Electrical	103.6	97.7	100.7	95.8	97.6	96.7	103.4	106.5	104.9	94.1	81.8	88.1	103.4	100.6	102.0
E	Equipment & Furnishings	100.0	100.9	100.1	100.0	80.2	98.8	100.0	101.4	100.1	100.0	79.4	98.7	100.0	101.1	100.1
G	Site Work	93.8	115.7	109.3	107.7	89.2	94.6	97.5	113.4	108.8	108.5	89.2	94.8	96.3	115.7	110.1
A-G	WEIGHTED AVERAGE	103.6	101.0	102.4	105.0	86.9	96.9	103.8	104.7	104.2	105.7	83.3	95.8	103.5	101.4	102.6

WASHINGTON / WEST VIRGINIA

DIV. NO.	BUILDING SYSTEMS	VANCOUVER			YAKIMA			CHARLESTON			HUNTINGTON			PARKERSBURG		
		MAT.	INST.	TOTAL	MAT.	INST.	TOTAL	MAT.	INST.	TOTAL	MAT.	INST.	TOTAL	MAT.	INST.	TOTAL
A	Substructure	108.1	98.4	102.1	103.8	95.8	98.9	96.7	94.5	95.3	100.7	96.3	98.0	96.0	93.1	94.2
B10	Shell: Superstructure	105.1	93.2	100.3	105.3	87.0	98.0	100.8	97.8	99.6	101.4	98.4	100.2	98.6	97.3	98.1
B20	Exterior Closure	115.0	99.5	107.4	107.6	70.9	89.7	97.3	92.3	94.9	98.2	93.0	95.7	96.1	89.5	92.9
B30	Roofing	106.9	89.6	100.2	104.3	81.2	95.3	101.2	91.1	97.3	101.5	92.0	97.8	101.6	88.5	96.5
C	Interior Construction	102.1	90.8	97.3	104.9	89.8	98.5	99.4	93.5	96.9	98.0	94.0	96.3	97.0	93.0	95.3
D10	Services: Conveying	100.0	78.1	93.7	100.0	98.2	99.5	100.0	95.7	98.8	100.0	96.8	99.1	100.0	95.6	98.8
D20 - 40	Mechanical	100.1	104.0	101.7	100.0	97.6	99.0	100.1	84.9	93.8	100.1	88.4	95.2	100.1	89.7	95.8
D50	Electrical	114.0	103.1	108.7	106.1	97.6	102.0	97.7	87.3	92.6	97.7	95.8	96.8	98.2	96.9	97.5
E	Equipment & Furnishings	100.0	97.4	99.8	100.0	100.2	100.0	100.0	90.3	99.4	100.0	92.2	99.5	100.0	90.9	99.4
G	Site Work	107.6	100.7	102.7	99.8	114.2	110.0	102.2	85.5	90.3	104.0	86.3	91.4	108.3	85.5	92.1
A-G	WEIGHTED AVERAGE	105.4	97.7	102.0	103.7	91.8	98.4	99.5	91.0	95.7	99.6	93.4	96.9	98.7	92.6	96.0

WEST VIRGINIA / WISCONSIN

DIV. NO.	BUILDING SYSTEMS	WHEELING			EAU CLAIRE			GREEN BAY			KENOSHA			LA CROSSE		
		MAT.	INST.	TOTAL	MAT.	INST.	TOTAL	MAT.	INST.	TOTAL	MAT.	INST.	TOTAL	MAT.	INST.	TOTAL
A	Substructure	96.0	93.2	94.3	96.9	100.9	99.3	100.4	98.4	99.2	107.7	104.4	105.6	88.7	98.7	94.9
B10	Shell: Superstructure	98.8	98.1	98.5	93.4	102.8	97.2	96.7	99.0	97.6	98.9	104.1	101.0	90.9	98.3	93.9
B20	Exterior Closure	107.2	91.7	99.6	93.9	100.0	96.9	108.5	97.5	103.1	96.9	110.6	103.6	90.8	97.4	94.0
B30	Roofing	101.7	88.7	96.7	97.5	89.7	94.5	99.5	87.5	94.9	96.4	102.8	98.9	96.9	88.7	93.7
C	Interior Construction	97.6	93.5	95.8	100.2	101.0	100.6	99.9	98.6	99.3	96.3	108.6	101.5	98.0	98.2	98.1
D10	Services: Conveying	100.0	104.3	101.2	100.0	94.3	98.4	100.0	92.4	97.8	100.0	99.9	100.0	100.0	94.3	98.4
D20 - 40	Mechanical	100.1	94.6	97.8	100.2	90.9	96.3	100.5	90.1	96.2	100.0	98.2	99.3	100.2	90.6	96.2
D50	Electrical	95.6	96.4	95.9	101.5	83.1	92.6	96.6	90.8	93.8	96.0	96.0	96.0	101.8	91.7	96.9
E	Equipment & Furnishings	100.0	90.7	99.4	100.0	97.9	99.9	100.0	98.3	99.9	100.0	108.1	100.5	100.0	98.3	99.9
G	Site Work	108.9	85.1	92.0	87.0	102.8	98.2	88.8	98.7	95.8	96.9	102.2	100.6	81.1	102.7	96.4
A-G	WEIGHTED AVERAGE	99.7	94.2	97.2	97.8	96.2	97.1	99.7	95.2	97.7	98.5	103.1	100.5	96.1	95.7	95.9

WISCONSIN / WYOMING

DIV. NO.	BUILDING SYSTEMS	MADISON			MILWAUKEE			RACINE			CASPER			CHEYENNE		
		MAT.	INST.	TOTAL	MAT.	INST.	TOTAL	MAT.	INST.	TOTAL	MAT.	INST.	TOTAL	MAT.	INST.	TOTAL
A	Substructure	99.1	100.4	99.9	99.4	105.8	103.4	99.1	105.2	102.9	99.2	67.0	79.4	97.9	67.2	79.0
B10	Shell: Superstructure	97.3	98.0	97.6	98.8	101.6	99.9	97.3	104.0	100.0	97.9	58.3	82.0	98.2	58.6	82.3
B20	Exterior Closure	98.7	102.8	100.7	98.8	116.3	107.3	98.5	110.6	104.4	105.6	40.2	73.6	104.5	42.0	74.0
B30	Roofing	95.7	96.4	96.0	94.9	111.6	101.4	96.3	102.3	98.6	99.6	56.2	82.7	99.6	56.3	82.8
C	Interior Construction	98.2	100.3	99.1	100.5	114.9	106.6	98.2	107.8	102.3	94.8	42.3	72.5	95.0	46.5	74.4
D10	Services: Conveying	100.0	92.5	97.9	100.0	102.2	100.6	100.0	99.9	100.0	100.0	85.0	95.7	100.0	85.0	95.7
D20 - 40	Mechanical	99.8	93.2	97.0	99.8	103.2	101.2	99.8	95.4	97.9	100.1	65.9	85.9	100.1	60.0	83.4
D50	Electrical	97.5	95.3	96.4	97.1	103.1	100.0	95.8	98.0	96.9	90.6	63.3	77.3	90.6	74.4	82.7
E	Equipment & Furnishings	100.0	98.5	99.9	100.0	113.0	100.8	100.0	108.2	100.5	100.0	41.3	96.4	100.0	41.8	96.5
G	Site Work	91.6	105.5	101.5	93.0	94.0	93.7	92.0	106.0	101.9	80.2	100.0	94.2	79.8	100.0	94.1
A-G	WEIGHTED AVERAGE	98.3	98.1	98.2	99.0	106.2	102.2	98.1	102.9	100.2	97.7	59.9	80.9	97.6	61.2	81.4

WYOMING / CANADA

DIV. NO.	BUILDING SYSTEMS	ROCK SPRINGS			CALGARY, ALBERTA			EDMONTON, ALBERTA			HALIFAX, NOVA SCOTIA			HAMILTON, ONTARIO		
		MAT.	INST.	TOTAL	MAT.	INST.	TOTAL	MAT.	INST.	TOTAL	MAT.	INST.	TOTAL	MAT.	INST.	TOTAL
A	Substructure	100.0	53.7	71.5	171.0	83.2	116.9	180.5	83.2	120.6	156.5	70.4	103.5	152.2	91.1	114.6
B10	Shell: Superstructure	96.6	50.0	77.8	143.8	79.7	118.0	146.2	79.7	119.5	128.1	66.7	103.4	142.7	86.6	120.1
B20	Exterior Closure	137.0	34.2	86.8	142.5	71.8	107.9	143.1	71.8	108.2	137.9	66.7	103.1	136.4	87.9	112.7
B30	Roofing	100.0	45.4	78.8	117.4	78.6	102.4	117.7	78.6	102.5	116.6	65.7	89.5	109.8	85.6	100.4
C	Interior Construction	98.6	31.7	70.2	108.7	75.0	94.4	111.0	75.0	95.7	101.6	62.2	84.9	112.2	85.4	100.8
D10	Services: Conveying	100.0	50.0	85.7	140.0	80.7	123.1	140.0	80.7	123.1	140.0	57.3	116.3	140.0	85.6	124.4
D20 - 40	Mechanical	100.1	54.9	81.3	101.4	74.7	90.3	101.5	74.7	90.4	101.0	66.4	86.7	101.5	82.4	93.6
D50	Electrical	89.9	62.7	76.7	117.6	84.2	101.4	118.3	84.2	101.8	127.4	68.3	98.7	130.4	91.7	111.6
E	Equipment & Furnishings	100.0	29.9	95.7	140.0	74.1	136.0	140.0	74.1	136.0	140.0	63.7	135.4	140.0	83.3	136.6
G	Site Work	83.6	99.5	94.9	119.9	104.7	109.1	123.8	104.7	110.2	103.7	95.4	97.8	116.0	103.3	107.0
A-G	WEIGHTED AVERAGE	101.5	51.4	79.2	124.3	79.6	104.4	125.9	79.6	105.3	119.3	68.2	96.6	124.5	87.9	108.2

COST INDEXES

DIV. NO.	BUILDING SYSTEMS	CANADA														
		KITCHENER, ONTARIO			LAVAL, QUEBEC			LONDON, ONTARIO			MONTREAL, QUEBEC			OSHAWA, ONTARIO		
		MAT.	INST.	TOTAL	MAT.	INST.	TOTAL	MAT.	INST.	TOTAL	MAT.	INST.	TOTAL	MAT.	INST.	TOTAL
A	Substructure	131.0	83.7	101.8	136.6	85.6	105.2	144.5	88.7	110.1	135.6	86.2	105.2	159.9	86.4	114.7
B10	Shell: Superstructure	120.2	82.1	104.8	114.4	82.3	101.4	129.6	84.9	111.6	129.1	82.8	110.5	119.6	83.9	105.2
B20	Exterior Closure	125.3	84.0	105.1	133.2	76.7	105.5	137.1	85.2	111.8	134.7	76.7	106.3	134.8	87.1	111.5
B30	Roofing	108.4	80.9	97.7	104.5	84.1	96.6	112.5	83.9	101.4	105.1	84.8	97.2	109.5	81.2	98.5
C	Interior Construction	104.3	79.5	93.8	106.1	80.7	95.3	112.8	82.2	99.8	108.0	80.9	96.5	107.8	83.6	97.6
D10	Services: Conveying	140.0	84.4	124.1	140.0	73.2	120.9	140.0	84.9	124.3	140.0	74.0	121.1	140.0	85.3	124.4
D20 - 40	Mechanical	101.0	78.8	91.8	101.0	71.8	88.9	101.6	79.0	92.2	101.6	71.9	89.3	101.0	80.4	92.5
D50	Electrical	125.2	88.5	107.4	122.1	71.6	97.5	130.7	88.7	110.3	121.1	71.6	97.0	126.3	89.2	108.3
E	Equipment & Furnishings	140.0	76.0	136.1	140.0	78.3	136.2	140.0	78.6	136.3	140.0	78.3	136.3	140.0	79.4	136.3
G	Site Work	104.9	102.7	103.3	98.0	96.4	96.9	115.9	103.0	106.8	97.7	96.7	97.0	117.8	102.9	107.2
A-G	**WEIGHTED AVERAGE**	115.6	83.7	101.4	115.2	78.7	99.0	121.9	85.3	105.6	118.5	79.0	101.0	118.7	85.7	104.1

DIV. NO.	BUILDING SYSTEMS	CANADA														
		OTTAWA, ONTARIO			QUEBEC, QUEBEC			REGINA, SASKATCHEWAN			SASKATOON, SASKATCHEWAN			ST CATHARINES, ONTARIO		
		MAT.	INST.	TOTAL	MAT.	INST.	TOTAL	MAT.	INST.	TOTAL	MAT.	INST.	TOTAL	MAT.	INST.	TOTAL
A	Substructure	153.2	88.9	113.6	141.9	86.4	107.8	143.7	68.0	97.1	133.0	68.0	93.0	126.7	84.5	100.7
B10	Shell: Superstructure	125.2	85.1	109.0	127.2	83.1	109.4	113.8	65.0	94.1	111.5	64.9	92.7	111.6	83.3	100.2
B20	Exterior Closure	136.4	85.0	111.2	135.1	79.0	107.7	128.7	58.6	94.4	128.4	58.6	94.3	124.9	82.2	104.1
B30	Roofing	110.0	82.2	99.2	104.7	85.0	97.1	105.3	60.1	87.8	104.1	59.1	86.6	108.4	82.0	98.1
C	Interior Construction	114.7	82.2	100.9	110.1	80.9	97.7	107.5	54.7	85.1	105.3	53.8	83.4	101.6	81.4	93.0
D10	Services: Conveying	140.0	82.6	123.6	140.0	74.2	121.2	140.0	56.5	116.1	140.0	56.5	116.1	140.0	61.0	117.4
D20 - 40	Mechanical	101.6	79.3	92.4	101.5	71.9	89.2	101.3	59.7	84.0	101.2	59.7	84.0	101.0	77.3	91.2
D50	Electrical	122.1	89.7	106.3	125.9	71.6	99.5	126.4	61.9	95.0	126.5	61.9	95.1	128.2	89.7	109.5
E	Equipment & Furnishings	140.0	80.1	136.4	140.0	78.4	136.3	140.0	53.0	134.7	140.0	53.1	134.7	140.0	79.7	136.3
G	Site Work	116.3	102.9	106.7	99.8	96.8	97.6	116.4	94.4	100.8	110.7	94.5	99.2	105.7	98.5	100.6
A-G	**WEIGHTED AVERAGE**	120.5	85.5	105.0	119.4	79.4	101.6	116.1	62.9	92.5	114.6	62.8	91.6	113.6	83.1	100.0

DIV. NO.	BUILDING SYSTEMS	CANADA														
		ST JOHNS, NEWFOUNDLAND			THUNDER BAY, ONTARIO			TORONTO, ONTARIO			VANCOUVER, B C			WINDSOR, ONTARIO		
		MAT.	INST.	TOTAL	MAT.	INST.	TOTAL	MAT.	INST.	TOTAL	MAT.	INST.	TOTAL	MAT.	INST.	TOTAL
A	Substructure	161.4	66.2	102.8	134.3	87.8	105.7	155.2	95.4	118.4	150.1	90.5	113.4	129.1	88.2	103.9
B10	Shell: Superstructure	119.7	62.9	96.8	113.5	84.5	101.9	136.5	90.2	117.9	143.1	86.5	120.3	112.6	85.3	101.6
B20	Exterior Closure	148.7	58.2	104.5	128.4	82.6	106.0	144.3	95.1	120.3	145.2	85.6	116.1	124.7	87.8	106.7
B30	Roofing	108.0	57.9	88.5	108.7	81.5	98.1	110.0	92.1	103.1	116.6	87.7	105.4	108.5	85.0	99.4
C	Interior Construction	109.7	54.0	86.1	104.0	76.9	92.5	112.1	91.1	103.2	110.4	82.5	98.6	103.1	82.3	94.3
D10	Services: Conveying	140.0	57.1	116.3	140.0	62.0	117.7	140.0	87.2	124.9	140.0	83.7	123.9	140.0	63.5	118.1
D20 - 40	Mechanical	101.4	56.5	82.8	101.0	78.0	91.5	101.5	88.0	95.9	101.5	81.4	93.2	101.0	80.5	92.5
D50	Electrical	121.0	57.4	90.1	125.2	87.9	107.1	129.8	92.3	111.6	132.9	80.8	107.5	133.4	89.7	112.1
E	Equipment & Furnishings	140.0	54.0	134.8	140.0	80.0	136.4	140.0	89.1	136.9	140.0	80.2	136.4	140.0	79.0	136.3
G	Site Work	119.5	94.1	101.5	111.4	98.6	102.3	117.2	103.7	107.6	120.2	107.1	110.9	101.2	98.8	99.5
A-G	**WEIGHTED AVERAGE**	120.1	60.9	93.8	114.8	82.8	100.6	124.2	92.1	109.9	125.7	85.6	107.9	114.6	85.4	101.6

DIV. NO.	BUILDING SYSTEMS	CANADA		
		WINNIPEG, MANITOBA		
		MAT.	INST.	TOTAL
A	Substructure	149.4	73.8	102.9
B10	Shell: Superstructure	142.0	69.8	112.9
B20	Exterior Closure	141.0	61.2	102.0
B30	Roofing	105.1	69.4	91.3
C	Interior Construction	108.5	63.6	89.4
D10	Services: Conveying	140.0	58.8	116.8
D20 - 40	Mechanical	101.5	65.9	86.7
D50	Electrical	131.9	68.6	101.1
E	Equipment & Furnishings	140.0	65.4	135.5
G	Site Work	115.8	99.9	104.6
A-G	**WEIGHTED AVERAGE**	124.1	68.9	99.6

Costs shown in *Means cost data publications* are based on National Averages for materials and installation. To adjust these costs to a specific location, simply multiply the base cost by the factor and divide by 100 for that city. The data is arranged alphabetically by state and postal zip code numbers. For a city not listed, use the factor for a nearby city with similar economic characteristics.

STATE/ZIP	CITY	MAT.	INST.	TOTAL
ALABAMA				
350-352	Birmingham	96.5	75.2	87.0
354	Tuscaloosa	96.3	55.4	78.1
355	Jasper	96.5	52.1	76.8
356	Decatur	96.2	55.9	78.3
357-358	Huntsville	96.3	72.0	85.5
359	Gadsden	96.1	58.8	79.6
360-361	Montgomery	97.0	57.1	79.3
362	Anniston	95.7	46.5	73.9
363	Dothan	96.3	49.0	75.3
364	Evergreen	95.7	50.2	75.5
365-366	Mobile	97.2	61.1	81.2
367	Selma	95.9	51.6	76.2
368	Phenix City	96.6	54.9	78.1
369	Butler	96.1	49.0	75.2
ALASKA				
995-996	Anchorage	133.7	114.7	125.3
997	Fairbanks	130.3	117.6	124.6
998	Juneau	132.5	114.0	124.3
999	Ketchikan	141.9	113.6	129.3
ARIZONA				
850,853	Phoenix	98.8	73.6	87.6
852	Mesa/Tempe	97.8	68.9	85.0
855	Globe	98.1	65.5	83.6
856-857	Tucson	97.1	71.2	85.6
859	Show Low	98.2	68.0	84.8
860	Flagstaff	100.4	72.3	87.9
863	Prescott	98.3	66.8	84.3
864	Kingman	96.7	67.2	83.6
865	Chambers	96.8	68.3	84.1
ARKANSAS				
716	Pine Bluff	95.2	60.9	80.0
717	Camden	93.2	38.9	69.1
718	Texarkana	94.6	44.0	72.1
719	Hot Springs	92.4	38.4	68.4
720-722	Little Rock	95.1	63.8	81.2
723	West Memphis	94.6	53.4	76.3
724	Jonesboro	95.2	53.4	76.6
725	Batesville	93.3	48.4	73.3
726	Harrison	94.5	48.4	74.0
727	Fayetteville	91.9	46.7	71.8
728	Russellville	93.2	46.6	72.5
729	Fort Smith	95.9	56.9	78.6
CALIFORNIA				
900-902	Los Angeles	102.5	112.1	106.8
903-905	Inglewood	98.2	109.0	103.0
906-908	Long Beach	99.6	109.0	103.8
910-912	Pasadena	100.4	109.3	104.4
913-916	Van Nuys	103.5	109.1	106.0
917-918	Alhambra	102.7	109.1	105.5
919-921	San Diego	104.0	103.9	104.0
922	Palm Springs	100.1	107.2	103.2
923-924	San Bernardino	97.9	108.1	102.4
925	Riverside	102.2	109.5	105.5
926-927	Santa Ana	99.9	108.2	103.5
928	Anaheim	102.6	110.6	106.1
930	Oxnard	103.8	110.0	106.5
931	Santa Barbara	102.6	110.5	106.1
932-933	Bakersfield	102.3	105.9	103.9
934	San Luis Obispo	103.4	107.9	105.4
935	Mojave	100.5	103.1	101.7
936-938	Fresno	104.1	110.9	107.1
939	Salinas	104.1	116.4	109.6
940-941	San Francisco	112.8	132.9	121.7
942,956-958	Sacramento	106.6	113.0	109.4
943	Palo Alto	105.5	127.2	115.1
944	San Mateo	108.4	127.8	117.0
945	Vallejo	105.5	122.2	112.9
946	Oakland	110.7	127.1	118.0
947	Berkeley	110.1	126.7	117.5
948	Richmond	109.5	126.1	116.9
949	San Rafael	110.2	122.8	115.8
950	Santa Cruz	109.4	116.4	112.5

STATE/ZIP	CITY	MAT.	INST.	TOTAL
CALIFORNIA (CONT'D)				
951	San Jose	107.3	127.5	116.3
952	Stockton	104.4	113.5	108.4
953	Modesto	104.3	111.0	107.3
954	Santa Rosa	104.1	122.0	112.1
955	Eureka	105.9	101.1	103.8
959	Marysville	105.0	112.5	108.3
960	Redding	106.6	110.3	108.3
961	Susanville	105.7	110.7	107.9
COLORADO				
800-802	Denver	100.6	89.9	95.8
803	Boulder	97.7	86.5	92.7
804	Golden	99.9	85.8	93.6
805	Fort Collins	101.1	82.4	92.8
806	Greeley	98.7	70.4	86.1
807	Fort Morgan	98.4	85.8	92.8
808-809	Colorado Springs	100.4	85.7	93.9
810	Pueblo	100.0	83.1	92.5
811	Alamosa	101.3	79.5	91.7
812	Salida	101.3	80.2	91.9
813	Durango	101.7	81.4	92.7
814	Montrose	100.3	79.6	91.1
815	Grand Junction	103.6	76.8	91.7
816	Glenwood Springs	101.1	83.3	93.2
CONNECTICUT				
060	New Britain	101.1	116.6	108.0
061	Hartford	102.3	117.3	109.0
062	Willimantic	101.8	117.2	108.7
063	New London	98.1	117.3	106.7
064	Meriden	100.1	117.8	108.0
065	New Haven	102.8	117.8	109.4
066	Bridgeport	102.4	117.8	109.2
067	Waterbury	101.9	117.7	108.9
068	Norwalk	101.9	118.2	109.1
069	Stamford	102.0	124.0	111.8
D.C.				
200-205	Washington	102.5	91.2	97.5
DELAWARE				
197	Newark	100.7	103.9	102.1
198	Wilmington	99.7	103.9	101.5
199	Dover	100.7	103.9	102.1
FLORIDA				
320,322	Jacksonville	98.3	59.6	81.1
321	Daytona Beach	98.3	72.8	87.0
323	Tallahassee	97.6	48.2	75.6
324	Panama City	99.4	36.8	71.6
325	Pensacola	99.1	57.2	80.5
326,344	Gainesville	99.6	57.7	81.0
327-328,347	Orlando	99.7	68.6	85.9
329	Melbourne	100.8	77.8	90.6
330-332,340	Miami	98.7	71.4	86.6
333	Fort Lauderdale	97.7	70.3	85.5
334,349	West Palm Beach	96.6	66.6	83.3
335-336,346	Tampa	99.4	70.8	86.7
337	St. Petersburg	101.6	54.2	80.5
338	Lakeland	98.4	70.5	86.0
339,341	Fort Myers	97.7	62.8	82.2
342	Sarasota	99.5	65.1	84.2
GEORGIA				
300-303,399	Atlanta	97.7	79.0	89.4
304	Statesboro	97.1	43.6	73.4
305	Gainesville	96.0	55.7	78.1
306	Athens	95.3	62.2	80.6
307	Dalton	97.4	49.4	76.1
308-309	Augusta	96.3	57.5	79.0
310-312	Macon	95.7	59.1	79.4
313-314	Savannah	97.0	57.7	79.6
315	Waycross	97.0	50.0	76.1
316	Valdosta	96.9	47.6	75.0
317,398	Albany	97.2	56.3	79.1
318-319	Columbus	97.2	60.4	80.8

STATE/ZIP	CITY	MAT.	INST.	TOTAL
HAWAII				
967	Hilo	115.0	124.8	119.3
968	Honolulu	120.6	124.8	122.5
STATES & POSS.				
969	Guam	196.4	52.7	132.5
IDAHO				
832	Pocatello	100.3	77.3	90.1
833	Twin Falls	100.8	57.7	81.7
834	Idaho Falls	98.3	58.8	80.8
835	Lewiston	107.4	83.6	96.8
836-837	Boise	100.2	78.9	90.7
838	Coeur d'Alene	106.6	75.9	93.0
ILLINOIS				
600-603	North Suburban	99.5	125.3	111.0
604	Joliet	99.4	118.5	107.9
605	South Suburban	99.5	125.3	111.0
606-608	Chicago	100.1	126.3	111.7
609	Kankakee	95.7	104.1	99.4
610-611	Rockford	98.1	107.9	102.4
612	Rock Island	95.4	97.0	96.1
613	La Salle	96.7	102.6	99.3
614	Galesburg	96.5	98.9	97.6
615-616	Peoria	99.1	99.9	99.4
617	Bloomington	95.8	99.2	97.3
618-619	Champaign	99.2	100.8	99.9
620-622	East St. Louis	94.5	102.0	97.8
623	Quincy	95.7	94.6	95.2
624	Effingham	95.1	97.1	96.0
625	Decatur	97.0	101.1	98.8
626-627	Springfield	97.3	99.8	98.4
628	Centralia	93.3	101.1	96.8
629	Carbondale	92.9	97.3	94.9
INDIANA				
460	Anderson	94.9	84.8	90.4
461-462	Indianapolis	98.8	87.2	93.7
463-464	Gary	96.2	104.3	99.8
465-466	South Bend	95.3	86.8	91.5
467-468	Fort Wayne	95.4	83.1	89.9
469	Kokomo	93.0	85.7	89.8
470	Lawrenceburg	92.0	81.8	87.5
471	New Albany	93.6	77.5	86.5
472	Columbus	96.1	84.0	90.7
473	Muncie	96.2	82.4	90.1
474	Bloomington	98.2	84.1	91.9
475	Washington	94.0	88.4	91.5
476-477	Evansville	95.3	90.0	93.0
478	Terre Haute	96.0	88.3	92.6
479	Lafayette	95.7	82.4	89.8
IOWA				
500-503,509	Des Moines	97.8	82.2	90.9
504	Mason City	96.0	62.0	80.9
505	Fort Dodge	96.2	58.1	79.3
506-507	Waterloo	97.8	62.0	81.9
508	Creston	96.4	66.0	82.9
510-511	Sioux City	98.4	74.9	88.0
512	Sibley	97.0	51.5	76.8
513	Spencer	98.8	49.9	77.1
514	Carroll	95.8	55.7	78.0
515	Council Bluffs	99.3	75.2	88.6
516	Shenandoah	96.2	52.9	77.0
520	Dubuque	97.8	77.2	88.7
521	Decorah	96.8	52.8	77.3
522-524	Cedar Rapids	98.9	84.4	92.5
525	Ottumwa	96.8	71.1	85.4
526	Burlington	96.0	75.5	86.9
527-528	Davenport	97.9	94.9	96.5
KANSAS				
660-662	Kansas City	98.9	96.1	97.7
664-666	Topeka	98.7	69.3	85.6
667	Fort Scott	97.6	72.3	86.4
668	Emporia	97.4	57.7	79.8
669	Belleville	99.2	63.7	83.5
670-672	Wichita	97.9	69.5	85.3
673	Independence	99.0	67.7	85.1
674	Salina	99.2	63.9	83.5
675	Hutchinson	94.5	49.5	74.5
676	Hays	98.5	67.0	84.5
677	Colby	99.2	53.1	78.8

STATE/ZIP	CITY	MAT.	INST.	TOTAL
KANSAS (CONT'D)				
678	Dodge City	100.4	68.4	86.2
679	Liberal	98.3	44.2	74.3
KENTUCKY				
400-402	Louisville	96.4	85.4	91.5
403-405	Lexington	96.0	65.7	82.5
406	Frankfort	96.3	66.7	83.1
407-409	Corbin	93.3	43.3	71.1
410	Covington	93.8	91.9	93.0
411-412	Ashland	92.5	96.9	94.5
413-414	Campton	94.1	43.7	71.7
415-416	Pikeville	95.0	60.3	79.6
417-418	Hazard	93.5	43.4	71.3
420	Paducah	92.2	88.4	90.5
421-422	Bowling Green	94.6	82.7	89.3
423	Owensboro	94.4	73.5	85.1
424	Henderson	92.0	90.4	91.3
425-426	Somerset	91.6	45.1	70.9
427	Elizabethtown	91.1	84.5	88.2
LOUISIANA				
700-701	New Orleans	99.8	69.7	86.4
703	Thibodaux	97.7	67.2	84.2
704	Hammond	95.0	62.7	80.7
705	Lafayette	97.4	62.3	81.8
706	Lake Charles	97.5	64.1	82.7
707-708	Baton Rouge	98.1	62.9	82.5
710-711	Shreveport	94.6	61.3	79.8
712	Monroe	95.3	57.9	78.7
713-714	Alexandria	95.4	59.5	79.5
MAINE				
039	Kittery	95.2	73.4	85.5
040-041	Portland	98.5	79.8	90.2
042	Lewiston	98.9	79.8	90.4
043	Augusta	97.4	73.2	86.7
044	Bangor	97.7	79.8	89.8
045	Bath	96.3	73.4	86.1
046	Machias	95.9	72.8	85.6
047	Houlton	96.0	76.4	87.3
048	Rockland	95.1	73.4	85.4
049	Waterville	96.3	73.3	86.1
MARYLAND				
206	Waldorf	98.3	72.6	86.9
207-208	College Park	98.4	81.4	90.8
209	Silver Spring	97.6	77.9	88.9
210-212	Baltimore	99.5	84.7	92.9
214	Annapolis	98.9	79.2	90.1
215	Cumberland	95.4	81.2	89.1
216	Easton	96.8	44.8	73.7
217	Hagerstown	96.2	81.9	89.8
218	Salisbury	97.3	53.5	77.8
219	Elkton	94.6	66.7	82.2
MASSACHUSETTS				
010-011	Springfield	100.4	105.9	102.8
012	Pittsfield	99.9	103.0	101.3
013	Greenfield	97.9	105.3	101.2
014	Fitchburg	96.6	119.8	106.9
015-016	Worcester	100.1	120.0	109.0
017	Framingham	96.1	123.8	108.4
018	Lowell	99.7	125.0	111.0
019	Lawrence	100.9	122.6	110.5
020-022, 024	Boston	102.5	132.0	115.6
023	Brockton	101.1	121.0	110.0
025	Buzzards Bay	95.5	119.0	105.9
026	Hyannis	98.1	119.0	107.4
027	New Bedford	100.8	119.1	109.0
MICHIGAN				
480,483	Royal Oak	95.3	107.9	100.9
481	Ann Arbor	97.4	105.9	101.2
482	Detroit	99.1	117.6	107.3
484-485	Flint	97.2	99.9	98.4
486	Saginaw	97.0	93.4	95.4
487	Bay City	96.9	93.5	95.4
488-489	Lansing	97.2	95.2	96.3
490	Battle Creek	95.5	87.9	92.1
491	Kalamazoo	95.8	85.8	91.4
492	Jackson	93.8	92.3	93.2
493,495	Grand Rapids	96.0	68.1	83.6
494	Muskegon	94.4	83.6	89.6

STATE/ZIP	CITY	MAT.	INST.	TOTAL
MICHIGAN (CONT'D)				
496	Traverse City	93.3	73.1	84.3
497	Gaylord	94.5	76.9	86.7
498-499	Iron Mountain	96.3	87.7	92.5
MINNESOTA				
550-551	Saint Paul	99.7	124.5	110.7
553-555	Minneapolis	100.5	126.7	112.2
556-558	Duluth	99.5	114.5	106.2
559	Rochester	99.2	108.5	103.4
560	Mankato	96.3	105.9	100.6
561	Windom	95.1	80.3	88.5
562	Willmar	94.4	87.4	91.3
563	St. Cloud	95.7	123.3	108.0
564	Brainerd	96.0	105.1	100.0
565	Detroit Lakes	97.9	100.6	99.1
566	Bemidji	97.3	101.2	99.0
567	Thief River Falls	96.8	96.1	96.5
MISSISSIPPI				
386	Clarksdale	95.9	30.8	67.0
387	Greenville	99.3	44.6	75.0
388	Tupelo	97.3	37.3	70.6
389	Greenwood	97.2	33.3	68.8
390-392	Jackson	97.8	46.1	74.8
393	Meridian	96.0	46.7	74.1
394	Laurel	97.1	33.4	68.8
395	Biloxi	97.8	55.5	79.0
396	McComb	95.7	52.1	76.4
397	Columbus	97.1	36.7	70.3
MISSOURI				
630-631	St. Louis	97.0	109.0	102.3
633	Bowling Green	95.7	95.8	95.7
634	Hannibal	94.5	86.3	90.9
635	Kirksville	96.8	79.2	89.0
636	Flat River	96.7	97.0	96.8
637	Cape Girardeau	96.4	92.3	94.6
638	Sikeston	94.5	83.2	89.5
639	Poplar Bluff	94.0	83.8	89.5
640-641	Kansas City	100.9	106.8	103.5
644-645	St. Joseph	99.9	93.3	97.0
646	Chillicothe	96.7	71.2	85.4
647	Harrisonville	96.4	101.3	98.6
648	Joplin	98.6	70.5	86.1
650-651	Jefferson City	96.1	89.7	93.3
652	Columbia	97.1	92.4	95.0
653	Sedalia	96.1	84.9	91.2
654-655	Rolla	95.0	78.6	87.7
656-658	Springfield	98.4	78.7	89.6
MONTANA				
590-591	Billings	102.1	74.5	89.8
592	Wolf Point	100.8	68.7	86.6
593	Miles City	98.8	70.0	86.0
594	Great Falls	102.7	73.8	89.9
595	Havre	99.6	68.9	86.0
596	Helena	102.5	71.5	88.7
597	Butte	101.2	70.9	87.7
598	Missoula	98.9	70.1	86.1
599	Kalispell	97.8	69.4	85.2
NEBRASKA				
680-681	Omaha	99.3	78.2	89.9
683-685	Lincoln	98.2	67.1	84.4
686	Columbus	96.1	49.6	75.5
687	Norfolk	97.7	60.6	81.3
688	Grand Island	97.5	65.6	83.3
689	Hastings	97.1	56.7	79.2
690	Mccook	96.9	48.2	75.3
691	North Platte	97.1	58.8	80.1
692	Valentine	99.2	38.5	72.3
693	Alliance	99.0	35.9	71.0
NEVADA				
889-891	Las Vegas	98.3	107.1	102.2
893	Ely	98.7	79.0	90.0
894-895	Reno	98.9	98.2	98.6
897	Carson City	98.6	97.9	98.3
898	Elko	97.6	85.4	92.2
NEW HAMPSHIRE				
030	Nashua	100.6	82.4	92.5
031	Manchester	100.9	82.4	92.7

STATE/ZIP	CITY	MAT.	INST.	TOTAL
NEW HAMPSHIRE (CONT'D)				
032-033	Concord	98.4	82.4	91.3
034	Keene	97.2	50.9	76.6
035	Littleton	97.4	61.6	81.5
036	Charleston	96.6	47.9	75.0
037	Claremont	95.9	47.9	74.6
038	Portsmouth	97.8	78.0	89.0
NEW JERSEY				
070-071	Newark	102.1	124.6	112.1
072	Elizabeth	100.8	123.6	110.9
073	Jersey City	99.8	124.5	110.8
074-075	Paterson	101.4	124.2	111.5
076	Hackensack	99.5	124.5	110.6
077	Long Branch	99.1	125.3	110.8
078	Dover	99.7	124.5	110.7
079	Summit	99.8	123.6	110.4
080,083	Vineland	97.6	117.3	106.3
081	Camden	99.6	116.8	107.2
082,084	Atlantic City	98.4	116.5	106.4
085-086	Trenton	99.6	122.6	109.8
087	Point Pleasant	99.5	122.3	109.6
088-089	New Brunswick	100.0	123.4	110.4
NEW MEXICO				
870-872	Albuquerque	98.5	76.2	88.6
873	Gallup	98.3	76.2	88.5
874	Farmington	98.7	76.2	88.7
875	Santa Fe	97.5	76.2	88.0
877	Las Vegas	96.9	76.2	87.7
878	Socorro	96.5	76.2	87.5
879	Truth/Consequences	96.9	70.8	85.3
880	Las Cruces	95.8	67.9	83.4
881	Clovis	97.0	75.2	87.3
882	Roswell	98.7	75.3	88.3
883	Carrizozo	99.0	76.2	88.9
884	Tucumcari	97.9	75.2	87.8
NEW YORK				
100-102	New York	106.0	164.4	131.9
103	Staten Island	101.6	159.9	127.5
104	Bronx	99.8	159.9	126.5
105	Mount Vernon	99.5	138.5	116.8
106	White Plains	99.5	138.2	116.7
107	Yonkers	104.5	138.5	119.6
108	New Rochelle	100.0	138.0	116.9
109	Suffern	99.8	126.2	111.5
110	Queens	101.4	158.8	126.9
111	Long Island City	103.1	158.8	127.8
112	Brooklyn	103.3	159.1	128.1
113	Flushing	103.3	158.8	128.0
114	Jamaica	101.5	158.8	126.9
115,117,118	Hicksville	101.2	149.5	122.7
116	Far Rockaway	103.4	158.8	128.0
119	Riverhead	102.2	149.5	123.2
120-122	Albany	97.8	96.4	97.2
123	Schenectady	98.1	95.9	97.1
124	Kingston	100.9	117.3	108.2
125-126	Poughkeepsie	100.1	119.4	108.7
127	Monticello	99.5	119.9	108.5
128	Glens Falls	92.7	91.3	92.1
129	Plattsburgh	96.8	88.4	93.1
130-132	Syracuse	99.0	94.7	97.1
133-135	Utica	96.9	90.9	94.2
136	Watertown	98.5	88.5	94.1
137-139	Binghamton	98.5	88.8	94.2
140-142	Buffalo	99.8	105.7	102.4
143	Niagara Falls	97.3	104.7	100.6
144-146	Rochester	99.8	99.7	99.8
147	Jamestown	96.3	87.9	92.6
148-149	Elmira	96.3	85.4	91.4
NORTH CAROLINA				
270,272-274	Greensboro	98.5	50.3	77.1
271	Winston-Salem	98.2	49.6	76.6
275-276	Raleigh	98.4	50.7	77.2
277	Durham	98.4	50.7	77.2
278	Rocky Mount	95.7	35.5	69.0
279	Elizabeth City	96.4	37.8	70.4
280	Gastonia	97.7	48.9	76.0
281-282	Charlotte	98.6	48.9	76.5
283	Fayetteville	97.1	50.8	76.5
284	Wilmington	96.4	49.9	75.7
285	Kinston	94.4	35.1	68.1

LOCATION FACTORS

STATE/ZIP	CITY	MAT.	INST.	TOTAL
NORTH CAROLINA (CONT'D)				
286	Hickory	94.8	33.9	67.8
287-288	Asheville	96.8	48.9	75.5
289	Murphy	95.7	33.0	67.9
NORTH DAKOTA				
580-581	Fargo	100.9	65.2	85.1
582	Grand Forks	100.5	59.2	82.2
583	Devils Lake	99.9	60.8	82.6
584	Jamestown	100.0	52.6	78.9
585	Bismarck	99.3	66.6	84.8
586	Dickinson	100.7	60.5	82.9
587	Minot	100.6	67.8	86.1
588	Williston	99.3	60.2	81.9
OHIO				
430-432	Columbus	97.3	89.6	93.9
433	Marion	93.9	88.4	91.5
434-436	Toledo	97.6	100.0	98.7
437-438	Zanesville	94.3	86.1	90.6
439	Steubenville	95.2	95.1	95.1
440	Lorain	96.9	97.1	97.0
441	Cleveland	97.2	104.3	100.4
442-443	Akron	97.8	98.0	97.9
444-445	Youngstown	97.2	93.3	95.5
446-447	Canton	97.3	88.4	93.4
448-449	Mansfield	94.6	93.8	94.2
450	Hamilton	94.5	89.6	92.3
451-452	Cincinnati	94.9	90.2	92.8
453-454	Dayton	94.6	87.0	91.2
455	Springfield	94.6	87.9	91.6
456	Chillicothe	93.5	94.2	93.8
457	Athens	96.3	79.9	89.0
458	Lima	96.8	87.9	92.8
OKLAHOMA				
730-731	Oklahoma City	97.1	64.3	82.6
734	Ardmore	94.0	63.7	80.5
735	Lawton	96.4	65.3	82.6
736	Clinton	95.3	62.0	80.6
737	Enid	96.0	62.0	80.9
738	Woodward	94.2	62.1	79.9
739	Guymon	95.2	32.9	67.5
740-741	Tulsa	96.2	59.9	80.1
743	Miami	93.0	67.5	81.7
744	Muskogee	95.6	44.0	72.7
745	Mcalester	92.6	55.1	75.9
746	Ponca City	93.1	62.2	79.4
747	Durant	93.1	61.7	79.2
748	Shawnee	94.6	60.2	79.3
749	Poteau	92.3	64.7	80.1
OREGON				
970-972	Portland	102.2	104.6	103.3
973	Salem	102.2	103.2	102.6
974	Eugene	101.7	102.8	102.2
975	Medford	103.3	101.1	102.3
976	Klamath Falls	103.4	101.1	102.4
977	Bend	102.2	102.9	102.5
978	Pendleton	96.5	102.6	99.2
979	Vale	94.4	93.2	93.9
PENNSYLVANIA				
150-152	Pittsburgh	97.0	104.2	100.2
153	Washington	94.0	102.8	98.0
154	Uniontown	94.2	101.2	97.3
155	Bedford	95.2	92.6	94.0
156	Greensburg	95.3	102.2	98.3
157	Indiana	94.1	100.7	97.0
158	Dubois	95.4	97.3	96.2
159	Johnstown	95.1	98.1	96.5
160	Butler	92.3	102.3	96.7
161	New Castle	92.3	101.0	96.2
162	Kittanning	92.8	103.6	97.6
163	Oil City	92.3	98.1	94.9
164-165	Erie	94.4	96.7	95.4
166	Altoona	94.4	94.5	94.5
167	Bradford	95.7	96.6	96.1
168	State College	95.4	93.9	94.7
169	Wellsboro	96.4	94.2	95.4
170-171	Harrisburg	98.2	95.3	96.9
172	Chambersburg	95.9	89.9	93.3
173-174	York	96.2	95.3	95.8
175-176	Lancaster	94.8	90.6	92.9

STATE/ZIP	CITY	MAT.	INST.	TOTAL
PENNSYLVANIA (CONT'D)				
177	Williamsport	93.4	82.4	88.5
178	Sunbury	95.4	94.4	95.0
179	Pottsville	94.5	94.0	94.3
180	Lehigh Valley	95.9	114.4	104.1
181	Allentown	98.0	107.4	102.2
182	Hazleton	95.3	96.6	95.9
183	Stroudsburg	95.2	101.1	97.8
184-185	Scranton	98.8	96.3	97.7
186-187	Wilkes-Barre	95.2	94.9	95.1
188	Montrose	94.8	95.5	95.1
189	Doylestown	94.7	119.7	105.8
190-191	Philadelphia	101.0	131.1	114.4
193	Westchester	97.6	122.0	108.5
194	Norristown	96.6	126.0	109.7
195-196	Reading	98.8	100.9	99.7
PUERTO RICO				
009	San Juan	135.2	25.5	86.5
RHODE ISLAND				
028	Newport	99.6	112.7	105.4
029	Providence	99.9	112.7	105.6
SOUTH CAROLINA				
290-292	Columbia	97.1	47.3	75.0
293	Spartanburg	95.9	45.4	73.5
294	Charleston	97.3	47.9	75.3
295	Florence	95.4	45.2	73.1
296	Greenville	95.7	45.4	73.4
297	Rock Hill	95.3	33.0	67.6
298	Aiken	96.1	72.0	85.4
299	Beaufort	97.0	36.2	70.0
SOUTH DAKOTA				
570-571	Sioux Falls	99.2	57.3	80.6
572	Watertown	97.9	51.0	77.1
573	Mitchell	96.9	50.8	76.4
574	Aberdeen	99.5	53.1	78.9
575	Pierre	98.8	53.5	78.7
576	Mobridge	97.5	50.4	76.6
577	Rapid City	99.4	52.2	78.4
TENNESSEE				
370-372	Nashville	97.5	73.7	86.9
373-374	Chattanooga	98.6	54.1	78.8
375,380-381	Memphis	96.6	76.2	87.5
376	Johnson City	97.3	56.4	79.1
377-379	Knoxville	94.7	57.5	78.2
382	Mckenzie	96.4	59.4	80.0
383	Jackson	98.3	51.7	77.6
384	Columbia	94.9	55.2	77.3
385	Cookeville	96.2	49.9	75.6
TEXAS				
750	Mckinney	97.6	55.5	78.9
751	Waxahackie	97.5	56.3	79.2
752-753	Dallas	98.3	67.1	84.4
754	Greenville	97.7	40.7	72.4
755	Texarkana	96.6	54.0	77.7
756	Longview	96.9	42.6	72.8
757	Tyler	97.4	56.2	79.1
758	Palestine	93.6	43.0	71.1
759	Lufkin	94.5	46.7	73.3
760-761	Fort Worth	96.2	62.8	81.4
762	Denton	96.5	51.6	76.6
763	Wichita Falls	97.1	57.9	79.7
764	Eastland	95.6	42.0	71.8
765	Temple	94.5	50.3	74.9
766-767	Waco	96.4	57.5	79.1
768	Brownwood	96.5	39.5	71.2
769	San Angelo	96.2	47.7	74.7
770-772	Houston	99.6	73.0	87.7
773	Huntsville	97.7	40.3	72.2
774	Wharton	99.1	45.1	75.1
775	Galveston	97.3	70.6	85.4
776-777	Beaumont	97.1	64.1	82.4
778	Bryan	94.7	64.5	81.3
779	Victoria	99.2	47.6	76.3
780	Laredo	94.6	53.7	76.5
781-782	San Antonio	95.0	67.0	82.5
783-784	Corpus Christi	97.3	53.1	77.7
785	Mc Allen	97.2	47.4	75.1
786-787	Austin	94.6	61.7	80.0

LOCATION FACTORS

STATE/ZIP	CITY	MAT.	INST.	TOTAL
TEXAS (CONT'D)				
88	Del Rio	96.5	33.2	68.4
89	Giddings	94.0	42.1	71.0
90-791	Amarillo	97.2	58.0	79.8
92	Childress	96.1	52.5	76.7
93-794	Lubbock	98.6	53.5	78.6
95-796	Abilene	96.6	53.3	77.4
97	Midland	98.7	49.9	77.0
98-799,885	El Paso	96.6	51.6	76.6
UTAH				
840-841	Salt Lake City	101.3	70.7	87.7
842,844	Ogden	97.4	70.5	85.5
843	Logan	99.1	70.5	86.4
845	Price	99.6	50.1	77.6
846-847	Provo	99.6	70.5	86.7
VERMONT				
50	White River Jct.	98.6	46.5	75.4
51	Bellows Falls	97.1	48.8	75.6
52	Bennington	97.4	49.1	76.0
53	Brattleboro	97.9	49.3	76.3
54	Burlington	100.5	64.9	84.7
56	Montpelier	97.3	64.9	82.9
57	Rutland	99.5	64.9	84.1
58	St. Johnsbury	98.8	49.1	76.7
59	Guildhall	97.4	48.6	75.7
VIRGINIA				
220-221	Fairfax	99.3	80.6	91.0
22	Arlington	100.5	79.3	91.1
23	Alexandria	99.9	84.9	93.3
24-225	Fredericksburg	98.0	68.4	84.9
26	Winchester	98.6	53.8	78.7
27	Culpeper	98.5	55.3	79.3
28	Harrisonburg	98.8	47.1	75.8
29	Charlottesville	99.1	61.5	82.4
230-232	Richmond	100.3	66.7	85.4
233-235	Norfolk	100.0	65.8	84.8
36	Newport News	99.6	64.7	84.1
37	Portsmouth	98.9	65.5	84.0
38	Petersburg	98.9	66.7	84.6
39	Farmville	98.3	41.1	72.9
240-241	Roanoke	100.2	48.1	77.1
42	Bristol	97.9	49.0	76.2
43	Pulaski	97.7	42.4	73.1
44	Staunton	98.4	45.6	75.0
45	Lynchburg	98.6	52.2	78.0
46	Grundy	98.0	44.4	74.2
WASHINGTON				
980-981,987	Seattle	103.8	104.7	104.2
82	Everett	103.6	101.0	102.4
83-984	Tacoma	103.5	101.4	102.6
85	Olympia	102.3	101.3	101.9
86	Vancouver	105.4	97.7	102.0
88	Wenatchee	103.5	86.6	96.0
89	Yakima	103.7	91.8	98.4
990-992	Spokane	105.7	83.3	95.8
93	Richland	105.0	86.9	96.9
94	Clarkston	103.9	84.3	95.2
WEST VIRGINIA				
247-248	Bluefield	96.8	79.4	89.1
49	Lewisburg	98.3	83.1	91.6
250-253	Charleston	99.5	91.0	95.7
54	Martinsburg	98.1	80.0	90.1
255-257	Huntington	99.6	93.4	96.9
258-259	Beckley	96.6	88.8	93.1
60	Wheeling	99.7	94.2	97.2
61	Parkersburg	98.7	92.6	96.0
62	Buckhannon	98.1	93.4	96.0
263-264	Clarksburg	98.5	93.4	96.2
65	Morgantown	98.6	93.5	96.4
66	Gassaway	97.9	92.2	95.4
67	Romney	97.9	86.5	92.9
68	Petersburg	97.9	89.3	94.1
WISCONSIN				
530,532	Milwaukee	99.0	106.2	102.2
31	Kenosha	98.5	103.1	100.5
34	Racine	98.1	102.9	100.2
35	Beloit	98.0	98.4	98.2
37	Madison	98.3	98.1	98.2

STATE/ZIP	CITY	MAT.	INST.	TOTAL
WISCONSIN (CONT'D)				
538	Lancaster	95.8	94.0	95.0
539	Portage	94.4	97.7	95.9
540	New Richmond	95.5	98.5	96.8
541-543	Green Bay	99.7	95.2	97.7
544	Wausau	94.9	94.3	94.6
545	Rhinelander	98.0	93.7	96.1
546	La Crosse	96.1	95.7	95.9
547	Eau Claire	97.8	96.2	97.1
548	Superior	95.4	102.8	98.7
549	Oshkosh	95.5	95.2	95.4
WYOMING				
820	Cheyenne	97.6	61.2	81.4
821	Yellowstone Nat'l Park	96.8	54.6	78.0
822	Wheatland	98.1	56.0	79.4
823	Rawlins	99.4	53.1	78.8
824	Worland	97.4	51.4	77.0
825	Riverton	98.4	53.1	78.3
826	Casper	97.7	59.9	80.9
827	Newcastle	97.2	53.1	77.6
828	Sheridan	98.1	58.7	80.6
829-831	Rock Springs	101.5	51.4	79.2
CANADIAN FACTORS (reflect Canadian currency)				
ALBERTA				
	Calgary	124.3	79.6	104.4
	Edmonton	125.9	79.6	105.3
	Fort McMurray	116.6	79.1	100.0
	Lethbridge	117.6	78.6	100.3
	Lloydminster	116.7	79.1	100.0
	Medicine Hat	116.7	78.6	99.8
	Red Deer	117.1	78.6	100.0
BRITISH COLUMBIA				
	Kamloops	117.8	81.8	101.8
	Prince George	119.1	81.8	102.5
	Vancouver	125.7	85.6	107.9
	Victoria	119.1	82.3	102.8
MANITOBA				
	Brandon	117.0	68.5	95.5
	Portage la Prairie	117.0	68.5	95.5
	Winnipeg	124.1	68.9	99.6
NEW BRUNSWICK				
	Bathurst	115.1	60.9	91.1
	Dalhousie	115.1	60.9	91.1
	Fredericton	117.0	64.9	93.8
	Moncton	115.4	60.9	91.2
	Newcastle	115.1	60.9	91.1
	Saint John	118.1	64.9	94.5
NEWFOUNDLAND				
	Corner Brook	119.8	60.9	93.7
	St. John's	120.1	60.9	93.8
NORTHWEST TERRITORIES				
	Yellowknife	120.4	77.4	101.3
NOVA SCOTIA				
	Dartmouth	117.8	67.1	95.3
	Halifax	119.3	68.2	96.6
	New Glasgow	115.8	67.1	94.2
	Sydney	113.3	67.1	92.8
	Yarmouth	115.6	67.1	94.1
ONTARIO				
	Barrie	119.9	84.1	104.0
	Brantford	119.1	87.6	105.1
	Cornwall	119.1	84.3	103.6
	Hamilton	124.5	87.9	108.2
	Kingston	120.1	84.7	104.4
	Kitchener	115.6	83.7	101.4
	London	121.9	85.3	105.6
	North Bay	119.1	82.4	102.8
	Oshawa	118.7	85.7	104.1
	Ottawa	120.5	85.5	105.0
	Owen Sound	120.1	82.5	103.4
	Peterborough	119.1	84.1	103.5
	Sarnia	119.3	87.9	105.4
	St. Catharines	113.6	83.1	100.0
	Sudbury	113.7	82.7	99.9

LOCATION FACTORS

533

STATE/ZIP	CITY	MAT.	INST.	TOTAL
ONTARIO (CONT'D)	Thunder Bay	114.8	82.8	100.6
	Toronto	124.2	92.1	109.9
	Windsor	114.6	85.4	101.6
PRINCE EDWARD ISLAND	Charlottetown	117.7	56.9	90.7
	Summerside	117.3	56.9	90.5
QUEBEC				
	Cap-de-la-Madeleine	115.9	78.9	99.5
	Charlesbourg	115.9	78.9	99.5
	Chicoutimi	115.0	78.6	98.8
	Gatineau	115.3	78.7	99.1
	Laval	115.2	78.7	99.0
	Montreal	118.5	79.0	101.0
	Quebec	119.4	79.4	101.6
	Sherbrooke	115.7	78.7	99.3
	Trois Rivieres	116.2	78.9	99.6
SASKATCHEWAN				
	Moose Jaw	114.2	62.9	91.4
	Prince Albert	113.7	62.7	91.1
	Regina	116.1	62.9	92.5
	Saskatoon	114.6	62.8	91.6
YUKON				
	Whitehorse	114.0	61.9	90.8

Square Foot and Cubic Foot Building Costs

The cost figures in division K1010 were derived from more than 11,200 projects contained in the Means Data Bank of Construction Costs, and include the contractor's overhead and profit, but do not include architectural fees or land costs. The figures have been adjusted to January of the current year. New projects are added to our files each year, and outdated projects are discarded. For this reason, certain costs may not show a uniform annual progression. In no case are all subdivisions of a project listed.

These projects were located throughout the U.S. and reflect a tremendous variation in S.F. and C.F. costs. This is due to differences, not only in labor and material costs, but also in individual owner's requirements. For instance, a bank in a large city would have different features than one in a rural area. This is true of all the different types of buildings analyzed. Therefore, caution should be exercised when using Division K1010 costs. For example, for court houses, costs in the data bank are local court house costs and will not apply to the larger, more elaborate federal court houses. As a general rule, the projects in the 1/4 column do not include any site work or equipment, while the projects in the 3/4 column may include both equipment and site work. The median figures do not generally include site work.

None of the figures "go with" any others. All individual cost items were computed and tabulated separately. Thus the sum of the median figures for Plumbing, HVAC and Electrical will not normally total up to the total Mechanical and Electrical costs arrived at by separate analysis and tabulation of the projects.

Each building was analyzed as to total and component costs and percentages. The figures were arranged in ascending order with the results tabulated as shown. The 1/4 column shows that 25% of the projects had lower costs, 75% higher. The 3/4 column shows that 75% of the projects had lower costs,

25% had higher. The median column shows that 50% of the projects had lower costs, 50% had higher.

There are two times when square foot costs are useful. The first is in the conceptual stage when no details are available. Then square foot costs make a useful starting point. The second is after the bids are in and the costs can be worked back into their appropriate units for information purposes. As soon as details become available in the project design, the square foot approach should be discontinued and the project priced as to its particular components. When more precision is required or for estimating the replacement cost of specific buildings, the current edition of **Means Square Foot Costs** should be used.

In using the figures in division K1010, it is recommended that the median column be used for preliminary figures if no additional information is available. The median figures, when multiplied by the total city construction cost index figures (see City Cost Indexes) and then multiplied by the project size modifier in RK1010-050, should present a fairly accurate base figure, which would then have to be adjusted in view of the estimator's experience, local economic conditions, code requirements and the owner's particular requirements. There is no need to factor the percentage figures as these should remain constant from city to city. All tabulations mentioning air conditioning had at least partial air conditioning.

The editors of this book would greatly appreciate receiving cost figures on one or more of your recent projects which would then be included in the averages for next year. All cost figures received will be kept confidential except that they will be averaged with other similar projects to arrive at S.F. and C.F. cost figures for next year's book. See the last page of the book for details and the discount available for submitting one or more of your projects.

Table K1010-031   Unit Gross Area Requirements

The figures in the table below indicate typical ranges in square feet as a function of the "occupant" unit. This table is best used in the preliminary design stages to help determine the probable size

requirement for the total project. See RK1010-050 for the typical total size ranges for various types of buildings.

Building Type	Unit	Gross Area in S.F.		
		1/4	Median	3/4
Apartments	Unit	660	860	1,100
Auditorium & Play Theaters	Seat	18	25	38
Bowling Alleys	Lane		940	
Churches & Synagogues	Seat	20	28	39
Dormitories	Bed	200	230	275
Fraternity & Sorority Houses	Bed	220	315	370
Garages, Parking	Car	325	355	385
Hospitals	Bed	685	850	1,075
Hotels	Rental Unit	475	600	710
Housing for the elderly	Unit	515	635	755
Housing, Public	Unit	700	875	1,030
Ice Skating Rinks	Total	27,000	30,000	36,000
Motels	Rental Unit	360	465	620
Nursing Homes	Bed	290	350	450
Restaurants	Seat	23	29	39
Schools, Elementary	Pupil	65	77	90
Junior High & Middle		85	110	129
Senior High		102	130	145
Vocational		110	135	195
Shooting Ranges	Point		450	
Theaters & Movies	Seat		15	

SQUARE FOOT

535

RK1010-050   Square Foot Project Size Modifier

One factor that affects the S.F. cost of a particular building is the size. In general, for buildings built to the same specifications in the same locality, the larger building will have the lower S.F. cost. This is due mainly to the decreasing contribution of the exterior walls plus the economy of scale usually achievable in larger buildings. The Area Conversion Scale shown below will give a factor to convert costs for the typical size building to an adjusted cost for the particular project.

The Square Foot Base Size lists the median costs, most typical project size in our accumulated data and the range in size of the projects.

The Size Factor for your project is determined by dividing your project area in S.F. by the typical project size for the particular Building Type. With this factor, enter the Area Conversion Scale at the appropriate Size Factor and determine the appropriate cost multiplier for your building size.

Example: Determine the cost per S.F. for a 100,000 S.F. Mid-rise apartment building.

$$\frac{\text{Proposed building area} = 100,000 \text{ S.F.}}{\text{Typical size from below} = 50,000 \text{ S.F.}} = 2.00$$

Enter Area Conversion scale at 2.0, intersect curve, read horizontally the appropriate cost multiplier of 0.94. Size adjusted cost becomes .94 x $90.00 = $84.60 based on national average costs.

Note: For Size Factors less than .50, the Cost Multiplier is 1.1
For Size Factors greater than 3.5, the Cost Multiplier is .90

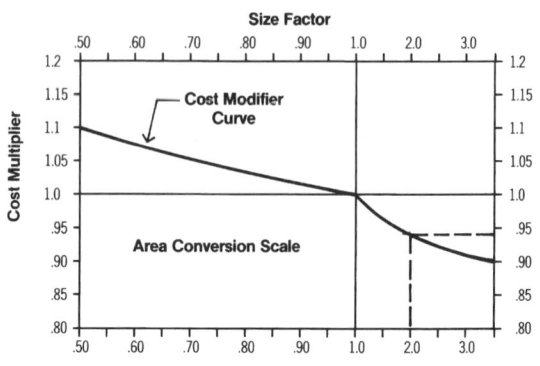

Square Foot Base Size

Building Type	Median Cost per S.F.	Typical Size Gross S.F.	Typical Range Gross S.F.	Building Type	Median Cost per S.F.	Typical Size Gross S.F.	Typical Range Gross S.F.
Apartments, Low Rise	$ 71.00	21,000	9,700 - 37,200	Jails	$216.00	40,000	5,500 - 145,000
Apartments, Mid Rise	90.00	50,000	32,000 - 100,000	Libraries	132.00	12,000	7,000 - 31,000
Apartments, High Rise	102.00	145,000	95,000 - 600,000	Living, Assisted	115.00	32,300	23,500 - 50,300
Auditoriums	119.00	25,000	7,600 - 39,000	Medical Clinics	122.00	7,200	4,200 - 15,700
Auto Sales	88.00	20,000	10,800 - 28,600	Medical Offices	115.00	6,000	4,000 - 15,000
Banks	158.00	4,200	2,500 - 7,500	Motels	87.00	40,000	15,800 - 120,000
Churches	109.00	17,000	2,000 - 42,000	Nursing Homes	119.00	23,000	15,000 - 37,000
Clubs, Country	111.00	6,500	4,500 - 15,000	Offices, Low Rise	99.50	20,000	5,000 - 80,000
Clubs, Social	105.00	10,000	6,000 - 13,500	Offices, Mid Rise	100.00	120,000	20,000 - 300,000
Clubs, YMCA	125.00	28,300	12,800 - 39,400	Offices, High Rise	128.00	260,000	120,000 - 800,000
Colleges (Class)	136.00	50,000	15,000 - 150,000	Police Stations	156.00	10,500	4,000 - 19,000
Colleges (Science Lab)	203.00	45,600	16,600 - 80,000	Post Offices	118.00	12,400	6,800 - 30,000
College (Student Union)	155.00	33,400	16,000 - 85,000	Power Plants	880.00	7,500	1,000 - 20,000
Community Center	113.00	9,400	5,300 - 16,700	Religious Education	101.00	9,000	6,000 - 12,000
Court Houses	151.00	32,400	17,800 - 106,000	Research	166.00	19,000	6,300 - 45,000
Dept. Stores	66.00	90,000	44,000 - 122,000	Restaurants	143.00	4,400	2,800 - 6,000
Dormitories, Low Rise	117.00	25,000	10,000 - 95,000	Retail Stores	70.50	7,200	4,000 - 17,600
Dormitories, Mid Rise	148.00	85,000	20,000 - 200,000	Schools, Elementary	104.00	41,000	24,500 - 55,000
Factories	64.00	26,400	12,900 - 50,000	Schools, Jr. High	108.00	92,000	52,000 - 119,000
Fire Stations	116.00	5,800	4,000 - 8,700	Schools, Sr. High	110.00	101,000	50,500 - 175,000
Fraternity Houses	110.00	12,500	8,200 - 14,800	Schools, Vocational	105.00	37,000	20,500 - 82,000
Funeral Homes	122.00	10,000	4,000 - 20,000	Sports Arenas	86.50	15,000	5,000 - 40,000
Garages, Commercial	80.00	9,300	5,000 - 13,600	Supermarkets	69.50	44,000	12,000 - 60,000
Garages, Municipal	99.50	8,300	4,500 - 12,600	Swimming Pools	163.00	20,000	10,000 - 32,000
Garages, Parking	42.50	163,000	76,400 - 225,300	Telephone Exchange	189.00	4,500	1,200 - 10,600
Gymnasiums	107.00	19,200	11,600 - 41,000	Theaters	101.00	10,500	8,800 - 17,500
Hospitals	190.00	55,000	27,200 - 125,000	Town Halls	121.00	10,800	4,800 - 23,400
House (Elderly)	96.50	37,000	21,000 - 66,000	Warehouses	51.00	25,000	8,000 - 72,000
Housing (Public)	89.00	36,000	14,400 - 74,400	Warehouse & Office	55.50	25,000	8,000 - 72,000
Ice Rinks	128.00	29,000	27,200 - 33,600				

SQUARE FOOT

			UNIT COSTS			% OF TOTAL			
K1010 | S.F. Costs		UNIT	1/4	MEDIAN	3/4	1/4	MEDIAN	3/4	
01	0010 **APARTMENTS** Low Rise (1 to 3 story)	S.F.	56.50	71	94				01
	0020 Total project cost	C.F.	5.05	6.70	8.25				
	0100 Site work	S.F.	4.81	6.55	11.55	6.05%	10.55%	14.05%	
	0500 Masonry		1.11	2.57	4.45	1.54%	3.67%	6.35%	
	1500 Finishes		5.95	8.20	10.15	9.05%	10.75%	12.85%	
	1800 Equipment		1.83	2.79	4.14	2.73%	4.03%	5.95%	
	2720 Plumbing		4.38	5.65	7.15	6.65%	8.95%	10.05%	
	2770 Heating, ventilating, air conditioning		2.79	3.44	5.05	4.20%	5.60%	7.60%	
	2900 Electrical		3.25	4.32	5.80	5.20%	6.65%	8.40%	
	3100 Total: Mechanical & Electrical		11.25	14.35	17.90	15.90%	18.05%	23%	
	9000 Per apartment unit, total cost	Apt.	52,500	80,000	118,000				
	9500 Total: Mechanical & Electrical	"	9,900	15,600	20,400				
02	0010 **APARTMENTS** Mid Rise (4 to 7 story)	S.F.	74.50	90	110				02
	0020 Total project costs	C.F.	5.80	8.05	11				
	0100 Site work	S.F.	2.98	5.90	11.65	5.25%	6.70%	9.20%	
	0500 Masonry		4.96	6.85	9.75	5.15%	7.40%	10.50%	
	1500 Finishes		9.40	13.10	15.20	10.55%	13.45%	17.70%	
	1800 Equipment		2.45	3.49	4.48	2.55%	3.48%	4.31%	
	2500 Conveying equipment		1.69	2.08	2.48	2.05%	2.29%	2.69%	
	2720 Plumbing		4.37	7	7.75	5.70%	7.20%	8.95%	
	2900 Electrical		4.92	7	8.20	6.65%	7.20%	8.95%	
	3100 Total: Mechanical & Electrical		15.75	19.70	24.50	18.85%	21%	25%	
	9000 Per apartment unit, total cost	Apt.	84,500	99,500	165,000				
	9500 Total: Mechanical & Electrical	"	15,900	18,400	23,300				
03	0010 **APARTMENTS** High Rise (8 to 24 story)	S.F.	84.50	102	117				03
	0020 Total project costs	C.F.	8.20	10.05	12.20				
	0100 Site work	S.F.	3.07	4.96	6.95	2.58%	4.84%	6.15%	
	0500 Masonry		4.89	8.90	11.05	4.74%	9.65%	11.05%	
	1500 Finishes		9.40	11.75	13.85	9.75%	11.80%	13.70%	
	1800 Equipment		2.72	3.34	4.43	2.78%	3.49%	4.35%	
	2500 Conveying equipment		1.92	2.92	4.16	2.23%	2.78%	3.37%	
	2720 Plumbing		6.25	7.35	10.30	6.80%	7.20%	10.45%	
	2900 Electrical		5.80	7.35	9.90	6.45%	7.65%	8.80%	
	3100 Total: Mechanical & Electrical		17.40	22	26.50	17.95%	22.50%	24.50%	
	9000 Per apartment unit, total cost	Apt.	88,000	97,000	134,500				
	9500 Total: Mechanical & Electrical	"	19,000	21,700	23,000				
04	0010 **AUDITORIUMS**	S.F.	87	119	170				04
	0020 Total project costs	C.F.	5.50	7.65	11.50				
	2720 Plumbing	S.F.	5.55	7.70	9.75	5.85%	7.20%	8.70%	
	2900 Electrical		7.10	9.85	13.20	6.80%	8.95%	11.30%	
	3100 Total: Mechanical & Electrical		13.60	19.30	39	24.50%	30.50%	31.50%	
05	0010 **AUTOMOTIVE SALES**	S.F.	63.50	88	108				05
	0020 Total project costs	C.F.	4.28	5.15	6.65				
	2720 Plumbing	S.F.	2.97	5.15	5.60	2.89%	6.05%	6.50%	
	2770 Heating, ventilating, air conditioning		4.58	7	7.55	4.61%	10%	10.35%	
	2900 Electrical		5.25	8.05	11.15	7.40%	9.95%	12.40%	
	3100 Total: Mechanical & Electrical		14.60	20.50	26	19.15%	20.50%	26%	
06	0010 **BANKS**	S.F.	127	158	200				06
	0020 Total project costs	C.F.	9.10	12.40	16.35				
	0100 Site work	S.F.	14.55	22	33	7.85%	12.95%	17%	
	0500 Masonry		6.65	12.45	23	3.36%	6.95%	10.35%	
	1500 Finishes		11.30	15.45	19.85	5.85%	8.45%	11.25%	
	1800 Equipment		5.10	10.50	22.50	1.34%	5.95%	10.65%	
	2720 Plumbing		4	5.70	8.35	2.82%	3.90%	4.93%	
	2770 Heating, ventilating, air conditioning		7.60	10.15	13.50	4.86%	7.15%	8.50%	
	2900 Electrical		12.05	16.10	21	8.20%	10.20%	12.20%	
	3100 Total: Mechanical & Electrical		29	38	46	16.55%	19.45%	23%	
	3500 See also division 11020 & 11030								

SQUARE FOOT

		K1010 | S.F. Costs		UNIT COSTS			% OF TOTAL			
			UNIT	1/4	MEDIAN	3/4	1/4	MEDIAN	3/4	
13	0010	**CHURCHES**	S.F.	86	109	141				13
	0020	Total project costs	C.F.	5.30	6.70	8.85				
	1800	Equipment	S.F.	1.09	2.46	5.20	1.03%	2.24%	4.50%	
	2720	Plumbing		3.34	4.67	6.90	3.51%	4.96%	6.25%	
	2770	Heating, ventilating, air conditioning		7.80	10.20	14.45	7.50%	10%	12%	
	2900	Electrical		7.20	9.90	13.25	7.30%	8.75%	10.90%	
	3100	Total: Mechanical & Electrical	↓	22	29	40	18.25%	22%	24%	
	3500	See also division 11040								
15	0010	**CLUBS, COUNTRY**	S.F.	92	111	140				15
	0020	Total project costs	C.F.	7.40	9.05	12.50				
	2720	Plumbing	S.F.	5.90	8.25	18.80	5.60%	7.90%	10%	
	2900	Electrical		7.25	10.35	13.65	7%	8.95%	11%	
	3100	Total: Mechanical & Electrical	↓	22	38.50	48	19%	26.50%	29.50%	
17	0010	**CLUBS, SOCIAL** Fraternal	S.F.	73.50	105	142				17
	0020	Total project costs	C.F.	4.59	6.95	8.30				
	2720	Plumbing	S.F.	4.62	5.75	8.70	5.60%	6.90%	8.55%	
	2770	Heating, ventilating, air conditioning		6.65	8.05	10.35	8.20%	9.25%	14.40%	
	2900	Electrical		5.85	9.10	11	6.50%	9.50%	10.55%	
	3100	Total: Mechanical & Electrical	↓	16.30	31	39.50	21%	23%	23.50%	
18	0010	**CLUBS, Y.M.C.A.**	S.F.	92.50	125	152				18
	0020	Total project costs	C.F.	4.25	7.10	10.60				
	2720	Plumbing	S.F.	5.85	11.60	13	5.65%	7.60%	10.85%	
	2900	Electrical		7	9.55	13.60	6.05%	7.80%	10.20%	
	3100	Total: Mechanical & Electrical	↓	26	31.50	35	18.40%	21.50%	28.50%	
19	0010	**COLLEGES** Classrooms & Administration	S.F.	103	136	184				19
	0020	Total project costs	C.F.	7.45	10.60	16.70				
	0500	Masonry	S.F.	6.85	12.90	15.85	5.65%	8.25%	10.50%	
	2720	Plumbing		5.05	10.10	18.15	5.10%	6.60%	8.95%	
	2900	Electrical		8.45	12.95	16.10	7.70%	9.85%	12%	
	3100	Total: Mechanical & Electrical	↓	31	43.50	51.50	24%	28%	31.50%	
21	0010	**COLLEGES** Science, Engineering, Laboratories	S.F.	173	203	247				21
	0020	Total project costs	C.F.	9.95	14.50	16.45				
	1800	Equipment	S.F.	9.65	22	24	2%	6.45%	12.65%	
	2900	Electrical		14.30	19.50	31	7.10%	9.40%	12.10%	
	3100	Total: Mechanical & Electrical	↓	53	63	97.50	28.50%	31.50%	41%	
	3500	See also division 11600								
23	0010	**COLLEGES** Student Unions	S.F.	110	155	182				23
	0020	Total project costs	C.F.	6.15	8.10	10.40				
	3100	Total: Mechanical & Electrical	S.F.	29	45	53	23.50%	26%	29%	
25	0010	**COMMUNITY CENTERS**	"	90	113	152				25
	0020	Total project costs	C.F.	5.95	8.50	11.05				
	1800	Equipment	S.F.	2.29	3.90	6.15	1.87%	3.12%	6%	
	2720	Plumbing		4.43	7.55	11.05	4.94%	7%	9.10%	
	2770	Heating, ventilating, air conditioning		7.30	10.50	14.45	6.95%	10.65%	13.05%	
	2900	Electrical		7.75	9.95	14.50	7.35%	9.10%	10.85%	
	3100	Total: Mechanical & Electrical	↓	27	32.50	46.50	22.50%	26.50%	32.50%	
28	0010	**COURT HOUSES**	S.F.	131	151	176				28
	0020	Total project costs	C.F.	10.10	12.20	15.20				
	2720	Plumbing	S.F.	6.25	8.75	12.55	5.95%	7.45%	8.20%	
	2900	Electrical		12.90	15.25	18.40	8.55%	9.95%	11.50%	
	3100	Total: Mechanical & Electrical	↓	35.50	41	51.50	22.50%	29.50%	30.50%	
30	0010	**DEPARTMENT STORES**	S.F.	48.50	66	83				30
	0020	Total project costs	C.F.	2.61	3.54	4.59				
	2720	Plumbing	S.F.	1.59	1.92	2.91	1.82%	4.21%	5.90%	
	2770	Heating, ventilating, air conditioning	↓	4.43	6.85	10.30	8.20%	9.10%	14.80%	

SQUARE FOOT

K1010 | S.F. Costs		UNIT	UNIT COSTS			% OF TOTAL			
			1/4	MEDIAN	3/4	1/4	MEDIAN	3/4	
0	2900	Electrical	S.F.	5.60	7.65	9.05	9.05%	12.15%	14.95%	30
	3100	Total: Mechanical & Electrical	↓	9.85	12.55	22	13.20%	21.50%	50%	
1	0010	**DORMITORIES** Low Rise (1 to 3 story)	S.F.	83	117	146				31
	0020	Total project costs	C.F.	5.50	8.45	12.65				
	2720	Plumbing	S.F.	5.55	7.45	9.40	8.05%	9%	9.65%	
	2770	Heating, ventilating, air conditioning	⃒	5.85	7.05	9.35	4.61%	8.05%	10%	
	2900	Electrical		5.95	8.95	12.45	6.55%	8.90%	9.55%	
	3100	Total: Mechanical & Electrical	↓	31	32.50	34	22%	26%	29%	
	9000	Per bed, total cost	Bed	39,100	43,500	93,000				
2	0010	**DORMITORIES** Mid Rise (4 to 8 story)	S.F.	113	148	181				32
	0020	Total project costs	C.F.	12.50	13.70	16.45				
	2900	Electrical	S.F.	12.05	13.70	17.80	8.20%	10.20%	11.10%	
	3100	Total: Mechanical & Electrical	"	31	35	68	19.50%	30.50%	37.50%	
	9000	Per bed, total cost	Bed	16,100	36,800	76,000				
4	0010	**FACTORIES**	S.F.	43	64	98.50				34
	0020	Total project costs	C.F.	2.75	4.11	6.80				
	0100	Site work	S.F.	4.90	8.95	14.15	6.95%	11.45%	17.95%	
	2720	Plumbing	⃒	2.31	4.31	7.10	3.73%	6.05%	8.10%	
	2770	Heating, ventilating, air conditioning		4.50	6.45	8.70	5.25%	8.45%	11.35%	
	2900	Electrical		5.35	8.45	12.85	8.10%	10.50%	14.20%	
	3100	Total: Mechanical & Electrical	↓	12.70	20.50	31	21%	28.50%	35.50%	
6	0010	**FIRE STATIONS**	S.F.	85	116	151				36
	0020	Total project costs	C.F.	4.98	6.85	9.10				
	0500	Masonry	S.F.	12.50	22.50	29.50	8.60%	11.65%	16.45%	
	1140	Roofing	⃒	2.78	7.50	8.55	1.90%	4.94%	5.05%	
	1580	Painting		2.28	3.22	3.29	1.37%	1.57%	2.07%	
	1800	Equipment		1.80	2.55	6.15	.90%	2.50%	4.07%	
	2720	Plumbing		5.40	7.95	11.70	5.85%	7.35%	9.50%	
	2770	Heating, ventilating, air conditioning		4.64	7.40	11.50	4.86%	7.25%	9.25%	
	2900	Electrical		6.05	10.40	13.70	6.80%	8.75%	10.65%	
	3100	Total: Mechanical & Electrical	↓	30	35	41.50	19.60%	23%	27%	
7	0010	**FRATERNITY HOUSES** and Sorority Houses	S.F.	85	110	150				37
	0020	Total project costs	C.F.	8.50	8.85	11.35				
	2720	Plumbing	S.F.	6.45	7.35	13.50	6.80%	8%	10.85%	
	2900	Electrical	⃒	5.60	12.15	14.85	6.60%	9.90%	10.65%	
	3100	Total: Mechanical & Electrical	↓	15	21.50	26		15.10%	15.90%	
8	0010	**FUNERAL HOMES**	S.F.	90	122	222				38
	0020	Total project costs	C.F.	9.15	10.20	19.60				
	2900	Electrical	S.F.	3.97	7.25	8.60	3.58%	4.44%	5.95%	
	3100	Total: Mechanical & Electrical	"	14.10	21	28.50	12.90%	12.90%	12.90%	
9	0010	**GARAGES, COMMERCIAL** (Service)	S.F.	53	80	109				39
	0020	Total project costs	C.F.	3.42	4.95	7.20				
	1800	Equipment	S.F.	2.87	6.45	10	2.69%	4.62%	6.80%	
	2720	Plumbing	⃒	3.52	5.45	9.90	5.45%	7.85%	10.65%	
	2730	Heating & ventilating		4.63	6.30	8.70	5.25%	6.85%	8.20%	
	2900	Electrical		4.90	7.50	10.70	7.15%	9.25%	10.85%	
	3100	Total: Mechanical & Electrical	↓	10.95	20.50	30.50	13.60%	17.40%	27%	
0	0010	**GARAGES, MUNICIPAL** (Repair)	S.F.	74.50	99.50	140				40
	0020	Total project costs	C.F.	4.67	5.90	10.20				
	0500	Masonry	S.F.	7	13.70	21.50	5.60%	9.15%	12.50%	
	2720	Plumbing	⃒	3.35	6.45	12.15	3.59%	6.70%	7.95%	
	2730	Heating & ventilating		5.75	8.30	16	6.15%	7.45%	13.50%	
	2900	Electrical		5.55	8.70	12.50	6.65%	8.15%	11.15%	
	3100	Total: Mechanical & Electrical	↓	17.75	35.50	51.50	21.50%	25.50%	28.50%	

SQUARE FOOT

		K1010 | S.F. Costs	UNIT	UNIT COSTS			% OF TOTAL		
				1/4	MEDIAN	3/4	1/4	MEDIAN	3/4
41	0010	**GARAGES, PARKING**	S.F.	29	42.50	73			
	0020	Total project costs	C.F.	2.73	3.71	5.40			
	2720	Plumbing	S.F.	.82	1.28	1.97	1.72%	2.70%	3.85%
	2900	Electrical		1.59	1.96	3.07	4.33%	5.20%	6.30%
	3100	Total: Mechanical & Electrical	▼	3.26	4.54	5.95	7%	8.90%	11.05%
	3200								
	9000	Per car, total cost	Car	12,300	15,400	19,700			
43	0010	**GYMNASIUMS**	S.F.	80.50	107	138			
	0020	Total project costs	C.F.	4.03	5.45	6.70			
	1800	Equipment	S.F.	1.92	3.60	6.90	2.07%	3.35%	6.70%
	2720	Plumbing		5.10	6.30	7.85	4.95%	6.75%	7.75%
	2770	Heating, ventilating, air conditioning		5.50	8.40	16.85	5.80%	9.80%	11.10%
	2900	Electrical		6.15	8.20	10.40	6.60%	8.30%	10.30%
	3100	Total: Mechanical & Electrical	▼	22	31	36.50	20.50%	26%	29.50%
	3500	See also division 11480							
46	0010	**HOSPITALS**	S.F.	155	190	288			
	0020	Total project costs	C.F.	11.75	14.65	21			
	1800	Equipment	S.F.	3.94	7.55	13.05	1.10%	2.68%	5%
	2720	Plumbing		13.35	18.70	24	7.60%	9.10%	10.85%
	2770	Heating, ventilating, air conditioning		19.60	25	34	7.80%	12.95%	16.65%
	2900	Electrical		16.95	22	34	9.85%	11.75%	14%
	3100	Total: Mechanical & Electrical	▼	47.50	63	103	26.50%	33%	36.50%
	9000	Per bed or person, total cost	Bed	127,000	203,500	272,000			
	9900	See also division 11700							
48	0010	**HOUSING** For the Elderly	S.F.	76.50	96.50	119			
	0020	Total project costs	C.F.	5.45	7.55	9.65			
	0100	Site work	S.F.	5.65	8.40	12.10	5.05%	7.90%	12.10%
	0500	Masonry		2.33	8.70	12.70	1.30%	6.05%	11%
	1800	Equipment		1.84	2.54	4.04	1.88%	3.23%	4.43%
	2510	Conveying systems		1.86	2.50	3.38	1.78%	2.20%	2.81%
	2720	Plumbing		5.65	7.25	9.45	8.15%	9.55%	10.50%
	2730	Heating, ventilating, air conditioning		2.91	4.12	6.15	3.30%	5.60%	7.25%
	2900	Electrical		5.70	7.70	9.90	7.30%	8.50%	10.25%
	3100	Total: Mechanical & Electrical	▼	19.60	23.50	31	18.10%	22.50%	29%
	9000	Per rental unit, total cost	Unit	71,000	83,000	92,500			
	9500	Total: Mechanical & Electrical	"	15,800	18,200	21,200			
50	0010	**HOUSING** Public (Low Rise)	S.F.	64	89	116			
	0020	Total project costs	C.F.	5.70	7.10	8.85			
	0100	Site work	S.F.	8.15	11.75	19.05	8.35%	11.75%	16.50%
	1800	Equipment		1.75	2.85	4.34	2.26%	3.03%	4.24%
	2720	Plumbing		4.63	6.10	7.75	7.15%	9.05%	11.60%
	2730	Heating, ventilating, air conditioning		2.33	4.51	4.94	4.26%	6.05%	6.45%
	2900	Electrical		3.88	5.80	8	5.10%	6.55%	8.25%
	3100	Total: Mechanical & Electrical	▼	18.40	23.50	26.50	14.50%	17.55%	26.50%
	9000	Per apartment, total cost	Apt.	70,500	80,000	100,500			
	9500	Total: Mechanical & Electrical	"	15,000	18,500	20,500			
51	0010	**ICE SKATING RINKS**	S.F.	57	128	141			
	0020	Total project costs	C.F.	4.03	4.12	4.75			
	2720	Plumbing	S.F.	2.05	3.84	3.93	3.12%	3.23%	5.65%
	2900	Electrical		5.85	9	9.55	6.30%	10.15%	15.05%
	3100	Total: Mechanical & Electrical	▼	9.80	13.80	17.25	18.95%	18.95%	18.95%
52	0010	**JAILS**	S.F.	167	216	278			
	0020	Total project costs	C.F.	15.75	21	26			
	1800	Equipment	S.F.	6.50	19.25	32.50	2.80%	5.55%	11.90%
	2720	Plumbing		17	21.50	28.50	7%	8.90%	13.35%
	2770	Heating, ventilating, air conditioning		15.05	20	39	7.50%	9.45%	17.75%
	2900	Electrical	▼	17.80	23	28.50	8.20%	11.55%	14.70%

SQUARE FOOT

K1010 | Square Foot Costs

SQUARE FOOT

		K1010 | S.F. Costs	UNIT	UNIT COSTS			% OF TOTAL			
				1/4	MEDIAN	3/4	1/4	MEDIAN	3/4	
2	3100	Total: Mechanical & Electrical	S.F.	44.50	83	98.50	27.50%	30%	34%	52
3	0010	**LIBRARIES**	S.F.	106	132	174				53
	0020	Total project costs	C.F.	7.20	9.05	11.55				
	0500	Masonry	S.F.	8.25	14.60	24.50	5.80%	8%	12.35%	
	1800	Equipment		1.44	3.88	6.05	.41%	1.50%	4.16%	
	2720	Plumbing		3.90	5.65	7.70	3.43%	4.60%	5.70%	
	2770	Heating, ventilating, air conditioning		8.65	14.60	19	7.80%	10.95%	12.80%	
	2900	Electrical		10.75	13.85	17.50	8.35%	10.40%	11.95%	
	3100	Total: Mechanical & Electrical		32	40	50	20.50%	23%	26.50%	
4	0010	**LIVING, ASSISTED**	S.F.	97.50	115	135				54
	0020	Total project costs	C.F.	8.20	9.60	10.90				
	0500	Masonry	S.F.	2.87	3.42	4.02	2.37%	3.16%	3.86%	
	1800	Equipment		2.22	2.58	3.30	2.12%	2.45%	2.66%	
	2720	Plumbing		8.20	10.95	11.35	6.05%	8.15%	10.60%	
	2770	Heating, ventilating, air conditioning		9.70	10.15	11.10	7.95%	9.35%	9.70%	
	2900	Electrical		9.55	10.55	12.20	9%	10%	10.70%	
	3100	Total: Mechanical & Electrical		27.50	32	36	26%	29%	31.50%	
5	0010	**MEDICAL CLINICS**	S.F.	99	122	155				55
	0020	Total project costs	C.F.	7.30	9.85	12.50				
	1800	Equipment	S.F.	2.69	5.65	8.80	1.10%	3.13%	6.35%	
	2720	Plumbing		6.60	9.30	12.45	6.15%	8.40%	10.10%	
	2770	Heating, ventilating, air conditioning		7.85	10.30	15.15	6.65%	8.85%	11.35%	
	2900	Electrical		8.35	12.10	15.80	8.10%	10%	12.20%	
	3100	Total: Mechanical & Electrical		27	37	51	22.50%	27%	33.50%	
	3500	See also division 11700								
7	0010	**MEDICAL OFFICES**	S.F.	93.50	115	141				57
	0020	Total project costs	C.F.	6.95	9.45	12.80				
	1800	Equipment	S.F.	3.25	6.10	8.65	.98%	5.10%	7.05%	
	2720	Plumbing		5.15	7.95	10.75	5.60%	6.80%	8.50%	
	2770	Heating, ventilating, air conditioning		6.25	9.15	11.90	6.15%	8.05%	9.70%	
	2900	Electrical		7.50	10.85	15.20	7.65%	9.80%	11.70%	
	3100	Total: Mechanical & Electrical		20	29	43	19.35%	23%	30.50%	
9	0010	**MOTELS**	S.F.	59	87	113				59
	0020	Total project costs	C.F.	5.25	7.05	11.50				
	2720	Plumbing	S.F.	6	7.60	9.10	9.45%	10.60%	12.55%	
	2770	Heating, ventilating, air conditioning		3.64	5.45	9.75	5.60%	5.60%	10%	
	2900	Electrical		5.60	7.15	9.10	7.45%	9.20%	10.80%	
	3100	Total: Mechanical & Electrical		18.95	23.50	40.50	18.50%	21%	25.50%	
	5000									
	9000	Per rental unit, total cost	Unit	30,000	57,000	61,500				
	9500	Total: Mechanical & Electrical	"	5,850	8,850	10,300				
0	0010	**NURSING HOMES**	S.F.	93.50	119	146				60
	0020	Total project costs	C.F.	7.25	9.10	12.40				
	1800	Equipment	S.F.	2.91	3.86	6.45	2.02%	3.62%	4.99%	
	2720	Plumbing		7.95	12	14.45	8.75%	10.10%	12.70%	
	2770	Heating, ventilating, air conditioning		8.35	12.65	16.80	9.70%	11.45%	11.80%	
	2900	Electrical		9.30	11.50	15.60	9.50%	10.60%	12.50%	
	3100	Total: Mechanical & Electrical		22	30.50	51	26%	29.50%	30.50%	
	9000	Per bed or person, total cost	Bed	41,000	51,500	66,000				
1	0010	**OFFICES** Low Rise (1 to 4 story)	S.F.	77.50	99.50	130				61
	0020	Total project costs	C.F.	5.55	7.65	10.10				
	0100	Site work	S.F.	6.20	10.50	15.35	6.20%	9.70%	13.60%	
	0500	Masonry	"	2.59	5.85	10.60	2.55%	5.40%	8%	
	1800	Equipment	S.F.	.82	1.62	4.42	.77%	1.50%	3.69%	
	2720	Plumbing		2.78	4.30	6.25	3.66%	4.50%	6.10%	
	2770	Heating, ventilating, air conditioning		6.15	8.70	12.55	7.20%	10.40%	11.70%	
	2900	Electrical		6.35	9.05	12.75	7.45%	9.65%	11.40%	

541

		K1010	S.F. Costs	UNIT	UNIT COSTS			% OF TOTAL		
					1/4	MEDIAN	3/4	1/4	MEDIAN	3/4
61	3100	Total: Mechanical & Electrical		S.F.	16.50	23	34.50	18.25%	22.50%	27%
62	0010	**OFFICES** Mid Rise (5 to 10 story)		S.F.	82.50	100	132			
	0020	Total project costs		C.F.	5.85	7.45	10.55			
	2720	Plumbing		S.F.	2.49	3.86	5.55	2.83%	3.74%	4.50%
	2770	Heating, ventilating, air conditioning			6.25	8.95	14.30	7.65%	9.40%	11%
	2900	Electrical			6.10	7.85	10.85	6.35%	7.80%	10%
	3100	Total: Mechanical & Electrical			15.65	19.95	39	18.95%	21%	27.50%
63	0010	**OFFICES** High Rise (11 to 20 story)		S.F.	101	128	157			
	0020	Total project costs		C.F.	7.10	8.85	12.70			
	2900	Electrical		S.F.	6.15	7.50	11.15	5.80%	7.85%	10.50%
	3100	Total: Mechanical & Electrical		"	19.80	26.50	45	16.90%	23.50%	34%
64	0010	**POLICE STATIONS**		S.F.	122	156	199			
	0020	Total project costs		C.F.	9.65	11.85	16.20			
	0500	Masonry		S.F.	11.40	20	26.50	6.70%	10.55%	11.35%
	1800	Equipment			1.93	8.50	13.45	1.43%	4.07%	6.70%
	2720	Plumbing			7	13.55	16.85	5.65%	6.90%	10.75%
	2770	Heating, ventilating, air conditioning			10.60	14.10	19.15	5.85%	10.55%	11.70%
	2900	Electrical			13.25	19.80	25	9.80%	11.85%	14.80%
	3100	Total: Mechanical & Electrical			43.50	52.50	71	28.50%	32%	32.50%
65	0010	**POST OFFICES**		S.F.	96	118	151			
	0020	Total project costs		C.F.	5.75	7.30	8.65			
	2720	Plumbing		S.F.	4.32	5.35	6.75	4.24%	5.30%	5.60%
	2770	Heating, ventilating, air conditioning			6.75	8.35	9.25	6.65%	7.15%	9.35%
	2900	Electrical			7.90	11.15	13.20	7.25%	9%	11%
	3100	Total: Mechanical & Electrical			23	30	34	16.25%	18.80%	22%
66	0010	**POWER PLANTS**		S.F.	665	880	1,625			
	0020	Total project costs		C.F.	18.30	40	85.50			
	2900	Electrical		S.F.	47	99.50	149	9.30%	12.75%	21.50%
	8100	Total: Mechanical & Electrical		"	117	380	855	32.50%	32.50%	52.50%
67	0010	**RELIGIOUS EDUCATION**		S.F.	77.50	101	119			
	0020	Total project costs		C.F.	4.29	6.15	7.70			
	2720	Plumbing		S.F.	3.23	4.58	6.45	4.40%	5.30%	7.10%
	2770	Heating, ventilating, air conditioning			8.15	9.25	13.05	10.05%	11.45%	12.35%
	2900	Electrical			6.15	8.65	11.35	7.60%	9.10%	10.35%
	3100	Total: Mechanical & Electrical			25	32.50	38.50	22%	23%	26%
69	0010	**RESEARCH** Laboratories and facilities		S.F.	115	166	242			
	0020	Total project costs		C.F.	9.05	17.40	20.50			
	1800	Equipment		S.F.	5.15	10.10	24.50	.90%	4.58%	8.80%
	2720	Plumbing			11.60	14.85	24	6.15%	8.30%	10.80%
	2770	Heating, ventilating, air conditioning			10.40	35	41.50	7.25%	16.50%	17.50%
	2900	Electrical			13.30	22.50	38	9.45%	11.15%	15.40%
	3100	Total: Mechanical & Electrical			41	78	111	29.50%	37%	45.50%
70	0010	**RESTAURANTS**		S.F.	111	143	188			
	0020	Total project costs		C.F.	9.45	12.35	16.25			
	1800	Equipment		S.F.	7.25	17.80	27	6.10%	13%	15.65%
	2720	Plumbing			8.90	10.75	14.15	6.10%	8.15%	9%
	2770	Heating, ventilating, air conditioning			11.30	15.60	20.50	9.20%	12%	12.40%
	2900	Electrical			11.85	14.60	19	8.35%	10.55%	11.55%
	3100	Total: Mechanical & Electrical			36.50	38.50	50	19.25%	24%	29.50%
	9000	Per seat unit, total cost		Seat	4,125	5,475	6,500			
	9500	Total: Mechanical & Electrical		"	1,025	1,375	1,625			
72	0010	**RETAIL STORES**		S.F.	52	70.50	92.50			
	0020	Total project costs		C.F.	3.55	5.05	7.05			
	2720	Plumbing		S.F.	1.90	3.17	5.40	3.26%	4.60%	6.80%
	2770	Heating, ventilating, air conditioning			4.10	5.60	8.45	6.75%	8.75%	10.15%
	2900	Electrical			4.70	6.45	9.30	7.25%	9.90%	11.65%
	3100	Total: Mechanical & Electrical			12.05	16.05	21.50	17.05%	21%	23.50%

SQUARE FOOT

				UNIT COSTS			% OF TOTAL			
K1010 | S.F. Costs			UNIT	1/4	MEDIAN	3/4	1/4	MEDIAN	3/4	
74	0010	**SCHOOLS** Elementary	S.F.	84.50	104	128				74
	0020	Total project costs	C.F.	5.60	7.15	9.25				
	0500	Masonry	S.F.	7.55	12.50	18.65	5.45%	10.65%	14.70%	
	1800	Equipment		2.49	4.23	7.80	1.90%	3.33%	5%	
	2720	Plumbing		4.90	6.95	9.30	5.70%	7.15%	9.35%	
	2730	Heating, ventilating, air conditioning		7.35	11.75	16.40	8.15%	10.80%	15.20%	
	2900	Electrical		7.95	10.40	13.05	8.40%	10.05%	11.70%	
	3100	Total: Mechanical & Electrical		28	35	42.50	24%	27.50%	30%	
	9000	Per pupil, total cost	Ea.	9,825	14,600	43,200				
	9500	Total: Mechanical & Electrical	"	2,775	3,500	12,500				
76	0010	**SCHOOLS** Junior High & Middle	S.F.	86	108	126				76
	0020	Total project costs	C.F.	5.60	7.25	8.10				
	0500	Masonry	S.F.	9.45	13.45	16.20	8%	11.60%	14.30%	
	1800	Equipment		2.81	4.53	7	1.81%	3.26%	5.15%	
	2720	Plumbing		5.60	6.35	8.20	5.50%	6.90%	8.15%	
	2770	Heating, ventilating, air conditioning		6.45	12.40	17.30	8.75%	12.75%	17.45%	
	2900	Electrical		8.50	10.35	13.30	7.90%	9.25%	10.60%	
	3100	Total: Mechanical & Electrical		25.50	36	44	22.50%	25.50%	29.50%	
	9000	Per pupil, total cost	Ea.	11,200	15,100	19,700				
78	0010	**SCHOOLS** Senior High	S.F.	90	110	139				78
	0020	Total project costs	C.F.	5.80	8.15	13.50				
	1800	Equipment	S.F.	2.41	5.70	8.40	1.88%	3.29%	4.80%	
	2720	Plumbing		5.25	7.80	14.10	5.70%	7%	8.35%	
	2770	Heating, ventilating, air conditioning		10.50	12.05	23	8.95%	11.60%	15%	
	2900	Electrical		9.10	11.70	18.75	8.35%	10.10%	12.35%	
	3100	Total: Mechanical & Electrical		31	35.50	60	23%	26.50%	28.50%	
	9000	Per pupil, total cost	Ea.	8,650	17,600	22,000				
80	0010	**SCHOOLS** Vocational	S.F.	74	105	132				80
	0020	Total project costs	C.F.	4.61	6.60	9.15				
	0500	Masonry	S.F.	4.36	10.75	16.45	3.53%	4.61%	10.95%	
	1800	Equipment	"	2.22	3.15	8	1.24%	3.13%	4.68%	
	2720	Plumbing	S.F.	4.75	7.25	10.85	5.40%	6.90%	8.55%	
	2770	Heating, ventilating, air conditioning		6.65	12.35	20.50	8.60%	11.90%	14.65%	
	2900	Electrical		7.75	10.55	14.90	8.45%	10.95%	13.20%	
	3100	Total: Mechanical & Electrical		27	29.50	51	23.50%	29.50%	31%	
	9000	Per pupil, total cost	Ea.	10,300	27,700	41,300				
83	0010	**SPORTS ARENAS**	S.F.	65	86.50	134				83
	0020	Total project costs	C.F.	3.52	6.30	8.15				
	2720	Plumbing	S.F.	3.76	5.70	12.05	4.35%	6.35%	9.40%	
	2770	Heating, ventilating, air conditioning		8.10	9.60	13.30	8.80%	10.20%	13.55%	
	2900	Electrical		6.75	9.15	11.85	8.60%	9.90%	12.25%	
	3100	Total: Mechanical & Electrical		16.80	30	39	21.50%	25%	27.50%	
85	0010	**SUPERMARKETS**	S.F.	60	69.50	84				85
	0020	Total project costs	C.F.	3.34	4.03	6.10				
	2720	Plumbing	S.F.	3.35	4.21	4.91	5.40%	6%	7.45%	
	2770	Heating, ventilating, air conditioning		4.92	6.55	7.95	8.60%	8.65%	9.60%	
	2900	Electrical		7.50	8.60	10.20	10.40%	12.45%	13.60%	
	3100	Total: Mechanical & Electrical		19.25	21	29	20.50%	26.50%	31%	
86	0010	**SWIMMING POOLS**	S.F.	97	163	345				86
	0020	Total project costs	C.F.	7.80	9.70	10.55				
	2720	Plumbing	S.F.	8.95	10.25	14.20	4.80%	9.70%	20.50%	
	2900	Electrical		7.30	11.80	17.20	6.50%	7.25%	7.60%	
	3100	Total: Mechanical & Electrical		17.80	45	61.50	11.15%	14.10%	23.50%	
87	0010	**TELEPHONE EXCHANGES**	S.F.	129	189	239				87
	0020	Total project costs	C.F.	8	12.85	17.70				
	2720	Plumbing	S.F.	5.45	8.60	12.30	4.52%	5.80%	6.90%	
	2770	Heating, ventilating, air conditioning		12.60	25.50	31.50	11.80%	16.05%	18.40%	

		K1010 | S.F. Costs	UNIT	UNIT COSTS			% OF TOTAL			
				1/4	MEDIAN	3/4	1/4	MEDIAN	3/4	
87	2900	Electrical	S.F.	13.10	21	37	10.90%	14%	17.85%	87
	3100	Total: Mechanical & Electrical	↓	38.50	73.50	104	29.50%	33.50%	44.50%	
91	0010	**THEATERS**	S.F.	81	101	153				91
	0020	Total project costs	C.F.	3.74	5.55	8.15				
	2720	Plumbing	S.F.	2.70	2.93	11.95	2.92%	4.70%	6.80%	
	2770	Heating, ventilating, air conditioning		7.85	9.55	11.80	8%	12.25%	13.40%	
	2900	Electrical		7.10	9.55	19.45	8.05%	9.95%	12.25%	
	3100	Total: Mechanical & Electrical	↓	18.25	27	56	23%	26.50%	27.50%	
94	0010	**TOWN HALLS** City Halls & Municipal Buildings	S.F.	95	121	153				94
	0020	Total project costs	C.F.	8.25	10.15	13.90				
	2720	Plumbing	S.F.	3.77	7.05	13	4.31%	5.95%	7.95%	
	2770	Heating, ventilating, air conditioning		6.85	13.55	19.80	7.05%	9.05%	13.45%	
	2900	Electrical		8.60	12.55	16.75	8.05%	9.50%	12.05%	
	3100	Total: Mechanical & Electrical	↓	29	34	45	22%	26.50%	31%	
97	0010	**WAREHOUSES** And Storage Buildings	S.F.	35.50	51	72				97
	0020	Total project costs	C.F.	1.94	2.77	4.57				
	0100	Site work	S.F.	3.48	6.90	10.45	6.05%	12.95%	19.85%	
	0500	Masonry		2.10	4.79	10.35	3.73%	7.40%	12.30%	
	1800	Equipment		.54	1.17	6.55	.91%	1.82%	5.55%	
	2720	Plumbing		1.12	2.02	3.77	2.90%	4.80%	6.55%	
	2730	Heating, ventilating, air conditioning		1.35	3.62	4.86	2.41%	5%	8.90%	
	2900	Electrical		2.05	3.76	6.25	5.15%	7.20%	10.10%	
	3100	Total: Mechanical & Electrical	↓	5.55	8.55	18.70	12.75%	18.90%	26%	
99	0010	**WAREHOUSE & OFFICES** Combination	S.F.	41.50	55.50	74.50				99
	0020	Total project costs	C.F.	2.12	3.08	4.56				
	1800	Equipment	S.F.	.72	1.39	2.15	.52%	1.21%	2.40%	
	2720	Plumbing		1.60	2.84	4.16	3.74%	4.76%	6.30%	
	2770	Heating, ventilating, air conditioning		2.53	3.96	5.55	5%	5.65%	10.05%	
	2900	Electrical		2.78	4.13	6.50	5.85%	8%	10%	
	3100	Total: Mechanical & Electrical	↓	7.75	11.80	18.85	14.40%	19.95%	24.50%	

SQUARE FOOT

Reference Aids — RL1010-100 — Live Loads

1010-101 Minimum Design Live Loads in Pounds per S.F. for Various Building Codes

Occupancy	Description	BOCA	IBC	UBC
		Minimum Live Loads, Pounds per S.F.		
Access Floors	Office use		50	50
	Computer use		100	100
Armories		150	150	150
Assembly	Fixed seats	60	60	50
	Movable seats	100	100	100
	Platforms or stage floors	125	125	125
Commercial & Industrial	Light manufacturing	125	125	75
	Heavy manufacturing	250	250	125
	Light storage	125	125	125
	Heavy storage	250	250	250
	Stores, retail, first floor	100	100	100
	Stores, retail, upper floors	75	75	100
	Stores, wholesale	125	125	100
Court rooms		100		
Dance halls	Ballrooms	100	100	
Dining rooms	Restaurants	100	100	
Fire escapes	Other than below	100	100	100
	Single family residential	40	40	100
Garages	Passenger cars only	50	50	50
Gymnasiums	Main floors and balconies	100	100	
Hospitals	Operating rooms, laboratories	60	60	
	Private room	40	40	40
	Wards	40	40	40
	Corridors, above first floor	80	80	
Libraries	Reading rooms	60	60	60
	Stack rooms	150	150	125
	Corridors, above first floor	80	80	
Marquees		75	75	
Office Buildings	Offices	50	50	50
	Lobbies	100	100	
	Corridors, above first floor	80	80	
Residential	Multi family private apartments	40	40	40
	Multi family, public rooms	100	100	
	Multi family, corridors	100	100	100
	Dwellings, first floor	40	40	40
	Dwellings, second floor & habitable attics	30	30	
	Dwellings, uninhabitable attics	20	20	
	Hotels, guest rooms	40	40	40
	Hotels, public rooms	100	100	
	Hotels, corridors serving public rooms	100	100	
	Hotels, corridors	40	100	
Schools	Classrooms	40	40	40
	Corridors	80	100	
Sidewalks	Driveways, etc. subject to trucking	250	250	250
Stairs	Exits	100	100	100
Theaters	Aisles, corridors and lobbies	100		
	Orchestra floors	60		
	Balconies	60		
	Stage floors	125	125	
Yards	Terrace, pedestrian	100	100	

BOCA = Building Officials & Code Administration International, Inc. National Building Code
IBC = International Building Code, International Code Council, Inc.
UBC = Uniform Building Code, International Conference of Building Officials

Table L1010-201   Design Weight Per S.F. for Walls and Partitions

Type	Wall Thickness	Description	Weight Per S.F.
Brick	4"	Clay brick, high absorption	34 lb.
		Clay brick, medium absorption	39
		Clay brick, low absorption	46
		Sand-lime brick	38
		Concrete brick, heavy aggregate	46
		Concrete brick, light aggregate	33
	8"	Clay brick, high absorption	69
		Clay brick, medium absorption	79
		Clay brick, low absorption	89
		Sand-lime brick	74
		Concrete brick, heavy aggregate	89
		Concrete brick, light aggregate	68
	12"	Common brick	120
		Pressed brick	130
		Sand-lime brick	105
		Concrete brick, heavy aggregate	130
		Concrete brick, light aggregate	98
	16"	Clay brick, high absorption	134
		Clay brick, medium absorption	155
		Clay brick, low absorption	173
		Sand-lime brick	138
		Concrete brick, heavy aggregate	174
		Concrete brick, light aggregate	130
Concrete block	4"	Solid conc. block, stone aggregate	45
		Solid conc. block, lightweight	34
		Hollow conc. block, stone aggregate	30
		Hollow conc. block, lightweight	20
	6"	Solid conc. block, stone aggregate	50
		Solid conc. block, lightweight	37
		Hollow conc. block, stone aggregate	42
		Hollow conc. block, lightweight	30
	8"	Solid conc. block, stone aggregate	67
		Solid conc. block, lightweight	48
		Hollow conc. block, stone aggregate	55
		Hollow conc. block, lightweight	38
	10"	Solid conc. block, stone aggregate	84
		Solid conc. block, lightweight	62
		Hollow conc. block, stone aggregate	55
		Hollow conc. block, lightweight	38
	12"	Solid conc. block, stone aggregate	108
		Solid conc. block, lightweight	72
		Hollow conc. block, stone aggregate	85
		Hollow conc. block, lightweight	55
Drywall	6"	Drywall on wood studs	10

Type	Wall Thickness	Description	Weight Per S.F.
Clay tile	2"	Split terra cotta furring	10 lb.
		Non load bearing clay tile	11
	3"	Split terra cotta furring	12
		Non load bearing clay tile	18
	4"	Non load bearing clay tile	20
		Load bearing clay tile	24
	6"	Non load bearing clay tile	30
		Load bearing clay tile	36
	8"	Non load bearing clay tile	36
		Load bearing clay tile	42
	12"	Non load bearing clay tile	46
		Load bearing clay tile	58
Gypsum block	2"	Hollow gypsum block	9.5
		Solid gypsum block	12
	3"	Hollow gypsum block	10
		Solid gypsum block	18
	4"	Hollow gypsum block	15
		Solid gypsum block	24
	5"	Hollow gypsum block	18
	6"	Hollow gypsum block	24
Structural facing tile	2"	Facing tile	15
	4"	Facing tile	25
	6"	Facing tile	38
Glass	4"	Glass block	18
	1"	Structural glass	15
Plaster	1"	Gypsum plaster (1 side)	5
		Cement plaster (1 side)	10
		Gypsum plaster on lath	8
		Cement plaster on lath	13
Plaster partition (2 finished faces)	2"	Solid gypsum on metal lath	18
		Solid cement on metal lath	25
		Solid gypsum on gypsum lath	18
		Gypsum on lath & metal studs	18
	3"	Gypsum on lath & metal studs	19
	4"	Gypsum on lath & metal studs	20
	6"	Gypsum on lath & wood studs	18
Concrete	6"	Reinf concrete, stone aggregate	75
		Reinf. concrete, lightweight	36-60
	8"	Reinf. concrete, stone aggregate	100
		Reinf. concrete, lightweight	48-80
	10"	Reinf. concrete, stone aggregate	125
		Reinf. concrete, lightweight	60-100
	12"	Reinf concrete stone aggregate	150
		Reinf concrete, lightweight	72-120

Table L1010-202   Design Weight per S.F. for Roof Coverings

Type		Description	Weight Per S.F.
Sheathing	Gypsum	1" thick	4
	Wood	¾" thick	3
Insulation	per 1"	Loose	.5
		Poured in place	2
		Rigid	1.5
Built-up	Tar & gravel	3 ply felt	5.5
		5 ply felt	

Type	Wall Thickness	Description	Weight Per S.F.
Metal	Aluminum	Corr. & ribbed, .024" to .040"	.4-.8
	Copper	or tin	1.0
	Steel	Corrugated, 29 ga. to 12 ga.	.6-5.0
Shingles	Asphalt	Strip shingles	3
	Clay	Tile	9-20
	Slate	¼" thick	10
	Wood		2

Table L1010-203 Design Weight per Square Foot for Floor Fills and Finishes

Type		Description	Weight per S.F.	Type		Description	Weight per S.F.
Floor fill	per 1"	Cinder fill	5	Wood	Single 7/8"	On sleepers, light concrete fill	16
		Cinder concrete	9			On sleepers, stone concrete fill	25
		Lightweight concrete	3-9		Double 7/8"	On sleepers, light concrete fill	19
		Stone concrete	12			On sleepers, stone concrete fill	28
		Sand	8		3"	Wood block on mastic, no fill	15
		Gypsum	6			Wood block on ½" mortar	16
Terrazzo	1"	Terrazzo, 2" stone concrete	25		per 1"	Hardwood flooring (25/32")	4
Marble	and mortar	on stone concrete fill	33			Underlayment (Plywood per 1")	3
Resilient	1/16"-1/4"	Linoleum, asphalt, vinyl tile	2	Asphalt	1-1/2"	Mastic flooring	18
Tile	3/4"	Ceramic or quarry	10		2"	Block on ½" mortar	30

Table L1010-204 Design Weight Per Cubic Foot for Miscellaneous Materials

Type		Description	Weight per C.F.	Type		Description	Weight per C.F.
Bituminous	Coal, piled	Anthracite	47-58	Masonry	Ashlar	Granite	165
		Bituminous	40-54			Limestone, crystalline	165
		Peat, turf, dry	47			Limestone, oolitic	135
		Coke	75			Marble	173
	Petroleum	Unrefined	54			Sandstone	144
		Refined	50		Rubble, in mortar	Granite	155
		Gasoline	42			Limestone, crystalline	147
	Pitch		69			Limestone, oolitic	138
	Tar	Bituminous	75			Marble	156
Concrete	Plain	Stone aggregate	144			Sandstone & Bluestone	130
		Slag aggregate	132		Brick	Pressed	140
		Expanded slag aggregate	100			Common	120
		Haydite (burned clay agg.)	90			Soft	100
		Vermiculite & perlite, load bearing	70-105		Cement	Portland, loose	90
		Vermiculite & perlite, non load bear	35-50			Portland set	183
	Reinforced	Stone aggregate	150		Lime	Gypsum, loose	53-64
		Slag aggregate	138		Mortar	Set	103
		Lightweight aggregates	30-120	Metals	Aluminum	Cast, hammered	165
Earth	Clay	Dry	63		Brass	Cast, rolled	534
		Damp, plastic	110		Bronze	7.9 to 14% Sn	509
		and gravel, dry	100		Copper	Cast, rolled	556
	Dry	Loose	76		Iron	Cast, pig	450
		Packed	95			Wrought	480
	Moist	Loose	78		Lead		710
		Packed	96		Monel		556
	Mud	Flowing	108		Steel	Rolled	490
		Packed	115		Tin	Cast, hammered	459
	Riprap	Limestone	80-85		Zinc	Cast rolled	440
		Sandstone	90	Timber	Cedar	White or red	24.2
		Shale	105		Fir	Douglas	23.7
	Sand & gravel	Dry, loose	90-105			Eastern	25
		Dry, packed	100-120		Maple	Hard	44.5
		Wet	118-120			White	33
Gases	Air	0C., 760 mm.	.0807		Oak	Red or Black	47.3
	Gas	Natural	.0385			White	47.3
Liquids	Alcohol	100%	49		Pine	White	26
	Water	4°C., maximum density	62.5			Yellow, long leaf	44
		Ice	56			Yellow, short leaf	38
		Snow, fresh fallen	8		Redwood	California	26
		Sea water	64		Spruce	White or black	27

REFERENCE AIDS

Table L1010-225 Design Weight Per S.F. for Structural Floor and Roof Systems

Type Slab		Description	Weight in Pounds per S.F. — Slab Depth in Inches												
Concrete Slab	Reinforced	Stone aggregate	1"	12.5	2"	25	3"	37.5	4"	50	5"	62.5	6"	75
		Lightweight sand aggregate		9.5		19		28.5		38		47.5		57
		All lightweight aggregate		9.0		18		27.0		36		45.0		54
	Plain, nonreinforced	Stone aggregate		12.0		24		36.0		48		60.0		72
		Lightweight sand aggregate		9.0		18		27.0		36		45.0		54
		All lightweight aggregate		8.5		17		25.5		34		42.5		51
Concrete Waffle	19" x 19"	5" wide ribs @ 24" O.C.	6+3	77	8+3	92	10+3	100	12+3	118				
			6+4½	96	8+4½	110	10+4½	119	12+4½	136				
	30" x 30"	6" wide ribs @ 36" O.C.	8+3	83	10+3	95	12+3	109	14+3	118	16+3	130	20+3	155
			8+4½	101	10+4½	113	12+4½	126	14+4½	137	16+4½	149	20+4½	173
Concrete Joist	20" wide form	5" wide rib	8+3	60	10+3	67	12+3	74	14+3	81				
		6" wide rib		63		70		78		86	16+3	94	20+3	111
		7" wide rib										99		118
		5" wide rib	8+4½	79	10+4½	85	12+4½	92	14+4½	99				
		6" wide rib		82		89		97		104	16+4½	113	20+4½	130
		7" wide rib										118		136
	30" wide form	5" wide rib	8+3	54	10+3	58	12+3	63	14+3	68				
		6" wide rib		56		61		67		72	16+3	78	20+3	91
		7" wide rib										83		96
		5" wide rib	8+4½	72	10+4½	77	12+4½	82	14+4½	87				
		6" wide rib		75		80		85		91	16+4½	97	20+4½	109
		7" wide rib										101		115
Wood Joists	Incl. Subfloor	12" O.C.	2x6	6	2x8	6	2x10	7	2x12	8	3x8	8	3x12	11
		16" O.C.		5		6		6		7		7		9

Table L1010-226 Superimposed Dead Load Ranges

Component	Load Range (PSF)
Ceiling	5-10
Partitions	20-30
Mechanical	4-8

Table L1010-301 Design Loads for Structures for Wind Load

Wind Loads: Structures are designed to resist the wind force from any direction. Usually 2/3 is assumed to act on the windward side, 1/3 on the leeward side.

For more than 1/3 openings, add 10 psf for internal wind pressure or 5 psf for suction, whichever is critical.

For buildings and structures, use psf values from Table L1010-301.

For glass over 4 S.F. use values in Table L1010-302 after determining 30' wind velocity from Table L1010-303.

Type Structure	Height Above Grade	Horizontal Load in Lb. per S.F.
Buildings	Up to 50 ft.	15
	50 to 100 ft.	20
	Over 100 ft.	20 + .025 per ft.
Ground signs & towers	Up to 50 ft.	15
	Over 50 ft.	20
Roof structures		30
Glass	See Table below	

Table L1010-302 Design Wind Load in PSF for Glass at Various Elevations (BOCA Code)

Height From Grade	Velocity in Miles per Hour and Design Load in Pounds per S.F.																			
	Vel.	PSF	Vel.	PSF	Vel.	PSF	Vel.	PSF	Vel.	PSF	Vel.	PSF	Vel.	PSF	Vel.	PSF	Vel.	PSF	Vel.	PSF
To 10 ft.	42	6	46	7	49	8	52	9	55	10	59	11	62	12	66	14	69	15	76	19	83	22
10-20	52	9	58	11	61	11	65	14	70	16	74	18	79	20	83	22	87	24	96	30	105	35
20-30*	60	12	67	14	70	16	75	18	80	20	85	23	90	26	95	29	100	32	110	39	120	46
30-60	66	14	74	18	77	19	83	22	88	25	94	28	99	31	104	35	110	39	121	47	132	56
60-120	73	17	82	21	85	12	92	27	98	31	104	35	110	39	116	43	122	48	134	57	146	68
120-140	81	21	91	26	95	29	101	33	108	37	115	42	122	48	128	52	135	48	149	71	162	84
240-480	90	26	100	32	104	35	112	40	119	45	127	51	134	57	142	65	149	71	164	86	179	102
480-960	98	31	110	39	115	42	123	49	131	55	139	62	148	70	156	78	164	86	180	104	197	124
Over 960	98	31	110	39	115	42	123	49	131	55	139	62	148	70	156	78	164	86	180	104	197	124

Determine appropriate wind at 30' elevation Fig. L1010-303 below.

Table L1010-303 Design Wind Velocity at 30 Ft. Above Ground

SPEEDS ARE FOR NORMAL EXPOSURE WHERE SURFACE FRICTION IS RELATIVELY UNIFORM FOR A FETCH OF ABOUT 25 MILES. IF THE EXPOSURE IS ELEVATED, SUBJECT TO CHANNELING, OR OTHER SPECIAL CONDITIONS AFFECTING THE EXTREME WIND SPEEDS, ADJUSTMENTS MUST BE MADE TO THE MAP VALUES.

Table L1010-401   Snow Load in Pounds Per Square Foot on the Ground

Based on 50 year
storm for INDUSTRIAL
BUSINESS, MERCANTILE,
& RESIDENTIAL

Table L1010-402   Snow Load in Pounds Per Square Foot on the Ground

Based on 100 year
storm for ASSEMBLY
INSTITUTIONAL, HIGH HAZARD
MOTELS

Table L1010-403   Snow Loads

To convert the ground snow loads on the previous page to roof snow loads, the ground snow loads should be multiplied by the following factors depending upon the roof characteristics.

Note, in all cases $\dfrac{\alpha - 30}{50}$ is valid only for > 30 degrees.

Description	Sketch	Formula	Angle	Conversion Factor = C_f	
				Sheltered	**Exposed**
Simple flat and shed roofs	Load Diagram	For $\alpha > 30°$ $C_f = 0.8 - \dfrac{\alpha - 30}{50}$	0° to 30° 40° 50° 60° 70° to 90°	0.8 0.6 0.4 0.2 0	0.6 0.45 0.3 0.15 0
				Case I	**Case II**
Simple gable and hip roofs	Case I Load Diagram Case II Load Diagram	$C_f = 0.8 - \dfrac{\alpha - 30}{50}$ $C_f = 1.25 \left(0.8 - \left(\dfrac{\alpha - 30}{50}\right)\right)$	10° 20° 30° 40° 50° 60°	0.8 0.8 0.8 0.6 0.4 0.2	— — 1.0 0.75 0.5 0.25
Valley areas of Two span roofs		$\beta = \dfrac{\alpha_1 + \alpha_2}{2}$ $C_f = 0.8 - \dfrac{\alpha - 30}{50}$	$\beta \leq 10°$ use Case I only $\beta > 10°$ $\beta < 20°$ use Case I & II $\beta \geq 20°$ use Case I, II & III		
Lower level of multi level roofs (or on an adjacent building not more then 15 ft. away)		$C_f = 15\,\dfrac{h}{g}$ h = difference in roof height in feet g = ground snow load in psf w = width of drift For h < 5, w = 10 h > 15, w = 30	When $15\,\dfrac{h}{g} < .8$, use 0.8 When $15\,\dfrac{h}{g} > 3.0$, use 3.0		

Summary of above:

1. For flat roofs or roofs up to 30°, use 0.8 x ground snow load for the roof snow load.
2. For roof pitches in excess of 30°, conversion factor becomes lower than 0.8.
3. For exposed roofs there is a further 25% reduction of conversion factor.
4. For steep roofs a more highly loaded half span must be considered.
5. For shallow roof valleys conversion factor is 0.8.
6. For moderate roof valleys, conversion factor is 1.0 for half the span.
7. For steep roof valleys, conversion factor is 1.5 for one quarter of the span.
8. For roofs adjoining vertical surfaces the conversion factor is up to 3.0 for part of the span.
9. If snow load is less than 30 psf, use water load on roof for clogged drain condition.

Table L1020-101   Floor Area Ratios

Table below lists commonly used gross to net area and net to gross area
ratios expressed in % for various building types.

Building Type	Gross to Net Ratio	Net to Gross Ratio	Building Type	Gross to Net Ratio	Net to Gross Ratio
Apartment	156	64	School Buildings (campus type)		
Bank	140	72	Administrative	150	67
Church	142	70	Auditorium	142	70
Courthouse	162	61	Biology	161	62
Department Store	123	81	Chemistry	170	59
Garage	118	85	Classroom	152	66
Hospital	183	55	Dining Hall	138	72
Hotel	158	63	Dormitory	154	65
Laboratory	171	58	Engineering	164	61
Library	132	76	Fraternity	160	63
Office	135	75	Gymnasium	142	70
Restaurant	141	70	Science	167	60
Warehouse	108	93	Service	120	83
			Student Union	172	59

The gross area of a building is the total floor area based on outside
dimensions.

The net area of a building is the usable floor area for the function intended
and excludes such items as stairways, corridors and mechanical rooms.

In the case of a commercial building, it might be considered as the "leasable
area."

Table L1020-201 Partition/Door Density

Building Type		Stories	Partition/Density	Doors	Description of Partition
Apartments		1 story	9 SF/LF	90 SF/door	Plaster, wood doors & trim
		2 story	8 SF/LF	80 SF/door	Drywall, wood studs, wood doors & trim
		3 story	9 SF/LF	90 SF/door	Plaster, wood studs, wood doors & trim
		5 story	9 SF/LF	90 SF/door	Plaster, metal studs, wood doors & trim
		6-15 story	8 SF/LF	80 SF/door	Drywall, metal studs, wood doors & trim
Bakery		1 story	50 SF/LF	500 SF/door	Conc. block, paint, door & drywall, wood studs
		2 story	50 SF/LF	500 SF/door	Conc. block, paint, door & drywall, wood studs
Bank		1 story	20 SF/LF	200 SF/door	Plaster, wood studs, wood doors & trim
		2-4 story	15 SF/LF	150 SF/door	Plaster, metal studs, wood doors & trim
Bottling Plant		1 story	50 SF/LF	500 SF/door	Conc. block, drywall, metal studs, wood trim
Bowling Alley		1 story	50 SF/LF	500 SF/door	Conc. block, wood & metal doors, wood trim
Bus Terminal		1 story	15 SF/LF	150 SF/door	Conc. block, ceramic tile, wood trim
Cannery		1 story	100 SF/LF	1000 SF/door	Drywall on metal studs
Car Wash		1 story	18 SF/LF	180 SF/door	Concrete block, painted & hollow metal door
Dairy Plant		1 story	30 SF/LF	300 SF/door	Concrete block, glazed tile, insulated cooler doors
Department Store		1 story	60 SF/LF	600 SF/door	Drywall, metal studs, wood doors & trim
		2-5 story	60 SF/LF	600 SF/door	30% concrete block, 70% drywall, wood studs
Dormitory		2 story	9 SF/LF	90 SF/door	Plaster, concrete block, wood doors & trim
		3-5 story	9 SF/LF	90 SF/door	Plaster, concrete block, wood doors & trim
		6-15 story	9 SF/LF	90 SF/door	Plaster, concrete block, wood doors & trim
Funeral Home		1 story	15 SF/LF	150 SF/door	Plaster on concrete block & wood studs, paneling
		2 story	14 SF/LF	140 SF/door	Plaster, wood studs, paneling & wood doors
Garage Sales & Service		1 story	30 SF/LF	300 SF/door	50% conc. block, 50% drywall, wood studs
Hotel		3-8 story	9 SF/LF	90 SF/door	Plaster, conc. block, wood doors & trim
		9-15 story	9 SF/LF	90 SF/door	Plaster, conc. block, wood doors & trim
Laundromat		1 story	25 SF/LF	250 SF/door	Drywall, wood studs, wood doors & trim
Medical Clinic		1 story	6 SF/LF	60 SF/door	Drywall, wood studs, wood doors & trim
		2-4 story	6 SF/LF	60 SF/door	Drywall, metal studs, wood doors & trim
Motel		1 story	7 SF/LF	70 SF/door	Drywall, wood studs, wood doors & trim
		2-3 story	7 SF/LF	70 SF/door	Concrete block, drywall on metal studs, wood paneling
Movie Theater	200-600 seats	1 story	18 SF/LF	180 SF/door	Concrete block, wood, metal, vinyl trim
	601-1400 seats		20 SF/LF	200 SF/door	Concrete block, wood, metal, vinyl trim
	1401-22000 seats		25 SF/LF	250 SF/door	Concrete block, wood, metal, vinyl trim
Nursing Home		1 story	8 SF/LF	80 SF/door	Drywall, metal studs, wood doors & trim
		2-4 story	8 SF/LF	80 SF/door	Drywall, metal studs, wood doors & trim
Office		1 story	20 SF/LF	200-500 SF/door	30% concrete block, 70% drywall on wood studs
		2 story	20 SF/LF	200-500 SF/door	30% concrete block, 70% drywall on metal studs
		3-5 story	20 SF/LF	200-500 SF/door	30% concrete block, 70% movable partitions
		6-10 story	20 SF/LF	200-500 SF/door	30% concrete block, 70% movable partitions
		11-20 story	20 SF/LF	200-500 SF/door	30% concrete block, 70% movable partitions
Parking Ramp (Open)		2-8 story	60 SF/LF	600 SF/door	Stair and elevator enclosures only
Parking garage		2-8 story	60 SF/LF	600 SF/door	Stair and elevator enclosures only
Pre-Engineered	Steel	1 story	0		
	Store	1 story	60 SF/LF	600 SF/door	Drywall on metal studs, wood doors & trim
	Office	1 story	15 SF/LF	150 SF/door	Concrete block, movable wood partitions
	Shop	1 story	15 SF/LF	150 SF/door	Movable wood partitions
	Warehouse	1 story	0		
Radio & TV Broadcasting		1 story	25 SF/LF	250 SF/door	Concrete block, metal and wood doors
& TV Transmitter		1 story	40 SF/LF	400 SF/door	Concrete block, metal and wood doors
Self Service Restaurant		1 story	15 SF/LF	150 SF/door	Concrete block, wood and aluminum trim
Cafe & Drive-in Restaurant		1 story	18 SF/LF	180 SF/door	Drywall, metal studs, ceramic & plastic trim
Restaurant with seating		1 story	25 SF/LF	250 SF/door	Concrete block, paneling, wood studs & trim
Supper Club		1 story	25 SF/LF	250 SF/door	Concrete block, paneling, wood studs & trim
Bar or Lounge		1 story	24 SF/LF	240 SF/door	Plaster or gypsum lath, wooded studs
Retail Store or Shop		1 story	60 SF/LF	600 SF/door	Drywall metal studs, wood doors & trim
Service Station	Masonry	1 story	15 SF/LF	150 SF/door	Concrete block, paint, door & drywall, wood studs
	Metal panel	1 story	15 SF/LF	150 SF/door	Concrete block, paint, door & drywall, wood studs
	Frame	1 story	15 SF/LF	150 SF/door	Drywall, wood studs, wood doors & trim
Shopping Center	(strip)	1 story	30 SF/LF	300 SF/door	Drywall, metal studs, wood doors & trim
	(group)	1 story	40 SF/LF	400 SF/door	50% concrete block, 50% drywall, wood studs
		2 story	40 SF/LF	400 SF/door	50% concrete block, 50% drywall, wood studs
Small Food Store		1 story	30 SF/LF	300 SF/door	Concrete block drywall, wood studs, wood trim
Store/Apt. above	Masonry	2 story	10 SF/LF	100 SF/door	Plaster, metal studs, wood doors & trim
	Frame	2 story	10 SF/LF	100 SF/door	Plaster, metal studs, wood doors & trim
	Frame	3 story	10 SF/LF	100 SF/door	Plaster, metal studs, wood doors & trim
Supermarkets		1 story	40 SF/LF	400 SF/door	Concrete block, paint, drywall & porcelain panel
Truck Terminal		1 story	0		
Warehouse		1 story	0		

Table L1020-301 Occupancy Determinations

Description		S.F. Required per Person		
		BOCA	SBC	UBC
Assembly Areas	Fixed Seats	**	6	7
	Movable Seats		15	15
	Concentrated	7		
	Unconcentrated	15		
	Standing Space	3		
Educational	Unclassified			
	Classrooms	20	40	20
	Shop Areas	50	100	50
Institutional	Unclassified		125	
	In-Patient Areas	240		
	Sleeping Areas	120		
Mercantile	Basement	30	30	20
	Ground Floor	30	30	30
	Upper Floors	60	60	50
Office		100	100	100

BOCA = Building Officials & Code Administrators' National Building Code
SBC = Standard Building Code
UBC = Uniform Building Code

** The occupancy load for assembly area with fixed seats shall be determined by the number of fixed seats installed.

Table L1020-302 Length of Exitway Access Travel (ft.)

Use Group	Without Fire Suppression System	With Fire Suppression System
Assembly	200	250
Business	200	250
Factory and industrial	200	250
High hazard	—	75
Institutional	150	200
Mercantile	200	250
Residential	200	250
Storage, low hazard	300	400
Storage, moderate hazard	200	250

Note: The maximum length of exitway access travel in unlimited area buildings shall be 400'.

Table L1020-303 Capacity per Unit Egress Width*

Use Group	Without Fire Suppression System (Inches per Person)*		With Fire Suppression System (Inches per Person)*	
	Stairways	Doors, Ramps and Corridors	Stairways	Doors, Ramps and Corridors
Assembly, Business, Educational, Factory Industrial, Mercantile, Residential, Storage	0.3	0.2	0.2	0.15
Institutional—1	0.4	0.2	0.2	0.2
Institutional—2	1.0	0.7	0.6	0.5
Institutional—3	0.3	0.2	0.3	0.2
High Hazard	—	—	0.3	0.2

* 1" = 25.4 mm

Table L1030-101 "U" Values for Type "A" Buildings

Type A buildings shall include:

A1 Detached one and two family dwellings

A2 All other residential buildings, three stories or less, including but not limited to:
multi-family dwellings, hotels and motels.

Table L1030-102 "U" Values for Type "B" Buildings

For all buildings Not Classified Type "A"

**Table L1030-201   Combinations of Wall and Single-Glazed Openings
(For Use With ASHRAE 90-75)**

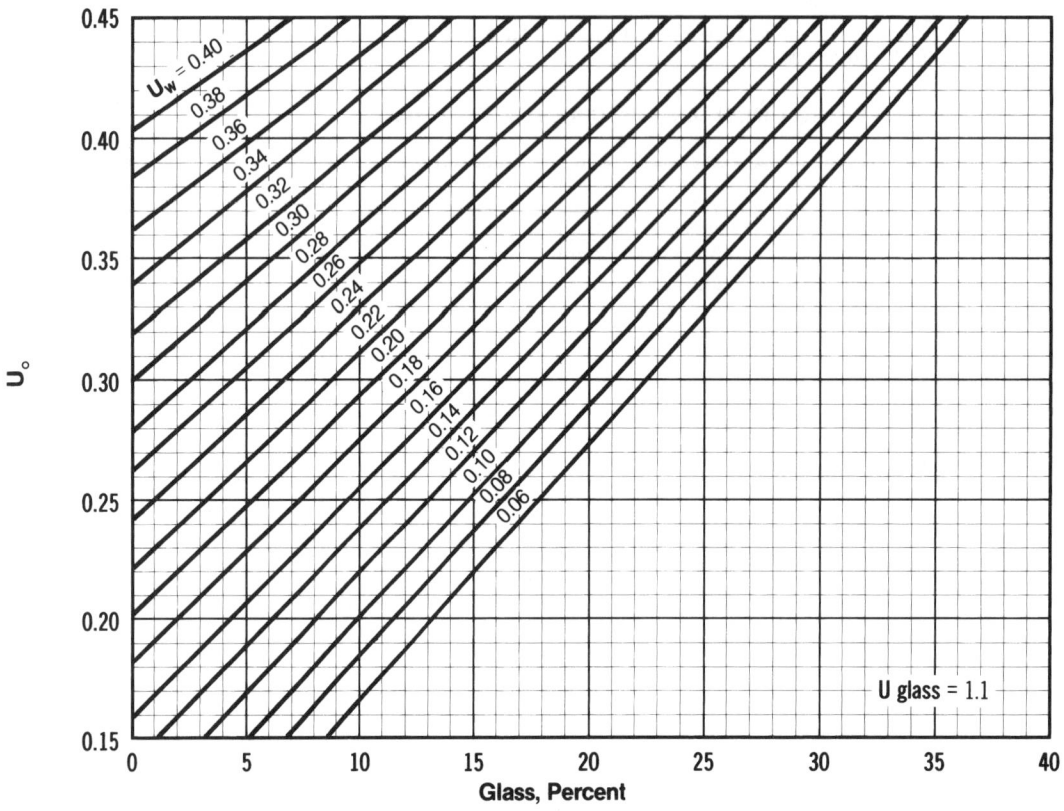

**Table L1030-202   Combinations of Wall and Double-Glazed Openings
(For Use With ASHRAE 90-75)**

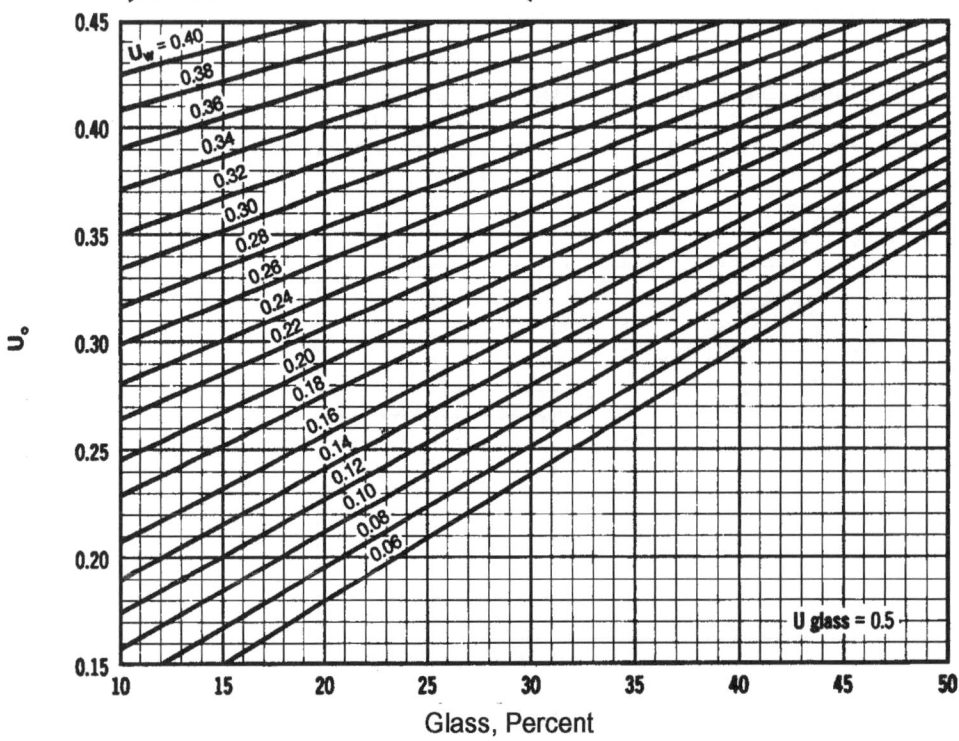

Table L1030-203   Influence of Wall and Roof Weight on "U" Value Correction for Heating Design

Effective "U" for walls: $Uw = Uw_{ss} \times M$ (similarly, Ur for roofs)

where Uw = effective thermal transmittance of opaque wall area BTU/h x Ft.² x °F

Uw_{ss} = steady state thermal transmittance of opaque wall area BTU/h x Ft.² x °F

(steady state "U" value)

M = weight correction factor

Example: Uw_{ss} = 0.20 with wall weight = 120 psf in Providence, R.I.
(6000 degree days)
Enter chart on bottom at 6000, go up to 120 psf curve, read to the left .88 M = .88
and $Uw = 0.20 \times 0.88 = 0.176$

Table L1030-204   "U" Values for Type "A" Buildings

One and two family dwellings and residential buildings of three stories or less shall have an overall U value as follows:

A. For roof assemblies in which the finished interior surface is essentially the underside of the roof deck, such as exposed concrete slabs, joist slabs, cored slabs, or cathedral ceilings or wood-beam construction.

$$U_{or} = 0.08 \text{ Btu/hr./ft.}^2/°F. \text{ (all degree days)}$$

B. For roof-ceiling assemblies, such as roofs with finished ceilings attached to or suspended below the roof deck.

$$U_{or} = 0.05 \text{ for below 8,000 degree days}$$
$$= 0.04 \text{ for 8,000 degree days and greater.}$$

Table L1030-205   "U" Values for Type "B" Buildings*

For all buildings not classified type "A".

*Minimum requirements for thermal design, ASHRAE Standard 90-75

Table L1030-301 "U" Values for Floors Over Unheated Spaces

Annual Celsius Heating Degree Days (18° C Base) (In Thousands)

U_o BTU/H-FT²-F

U_o — W/M²K

Annual Fahrenheit Heating Degree Days (65° F Base)(In Thousands)

Table L1030-302 "R" Values for Slabs on Grade

Annual Celsius Heating Degree Days (18° C Base) (In Thousands)

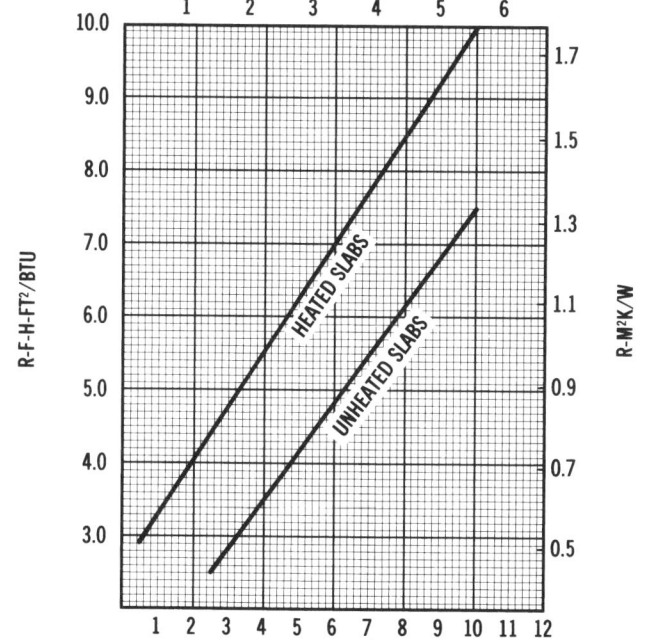

R-F-H-FT²/BTU

R-M²K/W

HEATED SLABS

UNHEATED SLABS

Annual Fahrenheit Heating Degree Days (65° F. Base) (In Thousands)

Table L1030-303 Insulation Requirements for Slabs on Grade for all Buildings

Insulation

24"

Section

24"

Insulation

Section (alternate)

Table L1030-401   Resistances ("R") of Building and Insulating Materials

Material	Wt./Lbs. per C.F.	R per Inch	R Listed Size
Air Spaces and Surfaces			
Enclosed non-reflective spaces, E=0.82			
50°F mean temp., 30°/10°F diff.			
.5"			.90/.91
.75"			.94/1.01
1.50"			.90/1.02
3.50"			.91/1.01
Inside vert. surface (still air)			0.68
Outside vert. surface (15 mph wind)			0.17
Building Boards			
Asbestos cement, 0.25" thick	120		0.06
Gypsum or plaster, 0.5" thick	50		0.45
Hardboard regular	50	1.37	
Tempered	63	1.00	
Laminated paper	30	2.00	
Particle board	37	1.85	
	50	1.06	
	63	0.85	
Plywood (Douglas Fir), 0.5" thick	34		0.62
Shingle backer, .375" thick	18		0.94
Sound deadening board, 0.5" thick	15		1.35
Tile and lay-in panels, plain or			
acoustical, 0.5" thick	18		1.25
Vegetable fiber, 0.5" thick	18		1.32
	25		1.14
Wood, hardwoods	48	0.91	
Softwoods	32	1.25	
Flooring Carpet with fibrous pad			2.08
With rubber pad			1.23
Cork tile, 1/8" thick			0.28
Terrazzo			0.08
Tile, resilient			0.05
Wood, hardwood, 0.75" thick			0.68
Subfloor, 0.75" thick			0.94
Glass			
Insulation, 0.50" air space			2.04
Single glass			0.91
Insulation Blanket or Batt, mineral, glass			
or rock fiber, approximate thickness			
3.0" to 3.5" thick			11
3.5" to 4.0" thick			13
6.0" to 6.5" thick			19
6.5" to 7.0" thick			22
8.5" to 9.0" thick			30
Boards			
Cellular glass	8.5	2.63	
Fiberboard, wet felted			
Acoustical tile	21	2.70	
Roof insulation	17	2.94	
Fiberboard, wet molded			
Acoustical tile	23	2.38	
Mineral fiber with resin binder	15	3.45	
Polystyrene, extruded,			
cut cell surface	1.8	4.00	
smooth skin surface	2.2	5.00	
	3.5	5.26	
Bead boards	1.0	3.57	
Polyurethane	1.5	6.25	
Wood or cane fiberboard, 0.5" thick			1.25

Material	Wt./Lbs. per C.F.	R per Inch	R Listed Size
Insulation Loose Fill			
Cellulose	2.3	3.13	
	3.2	3.70	
Mineral fiber, 3.75" to 5" thick	2-5		11
6.5" to 8.75" thick			19
7.5" to 10" thick			22
10.25" to 13.75" thick			30
Perlite	5-8	2.70	
Vermiculite	4-6	2.27	
Wood fiber	2-3.5	3.33	
Masonry Brick, Common	120	0.20	
Face	130	0.11	
Cement mortar	116	0.20	
Clay tile, hollow			
1 cell wide, 3" width			0.80
4" width			1.11
2 cells wide, 6" width			1.52
8" width			1.85
10" width			2.22
3 cells wide, 12" width			2.50
Concrete, gypsum fiber	51	0.60	
Lightweight	120	0.19	
	80	0.40	
	40	0.86	
Perlite	40	1.08	
Sand and gravel or stone	140	0.08	
Concrete block, lightweight			
3 cell units, 4"-15 lbs. ea.			1.68
6"-23 lbs. ea.			1.83
8"-28 lbs. ea.			2.12
12"-40 lbs. ea.			2.62
Sand and gravel aggregates,			
4"-20 lbs. ea.			1.17
6"-33 lbs. ea.			1.29
8"-38 lbs. ea.			1.46
12"-56 lbs. ea.			1.81
Plastering Cement Plaster,			
Sand aggregate	116	0.20	
Gypsum plaster, Perlite aggregate	45	0.67	
Sand aggregate	105	0.18	
Vermiculite aggregate	45	0.59	
Roofing			
Asphalt, felt, 15 lb.			0.06
Rolled roofing	70		0.15
Shingles	70		0.44
Built-up roofing .375" thick	70		0.33
Cement shingles	120		0.21
Vapor-permeable felt			0.06
Vapor seal, 2 layers of			
mopped 15 lb. felt			0.12
Wood, shingles 16"-7.5" exposure			0.87
Siding			
Aluminum or steel (hollow backed)			
oversheathing			0.61
With .375" insulating backer board			1.82
Foil backed			2.96
Wood siding, beveled, ½" x 8"			0.81

Table L1030-501 Weather Data and Design Conditions

City	Latitude (1) °	1'	Winter Temperatures (1) Med. of Annual Extremes	99%	97½%	Winter Degree Days (2)	Summer (Design Dry Bulb) Temperatures and Relative Humidity 1%	2½%	5%
UNITED STATES									
Albuquerque, NM	35	0	5.1	12	16	4,400	96/61	94/61	92/61
Atlanta, GA	33	4	11.9	17	22	3,000	94/74	92/74	90/73
Baltimore, MD	39	2	7	14	17	4,600	94/75	91/75	89/74
Birmingham, AL	33	3	13	17	21	2,600	96/74	94/75	92/74
Bismarck, ND	46	5	-32	-23	-19	8,800	95/68	91/68	88/67
Boise, ID	43	3	1	3	10	5,800	96/65	94/64	91/64
Boston, MA	42	2	-1	6	9	5,600	91/73	88/71	85/70
Burlington, VT	44	3	-17	-12	-7	8,200	88/72	85/70	82/69
Charleston, WV	38	2	3	7	11	4,400	92/74	90/73	87/72
Charlotte, NC	35	1	13	18	22	3,200	95/74	93/74	91/74
Casper, WY	42	5	-21	-11	-5	7,400	92/58	90/57	87/57
Chicago, IL	41	5	-8	-3	2	6,600	94/75	91/74	88/73
Cincinnati, OH	39	1	0	1	6	4,400	92/73	90/72	88/72
Cleveland, OH	41	2	-3	1	5	6,400	91/73	88/72	86/71
Columbia, SC	34	0	16	20	24	2,400	97/76	95/75	93/75
Dallas, TX	32	5	14	18	22	2,400	102/75	100/75	97/75
Denver, CO	39	5	-10	-5	1	6,200	93/59	91/59	89/59
Des Moines, IA	41	3	-14	-10	-5	6,600	94/75	91/74	88/73
Detroit, MI	42	2	-3	3	6	6,200	91/73	88/72	86/71
Great Falls, MT	47	3	-25	-21	-15	7,800	91/60	88/60	85/59
Hartford, CT	41	5	-4	3	7	6,200	91/74	88/73	85/72
Houston, TX	29	5	24	28	33	1,400	97/77	95/77	93/77
Indianapolis, IN	39	4	-7	-2	2	5,600	92/74	90/74	87/73
Jackson, MS	32	2	16	21	25	2,200	97/76	95/76	93/76
Kansas City, MO	39	1	-4	2	6	4,800	99/75	96/74	93/74
Las Vegas, NV	36	1	18	25	28	2,800	108/66	106/65	104/65
Lexington, KY	38	0	-1	3	8	4,600	93/73	91/73	88/72
Little Rock, AR	34	4	11	15	20	3,200	99/76	96/77	94/77
Los Angeles, CA	34	0	36	41	43	2,000	93/70	89/70	86/69
Memphis, TN	35	0	10	13	18	3,200	98/77	95/76	93/76
Miami, FL	25	5	39	44	47	200	91/77	90/77	89/77
Milwaukee, WI	43	0	-11	-8	-4	7,600	90/74	87/73	84/71
Minneapolis, MN	44	5	-22	-16	-12	8,400	92/75	89/73	86/71
New Orleans, LA	30	0	28	29	33	1,400	93/78	92/77	90/77
New York, NY	40	5	6	11	15	5,000	92/74	89/73	87/72
Norfolk, VA	36	5	15	20	22	3,400	93/77	91/76	89/76
Oklahoma City, OK	35	2	4	9	13	3,200	100/74	97/74	95/73
Omaha, NE	41	2	-13	-8	-3	6,600	94/76	91/75	88/74
Philadelphia, PA	39	5	6	10	14	4,400	93/75	90/74	87/72
Phoenix, AZ	33	3	27	31	34	1,800	109/71	107/71	105/71
Pittsburgh, PA	40	3	-1	3	7	6,000	91/72	88/71	86/70
Portland, ME	43	4	-10	-6	-1	7,600	87/72	84/71	81/69
Portland, OR	45	4	18	17	23	4,600	89/68	85/67	81/65
Portsmouth, NH	43	1	-8	-2	2	7,200	89/73	85/71	83/70
Providence, RI	41	4	-1	5	9	6,000	89/73	86/72	83/70
Rochester, NY	43	1	-5	1	5	6,800	91/73	88/71	85/70
Salt Lake City, UT	40	5	0	3	8	6,000	97/62	95/62	92/61
San Francisco, CA	37	5	36	38	40	3,000	74/63	71/62	69/61
Seattle, WA	47	4	22	22	27	5,200	85/68	82/66	78/65
Sioux Falls, SD	43	4	-21	-15	-11	7,800	94/73	91/72	88/71
St. Louis, MO	38	4	-3	3	8	5,000	98/75	94/75	91/75
Tampa, FL	28	0	32	36	40	680	92/77	91/77	90/76
Trenton, NJ	40	1	4	11	14	5,000	91/75	88/74	85/73
Washington, DC	38	5	7	14	17	4,200	93/75	91/74	89/74
Wichita, KS	37	4	-3	3	7	4,600	101/72	98/73	96/73
Wilmington, DE	39	4	5	10	14	5,000	92/74	89/74	87/73
ALASKA									
Anchorage	61	1	-29	-23	-18	10,800	71/59	68/58	66/56
Fairbanks	64	5	-59	-51	-47	14,280	82/62	78/60	75/59
CANADA									
Edmonton, Alta.	53	3	-30	-29	-25	11,000	85/66	82/65	79/63
Halifax, N.S.	44	4	-4	1	5	8,000	79/66	76/65	74/64
Montreal, Que.	45	3	-20	-16	-10	9,000	88/73	85/72	83/71
Saskatoon, Sask.	52	1	-35	-35	-31	11,000	89/68	86/66	83/65
St. John, Nwf.	47	4	1	3	7	8,600	77/66	75/65	73/64
Saint John, N.B.	45	2	-15	-12	-8	8,200	80/67	77/65	75/64
Toronto, Ont.	43	4	-10	-5	-1	7,000	90/73	87/72	85/71
Vancouver, B.C.	49	1	13	15	19	6,000	79/67	77/66	74/65
Winnipeg, Man.	49	5	-31	-30	-27	10,800	89/73	86/71	84/70

(1) Handbook of Fundamentals, ASHRAE, Inc., NY 1989
(2) Local Climatological Annual Survey, USDC Env. Science Services Administration, Asheville, NC

Table L1030-502   Maximum Depth of Frost Penetration in Inches

THIS MAP IS REASONABLY ACCURATE FOR MOST PARTS
OF THE UNITED STATES BUT IS NECESSARILY HIGHLY
GENERALIZED, AND CONSEQUENTLY NOT TOO ACCURATE IN
MOUNTAINOUS REGIONS, PARTICULARLY IN THE ROCKIES.

Table L1040-101 Fire-Resisting Ratings of Structural Elements (in hours)

Description of the Structural Element	No. 1 Fireproof		No. 2 Non Combustible			No. 3 Exterior Masonry Wall			No. 4 Frame	
			Protected		Unprotected	Heavy Timber	Ordinary		Protected	Unprotected
							Protected	Unprotected		
	1A	1B	2A	2B	2C	3A	3B	3C	4A	4B
Exterior, Bearing Walls	4	3	2	1½	1	2	2	2	1	1
Nonbearing Walls	2	2	1½	1	1	2	2	2	1	1
Interior Bearing Walls and Partitions	4	3	2	1	0	2	1	0	1	0
Fire Walls and Party Walls	4	3	2	2	2	2	2	2	2	2
Fire Enclosure of Exitways, Exit Hallways and Stairways	2	2	2	2	2	2	2	2	1	1
Shafts other than Exitways, Hallways and Stairways	2	2	2	2	2	2	2	2	1	1
Exitway access corridors and Vertical separation of tenant space	1	1	1	1	0	1	1	0	1	0
Columns, girders, trusses (other than roof trusses) and framing: Supporting more than one floor	4	3	2	1	0	—	1	0	1	0
Supporting one floor only	3	2	1½	1	0	—	1	0	1	0
Structural members supporting wall	3	2	1½	1	0	1	1	0	1	0
Floor construction including beams	3	2	1½	1	0	—	1	0	1	0
Roof construction including beams, trusses and framing arches and roof deck 15' or less in height to lowest member	2	1½	1	1	0	—	1	0	1	0

Note: a. Codes include special requirements and exceptions that are not included in the table above.

b. Each type of construction has been divided into sub-types which vary according to the degree of fire resistance required. Sub-types (A) requirements are more severe than those for sub-types (B).

c. Protected construction means all structural members are chemically treated, covered or protected so that the unit has the required fire resistance.

Type No. 1, Fireproof Construction — Buildings and structures of fireproof construction are those in which the walls, partitions, structural elements, floors, ceilings, and roofs, and the exitways are protected with approved noncombustible materials to afford the fire-resistance rating specified in Table L1040-101; except as otherwise specifically regulated. Fire-resistant treated wood may be used as specified.

Type No. 2, Noncombustible Construction — Buildings and structures of noncombustible construction are those in which the walls, partitions, structural elements, floors, ceilings, roofs and the exitways are approved noncombustible materials meeting the fire-resistance rating requirements specified in Table L1040-101; except as modified by the fire limit restrictions. Fire-retardant treated wood may be used as specified.

Type No. 3, Exterior Masonry Wall Construction — Buildings and structures of exterior masonry wall construction are those in which the exterior, fire and party walls are masonry or other approved noncombustible materials of the required fire-resistance rating and structural properties. The floors, roofs, and interior framing are wholly or partly wood or metal or other approved construction. The fire and party walls are ground-supported; except that girders and their supports, carrying walls of masonry shall be protected to afford the same degree of fire-resistance rating of the supported walls. All structural elements have the required fire-resistance rating specified in Table L1040-101.

Type No. 4, Frame Construction — Buildings and structures of frame construction are those in which the exterior walls, bearing walls, partitions, floor and roof construction are wholly or partly of wood stud and joist assemblies with a minimum nominal dimension of two inches or of other approved combustible materials. Fire stops are required at all vertical and horizontal draft openings in which the structural elements have required fire-resistance ratings specified in Table L1040-101.

Table L1040-201   Fire Grading of Building in Hours for Fire Walls and Fire Separation Walls

Fire Hazard for Fire Walls

The degree of fire hazard of buildings relating to their intended use is defined by "Fire Grading" the buildings. Such a grading system is listed in the Table L1040-201 below. This type of grading determines the requirements for fire walls and fire separation walls (exterior fire exposure). For mixed use occupancy, use the higher Fire Grading requirement of the components.

Building Type	Hours	Building Type	Hours
Businesses	2	Recreation Centers	2
Churches	1½	Residential Hotels	2
Factories	3	Residential, Multi-family Dwellings	1½
High Hazard*	4	Residential, One & Two Family Dwellings	¾
Industrial	3	Restaurants	2
Institutional, Incapacitated Occupants	2	Schools	1½
Institutional, Restrained Occupants	3	Storage, Low Hazard*	2
Lecture Halls	2	Storage, Moderate Hazard*	3
Mercantile	3	Terminals	2
Night Club	3	Theatres	3

Note: *The difference in "Fire Hazards" is determined by their occupancy and use.

High Hazard: Industrial and storage buildings in which the combustible contents might cause fires to be unusually intense or where explosives, combustible gases or flammable liquids are manufactured or stored.

Moderate Hazard: Mercantile buildings, industrial and storage buildings in which combustible contents might cause fires of moderate intensity.

Low Hazard: Business buildings that ordinarily do not burn rapidly.

Table L1040-202   Interior Finish Classification

Flame Spread for Interior Finishes

The flame spreadability of a material is the burning characteristic of the material relative to the fuel contributed by its combustion and the density of smoke developed. The flame spread classification of a material is based on a ten minute test on a scale of 0 to 100. Cement asbestos board is assigned a rating of 0 and select red oak flooring a rating of 100. The four classes are listed in Table L1040-202.

The flame spread ratings for interior finish walls and ceilings shall not be greater than the Class listed in the Table L1040-203 below.

Class of Material	Surface Burning Characteristics
I	0 to 25
II	26 to 75
III	76 to 200
IV	201 to 5000

Table L1040-203   Interior Finish Requirements by Class

Building Type	Vertical Exitways & Passageways	Corridors Providing Exitways	Rooms or Enclosed Spaces	Building Type	Vertical Exitways & Passageways	Corridors Providing Exitways	Rooms or Enclosed Space
Assembly Halls	I	I	II	Mercantile Walls	I	II	III
Businesses	I	II	III	Night Clubs	I	I	II
Churches	I	I	II	Residential Hotels	I	II	III
Factory	I	II	III	Residential, Multi-family	I	II	III
High Hazard	I	II	II	Residential, 1 & 2 Family	IV	IV	IV
Industrial	I	II	III	Restaurants	I	I	II
Institutional, Incapacitated	I	II	I	Storage, Low Hazard	I	II	III
Institutional, Restrained	I	I	I	Storage, Moderate Hazard	I	II	III
Mercantile Ceilings	I	II	II	Terminals	I	I	II

Description: This table is primarily for converting customary U.S. units in the left hand column to SI metric units in the right hand column.

In addition, conversion factors for some commonly encountered Canadian and non-SI metric units are included.

Table L1090-101 Metric Conversion Factors

If You Know		Multiply By		To Find
Length	Inches	x	25.4[a]	=	Millimeters
	Feet	x	0.3048[a]	=	Meters
	Yards	x	0.9144[a]	=	Meters
	Miles (statute)	x	1.609	=	Kilometers
Area	Square inches	x	645.2	=	Square millimeters
	Square feet	x	0.0929	=	Square meters
	Square yards	x	0.8361	=	Square meters
Volume (Capacity)	Cubic inches	x	16,387	=	Cubic millimeters
	Cubic feet	x	0.02832	=	Cubic meters
	Cubic yards	x	0.7646	=	Cubic meters
	Gallons (U.S. liquids)[b]	x	0.003785	=	Cubic meters[c]
	Gallons (Canadian liquid)[b]	x	0.004546	=	Cubic meters[c]
	Ounces (U.S. liquid)[b]	x	29.57	=	Milliliters[c, d]
	Quarts (U.S. liquid)[b]	x	0.9464	=	Liters[c, d]
	Gallons (U.S. liquid)[b]	x	3.785	=	Liters[c, d]
Force	Kilograms force[d]	x	9.807	=	Newtons
	Pounds force	x	4.448	=	Newtons
	Pounds force	x	0.4536	=	Kilograms force[d]
	Kips	x	4448	=	Newtons
	Kips	x	453.6	=	Kilograms force[d]
Pressure, Stress, Strength (Force per unit area)	Kilograms force per square centimeter[d]	x	0.09807	=	Megapascals
	Pounds force per square inch (psi)	x	0.006895	=	Megapascals
	Kips per square inch	x	6.895	=	Megapascals
	Pounds force per square inch (psi)	x	0.07031	=	Kilograms force per square centimeter[d]
	Pounds force per square foot	x	47.88	=	Pascals
	Pounds force per square foot	x	4.882	=	Kilograms force per square meter[d]
Bending Moment Or Torque	Inch-pounds force	x	0.01152	=	Meter-kilograms force[d]
	Inch-pounds force	x	0.1130	=	Newton-meters
	Foot-pounds force	x	0.1383	=	Meter-kilograms force[d]
	Foot-pounds force	x	1.356	=	Newton-meters
	Meter-kilograms force[d]	x	9.807	=	Newton-meters
Mass	Ounces (avoirdupois)	x	28.35	=	Grams
	Pounds (avoirdupois)	x	0.4536	=	Kilograms
	Tons (metric)	x	1000	=	Kilograms
	Tons, short (2000 pounds)	x	907.2	=	Kilograms
	Tons, short (2000 pounds)	x	0.9072	=	Megagrams[e]
Mass per Unit Volume	Pounds mass per cubic foot	x	16.02	=	Kilograms per cubic meter
	Pounds mass per cubic yard	x	0.5933	=	Kilograms per cubic meter
	Pounds mass per gallon (U.S. liquid)[b]	x	119.8	=	Kilograms per cubic meter
	Pounds mass per gallon (Canadian liquid)[b]	x	99.78	=	Kilograms per cubic meter
Temperature	Degrees Fahrenheit	(F-32)/1.8		=	Degrees Celsius
	Degrees Fahrenheit	(F+459.67)/1.8		=	Degrees Kelvin
	Degrees Celsius	C+273.15		=	Degrees Kelvin

[a]The factor given is exact
[b]One U.S. gallon = 0.8327 Canadian gallon
[c]1 liter = 1000 milliliters = 1000 cubic centimeters
  1 cubic decimeter = 0.001 cubic meter

[d]Metric but not SI unit
[e]Called "tonne" in England and
  "metric ton" in other metric countries

Table L1090-201   Metric Equivalents of Cement Content for Concrete Mixes

94 Pound Bags per Cubic Yard	Kilograms per Cubic Meter	94 Pound Bags per Cubic Yard	Kilograms per Cubic Meter
1.0	55.77	7.0	390.4
1.5	83.65	7.5	418.3
2.0	111.5	8.0	446.2
2.5	139.4	8.5	474.0
3.0	167.3	9.0	501.9
3.5	195.2	9.5	529.8
4.0	223.1	10.0	557.7
4.5	251.0	10.5	585.6
5.0	278.8	11.0	613.5
5.5	306.7	11.5	641.3
6.0	334.6	12.0	669.2
6.5	362.5	12.5	697.1

a. If you know the cement content in pounds per cubic yard,
   multiply by .5933 to obtain kilograms per cubic meter.

b. If you know the cement content in 94 pound bags per cubic yard,
   multiply by 55.77 to obtain kilograms per cubic meter.

Table L1090-202   Metric Equivalents of Common Concrete Strengths (to convert other psi values to megapascals, multiply by .006895)

U.S. Values psi	SI Value Megapascals	Non-SI Metric Value kgf/cm²*
2000	14	140
2500	17	175
3000	21	210
3500	24	245
4000	28	280
4500	31	315
5000	34	350
6000	41	420
7000	48	490
8000	55	560
9000	62	630
10,000	69	705

* kilograms force per square centimeter

Table L1090-203 Comparison of U.S. Customary Units and SI Units for Reinforcing Bars

Bar Designation No.[b]	Nominal Weight, lb/ft	Nominal Dimensions[a]			Deformation Requirements, in.		
		Diameter in.	Cross Sectional Area, in.²	Perimeter in.	Maximum Average Spacing	Minimum Average Height	Maximum Gap (Chord of 12-1/2% of Nominal Perimeter)
3	0.376	0.375	0.11	1.178	0.262	0.015	0.143
4	0.668	0.500	0.20	1.571	0.350	0.020	0.191
5	1.043	0.625	0.31	1.963	0.437	0.028	0.239
6	1.502	0.750	0.44	2.356	0.525	0.038	0.286
7	2.044	0.875	0.60	2.749	0.612	0.044	0.334
8	2.670	1.000	0.79	3.142	0.700	0.050	0.383
9	3.400	1.128	1.00	3.544	0.790	0.056	0.431
10	4.303	1.270	1.27	3.990	0.889	0.064	0.487
11	5.313	1.410	1.56	4.430	0.987	0.071	0.540
14	7.65	1.693	2.25	5.32	1.185	0.085	0.648
18	13.60	2.257	4.00	7.09	1.58	0.102	0.864

Bar Designation No.[b]	Nominal Weight kg/m	Nominal Dimensions[a]			Deformation Requirements, mm		
		Diameter, mm	Cross Sectional Area, cm²	Perimeter, mm	Maximum Average Spacing	Minimum Average Height	Maximum Gap (Chord of 12-1/2% of Nominal Perimeter)
3	0.560	9.52	0.71	29.9	6.7	0.38	3.5
4	0.994	12.70	1.29	39.9	8.9	0.51	4.9
5	1.552	15.88	2.00	49.9	11.1	0.71	6.1
6	2.235	19.05	2.84	59.8	13.3	0.96	7.3
7	3.042	22.22	3.87	69.8	15.5	1.11	8.5
8	3.973	25.40	5.10	79.8	17.8	1.27	9.7
9	5.059	28.65	6.45	90.0	20.1	1.42	10.9
10	6.403	32.26	8.19	101.4	22.6	1.62	11.4
11	7.906	35.81	10.06	112.5	25.1	1.80	13.6
14	11.384	43.00	14.52	135.1	30.1	2.16	16.5
18	20.238	57.33	25.81	180.1	40.1	2.59	21.9

[a]Nominal dimensions of a deformed bar are equivalent to those of a plain round bar having the same weight per foot as the deformed bar.
[b]Bar numbers are based on the number of eighths of an inch included in the nominal diameter of the bars.

Table L1090-204 Metric Rebar Specification - ASTM A615-81

Grade 300 (300 MPa* = **43,560 psi; +8.7% vs. Grade 40**)				
Grade 400 (400 MPa* = **58,000 psi; −3.4% vs. Grade 60**)				
Bar No.	Diameter mm	Area mm²	Equivalent in.²	Comparison with U.S. Customary Bars
10	11.3	100	.16	Between #3 & #4
15	16.0	200	.31	#5 (.31 in.²)
20	19.5	300	.47	#6 (.44 in.²)
25	25.2	500	.78	#8 (.79 in.²)
30	29.9	700	1.09	#9 (1.00 in.²)
35	35.7	1000	1.55	#11 (1.56 in.²)
45	43.7	1500	2.33	#14 (2.25 in.²)
55	56.4	2500	3.88	#18 (4.00 in.²)

* MPa = megapascals   Grade 300 bars are furnished only in sizes 10 through 35

PRELIMINARY ESTIMATE

PROJECT	TOTAL SITE AREA
BUILDING TYPE	OWNER
LOCATION	ARCHITECT
DATE OF CONSTRUCTION	ESTIMATED CONSTRUCTION PERIOD
BRIEF DESCRIPTION	

TYPE OF PLAN	TYPE OF CONSTRUCTION
QUALITY	BUILDING CAPACITY

Floor			Wall Areas				
Below Grade Levels			Foundation Walls	L.F.	Ht.		S.F.
Area		S.F.	Frost Walls	L.F.	Ht.		S.F.
Area		S.F.	Exterior Closure		Total		S.F.
Total Area		S.F.	Comment				
Ground Floor			Fenestration		%		S.F.
Area		S.F.			%		S.F.
Area		S.F.	Exterior Wall		%		S.F.
Total Area		S.F.			%		S.F.
Supported Levels			**Site Work**				
Area		S.F.	Parking		S.F. (For		Cars)
Area		S.F.	Access Roads		L.F. (X		Ft. Wide)
Area		S.F.	Sidewalk		L.F. (X		Ft. Wide)
Area		S.F.	Landscaping		S.F. (	% Unbuilt Site)	
Area		S.F.	**Building Codes**				
Total Area		S.F.	City		County		
Miscellaneous			National		Other		
Area		S.F.	**Loading**				
Area		S.F.	Roof	psf	Ground Floor		psf
Area		S.F.	Supported Floors	psf	Corridor		psf
Area		S.F.	Balcony	psf	Partition, allow		psf
Total Area		S.F.	Miscellaneous				psf
Net Finished Area		S.F.	Live Load Reduction				
Net Floor Area		S.F.	Wind				
Gross Floor Area		S.F.	Earthquake		Zone		
Roof			Comment				
Total Area		S.F.	Soil Type				
Comments			Bearing Capacity				K.S.F.
			Frost Depth				Ft.
Volume			**Frame**				
Depth of Floor System			Type		Bay Spacing		
Minimum		In.	Foundation, Standard				
Maximum		In.	Special				
Foundation Wall Height		Ft.	Substructure				
Floor to Floor Height		Ft.	Comment				
Floor to Ceiling Height		Ft.	Superstructure, Vertical		Horizontal		
Subgrade Volume		C.F.	Fireproofing	☐ Columns			Hrs.
Above Grade Volume		C.F.	☐ Girders	Hrs.	☐ Beams		Hrs.
Total Building Volume		C.F.	☐ Floor	Hrs.	☐ None		

ASSEMBLY NUMBER	DESCRIPTION	QTY	UNIT	TOTAL COST UNIT	TOTAL COST TOTAL	COST PER S.F.
A10	**Foundations**					
A20	Basement Construction					
B10	Superstructure					
B20	Exterior Closure					
B30	Roofing					

ESTIMATING FORMS

569

ASSEMBLY NUMBER	DESCRIPTION	QTY	UNIT	TOTAL COST		COST PER S.F.
				UNIT	TOTAL	
C	**Interior Construction**					
D10	**Conveying**					
D20	**Plumbing**					

ASSEMBLY NUMBER	DESCRIPTION	QTY	UNIT	TOTAL COST		COST PER S.F.
				UNIT	TOTAL	
D30	**HVAC**					
D40	**Fire Protection**					
D50	**Electrical**					

ESTIMATING FORMS

ASSEMBLY NUMBER	DESCRIPTION	QTY	UNIT	TOTAL COST UNIT	TOTAL COST TOTAL	COST PER S.F.
E	**Equipment & Furnishings**					
F	**Special Construction**					
G	**Sitework**					

Preliminary Estimate Cost Summary

PROJECT	TOTAL AREA	SHEET NO.
LOCATION	TOTAL VOLUME	ESTIMATE NO.
ARCHITECT	COST PER S.F.	DATE
OWNER	COST PER C.F.	NO OF STORIES
QUANTITIES BY	EXTENSIONS BY	CHECKED BY

DIV	DESCRIPTION	SUBTOTAL COST	COST/S.F.	PERCENTAGE
A	SUBSTRUCTURE			
B10	SHELL: SUPERSTRUCTURE			
B20	SHELL: EXTERIOR CLOSURE			
B30	SHELL: ROOFING			
C	INTERIOR CONSTRUCTION			
D10	SERVICES: CONVEYING			
D20	SERVICES: PLUMBING			
D30	SERVICES: HVAC			
D40	SERVICES: FIRE PROTECTION			
D50	SERVICES: ELECTRICAL			
E	EQUIPMENT & FURNISHINGS			
F	SPECIAL CONSTRUCTION			
G	SITEWORK			

BUILDING SUBTOTAL $ _____ -

Sales Tax __ % x Subtotal /2 .... $ _____ -

General Conditions ___% x Subtotal .... $ _____ -

Subtotal "A" $ _____ -

Overhead ___% x Subtotal "A" .... $ _____ -

Subtotal "B" $ _____ -

Profit ___ % x Subtotal "B" .... $ _____ -

Subtotal "C" $ _____ -

Location Factor ____ % x Subtotal "C"   Localized Cost $ _____ -
(Boston, MA)

Architects Fee___ x Localized Cost = ... $ _____ -
Contingency ____ x Localized Cost = ... $ _____ -

Project Total Cost $ _____ -

Square Foot Cost $ ____ / ____ S.F. =   **S.F. Cost** $ ____ -
Cubic Foot Cost $ ____ / ____ C.F. =   **C.F. Cost** $ ____ -

Abbreviations

A	Area Square Feet; Ampere	Cab.	Cabinet	Demob.	Demobilization
ABS	Acrylonitrile Butadiene Stryrene; Asbestos Bonded Steel	Cair.	Air Tool Laborer	d.f.u.	Drainage Fixture Units
A.C.	Alternating Current; Air-Conditioning; Asbestos Cement; Plywood Grade A & C	Calc	Calculated	D.H.	Double Hung
		Cap.	Capacity	DHW	Domestic Hot Water
		Carp.	Carpenter	Diag.	Diagonal
		C.B.	Circuit Breaker	Diam.	Diameter
		C.C.A.	Chromate Copper Arsenate	Distrib.	Distribution
A.C.I.	American Concrete Institute	C.C.F.	Hundred Cubic Feet	Dk.	Deck
AD	Plywood, Grade A & D	cd	Candela	D.L.	Dead Load; Diesel
Addit.	Additional	cd/sf	Candela per Square Foot	DLH	Deep Long Span Bar Joist
Adj.	Adjustable	CD	Grade of Plywood Face & Back	Do.	Ditto
af	Audio-frequency	CDX	Plywood, Grade C & D, exterior glue	Dp.	Depth
A.G.A.	American Gas Association			D.P.S.T.	Double Pole, Single Throw
Agg.	Aggregate	Cefi.	Cement Finisher	Dr.	Driver
A.H.	Ampere Hours	Cem.	Cement	Drink.	Drinking
A hr.	Ampere-hour	CF	Hundred Feet	D.S.	Double Strength
A.H.U.	Air Handling Unit	C.F.	Cubic Feet	D.S.A.	Double Strength A Grade
A.I.A.	American Institute of Architects	CFM	Cubic Feet per Minute	D.S.B.	Double Strength B Grade
AIC	Ampere Interrupting Capacity	c.g.	Center of Gravity	Dty.	Duty
Allow.	Allowance	CHW	Chilled Water; Commercial Hot Water	DWV	Drain Waste Vent
alt.	Altitude			DX	Deluxe White, Direct Expansion
Alum.	Aluminum	C.I.	Cast Iron	dyn	Dyne
a.m.	Ante Meridiem	C.I.P.	Cast in Place	e	Eccentricity
Amp.	Ampere	Circ.	Circuit	E	Equipment Only; East
Anod.	Anodized	C.L.	Carload Lot	Ea.	Each
Approx.	Approximate	Clab.	Common Laborer	E.B.	Encased Burial
Apt.	Apartment	Clam	Common maintenance laborer	Econ.	Economy
Asb.	Asbestos	C.L.F.	Hundred Linear Feet	E.C.Y	Embankment Cubic Yards
A.S.B.C.	American Standard Building Code	CLF	Current Limiting Fuse	EDP	Electronic Data Processing
Asbe.	Asbestos Worker	CLP	Cross Linked Polyethylene	EIFS	Exterior Insulation Finish System
A.S.H.R.A.E.	American Society of Heating, Refrig. & AC Engineers	cm	Centimeter	E.D.R.	Equiv. Direct Radiation
		CMP	Corr. Metal Pipe	Eq.	Equation
A.S.M.E.	American Society of Mechanical Engineers	C.M.U.	Concrete Masonry Unit	Elec.	Electrician; Electrical
		CN	Change Notice	Elev.	Elevator; Elevating
A.S.T.M.	American Society for Testing and Materials	Col.	Column	EMT	Electrical Metallic Conduit; Thin Wall Conduit
		CO₂	Carbon Dioxide		
Attchmt.	Attachment	Comb.	Combination	Eng.	Engine, Engineered
Avg.	Average	Compr.	Compressor	EPDM	Ethylene Propylene Diene Monomer
A.W.G.	American Wire Gauge	Conc.	Concrete		
AWWA	American Water Works Assoc.	Cont.	Continuous; Continued	EPS	Expanded Polystyrene
Bbl.	Barrel	Corr.	Corrugated	Eqhv.	Equip. Oper., Heavy
B&B	Grade B and Better; Balled & Burlapped	Cos	Cosine	Eqlt.	Equip. Oper., Light
		Cot	Cotangent	Eqmd.	Equip. Oper., Medium
B.&S.	Bell and Spigot	Cov.	Cover	Eqmm.	Equip. Oper., Master Mechanic
B.&W.	Black and White	C/P	Cedar on Paneling	Eqol.	Equip. Oper., Oilers
b.c.c.	Body-centered Cubic	CPA	Control Point Adjustment	Equip.	Equipment
B.C.Y.	Bank Cubic Yards	Cplg.	Coupling	ERW	Electric Resistance Welded
BE	Bevel End	C.P.M.	Critical Path Method	E.S.	Energy Saver
B.F.	Board Feet	CPVC	Chlorinated Polyvinyl Chloride	Est.	Estimated
Bg. cem.	Bag of Cement	C.Pr.	Hundred Pair	esu	Electrostatic Units
BHP	Boiler Horsepower; Brake Horsepower	CRC	Cold Rolled Channel	E.W.	Each Way
		Creos.	Creosote	EWT	Entering Water Temperature
B.I.	Black Iron	Crpt.	Carpet & Linoleum Layer	Excav.	Excavation
Bit.; Bitum.	Bituminous	CRT	Cathode-ray Tube	Exp.	Expansion, Exposure
Bk.	Backed	CS	Carbon Steel, Constant Shear Bar Joist	Ext.	Exterior
Bkrs.	Breakers			Extru.	Extrusion
Bldg.	Building	Csc	Cosecant	f.	Fiber stress
Blk.	Block	C.S.F.	Hundred Square Feet	F	Fahrenheit; Female; Fill
Bm.	Beam	CSI	Construction Specifications Institute	Fab.	Fabricated
Boil.	Boilermaker			FBGS	Fiberglass
B.P.M.	Blows per Minute	C.T.	Current Transformer	F.C.	Footcandles
BR	Bedroom	CTS	Copper Tube Size	f.c.c.	Face-centered Cubic
Brg.	Bearing	Cu	Copper, Cubic	f'c.	Compressive Stress in Concrete; Extreme Compressive Stress
Brhe.	Bricklayer Helper	Cu. Ft.	Cubic Foot		
Bric.	Bricklayer	cw	Continuous Wave	F.E.	Front End
Brk.	Brick	C.W.	Cool White; Cold Water	FEP	Fluorinated Ethylene Propylene (Teflon)
Brng.	Bearing	Cwt.	100 Pounds		
Brs.	Brass	C.W.X.	Cool White Deluxe	F.G.	Flat Grain
Brz.	Bronze	C.Y.	Cubic Yard (27 cubic feet)	F.H.A.	Federal Housing Administration
Bsn.	Basin	C.Y./Hr.	Cubic Yard per Hour	Fig.	Figure
Btr.	Better	Cyl.	Cylinder	Fin.	Finished
BTU	British Thermal Unit	d	Penny (nail size)	Fixt.	Fixture
BTUH	BTU per Hour	D	Deep; Depth; Discharge	Fl. Oz.	Fluid Ounces
B.U.R.	Built-up Roofing	Dis.;Disch.	Discharge	Flr.	Floor
BX	Interlocked Armored Cable	Db.	Decibel	F.M.	Frequency Modulation; Factory Mutual
c	Conductivity, Copper Sweat	Dbl.	Double		
C	Hundred; Centigrade	DC	Direct Current	Fmg.	Framing
C/C	Center to Center, Cedar on Cedar	DDC	Direct Digital Control	Fndtn.	Foundation

574

Fori.	Foreman, Inside	I.W.	Indirect Waste	M.C.F.	Thousand Cubic Feet
Foro.	Foreman, Outside	J	Joule	M.C.F.M.	Thousand Cubic Feet per Minute
Fount.	Fountain	J.I.C.	Joint Industrial Council	M.C.M.	Thousand Circular Mils
FPM	Feet per Minute	K	Thousand; Thousand Pounds;	M.C.P.	Motor Circuit Protector
FPT	Female Pipe Thread		Heavy Wall Copper Tubing, Kelvin	MD	Medium Duty
Fr.	Frame	K.A.H.	Thousand Amp. Hours	M.D.O.	Medium Density Overlaid
F.R.	Fire Rating	KCMIL	Thousand Circular Mils	Med.	Medium
FRK	Foil Reinforced Kraft	KD	Knock Down	MF	Thousand Feet
FRP	Fiberglass Reinforced Plastic	K.D.A.T.	Kiln Dried After Treatment	M.F.B.M.	Thousand Feet Board Measure
FS	Forged Steel	kg	Kilogram	Mfg.	Manufacturing
FSC	Cast Body; Cast Switch Box	kG	Kilogauss	Mfrs.	Manufacturers
Ft.	Foot; Feet	kgf	Kilogram Force	mg	Milligram
Ftng.	Fitting	kHz	Kilohertz	MGD	Million Gallons per Day
Ftg.	Footing	Kip.	1000 Pounds	MGPH	Thousand Gallons per Hour
Ft. Lb.	Foot Pound	KJ	Kiljoule	MH, M.H.	Manhole; Metal Halide; Man-Hour
Furn.	Furniture	K.L.	Effective Length Factor	MHz	Megahertz
FVNR	Full Voltage Non-Reversing	K.L.F.	Kips per Linear Foot	Mi.	Mile
FXM	Female by Male	Km	Kilometer	MI	Malleable Iron; Mineral Insulated
Fy.	Minimum Yield Stress of Steel	K.S.F.	Kips per Square Foot	mm	Millimeter
g	Gram	K.S.I.	Kips per Square Inch	Mill.	Millwright
G	Gauss	kV	Kilovolt	Min., min.	Minimum, minute
Ga.	Gauge	kVA	Kilovolt Ampere	Misc.	Miscellaneous
Gal.	Gallon	K.V.A.R.	Kilovar (Reactance)	ml	Milliliter, Mainline
Gal./Min.	Gallon per Minute	KW	Kilowatt	M.L.F.	Thousand Linear Feet
Galv.	Galvanized	KWh	Kilowatt-hour	Mo.	Month
Gen.	General	L	Labor Only; Length; Long;	Mobil.	Mobilization
G.F.I.	Ground Fault Interrupter		Medium Wall Copper Tubing	Mog.	Mogul Base
Glaz.	Glazier	Lab.	Labor	MPH	Miles per Hour
GPD	Gallons per Day	lat	Latitude	MPT	Male Pipe Thread
GPH	Gallons per Hour	Lath.	Lather	MRT	Mile Round Trip
GPM	Gallons per Minute	Lav.	Lavatory	ms	Millisecond
GR	Grade	lb.; #	Pound	M.S.F.	Thousand Square Feet
Gran.	Granular	L.B.	Load Bearing; L Conduit Body	Mstz.	Mosaic & Terrazzo Worker
Grnd.	Ground	L. & E.	Labor & Equipment	M.S.Y.	Thousand Square Yards
H	High; High Strength Bar Joist;	lb./hr.	Pounds per Hour	Mtd.	Mounted
	Henry	lb./L.F.	Pounds per Linear Foot	Mthe.	Mosaic & Terrazzo Helper
H.C.	High Capacity	lbf/sq.in.	Pound-force per Square Inch	Mtng.	Mounting
H.D.	Heavy Duty; High Density	L.C.L.	Less than Carload Lot	Mult.	Multi; Multiply
H.D.O.	High Density Overlaid	L.C.Y.	Loose Cubic Yard	M.V.A.	Million Volt Amperes
Hdr.	Header	Ld.	Load	M.V.A.R.	Million Volt Amperes Reactance
Hdwe.	Hardware	LE	Lead Equivalent	MV	Megavolt
Help.	Helper Average	LED	Light Emitting Diode	MW	Megawatt
HEPA	High Efficiency Particulate Air	L.F.	Linear Foot	MXM	Male by Male
	Filter	Lg.	Long; Length; Large	MYD	Thousand Yards
Hg	Mercury	L & H	Light and Heat	N	Natural; North
HIC	High Interrupting Capacity	LH	Long Span Bar Joist	nA	Nanoampere
HM	Hollow Metal	L.H.	Labor Hours	NA	Not Available; Not Applicable
H.O.	High Output	L.L.	Live Load	N.B.C.	National Building Code
Horiz.	Horizontal	L.L.D.	Lamp Lumen Depreciation	NC	Normally Closed
H.P.	Horsepower; High Pressure	lm	Lumen	N.E.M.A.	National Electrical Manufacturers
H.P.F.	High Power Factor	lm/sf	Lumen per Square Foot		Assoc.
Hr.	Hour	lm/W	Lumen per Watt	NEHB	Bolted Circuit Breaker to 600V.
Hrs./Day	Hours per Day	L.O.A.	Length Over All	N.L.B.	Non-Load-Bearing
HSC	High Short Circuit	log	Logarithm	NM	Non-Metallic Cable
Ht.	Height	L-O-L	Lateralolet	nm	Nanometer
Htg.	Heating	L.P.	Liquefied Petroleum; Low Pressure	No.	Number
Htrs.	Heaters	L.P.F.	Low Power Factor	NO	Normally Open
HVAC	Heating, Ventilation & Air-	LR	Long Radius	N.O.C.	Not Otherwise Classified
	Conditioning	L.S.	Lump Sum	Nose.	Nosing
Hvy.	Heavy	Lt.	Light	N.P.T.	National Pipe Thread
HW	Hot Water	Lt. Ga.	Light Gauge	NQOD	Combination Plug-on/Bolt on
Hyd.;Hydr.	Hydraulic	L.T.L.	Less than Truckload Lot		Circuit Breaker to 240V.
Hz.	Hertz (cycles)	Lt. Wt.	Lightweight	N.R.C.	Noise Reduction Coefficient
I.	Moment of Inertia	L.V.	Low Voltage	N.R.S.	Non Rising Stem
I.C.	Interrupting Capacity	M	Thousand; Material; Male;	ns	Nanosecond
ID	Inside Diameter		Light Wall Copper Tubing	nW	Nanowatt
I.D.	Inside Dimension; Identification	M²CA	Meters Squared Contact Area	OB	Opposing Blade
I.F.	Inside Frosted	m/hr; M.H.	Man-hour	OC	On Center
I.M.C.	Intermediate Metal Conduit	mA	Milliampere	OD	Outside Diameter
In.	Inch	Mach.	Machine	O.D.	Outside Dimension
Incan.	Incandescent	Mag. Str.	Magnetic Starter	ODS	Overhead Distribution System
Incl.	Included; Including	Maint.	Maintenance	O.G.	Ogee
Int.	Interior	Marb.	Marble Setter	O.H.	Overhead
Inst.	Installation	Mat; Mat'l.	Material	O&P	Overhead and Profit
Insul.	Insulation/Insulated	Max.	Maximum	Oper.	Operator
I.P.	Iron Pipe	MBF	Thousand Board Feet	Opng.	Opening
I.P.S.	Iron Pipe Size	MBH	Thousand BTU's per hr.	Orna.	Ornamental
I.P.T.	Iron Pipe Threaded	MC	Metal Clad Cable	OSB	Oriented Strand Board

O.S.&Y.	Outside Screw and Yoke	Rsr	Riser	Th.;Thk.	Thick
Ovhd.	Overhead	RT	Round Trip	Thn.	Thin
OWG	Oil, Water or Gas	S.	Suction; Single Entrance; South	Thrded	Threaded
Oz.	Ounce	SC	Screw Cover	Tilf.	Tile Layer, Floor
P.	Pole; Applied Load; Projection	SCFM	Standard Cubic Feet per Minute	Tilh.	Tile Layer, Helper
p.	Page	Scaf.	Scaffold	THHN	Nylon Jacketed Wire
Pape.	Paperhanger	Sch.; Sched.	Schedule	THW.	Insulated Strand Wire
P.A.P.R.	Powered Air Purifying Respirator	S.C.R.	Modular Brick	THWN;	Nylon Jacketed Wire
PAR	Parabolic Reflector	S.D.	Sound Deadening	T.L.	Truckload
Pc., Pcs.	Piece, Pieces	S.D.R.	Standard Dimension Ratio	T.M.	Track Mounted
P.C.	Portland Cement; Power Connector	S.E.	Surfaced Edge	Tot.	Total
P.C.F.	Pounds per Cubic Foot	Sel.	Select	T-O-L	Threadolet
P.C.M.	Phase Contrast Microscopy	S.E.R.; S.E.U.	Service Entrance Cable	T.S.	Trigger Start
P.E.	Professional Engineer;	S.F.	Square Foot	Tr.	Trade
	Porcelain Enamel;	S.F.C.A.	Square Foot Contact Area	Transf.	Transformer
	Polyethylene; Plain End	S.F. Flr.	Square Foot of Floor	Trhv.	Truck Driver, Heavy
Perf.	Perforated	S.F.G.	Square Foot of Ground	Trlr.	Trailer
Ph.	Phase	S.F. Hor.	Square Foot Horizontal	Trlt.	Truck Driver, Light
P.I.	Pressure Injected	S.F.R.	Square Feet of Radiation	TTY	Teletypewriter
Pile.	Pile Driver	S.F. Shlf.	Square Foot of Shelf	TV	Television
Pkg.	Package	S4S	Surface 4 Sides	T.W.	Thermoplastic Water Resistant
Pl.	Plate	Shee.	Sheet Metal Worker		Wire
Plah.	Plasterer Helper	Sin.	Sine	UCI	Uniform Construction Index
Plas.	Plasterer	Skwk.	Skilled Worker	UF	Underground Feeder
Pluh.	Plumbers Helper	SL	Saran Lined	UGND	Underground Feeder
Plum.	Plumber	S.L.	Slimline	U.H.F.	Ultra High Frequency
Ply.	Plywood	Sldr.	Solder	U.L.	Underwriters Laboratory
p.m.	Post Meridiem	SLH	Super Long Span Bar Joist	Unfin.	Unfinished
Pntd.	Painted	S.N.	Solid Neutral	URD	Underground Residential
Pord.	Painter, Ordinary	S-O-L	Socketolet		Distribution
pp	Pages	sp	Standpipe	US	United States
PP; PPL	Polypropylene	S.P.	Static Pressure; Single Pole; Self-	USP	United States Primed
P.P.M.	Parts per Million		Propelled	UTP	Unshielded Twisted Pair
Pr.	Pair	Spri.	Sprinkler Installer	V	Volt
P.E.S.B.	Pre-engineered Steel Building	spwg	Static Pressure Water Gauge	V.A.	Volt Amperes
Prefab.	Prefabricated	S.P.D.T.	Single Pole, Double Throw	V.C.T.	Vinyl Composition Tile
Prefin.	Prefinished	SPF	Spruce Pine Fir	VAV	Variable Air Volume
Prop.	Propelled	S.P.S.T.	Single Pole, Single Throw	VC	Veneer Core
PSF; psf	Pounds per Square Foot	SPT	Standard Pipe Thread	Vent.	Ventilation
PSI; psi	Pounds per Square Inch	Sq.	Square; 100 Square Feet	Vert.	Vertical
PSIG	Pounds per Square Inch Gauge	Sq. Hd.	Square Head	V.F.	Vinyl Faced
PSP	Plastic Sewer Pipe	Sq. In.	Square Inch	V.G.	Vertical Grain
Pspr.	Painter, Spray	S.S.	Single Strength; Stainless Steel	V.H.F.	Very High Frequency
Psst.	Painter, Structural Steel	S.S.B.	Single Strength B Grade	VHO	Very High Output
P.T.	Potential Transformer	sst	Stainless Steel	Vib.	Vibrating
P. & T.	Pressure & Temperature	Sswk.	Structural Steel Worker	V.L.F.	Vertical Linear Foot
Ptd.	Painted	Sswl.	Structural Steel Welder	Vol.	Volume
Ptns.	Partitions	St.;Stl.	Steel	VRP	Vinyl Reinforced Polyester
Pu	Ultimate Load	S.T.C.	Sound Transmission Coefficient	W	Wire; Watt; Wide; West
PVC	Polyvinyl Chloride	Std.	Standard	w/	With
Pvmt.	Pavement	STK	Select Tight Knot	W.C.	Water Column; Water Closet
Pwr.	Power	STP	Standard Temperature & Pressure	W.F.	Wide Flange
Q	Quantity Heat Flow	Stpi.	Steamfitter, Pipefitter	W.G.	Water Gauge
Quan.;Qty.	Quantity	Str.	Strength; Starter; Straight	Wldg.	Welding
Q.C.	Quick Coupling	Strd.	Stranded	W. Mile	Wire Mile
r	Radius of Gyration	Struct.	Structural	W-O-L	Weldolet
R	Resistance	Sty.	Story	W.R.	Water Resistant
R.C.P.	Reinforced Concrete Pipe	Subj.	Subject	Wrck.	Wrecker
Rect.	Rectangle	Subs.	Subcontractors	W.S.P.	Water, Steam, Petroleum
Reg.	Regular	Surf.	Surface	WT., Wt.	Weight
Reinf.	Reinforced	Sw.	Switch	WWF	Welded Wire Fabric
Req'd.	Required	Swbd.	Switchboard	XFER	Transfer
Res.	Resistant	S.Y.	Square Yard	XFMR	Transformer
Resi.	Residential	Syn.	Synthetic	XHD	Extra Heavy Duty
Rgh.	Rough	S.Y.P.	Southern Yellow Pine	XHHW; XLPE	Cross-Linked Polyethylene Wire
RGS	Rigid Galvanized Steel	Sys.	System		Insulation
R.H.W.	Rubber, Heat & Water Resistant;	t.	Thickness	XLP	Cross-linked Polyethylene
	Residential Hot Water	T	Temperature; Ton	Y	Wye
rms	Root Mean Square	Tan	Tangent	yd	Yard
Rnd.	Round	T.C.	Terra Cotta	yr	Year
Rodm.	Rodman	T & C	Threaded and Coupled	Δ	Delta
Rofc.	Roofer, Composition	T.D.	Temperature Difference	%	Percent
Rofp.	Roofer, Precast	Tdd	Telecommunications Device for	~	Approximately
Rohe.	Roofer Helpers (Composition)		the Deaf	Ø	Phase
Rots.	Roofer, Tile & Slate	T.E.M.	Transmission Electron Microscopy	@	At
R.O.W.	Right of Way	TFE	Tetrafluoroethylene (Teflon)	#	Pound; Number
RPM	Revolutions per Minute	T. & G.	Tongue & Groove;	<	Less Than
R.S.	Rapid Start		Tar & Gravel	>	Greater Than

Index

INDEX

Notes

581

Notes

Notes

588

Reed Construction Data, Inc.

Reed Construction Data, Inc., a leading worldwide provider of total construction information solutions, is comprised of three main product groups designed specifically to help construction professionals advance their businesses with timely, accurate and actionable project, product and cost data. Reed Construction Data is a division of Reed Business Information, a member of the Reed Elsevier plc group of companies.

The Project, Product and Cost & Estimating divisions offer a variety of innovative products and services designed for the full spectrum of design, construction and manufacturing professionals. Through its International companies, Reed Construction Data's reputation for quality construction market data is growing worldwide.

Cost Information
RSMeans, the undisputed market leader and authority on construction costs, publishes current cost and estimating information in annual cost books and on the *Means CostWorks* CD-ROM. RSMeans furnishes the construction industry with a rich library of corresponding reference books and a series of professional seminars that are designed to sharpen professional skills and maximize the effective use of cost estimating and management tools. RSMeans also provides construction cost consulting for owners, manufacturers, designers, and contractors.

Project Data
Reed Construction Data provides complete, accurate and relevant project information through all stages of construction. Customers are supplied industry data through leads, project reports, contact lists, plans and specifications surveys, market penetration analyses and sales evaluation reports. Any of these products can pinpoint a county, look at a state, or cover the country. Data is delivered via paper, e-mail, CD-ROM or the Internet.

Building Product Information
The First Source suite of products is the only integrated building product information system offered to the commercial construction industry for comparing and specifying building products. These print and online resources include *First Source*, CSI's SPEC-DATA™, CSI's MANU-SPEC™, First Source CAD, and Manufacturer Catalogs. Written by industry professionals and organized using CSI's MasterFormat™, construction professionals use this information to make better design decisions. FirstSourceONL.com combines Reed Construction Data's project, product and cost data with news and information from Reed Business Information's *Building Design & Construction* and *Consulting-Specifying Engineer magazines.* This industry-focused site offers easy and unlimited access to vital information for all construction professionals.

International
Reed Construction Data Canada serves the Canadian construction market with reliable and comprehensive project and product information services that cover all facets of construction. Core services include: *BuildSource, BuildSpec, BuildSelect,* product selection and specification tools available in print and on the Internet; Building Reports, a national construction project lead service; CanaData, statistical and forecasting information; *Daily Commercial News,* a construction newspaper reporting on news and projects in Ontario; and *Journal of Commerce,* reporting news in British Columbia and Alberta.

BIMSA/Mexico provides construction project news, product information, cost data, seminars and consulting services to construction professionals in Mexico. Its subsidiary, PRISMA, provides job costing software.

Byggfakta Scandinavia AB, founded in 1936, is the parent company for the leaders of customized construction market data for Denmark, Estonia, Finland, Norway and Sweden. Each company fully covers the local construction market and provides information across several platforms including via subscription, on an ad-hoc basis, electronically and on paper.

Cordell Building Information Services, with its complete range of project and cost and estimating services, is Australia's specialist in the construction information industry. Cordell provides in-depth and historical information on all aspects of construction projects and estimation, including several customized reports, construction and sales leads and detailed cost information.

For more information, please visit our Web site at www.reedconstructiondata.com.

Reed Construction Data, Inc., Corporate Office
30 Technology Parkway South
Norcross, GA 30092-2912
(800) 322-6996
(800) 895-8661 (fax)
info@reedbusiness.com
www.reedconstructiondata.com

 Reed Construction Data®

Means Project Cost Report

By filling out this report, your project data will contribute to the database that supports the Means Project Cost Square Foot Data. When you fill out this form, Means will provide a 30% discount off one of the Means products advertised in the following pages. Please complete the form including all items where you have cost data, and all the items marked (✔).

$30.00 Discount per product for each report you submit.

Project Description (No remodeling projects, please.)

✔ Building Use (Office, School...) _____

✔ Address (City, State) _____

_____

✔ Frame (Wood, Steel...) _____

✔ Exterior Wall (Brick, Tilt-up...) _____

✔ Basement: (check one)   ☐ Full ☐ Partial ☐ None

✔ Number Stories _____

✔ Floor-to-Floor Height _____

% Air Conditioned _____ Tons _____

Comments _____

Total Project Cost $ _____

Owner _____

Architect _____

General Contractor _____

✔ Bid Date _____

Typical Bay Size _____

✔ Labor Force: _____ % Union _____ % Non-Union

✔ Project Description (Circle one number in each line)

1. Economy 2. Average   3. Custom 4. Luxury
1. Square   2. Rectangular 3. Irregular 4. Very Irregular

A	✔ General Conditions	$
B	✔ Site Work	$
C	✔ Concrete	$
D	✔ Masonry	$
E	✔ Metals	$
F	✔ Wood & Plastics	$
G	✔ Thermal & Moisture Protection	
GR	Roofing & Flashing	$
H	✔ Doors and Windows	$
J	✔ Finishes	$
JP	Painting & Wall Covering	$

K	✔ Specialties	$
L	✔ Equipment	$
M	✔ Furnishings	$
N	✔ Special Construction	$
P	✔ Conveying Systems	$
Q	✔ Mechanical	$
QP	Plumbing	$
QB	HVAC	$
R	✔ Electrical	$
S	✔ Mech./Elec. Combined	$

Please specify the Means product you wish to receive. Complete the address information.

Product Name _____

Product Number _____

Your Name _____

Title _____

Company _____
☐ Company
☐ Home   Street Address _____

City, State, Zip _____

Method of Payment:

Credit Card # _____

Expiration Date _____

Check _____

Purchase Order _____

Reed Construction Data/RSMeans
Square Foot Costs Department
P.O. Box 800
Kingston, MA 02364-9988

Return by mail or Fax 888-492-6770.

For more information
visit Means Web site
at www.rsmeans.com

Reed Construction Data/RSMeans . . . a tradition of excellence in Construction Cost Information and Services since 1942.

Table of Contents

Book Selection Guide

The following table provides definitive information on the content of each cost data publication. The number of lines of data provided in each unit price or assemblies division, as well as the number of reference tables and crews is listed for each book. The presence of other elements such as an historical cost index, city cost indexes, square foot models or cross-referenced index is also indicated. You can use the table to help select the Means' book that has the quantity and type of information you most need in your work.

Unit Cost Divisions	Building Construction Costs	Mechanical	Electrical	Repair & Remodel.	Square Foot	Site Work Landsc.	Assemblies	Interior	Concrete Masonry	Open Shop	Heavy Construc.	Light Commercial	Facil. Construc.	Plumbing	Western Construction Costs	Residential
1	554	273	358	465		496		297	465	553	502	210	979	325	553	161
2	2560	1345	460	1457		8234		543	1383	2466	5287	695	4835	1593	2543	768
3	1393	112	99	721		1250		191	1774	1388	1395	229	1315	81	1388	262
4	854	18	0	666		660		576	1075	830	597	432	1076	0	840	352
5	1876	239	193	965		817		967	738	1844	1077	835	1856	316	1816	759
6	1487	82	78	1465		450		1385	317	1477	607	1603	1574	47	1824	1734
7	1252	159	71	1240		472		485	401	1251	355	963	1306	169	1252	760
8	1893	28	0	1965		313		1682	642	1870	51	1329	2095	0	1852	1263
9	1595	47	0	1425		232		1688	360	1547	157	1337	1811	47	1586	1217
10	871	47	25	508		193		710	170	873	0	404	917	233	871	217
11	1018	322	169	502		137		813	44	925	107	218	1173	291	925	104
12	313	0	0	46		213		1422	27	304	0	65	1442	0	304	63
13	1317	1016	385	659		451		884	75	1292	340	484	1931	929	1183	194
14	345	36	0	258		36		292	0	344	30	12	362	35	343	6
15	1976	13019	636	1752		1542		1192	58	1985	1729	1225	10787	9669	2005	829
16	1279	470	10114	1002		739		1108	55	1297	760	1083	9885	415	1228	554
Totals	20583	17213	12588	15096		16235		14235	7584	20246	12994	11124	43344	14150	20513	9243

Assembly Divisions	Building Construction Costs	Mechanical	Electrical	Repair & Remodel.	Square Foot	Site Work Landsc.	Assemblies	Interior	Concrete Masonry	Open Shop	Heavy Construc.	Light Commercial	Facil. Construc.	Plumbing	Western Construction Costs	Asm Div	Residential
A	612	19	0	192	150	540	612	0	550		542	149	24	0		1	374
B	5584	0	0	809	2479	0	5584	333	1914		0	2023	144	0		2	217
C	1208	0	0	635	853	0	1208	1555	132		0	757	238	0		3	588
D	2392	1014	780	693	1776	0	2392	752	0		0	1262	1010	890		4	867
E	294	0	0	85	257	0	294	5	0		0	257	5	0		5	393
F	124	0	0	0	123	0	124	0	0		0	123	3	0		6	357
G	584	465	172	330	111	1844	584	0	471		423	110	113	559		7	299
																8	760
																9	80
																10	0
																11	0
																12	0
Totals	10798	1498	952	2744	5749	2384	10798	2645	3067	0	965	4681	1537	1449			3935

Reference Section	Building Construction Costs	Mechanical	Electrical	Repair & Remodel.	Square Foot	Site Work Landsc.	Assemblies	Interior	Concrete Masonry	Open Shop	Heavy Construc.	Light Commercial	Facil. Construc.	Plumbing	Western Construction Costs	Residential
Tables	237	237	237	237	4	237	219	237	237	237	237	237	237	237	237	35
Models					105							46				32
Crews	452	452	452	434		452		452	452	431	452	431	434	452	452	431
City Cost Indexes	yes	yes	yes	yes	yes	yes	yes	yes	yes	yes	yes	yes	yes	yes	yes	yes
Historical Cost Indexes	yes	yes	yes	yes	yes	yes	yes	yes	yes	yes	yes	yes	yes	yes	yes	no

Annual Cost Guides

For more informati◌
visit Means Web sit◌
at www.rsmeans.co◌

Means Building Construction Cost Data 2006

Available in Both Softbound and Looseleaf Editions

The "Bible" of the industry comes in the standard softcover edition or the looseleaf edition.

Many customers enjoy the convenience and flexibility of the looseleaf binder, which increases the usefulness of Means *Building Construction Cost Data 2006* by making it easy to add and remove pages. You can insert your own cost information pages, so everything is in one place. Copying pages for faxing is easier also. Whichever edition you prefer, softbound or the convenient looseleaf edition, you'll be eligible to receive *The Change Notice* FREE. Current subscribers receive *The Change Notice* via e-mail.

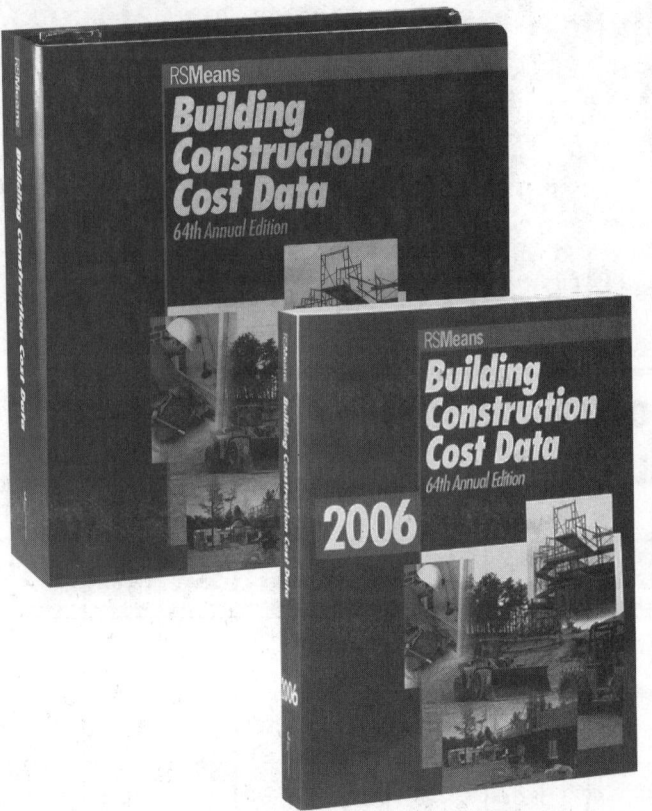

$157.95 per copy, Looseleaf
Catalog No. 61016

Means Building Construction Cost Data 2006

Offers you unchallenged unit price reliability in an easy-to-use arrangement. Whether used for complete, finished estimates or for periodic checks, it supplies more cost facts better and faster than any comparable source. Over 23,000 unit prices for 2006. The City Cost Indexes cover over 930 areas, for indexing to any project location in North America. Order and get *The Change Notice* FREE. You'll have year-long access to the Means Estimating **HOTLINE** FREE with your subscription. Expert assistance when using Means data is just a phone call away.

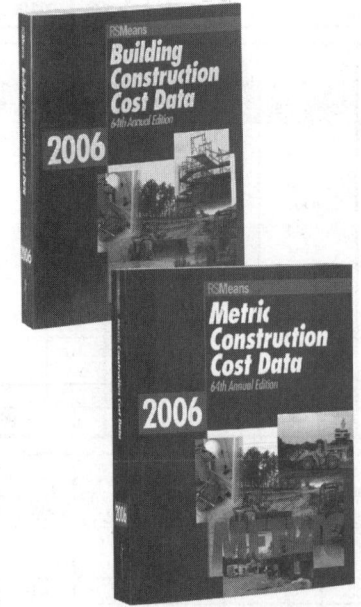

$126.95 per copy
Over 700 pages, illustrated, available Oct. 2005
Catalog No. 60016

Means Metric Construction Cost Data 2006

An 880+ page compendium of all the data from both the 2006 *Building Construction Cost Data* AND the *Heavy Construction Cost Data*, in **metric** format! Access all of this vital information from one complete source. It contains more than 600 pages of unit costs and 40 pages of assemblies costs. The Reference Section contains over 200 pages of tables, charts and other estimating aids. A great way to stay in step with today's construction trends and rapidly changing costs.

$149.95 per copy
Over 800 pages, illus., available Dec. 2005
Catalog No. 63016

For more information
visit Means Web site
www.rsmeans.com

Annual Cost Guides

Means Mechanical Cost Data 2006

HVAC   Controls

Total unit and systems price guidance for mechanical construction... materials, parts, fittings, and complete labor cost information. Includes prices for piping, heating, air conditioning, ventilation, and all related construction.

Plus new 2006 unit costs for:

- Over 2500 installed HVAC/controls assemblies
- "On Site" Location Factors for over 930 cities and towns in the U.S. and Canada
- Crews, labor, and equipment

$126.95 per copy
Over 600 pages, illustrated, available Oct. 2005
Catalog No. 60026

Means Electrical Cost Data 2006

Pricing information for every part of electrical cost planning. More than 15,000 unit and systems costs with design tables; clear specifications and drawings; engineering guides and illustrated estimating procedures; complete labor-hour and materials costs for better scheduling and procurement; and the latest electrical products and construction methods.

- A variety of special electrical systems including cathodic protection
- Costs for maintenance, demolition, HVAC/mechanical, specialties, equipment, and more

$126.95 per copy
Over 450 pages, illustrated, available Oct. 2005
Catalog No. 60036

Means Square Foot Costs 2006

Is Accurate and Easy To Use!

- **Updated 2006 price information**, based on nationwide figures from suppliers, estimators, labor experts, and contractors
- "How-to-Use" sections, with **clear examples** of commercial, residential, industrial, and institutional structures
- Realistic graphics, offering true-to-life illustrations of building projects
- Extensive information on using square foot cost data, including sample estimates and alternate pricing methods

$137.95 per copy
Over 450 pages, illustrated, available Nov. 2005
Catalog No. 60056

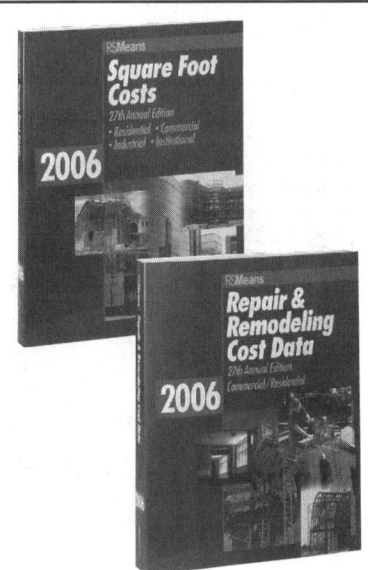

Means Plumbing Cost Data 2006

Comprehensive unit prices and assemblies for plumbing, irrigation systems, commercial and residential fire protection, point-of-use water heaters, and the latest approved materials. This publication and its companion, Means *Mechanical Cost Data*, provide full-range cost estimating coverage for all the mechanical trades.

$126.95 per copy
Over 550 pages, illustrated, available Oct. 2005
Catalog No. 60216

Means Electrical Change Order Cost Data 2006

Means *Electrical Change Order Cost Data* 2006 provides you with electrical unit prices exclusively for pricing change orders based on the recent, direct experience of contractors and suppliers. Analyze and check your own change order estimates against the experience others have had doing the same work. It also covers productivity analysis and change order cost justifications. With useful information for calculating the effects of change orders and dealing with their administration.

$126.95 per copy
Over 450 pages, available Oct. 2005
Catalog No. 60236

Means Repair & Remodeling Cost Data 2006

Commercial/Residential

Use this valuable tool to estimate commercial and residential renovation and remodeling.

Includes: New costs for hundreds of unique methods, materials, and conditions that only come up in repair and remodeling. PLUS:

- Unit costs for over 16,000 construction components
- Installed costs for over 90 assemblies
- Costs for 300+ construction crews
- Over 930 "On Site" localization factors for the U.S. and Canada.

$107.95 per copy
Over 650 pages, illustrated, available Nov. 2005
Catalog No. 60046

Annual Cost Guides

For more informatio
visit Means Web sit
at www.rsmeans.co

Means Facilities Construction Cost Data 2006

For the maintenance and construction of commercial, industrial, municipal, and institutional properties. Costs are shown for new and remodeling construction and are broken down into materials, labor, equipment, overhead, and profit. Special emphasis is given to sections on mechanical, electrical, furnishings, site work, building maintenance, finish work, and demolition.

More than 45,000 unit costs, plus assemblies and reference sections are included.

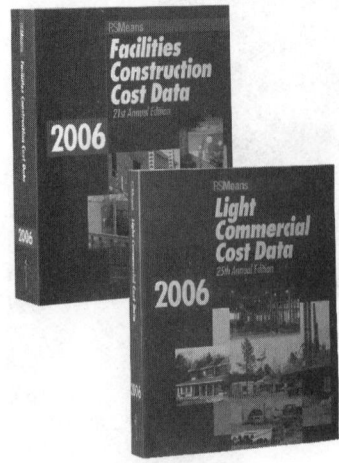

$299.95 per copy
Over 1200 pages, illustrated, available Nov. 2005
Catalog No. 60206

Means Light Commercial Cost Data 2006

Specifically addresses the light commercial market which is an increasingly specialized niche in the industry. Aids you, the owner/designer/contractor, in preparing all types of estimates, from budgets to detailed bids. Includes new advances in methods ar materials.

Assemblies section allows you to evaluate alternatives in the early stages of design/planning.

Over 13,000 unit costs for 2006 ensure you have th prices you need... when you need them.

$107.95 per copy
Over 650 pages, illustrated, available Nov. 20(
Catalog No. 60186

Means Residential Cost Data 2006

Contains square foot costs for 30 basic home models with the look of today, plus hundreds of custom additions and modifications you can quote right off the page. With costs for the 100 residential systems you're most likely to use in the year ahead. Complete with blank estimating forms, sample estimates and step-by-step instructions.

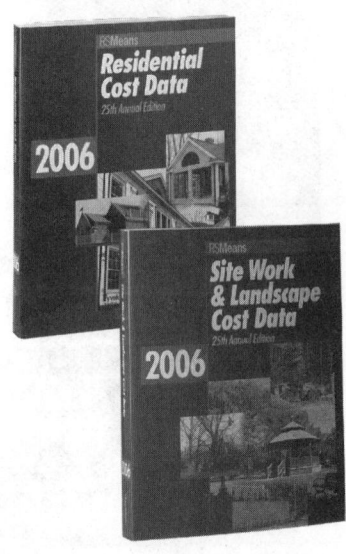

$107.95 per copy
Over 600 pages, illustrated, available Oct. 2005
Catalog No. 60176

Means Site Work & Landscape Cost Data 2006

Means *Site Work & Landscape Cost Data* 2006 is organized to assist you in all your estimating needs Hundreds of fact-filled pages help you make accura cost estimates efficiently.

Updated for 2006!
- Demolition features—including ceilings, doors, electrical, flooring, HVAC, millwork, plumbing, roofing, walls, and windows
- State-of-the-art segmental retaining walls
- Flywheel trenching costs and details
- Updated wells section
- Landscape materials, flowers, shrubs, and trees

$126.95 per copy
Over 600 pages, illustrated, available Nov. 20(
Catalog No. 60286

Means Assemblies Cost Data 2006

Means *Assemblies Cost Data* 2006 takes the guesswork out of preliminary or conceptual estimates. Now you don't have to try to calculate the assembled cost by working up individual components costs. We've done all the work for you.

Presents detailed illustrations, descriptions, specifications and costs for every conceivable building assembly—240 types in all—arranged in the easy-to-use UNIFORMAT II system. Each illustrated "assembled" cost includes a complete grouping of materials and associated installation costs, including the installing contractor's overhead and profit.

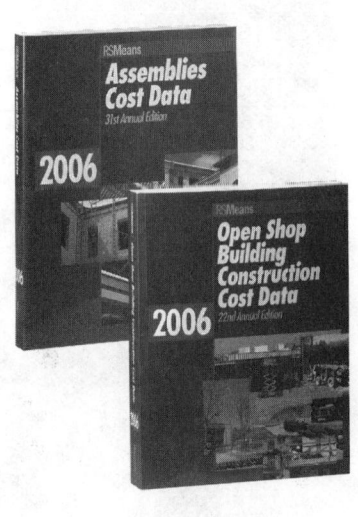

$206.95 per copy
Over 600 pages, illustrated, available Oct. 2005
Catalog No. 60066

Means Open Shop Building Construction Cost Data 2006

The latest costs for accurate budgeting and estimating of new commercial and residential construction... renovation work... change orders... cost engineering.

Means *Open Shop BCCD* will assist you to:
- Develop benchmark prices for change orders
- Plug gaps in preliminary estimates, and budgets
- Estimate complex projects
- Substantiate invoices on contracts
- Price ADA-related renovations

$126.95 per copy
Over 700 pages, illustrated, available Dec. 200
Catalog No. 60156

r more information
sit Means Web site
www.rsmeans.com

Annual Cost Guides

leans Building Construction ost Data 2006

estern Edition

is regional edition provides more precise cost ormation for western North America. Labor rates based on union rates from 13 western states and stern Canada. Included are western practices and terials not found in our national edition: tilt-up ncrete walls, glu-lam structural systems, specialized ber construction, seismic restraints, and landscape d irrigation systems.

26.95 per copy
ver 650 pages, illustrated, available Dec. 2005
atalog No. 60226

Means Heavy Construction Cost Data 2006

A comprehensive guide to heavy construction costs. Includes costs for highly specialized projects such as tunnels, dams, highways, airports, and waterways. Information on different labor rates, equipment, and material costs is included. Features unit price costs, systems costs, and numerous reference tables for costs and design. Valuable not only to contractors and civil engineers, but also to government agencies and city/town engineers.

$126.95 per copy, Over 450 pages,
illustrated, available Nov. 2005
Catalog No. 60166

leans Construction Cost dexes 2006

o knows what 2006 holds? What materials d labor costs will change unexpectedly? By how ch?
Breakdowns for 316 major cities
National averages for 30 key cities
Expanded five major city indexes
Historical construction cost indexes

72.00 per year
8.00 individual quarters
atalog No. 60146 A,B,C,D

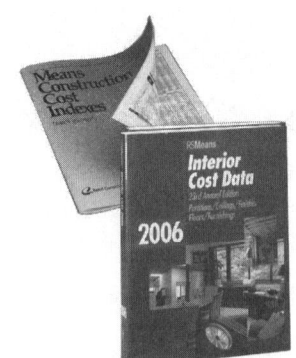

Means Interior Cost Data 2006

Provides you with prices and guidance needed to make accurate interior work estimates. Contains costs on materials, equipment, hardware, custom installations, furnishings, labor costs... every cost factor for new and remodel commercial and industrial interior construction, including updated information on office furnishings, plus more than 50 reference tables. For contractors, facility managers, and owners.

$126.95 per copy, Over 600 pages,
illustrated, available Oct. 2005
Catalog No. 60096

leans Concrete & asonry Cost Data 2006

ovides you with cost facts for virtually all ncrete/masonry estimating needs, from mplicated formwork to various sizes and face ishes of brick and block, all in great detail. The mprehensive unit cost section contains more an 8,500 selected entries. Also contains an semblies cost section, and a detailed reference ction that supplements the cost data.

14.95 per copy
ver 450 pages, illustrated, available Dec. 2005
atalog No. 60116

Means Labor Rates for the Construction Industry 2006

Complete information for estimating labor costs, making comparisons, and negotiating wage rates by trade for over 300 U.S. and Canadian cities. With 46 construction trades listed by local union number in each city, and historical wage rates included for comparison. No similar book is available through the trade.

Each city chart lists the county and is alphabetically arranged with handy visual flip tabs for quick reference.

$274.95 per copy
Over 300 pages, available Dec. 2005
Catalog No. 60126

leans Facilities Maintenance & epair Cost Data 2006

eans *Facilities Maintenance & Repair Cost Data* gives you a complete system to anage and plan your facility repair and maintenance costs and budget efficiently. uidelines for auditing a facility and developing an annual maintenance plan. Budgeting included, along with reference tables on cost and management and information on equency and productivity of maintenance operations.

he only nationally recognized source of maintenance and repair costs. Developed in operation with the Army Corps of Engineers.

74.95 per copy
ver 600 pages, illustrated, available Dec. 2005
atalog No. 60306

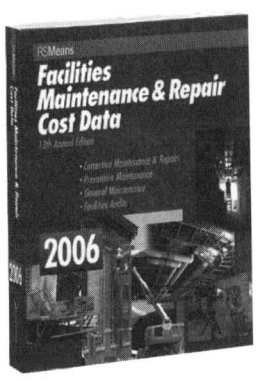

Reference Books

For more informati‹
visit Means Web si‹
at www.rsmeans.co

Builder's Essentials: Estimating Building Costs for the Residential & Light Commercial Contractor

By Wayne J. DelPico

Step-by-step estimating methods for residential and light commercial contractors. Includes a detailed look at every construction specialty—explaining all the components, takeoff units, and labor needed for well-organized, complete estimates. Covers:

- Correctly interpreting plans and specifications
- Developing accurate and complete labor and material costs
- Understanding direct and indirect overhead costs... and accounting for time-sensitive costs
- Using historical cost data to generate new project budgets
- Allocating the right amount for profit and contingencies

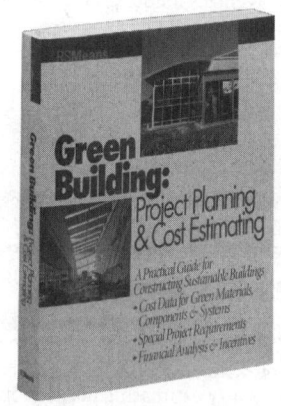

$29.95 per copy
Over 400 pages, illustrated, Softcover
Catalog No. 67343

Green Building: Project Planning & Cost Estimating

By RSMeans and Contributing Authors

Written by a team of leading experts in sustainable design, this book is a complete guide to planning and estimating green building projects, a growing trend in building design and construction. It explains:

- All the different criteria for "green-ness"
- What criteria your building needs to meet to get a LEED, ENERGYSTAR®, or other recognized rating for green buildings
- How the project team works differently on a green versus a traditional building project
- How to select and specify green products
- How to evaluate the cost and value of green products versus conventional ones—not only for their initial installation cost, but their cost over time in maintenance and operation

Features an extensive Green Building Cost Data section

$89.95 per copy
350 pages, illustrated, Hardcover
Catalog No. 67338

Means ADA Compliance Pricing Guide

New Second Edition

By Adaptive Environments and RSMeans

Completely updated and revised to the new 2004 *Americans with Disabilities Act Accessibility Guidelines*, this book features more than 70 of the most commonly needed modifications for ADA compliance—their design requirements, suggestions, and final cost. Projects range from installing ramps and walkways, widening doorways and entryways, and installing and refitting elevators, to relocating light switches and signage, and remodeling bathrooms and kitchens. Also provided are:

- Detailed cost estimates for budgeting modification projects, including estimates for each of 260 alternates.
- An assembly estimate for every project, with detailed cost breakdown including materials, labor-hours, and contractor's overhead.
- 3,000 additional ADA compliance-related unit cost line items.
- Costs that are easily adjusted to over 900 cities and towns.

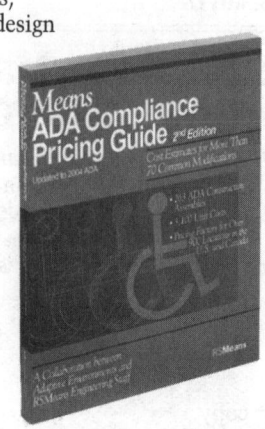

$79.95 per copy
Over 350 pages, illustrated, Softcover
Catalog No. 67310A

For more information
visit Means Web site
www.rsmeans.com

Reference Books

Value Engineering: Practical Applications . . . For Design, Construction, Maintenance & Operations

By Alphonse Dell'Isola, PE

A tool for immediate application—for engineers, architects, facility managers, owners, and contractors. Includes: making the case for VE—the management briefing, integrating VE into planning, budgeting, and design, conducting life cycle costing, using VE methodology in design review and consultant selection, case studies, VE workbook, and a life cycle costing program on disk.

$79.95 per copy
Over 450 pages, illustrated, Softcover
Catalog No. 67319

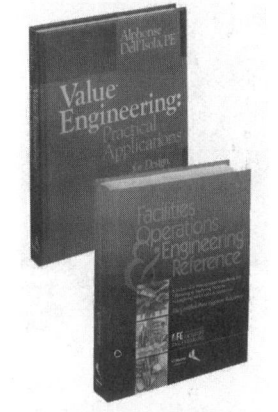

Facilities Operations & Engineering Reference

By the Association for Facilities Engineering and RSMeans

An all-in-one technical referance for planning and managing facility projects and solving day-to-day operations problems. Selected as the official Certified Plant Engineer reference, this handbook covers financial analysis, maintenance, HVAC and energy efficiency, and more.

$109.95 per copy
Over 700 pages, illustrated, Hardcover
Catalog No. 67318

The Building Professional's Guide to Contract Documents

3rd Edition

By Waller S. Poage, AIA, CSI, CVS

A comprehensive reference for owners, design professionals, contractors, and students.
Structure your documents for maximum efficiency.
Effectively communicate construction requirements.
Understand the roles and responsibilities of construction professionals.
Improve methods of project delivery.

$62.48 per copy, 400 pages
Diagrams and construction forms, Hardcover
Catalog No. 67261A

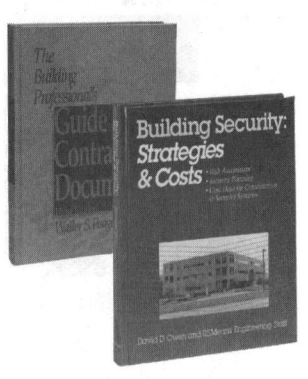

Building Security: Strategies & Costs

By David Owen

This comprehensive resource will help you evaluate your facility's security needs, and design and budget for the materials and devices needed to fulfill them.

Includes over 130 pages of Means cost data for installation of security systems and materials, plus a review of more than 50 security devices and construction solutions.

$44.98 per copy
350 pages, illustrated, Hardcover
Catalog No. 67339

Cost Planning & Estimating for Facilities Maintenance

In this unique book, a team of facilities management authorities shares their expertise at:
Evaluating and budgeting maintenance operations
Maintaining and repairing key building components
Applying Means *Facilities Maintenance & Repair Cost Data* to your estimating

Covers special maintenance requirements of the ten major building types.

$89.95 per copy
Over 475 pages, Hardcover
Catalog No. 67314

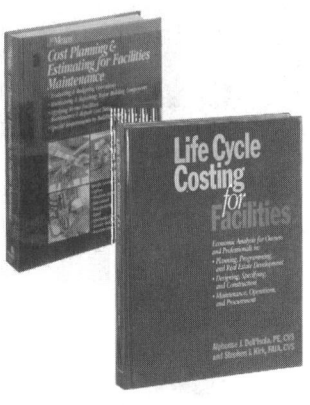

Life Cycle Costing for Facilities

By Alphonse Dell'Isola and Dr. Steven Kirk

Guidance for achieving higher quality design and construction projects at lower costs! Cost-cutting efforts often sacrifice quality to yield the cheapest product. Life cycle costing enables building designers and owners to achieve both. The authors of this book show how LCC can work for a variety of projects – from roads to HVAC upgrades to different types of buildings.

$99.95 per copy
450 pages, Hardcover
Catalog No. 67341

Building & Renovating Schools

This all-inclusive guide covers every step of the school construction process—from initial planning, needs assessment, and design, right through moving into the new facility. A must-have resource for anyone concerned with new school construction or renovation, including architects and engineers, contractors and project managers, facility managers, school administrators and school board members, building committees, community leaders, and anyone else who wants to ensure that the project meets schools' needs in a cost-effective, timely manner. With square foot cost models for elementary, middle, and high school facilities and real-life case studies of recently completed school projects.

The contributors to this book—architects, construction project managers, contractors, and estimators who specialize in school construction—provide start-to-finish, expert guidance on the process.

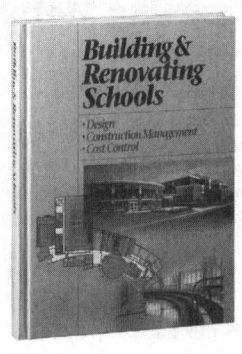

$99.95 per copy
Over 425 pages, Hardcover
Catalog No. 67342

Reference Books

For more informatic
visit Means Web sit
at www.rsmeans.co

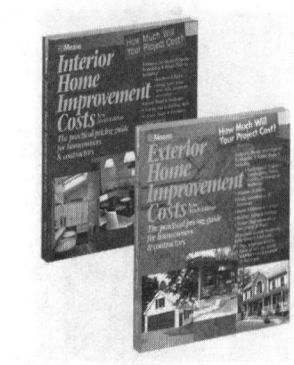

Interior Home Improvement Costs, New 9th Edition

Estimates for the most popular remodeling and repair projects—from small, do-it-yourself jobs to major renovations and new construction. Includes: Kitchens & Baths; New Living Space from your Attic, Basement, or Garage; New Floors, Paint, and Wallpaper; Tearing Out or Building New Walls; Closets, Stairs, and Fireplaces; New Energy-Saving Improvements, Home Theaters, and More!

$24.95 per copy
250 pages, illustrated, Softcover
Catalog No. 67308E

Exterior Home Improvement Costs New 9th Edition

Estimates for the most popular remodeling and repair projects—from small, do-it-yourself jobs, to major renovations and new construction. Includes: Curb Appeal Projects—Landscaping, Patios, Porches, Driveways, and Walkways; New Windows and Doors; Decks, Greenhouses, and Sunrooms; Room Additions and Garages; Roofing, Siding, and Painting; "Green" Improvements to Save Energ & Water.

$24.95 per copy
Over 275 pages, illustrated, Softcover
Catalog No. 67309E

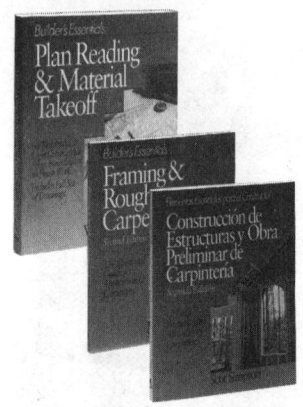

Builder's Essentials: Plan Reading & Material Takeoff

By Wayne J. DelPico

For Residential and Light Commercial Construction

A valuable tool for understanding plans and specs, and accurately calculating material quantities. Step-by-step instructions and takeoff procedures based on a full set of working drawings.

$35.95 per copy
Over 420 pages, Softcover
Catalog No. 67307

Builder's Essentials: Framing & Rough Carpentry, 2nd Edition

By Scot Simpson

Develop and improve your skills with easy-to-follow instructions and illustrations. Learn proven techniques for framing walls, floors, roofs, stairs, doors, and windows. Updated guidance on standards, building codes, safety requirements, and more. Also available in Spanish!

$24.95 per copy
Over 150 pages, Softcover
Catalog No. 67298A
Spanish Catalog No. 67298AS

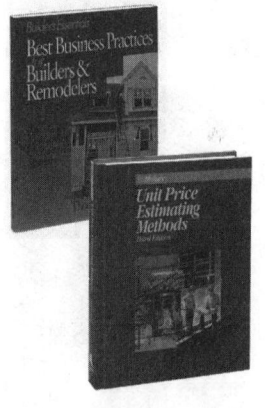

Builder's Essentials: Best Business Practices for Builders & Remodelers

An Easy-to-Use Checklist System

By Thomas N. Frisby

A comprehensive guide covering all aspects of running a construction business, with more than 40 user-friendly checklists. Provides expert guidance on: increasing your revenue and keeping more of your profit, planning for long-term growth, keeping good employees, and managing subcontractors.

$29.95 per copy
Over 220 pages, Softcover
Catalog No. 67329

Unit Price Estimating Methods

New 3rd Edition

This new edition includes up-to-date cost data and estimating examples, updated to reflect changes to the CSI numbering system and new features of Means cost data. It describes the most productive, universally accepted ways to estimate, and uses checklists and forms to illustrate shortcuts and timesavers. A model estimate demonstrates procedures. A new chapter explores computer estimating alternatives.

$59.95 per copy
Over 350 pages, illustrated, Hardcover
Catalog No. 67303A

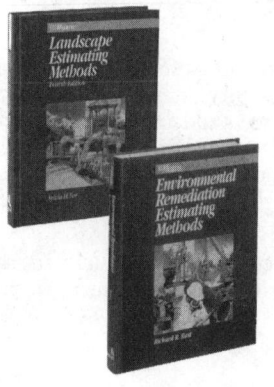

Means Landscape Estimating Methods

4th Edition

By Sylvia H. Fee

This revised edition offers expert guidance for preparing accurate estimates for new landscape construction and grounds maintenance. Includes a complete project estimate featuring the latest equipment and methods, and chapters on *Life Cycle Costing* and *Landscape Maintenance Estimating*.

$62.95 per copy
Over 300 pages, illustrated, Hardcover
Catalog No. 67295B

Means Environmental Remediation Estimating Methods, 2nd Edition

By Richard R. Rast

Guidelines for estimating 50 standard remediation technologies. Use it to prepare preliminary budgets, develop estimates, compare costs and solutions, estimate liability, review quotes, negotiate settlements.

A valuable support tool for Means Environmental Remediation Unit Price and Assemblies books.

$99.95 per copy
Over 750 pages, illustrated, Hardcover
Catalog No. 64777A

or more information
isit Means Web site
: www.rsmeans.com

Reference Books

Means Illustrated Construction Dictionary, Condensed, 2nd Edition

y RSMeans

ecognized in the industry as the best resource of its kind.
his essential tool has been further enhanced with updates
existing terms and the addition of hundreds of new terms
d illustrations—in keeping with recent developments. For
ntractors, architects, insurance and real estate personnel,
meowners, and anyone who needs quick, clear definitions
r construction terms.

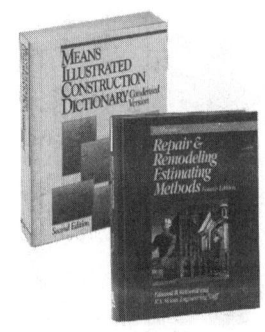

59.95 per copy
ver 500 pages, Softcover
atalog No. 67282A

Means Repair & Remodeling Estimating New 4th Edition

By Edward B. Wetherill & RSMeans

This important reference focuses on the unique problems
of estimating renovations of existing structures and helps
you determine the true costs of remodeling through careful
evaluation of architectural details and a site visit.

New section on disaster restoration costs.

$69.95 per copy
Over 450 pages, illustrated, Hardcover
Catalog No. 67265B

Facilities Planning & Relocation

New, lower price and user-friendly format.

y David D. Owen

complete system for planning space needs and managing
locations. Includes step-by-step manual, over 50 forms,
d extensive reference section on materials and
rnishings.

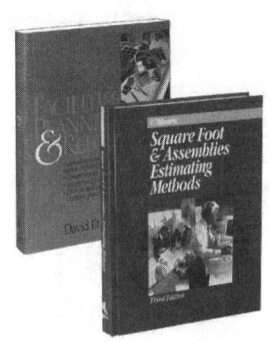

89.95 per copy
ver 450 pages, Softcover
atalog No. 67301

Means Square Foot & Assemblies Estimating Methods, 3rd Edition

Develop realistic square foot and assemblies costs for
budgeting and construction funding. The new edition
features updated guidance on square foot and assemblies
estimating using UNIFORMAT II. An essential reference for
anyone who performs conceptual estimates.

$69.95 per copy
Over 300 pages, illustrated, Hardcover
Catalog No. 67145B

Means Electrical Estimating Methods 3rd Edition

xpanded new edition includes sample estimates and cost
formation in keeping with the latest version of the CSI
asterFormat and UNIFORMAT II. Complete coverage
fiber optic and uninterruptible power supply electrical
stems, broken down by components and explained in
etail. Includes a new chapter on computerized estimating
ethods. A practical companion to Means *Electrical Cost
ata.*

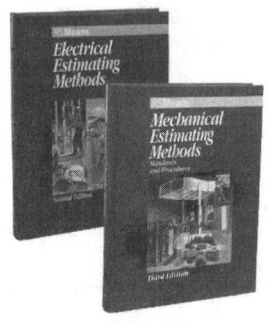

64.95 per copy
ver 325 pages, Hardcover
atalog No. 67230B

Means Mechanical Estimating Methods 3rd Edition

This guide assists you in making a review of plans, specs, and
bid packages with suggestions for takeoff procedures, listings,
substitutions and pre-bid scheduling. Includes suggestions for
budgeting labor and equipment usage. Compares materials
and construction methods to allow you to select the best
option.

$64.95 per copy
Over 350 pages, illustrated, Hardcover
Catalog No. 67294A

Means Spanish/English Construction Dictionary

y RSMeans, The International Conference of Building Officials (ICBO), nd Rolf Jensen & Associates (RJA)

esigned to facilitate communication among Spanish-
nd English-speaking construction personnel—improving
erformance and job-site safety. Features the most common
ords and phrases used in the construction industry, with
asy-to-follow pronunciations. Includes extensive building
ystems and tools illustrations.

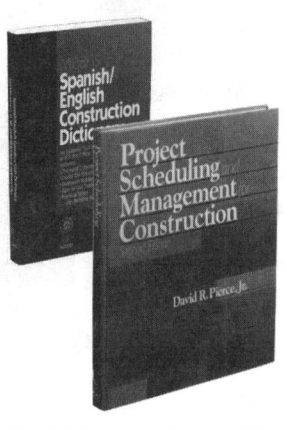

22.95 per copy
50 pages, illustrated, Softcover
atalog No. 67327

Project Scheduling & Management for Construction

New 3rd Edition
By David R. Pierce, Jr.

A comprehensive yet easy-to-follow guide to construction
project scheduling and control—from vital project
management principles through the latest scheduling,
tracking, and controlling techniques. The author is a
leading authority on scheduling with years of field and
teaching experience at leading academic institutions. Spend
a few hours with this book and come away with a solid
understanding of this essential management topic.

$64.95 per copy
Over 300 pages, illustrated, Hardcover
Catalog No. 67247B

Reference Books

For more informatio
visit Means Web sit
at www.rsmeans.co

Concrete Repair and Maintenance Illustrated
By Peter Emmons

Hundreds of illustrations show users how to analyze, repair, clean, and maintain concrete structures for optimal performance and cost effectiveness. From parking garages to roads and bridges to structural concrete, this comprehensive book describes the causes, effects, and remedies for concrete wear and failure. Invaluable for planning jobs, selecting materials, and training employees, this book is a must-have for concrete specialists, general contractors, facility managers, civil and structural engineers, and architects.

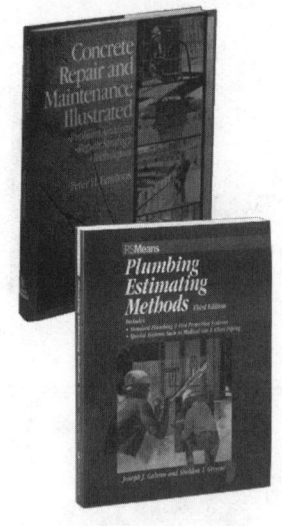

$34.98 per copy
300 pages, illustrated, Softcover
Catalog No. 67146

Plumbing Estimating Methods 3rd Edition
By Joseph Galeno and Sheldon Greene

Updated and revised! This practical guide walks you through a plumbing estimate, from basic materials and installation methods through change order analysis. *Plumbing Estimating Methods* covers residential, commercial, industrial, and medical systems, and features sample takeoff and estimate forms and detaile illustrations of systems and components.

$29.98 per copy
330+ pages, Softcover
Catalog No. 67283B

Building Spec Homes Profitably
By Kenneth V. Johnson

Professional guidance from a veteran spec home builder... This complete system covers all aspects of spec home building—from market research and site selection to financing and design to scheduling and supervision, and more!

You'll learn to reduce risk and ensure profits in spec home building, no matter what the economic climate.

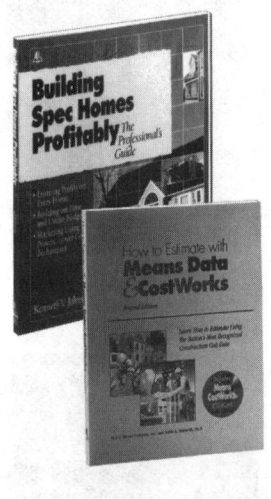

$29.95 per copy
200 pages, Softcover
Catalog No. 67312

How to Estimate with Means Data & CostWorks
New 2nd Edition
By RSMeans and Saleh Mubara

Learn estimating techniques using Means cost data. Includes an instructional version of Means *CostWorks* CD–ROM with sample building plans. The step-by-step guide takes you through all the major construction items. Over 300 sample estimating problems are included.

$59.95 per copy
Over 190 pages, Softcover
Catalog No. 67324A

Means Productivity Standards for Construction
3rd Edition
By RSMeans

Completely updated, with well over 3,000 new work items added to cover topics in current demand! This comprehensive reference for the productivity of construction crews, equipment, and labor includes precise descriptions of each construction task organized according to the Construction Specifications Institute's MasterFormat.

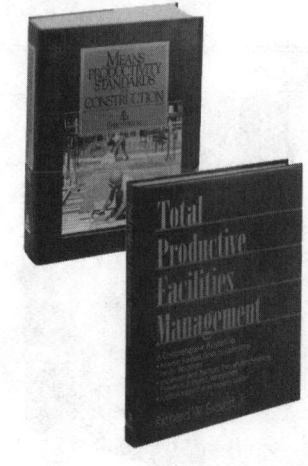

$49.98 per copy
800+ pages, Hardcover
Catalog No. 67236A

Total Productive Facilities Management
By Richard W. Sievert, Jr.

Today, facilities are viewed as strategic resources... elevating the facility manager to the role of asset manager supporting the organization's overall business goals. Now, Richard Sievert Jr., in this well-articulated guidebook, sets forth a new operational standard for the facility manager's emerging role... a comprehensive program for managing facilities as a true profit center.

$29.98 per copy
275 pages, Softcover
Catalog No. 67321

For more information
visit Means Web site
at www.rsmeans.com

Reference Books

Preventive Maintenance Guidelines for School Facilities

By John C. Maciha

A complete PM program for K-12 schools that ensures sustained security, safety, property integrity, user satisfaction, and reasonable ongoing expenditures.

Includes schedules for weekly, monthly, semiannual, and annual maintenance with hard copy and electronic forms.

$149.95 per copy
Over 225 pages, Hardcover
Catalog No. 67326

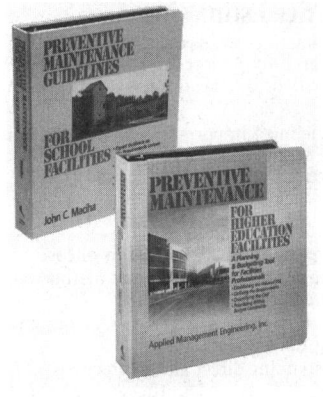

Preventive Maintenance for Higher Education Facilities

By Applied Management Engineering, Inc.

An easy-to-use system to help facilities professionals establish the value of PM, and to develop and budget for an appropriate PM program for their college or university. Features interactive campus building models typical of those found in different-sized higher education facilities, and PM checklists linked to each piece of equipment or system in hard copy and electronic format.

$149.95 per copy
150 pages, Hardcover
Catalog No. 67337

Historic Preservation: Project Planning & Estimating

By Swanke Hayden Connell Architects

Expert guidance on managing historic restoration, rehabilitation, and preservation building projects and determining and controlling their costs. Includes:
 How to determine whether a structure qualifies as historic
 Where to obtain funding and other assistance
 How to evaluate and repair more than 75 historic building materials

$99.95 per copy
Over 675 pages, Hardcover
Catalog No. 67323

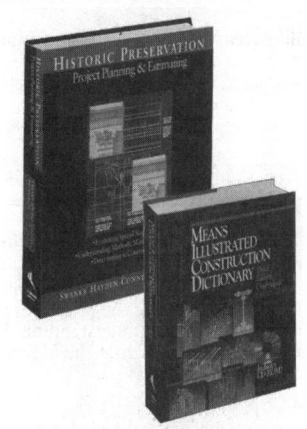

Means Illustrated Construction Dictionary

Unabridged 3rd Edition

Long regarded as the Industry's finest, the Means Illustrated Construction Dictionary is now even better. With the addition of over 1,000 new terms and hundreds of new illustrations, it is the clear choice for the most comprehensive and current information. The companion CD-ROM that comes with this new edition adds many extra features: larger graphics, expanded definitions, and links to both CSI MasterFormat numbers and product information.

$99.95 per copy
Over 790 pages, illustrated, Hardcover
Catalog No. 67292A

Designing & Building with the IBC, 2nd Edition

By Rolf Jensen & Associates, Inc.

This updated comprehensive guide helps building professionals make the transition to the 2003 International Building Code®. Includes a side-by-side code comparison of the IBC 2003 to the IBC 2000 and the three primary model codes, a quick-find index, and professional code commentary. With illustrations, abbreviations key, and an extensive Resource section.

$99.95 per copy
Over 875 pages, Softcover
Catalog No. 67328A

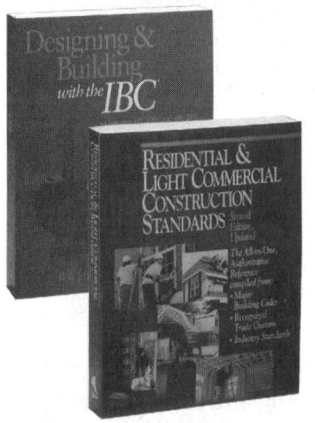

Residential & Light Commercial Construction Standards, 2nd Edition

By RSMeans and Contributing Authors

For contractors, subcontractors, owners, developers, architects, engineers, attorneys, and insurance personnel, this book provides authoritative requirements and recommendations compiled from the nation's leading professional associations, industry publications, and building code organizations.

$59.95 per copy
600 pages, illustrated, Softcover
Catalog No. 67322A

Builder's Essentials: Advanced Framing Methods

By Scot Simpson

A highly illustrated, "framer-friendly" approach to advanced framing elements. Provides expert, but easy to interpret, instruction for laying out and framing complex walls, roofs, and stairs, and special requirements for earthquake and hurricane protection. Also helps bring framers up to date on the latest building code changes, and provides tips on the lead framer's role and responsibilities, how to prepare for a job, and how to get the crew started.

$24.95 per copy
150 pages, illustrated, Softcover
Catalog No. 67330

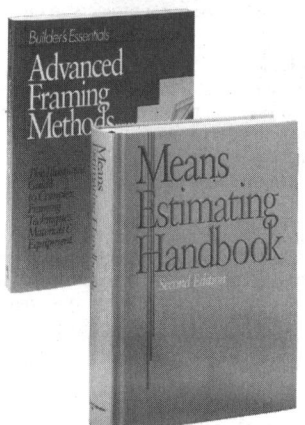

Means Estimating Handbook
2nd Edition

By RSMeans

Updated new Second Edition answers virtually any estimating technical question—all organized by CSI MasterFormat. This comprehensive reference covers the full spectrum of technical data required to estimate construction costs. The book includes information on sizing, productivity, equipment requirements, code-mandated specifications, design standards, and engineering factors.

$99.95 per copy
Over 900 pages, Hardcover
Catalog No. 67276A

For more information
visit Means Web site
at www.rsmeans.com

Seminars

Means CostWorks Training

This one-day seminar course has been designed with the intention of assisting both new and existing users to become more familiar with the *CostWorks* program. The class is broken into two unique sections: (1) A one-half day presentation on the function of each icon; and each student will be shown how to use the software to develop a cost estimate. (2) Hands-on estimating exercises that will ensure that each student thoroughly understands how to use *CostWorks*. You must bring your own laptop computer to this course.

CostWorks Benefits/Features:
• Estimate in your own spreadsheet format
• Power of Means National Database
• Database automatically regionalized
• Save time with keyword searches
• Save time by establishing common estimate items in "Bookmark" files
• Customize your spreadsheet template
• Hot key to Product Manufacturers' listings and specs
• Merge capability for networking environments
• View crews and assembly components
• AutoSave capability
• Enhanced sorting capability

Unit Price Estimating

This interactive two-day seminar teaches attendees how to interpret project information and process it into final, detailed estimates with the greatest accuracy level.

The single most important credential an estimator can take to the job is the ability to visualize construction in the mind's eye, and thereby estimate accurately.

Some Of What You'll Learn:
• Interpreting the design in terms of cost
• The most detailed, time-tested methodology for accurate "pricing"
• Key cost drivers—material, labor, equipment, staging, and subcontracts
• Understanding direct and indirect costs for accurate job cost accounting and change order management

Who Should Attend: Corporate and government estimators and purchasers, architects, engineers... and others needing to produce accurate project estimates.

Square Foot and Assemblies Cost Estimating

This two-day course teaches attendees how to quickly deliver accurate square foot estimates using limited budget and design information.

Some Of What You'll Learn:
• How square foot costing gets the estimate done faster
• Taking advantage of a "systems" or "assemblies" format
• The Means "building assemblies/square foot cost approach"
• How to create a very reliable preliminary and systems estimate using bare-bones design information

Who Should Attend: Facilities managers, facilities engineers, estimators, planners, developers, construction finance professionals... and others needing to make quick, accurate construction cost estimates at commercial, government, educational, and medical facilities.

Repair and Remodeling Estimating

This two-day seminar emphasizes all the underlying considerations unique to repair/remodeling estimating and presents the correct methods for generating accurate, reliable R&R project costs using the unit price and assemblies methods.

Some Of What You'll Learn:
• Estimating considerations—like labor-hours, building code compliance, working within existing structures, purchasing materials in smaller quantities, unforeseen deficiencies
• Identify problems and provide solutions to estimating building alterations
• Rules for factoring in minimum labor costs, accurate productivity estimates and allowances for project contingencies
• R&R estimating examples are calculated using unit prices and assemblies data

Who Should Attend: Facilities managers, plant engineers, architects, contractors, estimators, builders... and others who are concerned with the proper preparation and/or evaluation of repair and remodeling estimates.

Mechanical and Electrical Estimating

This two-day course teaches attendees how to prepare more accurate and complete mechanical/electrical estimates, avoiding the pitfalls of omission and double-counting, while understanding the composition and rationale within the Means Mechanical/Electrical database.

Some Of What You'll Learn:
• The unique way mechanical and electrical systems are interrelated
• M&E estimates, conceptual, planning, budgeting, and bidding stages
• Order of magnitude, square foot, assemblies, and unit price estimating
• Comparative cost analysis of equipment and design alternatives

Who Should Attend: Architects, engineers, facilities managers, mechanical and electrical contractors... and others needing a highly reliable method for developing, understanding, and evaluating mechanical and electrical contracts.

Plan Reading and Material Takeoff

This two-day program teaches attendees to read and understand construction documents and to use them in the preparation of material takeoffs.

Some of What You'll Learn:
• Skills necessary to read and understand typical contract documents—blueprints and specifications
• Details and symbols used by architects and engineers
• Construction specifications' importance in conjunction with blueprints
• Accurate takeoff of construction materials and industry-accepted takeoff methods

Who Should Attend: Facilities managers, construction supervisors, office managers... and others responsible for the execution and administration of a construction project including government, medical, commercial, educational, or retail facilities.

Facilities Maintenance and Repair Estimating

This two-day course teaches attendees how to plan, budget, and estimate the cost of ongoing and preventive maintenance and repair for existing buildings and grounds.

Some Of What You'll Learn:
• The most financially favorable maintenance, repair, and replacement scheduling and estimating
• Auditing and value engineering facilities
• Preventive planning and facilities upgrading
• Determining both in-house and contract-out service costs; annual, asset-protecting M&R plan

Who Should Attend: Facility managers, maintenance supervisors, buildings and grounds superintendents, plant managers, planners, estimators... and others involved in facilities planning and budgeting.

Scheduling and Project Management

This two-day course teaches attendees the most current and proven scheduling and management techniques needed to bring projects in on time and on budget.

Some Of What You'll Learn:
• Crucial phases of planning and scheduling
• How to establish project priorities and develop realistic schedules and management techniques
• Critical Path and Precedence Methods
• Special emphasis on cost control

Who Should Attend: Construction project managers, supervisors, engineers, estimators, contractors... and others who want to improve their project planning, scheduling, and management skills.

Advanced Project Management

This two-day seminar will teach you how to effectively manage and control the entire design-build process and allow you to take home tangible skills that will be immediately applicable on existing projects.

Some Of What You'll Learn:
• Value engineering, bonding, fast-tracking, and bid package creation
• How estimates and schedules can be integrated to provide advanced project management tools
• Cost engineering, quality control, productivity measurement, and improvement
• Front loading a project and predicting its cash flow

Who Should Attend: Owners, project managers, architectural and engineering managers, construction managers, contractors... and anyone else who is responsible for the timely design and completion of construction projects.

Seminars

2006 Means Seminar Schedule

Note: Call for exact dates and details.

Location	Dates
Las Vegas, NV	March
Washington, DC	April
Phoenix, AZ	April
Denver, CO	May
San Francisco, CA	June
Philadelphia, PA	June
Washington, DC	September
Dallas, TX	September
Las Vegas, NV	October
Orlando, FL	November
Atlantic City, NJ	November
San Diego, CA	December

Registration Information

Register Early... Save up to $100! Register 30 days before the start date of seminar and save $100 off your total fee. *Note: This discount can be applied only once per order. It cannot be applied to team discount registrations or any other special offer.*

How to Register Register by phone today! Means' toll-free number for making reservations is: **1-800-334-3509.**

Individual Seminar Registration Fee $895. Individual CostWorks Training Registration Fee $349. To register by mail, complete the registration form and return with your full fee to: Seminar Division, Reed Construction Data, RSMeans Seminars, 63 Smiths Lane, Kingston, MA 02364.

Federal Government Pricing All federal government employees save 25% of regular seminar price. Other promotional discounts cannot be combined with Federal Government discount.

Team Discount Program Two to four seminar registrations: Call for pricing.

Multiple Course Discounts When signing up for two or more courses, call for pricing.

Refund Policy Cancellations will be accepted up to ten days prior to the seminar start. There are no refunds for cancellations received later than ten working days prior to the first day of the seminar. A $150 processing fee will be applied for all cancellations. Written notice of cancellation is required. Substitutions can be made at any time before the session starts. **No-shows are subject to the full seminar fee.**

AACE Approved Courses The RSMeans Construction Estimating and Management Seminars described and offered to you here have each been approved for 14 hours (1.4 recertification credits) of credit by the AACE International Certification Board toward meeting the continuing education requirements for recertification as a Certified Cost Engineer/Certified Cost Consultant.

AIA Continuing Education We are registered with the AIA Continuing Education System (AIA/CES) and are committed to developing quality learning activities in accordance with the CES criteria. RSMeans seminars meet the AIA/CES criteria for Quality Level 2. AIA members will receive (14) learning units (LUs) for each two-day RSMeans Course.

NASBA CPE Sponsor Credits We are part of the National Registry of CPE Sponsors. Attendees may be eligible for (16) CPE credits.

Daily Course Schedule The first day of each seminar session begins at 8:30 A.M. and ends at 4:30 P.M. The second day is 8:00 A.M.–4:00 P.M. Participants are urged to bring a hand-held calculator since many actual problems will be worked out in each session.

Continental Breakfast Your registration includes the cost of a continental breakfast, a morning coffee break, and an afternoon break. These informal segments will allow you to discuss topics of mutual interest with other members of the seminar. (You are free to make your own lunch and dinner arrangements.)

Hotel/Transportation Arrangements RSMeans has arranged to hold a block of rooms at each hotel hosting a seminar. To take advantage of special group rates when making your reservation, be sure to mention that you are attending the Means Seminar. You are, of course, free to stay at the lodging place of your choice. **(Hotel reservations and transportation arrangements should be made directly by seminar attendees.)**

Important Class sizes are limited, so please register as soon as possible.

Note: Pricing subject to change.

Registration Form   Call 1-800-334-3509 to register or FAX 1-800-632-6732. Visit our Web site www.rsmeans.com

Please register the following people for the Means Construction Seminars as shown here. Full payment or deposit is enclosed, and we understand that we must make our own hotel reservations if overnight stays are necessary.

Full payment of $_____ enclosed.

Bill me

Name of Registrant(s)
(to appear on certificate of completion) _____

_____

_____

P.O. #: _____
GOVERNMENT AGENCIES MUST SUPPLY PURCHASE ORDER NUMBER

Firm Name _____

Address _____

City/State/Zip _____

Telephone No. Fax No. _____

E-mail Address _____

Charge our registration(s) to: ☐ MasterCard ☐ VISA ☐ American Express ☐ Discover

Account No. _____ Exp. Date _____

Cardholder's Signature _____

Seminar Name	City	Dates

Please mail check to: Seminar Division, Reed Construction Data, RSMeans Seminars, 63 Smiths Lane, P.O. Box 800, Kingston, MA 02364 USA

MeansData™

CONSTRUCTION COSTS FOR SOFTWARE APPLICATIONS
Your construction estimating software is only as good as your cost data.

A proven construction cost database is a mandatory part of any estimating package. The following list of software providers can offer you MeansData™ as an added feature for their estimating systems. See the table below for what types of products and services they offer (match their numbers). Visit online at www.rsmeans.com/demosource/ for more information and free demos. Or call their numbers listed below.

1. **3D International**
713-871-7000
venegas@3di.com

2. **4Clicks-Solutions, LLC**
719-574-7721
mbrown@4clicks-solutions.com

3. **Aepco, Inc.**
301-670-4642
blueworks@aepco.com

4. **Applied Flow Technology**
800-589-4943
info@aft.com

5. **ArenaSoft Estimating**
888-370-8806
info@arenasoft.com

6. **BSD - Building Systems Design, Inc.**
888-273-7638
bsd@bsdsoftlink.com

7. **CMS - Computerized Micro Solutions**
800-255-7407
cms@proest.com

8. **Corecon Technologies, Inc.**
714-895-7222
sales@corecon.com

9. **CorVet Systems**
301-622-9069
sales@corvetsys.com

10. **Estimating Systems, Inc.**
800-967-8572
esipulsar@adelphia.net

11. **MC² - Management Computer**
800-225-5622
vkeys@mc2-ice.com

12. **Maximus Asset Solutions**
800-659-9001
assetsolutions@maximus.com

13. **Shaw Beneco Enterprises, Inc.**
877-719-4748
inquire@beneco.com

14. **Timberline Software Corp.**
800-628-6583
product.info@timberline.com

15. **US Cost, Inc.**
800-372-4003
sales@uscost.com

16. **Vanderweil Facility Advisors**
617-451-5100
info@VFA.com

17. **WinEstimator, Inc.**
800-950-2374
sales@winest.com

TYPE	1	2	3	4	5	6	7	8	9	10	11	12	13	14	15	16	17
BID					•	•	•		•		•			•			•
Estimating		•			•	•	•	•	•	•	•		•	•	•		•
DOC/JOC/SABER		•			•	•			•	•		•	•	•			•
ID/IQ		•							•	•		•	•				•
Asset Mgmt.												•				•	•
Facility Mgmt.	•	•	•									•				•	
Project Mgmt.	•	•						•	•				•	•			
TAKE-OFF					•		•	•	•		•			•	•		
EARTHWORK								•			•						
Pipe Flow				•													
HVAC/Plumbing					•		•		•								
Roofing					•		•		•								
Design	•				•								•	•			•
Other Offers Links:																	
Accounting/HR		•			•								•	•			
Scheduling					•									•	•		•
CAD														•	•		•
PDA														•	•		•
Lt. Versions		•					•							•	•		•
Consulting	•	•			•		•		•			•		•	•	•	•
Training		•			•	•	•	•	•	•		•	•	•	•	•	•

Qualified re-seller applications now being accepted. Call Carol Polio, Ext. 5107.

FOR MORE INFORMATION
CALL 1-800-448-8182, EXT. 5107 OR FAX 1-800-632-6732

For more information
visit Means Web site
at www.rsmeans.com

Preventive Maintenance for Multi-Family Housing
by John C. Maciha

With 24/7 occupancy, multi-family buildings can be some of the toughest to maintain. Prepared by one of the nation's leading experts on multi-family housing, Preventive Maintenance for Multi-Family Housing puts easy-to-use guidelines right at your fingertips for the what, when, why, and how much of multi-family preventive maintenance.

This complete PM system for apartment and condominium communities features expert guidance, checklists for buildings and grounds maintenance tasks and their frequencies, a reusable wall chart to track maintenance, and a dedicated Web site featuring customizable electronic forms. A must-have for anyone involved with multi-family housing maintenance and upkeep.

$89.95 per copy
325 pages
Catalog No. 67346

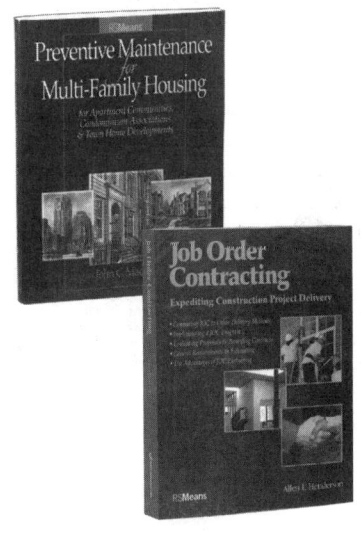

Job Order Contracting
Expediting Construction Project Delivery
by Allen Henderson

Expert guidance to help you implement JOC—fast becoming the preferred project delivery method for repair and renovation, minor new construction, and maintenance projects in the public sector and in many states and municipalities. The author, a leading JOC expert and practitioner, shows how to:
• Establish a JOC program
• Evaluate proposals and award contracts
• Handle general requirements and estimating
• Partner for maximum benefits

This book will quickly bring you up to speed on JOC (also known as Delivery Order Contracting and SABER) so you can achieve faster project starts with fewer delays, cost overruns, and quality disputes.

$89.95 oer copy
192 pages, illustrated, Hardcover
Catalog No. 67348

Kitchen & Bath Project Costs: Planning & Estimating Successful Projects

Project estimates for 35 of the most popular kitchen and bath renovations... from replacing a single fixture to whole-room remodels. Each estimate includes:
All materials needed for the project
Labor-hours to install (and demolish/remove) each item
Subcontractor costs for certain trades and services
An allocation for overhead and profit

PLUS! Takeoff and pricing worksheets—forms you can photocopy or access electronically from the book's Web site; alternate materials—unit costs for different finishes and fixtures, location factors—easy multipliers to adjust the costs to your location; and expert guidance on estimating methods, project design, contracts, marketing, working with homeowners, and tips for each of the estimated projects.

$29.95 per copy
Over 175 pages
Catalog No. 67347

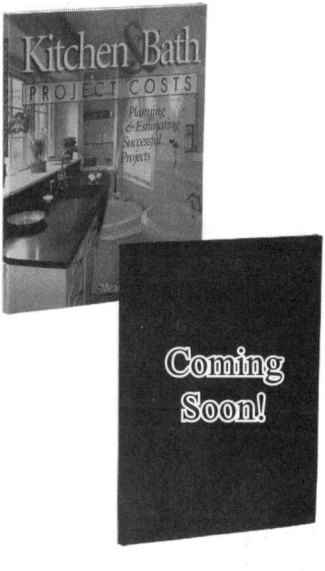

Coming Soon!

Home Addition & Renovation Costs

New! This essential home remodeling reference gives you 35 project estimates and guidance for some of the most popular home renovation and addition projects... from opening up a simple interior wall to adding an entire second story. Each estimate includes a floor plan, color photos, and detailed costs. Use the project estimates as backup for pricing, to check your own estimates, or as a cost reference for preliminary discussion with homeowners.

Includes:
• Case studies—with creative solutions and design ideas.
• Alternate materials costs—so you can match the estimates to the particulars of your projects.
• Location factors—easy multipliers to adjust the book's costs to your own location.

$29.95 per copy
Over 200 pages, illustrated, Softcover
Catalog No. 67349

The Practice of Cost Segregation Analysis
by Bruce A. Desrosiers and Wayne J. DelPico

This expert guide walks you through the practice of cost segregations analysis, which enables property owners to defer taxes and benefit from "accelerated cost recovery" through depreciation deductions on assets that are properly identified and classified. Cost segregation practice requires knowledge of both tax law and the construction process. In this book, the authors share their expertise in these areas with tax and accounting professionals, cost segregation consultants, facility owners, architects, and general contractors—providing guidance on major aspects of a professional, defensible cost segregation study.

With a glossary of terms, sample cost segregation estimates for various building types, key information resources, and updates via a dedicated Web site, this book is a critical resource for anyone involved in cost segregation analysis.

$99.95 per copy
Over 225 pages
Catalog No. 67345

FOR REFERENCE

NOT TO BE TAKEN FROM THIS ROOM

2006 Order Form

	COST ESTIMATING BOOKS	Unit Price	Total
	Assemblies Cost Data 2006	$206.95	
0016	Building Construction Cost Data 2006	126.95	
61016	Building Const. Cost Data–Looseleaf Ed. 2006	157.95	
60226	Building Const. Cost Data–Western Ed. 2006	126.95	
60116	Concrete & Masonry Cost Data 2006	114.95	
60146	Construction Cost Indexes 2006	272.00	
60146A	Construction Cost Index–January 2006	68.00	
60146B	Construction Cost Index–April 2006	68.00	
60146C	Construction Cost Index–July 2006	68.00	
60146D	Construction Cost Index–October 2006	68.00	
60346	Contr. Pricing Guide: Resid. R & R Costs 2006	39.95	
60336	Contr. Pricing Guide: Resid. Detailed 2006	39.95	
60326	Contr. Pricing Guide: Resid. Sq. Ft. 2006	39.95	
60236	Electrical Change Order Cost Data 2006	126.95	
60036	Electrical Cost Data 2006	126.95	
60206	Facilities Construction Cost Data 2006	299.95	
60306	Facilities Maintenance & Repair Cost Data 2006	274.95	
60166	Heavy Construction Cost Data 2006	126.95	
60096	Interior Cost Data 2006	126.95	
60126	Labor Rates for the Const. Industry 2006	274.95	
60186	Light Commercial Cost Data 2006	107.95	
60026	Mechanical Cost Data 2006	126.95	
63016	Metric Construction Cost Data 2006	149.95	
60156	Open Shop Building Const. Cost Data 2006	126.95	
60216	Plumbing Cost Data 2006	126.95	
60046	Repair and Remodeling Cost Data 2006	107.95	
60176	Residential Cost Data 2006	107.95	
60286	Site Work & Landscape Cost Data 2006	126.95	
60056	Square Foot Costs 2006	137.95	
	REFERENCE BOOKS		
67310A	ADA Compliance Pricing Guide, 2nd Ed.	79.95	
67273	Basics for Builders: How to Survive and Prosper	34.95	
67330	Bldrs Essentials: Adv. Framing Methods	24.95	
67329	Bldrs Essentials: Best Bus. Practices for Bldrs	29.95	
67298A	Bldrs Essentials: Framing/Carpentry 2nd Ed.	24.95	
67298AS	Bldrs Essentials: Framing/Carpentry Spanish	24.95	
67307	Bldrs Essentials: Plan Reading & Takeoff	35.95	
67261A	Bldg. Prof. Guide to Contract Documents 3rd Ed.	32.48	
67342	Building & Renovating Schools	99.95	
67339	Building Security: Strategies & Costs	44.98	
67312	Building Spec Homes Profitably	29.95	
67146	Concrete Repair & Maintenance Illustrated	34.98	
67314	Cost Planning & Est. for Facil. Maint.	89.95	
67317A	Cyberplaces: The Internet Guide 2nd Ed.	59.95	
67328A	Designing & Building with the IBC, 2nd Ed.	99.95	
67230B	Electrical Estimating Methods 3rd Ed.	64.95	
64777A	Environmental Remediation Est. Methods 2nd Ed.	$ 99.95	
67343	Estimating Bldg. Costs for Resi. & Lt. Comm.	29.95	
67160	Estimating for Contractors	17.98	
67276A	Estimating Handbook 2nd Ed.	99.95	
67249	Facilities Maintenance Management	86.95	
67246	Facilities Maintenance Standards	79.95	

Qty.	Book No.	REFERENCE BOOKS (Cont.)	Unit Price	Total
	67318	Facilities Operations & Engineering Reference	109.95	
	67301	Facilities Planning & Relocation	89.95	
	67231	Forms for Building Const. Professional	47.48	
	67260	Fundamentals of the Construction Process	25.98	
	67338	Green Building: Proj. Planning & Cost Est.	89.95	
	67323	Historic Preservation: Proj. Planning & Est.	99.95	
	67308E	Home Improvement Costs–Int. Projects 9th Ed.	24.95	
	67309E	Home Improvement Costs–Ext. Projects 9th Ed.	24.95	
	67324A	How to Est. w/Means Data & CostWorks 2nd Ed.	59.95	
	67304	How to Estimate with Metric Units	9.98	
	67306	HVAC: Design Criteria, Options, Select. 2nd Ed.	84.95	
	67281	HVAC Systems Evaluation	84.95	
	67282A	Illustrated Construction Dictionary, Condensed	59.95	
	67292A	Illustrated Construction Dictionary, w/CD-ROM	99.95	
	67348	Job Order Contracting	89.95	
	67347	Kitchen & Bath Project Costs	29.95	
	67295B	Landscape Estimating 4th Ed.	62.95	
	67341	Life Cycle Costing	99.95	
	67302	Managing Construction Purchasing	19.98	
	67294A	Mechanical Estimating 3rd Ed.	64.95	
	67283B	Plumbing Estimating Methods 3rd Ed.	29.98	
	67345	Pract. of Cost Segregation Analysis	99.95	
	67326	Preventive Maint. Guidelines for School Facil.	149.95	
	67337	Preventive Maint. for Higher Education Facilities	149.95	
	67346	Preventive Maint. for Multi-Family Housing	89.95	
	67236A	Productivity Standards for Constr.–3rd Ed.	49.98	
	67247B	Project Scheduling & Management for Constr. 3rd Ed.	64.95	
	67265B	Repair & Remodeling Estimating 4th Ed.	69.95	
	67322A	Resi. & Light Commercial Const. Stds. 2nd Ed.	59.95	
	67327	Spanish/English Construction Dictionary	22.95	
	67145B	Sq. Ft. & Assem. Estimating Methods 3rd Ed.	69.95	
	67287	Successful Estimating Methods	32.48	
	67313	Successful Interior Projects	24.98	
	67233	Superintending for Contractors	35.95	
	67321	Total Productive Facilities Management	29.98	
	67284	Understanding Building Automation Systems	29.98	
	67303A	Unit Price Estimating Methods 3rd Ed.	59.95	
	67319	Value Engineering: Practical Applications	79.95	

MA residents add 5% state sales tax		
Shipping & Handling**		
Total (U.S. Funds)*		

Prices are subject to change and are for U.S. delivery only. *Canadian customers may call for current prices. **Shipping & handling charges: Add 7% of total order for check and credit card payments. Add 9% of total order for invoiced orders.

Send Order To: **ADDV-1001**

Name (Please Print) _____

Company _____

☐ **Company**
☐ **Home**   Address _____

City/State/Zip _____

Phone # _____ P.O. # _____

(Must accompany all orders being billed)

Mail To: **RSMeans** P.O. Box 800, Kingston, MA 02364-0800